Anionic Annulations in Organic Synthesis

A Versatile and Prolific Class of Ring-forming Reactions

Anionic Annulations in Organic Synthesis

A Versatile and Prolific Class of Ring-forming Reactions

Dipakranjan Mal
Former Professor
Department of Chemistry
Indian Institute of Technology Kharagpur
Kharagpur, India

ELSEVIER

Elsevier
Radarweg 29, PO Box 211, 1000 AE Amsterdam, Netherlands
The Boulevard, Langford Lane, Kidlington, Oxford OX5 1GB, United Kingdom
50 Hampshire Street, 5th Floor, Cambridge, MA 02139, United States

Library of Congress Cataloging-in-Publication Data
A catalog record for this book is available from the Library of Congress

British Library Cataloguing-in-Publication Data
A catalogue record for this book is available from the British Library

ISBN: 978-0-12-813800-7

Publisher: Susan Dennis
Acquisition Editor: Emily M. McCloskey
Editorial Project Manager: Michelle W. Fisher
Production Project Manager: Maria Bernard
Cover Designer: Greg Harris

Typeset by SPi Global, India

Dedication

Dedicated to my parents and my teachers. (Professors Frank M. Hauser, Layton L. McCoy, Jerry R. Dias, (Late) Asima Chatterjee, (Late) Achintya K. Sen, Priyalal Majumdar, (Late) Suresh C. Sengupta, and Bimal K. Mahapatra and many more).

Contents

10. Applications of [4+2] Anionic Annulations in the Total Synthesis

11. Miscellaneous Anionic Annulations

12. Concluding Remarks

Preface

Annulation, another name for the ring formation with two new bonds of the ring, is a powerful tool in the synthesis of organic molecules with a cyclic structure. The term gained prominence with chemists when Robinson invented his annulation in 1935. Since then, progress in this area has continued unabated, and the topic enticed organic chemists into exploring new dimensions. However, the concept largely remained scattered and random. This book is intended to reinforce the fact that the annulation is a commonly used term, and it itself forms a class of reactions. Annulations are of fundamental importance due to the occurrence of cyclic and polycyclic motifs in a great majority of organic molecules. These motifs can be constructed successfully by the applications of annulations. More specifically, annulations are found useful for the synthesis of polyoxygenated polycyclic compounds, which include tetracyclines, anthracyclines, angucyclines, uncialamycins, carbazoles, lignans, etc.

The landmark in the post-Robinson era is probably the Hauser annulation, discovered in 1978. The underlying concept enunciated by this annulation has emerged to be either competitive with or complementary to the famous Diels-Alder reaction. It may be termed as an aromatic Diels-Alder reaction. As a follow-up of our review in 2007, and the contributions made by the Myers group in the field, I decided to write a book on annulations. The motivation for writing this book is to encourage the use of annulations in the synthesis of natural products and drugs by researchers working on organic synthesis.

The first chapter focuses on the definition of the term "annulation." It also illustrates the term with an overview of the development of the Robinson annulation and the Darzens [2 + 1] condensation over the past 80 years. The primary objective of the book is to document the Hauser annulation, which is presented in Chapters 2–4. Chapters 5–10 are devoted to the allied [4 + 2] annulations. Chapters 11 and 12 deal with miscellaneous anionic annulations, which are classified as [2 + 1], [3 + 2], and [3 + 3] annulations. The final part of the book may be considered as an appendix, consisting of a few short subchapters briefly describing a series of related topics. The annulations consisting in two discrete reactions conducted in two steps are not included in this book.

Annulations, in general, are riskier than cyclizations, since the former demand simultaneous tuning for two ends of a cycle to be formed, and this involves overcoming the entropy of an intermolecular reaction as opposed to an intramolecular reaction. To give a comprehensive view of annulations, this book—the first of its kind—was conceptualized a few years ago. It is expected to benefit senior undergraduate and graduate students. It is likely to be equally useful to practicing synthetic organic chemists of academia and industries. Most of the reaction schemes of the book refer to the abbreviations of journals in the format [Year of publication—Journal abbreviation—Initial page number—Last name of the corresponding author followed by their initials].

It is possible that I might have missed many important reactions which deserved a place in this book. These omissions, along with suggestions for improvement of the book or inclusions of any relevant work, may be communicated to me at dipak.mal@gmail.com/dmal@chem.iitkgp.ac.in.

Dipakranjan Mal

Acknowledgments

I express my gratitude to all those who have helped me for completing the task of writing the manuscript. It is a pleasure to acknowledge the fine contributions of both past and present group members, many of whose names appear in the references related to the annulations. All of them played a major role in shaping my research career. Some went through the manuscript of the book. Special thanks go to my group members Dr. Supriti Jana, Dr. Shyam Basak, Dr. Soumen Chakraborty, Bidhan Ghosh, and Biswajit Roy. I am indebted to Professor Samik Nanda of IIT Kharagpur for his continuing collegial interest and support. Cordial thanks are due to Mr. Subhasis Biswas for his involvement in the production of the schemes and figures. I would also like to thank the reviewers of the original proposals of the book for their valuable suggestions.

I am thankful to my wife Sobhana, my sons Rudrajit and Indrajit for their forbearance and relentless support.

Abbreviations

JOURNALS

ACIE	Angewandte Chemie International Edition
AJC	Australian Journal of Chemistry
BJOC	Beilstein Journal of Organic Chemistry
BMC	Biorganic Medicinal Chemistry
CEJ	Chemistry: A European Journal
CLS	Chemistry Letters
CRS	Chemical Reviews
CSE	Chemical Science
CST	ChemistrySelect
COS	Current Organic Synthesis
HCA	Helvetica Chimica Acta
HCS	Heterocycles
IJCY	Indian Journal of Chemistry
JABS	Journal of Antibiotics
JCE	Journal of Chemical Education
JCRS	Journal of Chemical Research (S)
JCSCCS	Journal of Chemical Society, Chemical Communications
JCSDT	Journal of Chemical Society, Dalton Transactions
JCSPT1	Journal of Chemical Society, Perkin Transactions 1
JHC	Journal of Heterocyclic Chemistry
JOC	Journal of Organic Chemistry
JEMC	Journal of Enzyme Inhibition and Medicinal Chemistry
JACS	Journal of the American Chemical Society
JKCS	Journal of the Korean Chemical Society
LAN	Liebigs Annalen
NCY	Nature Chemistry
NCS	Nature Communications
NPC	Natural Product Communication
OPRD	Organic Process Research & Development
SCE	Science
SYS	Synthesis
SYT	Synlett
THN	Tetrahedron
THS	Tetrahedron Letters
TCC	Topics in Current Chemistry

REAGENTS AND CONDITIONS

Aq	aqueous
B_2Pin_2	bis(pinacollato)diboron
BINOL	1,1′-bi-2-naphthol
CAN	ceric ammonium nitrate
DBU	1,8-diazabicyclo[5.4.0]undec-7-ene
DCM	dichloromethane
DDQ	2,3,-dichloro-5,6-dicyano-1,4-benzoquinone
DIBALH	diisobutylaluminium hydride
DMF	*N,N*-dimethylformamide
DMSO	dimethyl sulfoxide
equiv	equivalent
h	hour
dppf	1,1′-bis(diphenylphosphino)ferrocene
HMDS	hexamethyldisilazane
KHMDS	potassium bis(trimethylsilyl)amide
KTB	potassium *tert*-butoxide
LAH	lithium aluminium hydride
LDA	lithium diisopropylamide
LHMDS	lithium hexamethyldisilazide
LTB	lithium *tert*-butoxide
LTMP	lithium tetramethylpiperidide
MCPBA	*meta*-chloroperoxybenzoic acid
NBS	*N*-bromosuccinimide
NMO	4-methylmorpholine *N*-oxide
PCC	pyridinium chlorochromate
PDC	pyridinium dichromate
PivCl	pivaloyl chloride
PTSA	*p*-toluenesulfonic acid
Py	pyridine
rt	room temperature
TBAF	tetrabutylammonium fluoride
TFA	trifluoroacetic acid
THF	tetrahydrofuran
TBHP	*tert*-butyl hydroperoxide
TMEDA	*N,N,N′,N′*-tetramethylethylenediamine

Chapter 1

Introduction

1.1 ANNULATIONS: HISTORY AND PERSPECTIVES

Organic compounds abound in the chemical literature. Their chemistry has emerged as a separate and central discipline of science. Many of the molecules contribute to the direct benefit of mankind, and have become part and parcel of human existence. An overwhelming number of them are medicinally important and bioactive compounds, and occur in different structures. A vast majority of organic molecules display cyclic and polycyclic motifs as the distinguishing structural features in addition to the multiple bonds, heteroatoms, stereochemistry, and their combinations. As evident from the article by Njardarson, over 90% of the bestselling medicines feature cyclic structures.[1] Consequently, the synthesis of cyclic or polycyclic has become a central and vitally important area of research interest, and constitutes a subdiscipline of organic chemistry.

There are five commonly used processes for construction of cyclic compounds: (i) annulation, (ii) cyclization, (iii) cycloaddition, (iv) cycloisomerization, and (v) fragmentation. Since the cyclic motifs vary in their sizes and numbers, the degree of the synthetic challenges varies from molecule to molecule, owing to the varying ring strains and stereoelectronic effects, and the occurrence of sensitive functional groups. Furthermore, the presence of the heteroatoms in cyclic scaffolds poses additional synthetic challenges, and demands innovative solutions. Consequently, heterocycles constitute an entirely distinct subclass of organic compounds.

Although the importance of the construction of the cyclic compounds has been recognized very early in the development of organic chemistry, few textbooks have covered this topic. Two books are *Cyclization Reactions* by Thebtaranonth and Thebtaranonth, and *Name Reactions for Carbocyclic Ring Formations* by Jie Jack Li.[2] Likewise, annulation, a way of building cyclic motifs in a convergent manner, has not found a place in a textbook.[3] It is discussed only in passing as the benzannulation in the chapter "Anionic cyclization" of the book by Thebtaranonth and Thebtaranonth.

1.1.1 What is an Annulation?

According to the IUPAC, an "annulation" is defined as "a transformation involving fusion of a new ring to a molecule via two new bonds." According

Anionic Annulations in Organic Synthesis. https://doi.org/10.1016/B978-0-12-813800-7.00001-4

to Danhesier, annulations are ring-forming processes in which two molecular fragments are united with the formation of two new bonds.[4] The word "annulation" is derived from the Latin word *annulatus*, meaning "the formation of rings." In organic chemistry, this term means the process of ring-forming onto a preexisting cyclic or acyclic system. In the older literature, the term "annelation" was used in the place of annulation. However, the recent literature tends to prefer "annulation." Historically, the concept of the annulation was seeded in Darzens reaction as early as 1902. The term "annulation" was first pioneered by Rapson and Robinson in 1935 by their discovery of the annulation, consisting of a tandem Michael-aldol reaction.[5] The next landmark in the field was the McCoy cyclopropanation, discovered in 1958, which is a Michael-initiated intramolecular substitution for three-membered ring annulation.[6] In 1965, Schmid et al. introduced the concept of Michael-initiated Dieckmann condensation for six-membered ring annulation in the synthesis of tetralones.[7] The last breakthroughs were due to Hauser, Kraus, Sammes, van Leusen, and Staunton's groups in 1978 for the synthesis of benzenoid carbocycles.[8] In 1980, Little had attempted to categorize the anionic cyclizations under the term "Michael initiated ring closure reactions (MIRC)."[9] This term is illustrated by reactions 1–3 in Scheme 1.1. The third reaction is distinctly different from the first two reactions. Reactions 1 and 2 involve formation of only one new bond of the new ring, whereas reaction 3 involves two bonds of the new cyclopropane ring. Thus, the McCoy cyclopropanation (reaction 3) should be viewed as an anionic annulation. It can also be considered as a Michael-initiated [2+1] annulation.

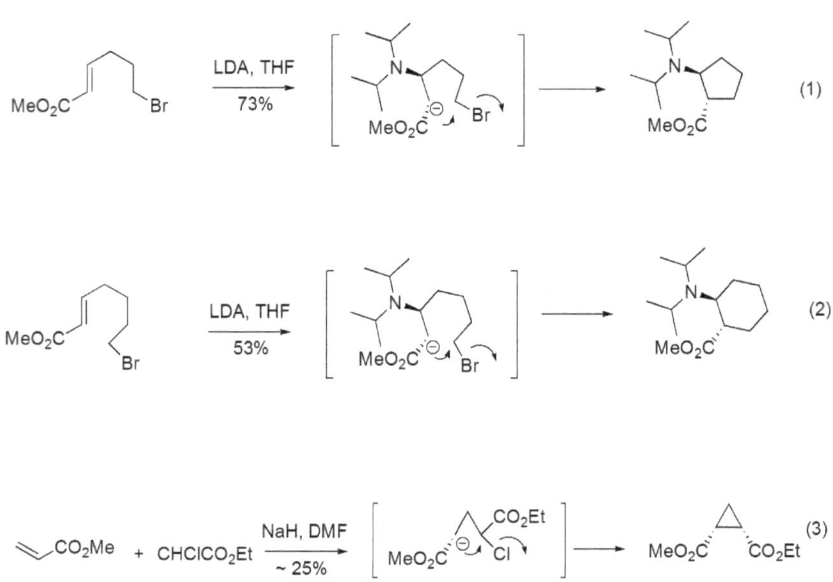

SCHEME 1.1 Michael-initiated ring closure and annulation reactions [1980 THS 2609 Little RD; 1958 JACS 6568 McCoy LL].

As yet, the key concept of the reactions implied by “annulations” has not been appreciated by the students of chemistry. It is shadowed by the broader titles like domino reactions, which essentially cover the multistep construction of cyclic compounds in one pot. Nevertheless, there is a clear distinction between a domino reaction and an annulation. The domino reactions are often unimolecular and the annulations are bimolecular.

It is imperative to acquire comprehensive and systematic knowledge of the successful but scattered annulations for better understanding of the concept of annulations. In Scheme 1.2, five commonly used processes for the construction of cyclic compounds are illustrated. Entry 1 shows a classic example of an annulation, i.e., Robinson annulation. Methyl vinyl ketone (**1**) undergoes an annulation with cyclohexanone (**2**) in presence of a base to give decalenone (**3**) in a single operation. Of the two new bonds of the new ring of the decalenone, the first one is formed in the initiation step, i.e., Michael addition, and the second new bond in the termination step, which is an intramolecular aldol

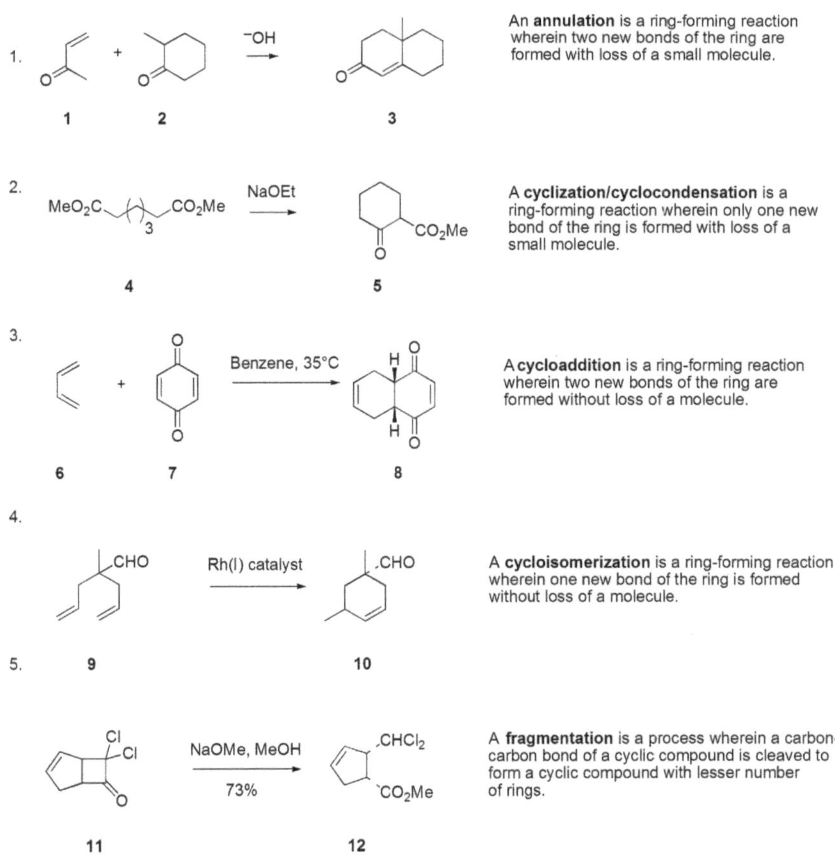

SCHEME 1.2 Representative classes of reactions for the formation of cyclic compounds.

condensation with the loss of a water molecule. Reaction 2 is a cyclization, in which dimethyl pimelate (**4**) gives 2-oxo-cyclohexane carboxylate **5** with loss of methanol in the presence of NaOEt. The key step is an intramolecular reaction forming only one new bond of the ring. Reaction 3 is a [4+2] cycloaddition between butadiene (**6**) and *p*-benzoquinone (**7**) to give decalenedione **8**. Reaction 4 is a rhodium-catalyzed cycloisomerization of aldehyde (**9**) to cyclohexene (**10**). Fragmentation (reaction 5, Scheme 1.2) also provides a cyclic structure, provided it begins with a bicyclic or a polycyclic compound. In the NaOMe promoted reaction of compound (**11**), the four-membered ring is cleaved to give cyclopentene derivative (**12**). No new cycle is formed.

In a cycloaddition (reaction 3) or a cycloisomerization (reaction 4), no small molecule is eliminated in course of the reaction, whereas in annulations (reaction 1) and cyclizations (reaction 2), a small molecule is often eliminated. This book has dealt with the annulations entailing a loss of a small molecule as shown in Scheme 1.2. Therefore, an annulation can be viewed as a cyclization proceeded by an intermolecular reaction.

In organic chemistry, cyclizations are well-studied and well-understood reactions. For the last 40 years, Baldwin rules[10] have provided valuable guidelines for cyclization reactions in terms of stereoelectronic factors. The effect of ring sizes on the cyclizations is somewhat foreseeable.[11] However, annulations are much more complex to predict, compared to the cyclizations, since they consist of two discrete steps: an intermolecular reaction and then an intramolecular reaction. In general, the intermolecular steps are less favorable than their intramolecular counterparts. This is due to the fact that for the formation of transition states in the former, two molecular species join to form one species, resulting in the loss of translation freedom. Nevertheless, reactions such as electron transfer reactions, halogen metal exchange, deprotonation of molecules with active hydrogens, and Michael additions are qualitatively ranked among the very fast intermolecular organic reactions.

Annulations must be differentiated from cascade/domino reactions, which are often unimolecular. The former entails at least two fragments. Moreover, at least two new bonds of the newly formed ring should be formed in the process of annulations. A domino reaction is defined as a reaction in which two or more bond-forming transformations take place under the same reaction conditions, without additional reagents and catalysts, and the first step is responsible for the succeeding steps.[12] The Wiemer domino reaction[13] for the synthesis angelichalcone (**13**) from epoxide (**14**) is presented in Scheme 1.3 to illustrate that it is not an annulation, although two new rings and three new bonds are formed in a one-pot reaction. No two new bonds belong to a particular ring.

A Diels-Alder reaction (Scheme 1.2, reaction 3) or Pauson-Khand reaction can, arguably, be classified as an annulation by IUPAC definition. But, in this book, such reactions are classified as cycloadditions. Moreover, Diels-Alder reactions/Pauson-Khand reactions are very rarely catalyzed by a base. Thus, they

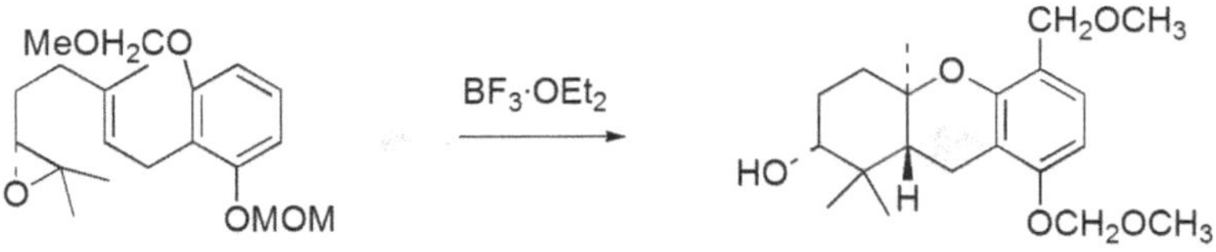

14 **13**

SCHEME 1.3 Wiemer domino reaction [2009 JACS 14630 Wiemer DF].

are not classified as anionic reactions. Annulation processes that seemingly take place in stepwise manner under basic conditions are covered in this book.

1.1.2 Definition and Classification of Anionic Annulations

Like cyclization processes, annulations can broadly be classified into five categories based on their mechanisms and the types of reactive intermediates in the initiating/triggering steps. These are (i) anionic annulations, (ii) cationic annulations, (iii) transition metal-catalyzed annulations, (iv) radical annulations, and (v) pericyclic annulations.

To clarify what is meant by "anionic annulations," we define them as the ones entailing a reactive anionic species in the initial steps of the mechanisms. In other words, anionic annulations are those annulations which are promoted by alkali/alkaline earth metal bases. The fragments or the synthon equivalents that generate the initial anionic species would be termed as donors, and those which undergo reaction with the donors are called acceptors. An annulation can further be classified as [4+2], [3+3], [3+2], [5+1], etc. based upon the total numbers of constituent atoms derived from a donor and an acceptor for the new ring. If the ring is a benzenoid, the process is termed as benzannulation. Likewise, an annulation can be termed "heteroannulation" if it involves formation of a heterocycle. The following annulation discovered by Njardarson et al. is illustrative of the above classifications.[14] As shown in Scheme 1.4, the reaction bromo ester (**15**) with sulfoxide imine (**16**) in the presence of excess LDA furnished pyrrolidine (**17**) in 63% yield. The bromo ester (**15**) provided three atoms and the imine (**16**) two atoms of the pyrrolidine heterocycle (**17**). Therefore, it is a [3+2] annulation. The reaction supposedly proceeds through the anionic intermediates (**18a**, **18b**). Therefore, the reaction in Scheme 1.4 is classified as an anionic [3+2] heteroannulation.

1.1.3 Why are the Anionic Annulations Chosen?

First reason: Among the five types of annulations, the anionic annulations are most useful in organic synthesis in line with the general trends with anionic reactions as a whole. It is pertinent to note that a vast majority of the industrial reactions are promoted by bases. According to Ripin et al.,[15] anionic reactions for the CC bond

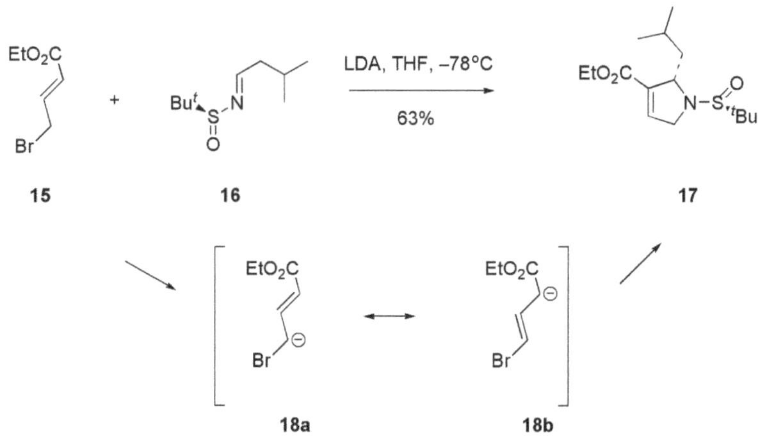

SCHEME 1.4 Njardarson [3+2] heteroannulation [2015 ACIE 13706 Njardarson JT].

formation were more than 65% in a survey for the period 1997–2002. The same review records that "The use of anionic chemistry and reductions remained about four times as frequent as the use of Lewis acidic chemistry and oxidations."

Similar remarks are echoed by Thebtaranonth and Thebtaranonth in their book[2a] as "Not withstanding the newly developed radical cyclizations discussed in the previous chapter, it would seem that the anionic cyclization is still, so to speak, the bread and butter of synthetic chemists in the business of ring construction. Development in this area is also fast taking place, particularly strong quarters to watch out for being tandem cyclization in the construction of polycyclic compounds and the *ortho*-metalation of arene rings and their subsequent reactions."

Second reason: Annulations are hardly covered in the literature.[3] They are minimally covered in Chapter 4 of the book by Thebtaranonth and Thebtaranonth.[2a] Twenty-two years have elapsed since that book was first published. Annulation chemistry has matured and fully grown over those years. It is being routinely used in the total syntheses of very complex molecules. Lately, this concept has been extended to transitional metal chemistry and heterocyclic chemistry. Therefore, it is felt that the science behind the anionic cyclocondensations should be reanalyzed. Emphasis will placed on the Hauser [4+2] annulation.[16] Chemoselectivity and regioselectivity have made annulation useful in organic synthesis. Moreover, the value of directed metalation/benzylic CH functionalization in the synthesis of complex aromatic and hydroaromatic systems will be highlighted, because it has provided the chemical community with immensely powerful tools for producing polycyclic aromatic compounds, in both academic laboratories and industrial plants. In general, this volume will, to a large extent, focus on the synthesis of carbocycles or polycyclic compounds. Azacycles and oxacycles are minimally covered in a latter chapter under the heading "heteroannulation."

Third reason: The Diels-Alder reaction, formally considered as an annulation, is treated as an [4+2] cycloaddition. According to a recent highlight[17] by Svenda et al., it is apparent that [4+2] anionic annulations can now complement the most successful cycloaddition, i.e., Diels-Alder (DA) reactions. A case in point is the Bach total synthesis[18] of fredericamycin in 1994, where the DA reaction failed, but the anionic annulations succeeded.

Fourth reason: The concept of convergency is vital for making a multistep synthesis more efficient. It is more so when it is brought into play at a later stage of a synthesis. In a recent review[17] on such a topic by Svenda et al., the importance of convergency was reiterated and emphasized. A glimpse at the review would make it apparent that the anionic annulations are endowed with features of convergency.

Fifth reason: Anionic annulations are hardly practiced as a step in the retrosynthesis of a target molecule. Retrosynthesis is a pathway of planning a multistep synthesis by sequential hypothetical fragmentations of the target molecules to identify the starting materials. Since the random fragmentations of targets by cleavage of one bond at a time can lead to enormous possibilities, a number of guidelines are prescribed in the textbooks[3] to narrow down the possibilities. For the cyclic or polycyclic targets, retrosyntheses based on cycloadditions like Diels-Alder reaction and [2+2] cycloadditions are frequently employed. On the contrary, anionic annulations are not considered. However, analyses based on two-bond cleavages in a retrosynthesis are quite prevalent in heterocyclic chemistry. The presence of a heteroatom in the targets dictates two bond cleavages at a time. The famous heteroannulations, namely the Paal-Knorr reaction (Scheme 1.5)[19] and Krohnke reaction (Scheme 1.6)[20] are presented here as illustrations. Two CN bonds of compound (**19**) are formed in single step from the intermediate (**20**). Likewise, pyridine derivative (**21**) is synthesized from 1,5-diketo intermediate (**22**) by simultaneous formation of two CN bonds.

Similarly, other well-known reactions, including Fischer indole synthesis, Skraup synthesis, and Pictet-Spengler synthesis, are categorized into cationic annulations and utilized in heterocyclic chemistry. Among the 20 most powerful transforms listed by Smith, Fischer indole synthesis and Knorr pyrrole synthesis are identified as two heteroannulations. The only anionic annulation that has figured in the list is Robinson annulation.[3d]

Nevertheless, a complete coverage of anionic annulation in general with finer details is warranted. This book is intended to cover major anionic annulations

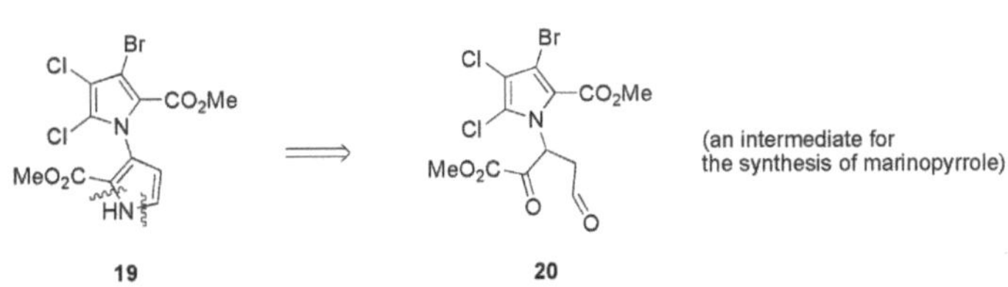

SCHEME 1.5 Classical [4+1] heteroannulation: Paal-Knorr synthesis [2013 CCS 558 Clive DLJ].

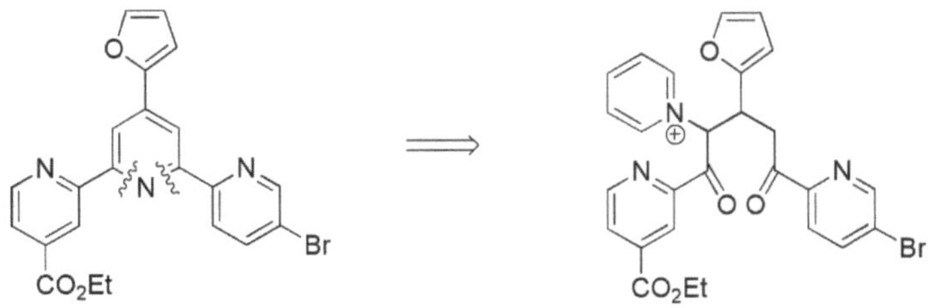

21 22

SCHEME 1.6 Krohnke [5 + 1] heteroannulation [2016 SYS 1974 Sasaki I].

leading to carbocycles and heterocycles. In particular, it will sketch the development of [4+2] annulations and so on in organic synthesis during the past 50 years. The underlying motivation is to promote the annulations in the synthesis of natural products, the relevance of which prominently persists even today. Such endeavors of the chemical synthesis have led to the discovery of truly new reactions as well as meeting new challenges of chemical synthesis. The annulations that build up new aromatic rings are termed as benzannulations. In this book, one such annulation, i.e. the Hauser annulation, will be presented in details as a powerful ring-forming anionic reaction which, in certain instances, rivals the famous Diels-Alder reactions.

1.2 ROBINSON [4+2] ANNULATION

Robinson annulation is one of the most famous ring-forming reactions. It is, therfeore, imperative that a prelude to this annulation is included herein. Although it is introduced in the syllabus of undergraduate chemistry, the present chapter is intended to pave the way to learning and mastering the concepts of the anionic annulations in organic synthesis. Robinson annulation was discovered by Sir Robert Robinson in 1935. Since then, it has been a cornerstone in the field of organic synthesis. The annulation was developed as a way to synthesize sterols and their derivatives. Total synthesis of many intriguing natural products has used this annulation as a key step.

Scheme 1.7 presents an example of Robinson annulation. By convention, it should be noted as anionic [4+2] annulation. The reaction of keto ester (**23**) with methyl vinyl ketone (**1**) in the presence of potassium *tert*-butoxide in refluxing ethanol gives decalenone (**24**).[21] Of the six carbon carbons of the new ring, i.e., A ring of (**24**), two carbons are derived from the keto ester (**23**) and four from the acceptor (**1**). Formally, (**23**) is considered a 1,2-dipole (**25a**) equivalent and (**1**) a 1,4-dipole (**25b**).

The process comprises four tandem steps (Scheme 1.8): (i) enolate formation (**23–26**), (ii) Michael addition of the enolate nucleophile to α,β-unsaturated ketone (**26–1**), (iii) aldol cyclization (**27–28**), and (iv) dehydration of (**29**) to

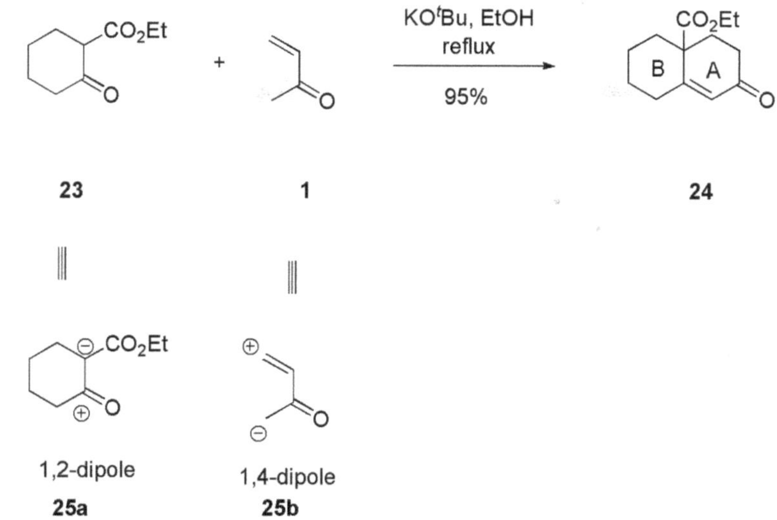

SCHEME 1.7 An example of Robinson annulation [1996 *JCSPT1* 1523 Ley SV].

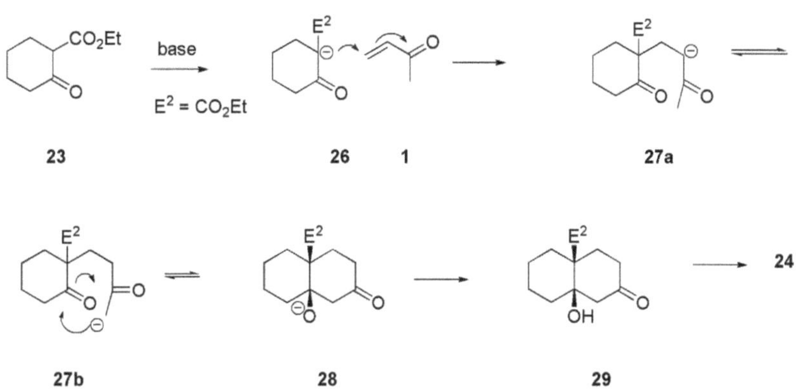

SCHEME 1.8 Mechanism of the formation of (**24**) by Robinson annulation.

afford the decalenone (**24**). These reactions are usually carried out under basic conditions. The very first step , i.e. selective deprotonation of **23** forming intermediate **26** is important for the success of the annulation.

Although the annulation primarily provides nonaromatic six-membered carbocycles, it can be modified to prepare aromatics such as phenolics. When sulfoxide-activated cyclohexanones are submitted to Robinson annulation, a cascade of reactions ensues to produce tetrahydro-β-naphthols in one step. For example, the reaction of sulfoxide (**30**) with vinyl methyl ketone (**31**) in the presence of NaOMe afforded tetrahydronaphthol (**32**) in 43% yield (Scheme 1.9). This reaction is quite general for the preparation of substituted β-naphthols.

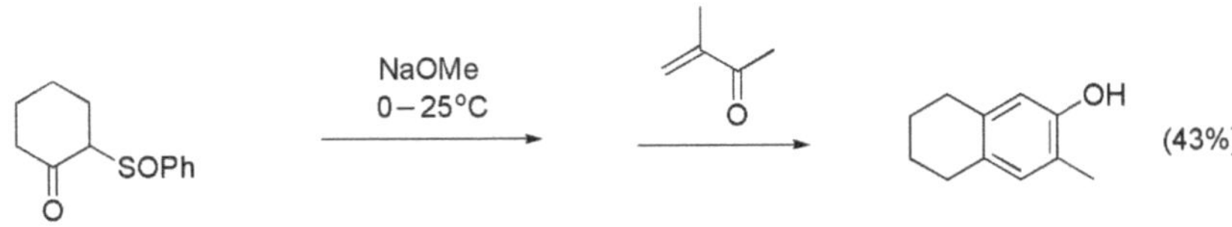

30 31 32

SCHEME 1.9 Boger phenol synthesis [1980 JOC 5002 Boger DL].

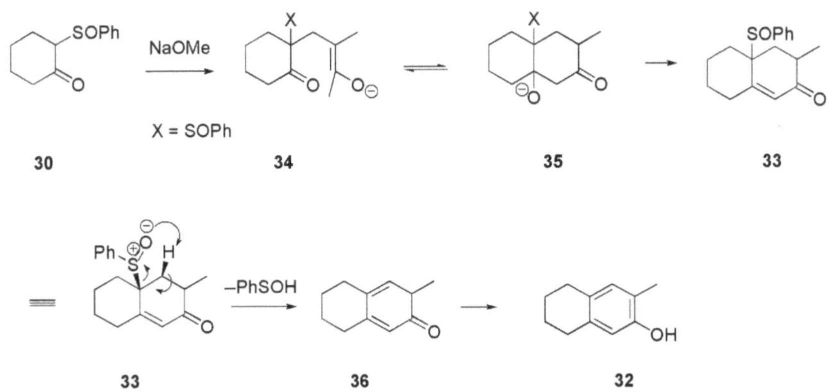

SCHEME 1.10 Representative mechanism for Boger phenol synthesis.

The reaction follows the same mechanism as one in Scheme 1.8 to produce (**33**) via inter mediates (**34** and **35**). Bicyclic sulfoxide (**33**) undergoes pericyclic thermal elimination of phenylsulfenic acid to give (**36**). Tautomerization of (**36**) leads to the formation of phenol (**32**) (Scheme 1.10).

A similar phenol synthesis has been developed by Takaki et al. This approach has utilized the phenylthio group as a nucleofuge.[22]

Asymmetric Robinson annulation: Robinson annulation is largely known for its applications in the synthesis of decalone/octalone building blocks in racemic forms. In recent times, it has been used for accessing Hajos-Parrish (H-P) ketones (**37**) and Wieland-Miescher (W-M) ketones (**38**) in enantiomeric forms, which are useful for constructing natural products containing 6–5 fused carbocycles and 6–6 fused carbocycles, respectively. They have been the key starting building blocks for total synthesis of complex target molecules. For example, Dethe et al. have achieved the total synthesis of (−)-mycoleptodiscin A, a hybrid of an indole and a terpenoid from the W-M ketone.[23]

Historically, L-proline-catalyzed Robison annulation has long been known in the literature. It is probably one of the early examples of organocatalyzed reactions. It can be run in two modes. The 1,3-diketones (**39** and **40**) can be converted to the bicyclic enones (**37** and **38**) in one step or two steps via the initial Michael adducts (**41** and **42**), and the initial annulation products (**43** and **44**) (Scheme 1.11).[24,25]

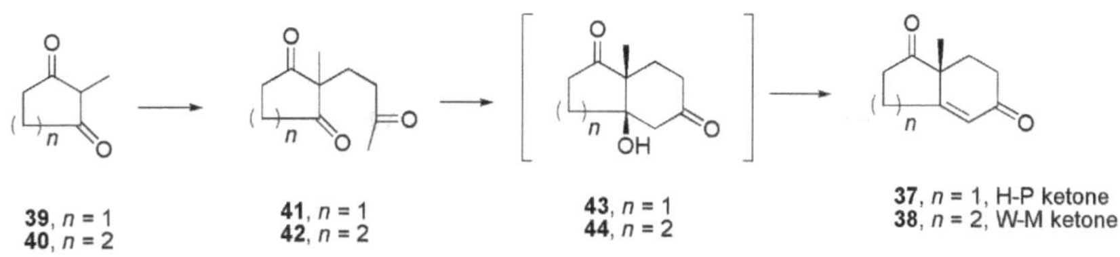

SCHEME 1.11 Asymmetric Robinson annulation.

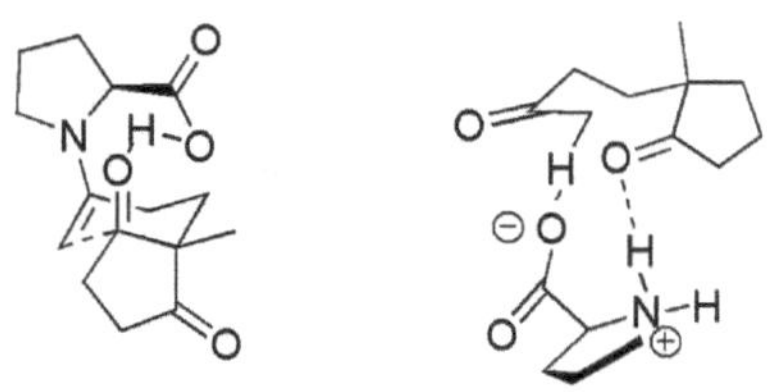

45 46

FIG. 1.1 Transition state models for asymmetric Robinson annulation.

As to the mechanism (Fig. 1.1), it is believed that the asymmetric induction takes place in the second step, i.e., conversion of the Michael adducts (**41** and **42**) to the initial annulation product (**43** and **44**). There are four models for the transition state of the step. Presented below is the structure (**45**) proposed by Houk et al.[26] However, the Swaminathan model did not involve any enamine. It involves zwitter ionic structure (**46**) as evident from the NMR analysis.[27]

Asymmetric synthesis of W-M ketone (**38**) was realized by D-proline catalyzed Robinson annulation of cyclohexane-1,3-dione (**40**) and methyl vinyl ketone (**1**) in two steps: Michael addition and intramolecular aldol reaction of (**42**) (Scheme 1.12).[28]

Hagiwara group has demonstrated that the second step, i.e., the asymmetric cyclization step (**42**→**38**) of Robinson annulation can be effected by optically active phenylalanine to provide either enantiomer of the product (**38**) in high optical and chemical yields. The reaction of (**40**) with (**47**) gives Michael adduct (**48**) which, upon reaction with L-*β*-phenylalanine and D-camphorsulfonic acid in DMF, gives decalenone (**49**). The use of D-*β*-phenylalanine in the second step furnishes the other optical antipode (Scheme 1.13).[29]

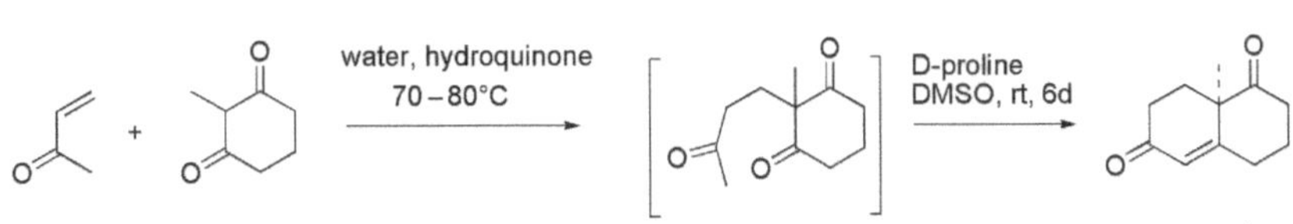

1 40 42 38

SCHEME 1.12 Synthesis of Wieland-Miescher ketone by asymmetric Robinson annulation [1990 SYS 53, Harada N].

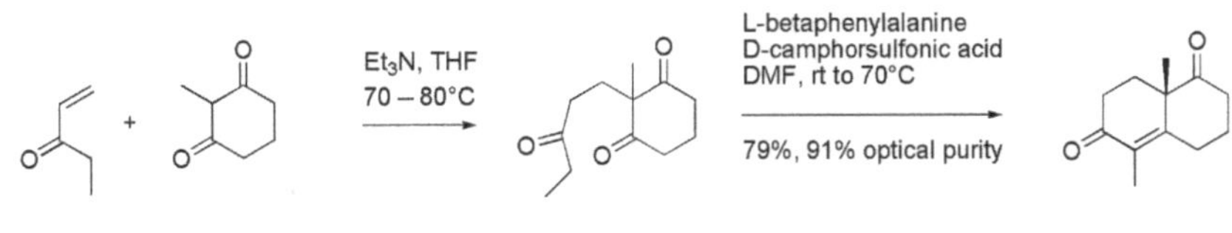

47 **40** **48** **49**

SCHEME 1.13 Two-step asymmetric Robinson annulation [1988 JOC 2308 Hagiwara H].

The synthesis of magellanine type of lycopodium alkaloids necessitated Robinson annulation of a pyridylmethyl substituted cyclopentane-1,3-dione. The reaction of diketone (**50**) with methyl vinyl ketone (**1**) in the presence of L-proline followed by acid catalyzed dehydration afforded the annulation product (**51**) in 50% yield (Scheme 1.14). This study revealed that the higher basicity of the pyridine nucleus impedes the yield of the annulation.[30]

Although both pyrolidine nucleus and the carboxylic acid group are necessary for an asymmetric Robinson annulation, Luo et al. showed that amino acid-derived chiral primary amines in conjunction with *m*-nitrobenzoic acid are equally effective for Robinson annulation (Scheme 1.15). Both H-J and W-M ketones are obtained in good yield from triketone (**41**) and (**42**), respectively.[31]

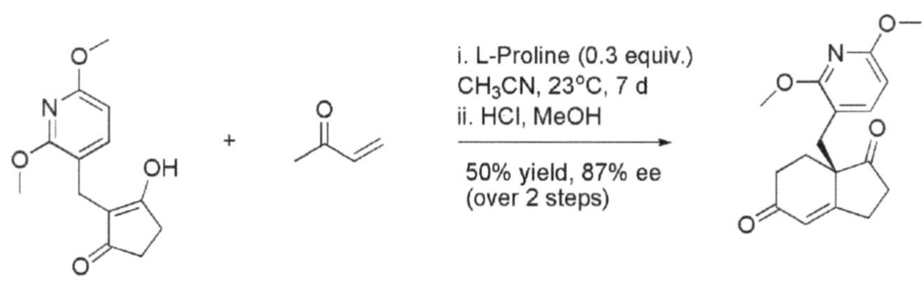

50 **1** **51**

SCHEME 1.14 Robinson annulation to lycopodium alkaloids [2012 ORS 632 Sarpong R].

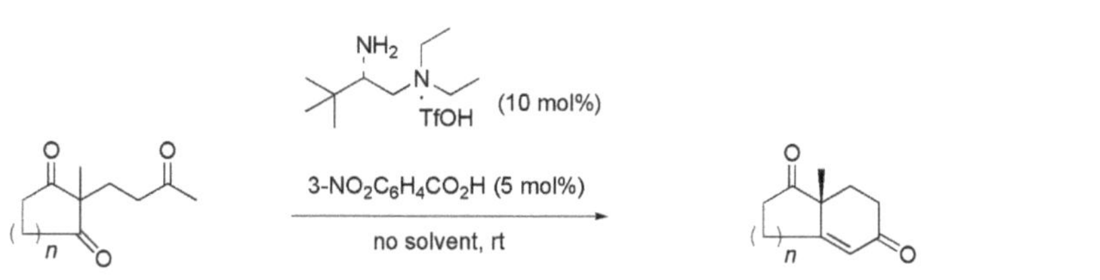

41, *n* = 1
42, *n* = 2

37, *n* = 1, H-P ketone, 96% yield, 96% ee
38, *n* = 2, W-M ketone, 98% yield, 91% ee

SCHEME 1.15 Amine catalyzed asymmetric Robinson annulation [2012 JOC 2526 Luo S].

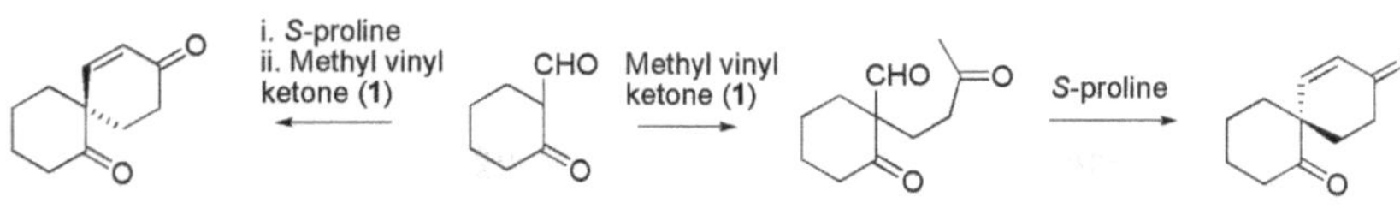

52 54 55 53

SCHEME 1.16 Robinson annulation for spirocycles [1990 IJCY 401 Swaminathan S].

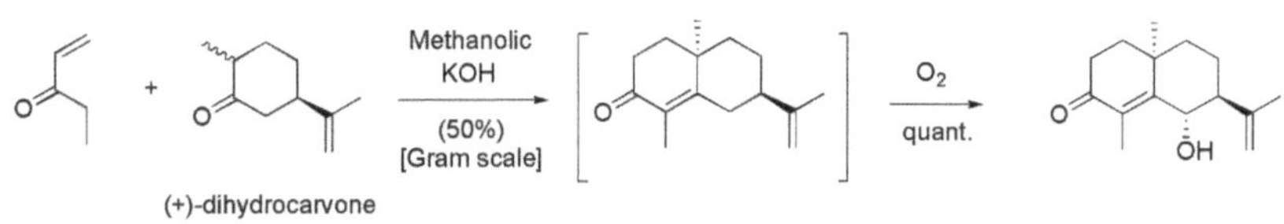

47 56 57 58

SCHEME 1.17 Chiral pool approach applied to Robinson annulation [2001 THN 2171 Aladro FJ].

Robinson annulation is also used for synthesizing spirocyclic compounds. Quite interestingly, both the enantiomers (**52** and **53**) can be obtained from the same starting material (**54**), using the same chiral catalyst, S-proline (Scheme 1.16), by changing the sequence of addition of reagents and catalysts.[32] For the preparation of ketone (**52**), proline followed by methyl vinyl ketone is used. For (**53**), the sequence is inverse, and the Michael adduct (**55**) is first isolated.

Optically pure carvone derivatives have been the subject of Robinson annulation to prepare enantiopure W-M ketones. The reaction of (**56**) with (**47**) in the presence of KOH gave annulation product (**57**), which upon exposure to oxygen furnished optically active (**58**) in a gram-scale (Scheme 1.17).[33]

1.2.1 Applications in the Total Synthesis

The merits of the Robinson annulation lie in its applicability in the total synthesis of complex molecules, and it continues to be the strategy of choice for the synthesis of naturally occurring steroids, alkaloids, and terpenoids. In the following paragraphs, synthesis of natural products, namely, cortisone (**59**)[34] and morphine (**60**),[35] is presented.

Cortisone synthesis: A landmark moment in steroid research was the announcement in 1949 by the Mayo Clinic that cortisone (**59**) had a dramatic effect on patients suffering from rheumatoid arthritis and other inflammatory diseases. In the early 1950s, both Merck (US) and Ciba (Swiss) made cortisone available through synthesis. After Sir Robert Robinson established his annulation reaction, cortisone (**59**) became a relevant target. In Scheme 1.18, Woodward total synthesis of cortisone (**59**) is highlighted. A key step of the synthesis is Robinson annulation. Benzoquinone (**61**) was reacted with butadiene (**6**) in a sealed tube to give the Diels-Alder adduct (**62**), which was transformed to decalenone (**63**) via two reductions. Anionic formylation of (**63**) furnished formyl derivative (**64**). Robinson annulation of (**65**) followed by deformylation

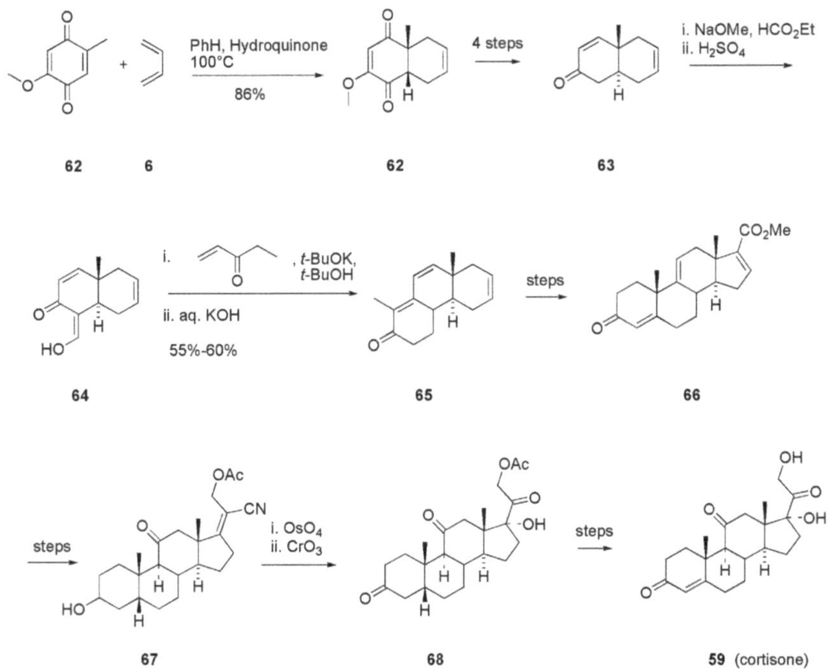

SCHEME 1.18 Highlights of the synthesis of cortisone (**59**).

afforded tricyclic intermediate (**65**). Further elaboration of (**65**) through (**66**), (**67**), and (**68**) provided cortisone (**59**).

Synthesis of morphine (60): Morphine (**60**), the first alkaloid isolated from the opium poppy, is well known for its use in medication for pain relief. Over the years, extensive synthetic studies of (**60**) and related natural products have been made, resulting in more than 30 total syntheses highlighting various CC bond-forming reactions. The synthetic prowess of Robinson annulation was also established through the synthesis of morphine (**60**). It is found in the resin on *Papaver somniferum* (opium poppy). From a chemistry perspective, morphine consists of a benzylisoquinoline alkaloid backbone with two additional rings. Mulzer synthesis had recourse to tetralone (**69**) and (**70**). Robinson annulation of (**71**) with methyl vinyl ketone (**1**) provided angular tricyclic intermediate (**74**), which was converted into intermediate (**72**) by an alkaline deformylation. Further transformations with more than 10 steps yielded morphine (**60**) (Scheme 1.19).

Pallambins (**73** and **74**) are natural products with extraordinary structural complexity and no significant bioactivity. Baran synthesis of pallambins strategically relied upon Robinson annulation.[36] The key intermediate (**75**) was prepared by Robinson annulation of (76) with (**47**). Furfuryl alcohol (**77**) was submitted to Eschenmoser-Claisen rearrangement to produce amide (**78**), which was reduced with TMDS (tetramethyldisiloxane) to give aldehyde (**76**). Robinson annulation

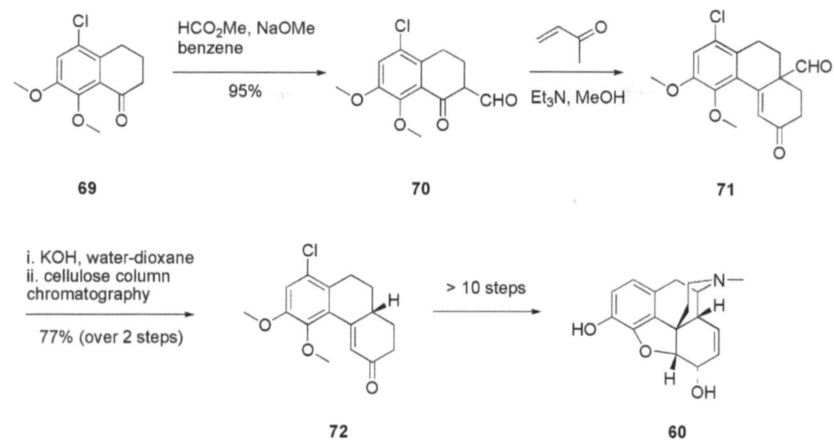

SCHEME 1.19 Synthesis of morphine (**60**) [1997 ACIE 2830 Mulzer J].

was then applied to aldehyde (**76**) and ethyl vinyl ketone (**47**) to furnish furan substituted cyclohexanone (**78**). It was next elaborated to pallambins in nine steps including a furan photooxygenation step (Scheme 1.20).

Limonin (**79**) is a flagship member of several hundred limonoids. It has attracted immense interest from the synthesis community. In a recent yet first total synthesis of (±)-limonin, Robinson annulation formed the key step (Scheme 1.21).[37] Snider's manganese(III)-mediated tandem cyclization applied to the acyclic precursor (**80**) furnished the tricyclic intermediate (**81**) with C13α configuration as the major product via radical (**82**). The necessary carbanion for Robinson annulation was generated by reductive removal of chlorine in (**81**), and it was allowed to undergo Robinson annulation with methyl vinyl ketone (**24**) in the presence of potassium *tert*-butoxide (cat.) to

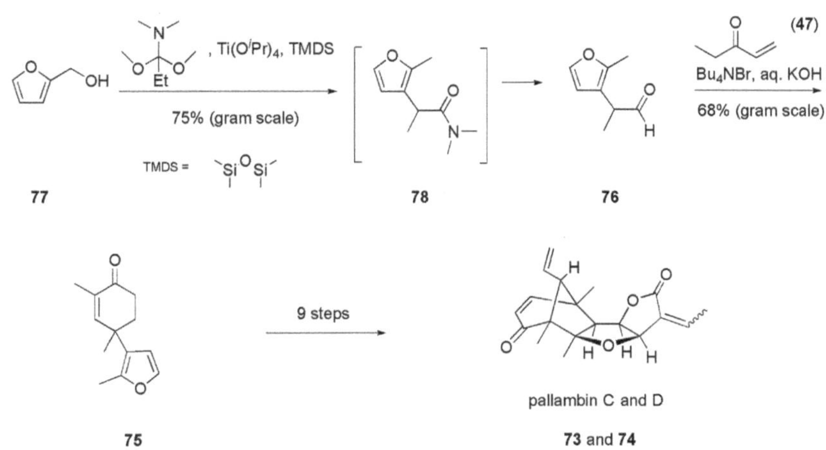

SCHEME 1.20 Baran synthesis of pallambins [2016 JACS 7536 Baran PS].

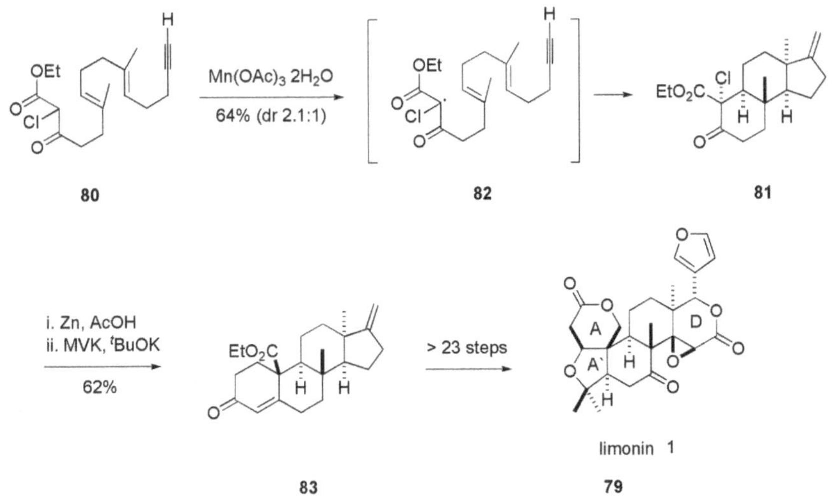

SCHEME 1.21 Total synthesis of (±)-limonin [2015 ACIE 8538 Yamashita S].

furnish tetracyclic system (**83**), which was carried over to complete the total synthesis of (±)-limonin (**79**) in a total of 35 steps from geraniol.

Intramolecular Robinson annulation, where two rings are formed in one pot, is not frequently discussed in the literature. Watanabe synthesis of (−)- tanabalin (**84**) is a classic example of such a process (Scheme 1.22).[38] Two acyclic fragments (**85** and **86**) were reacted in the presence of NaOMe to afford decalenone (**87**) as a single isomer in 82% yield. This tandem reaction is proposed to take place via intermediates (**88** and **89**) in a sequence of intermolecular alkylation followed by intramolecular Robinson annulation. The decalin intermediate (**87**) was elaborated to tanabalin (**84**) in nine more steps.

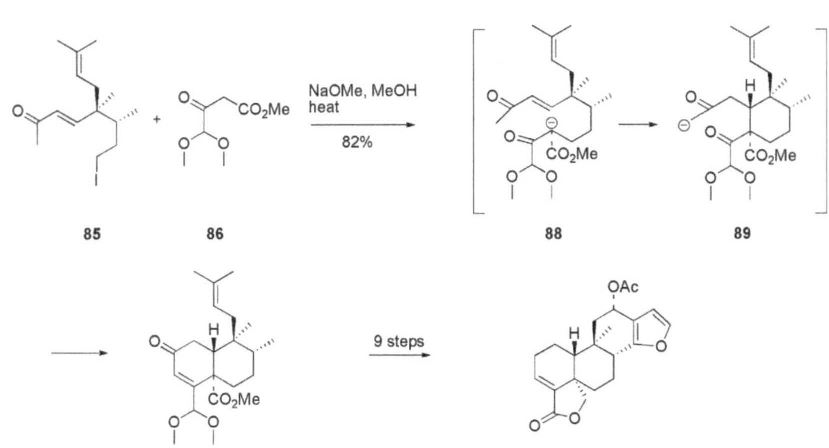

SCHEME 1.22 Intramolecular Robinson annulation for decalenone synthesis.

Intramolecular Robinson annulation culminated successfully in a formal total synthesis of platensimycin (**90**) (Scheme 1.23).[39] It was discovered by scientists of Merck. Its cage-like tetracyclic core with several stereogenic centers was appealing as a target for chemical synthesis. Presented below is the synthesis by Yamamoto et al. The bicyclic ketone (**91**), available from Bronsted acid-assisted chiral Lewis acid-catalyzed enantioselective Diels-Alder reaction between methylcyclopentadiene and methyl acrylate, was the key starting material. Vinyl cuprate reagent prepared in situ, when reacted with lactone (**92**), gave a carboxylic acid, which was treated with a catalytic amount of trifluoromethanesulfonimide to give vinyl lactone (**93**) as a diastereomeric mixture. Cyanide (**94**) was prepared by DIBAL-H reduction of (**93**), followed by Lewis acid-mediated cyanation. It was converted to aldehyde (**95**) by DIBAL-H/^{n}BuLi reduction. The Wadsworth-Emmons reaction of (**95**) delivered enone (**96**). Ruthenium-catalyzed oxidative degradation of (**96**) gave the precursor (**97**) for Robinson annulation. The late-stage annulation was performed in one pot using L-proline and sodium hydroxide to furnish tetracyclic core structure (**98**) as the major product and (**99**) as the minor product.

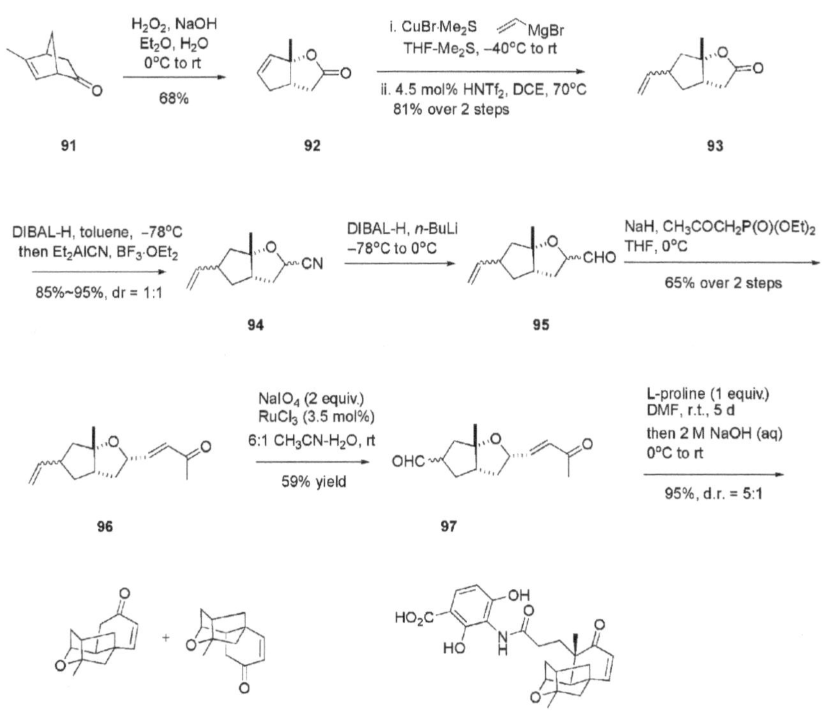

SCHEME 1.23 Total synthesis of platensimycin via intramolecular Robinson annulation [2007 JACS 9534 Yamamoto H].

Atypical Robinson annulation: Since Robinson annulation is considered a formal [4+2] condensation reaction, the new rings formed by the reaction are always a six-membered ring. Over the years, Robinson annulation has been broadened to encompass [3+3] annulations. In general, all ring-forming cascades involving the sequence of Michael and intramolecular aldol reaction are classified as Robinson annulation. The one-pot synthesis of cyclohexanone (**100**) from ethyl acetoacetate (**101**) and crotonaldehyde (**102**) is one such example (Scheme 1.24).[40] In the presence of potassium carbonate and benzyl triethyl ammonium chloride in toluene at r.t., the annulation between (**101**) and (**102**) gave cyclohexanone (**100**). The reaction proceeds through Michael addition of enolate (**103**) to crotonaldehyde (**102**) to form adduct (**104**), which remains in equilibrium with its proton exchanged form (**105**). Under the basic conditions, it undergoes aldol condensation to give (**106**), which upon epimerization gives product (**100**).

However, the key concept, i.e., Michael addition followed by aldol condensation, can be innovated to construct a five-membered ring in two steps through a formal [3+2] condensation mode. As shown in Scheme 1.25, symmetrical diketodiester (**107**) underwent Michael addition with methyl vinyl ketone (**1**) to give adduct (**108**). Pyrrolidine catalyzed intramolecular aldol reaction of (**108**) furnished 5-6-5 tricyclic compound (**109**) in 25% yield.[41a]

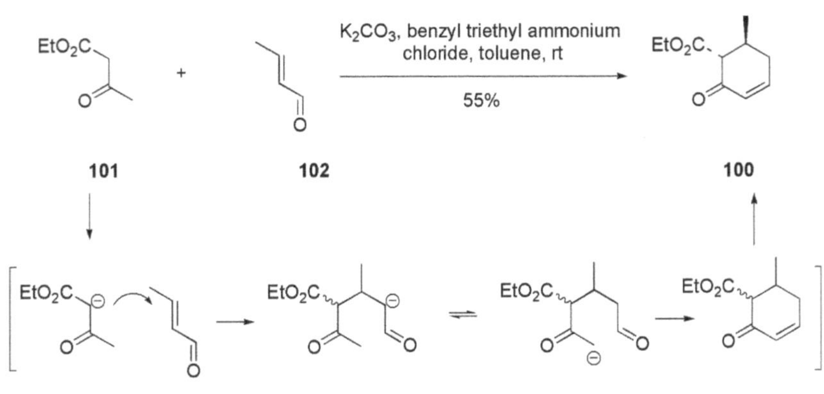

SCHEME 1.24 Robinson [3+3] annulation [1988 JCE 637 Soriano DS].

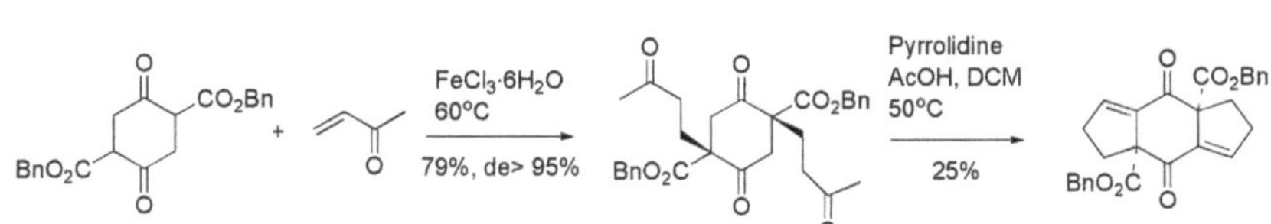

SCHEME 1.25 Robinson annulation for five-membered carbocycles [2006 SYT 624 Christoffers J].

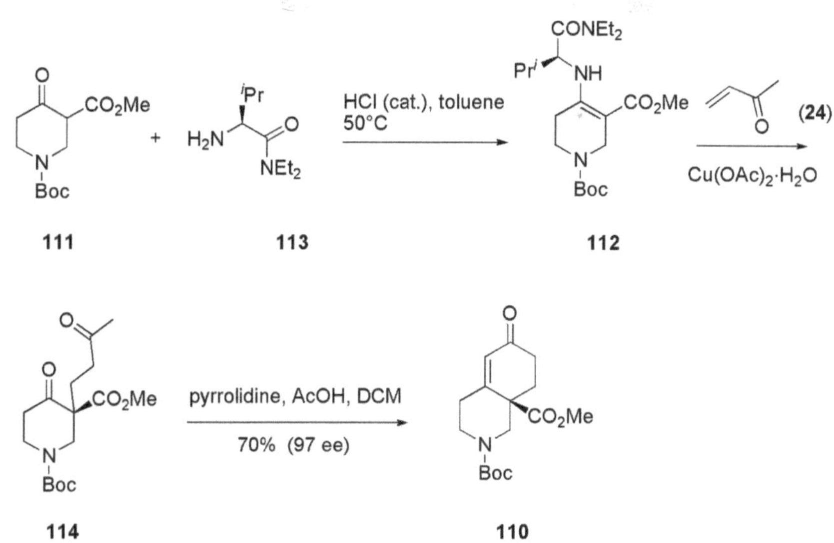

SCHEME 1.26 Asymmetric synthesis of an isoquinolone by Robinson-type annulation.

In conclusion, innovations in Robinson annulation and allied reactions are manifold. They are expected to develop continually in newer directions. For example, the asymmetric synthesis of octahydroisoquinolone (**110**) from keto ester (**111**) was achieved in two steps via enamine (**112**). The enamine (**112**) was prepared by the reaction of keto ester (**110**) with L-valine diethyl amide (**113**) in the presence of HCl (cat.). Treatment of the enamine (**112**) with methyl vinyl ketone (**1**) and a catalytic amount of $Cu(OAc)_2$ gave optically pure 1,5-diketo product (**114**). Intramolecular cyclization of (**114**) with pyrrolidine in acetic acid afforded hexahydroisoquinolone (**110**) with 97% ee (Scheme 1.26).[41b]

1.3 DARZENS [2+1] CONDENSATION

Darzens condensation, a [2+1] annulation, also known as Darzens glycidic ester condensation and discovered in 1911, is one of the oldest annulations. It employs α-halogenated carbonyl compounds and aldehydes or ketones to form directly α,β-epoxy carbonyl compounds. It can be classified as a hetero-annulation, because a new heterocyclic ring is formed in the process. Due to its general utility in organic synthesis and its impact in the discovery of other annulations, namely McCoy cyclopropanation, allowing formation of three-membered rings, its coverage in an early chapter was deemed prudent.

A general Darzens condensation is presented in Scheme 1.27.[42] It is a non-oxidative method for the preparation of epoxides (**115**) from donors (**116**) and acceptors (**117**). The donors (**116**) usually have an electron withdrawing group and a nucleofuge. The electron withdrawing groups are carboxylic acid derivatives, such as esters, thiol esters, amides and nitriles, and ketones. Schiff's bases,

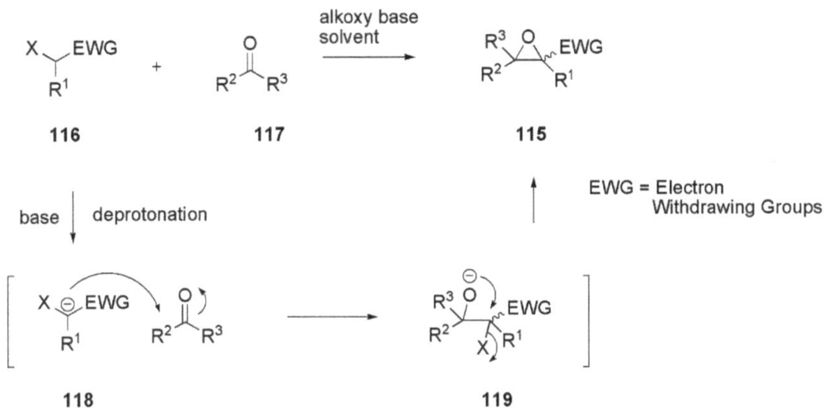

SCHEME 1.27 General scheme for Darzens condensations.

sulfones, sulfoxides, and sulfoximines have also been used. A chlorine atom is commonly used as a nucleofuge. In certain cases, bromine and iodine have been used as nucleofuges. The acceptors (**117**) are aldehydes and ketones, including α,β-unsaturated ones. However, aliphatic aldehydes are not considered good acceptors. The reagents for effecting the condensation are a base in an anhydrous aprotic nonpolar solvent. The bases include NaH, LDA, KO^{t}Bu, NaOEt, etc. In some cases, the reactions are conducted under phase transfer conditions with crown ethers and quaternary salts.

The mechanism of the reaction is formulated on the initial formation of the carbanions (**118**) by α-deprotonation of halogenated donors (**116**). Like an aldol reaction, the carbanions add to the carbonyl group of (**117**). The resulting oxy anions (**119**) then undergo intramolecular substitution to form the epoxides **113**. The stereochemistry of the products is often the consequence of steric effects in the transition states.

The applications of the Darzens reaction are not limited to the synthesis of epoxides. Traditionally, the glycidic esters obtained from the Darzens reactions are used for one-carbon homologation of aldehydes and ketones. For example, a robust process for the homologation has been prescribed by Gallou et al. As shown in Scheme 1.28, *p*-substituted aldehydes (**120**), on reaction with methyl chloropropionate in the presence of NaOMe in toluene at 0°C followed by hydrolysis (aq NaOH in toluene) and decarboxylative rearrangement (aq HCl in toluene, give ketones (**121**) in good yields.[43] Epoxides (**122**), formed as intermediates, undergo decarboxylative rearrangement via (**123**) and (**124**). For substrates with electron withdrawing groups, the yields of the products are not satisfactory.

Watt et al. developed a two-step homologation of ketones to α-hydroxy aldehydes based upon Darzens condensations (Scheme 1.29).[44] Darzens reaction of ketones (**125**) with chloroemethyl phenyl sulfone (**126**) affords α,β-epoxy sulfones (**127**). Treatment of the epoxides (**127**) with hydroxide generated

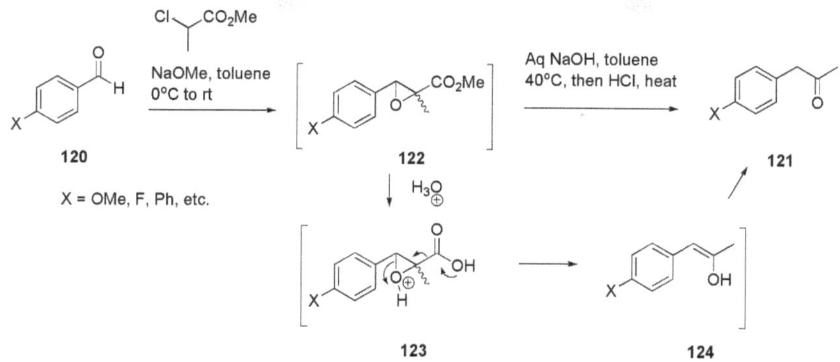

SCHEME 1.28 Darzens condensation for homologation of aldehydes [2011 SYS 1221 Gallou F].

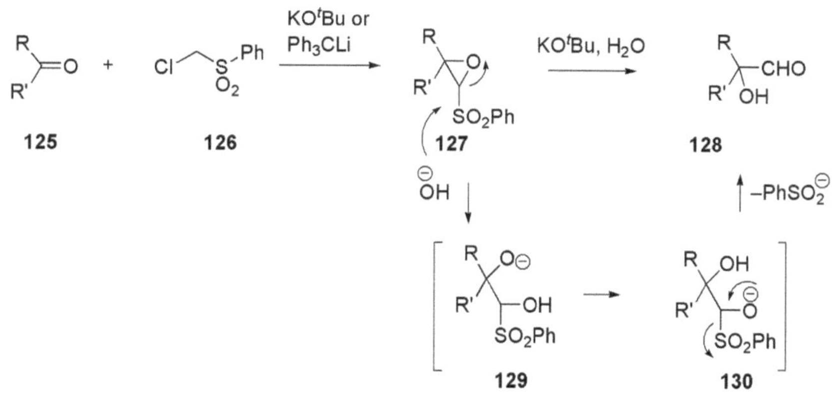

SCHEME 1.29 Watt homologation [1984 JOC 1378W DS].

by adding water to potassium *tert*-butoxide (Gassman's procedure) affords hydroxyl aldehydes (**128**) in good yields with high stereoselectivity. The rearrangement of the epoxides (**127**) to hydroxy aldehydes (**128**) proceeds via epoxides-opening to (**129**) and (**130**), which collapse to aldehydes (**128**) with the loss of benzenesulfinate.

Darzens condensation is also chemoselective. For anthraquinone carboxaldehyde (**129**), the Darzens reaction with methyl chloroacetate takes place selectively at its aldehyde function to give anthraquinone (**132**) (Scheme 1.30).[45]

The Darzens condensation has been used as one of the key steps in the total synthesis of berkeleyamide D (**133**), which is accomplished in the presence of Et_3N, a mild base (Scheme 1.31).[46] The reaction between (**134**) and (**135**) under the conditions afforded exclusively one diastereomer, i.e., (**136**) in excellent yield. The selectivity is explained by steric interaction between dioxolane-containing side chain and isovaleryl group in the respective transition state. Treatment of the epoxide (**136**) with aqueous $HClO_4$ in DCM afforded berkeleyamide D (**133**) and pyridone derivative (**137**).

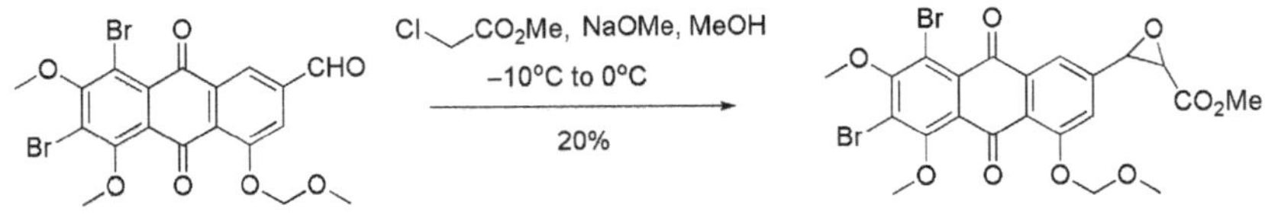

SCHEME 1.30 Selective Darzens condensation [2016 BJOC 531 Mal D].

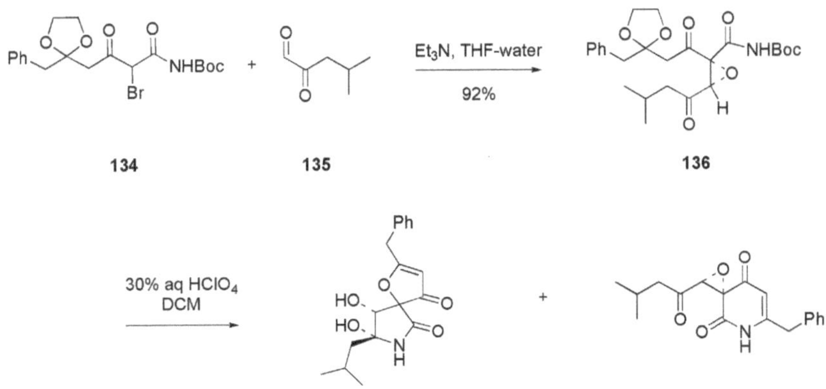

SCHEME 1.31 Darzens condensation in the total synthesis of berkeleyamide D [2016 THN 6640 Kuramochi K].

1.3.1 Vinylogous Darzens Condensation

When Darzens condensation was applied to 4-bromocrotontane (**138**) and piperonal (**139**), it resulted in vinylogous condensation product (**140**) (Scheme 1.32). The condensation is proposed to take place via intermediate (**141**). Alkenyl epoxide-dihydrofuran rearrangement followed by a Lewis acid-mediated reaction with dimethyl acetal of piperonal furnished (+/−) epiasarinin (**142**).[47]

Similarly, the condensation of α-bromocrotonates (**143**) with aldehydes (**144**) gives vinyl substituted epoxides (**145**) via α-carbanionic species (**146** and **147**) (Scheme 1.33).[48]

The asymmetric version of Darzens condensation is moderately successful. In their early work, Shioiri et al. showed that enantioselective synthesis of α,β-epoxy sulfones (**148**) can be achieved by the reaction of chloromethyl phenyl sulfone (**126**) with various aromatic aldehydes (**149**) in the presence of chiral quaternary ammonium salts derived from quinine with KOH at room temperature with ee up to 81%.[49] Jeong et al. developed an improved procedure (Scheme 1.34).[50]

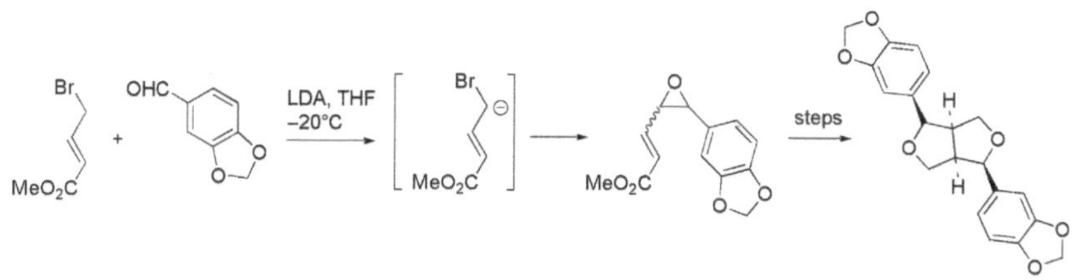

138 139 141 140 142

SCHEME 1.32 Vinylogous Darzens condensation of γ-bromocrotonate [2002 ORS 1159 Steel PG].

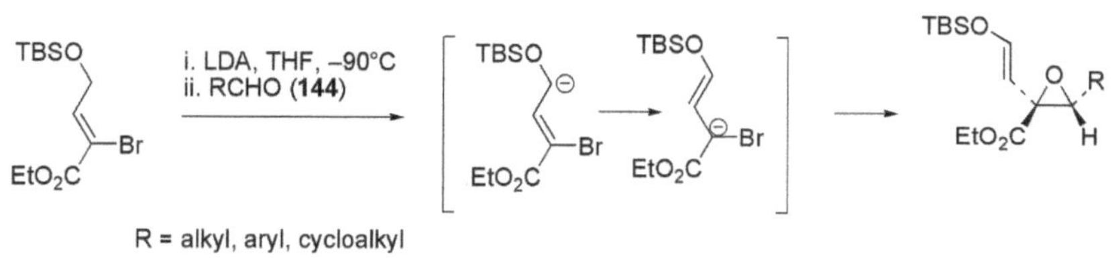

143 146 147 145

SCHEME 1.33 Vinylogous Darzens condensation of α-bromocrotonate [1991 JOC 4598 Hudlicky T and 2015 ORS 5998 Rizzacasa MA].

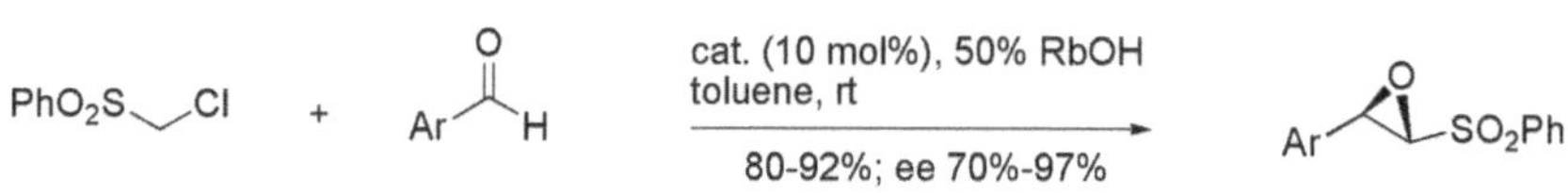

126 149 148

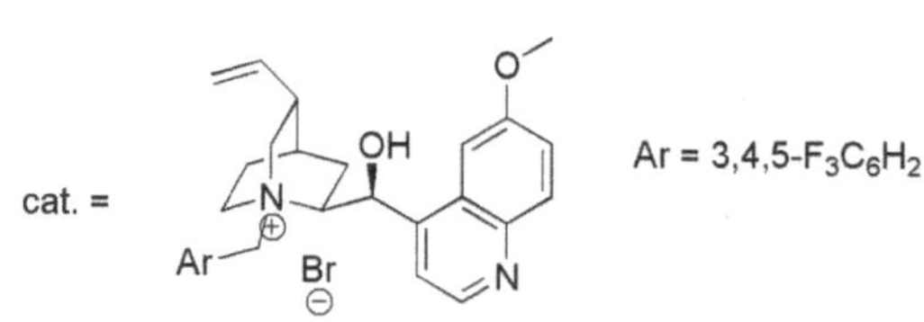

X

SCHEME 1.34 Asymmetric Darzens reaction of chloromethyl phenyl sulfone [2007 THN 8099 Jeong B-S].

In a recent publication, Deng et al. demonstrated the similar effect of cupreinium salts for the enantioselective Darzens reaction of aliphatic and aromatic aldehydes (Scheme 1.35).[51] Chloroketones (**150**) undergo condensation with aryl aldehydes (**151**) in the presence of LiOH and a catalyst give epoxides (**152**). The process is also useful for the synthesis of trisubstituted epoxides. In general, ee is in the range of 90%–99%.

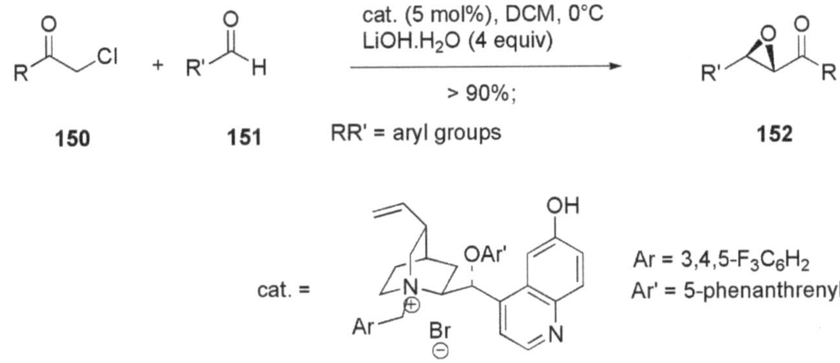

SCHEME 1.35 Deng asymmetric Darzens reaction [2011 CSE 1301 Deng L].

Jorgensen et al. found that Michael addition can be combined with Darzens condensation to give functionalized complex epoxycyclohexanone derivatives (**153**).[52] With the application of Hayashi-Jorgensen catalyst, this process leads to creation of four chiral centers of the cychexanones with high diastereo- and enantioselectivities.

As shown in Scheme 1.36, the tandem reaction of (**154**) with (**155**) in the presence of catalyst (**156**) and K_2CO_3 gives cyclohexane epoxides (**153**).

The mechanism of the Jorgensen reaction is given in Scheme 1.37. Initially the catalyst (**156**) reacts with the aldehydes (**155**) to form iminium salts (**157**). The ketoesters (**154**) approach the planar iminium ion from the *Re*-face due to the steric hindrance exerted by the bulky groups of the pyrrolidine ring of catalyst (**156**). Hydrolysis of the enamine (**158**) furnishes the Michael adducts (**159**) via (**160**). Sodium acetate-promoted aldol reaction of (**159**) gives (**161**), which undergo Darzens reaction under the influence of K_2CO_3, leading to cyclohexanone epoxides (**153**).

Aza-Darzens annulation: The Darzens condensation has witnessed many modifications. Among them, the most useful variant is the aza-Darzens reaction forming aziridine heterocycles.[53] It has been used to synthesize a diverse range of substituted aziridines. The process involves one-step reaction of aldimines/ketimines (**162**) with precursors (**163**) of stabilized carbanions with α-leaving

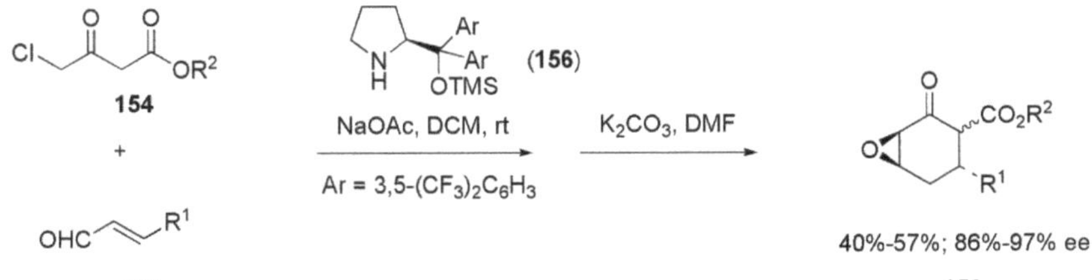

SCHEME 1.36 Asymmetric tandem Michael-Darzens reaction [2006 JACS 5475 Jorgensen KA].

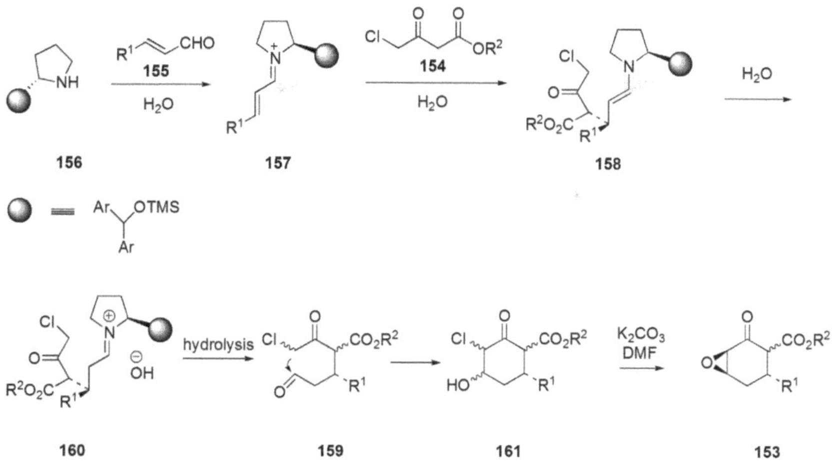

SCHEME 1.37 Mechanism of the Michael-Darzens reaction [2006 JACS 5475 Jorgensen KA].

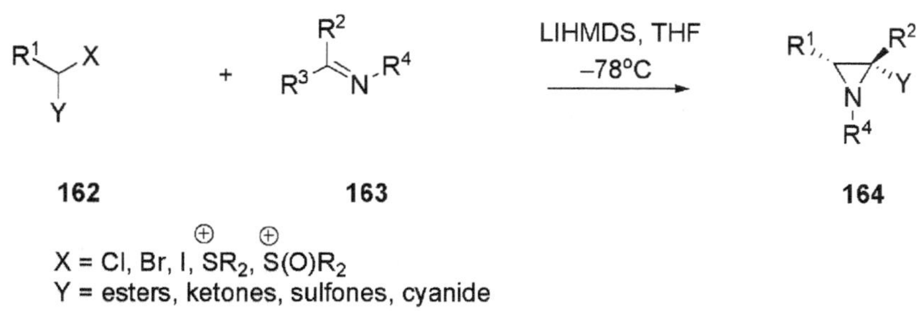

SCHEME 1.38 General aza-Darzens reaction.

groups in the presence of LHMDS in THF (Scheme 1.38) to give (**164**). A broad range of anion-stabilizing groups allows the process. Generally, *trans* isomers of (**164**) are the major products.

The salient feature of the aza-Darzens reaction is that a good degree of enantioselectivity can be achieved by employing optically active sulfinyl imines. The reaction of (**165**) with (**166**) in the presence of LHMDS gave (**167**). The selectivity is explained by the preferential formation of Zimmerman-Traxler transition state (**168**) from the intermediate (**169**) and the sulfinyl imine (**166**) (Scheme 1.39).[54]

The vinylogous aza-Darzens reaction of substituted α-bromocrotonate esters with chiral mesitylsulfinimines in the presence of LHMDS at −78°C provides a range of vinyl substituted aziridines in good yields and excellent stereoselectivity. For example, crotonate (**170**) reacted with imine (**171**) in the presence of LHMDS to give vinylaziridine (**172**). The initial γ-carbanion (**173**), derived from the crotonate, takes part in the annulation in its α-carbanion form (**174**) (Scheme 1.40).[55]

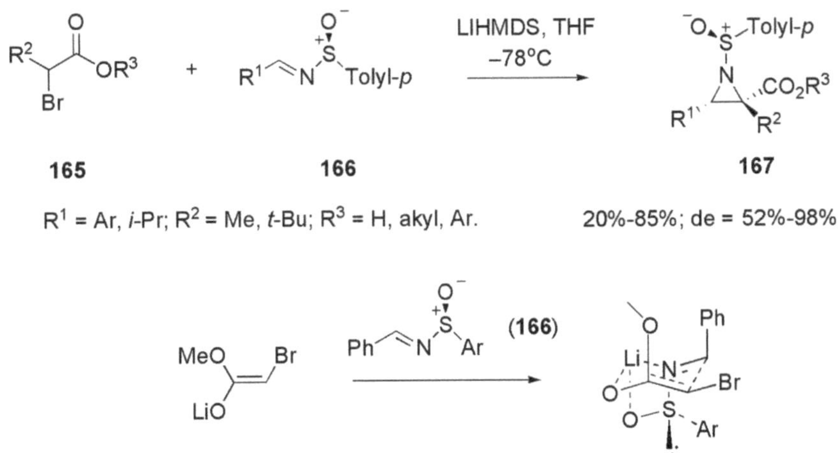

165 166 167 169 168

R[1] = Ar, *i*-Pr; R^2 = Me, *t*-Bu; R^3 = H, akyl, Ar. 20%-85%; de = 52%-98%

SCHEME 1.39 An example of stereoselective aza-Darzens [1999 JOC 7559 Davis FA] and [2011 OBC 5034 Stockman RA].

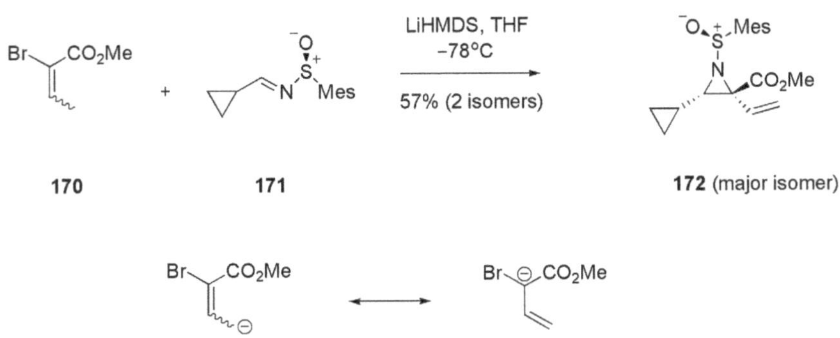

170 171 172 (major isomer)

173 174

SCHEME 1.40 An example of vinylogous aza-Darzens reactions [2014 ORS 6290 Stockman RA].

On the contrary, vinylogous aziridination of bromo lactone (**175**) in Scheme 1.41, with imine (**176**) took place at γ-position to give optically active aziridine (**177**) (Scheme 1.41).[56]

Only recently was a catalytic enantioselective version of aza-Darzens reaction developed by Trost et al. It was employed for the construction of complex trisubstituted aziridines. This methodology was based upon the use of Zn-ProPhenol catalysts. The reaction of achiral haloketones (**178**) with imines (**179**) in the presence of catalyst (**180**) and 2 equiv. of Cs_2CO_3 gave a series of chiral aziridines (**181**) in excellent yields (Scheme 1.42).[57] The noncyclized intermediates formed in the first step was isolated under controlled reaction conditions.

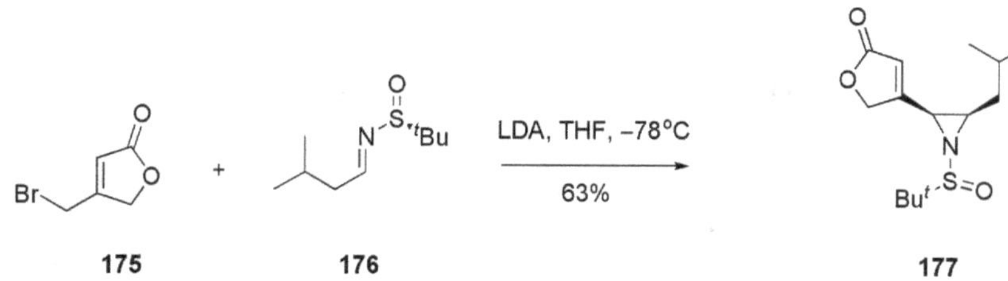

SCHEME 1.41 Njardarson asymmetric aza-Darzens reaction [ACIE 2015 13,706 Njardarson JT].

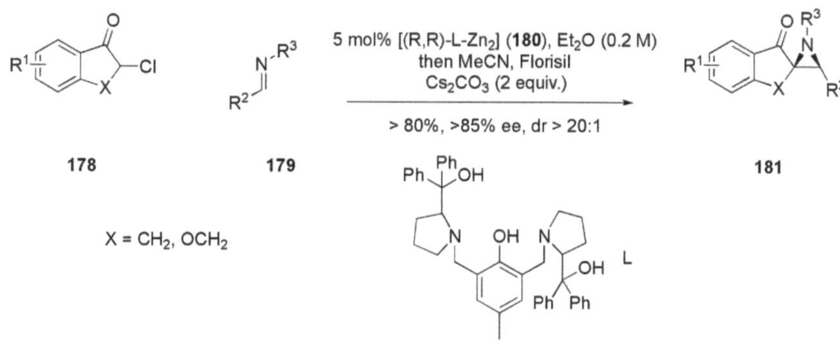

SCHEME 1.42 A catalytic enantioselective aza-Darzens reaction [2017 ACIE 2448 Trost BM].

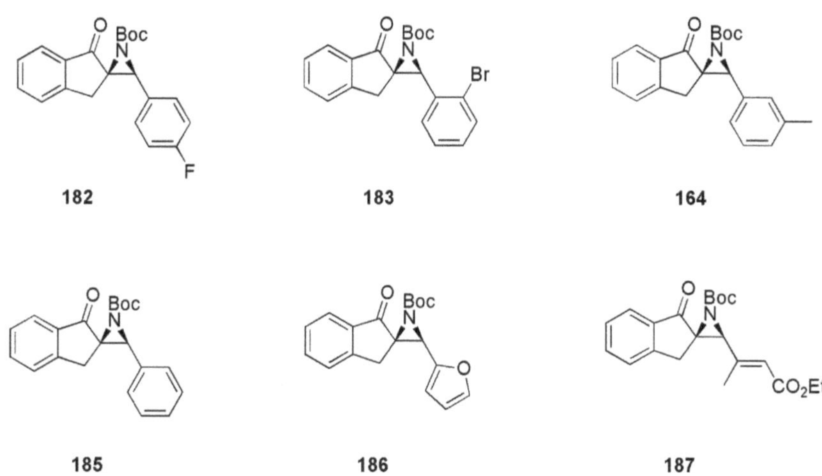

FIG. 1.2 Chiral aziridines synthesized by the Trost method.

Selected examples of prepared chiral aziridines such as (**182–187**) are displayed in Fig. 1.2.

In conclusion, the Darzens condensation, the simplest form of annulations, was discovered more than 100 years ago. Yet it appears to be more valuable than ever before in the synthesis of heterocycles.

REFERENCES

1. McGrath, N. A.; Brichacek, M.; Njardarson, J. T. *J. Chem. Educ.* **2010**, *87*, 1348.
2. (a) Thebtaranonth, C.; Thebtaranonth, Y. Cyclization Reactions; CRC Press Inc., 1994. (b) Li, J. J. Name Reactions for Carbocyclic Ring Formations; Wiley, 2010.
3. (a) Carruthers, W.; Coldham, I. Modern Methods of Organic Synthesis; Cambridge University Press, 2004. (b) Wyatt, P.; Warren, S. Organic Synthesis: Strategy and Control; John Wiley & Sons Ltd, 2007.(c)Zweifel, G. S.; Nantz, M. H. Modern Organic Synthesis; W.H. Freeman and Company, 2007. .(d)Smith, M. B. Organic Synthesis; Wavefunction Inc, 2010.(e)Clayden, J.; Greeves, N.; Warren, S. Organic Chemistry; Oxford University Press, 2012.
4. Danheiser, R. L.; Gee, S. K.; Sard, H. *J. Am. Chem. Soc.* **1982**, *104*, 7670.
5. Rapson, W. S.; Robinson, R. *J. Chem. Soc.* **1935**, 1285.
6. McCoy, L. L. *J. Am. Chem. Soc.* **1958**, *80*, 6568.
7. Eisenmuth, W.; Renfroe, H. B.; Schmid, H. *Helv. Chim. Acta* **1965**, *48*, 375.
8. (a) Hauser, F. M.; Rhee, R. P. *J. Org. Chem.* **1978**, *43*, 178. (b) Broom, N. J. P.; Sammes, P. G. *J. Chem. Soc. Chem. Commun.* **1978**, 162.(c)Kraus, G. A.; Sugimoto, H. *Tetrahedron Lett.* **1978**, *26*, 2263. .(d)Wildeman, J.; Borgen, P. C.; Pluim, H.; Rouwette, P.H.F.M.; van Leusen, A. M. *Tetrahedron Lett.* **1978**, *25*, 2213.(e)Evans, G. E.; Leeper, F. J.; Murphy, J. A.; Staunton, J. *J. Chem. Soc. Chem. Commun.* **1979**, 205. .(f)Dodd, J. H.; Weinreb, S. M. *Tetrahedron Lett.* **1979**, *38*, 3593.
9. Little, R. D.; Dawson, J. R. *Tetrahedron Lett.* **1980**, *21*, 2609.
10. Alabugin, I. V.; Gilmore, K. *Chem. Commun.* **2013**, 11246.
11. Galli, C.; Mandolini, L. *Eur. J. Org. Chem.* **2000**, 3117.
12. Tietze, L. F.; Brasche, G.; Gericke, K. M. Domino Reactions in Organic Synthesis; Wiley-VCH, 2006.
13. Topczewski, J. J.; Callahan, M. P.; Neighbors, J. D.; Wiemer, D. F. *J. Am. Chem. Soc.* **2009**, *131*, 14630.
14. Chogii, I.; Njardarson, J. T. *Angew. Chem. Int. Ed.* **2015**, *54*, 13706.
15. Dugger, R. W.; Ragan, J. A.; Ripin, D. H. B. *Org. Process Res. Dev.* **2005**, *9*, 253.
16. Mal, D.; Pahari, P. *Chem. Rev.* **2007**, *107*, 1892.
17. Hill, N.; Paruch, K.; Svenda, J. *Tehtrahedron* **2016**, *72*, 3345.
18. Wendt, J. A.; Gauvreau, P. J.; Bach, R. D. *J. Am. Chem. Soc.* **1994**, *116*, 9921.
19. Cheng, P.; Clive, D. L. J.; Fernandopulle, S.; Chen, Z. *Chem. Commun.* **2013**, *49*, 558.
20. Sasaki, I. *Synthesis* **2016**, *48*, 1974.
21. de la Puente, M. L.; Ley, S. V.; Simmonds, M. S.; Blaney, W. M. *J. Chem. Soc., Perkin Trans. 1* **1996**, 1523.
22. Takaki, K.; Okada, M.; Yamada, M.; Negoro, K. *J. Org. Chem.* **1982**, *47*, 1200.
23. Dethe, D. H.; Sau, S. K.; Mahapatra, S. *Org. Lett.* **2016**, *18*, 6392.
24. Hajos, Z. G.; Parrish, D. R. *J. Org. Chem.* **1974**, *39*, 1615.
25. Zhou, P.; Zhang, L.; Luo, S.; Cheng, J.-P. *J. Org. Chem.* **2012**, *77*, 2526.
26. Clemente, F. R.; Houk, K. N. *J. Am. Chem. Soc.* **2005**, *127*, 11294.
27. Rajagopal, D.; Narayanan, R.; Swaminathan, S. *Tetrahedron Lett.* **2001**, *42*, 4887.
28. Harada, N.; Sugioka, T.; Uda, H.; Kuriki, T. *Synthesis* **1990**, 53.
29. Hagiwara, H.; Uda, H. *J. Org. Chem.* **1988**, *53*, 2308.
30. Murphy, R. A.; Sarpong, R. *Org. Lett.* **2012**, *14*, 632.
31. Zhou, P.; Zhang, L.; Luo, S.; Cheng, J.-P. *J. Org. Chem.* **2012**, *77*, 2526.
32. Ramamurthi, N.; Swaminathan, S. *Ind. J. Chem.* **1990**, *29B*, 401.
33. Aladro, F. J.; Guerra, F. M.; Moreno-Dorado, F. J.; Bustamante, J. M.; Jorge, Z. D.; Massanet, G. M. *Tetrahedron* **2001**, *57*, 2171.

34. Woodward, R. B.; Sondheimer, F.; Taub, D. *J. Am. Chem. Soc.* **1951**, *73*, 4057.
35. Mulzer, J.; Duerner, G.; Trauner, D. *Angew. Chem. Int. Ed.* **1997**, *35*, 2830.
36. Martinez, L. P.; Umemiya, S.; Wengryniuk, S. E.; Baran, P. S. *J. Am. Chem. Soc.* **2016**, *138*, 7536.
37. Yamashita, S.; Naruko, A.; Nakazawa, Y.; Zhao, L.; Hayashi, Y.; Hirama, M. *Angew. Chem. Int. Ed.* **2015**, *54*, 8538.
38. Watanabe, H.; Onoda, T.; Kitahara, T. *Tetrahedron Lett.* **1999**, *40*, 2545.
39. Li, P.; Payette, J. N.; Yamamoto, H. *J. Am. Chem. Soc.* **2007**, *129*, 9534.
40. Soriano, D. S.; Lombardi, A. M.; Persichini, P. J.; Nalewajek, D. *J. Chem. Educ.* **1988**, *65*, 637.
41. (a) Christoffers, J.; Zhang, Y.; Frey, W.; Fischer, P. *Synlett* **2006**, 624. (b) Christoffers, J.; Scharl, H. *Eur. J. Org. Chem.* **2002**, 1505.
42. Bako, P.; Rapi, Z.; Keglevich, G. *Curr. Org. Synth.* **2014**, *11*, 361.
43. Zimbron, J. M.; Seeger-Weibel, M.; Hirt, H.; Gallou, F. *Synthesis* **2008**, 1221.
44. Adamczyk, M.; Dolence, E. K.; Watt, D. S.; Christy, M. R.; Reibenspies, J. H.; Anderson, O. P. *J. Org. Chem.* **1984**, *49*, 1378.
45. Roy, J.; Mal, T.; Jana, S.; Mal, D. *Beilstein J. Org. Chem.* **2016**, *12*, 531.
46. Mizutani, S.; Komori, K.; Kai, C.; Kuramochi, K.; Tsubaki, K. *Tetrahedron* **2016**, *72*, 6640.
47. Aldous, D. J.; Dalencon, A. J.; Steel, P. G. *Org. Lett.* **2002**, *4*, 1159.
48. Cameron, A.; Fisher, B.; Fisk, N.; Hummel, J.; White, J. M.; Rizzacasa, M. A.; Krenske, E. H. *Org. Lett.* **2015**, *17*, 5998.
49. Arai, S.; Ishida, T.; Shioiri, T. *Tetrahedron Lett.* **1998**, *39*, 8299.
50. Ku, J.-M.; Yoo, M.-S.; Park, H.-G.; Jew, S.-S.; Jeong, B.-S. *Tetrahedron* **2007**, *63*, 8099.
51. Liu, Y.; Provencher, B. A.; Bartelson, K. J.; Deng, L. *Chem. Sci.* **2011**, *2*, 1301.
52. Marigo, M.; Bertelsen, S.; Landa, A.; Jørgensen, K. A. *J. Am. Chem. Soc.* **2006**, *128*, 5475.
53. Sweeney, J. *Eur. J. Org. Chem.* **2009**, 491.
54. (a) Davis, F. A.; Liu, H.; Zhou, P.; Fang, T.; Reddy, G. V.; Zhang, Y. *J. Org. Chem.* **1999**, *64*, 7559. (b) Sola, T. M.; Churcher, I.; Lewis, W.; Stockman, R. A. *Org. Biomol. Chem.* **2011**, *9*, 5034.
55. Moragas, T.; Churcher, I.; Lewis, W.; Stockman, R. A. *Org. Lett.* **2014**, *16*, 6290.
56. Chogii, I.; Njardarson, J. T. *Angew. Chem. Int. Ed.* **2015**, *54*, 13706.
57. Trost, B. M.; Saget, T.; Hung, C.-I. *Angew. Chem. Int. Ed.* **2017**, *56*, 2440.

FURTHER READING

58. Boger, D. L.; Mullican, M. D. *J. Org. Chem.* **1980**, *45*, 5002.

Chapter 2

Hauser [4+2] Annulation: Definition, History, Mechanism, and Types of Phthalide Donors and Toluate Donors

The Hauser annulation is an anionic benzannulation reaction.[1] It provides a facile and useful synthesis of six-membered aromatics, especially oxygen-substituted benzenoids in regiospecific manner under basic conditions. In the literature, there are many more benzannulations, such as Dotz benzannulation, Danheiser benzannulation, Yamamoto annulation, etc.[2] They are developed owing to the enormous importance of aromatic rings in medicinal chemistry and material chemistry. The chemistry of aromatics, including the electrophilic substitutions and their regiochemistry, is thoroughly discussed in the undergraduate textbooks. With the knowledge of directing effects of substituents, the synthesis of a particular regioisomer is predictable. For a multistep synthesis of aromatics like benzenoids, naphthalenoids, and fused aromatics, elaboration of benzene derivatives is the first choice, because such compounds are readily and commercially available. Furthermore, the electrophilic substitutions are so well understood, it has been almost customary to utilize them to complete a synthesis. Step-economy, time economy, and separation problems are often not given serious consideration. In the synthesis of a disubstituted benzene from a monosubstituted benzene by electrophilic substitution reactions, three different isomeric disubstituted products are formed, because of the presence of three chemically nonequivalent nuclear carbon atoms in the starting benzene derivative. Similarly, a monosubstituted naphthalene would give rise to formation of seven different disubstituted isomers due to seven nonequivalent nuclear carbon atoms. Consequently, their separations become a difficult task. Moreover, the intrinsic orientation effect of a ring substituent is rarely modifiable.

Hauser et al. proposed to resolve the issue of regiochemistry at an early stage of a multistep synthesis. In pursuance of their research on pharmaceutically important anthracyclines and anthraquinones, they conceived of anionic benzannulations for regioselective construction of 1,4-dioxygenated naphthalenes. This topic is covered in passing in the book *Cyclization Reactions*.

Anionic Annulations in Organic Synthesis. https://doi.org/10.1016/B978-0-12-813800-7.00002-6

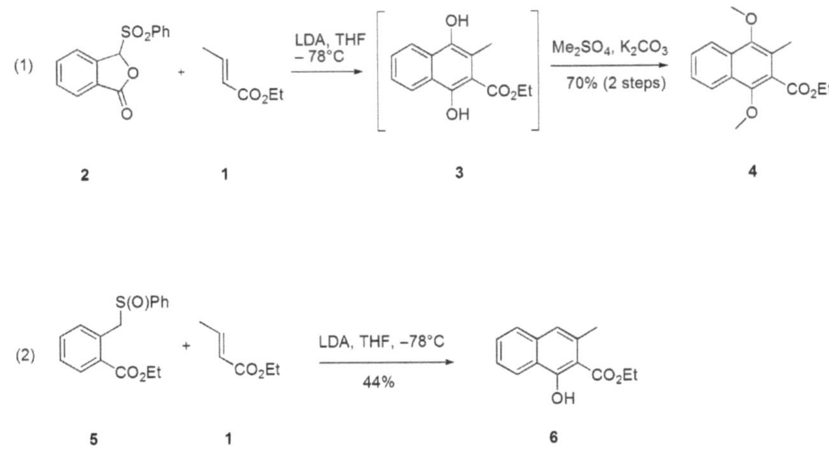

SCHEME 2.1 Two original annulations of Hauser and Rhee [1978 JOC 178 Hauser FM].

The seminal paper on benzannulation was first published in 1978 by Hauser and Rhee.[3a] It described two different modes of cyclocondensations, which can be termed Hauser phthalide annulation and Hauser toluate sulfoxide annulation (Scheme 2.1). Both the annulations share many common features. They were performed in the presence of LDA at −78°C on the same acceptor, i.e., ethyl crotonate (**1**). Both gave rise to a new fused phenol unit in the products. More importantly, the products are formed in a regiospecific manner. In the first case, condensation of 3-phenylsufonylphthalide (**2**) with (**1**) gave 1,4-naphthoquinol **3**, which was characterized as its dimethyl ether (**4**) prepared by reaction with methyl sulfate in the presence of K_2CO_3. The second reaction involves 1′-phenylsulfinyl *ortho* toluate (**5**) with ethyl crotonate (**1**), leading to α-naphthol (**6**) (Scheme 2.1).

Although the acceptors and the reaction conditions for both Hauser phthalide annulation (reaction 1) and Hauser toluate sulfoxide annulation (reaction 2) are very similar, the structures of the donors, i.e., (**2**) and (**5**) are different. As a result, the outcomes of the reactions markedly differ from each other with regard to oxidation levels. The former gives a 1,4-naphthoquinol (**3**) as the product, and the latter gives a 1-naphthol (**6**). Concerning the regiochemical outcomes, both prove to be robust. Between the above two Hauser annulations, the phthalide annulation has had more of an impact on organic synthesis than the Hauser toulate sulfoxide annulation on two counts. First, there is hardly any nonreductive method for directive synthesis of 1,4-naphthoquinol. Second, there are a good number of synthetic methods for the synthesis of 1-naphthols.

Hauser phthalide annulation is defined as the condensation reaction between C3 nucleofugal phthalides (**7**) with acceptors with polarized multiple bonds to give the respective benzannulated products (Scheme 2.2). Similarly, Hauser toluate annulation is referred to as the condensation between 1′-phenylsulfinyl *o*-toluates (**8**) and the above acceptors to give α-naphthols (Scheme 2.2).

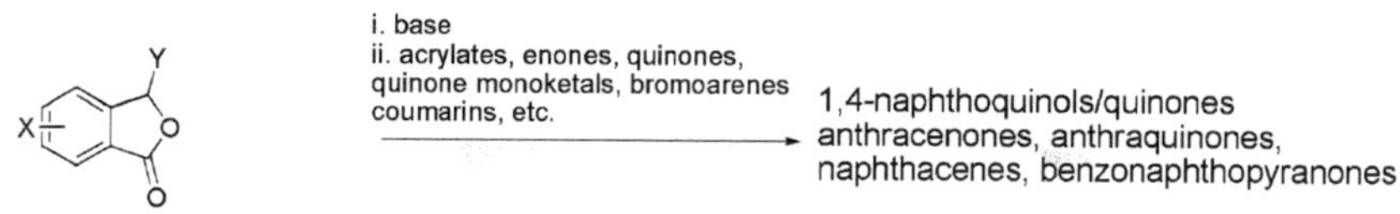

Y = SPh, SO_2Ph, CN, etc.

7

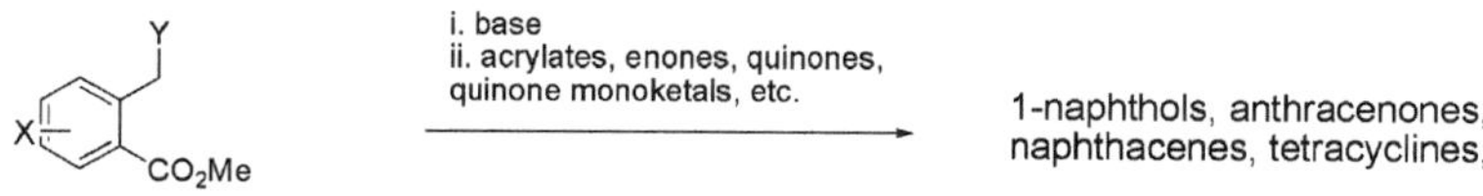

Y = SOPh, SO_2Ph, etc.

8

SCHEME 2.2 Generalized reactions of Hauser phthalides annulations and toluate annulations.

History: The discovery of Hauser annulation and the allied benzannulations deserve special mention with regard to the timings of their publications. Fascinatingly, reports of seven similar annulation reactions appeared in the same year, i.e., 1978. All of the reactions led to benzannulated products. Table 2.1 provides a summary of reports and demonstrates a mere coincidence/close competitions to a common objective of synthesizing oxygenated aromatic

TABLE 2.1 Summary of Phthalide Donors and Toluate Donors Reported in 1978

Entry	Structure of Donor	Date of Receipt of the Manuscript at Office	Date of the Publication	Group Reference
1	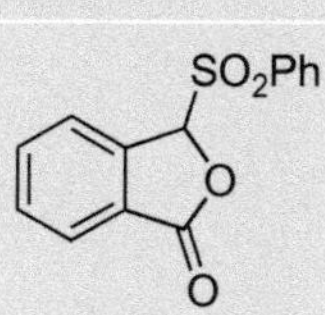 **2**	27 July 1977	1 January 1978	Hauser FM[3a]
2	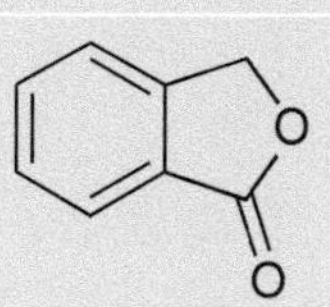 **9**	17 November 1977	February 1978	Sammes PG[3b]

Continued

TABLE 2.1 Summary of Phthalide Donors and Toluate Donors Reported in 1978—cont'd

Entry	Structure of Donor	Date of Receipt of the Manuscript at Office	Date of the Publication	Group Reference
3	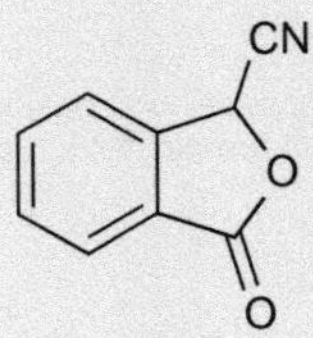 **10**	9 February 1978	2 May 1978	Kraus GA[3c]
4	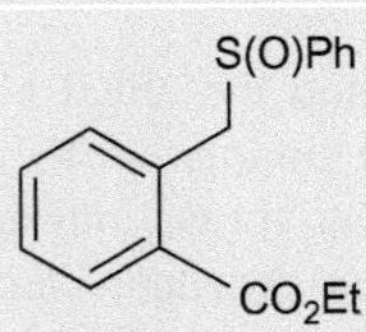 **11**	27 July 1977	1 January 1978	Hauser FM[3a]
5	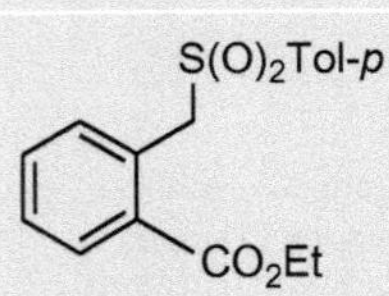 **12**	10 April 1978	27 April 1978	van Leusen AT[3d]
6	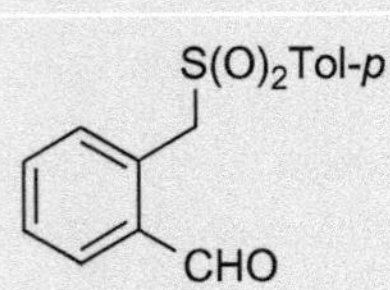 **13**	10 April 1978	27 April 1978	van Leusen AT[3d]
7	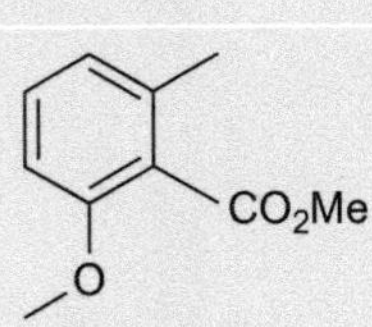 **14**	2 October 1978	March 1979	Staunton J[3e]

polykeyides. The Hauser group submitted their report on the annulation of 3-phenylsulfonyl substituted isobenzofuranone (**2**) with acceptors in mid-1977 (entry 1, Table 2.1).[3a] In the same year, the Sammes group submitted their related work on phthalide (**9**) (entry 2).[3b] After the publication of Hauser's work, Kraus et al. demonstrated that 3-cyanoisobenzofuranone (**10**) could be utilized as an annulating agent in the place of (**2**).[3c] More interestingly, Hauser and Rhee applied the above concept of annulations to sulfoxide activated *o*-toluate (**11**) for the synthesis of the corresponding α-naphthols, and reported the work in the same publication.[3a] Two more similar toluate/tolulaldehyde donors (**12** and **13**) were developed later in the same year by van Leusen et al.[3d] Immediately after the publication of these works, Staunton et al. reported their work on annulation of toluate donors **14** without an activating group, which is now known as Staunton-Weinreb annulation.[3e]

2.1 MECHANISM

There is no systematic study on elucidation of the mechanism of the Hauser reactions, although a great deal of synthetic aspects of the reaction is reported in literature over the past 40 years. That phthalide (**9**) undergoes deprotonation and carbonation at C-3 has been known since the 1960s. However, the respective carbanionic chemistry remained unexplored for many years. If substituted with phenylthio group at C-3, isobenzofuranone (**15**) can readily undergo deprotonation to form the corresponding carbanion which, in turn, manifests in *C*-alkylation. For instance, the reaction of phthalide (**15**) with ethyl crotonate (**1**) gave the corresponding Michael adducts (**16**) via carbanion (**17**) (Scheme 2.3).[4]

Similar mechanistic steps in Scheme 2.3 are accepted for Hauser phthalide annulation. The mechanism originally proposed by Hauser and Rhee is depicted in Scheme 2.4. The first step is selective deprotonation of (**2**) at C-3. The resulting carbanion (**18**) undergoes Michael addition to the acceptor (**19**), and then the incipient carbanion (**20**) undergoes Dieckmann cyclization/Claisen condensation at the lactone carbonyl group to give intermediates (**21**) and (22). The intermediate (**22**) collapses to annulated product (**23**) through expulsion of benzenesulfenate ions. Base-promoted tautomerization and aromatization of (**23**) finally give naphthoquinol dioxide (**24**). The five-step mechanism consisting of the sequence of (i) lateral deprotonation, (ii) intermolecular Michael addition, (iii) Dieckmann/Claisen condensation, (iv) elimination of benzenesulfenate anion, and (v) tautomerization is accepted in the literature.

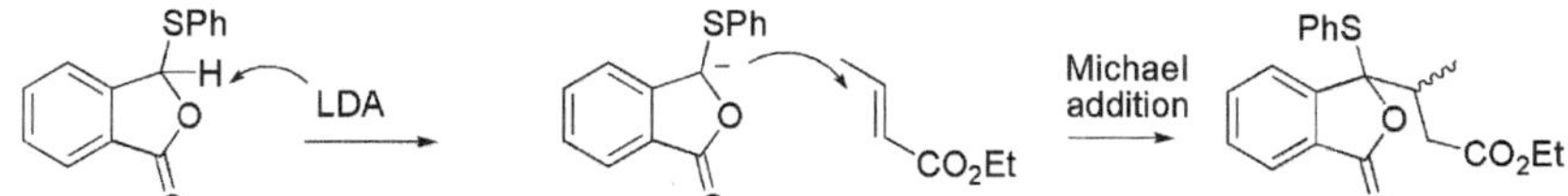

15 **17** **1** **16**

SCHEME 2.3 A Michael addition of 3-phenylthiophthalide (**15**).

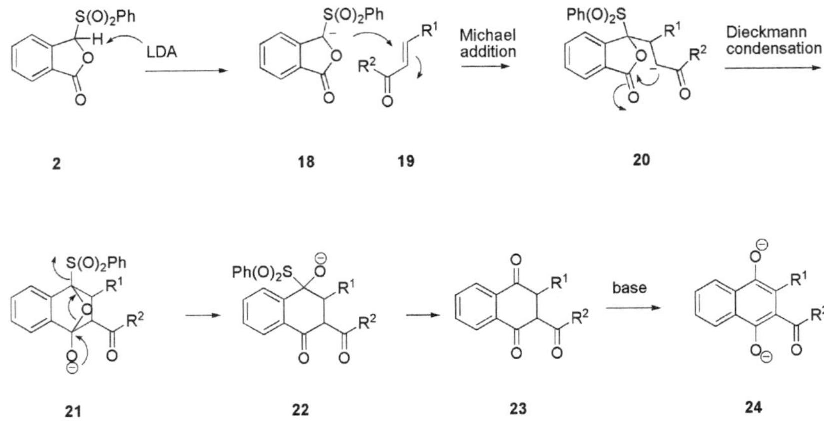

SCHEME 2.4 Mechanism of Hauser phthalide annulation.

The phenylsulfonyl group in (**2**) is proposed to serve two functions: (i) it increases acidity of the C3 hydrogen and thus, facilitates carbanion formation; and (ii) it acts as a good nucleofuge. In addition, it may be expected to assist Dieckmann/Claisen cyclization by contributing to the Thorpe-Ingold effect.[5]

In the synthon approach, Hauser phthalide annulation can be viewed as the coupling of two synthons (**25** and **26**) derived from their equivalents (**2** and **19**, respectively) (Scheme 2.5). In essence, the isobenzofuranone (**2**) is a synthetic equivalent of 1,4-dipolar synthon (**25**). The requirement of the annulation, therefore, is four-atom Hauser donors with two reactive centers of opposite polarity at two termini embedded in an aromatic nucleus. These 1,4-dipolar synthons equivalents are termed as Hauser donors. Under the influence of strong bases, these donors deliver reactive anions and initiate the annulation reaction. Therefore, Hauser annulations are classified as anionic annulations.

The supports for the stepwise mechanism presented in Scheme 2.4 stem from the reaction conditions and the structural elements that promote these

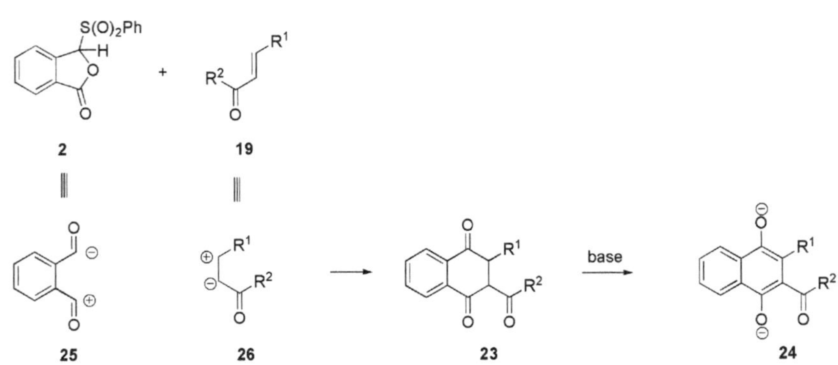

SCHEME 2.5 Mechanism of Hauser phthalide annulation based on synthons.

reactions. The mechanism is also corroborated by isolation of the initial Michael adducts in many instances. These intermediates, on treatment with bases, undergo ring-closure to give same annulated products.[6] Hauser phthalide annulation is therefore defined, in mechanistic terms, as the lateral metalation-initiated Michael-Dieckmann/Claisen condensations. These are commonly accomplished in a one-pot operation.

Similar to Hauser phthalide annulation (reaction 1, Scheme 2.1), the mechanism of the Hauser toluate annulation (reaction 2, Scheme 2.1) can be depicted in stepwise fashion (Scheme 2.6). The very first step is lateral lithiation/deprotonation at the side chain benzylic position of the sulfoxide (**5**) in the presence of LDA to form carbanion (**25**). In the second step, Michael addition of (**25**) to (**1**) takes place to carbanion (**26**), which then participates in Dieckmann condensation to form tetralone (**27**). Thermal *cis*-elimination of phenylsulfenic acid from (**27**) provides (**28**), which tautomerizes to naphthol (**29**) (Scheme 2.6).

The most remarkable feature of these annulations is the regiochemical outcome. Attendant with Michael additions, both the addition and the ring closure steps occur in a regiospecific manner with the Michael acceptors. If appropriately substituted phthalides/toluates and Michael acceptors are used, a single annulation product with defined regiochemistry is formed. No examples have been found where regiochemistry has been reversed formally or informally. In reactions with a few unsymmetrical benzynes and heteroarynes, the regioselectivity is not very pronounced.[7]

2.2 MECHANISTIC DICHOTOMY

Although he stepwise mechanisms based upon two key steps—Michael addition and Dieckmann/Claisen condensation are generally accepted for Hauser phthalide annulation—an alternative based on a synchronous [4+2] cycloaddition such as

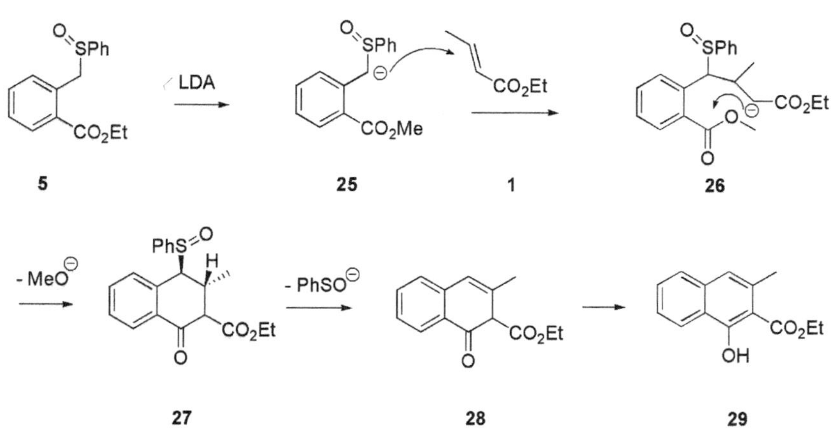

SCHEME 2.6 Mechanism of Hauser toluate annulation.

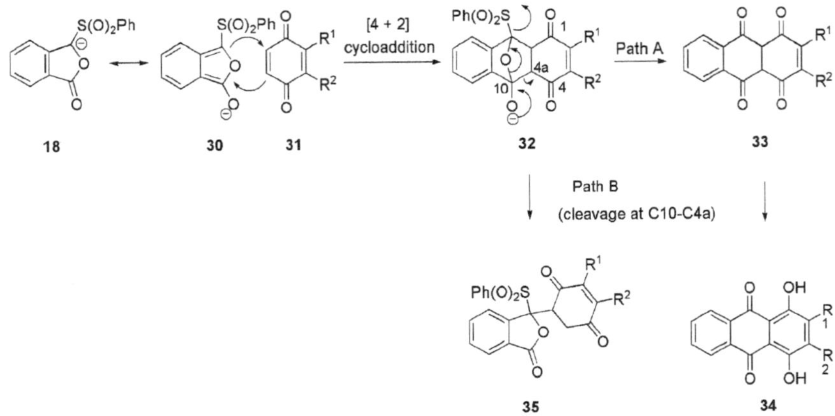

SCHEME 2.7 Mechanism of Hauser annulation based on cycloaddition-fragmentation.

one between isobenzofuranone oxy anion (**30**), the resonance structure of deprotonated (**18**), and the acceptor (**31**) is not ruled out. Fragmentation of the newly formed furan ring of (**32**) (Scheme 2.7), can occur in two different competing pathways. Path A involving cleavage of oxa-bridge would give the annulation product (**33**) and then product (**34**), and path B involving C4a–C10 would provide the 1,4-adduct (**35**) (Scheme 2.7).[8]

This mechanism is, in part, supported by the findings with TMS-protected isobenzofuranone (**36**), a structural analog of (**30**). The intermediate (**36**), prepared from cyanophthalide (**37**), gave expected TMS protected annulated product (**38**) on reaction with enone (**39**) (Scheme 2.8). Though the isobenzofuranone (**36**) seemed to be too unstable to be isolated, its formation was apparent from the analysis of its NMR spectrum. Likewise, the annulation product (**38**) displayed a TMS peak in ^{1}H NMR spectrum. Compound (**38**) was desilylated with silica gel to produce quinol (**40**), a model compound for the synthesis of natural product fredericamycin A.[9]

The fact that the Michael addition product (**41**) from the reaction of (**42**) with phthalide (**10**) in the presence of LDA could not be cyclized to give anthraquinone (**43**) can be interpreted, in principle, by the concerted mechanism (Scheme 2.9).[10]

However, the duality of the mechanism of Hauser phthalide annulation remains the subject of more scrutiny. The reaction of benzocycloheptadienone (**44**) with isobenzofuranone (**2**) yielded three different annulated products (**45**–**47**) (Scheme 2.10).[11] While formation of the first two products (**45** and **46**) can be explained with the suggested concerted mechanism (Scheme 2.7), that of the third one (**47**) should be attended with a different mechanism. It is a product of formal [4+4] cycloaddition, which is not thermally allowed. It must have been formed by stepwise manner, i.e., Michael addition followed by Dieckmann reaction. In view of the above results, both stepwise and concerted mechanism may be considered for further research on Hauser annulation.

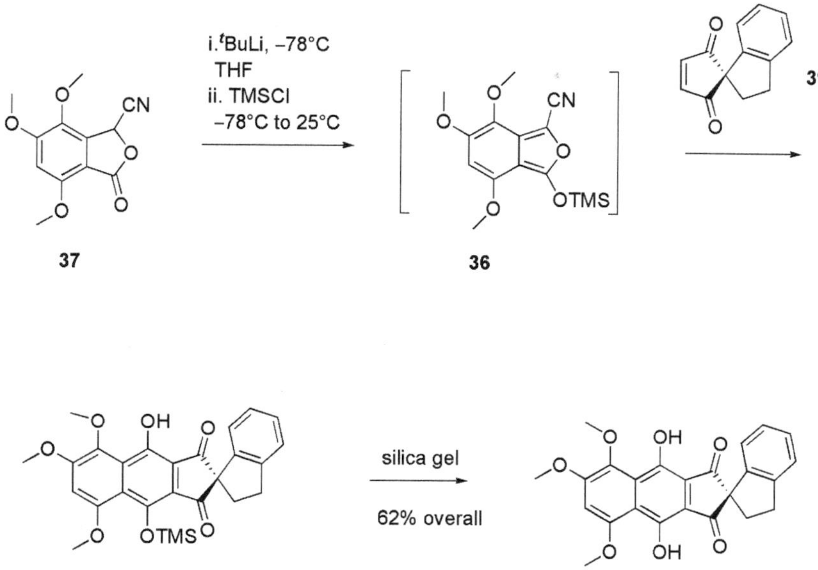

37 36 38 40

SCHEME 2.8 Synthesis of model compound related to fredericamycin A.

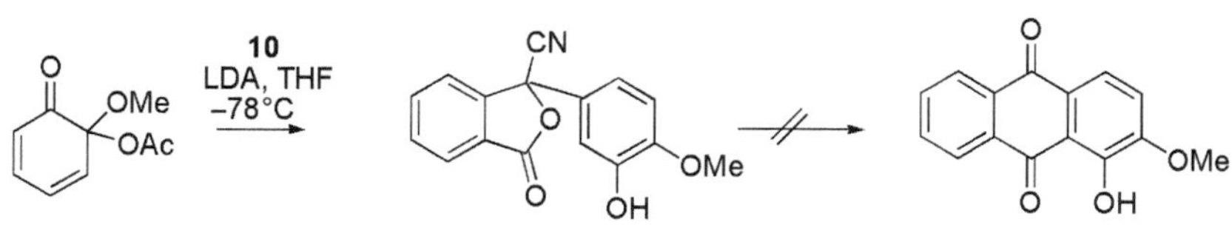

42 41 43

SCHEME 2.9 Isolation of an inert Michael adduct.

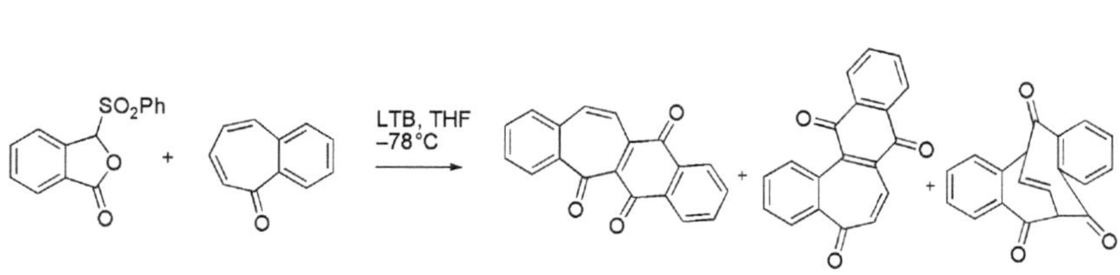

2 44 45 (15%) 46 (28%) 47(7%)

SCHEME 2.10 Hauser phthalide annulation with benzocycloheptadienone (**44**).

2.3 DONORS IN HAUSER ANNULATION

The concept underlined by Hauser annulation has had an enormous impact on the development of other related annulations. That is why most of the benzannulations involving lateral metalations and anionic [4+2] cycloadditions are included under the umbrella term "Hauser annulation."

The success of these reactions hinges on the strong tendency of the initial anions to add to acceptors in 1,4 mode rather than 1,2 mode. As is often the case in the Michael reactions, the addition is virtually regiospecific to give the respective Michael addition products which, upon warming, cyclize to give the annulation products. Thus the reactivity of the Hauser donors is critically important. On the basis of structures, the donors are classified into four categories: i) C3 nucleofugal phthalides, ii) C3 unactivated phthalides, iii) laterally activated *o*-toluates, and iv) unactivated *o*-toluates donors. All four of these classes share a common structural feature, i.e., a potential carbanionic center at the lateral positions. The annulation chemistry of these donors will be described in later chapters.

2.4 TYPES OF PHTHALIDE DONORS

Mechanistic considerations would predict a large number of nucleofuges for the synthesis of 3-nucleofugal phthalides. From the synthetic chemists' perspective, the ones that are readily synthesized and utilized in Hauser annulation are presented in Fig. 2.1. These are phthalides (**48–57**).[12] Further, there are examples of

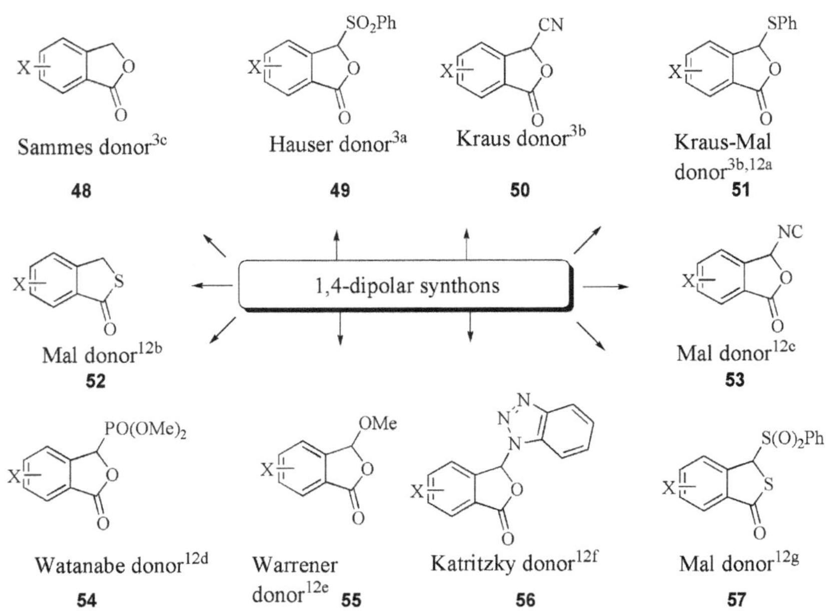

FIG. 2.1 Types of phthalide donors for benzannulations.

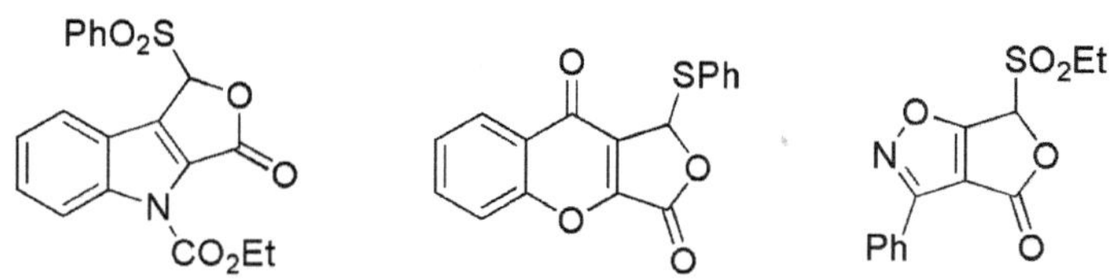

58 **59** **60**

FIG. 2.2 Examples of heterocyclic analogs of Hauser donors.

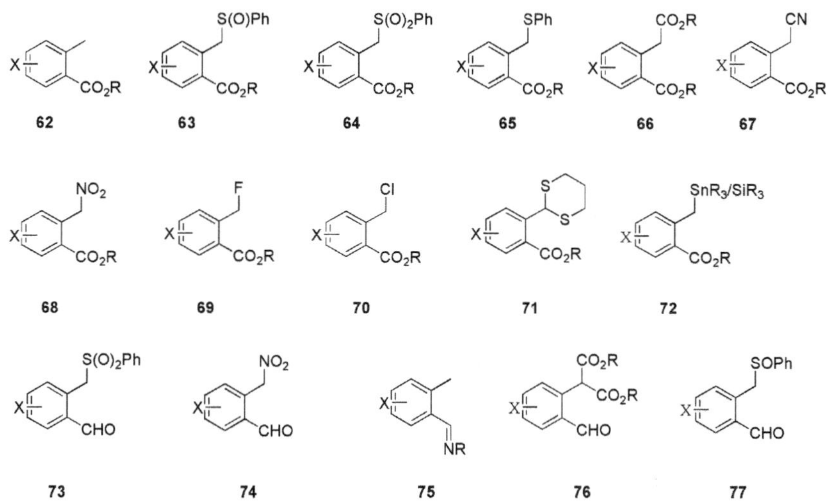

FIG. 2.3 Types of toluate donors.[14]

Hauser donors which result from replacement benzene ring of phthalides with a heterocycle. These donors have not been studied in detail. Analogs of the type (58),[13a] (59),[13b] and (60)[13c] (Fig. 2.2) appear in the literature.

Like the phthalides in Fig. 2.1, toluate donors are of mainly two types: laterally activated and laterally nonactivated. The former may be termed as Hauser-van Leusen toluate donors and the latter as Staunton-Weinreb-Myers donors. All toluates that have been studied till date are displayed in Fig. 2.3.[14]

REFERENCES

1. (a) Mal, D.; Pahari, P. *Chem. Rev.* **2007**, *107*, 1892. (b) Rathwell, K.; Brimble, M. A. *Synthesis* **2007**, *5*, 643. (c) Mitchell, A. S.; Russell, R. A. *Tetrahedron* **1995**, *51*, 5207.
2. Kotha, S.; Misra, S.; Halder, S. *Tetrahedron* **2008**, *64*, 10775.
3. (a) Hauser, F. M.; Rhee, R. P. *J. Org. Chem.* **1978**, *43*, 178. (b) Broom, N. J. P.; Sammes, P. G. *J. Chem. Soc. Chem. Commun.* **1978**, 162. (c) Kraus, G. A.; Sugimoto, H. *Tetrahedron Lett.*

1978, *26*, 2263. (d) Wildeman, J.; Borgen, P. C.; Pluim, H.; Rouwette, P.H.F.M.; van Leusen, A. M. *Tetrahedron Lett.* **1978**, *25*, 2213. (e) Evans, G. E.; Leeper, F. J.; Murphy, J. A.; Staunton, J. *J. Chem. Soc. Chem. Commun.* **1979**, 205. (f) Dodd, J. H.; Weinreb, S. M. *Tetrahedron Lett.* **1979**, *38*, 3593.

4. Kraus, G.; Sugimoto, H. *Synth. Commun.* **1977**, *7*, 505.
5. Jung, M. E.; Piizzi, G. *Chem. Rev.* **2005**, *105*, 1735.
6. Saint-Jalmes, L.; Lila, C.; Xu, J. Z.; Moreau, L.; Pfeiffer, B.; Eck, G.; Pelsez, L.; Rolando, C.; Julia, M. *Bull. Soc. Chim. Fr.* **1993**, *130*, 447.
7. Khanapure, S. P.; Reddy, R. T.; Biehl, E. R. *J. Org. Chem.* **1987**, *52*, 5685.
8. Mal, D.; Ray, S.; Sharma, I. *J. Org. Chem.* **2007**, *72*, 4981.
9. Evans, J. C.; Klix, R. C.; Bach, R. D. *J. Org. Chem.* **1988**, *53*, 5519.
10. Lebrasseur, N.; Fan, G.-J.; Oxoby, M.; Looney, M. A.; Quideau, S. *Tetrahedron* **2005**, *61*, 1551.
11. Hauser, F. M.; Yin, H. *Org. Lett.* **2000**, *2*, 1045.
12. (a) Ghorai, S. K.; Roy, H. N.; Bandopadhyay, M.; Mal, D. *J. Chem. Res. (S)* **1999**, 30. (b) Mal, D.; Pal, R.; Murty, K.V.S.N. *J. Chem. Soc. Chem. Commun.* **1992**, 821. (c) Mal, D.; Ghosh, K.; Chakraborty, S. *Synthesis* **2015**, *47*, 2473. (d) Watanabe, M.; Morimoto, H.; Nogami, K.; Ijichi, S.; Furukawa, S. *Chem. Pharm. Bull.* **1993**, *41*, 968. (e) Warrener, R. N.; Liu, L.; Russell, R. A.; Tiekink, E. R. T. *Synlett* **1998**, *4*, 387. (f) Katritzky, A. R.; Zhang, G.; Xie, L. *Synth. Commun.* **1997**, *27*, 3951. (g) Majumdar, G.; Pal, R.; Murty, K.V.S.N.; Mal, D. *J. Chem. Soc. Perkin Trans.* **1994**, *1*, 309.
13. (a) Mal, D.; Senapati, B.; Pahari, P. *Synlett* **2005**, 994. (b) Hauser, F. M.; Dorsch, W. A. *Org. Lett.* **2003**, *5*, 3753. (c) Alguacil, R.; Martin, M. V.; Paredes, M. C. *Heterocycles* **2000**, *53*, 1029.
14. Donner, C. D. *Tetrahedron* **2013**, *69*, 3747.

Chapter 3

Lateral Lithiations

The chemistry of aromatics is of enormous importance because of its applications in virtually every aspect of human life, which include agro-chemicals, dyes, food additives, cosmetics, functional materials, and pharmaceuticals. Consequently, the pursuit of newer methods for the synthesis of aromatics is a long-standing and major research area in organic synthesis. One of the major developments in this area is the growth of chemistry of organolithiums. It has emerged as an invaluable tool in the synthesis of aromatics, fused carbocycles, and heterocycles. Due to the ready availability of hundreds of organolithiums at reasonable costs, and oxygen- and moisture-free setups, organolithiums are routinely used without much difficulty.

Among many reactions, the lithiation/metalation of arenes in the proximity of an appropriate directing metalation group (DMG) by organolithiums has proven to be a powerful avenue for the regiodefined preparation of polyfunctional arenes. It has supplemented the three major strategies, namely, the Friedel-Crafts acylation, the Diels-Alder reaction, and cyclization of polyketides for the construction of polyketide aromatics.

In 1977, the foundation for the discovery of the Hauser annulation was laid by Hauser and Rhee in their synthesis of polyketide natural products.[1] It stemmed from lateral lithiations, a subclass of the organolithium chemistry applied to aromatics. Since it forms the founding step of the Hauser annulations presented in Chapter 2, an overview of lateral lithiation is presented in this chapter.

The directed organolithium chemistry is categorized into three major subareas: *ortho* lithiation, α-lithiation (or lithiation α to heteroatoms), and lateral lithiation.[2] In special cases, remote lithiations are observed. An example of α-lithiation is given in Scheme 3.1. Methoxyallene (**1**), on reaction with ^{n}BuLi, undergoes regioselective lithiation at position α to the methoxy group to form (**2**), which quantitatively adds to benzaldehyde, forming 1,2-adduct (**3**). The details of *ortho* lithiation are reviewed by Snieckus. In a recent review[3] by Donner, lateral lithiations have been highlighted in the context of Staunton-Weinreb annulation. In this section, the concept of the lateral lithiation, also known as benzylic lithiation, is introduced and its pertinent uses in general organic synthesis are described.

There are three general strategies by which organolithium compounds are commonly prepared. These are: (i) deprotonation reaction with an organolithium reagent, (ii) halogen-lithium exchange, and (iii) reduction of an organic species with metallic lithium. The first strategy is practiced most commonly. In addition,

Anionic Annulations in Organic Synthesis. https://doi.org/10.1016/B978-0-12-813800-7.00003-8

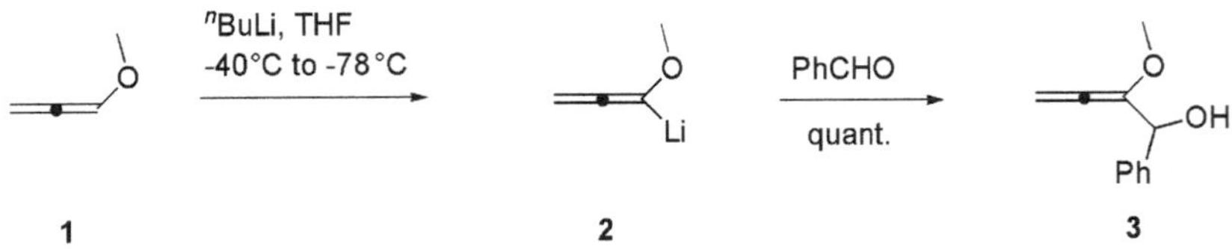

SCHEME 3.1 α-Lithiation of methoxyallene [2004 EJOC 2297 Reisssig H-U].

it is useful in achieving chemoselectivity, regioselectivity, and stereoselectivity in the formation of a new carbon–carbon or carbon-heteroatom bond.

In Scheme 3.2, a few typical examples are presented to give an overview of the preparative methods for the organolithiums. Strong bases like LDA and

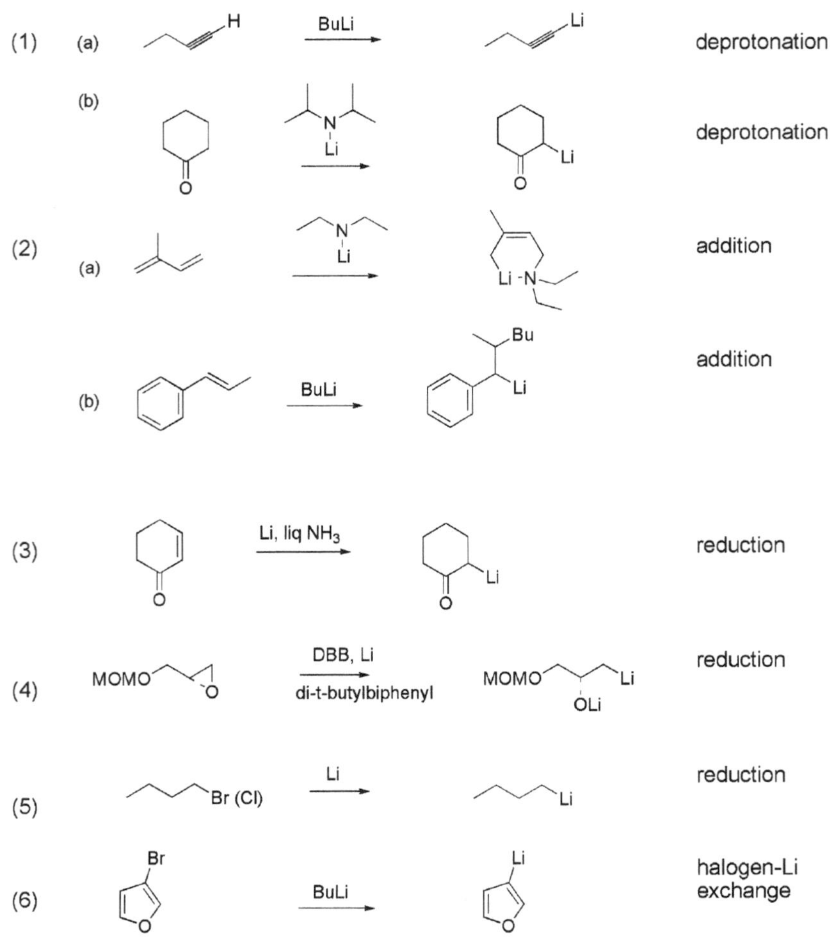

SCHEME 3.2 Highlights of the methods for the preparation of organolithium compounds.

BuLi react with CH acids to form the corresponding lithio species (reaction 1). Reaction 2a illustrates the generation of organolithiums by addition of an organolithium to a CC double bond. The driving force for such addition is likely to be chelation. For reaction 2b, the greater resonance stability of the resulting lithio species due to phenyl ring seems to be appropriate. Birch reduction, though limited in use, is used for the preparation of organolithiums in certain cases. Reaction 3 represents an example of Birch condition to produce organolithium species under reductive mode. There are instances where oxiranes can be used as precursors for organolithiums by reduction with DBB (di-*t*-butylbiphenyl) (reaction 4). Reaction 5 represents another reductive procedure. Reaction 6 is commonly categorized as halogen-Li exchange, and it is often kinetically favored over deprotonations. Similar exchange reactions are known for organotin substrates.

3.1 WHAT IS LATERAL LITHIATION?

Lateral lithiation is the lithiation of a benzylic position *ortho* to a heteroatom-containing substituent, often termed as the directing group (DG). Toluene derivative (**4**) with a directing group, when treated with either alkyllithiums or lithium dialkylamides, undergoes deprotonation at the benzylic position to give lithiated species (**5**) (Scheme 3.3). Such a process is termed as lateral lithiation. The directing groups are the groups that contain nitrogen, oxygen, or both. The commonly used groups are amides, amines, carbamates, carboxylates, ethers, oxazolines, sulfonamides, and ureas. However, there are instances where lateral lithiations can be achieved without an *ortho* directing group (cf. Scheme 3.3, entry 2b).

The driving force for lithiation is the stability of the lithiated species, derived from both resonance stabilization and chelation. For example, *ortho* toluate (**6**), when reacted with LDA, produces lithio species (**7**) (Scheme 3.4, reaction 1). For reaction 2, a temporal chelating group is used to induce lateral lithiation to produce (**8**) from (**7**) as well as to protect the aldehyde functionality from its reaction with BuLi. The lithiation of (**7**) is specifically directed to methyl group α to nitrogen to produce **8**. Likewise, phthalide (**9**) can be lithiated by reaction with LDA at −78°C to produce reactive 3-lithiophthalide (**10**). Although the metalation of phthalides has been known for a long time, it is now categorized as lateral lithiations, since it involves a formal benzylic position.

Perhaps the earliest known manifestation of lateral metalation occurred in the famous Madelung synthesis of indoles. In 1912, Madelung reported the

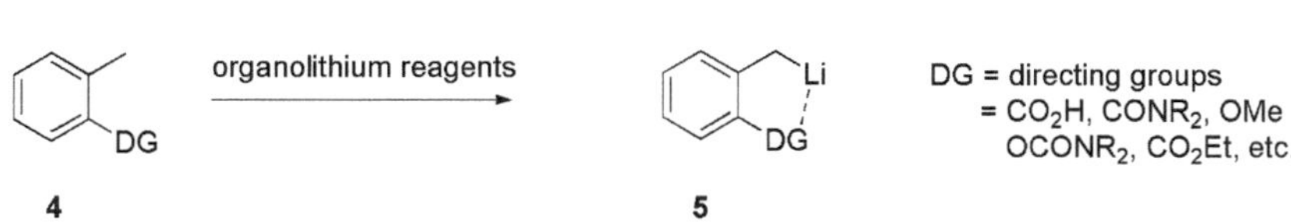

SCHEME 3.3 Lateral lithiation defined.

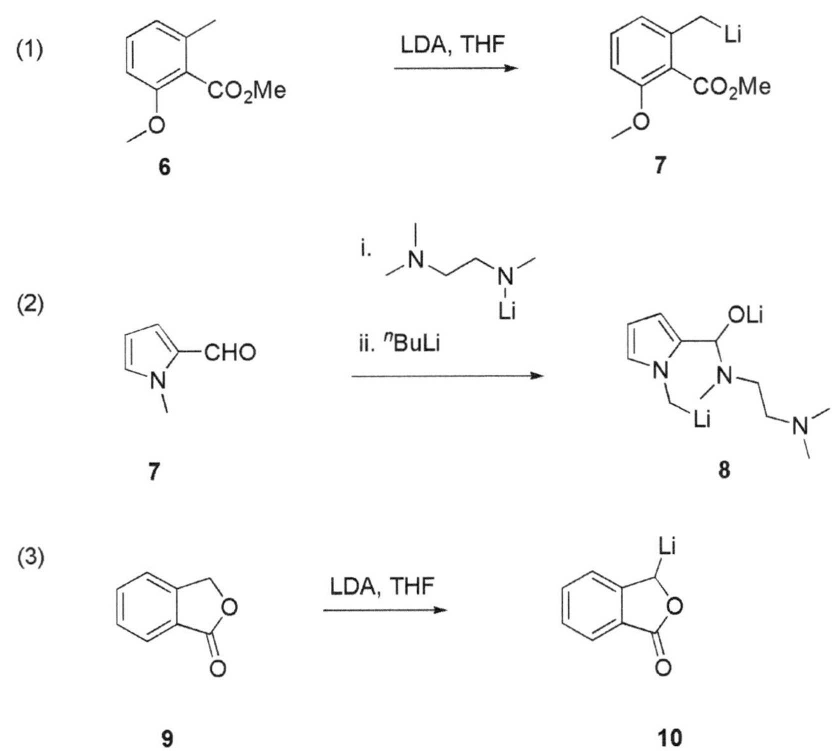

SCHEME 3.4 Typical examples of lateral lithiations.

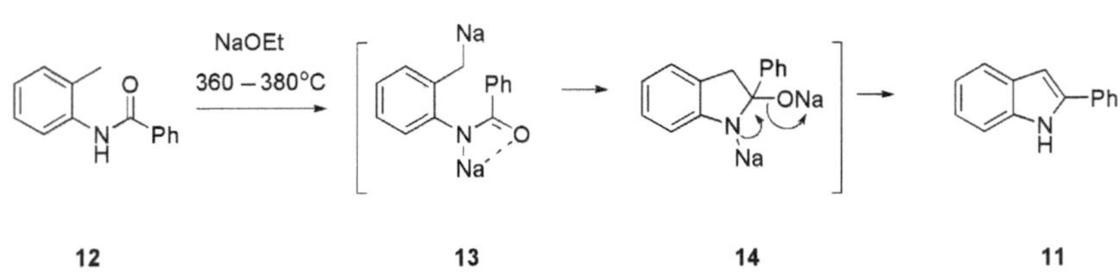

SCHEME 3.5 Madelung synthesis of 2-phenylindole.

synthesis of 2-phenylindole (**11**) by heating *N*-benzoyl-*o*-toluidine (**12**) with sodium ethoxide at 360–380°C (Scheme 3.5).[4] The structures (**13 and 14**) were proposed as the key intermediates of the transformation. The concept was utilized by Houlihan et al. to show that the procedure consisting in 2 or 3 equiv. of *n*-BuLi or LDA in tetrahydrofuran reduces the reaction temperatures to the range of −20°C to 25°C.[5] This finding has paved the way for the discovery of a variant, named Smith indole synthesis.[6] The support for the sodiation of (**12**) can be derived from Chiba's recent work, cited in Scheme 3.6.[7] The reaction of amide (**15**) with NaH-LiI in THF at 85°C furnished indanone (**16**) in 84% yield.

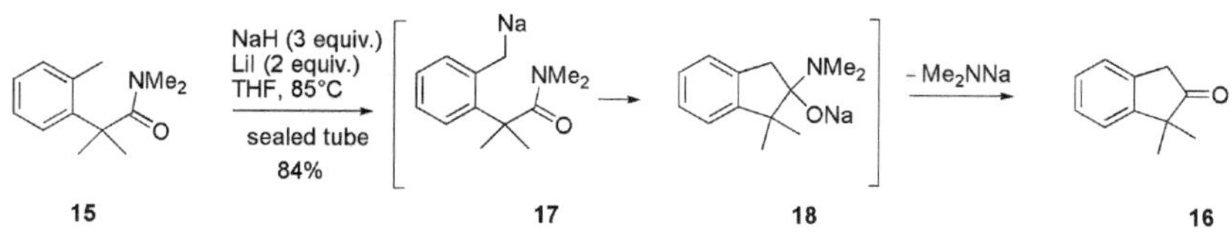

SCHEME 3.6 An example of lateral sodiation [2017 ACIE 6544 Chiba S].

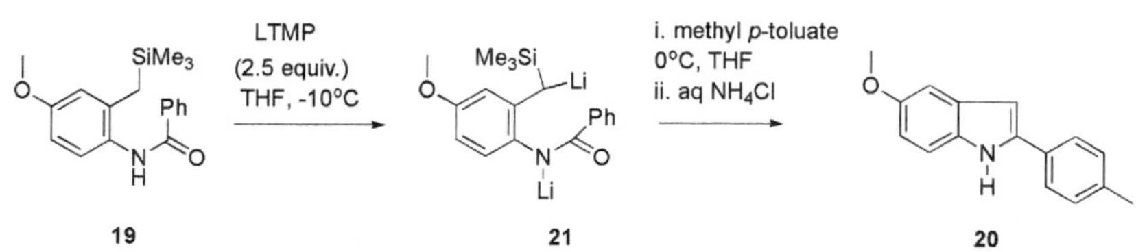

SCHEME 3.7 Variant of Madelung synthesis [1990 THN 1379 Bartoli G].

It is believed that the reaction proceeds through lateral sodio intermediate (**17**) and its cyclized form (**18**). The presence of an iodide additive is essential for the reaction.

In 1990, Bartoli et al. combined lateral lithiation with Peterson olefination for the synthesis of N-unsubstituted indoles (Scheme 3.7).[8] The reaction of 4-methoxy-2-trimethylsilylmethylbenzanilide (**19**) with LTMP (2.5 equiv.) at $-10°C$ followed by methyl *p*-toluate gave 5-methoxy-2-(*p*-tolyl)indole (**20**). The reaction is believed to take place via dilithiated species (**21**). Its reaction with methyl *p*-toluate gives keto intermediate (**22**), which undergoes intramolecular Peterson olefination via (**23**) to give *N*-benzoylindole (**24**). The lithium methoxide formed in the first stage knocks off the *N*-benzoyl group of the precursor (**24**) to produce (**20**) (Scheme 3.8).

One of the most commonly used directing groups for practical lithiations is the diethylamide group. It has been extensively used in preparing highly substituted aromatics with defined regiochemistry. The lithiation of amide (**25**) with LDA to (**26**) followed by silylation to silyl derivative (**27**) clearly indicates that the *ortho* selectivity is due to chelation of the intermediate (**26**) (Scheme 3.9).[9]

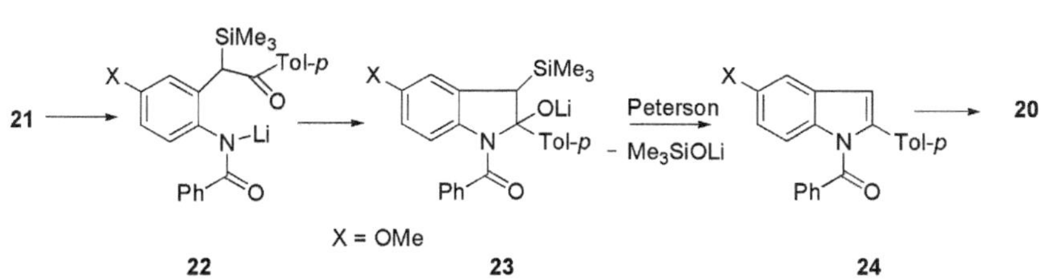

SCHEME 3.8 Mechanism of the Bartoli reaction [1990 THN 1379 Bartoli G].

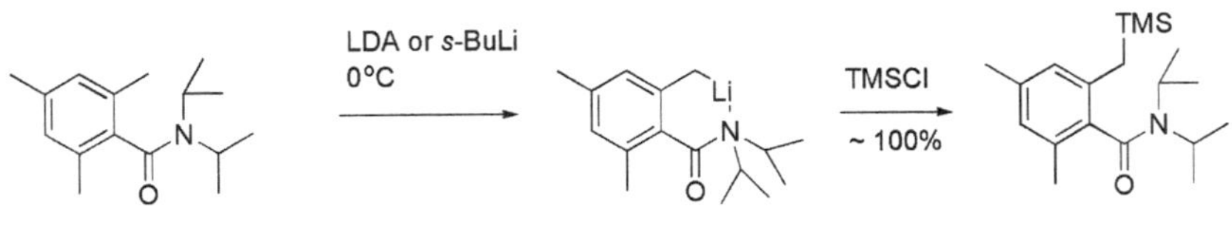

SCHEME 3.9 *Ortho* selective lateral lithiation [2001 JCSPT1 371 Clayden J].

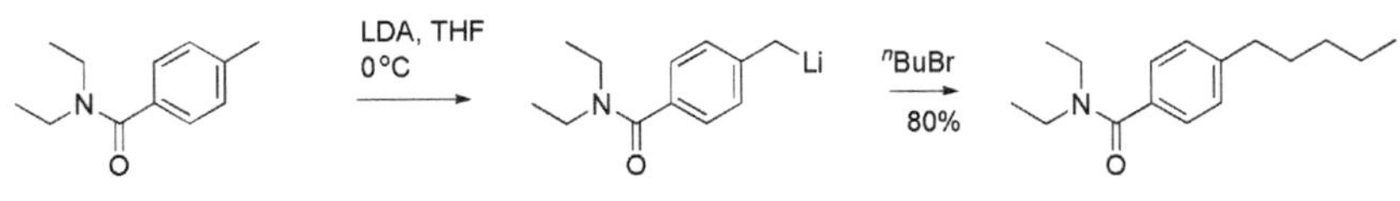

SCHEME 3.10 Lithiation of a *para* methyl group.

That the chelation is not essential is evident from the following example (Scheme 3.10). The tertiary carboxamide group can lead to lithiation of *para* or *meta* methyl groups. Under the influence of LDA in THF, toluamide (**28**) underwent lateral lithiation in the *para* methyl group to give lithio species (**29**), which was successfully butylated with butyl bromide to give homologated amide (**30**) (Scheme 3.10).

3.2 APPLICATIONS OF LATERAL LITHIATIONS

This section presents selected applications of lateral lithiations. It has been shown by Macos et al. that lateral lithiation can directly be applied to norditerpene derivatives to accede to the tricyclic diterpene natural products with abietane skeletons. Their approach relied upon the use of a carbamate directing group. The intermediate (**31**), derived from sclareol, a naturally occurring diterpenoid, was employed in the synthesis of (+)-ferruginol (**32**) (Scheme 3.11). The lateral methyl group in terpene derivative (**31**) was converted into an isopropyl group by two successive lateral lithiation-mediated methylations. Formation of both (**33**) and (**34**) was accomplished in excellent yields. The carbamate group in (**34**) was removed by the action of LAH to give enantiopure ferruginol (**32**).[10]

Aromatic methoxy groups well-known for directed *ortho* metalation are also capable of promoting lateral lithiations. Anisole (**35**), on reaction with BuLi, undergoes lithiation in the methyl group to give lithio species (**36**), an in situ reaction of which with cyclohexane carboxaldehyde (**37**) gives adduct (**38**) (Scheme 3.12).

Similar lithiation was applied to methyl ether (**39**). Its reaction with ^{s}BuLi, followed by treatment with *N*-formylpiperidine, resulted in benzyl formylated product (**40**) (Scheme 3.13) through lateral lithiated species (**41**).[11]

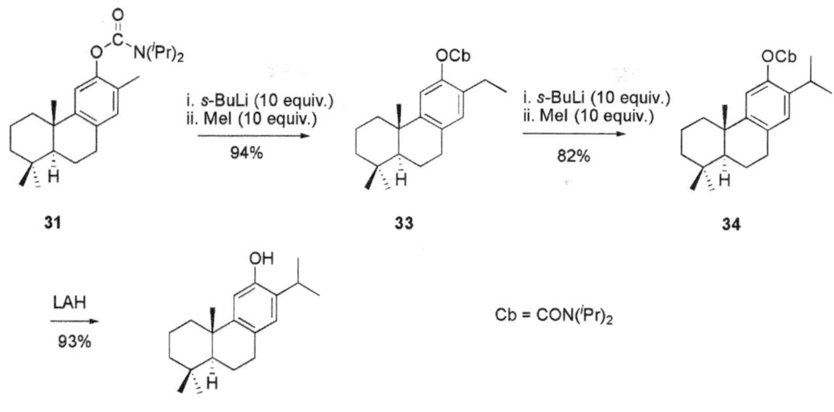

SCHEME 3.11 Carbamate, a directing group for a terpene synthesis [2010 THN 7773 Marcos IS].

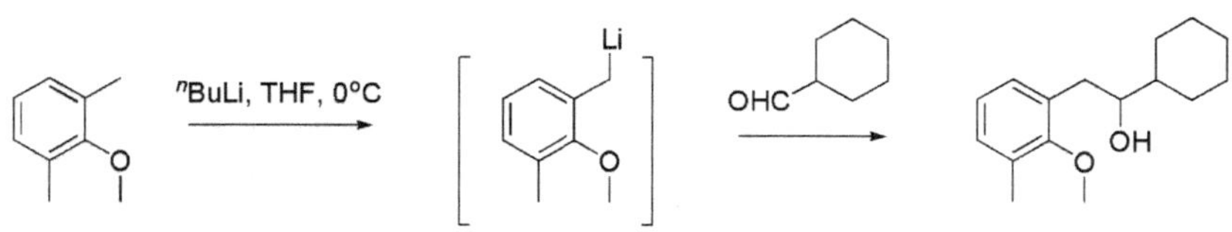

SCHEME 3.12 Lateral lithiation of dimethylanisole (**35**).

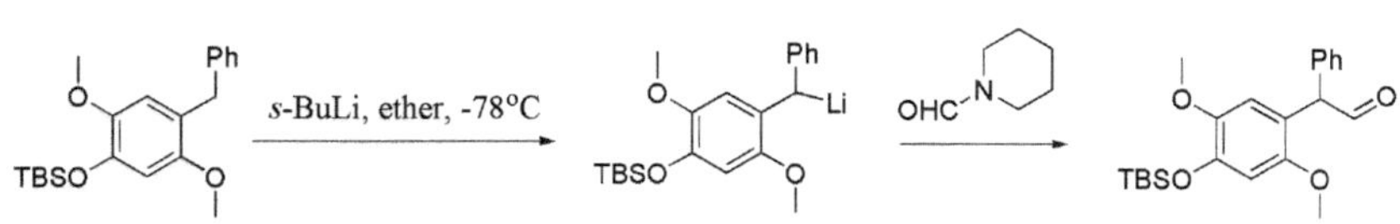

SCHEME 3.13 Formylation of a side chain in aromatics [2013 THS 1469 Wilkinson JA].

Methoxy assisted threefold lithiation was observed in the metalation of 4-hydroxy-6,7-dimethoxy-8-methyl-2-naphthoic acid (**42**) with sevenfold LTMP (Scheme 3.14).[12] It produced trianion (**43**). Interestingly, there was no sign of directed *ortho* lithiation, and the C7 methoxy group directed lateral lithiation resulting in the trianion (**43**). Subsequent methylation at C8 methide of the intermediate (**43**) yielded naphthoic acid (**44**) in excellent yield. The naphthoic acid (**44**) served as a precursor to a gossypol derivative.

Lateral lithiation has been advantageously utilized for the one-pot synthesis of 2-aryl-1,4-benzoxazin-3-ones via a cascade of reactions.[13] Tert-butyl lithium promoted tandem lateral lithiation—anionic cyclization—and the alkylation reaction of readily accessible N-Boc-O-benzyl derivatives (**45**), obtainable from

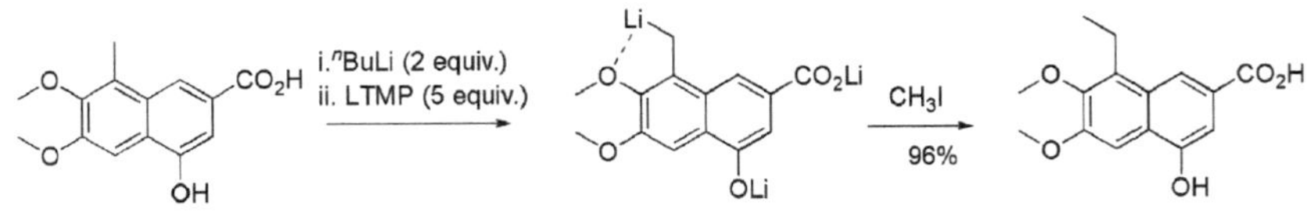

42 43 44

SCHEME 3.14 Methoxy directed lateral lithiation [2011 JOC 601 Mortier J].

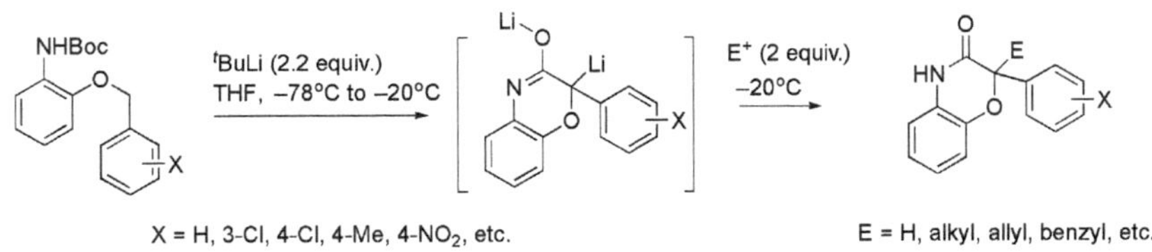

45 47 46

SCHEME 3.15 Lateral lithiation-initiated cyclization [2017 SYT 471 Spivey AC].

2-aminophenols, gave 2-aryl-1,4-benzoxazin-3-ones (**46**) (Scheme 3.15). The lithio oxazines (**47**) are considered the advanced reactive intermediates. The Boc protecting group in (**45**) plays dual roles. It assists lateral lithiation to form (**48**) from (**49**) as well as forming the reactive isocyanate (**50** and **51**) through loss of LiO^{t}Bu (Scheme 3.16). The released *tert*-butoxide is strong enough to effect the benzylic deprotonation of (**51**) to give (**52**), necessary for its subsequent alkylation to (**53**) and (**54**).

A heteroatom of a heterocycle can promote the lateral lithiations of aromatics. As shown in Scheme 3.17, the aziridine nucleus in (**55**) gave lateral lithiated species (**56**), which was alkylated with PrI to give propyl derivative (**57**).[14] During the lithiation process, the aziridine ring remained intact.

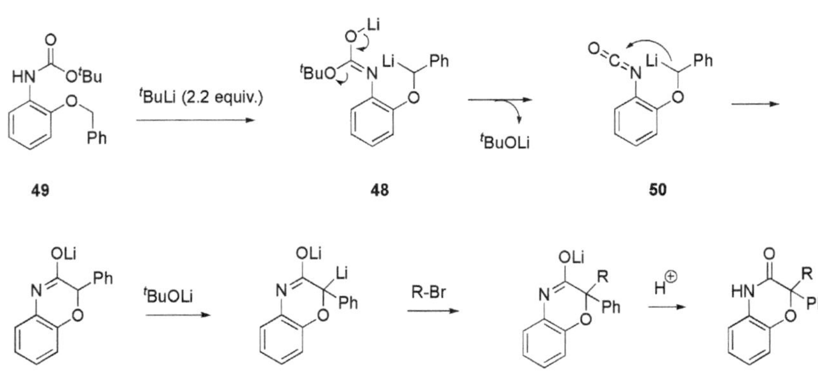

49 48 50

51 52 53 54

SCHEME 3.16 Possible mechanism of the oxazine formation.

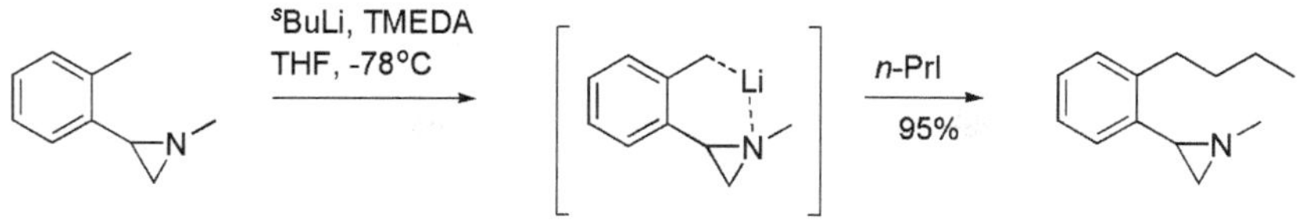

55 56 57

SCHEME 3.17 Aziridine-promoted lateral lithiation [2009 JOC 6319 Florio S].

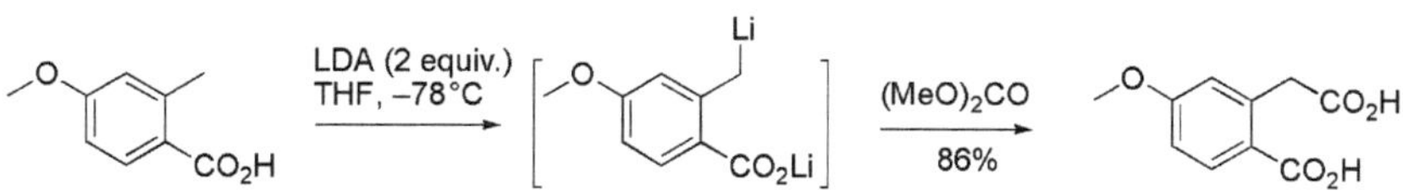

58 59 60

SCHEME 3.18 Lateral lithiation of an *o*-toluic acid [2002 ORS 2237 Hauser FM].

When there are two directing groups, namely OMe and CO_2H, in a substrate, the carboxyl group at the *ortho* position may exert its directing effect. As shown in Scheme 3.18, carboxy-assisted lateral lithiation wins over OMe directed *ortho* lithiation for toluic acid (**58**) to provide dilithio species (**59**) in the presence of LDA. Carboxylation of (**59**) with dimethyl carbonate gave homophthalic acid (**60**) after alkaline work-up.[15]

In the work of Shea et al., it is shown that an aromatic bromine substituent remains unaffected during lateral lithiation. Lithiation of bromotoluic acid (**61**) with LDA gave dilithiated species (**62**), which was selectively alkylated with allylic chloride (**63**) to give (**64**) (Scheme 3.19).[16]

The following reaction scheme illustrates the first example of extension of lateral lithiation to a naphthoic acid (Scheme 3.20).[1] The very first step of the scheme consists of a lateral lithiation-mediated carboxylation with diemthyl carbonate. Methoxybenzoic acid (**65**) was transformed to homophthalic acid (**66**) in good yield by lateral lithiation medication carboxylation (cf. Schemes 3.18 and 3.19). Acylation of (**66**) with acetic anhydride and pyridine gave acetohomophthalic anhydride (**67**). Alkaline hydrolysis of (**67**) with concomitant decarboxylation gave (**68**). Ring closure of the intermediate (**68**) using acetic

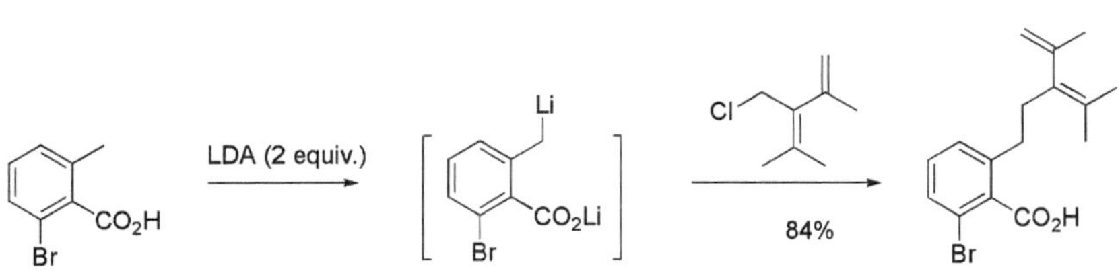

61 62 63 64

SCHEME 3.19 Lateral lithiation of a bromo *o*-toluic acid [1992 THN 7013 Shea KJ].

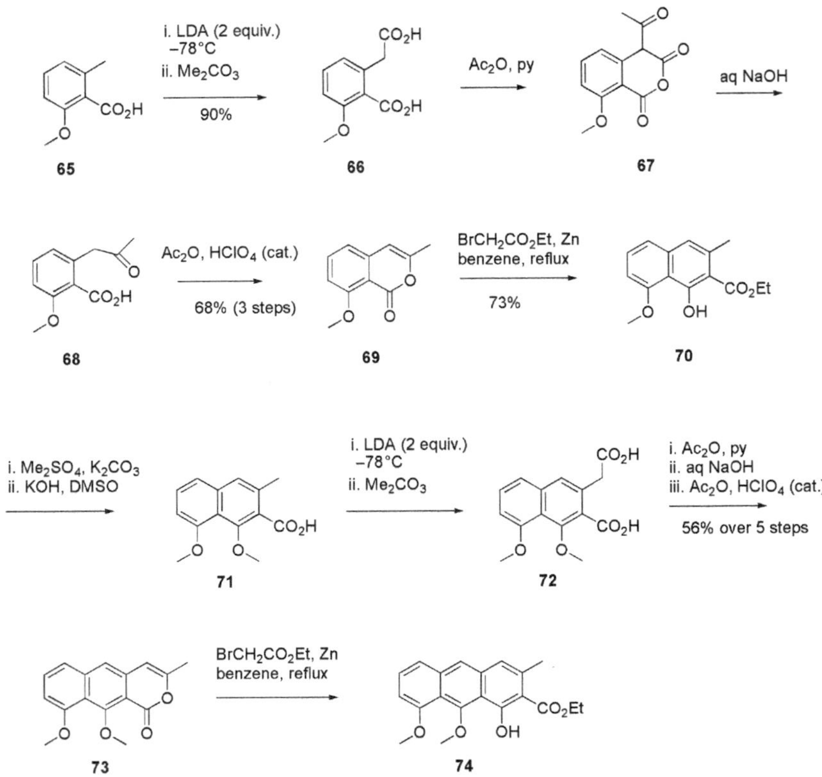

SCHEME 3.20 Hauser recursive lateral lithiation strategy for the synthesis of polynuclear phenolics [1977 JACS 4533 Hauser FM].

anhydride and perchloric acid yielded isocoumarin (**69**). This was transformed to naphthoate (**70**) in one step by Zn-mediated Reformatsky reaction with ethyl bromoacetate in refluxing benzene. *O*-Methylation of (**70**) with dimethyl sulfate and potassium carbonate followed by alkaline hydrolysis in dimethyl sulfoxide furnished naphthoic acid (**71**), a benzo analog of the starting toluic acid (**65**). The sequence applied to (**65**) was repeated with acid (**71**) to obtain through (**72**) naphtho[2,3-*c*]pyran (**73**), and finally 1-hydroxy-3-methylanthracenecarboxylate (**74**), embodying naturally occurring linear polynuclear aromatics.

Lateral lithiation under the influence of LDA is compatible with substituted ortho toluate ester. The resulting lithiated species can be phenylthiolated. With LDA (1 equiv.) and diphenyldisulfide (1.2 equiv.), monophenylthio derivative (**75**) is formed from toluate (**76**). Toluate (**76**) provides product (**77**) when the reaction is run with LDA (3 equiv.) and diphenyldisulfide (2.2 equiv.) (Scheme 3.21).[17] The above reactions served as the key steps for the synthesis of bisphthalide sulfone (**78**), a valuable intermediate of the first total synthesis of the 7,7′- linked bianthraquinone biphyscion via a one-pot double annulation

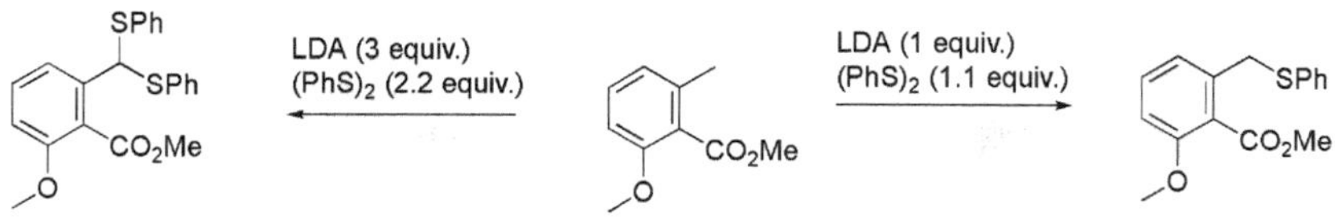

77 76 75

SCHEME 3.21 Lateral phenylthiolation of *ortho*-toluates [1980 SYS 72 Hauser FM].

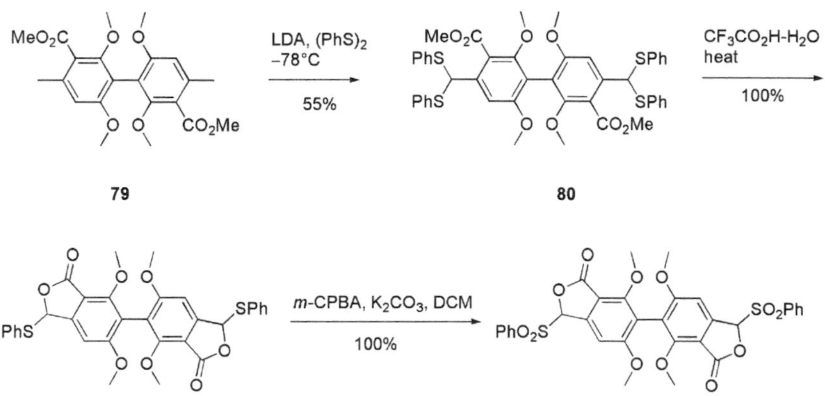

79 80

81 78

SCHEME 3.22 Synthesis of bis-phthalide sulfone donors [1999 ORS 671 Hauser FM].

(Scheme 3.22).[18] Treatment of (**79**) with 6 equiv. of LDA at −78°C and 4.4 equiv. of $(PhS)_2$ afforded tetraphenylthio product (**80**) in 55% yield. Reaction of (**80**) with CF_3CO_2H-H_2O afforded a quantitative yield of 3-phenylthiobisphthalide (**81**) as a mixture of diastereoisomers. It was oxidized with *m*-CPBA to give sulfone (**78**) in 100% yield. The sulfone (**78**) was successfully used for double annulation on the way to the total synthesis of biphyscion.

Although the aromatic carboxylic esters are known to induce lateral lithiations (Schemes 3.21 and 3.22), the *ortho* lithio species like (**82**) is unstable, and tends to dimerize via self-condensation. For the parent ester (**83**), the lithio species could be trapped as its silyl derivative (**84**) by reaction with trimethylsilyl chloride. The silyl derivative (**84**), however, on reaction with CsF, formed carbanion (**82**); this in turn reacted with 2-cyclohexenone to give adduct (**85**) (Scheme 3.23).[19]

Laterally lithiated species derived from *o*-toluate esters smoothly react with Weinreb amides to give the corresponding ketones. In Mohapatra synthesis of isocladosporin (**86**), ester (**87**) was coupled with Weinreb amide (**88**) to give ketone (**89**), which was converted to the target (**86**) in two simple steps: $NaBH_4$ reduction and acid catalyzed MOM deprotection (Scheme 3.24).[20] It may be noted that the condensation of the laterally lithiated toluates with Weinreb amides is a general reaction, and it has been applied to the Weinreb derived from methoxyacetic acid.

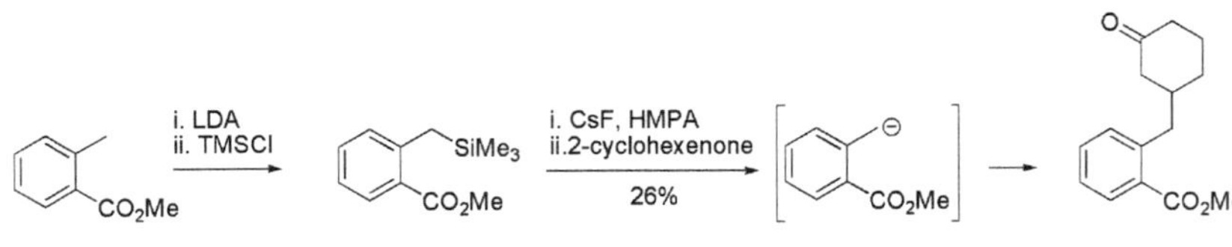

83 84 82 85

SCHEME 3.23 Lateral lithiation and cesiation [1985 CLS 339 Achiwa K].

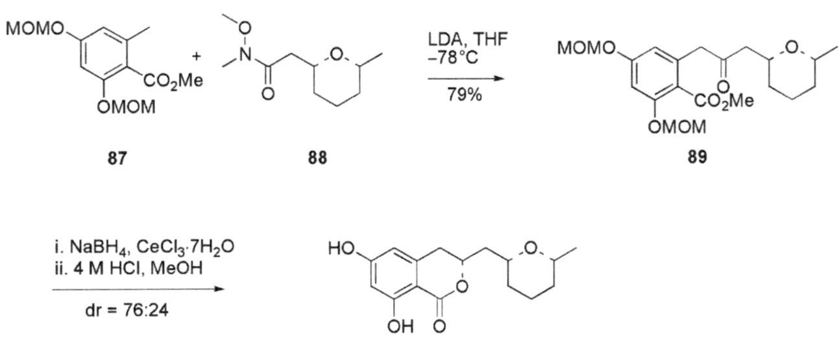

87 88 89

86 isocladosporin

SCHEME 3.24 Mohapatra synthesis of isocladosporin [2016 THS 53 Mohapatra DK].

Heteroatom assisted lateral lithiation has been achieved with methyl 5-methyl-2-furoate (**90**) by using bulky Lewis acid like aluminum tris(2,6-di-2-naphthylphenoxide) (ATNP). The vinylogous aldol reaction between (**90**) and aldehyde (**91**) proceeded smoothly in high yields in the presence of ATPN and LTMP to give (**92**). This type of chain lengthening is quite general with both aldehydes and ketones (Scheme 3.25).[21]

However, lateral lithiation fails for the sterically congested naphthoate such as (**93**). As shown in Scheme 3.26, the reaction of (**93**) with LDA/LTMP followed by ethyl formate/dimethylformamide returned the starting material.[22] There was no evidence for the formation of lithio species (**94**). Similar failures have been encountered in the Staunton-Weinreb annulation of methoxy-substituted toluates.

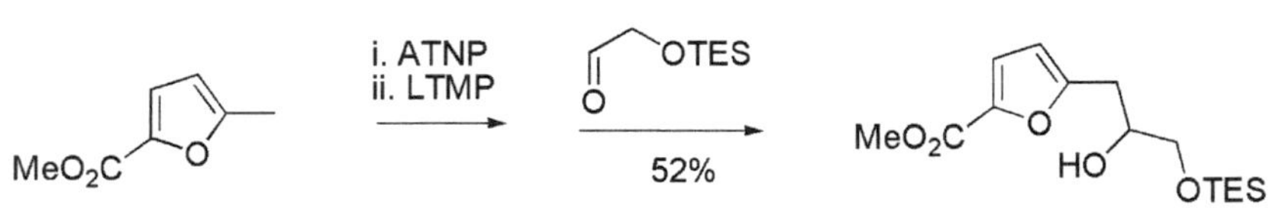

90 91 92

SCHEME 3.25 Lateral lithiation of furoates [2017 JOC 759 Sammakia T].

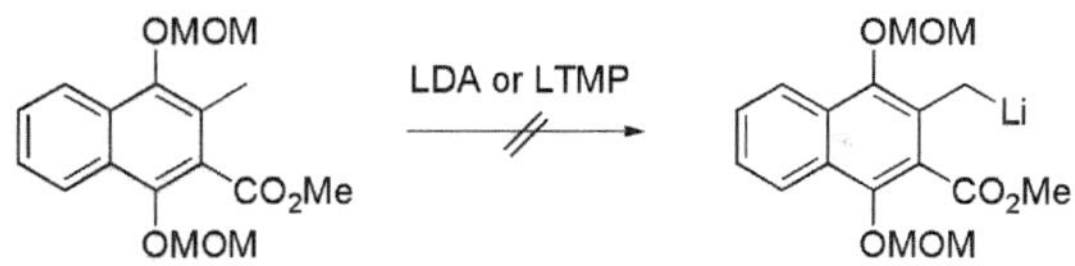

93 94

SCHEME 3.26 Failed lateral lithiation of a naphthoate [2015 NPC 1025 Kraus GA].

Like laterally lithiated toluate derived from (**76**), lithiated phthalides can also be carboxylated at C3. Reaction of phthalide (**9**) with LDA at −78°C gave (**95**), which was treated with carbon dioxide to furnish with acid (**96**). Further derivatization of (**96**) through the reaction of oxalyl chloride and *tert*-butanol produced *tert*-butyl ester (**97**) in 41% overall yield (Scheme 3.27).[23]

Similar chemistry was used in the Ciufolini synthesis of fredericamycin. It involved the addition of 3-lithiophthalide (**95**), prepared by deprotonation of phthalide (**9**), to aldehyde (**98**) to give the corresponding alcohol (**99**) en route to an alkylidene phthalide (Scheme 3.28).[24]

Like phthalides, their aza analogs, i.e., isoindolones (**100**), smoothly undergo lithiation at C3 upon treatment with LDA at −78°C. The resulting lithio species react with a variety of electrophilic reagents to give three-substituted isoindolone derivatives (**101**) (Scheme 3.29).[25]

Lateral lithiation has also been a key step in the synthesis of protoberberine alkaloids, namely (S)- and (*R*)-*O*-methylbharatamine (**102**). The reaction of the lateral carbanion, generated from (**103**) under the action of 1.1 equiv. of nBuLi at −72°C, with imine (**104**) gave two products: annulation product (**105**) and addition product (**106**) (Scheme 3.30).[26] The latter was converted to isoquinolone (**105**) upon treatment with equimolar ^{n}BuLi at −72°C. The chirality of the toluamide (**103**) was transferred completely to the products. LAH of

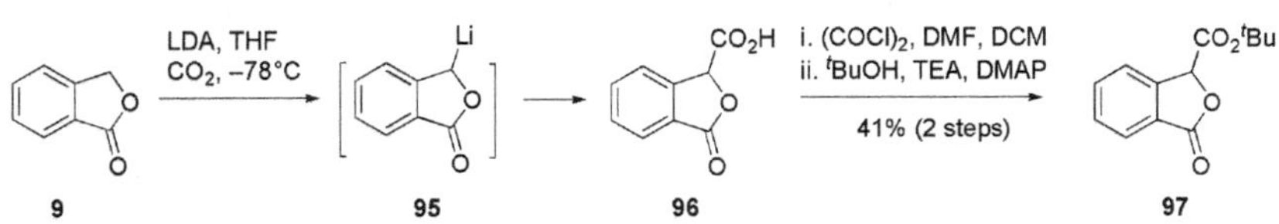

9 95 96 97

SCHEME 3.27 Carboxylation of phthalides [2012 CCS 4707 Lu Y].

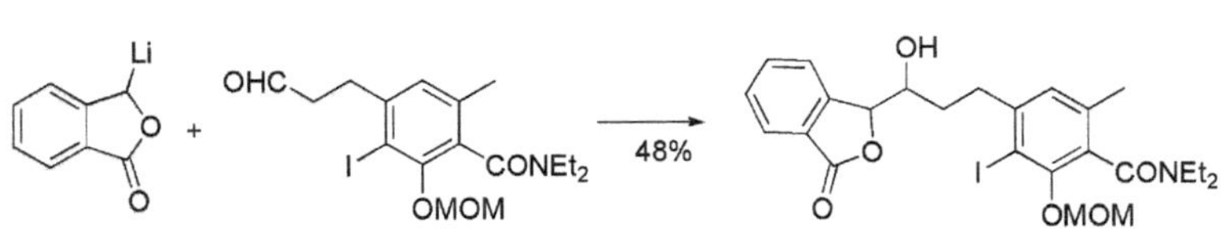

95 98 99

SCHEME 3.28 Reaction of 3-lithiophthalide with an aldehyde [1987 THS 171 Ciufolini MA].

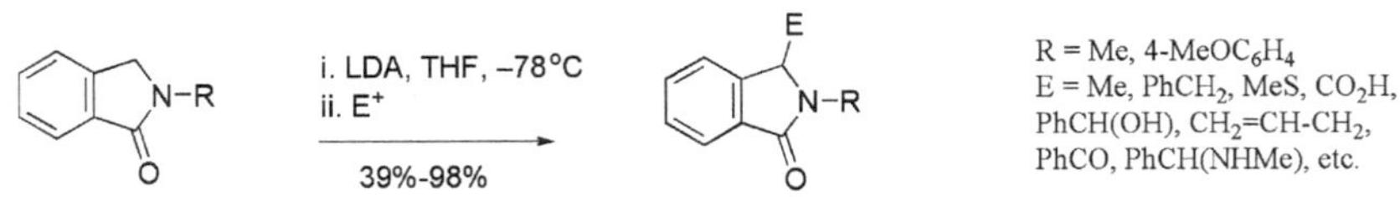

100 101

SCHEME 3.29 Lateral lithiation of isoindolinones [1998 THS 2319 Couture A].

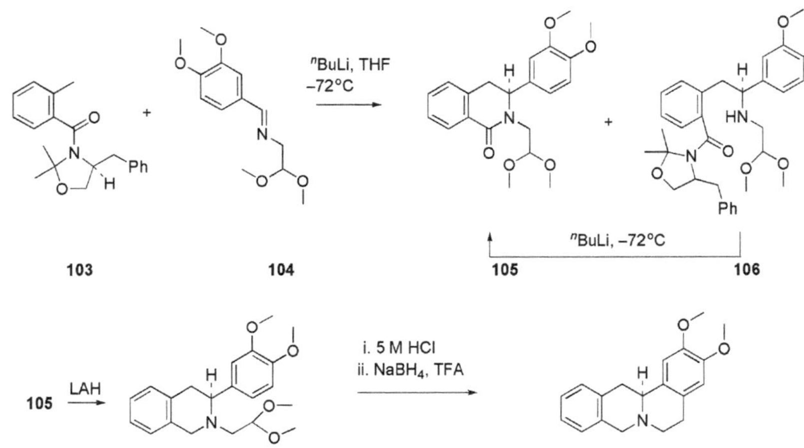

103 104 105 106

107

102 (S)-O-Methylbharatamine
overall yield 32%; 88% ee

SCHEME 3.30 Reactivity of toluamide containing an oxazolidine [2005 THY 2561 Rozwadowska MD].

isoquinolone (**105**) gave tetrahydroisoquinoline (**107**). The cyclization of (**107**) to (**102**) was affected with 88% ee by stirring in 5 M aqueous hydrochloric acid for 18 h at room temperature, followed by reduction with sodium borohydride in trifluoroacetic acid.

With benzonitriles, the reaction of lithiated toluamides results in hetero-annulation, leading to isoquinolones. For example, the reaction of (**108**) with (**109**) proceeded through (**110**) to give isoquinolone (**111**) (Scheme 3.31).[27]

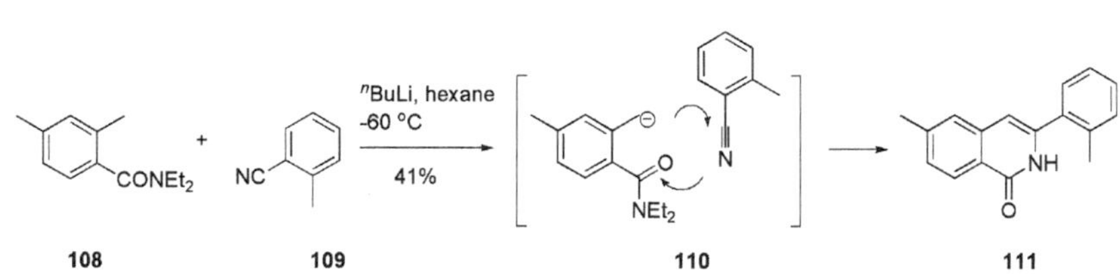

108 109 110 111

SCHEME 3.31 Toluamide-nitrile annulation [2015 EJMC 583 Cho W-J].

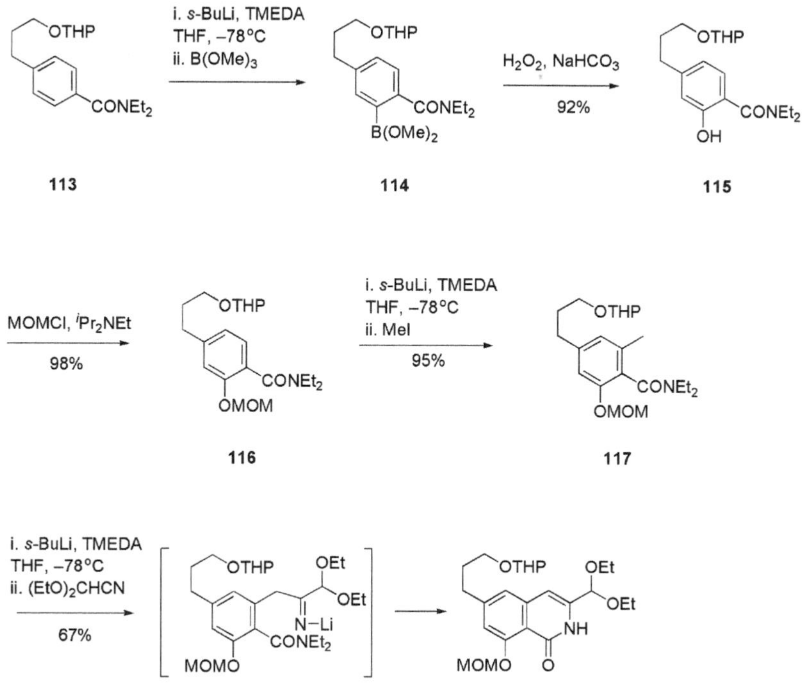

113 114 115 116 117 118 112

SCHEME 3.32 Aromatic amide group, a directing group [2011 BJOC 1475 Ciufolini MA].

Effectiveness of a diethylamide group in directed lithiations is appropriately illustrated by the multistep synthesis of (**112**), as a precursor used in the synthesis of fredericamycin A (Scheme 3.32).[28] *Para* substituted benzamide (**113**) was first converted into boronate (**114**) via directed *ortho* lithiation (*s*-BuLi, TMEDA), followed by trapping with methyl borate. The boronate (**114**) was oxidized with hydrogen peroxide to give phenolic compound (**115**). Protection of the phenolic OH group in (**115**) with MOMCl and Hunig base afforded the intermediate (**116**). Second, *ortho* lithiation of (**116**) followed by methylation with MeI selectively provided *ortho* toluate derivative (**117**), a substrate suited for lateral lithiation-initiated annulation. Treatment of toluate (**117**) with ^{s}BuLi-TMEDA at −78°C followed by treatment with diethoxyacetonitrile yielded heteroannulated product (**112**) via lithiated adduct (**118**). It is noteworthy that all the steps are high-yielding.

Likewise, Yang et al. synthesized stepholidine (**119**), a promising antipsychotic drug candidate, by an application of the lateral lithiation methodology. The necessary chiral induction was achieved by the use of (*S*)-*tert*-butylimine auxiliary. The reaction of tolunitrile (**120**) with enantiopure sulfoxide (**121**) in the presence of LDA gave the desired diastereomer (**122**) in 52% isolated yield

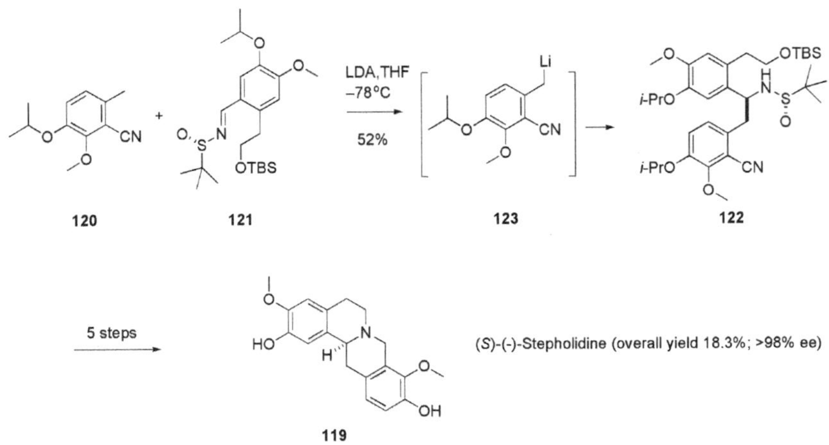

SCHEME 3.33 Yang synthesis of (*S*)-(−)-stepholidine [2010 HCS 2581 Yang Y].

through the lithiated species (**123**) (Scheme 3.33).[29] The transformation of the adduct (**122**) to alkaloid (**119**) required five more steps. These were (i) conc. HCl-catalyzed removal of the *N*-sulfinyl group and the TBS group, (ii) KOH-promoted lactamization, (iii) tosylation with TsCl followed by cyclization with NaH, (iv) LAH reduction, and (v) selective deprotection of isopropyl groups by BCl_3 treatment.

Both lateral lithiations and remote lithiations are induced by a remote amide group (Scheme 3.34).[30] As shown below, lithiation of a proximate methyl group in the adjacent benzene nucleus can be directed by an amide group. They are useful for the synthesis of phenanthrene natural products. Snieckus et al. combined both nuclear and remote lateral lithiation to achieve a total synthesis of gymnopusin (**124**). Differentially protected biphenyl carbamate (**125**), obtainable from benzene derivatives by the Suzuki coupling reaction, was treated with LDA followed MeI to give biphenyl amide (**126**) through remotely lithiated (**127**). This, on treatment with BuLi followed by methylation with MeI, gave (**128**) through (**129**) and (**130**). Selective deprotection of isopropyl groups of (**128**) furnished gymnopusin (**124**).

An early example of lateral metalation can be found in the synthesis of drug citalopram (**131**), as in Scheme 3.35. The reaction of (**132**) with 3-dimethylaminopropyl chloride afforded (**131**). Intrinsic acidity of C1 hydrogen, increased due to the *p*-cyano group, was sufficient to cause sodiation and subsequent alkylation.

Bodipy dyes, characterized as having 4,4-difluoro-4-bora-3a,4a-diaza-s-indacene core, are a well-known class of fluorophores. Their photophysical properties are sensitive to appended functional groups. Lateral lithiation has, conveniently, been utilized for selective introduction of the functional group at the *meso* methyl group.

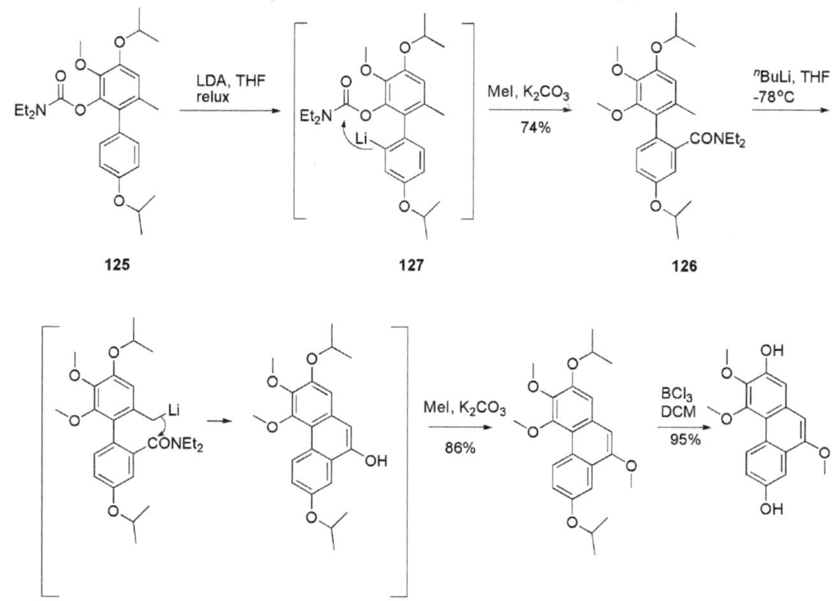

125 127 126

129 130 128 124

SCHEME 3.34 Directed remote lateral lithiation [2012 HCA 2680 Snieckus V].

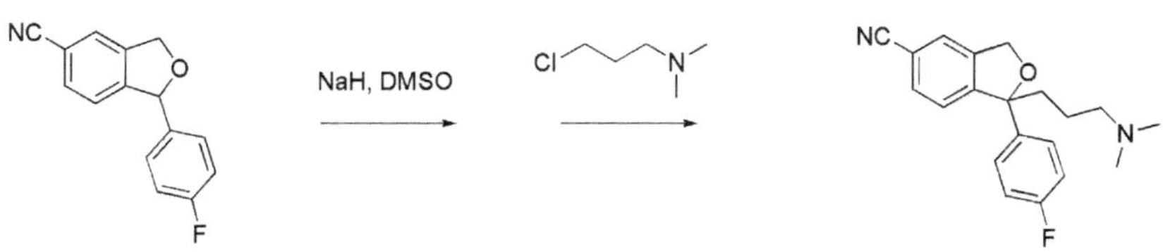

132 131

SCHEME 3.35 Alkylative process for citalopram.

As shown in Scheme 3.36, treatment of bodipy derivative (**133**) with LDA followed by electrophilic reagents results in selective functionalization of meso methyl group, giving derivatives (**134**) in good yields.[31] Apparently, there is no involvement of chelation control in the lithiation process.

In exceptional cases, benzene rings can facilitate lithiation of a methyl group in the adjacent benzene ring (Scheme 3.37).[32] Tetramethylbiphenyl (**135**) was laterally lithiated with ^{n}BuLi-sparteine to the dilithio species (**136**). It was characterized by its conversion into homologated alcohol (**137**) via carboxylation with CO_2 followed by LAH reduction of its methyl ester.

In summary, this chapter highlights the structural requirements for various selective lateral lithiations. For phthalides, there is no prerequisite of a directing group for necessary deprotonations. On the other hand, for *ortho* toluates, the nature of the directing group *ortho* to the benzylic positions is a crucial factor.

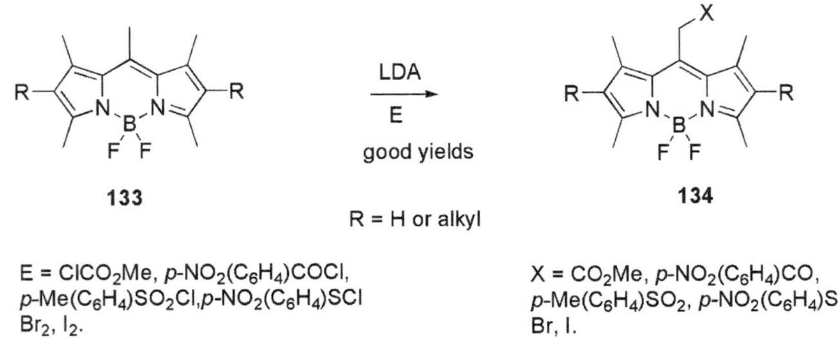

SCHEME 3.36 Selective lateral lithiation of a BODIPY derivative [2014 ORS 4364 Ortiz MJ].

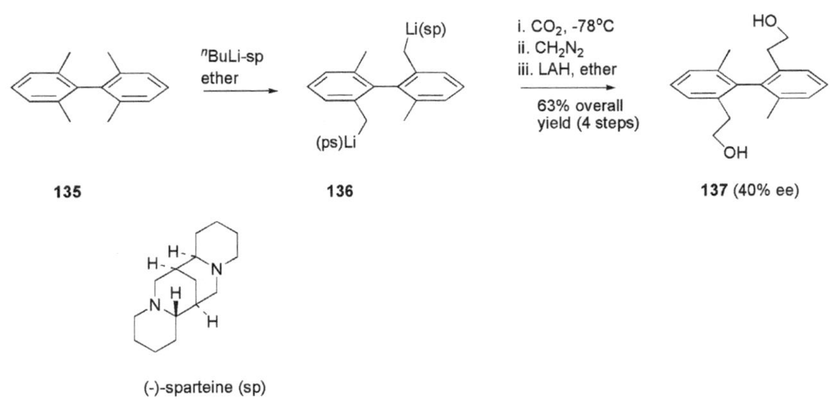

SCHEME 3.37 Lateral lithiation of 2,2′,6,6′-tetramethyl-1,1′-biphenyl [1988 JCSDT 2403 Raston CL].

REFERENCES

1. Hauser, F. M.; Rhee, R. *J. Am. Chem. Soc.* **1977**, *99*, 4533.
2. (a) Clark, R. D.; Jahangir, A. *Org. Reactions* **1995**, *47*. 419. (b) Clayden, J. Organolithiums: Selectivity for Synthesis; Elsevier Science Ltd, 2002.
3. Donner, C. D. *Tetrahedron* **2013**, *69*, 3747.
4. Madelung, W. *Ber.* **1912**, *45*, 1128.
5. Joulihan, H. W.; Parrino, V. A.; Uike, Y. *J. Org. Chem.* **1981**, *46*, 4511.
6. (a) Smith, A. B.; Visnick, M.; Haseltne, J. N.; Sprengeler, P. A. *Tetrahedron* **1986**, *42*, 2957. (b) Gribble, G. W. *J. Chem. Soc. Perkin Trans. 1* **2000**, 1045.
7. Huang, Y.; Chan, G. H.; Chiba, S. *Angew. Chem. Int. Ed.* **2017**, *56*, 6544.
8. Bartoli, G.; Palmieri, G.; Petrini, M. *Tetrahedron* **1990**, *46*, 1379.
9. Clayden, J.; Johnson, P.; Pink, J. H. *J. Chem. Soc. Perkin Trans.* **2001**, *1*, 371.
10. Marcos, I. S.; Beneitez, A.; Moro, R. F.; Basabe, P.; Diez, D.; Urones, J. G. *Tetrahedron* **2010**, *66*, 7773.

11. Slabu, I.; Rossington, S. B.; Killoran, P. M.; Hirst, N.; Wilkinson, J. A. *Tetrahedron Lett.* **2013**, *54*, 1489.
12. Le, T.-T.; Chau, N.T.-T.; Nguyen, T.-T.; Brien, J. N.; Thai, T.-T.; Nourry, A.; Castanet, A.-S.; Nguyen, K.P.-P.; Mortier, J. *J. Org. Chem.* **2011**, *76*, 601.
13. Bodero, O.; Spivey, A. C. *Synlett* **2017**, *28*, 471.
14. Dammacco, M.; Degennaro, L.; Florio, S.; Luisi, R.; Musio, B.; Altomare, A. *J. Org. Chem.* **2009**, *74*, 6319.
15. Hauser, F. M.; Dorsch, W. A.; Mal, D. *Org. Lett.* **2002**, *4*, 2237.
16. Jackson, R. W.; Higby, R. G.; Gilman, J. W.; Shea, K. *J. Tetrahedron* **1992**, *48*, 7013.
17. Hauser, F. M.; Rhee, R. P.; Prasanna, S.; Weinreb, S. M.; Dodd, J. H. *Synthesis* **1980**, 72.
18. Hauser, F. M.; Gauuan, P. J. F. *Org. Lett.* **1999**, *1*, 671.
19. Aono, M.; Terao, Y.; Achiwa, K. *Chem. Lett.* **1985**, 339.
20. Mohapatra, D. K.; Maity, S.; Banoth, S.; Gonnade, R. G.; Yadav, J. S. *Tetrahedron Lett.* **2016**, *57*, 53.
21. Hartwig, W. T.; Sammakia, T. *J. Org. Chem.* **2017**, *82*, 759.
22. Kraus, G. A.; Dong, P. *Nat. Prod. Commun.* **2015**, *10*, 1025.
23. Luo, J.; Wang, H.; Zhong, F.; Kwiatkowski, J.; Xu, L.-W.; Lu, Y. *Chem. Commun.* **2012**, *48*, 4707.
24. Ciufolini, M. A.; Browne, M. E. *Tetrahedron Lett.* **1987**, *28*, 171.
25. Couture, A.; Deniau, E.; Ionescu, D.; Grandclaudon, P. *Tetrahedron Lett.* **1998**, *39*, 2319.
26. Chrzanowska, M.; Dreas, A.; Rozwadowska, M. D. *Tetrahedron: Asymmetry* **2005**, *16*, 2954.
27. Khadka, D. B.; Woo, H.; Yang, S. H.; Zhao, C.; Jin, Y.; Le, T. N.; Kwon, Y.; Cho, W.-J. *Eur. J. Med. Chem.* **2015**, *92*, 583.
28. Turner, C. D.; Ciufolini, M. A. *Beilstein J. Org. Chem.* **2011**, *7*, 1475.
29. Cheng, J.; Fu, L.; Ling, C.; Yang, Y. *Heterocycles* **2010**, *81*, 2581.
30. Wang, X.; Fu, J.-m.; Snieckus, V. *Helv. Chim. Acta* **2012**, *95*, 2680.
31. Palao, E.; de la Moya, S.; Agarrabeitia, A. R.; Esnal, I.; Banuelos, J.; Lopez-Arbeloa, I.; Ortiz, M. *J. Org. Lett.* **2014**, *16*, 4364.
32. Engelhardt, L. M.; Leung, W. P.; Raston, C. L.; Salem, G.; Twiss, P.; White, A. H. *J. Chem. Soc., Dalton Trans.* **1988**, 2403.

Chapter 4

[4+2] Annulation of 3-Nuclofugal Phthalides (Hauser Annulation)

The Hauser [4+2] annulation is now defined as an annulation of a C3-nucleofugal phthalide (benzo[*c*]furanone) (**1a**) with an acceptor molecule having a polarized or otherwise activated (e.g., aryne) multiple bond to furnish a naphthalene-1,4-diol (**1b**) or 1,4-naphthoquinone (**1c**) (Scheme 4.1). This definition encompasses a wide variety of phthalides with different nucleofuges at C3 as summarized in Chapter 2, and their benzo analogs and heterocyclic analogs. In this book, we term these phthalides as the donors of the Hauser annulation. They largely differ in the nature of the nucleofuges, γ-lactone moiety of the phthalides being the common structural unit.[1]

The success of the annulations is crucially dependent on the reaction conditions as well as the nucleofuges of the donors. Among the reported donors (**2–9**) for Hauser annulation (Fig. 4.1), the most frequently used are Hauser donors (**2**), Kraus donors (**3**), and Kraus-Mal donors (**4**). Since the introduction of LiO^{t}Bu[2] as the base by Hauser and Mal, and LHMDS[3] by Tatsuta et al., the efficacies of the annulations have greatly improved. In certain cases, Cs_2CO_3[4] is found to be effective. Of late, Kraus donors (**3**) have shown greater promise and potential in their reactivity and applicability. However, one advantage of the Hauser donors (**2**) over the Kraus donors (**3**) is that they are more easily prepared; there are also a good number of routes available. The applications of phenylthiophthalides (**4**) in Hauser annulation have received renewed interests since Mal's reinvestigation on their reactions under different conditions.[5] The progress of the chemistry of phthalide-3-phosphonates (**5**) is, however, hindered due to the limited choice of their preparation.[6] The donors (**6–9**) are occasionally used in organic synthesis.

The typical acceptors for the Hauser annulation are acrylates, conjugated dienoates, allene carboxylates, acrylonitriles, enones, vinyl sulfones, imines, vinylphosphonates, nitroethenes, arynes, quinone monoketals, etc. Although the results of Hauser annulations are general in nature, they may vary with structures of the acceptors. In the following sections, the reactivities of various kinds of acceptors to the Hauser annulation are described.

Anionic Annulations in Organic Synthesis. https://doi.org/10.1016/B978-0-12-813800-7.00004-X

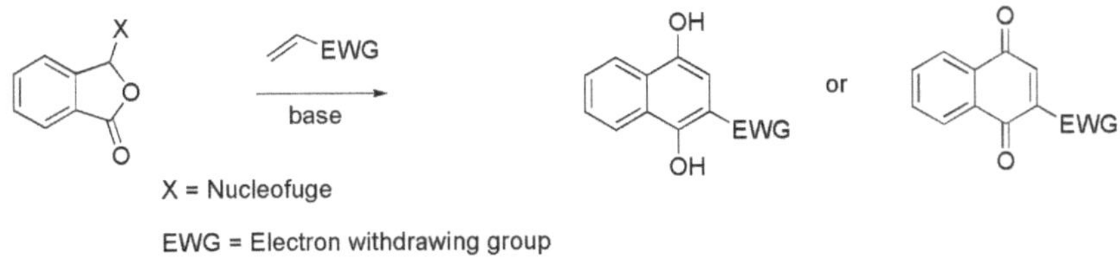

1a **1b** **1c**

SCHEME 4.1 Abbreviated form of the Hauser [4+2] phthalide annulation.

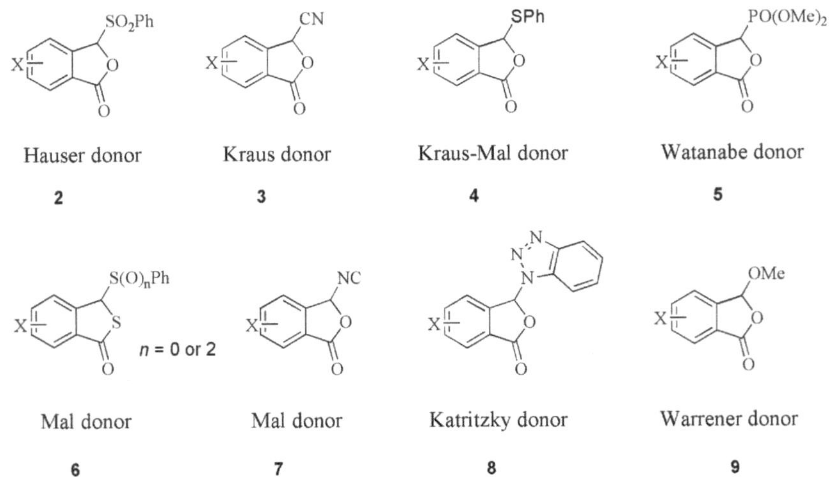

FIG. 4.1 Phthalides with nucleofuges at C3 for the Hauser annulation [X = a substituent like alkyl, alkoxy, bromo, cyano, fluoro, chloro, etc.]

4.1 CONJUGATED ESTERS, LACTONES, NITRILES, ALLENE CARBOXYLATES

Activated phthalide donors (**2–9**), differing in the nature of the nucleofuges at C3 (Fig. 4.1), give the same product in most cases, irrespective of the nucleofuges. The earliest example of Hauser annulation is the one with ethyl acrylate (**10**), which is often used as the test acceptor for a phthalide annulation. As shown in Scheme 4.2, the yields of the annulation vary with donors, bases, solvents, and conditions. The reaction of the acrylate (**10**) with phthalide sulfone (**11**) furnished quinol (**12**), which was, without isolation, *O*-methylated with Me_2SO_4-K_2CO_3 to give dimethyl ether (**13**) in 32% overall yield.[7] When the reaction was carried out with cyanophthalide (**14**) in the presence of LDA-HMPA, the product (**11**) was obtained in an improved yield.

In contrast to the cyanide (**14**), the corresponding isocyanide (**15**) gave a poorer yield for the same product, i.e., (**12**).[8] However, 3-cyano-7-methoxyphthalide underwent cyclocondensation with methyl acrylate to give

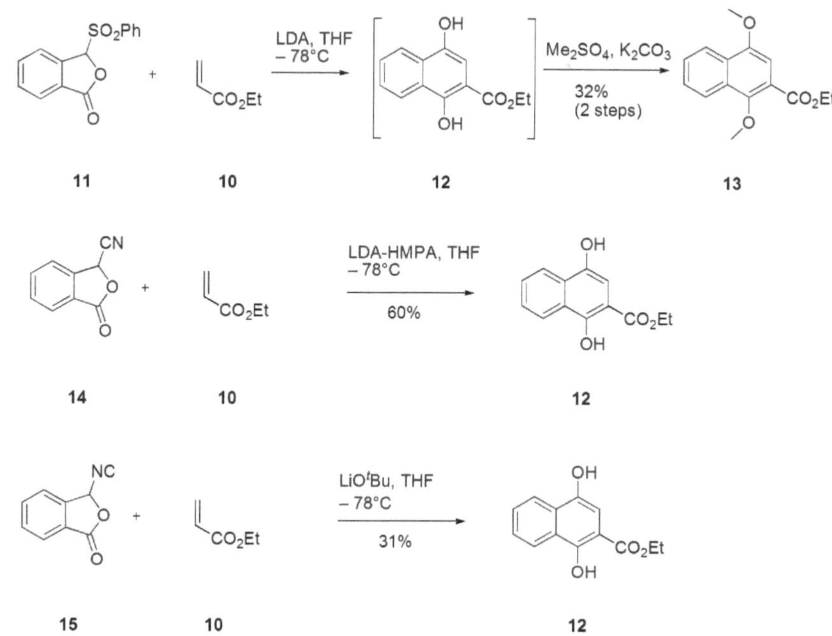

SCHEME 4.2 Annulation with ethyl acrylate with different donors [1978 JOC 178 Hauser FM; 1983 JOC 3439 Kraus GA; 2015 SYS 2473 Mal D].

the annulated product in 95% yield. In 2007, Brimble showed the highest ever yield of a Hauser annulation with an acrylate. The annulation of 3-cyano-4,6-dimethoxyphthalide (**16**) with ethyl acrylate afforded the naphthol (**17**) in 99% yield.[9] In contrast, the annulation of 3-cyano-5,7-dimethoxyphthalide (**18**) furnished the corresponding product (**19**) in 0 to 54% yield. On a large scale, the same reaction was irreproducible. *Tert*-butyllithium, among other bases used, was partially effective, giving the product (**19**) in a moderate yield. The marked difference in reactivities of the cyanophthalides (**16** and **18**) has been attributed to the effect of the methoxy substituents. For 5,7-dimethoxyphthalide (**18**), the methoxy groups are *para* and *ortho* to the carbonyl group. Consequently, they exert resonance effect on the carbonyl group, resulting in its reduced electrophilicity for the Dieckmann step in the intermediate (**20**) (Scheme 4.3).

Similar to the unsubstituted acrylates (e.g., **10**), simple crotonate ester (**21**) undergoes annulation with sulfone (**11**), cyanide (**14**), sulfide (**22**), and phosphonates (**5**) (X=H) to give the annulation product (**23**) in the range of 61%–89% yields. At low temperature under the conditions of Kraus et al., the Michael adduct derived from the C3 anion of phenylthiophthalide (**22**) was isolated. Mal et al., however, have shown that such an annulation in the presence of LiO^{t}Bu in the temperature range −60 to 0°C gives the expected product (**23**) in 61% yield (Scheme 4.4).[5]

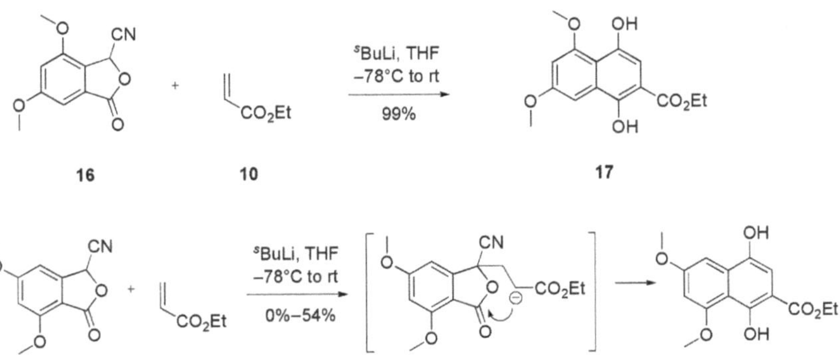

16 10 17

18 10 20 19

SCHEME 4.3 Substituent effects on the Hauser annulations [2007 THN 880 Brimble MA].

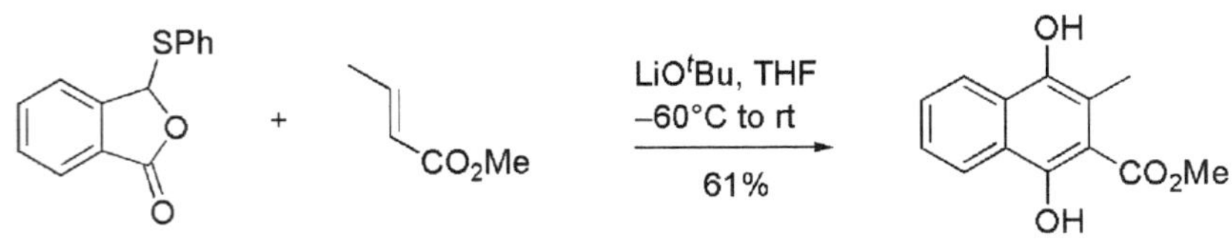

22 21 23

SCHEME 4.4 Annulation of phenylthiophthalide (**22**) with crotonate [1999 JCRS 30 Mal D].

In general, the yields of the annulation products from crotonates are higher than those with unsubstituted acrylates. The annulation of phthalide sulfone (**24**) with ethyl 2-hexenoate (**25**) is unique in that the yield of the product (**26**) is highest in an annulation with a crotonate in the presence of LDA (Scheme 4.5). The naphthoate (**26**) was degraded in a few steps to give the formyl derivative (**27**) for the synthesis of polycyclic compounds.[10]

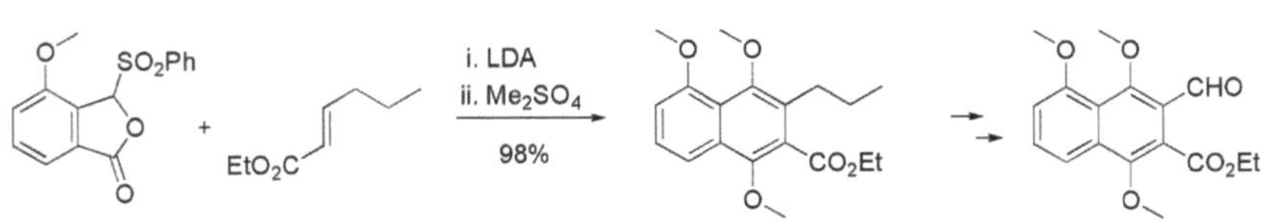

24 25 26 27

SCHEME 4.5 A high-yielding annulation of a phthalide sulfone [1981 JACS 6378 Hauser FM].

The efficacy of phenylthiophthalide (**22**) was demonstrated by its tandem annulation with crotonate derivative (**27**) to give furan-containing tricyclic compound (**29**) in 61% yield. Treatment of (**29**) with PTSA resulted in the elimination of ethanol, leading to a short synthesis of furomollugin (**30**) (Scheme 4.6).[11]

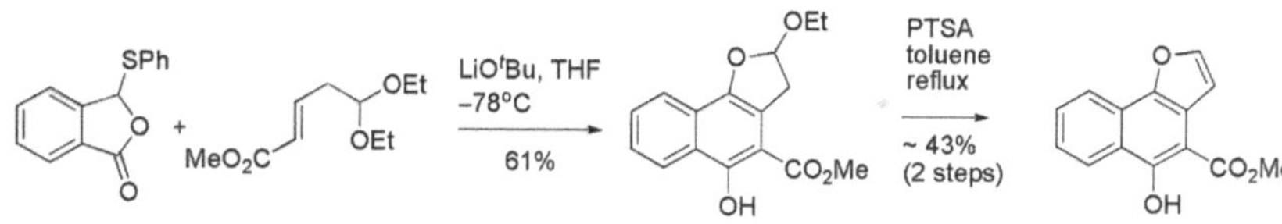

22 28 29 30

SCHEME 4.6 A short synthesis of furomollugin [2015 NPC 1025 Kraus GA].

It appears that the size of the γ-substituent of crotonates has little effect on the efficiency of the annulations. The annulation of sulfone phthalide (**31**) with crotonates (**32**) in the presence of LTB provided naphthoquinol (**32**) in 85% yield (Scheme 4.7).[12]

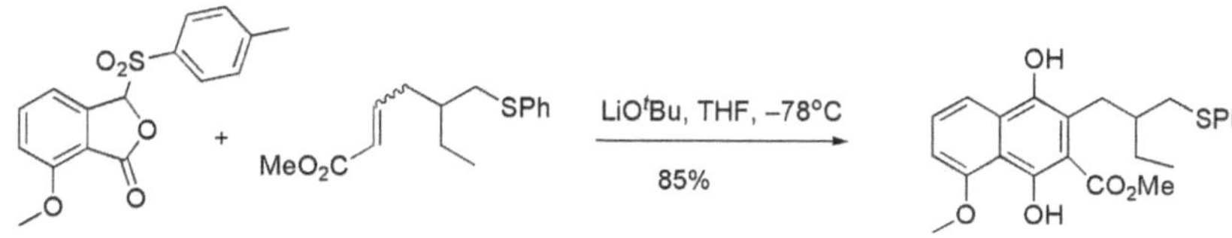

31 32 33

SCHEME 4.7 Annulation with substituted crotonate [1991 JKCS 756 Rho YS].

The annulation reactivity of isomeric fumarates and maleates has been little studied. Both dimethyl fumarate (**34a**) and dimethyl maleate (**34b**), when reacted with phthalide-3-phosphonate (**35**) in the presence of lithium *tert*-butoxide in THF, gave 1,4-dihydroxynaphthalenedicarboxylate (**36**). The yield of the product (**36**) was 98% with the fumarate, whereas it was 74% with dimethyl maleate.[6] Benzotriazolylphthalide (**7**) (X=H) gave the same product (**36**) in 40% yield in annulation with dimethyl fumarate (Scheme 4.8).

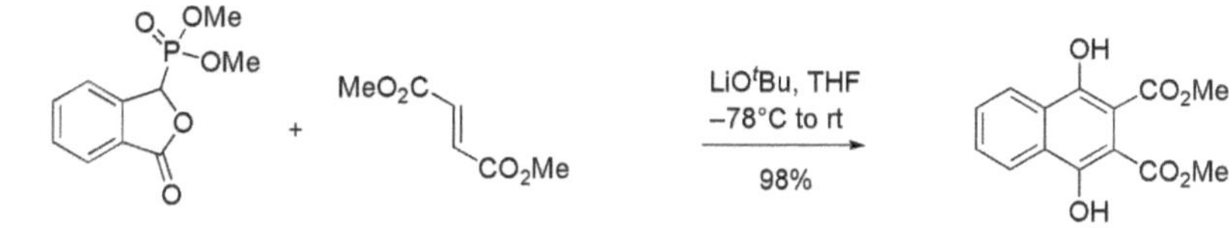

35 34a 36

SCHEME 4.8 Annulation of fumarate with phthalide phosphonate [1993 CPB 968 Watanabe M].

Cinnamate esters, expectedly, are sufficiently reactive to undergo annulation with various Hauser donors. Treatment of cyanophthalide (**37**) with LDA in THF followed by cinnamate (**38**) afforded naphthoquinol (**39**), which, without isolation, was air-oxidized in aq KOH-THF to give 1,4-naphthoquinone (**40**) (Scheme 4.9).[13] The naphthoquinone (**40**) was reductively methylated with

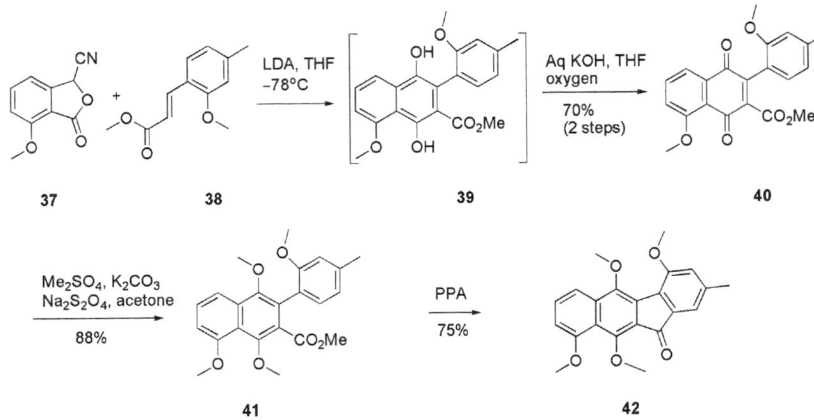

SCHEME 4.9 Hauser annulation with a cinnamate [1992 JOC 2774 Gould SJ].

Me_2SO_4-K_2CO_3-$Na_2S_2O_4$ to give quinol ether (**41**), which served as an intermediate for prekinamycin. Intramolecular Friedel-Crafts reaction of (**41**) with polyphosphoric acid (PPA) furnished benzindenone (**42**), featuring prekinamycin scaffold. The same sequence of reactions has successfully been used in the synthesis of ^{13}C labeled prekinamycins. It may be noted that the annulation of a cinnamate with the Katritzky synthon (**7**) (X = H) gives lower yield (~40%) of the annulation product.[14]

A similar scheme of reactions was used for the synthesis of phenanthroviridin aglycon (**43**) (Scheme 4.10). The Hauser annulation of the cyanophthalide (**37**) with 2-(trimethylsilyl)ethyl ester (**44**) gave naphthoquinone (**45**) via naphthoquinol (**46**). Reductive methylation of the quinone (**45**) with Me_2SO_4-K_2CO_3-$Na_2S_2O_4$ to dimethyl ether (**47**) followed by functional group interconversions involving desilylation, mixed anhydride formation, and aminolysis gave carboxamide (**48**). Nuclear lithiation of the bromo carboxamide (**48**) with ^{t}BuLi, followed by formylation of the resultant lithio intermediate with DMF, provided the precursor (**49**) of phenanthroviridin aglycon (**43**).

Compared to the earlier schemes (cf. Schemes 4.9 and 4.10), a different course of reactions ensues when cinnamate acceptors bear a functionalized *ortho* appendage. The concomitant formation of two six-membered rings via a Hauser-initiated tandem annulation is realized. Such reactions are possible when the appendages are an ester, a bromomethyl, or a cyano group. Thus, a series of benzo[*d*]naphtho[*b*]pyranones (**50**) were obtained in good yields. This tandem annulation is depicted in Scheme 4.11.[15] The reaction is quite general with regard to various substituents in both the donors (**51**) and the acceptors (**52**). The annulation is proposed to take place via initial Hauser annulation between (**51**) and (**52**), forming the trioxide (**53**).

As is evident from the following tandem annulation (Scheme 4.12), the *cis* geometry in the cinnamate acceptor (**54**) has little effect on the efficiency of the annulation. Furthermore, the lactone ring of the tetracycle (**55**) is directly derivable from a nitrile appendage. The reaction between the phthalide sulfone (**11**)

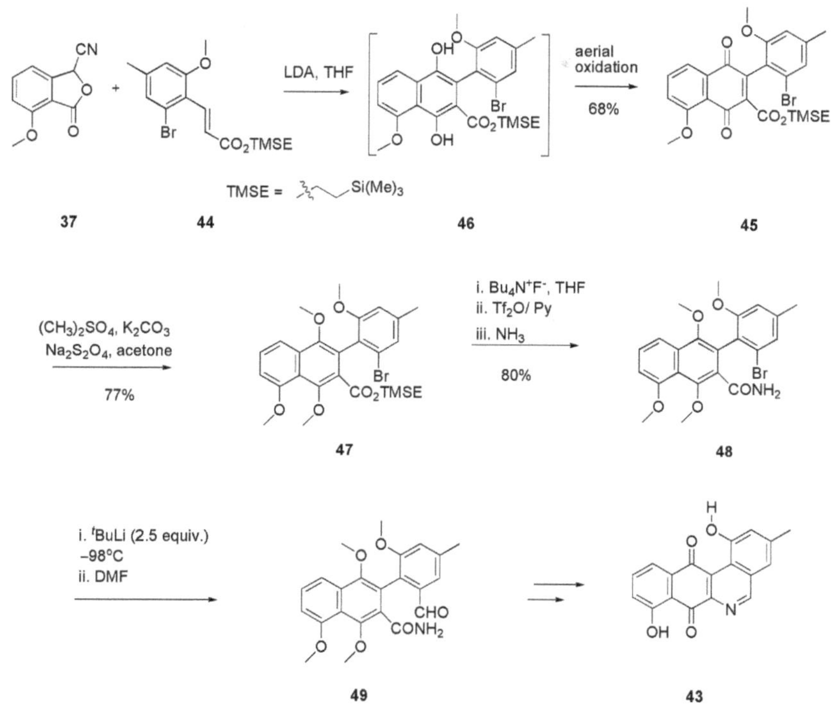

SCHEME 4.10 Hauser annulation route to the synthesis of phenanthroviridin [1992 JOC 2774 Gould SJ].

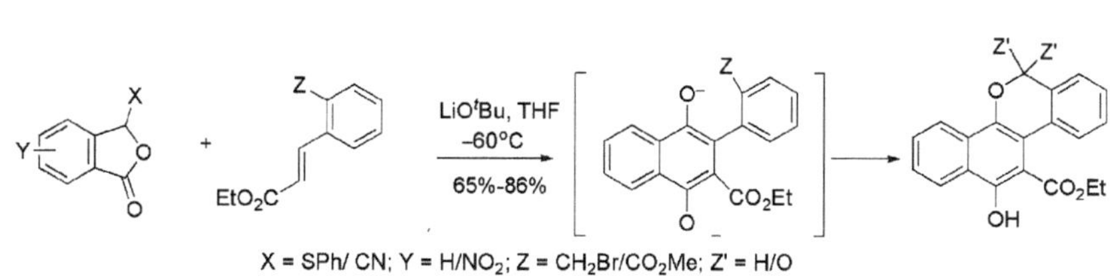

51 52 53 50

SCHEME 4.11 Hauser-initiated tandem annulation [2008 THN 3253 Mal D].

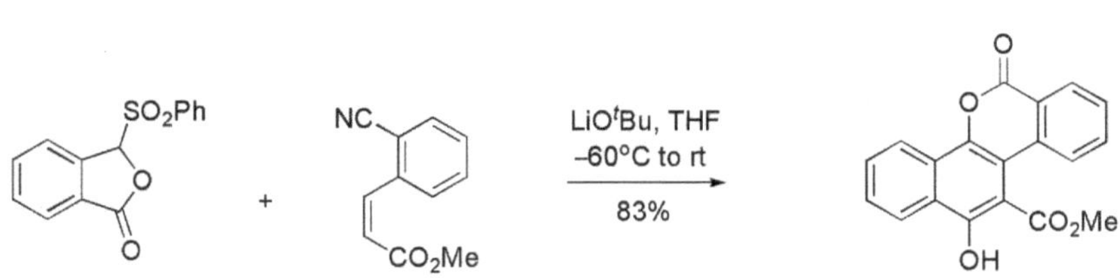

11 54 55

SCHEME 4.12 Tandem annulation with a *cis* cinnamate [2008 THN 3253 Mal D].

and the cinnamate (**54**) in the presence of LTB furnished the tetracyclic product (**55**) in 83% yield in a one-pot operation.

As a probable mechanism of the annulation, it is suggested that the Hauser annulation between (**54**) and anion of (**11**) affords quinol dioxide (**56**), which undergoes further cyclization to form an imidate anion (**57**) (Scheme 4.13). During the acidic work-up, the anion (**57**) is hydrolyzed via intermediates (**58** and **59**) to the tetracyclic product (**55**).

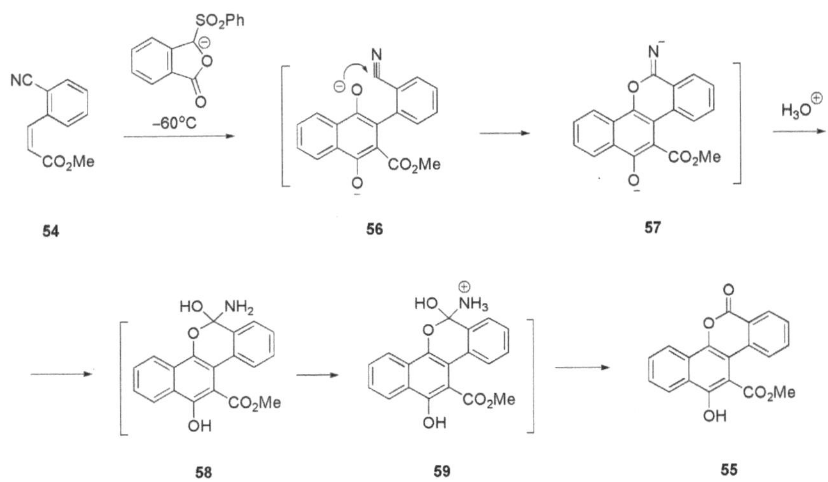

SCHEME 4.13 Mechanism of Hauser-initiated tandem annulation of a cyano-appended cinnamate.

The required *cis* cinnamate ester (**54**) was prepared through an uncommon fragmentation. Nitroso β-naphthol (**60**) was tosylated to give dearomatized product (**61**) (Scheme 4.14). Treatment of the intermediate (**61**) with alkali caused fragmentation to give the cinnamic acid (**62**). DBU-CH_3I promoted methylation**16** of (**62**) furnished the methyl ester (**54**) in excellent yield.

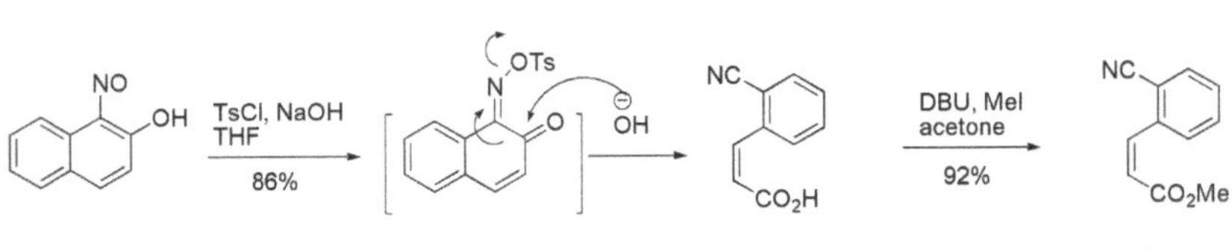

SCHEME 4.14 Preparation of the *cis* cinnamate (**54**) [2008 THN 3253 Mal D].

The Hauser annulation of functionalized cinnamates is impressively selective, as shown in Scheme 4.15. The reaction between the phthalide (**14**) and acceptor (**63**) afforded pentacycle (**64**) in a single step in excellent yield, although the acrylate appendage in (**63**) is flanked by two reactive ester units as *ortho* substituents. Evidently, the anticipated side reactions are slower than the initial annulation, to enable selective formation of trioxide (**65**). The trianion then undergoes two-fold lactonization to give fused benzopyranone (**64**).[15]

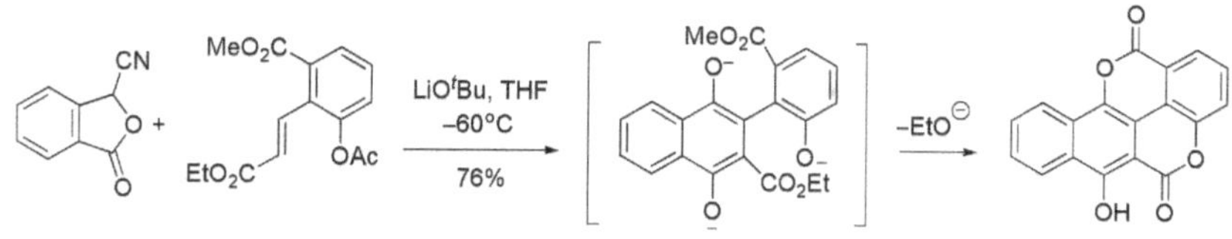

14 63 65 64

SCHEME 4.15 Tandem annulation [2008 THN 3253 Mal D].

Acrylates (**66–69**) with sugar moieties appended to their β-position are successfully annulated in the presence of LTB to give naphthoquinone/naphthoquinol-carbohydrate hybrids in excellent yields (Fig. 4.2). The sugar moieties do not apparently affect the efficiency of the annulations. Since the resulting naphthoquinols were highly air-sensitive, they were immediately methylated with dimethyl sulfate-potassium carbonate, and isolated as their dimethyl ethers (**70** and **71**). Some of the synthesized naphthoquinone-conjugated sugars were shown to be active against the human cervical cancer cell line (HeLa).[17]

When ethyl sorbate (**73**), a representative member of 2,4-dienoates, was reacted with phthalide sulfone (**11**) in the presence of LDA, the condensation took place at the C4 double bond of the sorbate, leaving the C2 double bond intact, and the resultant naphthol (**74**) was dimethylated with Me_2SO_4-K_2CO_3 to give naphthalene (**75**) as the sole product in 83% yield (Scheme 4.16).[10] Such a study has not been extended to other conjugated dienoate acceptors.

Acceptors that were successfully annulated:

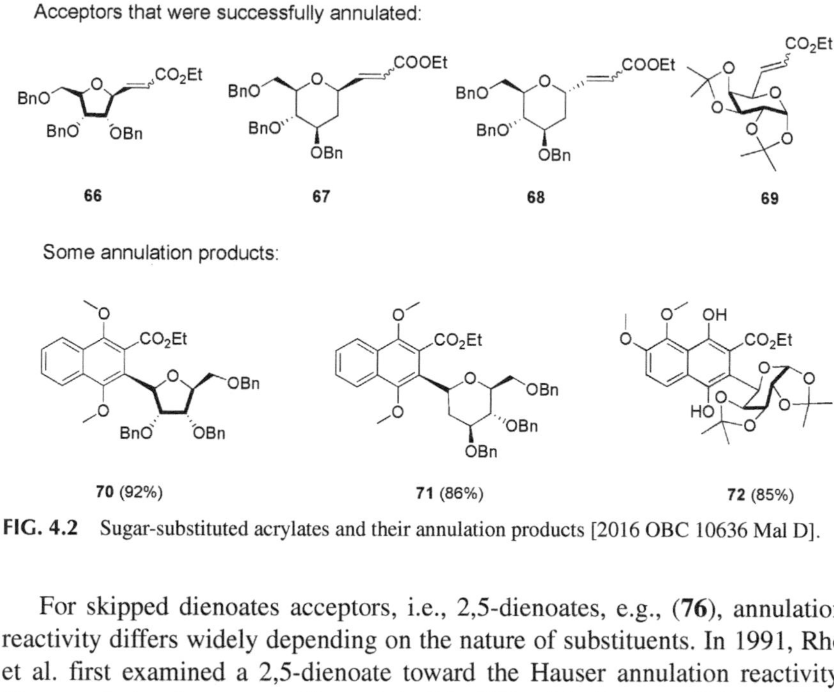

66 67 68 69

Some annulation products:

70 (92%) 71 (86%) 72 (85%)

FIG. 4.2 Sugar-substituted acrylates and their annulation products [2016 OBC 10636 Mal D].

For skipped dienoates acceptors, i.e., 2,5-dienoates, e.g., (**76**), annulation reactivity differs widely depending on the nature of substituents. In 1991, Rho et al. first examined a 2,5-dienoate toward the Hauser annulation reactivity. The annulation between sulfone phthalide (**31**) with dienoate (**76**) furnished

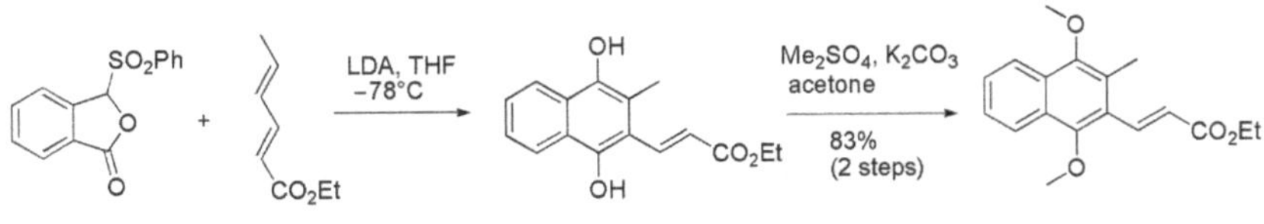

11 73 74 75

SCHEME 4.16 Annulation with a 2,4-pentadienoate [JACS 1981 6378 Hauser FM].

isomerized product (**77**) in 85% yield (Scheme 4.17). Rho et al. wrongly proposed the structure (**78**) as the product.[12] On careful scrutiny of their NMR data for the product (**78**), it was found to be incorrect. The structure (**77**) rightly corresponds to two singlets at δ 1.92 and 1.75, and one singlet at 6.17.

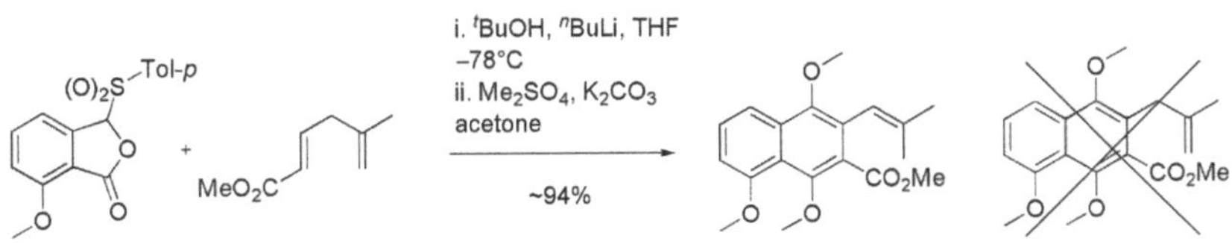

31 76 77 78

SCHEME 4.17 Hauser annulation with skipped dienoates under Rho conditions [1991 JKCS 756 Rho YS].

In recent years, the study on the skipped dienoates has been pursued thoroughly by Mal et al. Similar reactions of cyanophthalides with the skipped dienoates in the presence of LHMDS in a mixture of THF and MTBE furnished normal annulation products in good yields (Scheme 4.18).[18] These products did not undergo any isomerization under the conditions of the reaction. Cyanophthalides (**79**), on annulation 2,5-dienoates (**80**), furnished annulation products (**81**), after in situ dimethylation.

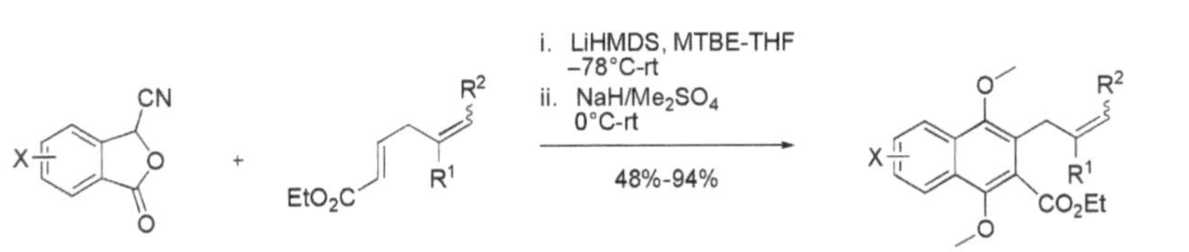

79 80 81

81: R^1 = Me, R^2 = H; R^1 = Et, R^2 = Me X = H, Br, methoxy, dimethoxy

SCHEME 4.18 Hauser annulation with skipped dienoates [2017 JOC 11035 Mal D].

For the unsubstituted 2,5-dienoate, i.e., (**82**), the results are somewhat unexpected. Under the above set of conditions, the reaction between the cyanophthalide (**14**) and the dienoate (**82**) gave a mixture of three products (**83–85**). It appears that the dienoate (**82**) is prone to undergo isomerization under the conditions, and the respective double bond isomerization of the dienoate (**82**) competes with the formation of product (**84**) (Scheme 4.19).[19]

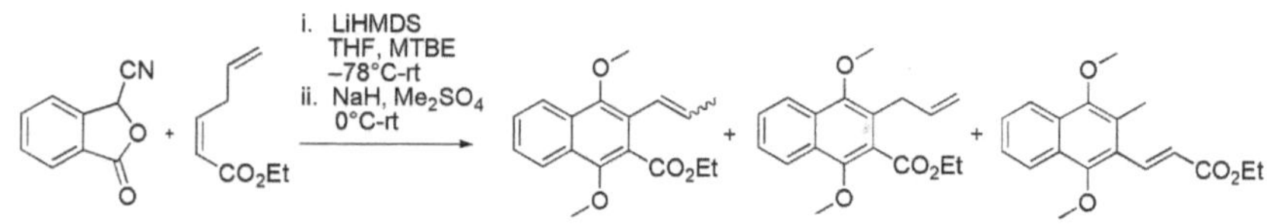

14 82 83 (52%) 84 (16%) 85 (11%)

SCHEME 4.19 Hauser annulation of an unsaturated 2,5-dienoate [2017 JOC 11035 Mal D].

A more interesting result was obtained for the prenyl-containing acceptor (**86**). The annulation of the cyanophthalide (**14**) with (**86**) followed by in situ methylation with NaH-Me_2SO_4 gave the hydroxylated product (**87**), contrary to the expected product (**88**). The unusual formation of (**87**) was explained by auto-oxidation of the initial annulation product (**88**) (Scheme 4.20).[19]

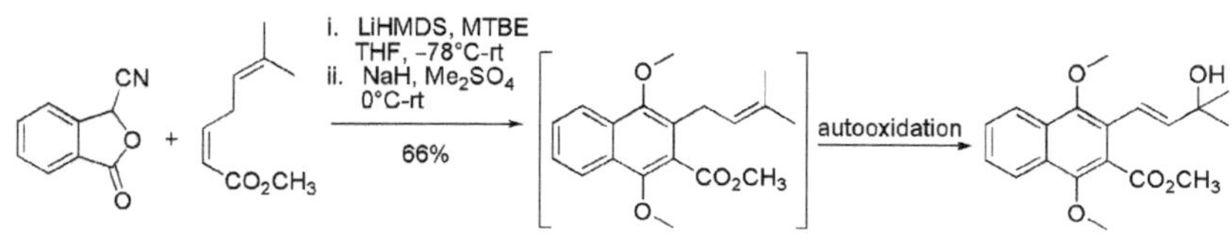

14 86 88 87

SCHEME 4.20 Annulation with a β-prenyl acrylate [2017 JOC 11035 Mal D].

Very recently, the Kesavan group examined various heavily substituted acrylates toward the Hauser annulation. Indolinone-embedded acrylates (**89**) were shown to react smoothly with the sulfone phthalide (**11**) at room temperature in the presence of Cs_2CO_3 in THF to give spirocyclic products (**90**) in good yields.[19] Since the annulation (**90**) products exist in equilibrium with their enol forms, they were isolated as Boc derivatives (**91**) (Scheme 4.21). More recently, the products have been characterized as their enolic forms.[20]

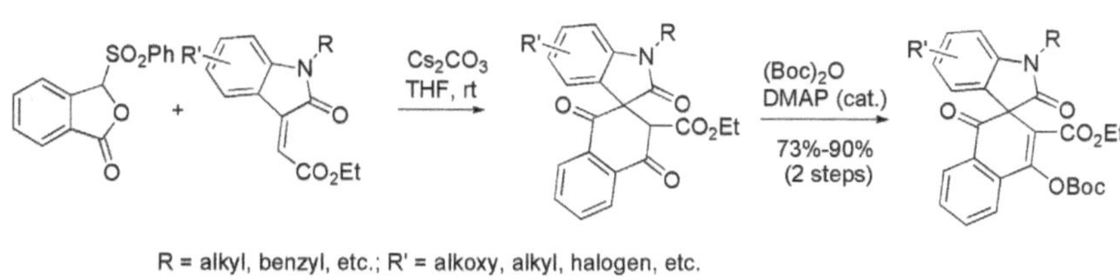

11 89 90 91

SCHEME 4.21 Annulation with methylene indolinones [2017 EJOC 5686 Kesavan V; 2017/2018 JOC 0000 Namboothiri INN].

4.1.1 Allenecarboxylates

Although allenecarboxylates are expected to resemble acrylates in their reactivity toward nucleophiles due to conjugation of one of the double bonds to the ester groups, the literature reports on their reactivity are confined to an insignificant number of publications. Likewise, the studies on their anionic annulation reactivity are marginal. The Mal group recently began to investigate their reactivity toward the Hauser annulation. The results with allenecarboxylates are diverse, as shown in the following schemes. Scheme 4.22 shows how the reaction of 3-cyanophthalides (**92**) with aryl substituted allenecarboxylates (**93**) in the presence of LiO^{t}Bu results in a tandem annulation furnishing naphtho[*b*] furanones (**94**) in good yields with no loss of carbon.[21] This process entails formation of two rings in a one-pot operation. There was no sign of formation of the expected quinone methide (**95**). Interestingly, the total number of carbon of the reactants is the same as that of the products.

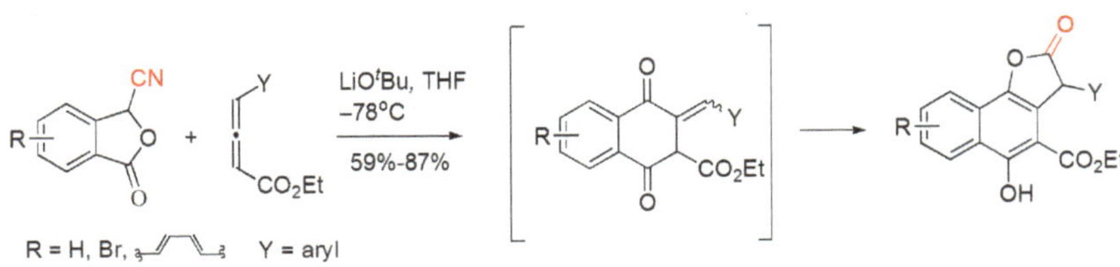

92 93 95 94

SCHEME 4.22 Mal naphtho[*b*]furanone synthesis [2018 JOC 4537 Mal D].

The carbon-economy and the one-pot formation of naphtho[*b*]furanones (**94**) are explained by a tandem process in which the initially expelled cyanide induces the second annulation (Scheme 4.23). For the representative reactants, the annulation cascade is initiated by LiO^{t}Bu-promoted generation of

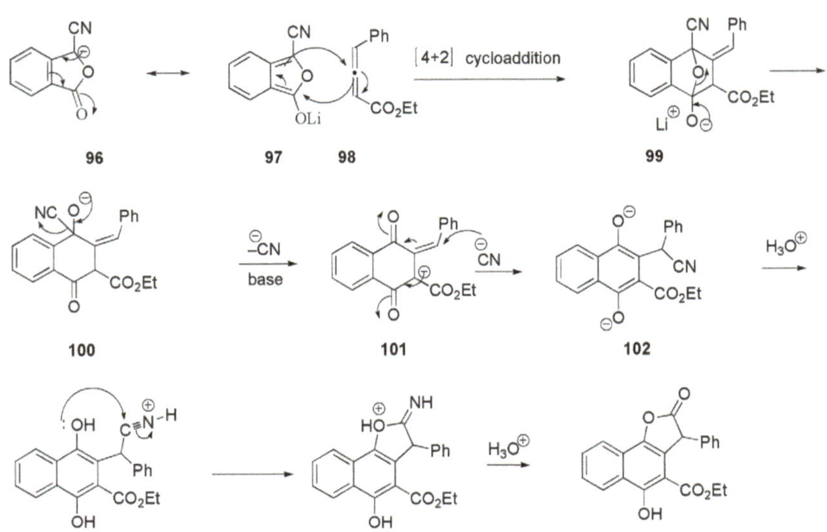

96 97 98 99 100 101 102 103 104 105

SCHEME 4.23 Probable mechanism of formation of naphtho[*b*]furanones (**94**).

3-lithiophthalide (**96**). Its diene form (**97**) undergoes cycloaddition with (**98**) to give first tricyclic intermediate (**99**) and then tetralone (**100**). As the cyanide group is a good leaving group, it is expelled from the intermediate (**100**) to give quinone methide (**101**) (cf. **95**). The expelled cyanide ion then adds to the double bond of the intermediate (**101**), forming the naphthoquinone dianion (**102**). Under the acidic work-up conditions, the dioxide and the cyanide groups in (**102**) are protonated to form (**103**). Intramolecular imidation to (**104**), followed by hydrolysis during acidic work-up conditions, gives the naphthofuranone (**105**).

The tandem annulation described in Scheme 4.23 is very sensitive to effects of substituents of phthalide donors. For the phthalide donors with a methoxy group at C4 or C6 like (**106**), the reaction diverts to form naphtho[*c*]furanones (**107**). For example, the annulation of 6,7-dimethoxycyanophthalide (**106a**) with the allene carboxylate (**93b**) under the influence of LTB at −78°C, furnished 1,4-dihydroxynaphthalide (**107a**) in 65% yield, contrary to the expected formation of a naphtho[*b*]furanone (cf. **105**). This trend is generalized with the conversions: (**106b** → **107b**), (**106c** → **107c**), and (**106d** → **107d**) (Scheme 4.24).

The product divergence is explained by preferential conjugate addition of water over a cyanide anion during acidic quench to the quinone methide intermediate (**108**) (cf. **101**), formed in the first annulation (Scheme 4.25). The hydroxy ester (**109**), under acidic conditions, lactonizes to naphthalide (**107**).

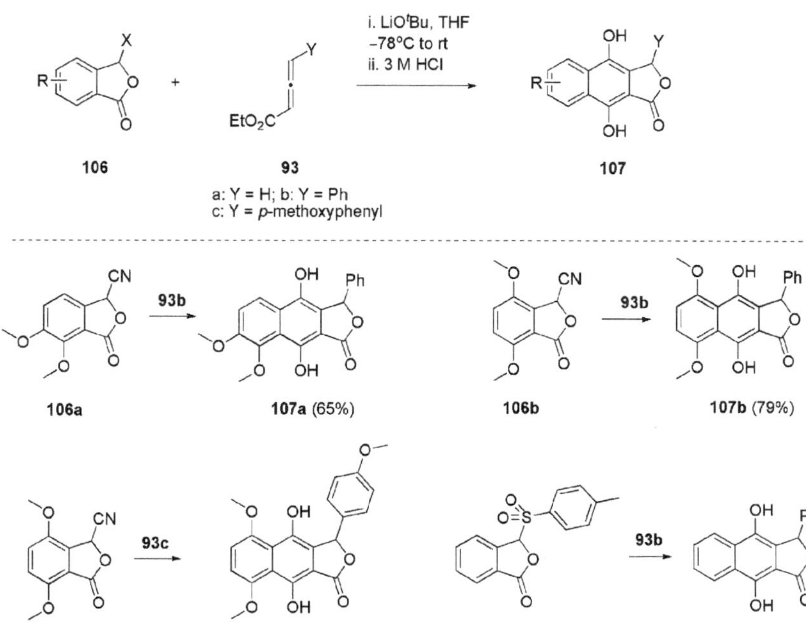

SCHEME 4.24 Scheme for formation of the annulation product (**107**) [2018 JOC 4537 Mal D].

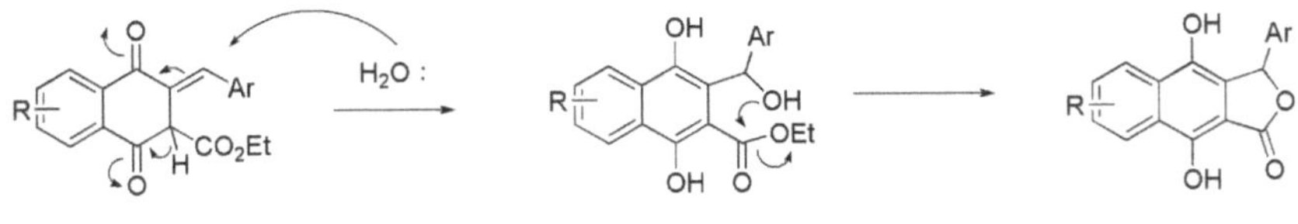

108 109 107

SCHEME 4.25 Mechanism of the formation of naphtho[*c*]furanones from allenecarboxylates.

The reaction of the phthalide sulfone (**11**) or cyanophthalide (**14**) with the parent allene carboxylate (**93a**) in the presence of LiO^tBu leads to new annulation and affords yet another class of product, e.g., (**110**) in an unprecedented manner (Scheme 4.26).[21]

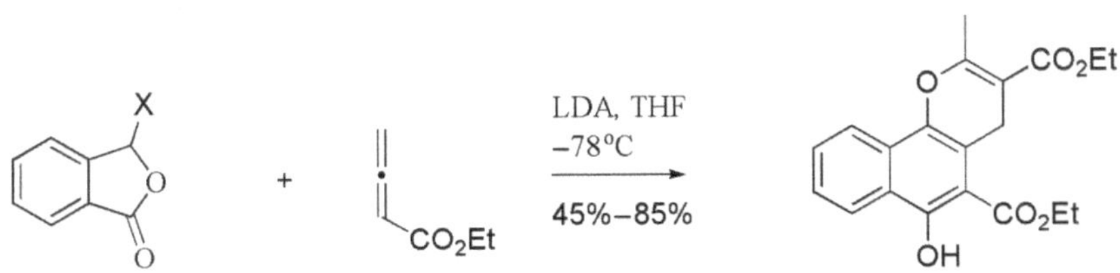

11: X = SO_2Ph, **14**: CN **93a** **110**

SCHEME 4.26 Annulation with unsubstituted allenecarboxylates [2018 PhD Thesis, IIT Kharagpur, Supriti Jana].

The mechanistic rationale behind the formation of the naphthopyran (**110**) is based upon [4+2] cycloaddition between the initial annulation product (**111**) and the allene acceptor (**93a**). The resulting adduct (**112**) then isomerizes to the final product, i.e., (**110**) under the basic condition of the reaction (Scheme 4.27).

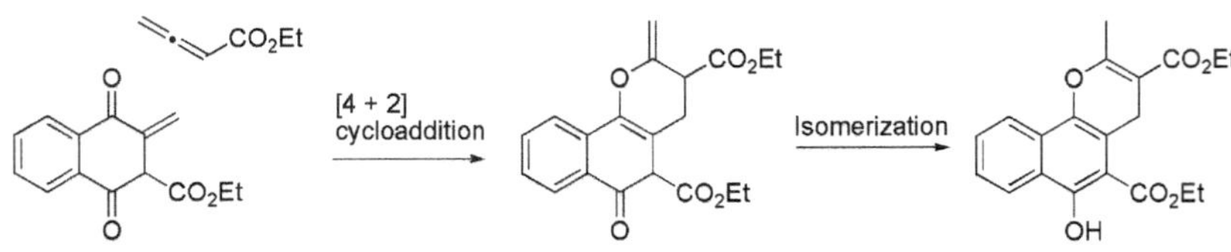

111 **112** **110**

SCHEME 4.27 Mechanistic interpretation of the formation of the naphthopyran (**110**).

4.1.2 Alpha-Substituted Acrylates

The Hauser annulation was discovered in 1978, and has been studied in depth over the years. However, the reactivity of the α-substituted acrylates toward the Hauser donors remained unexplored till 2015, when Mal et al. first disclosed their results. This large gap is perhaps due to the perception that the success of the annulations is derived from the enolate formation of the initial Michael-Dieckmann product, and the enolate formation is not logically permissible for substrates like α-substituted acrylates. Mal et al. reported that these acrylates undergo smooth annulation with 3-cyanophthalides in presence of bases and directly yield 1,4-naphthoquinones.[22] For example, the reaction of 3-cyanophthalide (**14**) with methyl methacrylate (**113**) in the presence of LTB followed by routine acidic quench furnished 2-methyl-1,4-naphthoquinone (**114**) in 64% yield (Scheme 4.28).

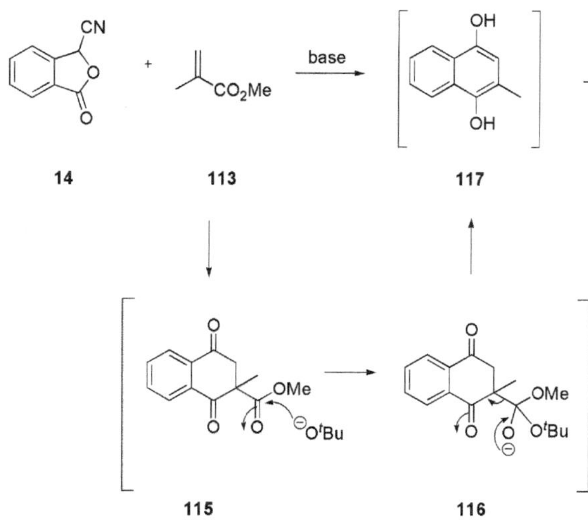

SCHEME 4.28 Mal demethoxycarbonylative annulation: synthesis of naphthoquinones [2015 ORS 5800 Mal D].

Mechanistically, the Mal annulation follows a demethoxycarbonylation of initial [4+2] annulation product (**115**) via intermediate (**116**) and gives 1,4-naphthoquinol (**117**). Aerial oxidation of (**117**) produces quinone (**114**). The support for the mechanism in Scheme 4.28 is derived from the studies, as discussed below, on more substituted acrylates, where the initial annulation products were isolated, and the dealkoxycarbonylations were carried out separately. This dealkoxycarbonylative annulation (i.e., **115–117**) is fairly general. It has been applied to the synthesis of a variety of naphthoquinones (**118–121**), accommodating alkyl and aryl groups in the quinone products (Fig. 4.3).

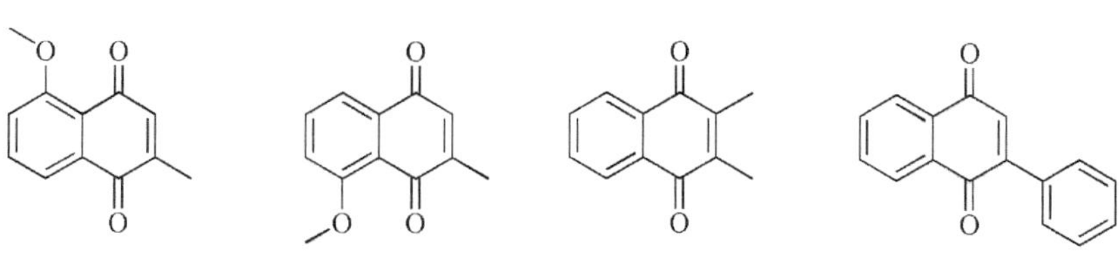

FIG. 4.3 Selected naphthoquinones obtained by dealkoxycarbonylative annulation.

However, a new rearrangement reaction ensues when α-prenyl acrylate (**122**) is subjected to the annulation with the phthalide (**14**). 4-Prenyloxynaphthoate (**123**) (Scheme 4.29) was formed in 83% yield.[22] The expected 2-prenylnaphthoquinone (**124**) was not formed. The formation of the extraordinary product (**123**) is explained by preferential Cope rearrangement of initial annulation product (**125**) to alkoxy tetralone (**126**), followed by retro-Wittig rearrangement of (**126**) and aromatization of naphthalenone (**127**).

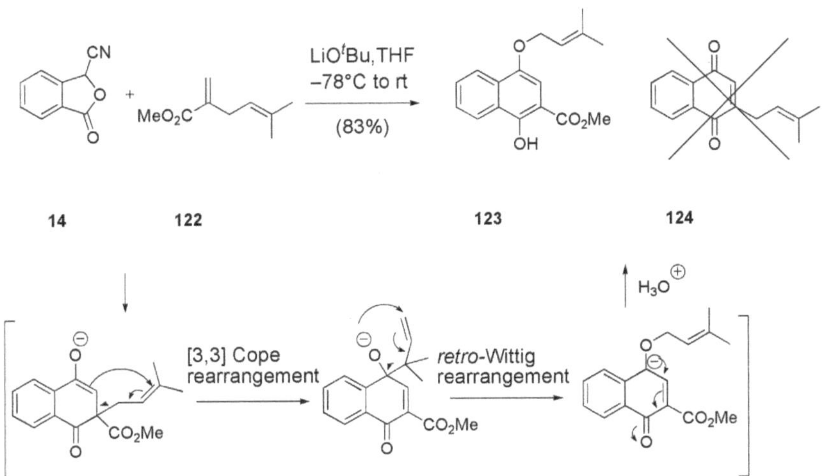

SCHEME 4.29 Annulation with α-prenylacrylate: a case of tandem rearrangement [2015 ORS 5800 Mal D].

This rearrangement is prevented when the acrylate acceptors contain both α and β substituents. However, the annulations with phthalides (**3**) proceed to give isolable 2,3-substituted naphthalene-1,4-diones. This finding was employed to accomplish a short total synthesis vitamin K_2 as presented in the following section.

Mal synthesis of vitamin K: Vitamin K, a family of natural products, is of importance in many key biochemical processes of living systems, including blood coagulation, bone metabolism, and cell growth. In addition, vitamin K is shown to inhibit the growth of a number of tumor cell lines. Although vitamin K molecules are made commercially available by chemical synthesis, the development of an environmental friendly process for their production on a commercial scale continues to be a challenging goal. The Mal group utilized their dealkoxycarbonylative benzannulation and developed a short synthesis of vitamin K molecules, e.g., vitamin K_2 (**128**) (Scheme 4.30).[22]

The required α-alkenylacrylate (**129**) was obtained in two steps from phosphonoacetate (**130**). Geranyl bromide (**131**) was reacted with methyl 2-(diethoxyphosphoryl)acetate (**130**) in the presence of KO^{t}Bu in dry DMF at 0°C to give phosphonoacetate (**132**) in 75% yield. Wittig-Horner reaction of (**132**) with acetaldehyde in the presence of NaH furnished α-alkenylacrylate

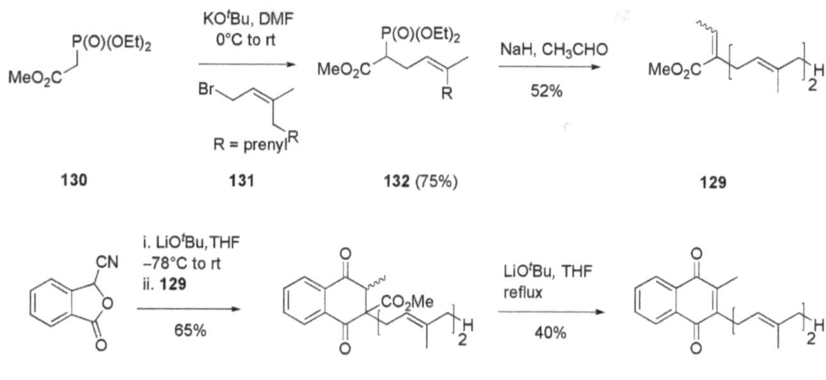

SCHEME 4.30 Mal synthesis of vitamin K_2 by demethoxycarbonylative annulation [2015 ORS 5800 Mal D].

(**129**) in good yield (Scheme 4.30). Treatment of the cyanophthalide (**14**) with LiO^{t}Bu followed by the acrylate (**129**) afforded annulation product (**133**) as a mixture of stereoisomers. Demethoxycarbonylation of the dione (**133**) by its reaction with LiO^{t}Bu in refluxing THF resulted in vitamin K_2 (**128**). This synthesis of vitamin K_2 is free from the stoichiometric use of transition metals, and the by-products are water-soluble.

The demethoxycarbonylative synthesis of the dialkyl naphthoquinones (Scheme 4.30) is, in principle, applicable for the synthesis of naphth[*d*]oxazoles or 2-amino-1,4-naphthoquinol/naphthoquinones, the key structural features of the natural products (**134–138**) (Fig. 4.4). Accordingly, Mal et al. exploited their demethoxycarbonylative annulation, and examined annulation reactivity of various α-acylamidoacrylates toward activated phthalides. Although such an approach incorporates a N atom in a 1,4-naphthoquinone moiety concomitant with the benzannulation, the structures of the products vary with the actual structure of the acrylates.

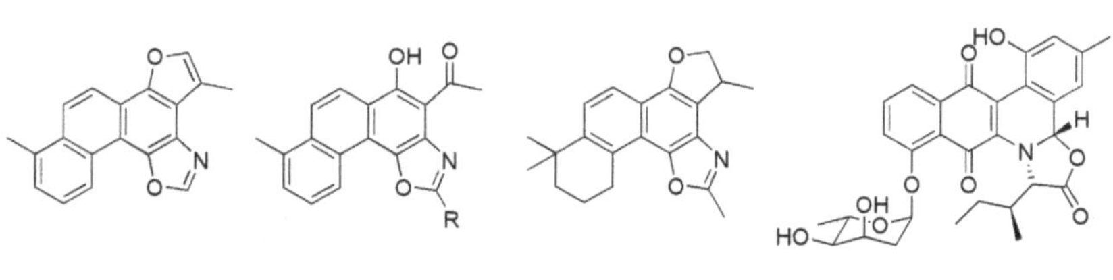

FIG. 4.4 Representative naphth[*d*]oxazole and aminonaphthoquinone natural products.

The reaction between the 3-cyanophthalide (**14**) with α-acylamidoacrylate (**139**) in the presence of LHMDS at −78°C afforded expected 2-acetamino-1,4-naphthoquinone (**140**) in 75% yield. All three steps, i.e., annulation to (**141**), its demethoxycarbonylation, and aerial oxidation (**142** → **140**), took place in one operation (Scheme 4.31).[23]

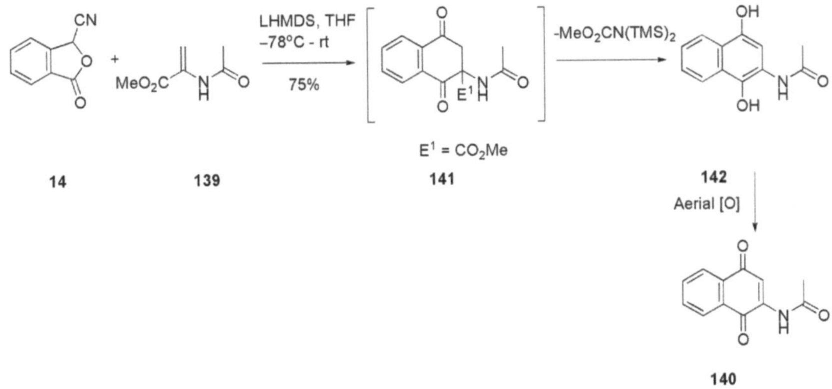

SCHEME 4.31 Annulation with an α-acetaminoacrylate [2016 SYS 1235 Mal D].

It is remarkable that the active hydrogen of NH group at C2 of the acceptor (**139**) interfered with neither the initial annulation to give (**141**) nor the subsequent demethoxycarbonylation of (**141**). More notable are the results with β-substituted α-acylamidoacrylates. Treatment of phthalides (**143**) with acrylates (**144**) under the above conditions afforded double annulation products, i.e., naphthoxazoles (**145**) in good yields[24] (Scheme 4.32).

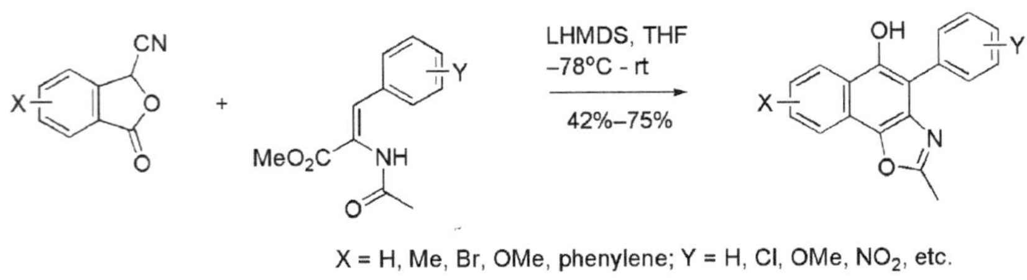

SCHEME 4.32 Demethoxycarbonylative double annulation with α-acetamido acrylates to naphthoxazoles [2016 SYS 1235 Mal D].

The formation of the naphthoxazoles (**145**) is explained by the mechanism shown in Scheme 4.33. It is believed to proceed via a five-step cascade consisting of lateral lithiation, Michael addition, Dieckmann condensation, demethoxycarbonylation, and heterocyclization. The initial Hauser annulation product (**146**), formed after the first three steps, undergoes LHMDS-promoted demethoxycarbonylation followed by enolization to produce dianion (**147**). Acidic work-up of (**147**) leads to formation of (**148**) first and then the oxazole ring of naphthoxazoles (**145**). Steric compression exerted by the aryl groups of the acrylates is perhaps the driving force for the facile oxazole annulation.

The similar reaction of the angular phthalide (**149**) with amidoacrylate (**150**) in the presence of LHMDS furnished amidophenanthroquinone (**151**) (cf. **140**) in 57% yield. The failure in the formation of the corresponding oxazole ring is ascribed to a *peri* effect (Scheme 4.34).

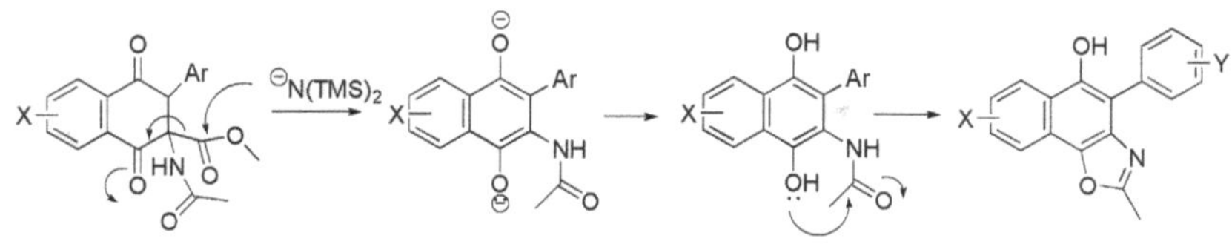

SCHEME 4.33 Mechanism of formation of naphthoxazoles (**144**) by annulation with α-acylamidoacrylates.

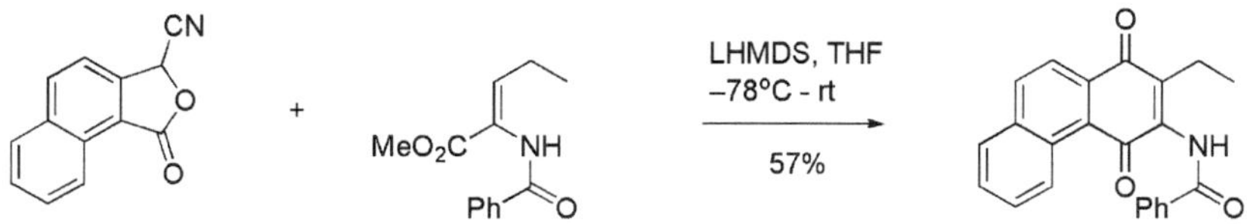

SCHEME 4.34 Annulation of an angular cyanophthalide with an α-acylamidoacrylate.

4.1.3 Morita-Baylis-Hilman Adducts (MBH)

Morita-Baylis-Hilman (MBH) adducts (**152**) can be viewed as α-substituted acrylates. Consequently, it is expected that they would undergo normal Mal dealkoxycarbonylative annulation with the phthalide (**14**) to give the quinols (**153**) or the corresponding air-oxidized products (**154**) (Scheme 4.35).[25]

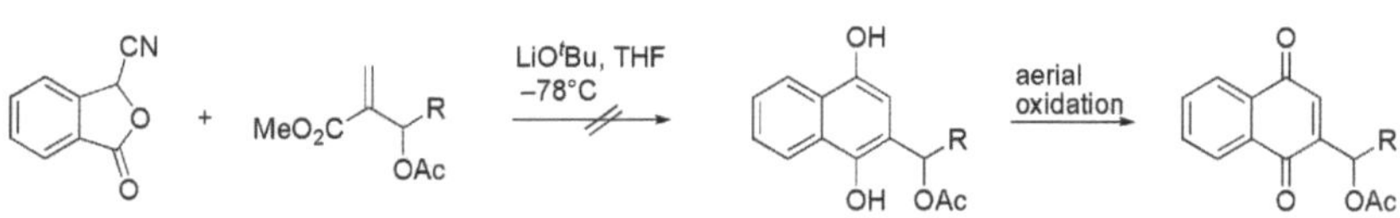

SCHEME 4.35 Proposed annulation with Morita-Baylis-Hilman adducts [Unpublished results Mal D].

Contrary to expectation, an entirely new reaction evolved when acrylates (**152**) were submitted to the reaction with the phthalide (**14**) in the presence of LTB, and indane-1,3-diones (**155**) (Scheme 4.36) were formed.[25] With *R*=2-furyl, phenyl, 2-bromophenyl and 4-cyclohexenyl, the respective indanones (**155a–155d**) were produced in good yields.

The probable mechanism of the formation of the indanones (**155**) is presented in Scheme 4.37. The first step of the cascade is believed to be lateral deprotonation of the phthalide (**14**) to form anion (**96**). The Michael addition of the anion (**96**) to MBH adducts (**152**) with concomitant loss of the acetoxy of the acceptors generates acrylates (**156**). Allylic deprotonation of (**156**) gives new carbanions (**157**) which undergo Dieckmann-type cyclization to give cycloalkoxide (**158**). Alkoxide-assisted elimination of the cyano group gives indanones (**155**).

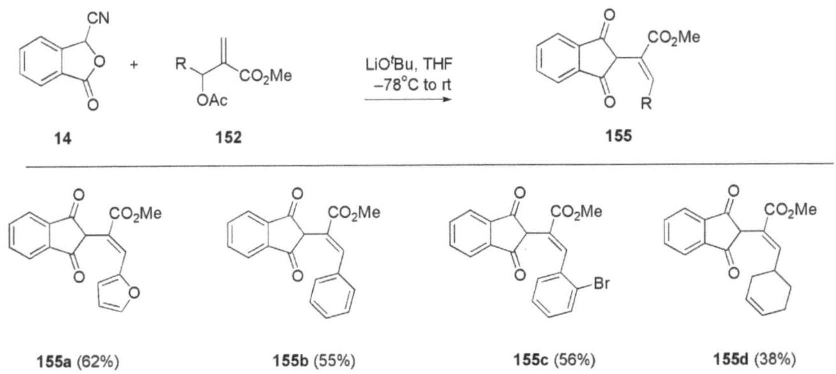

SCHEME 4.36 Annulation with Morita-Baylis-Hilman adducts: A case of [4+1] annulation [Unpublished results Mal D].

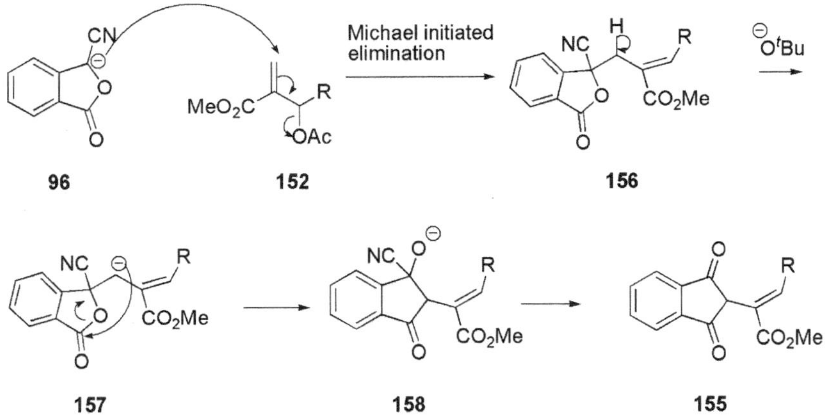

SCHEME 4.37 Probable mechanism of Mal 1,3-indanedione annulation.

4.1.4 Acrylonitriles

The Hauser annulation with acrylonitriles has scarcely been studied. It has been utilized only in few instances of total syntheses. Historically, the reaction of acrylonitriles (**159**) with the Hauser donor (**22**) gave Michael adduct (**160**) in the presence of LDA at −78°C to rt. (Scheme 4.38).[26]

When the same reaction was carried out in the presence of LiO^{t}Bu at −60°C to rt., it yielded quinol (**161**), which was *O*-methylated to give (**162**) in 76% overall yield. The Hauser annulation of acrylonitrile has been successfully applied to the total synthesis of optically active shikonin (**163**) (Scheme 4.39).[27]

Cyclocondensation of the phthalide sulfone (**164**) with acrylonitrile (**159**) in the presence of LDA followed by *O*-methylation of the annulation product gave tetraoxygenated naphthalene carbonitrile (**165**). DIBALH reduction of (**165**) gave carboxaldehyde (**166**), which upon asymmetric allylation with allyldiisopinocamphenyl borane gave alcohol (**167**). Following the protection of

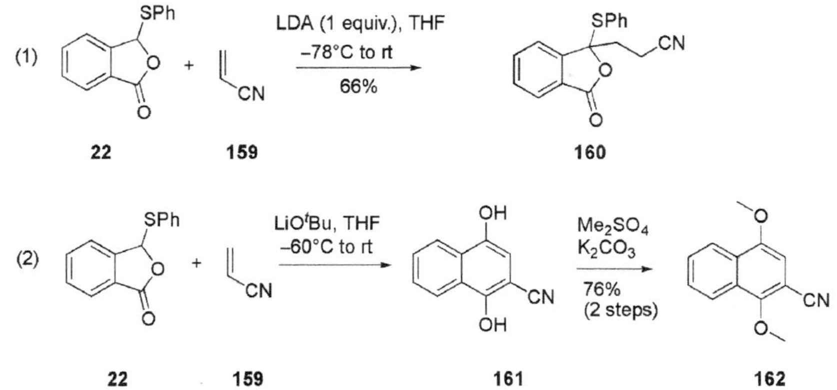

SCHEME 4.38 Annulation with acrylonitrile [1999 JCRS 30 Mal D].

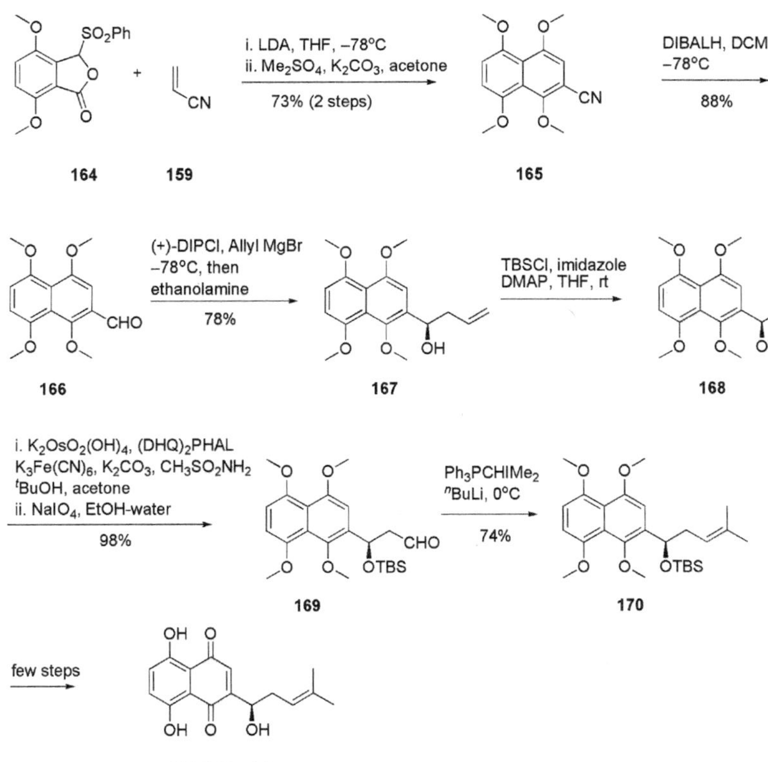

SCHEME 4.39 Asymmetric synthesis of shikonin [1997 THS 7263 Couladouros EA].

the OH group in (**167**) as TBS derivative (**168**) using TBSCl, imidazole, and DMAP, the terminal olefin in (**168**) was quantitatively cleaved to (**169**) in two steps: Sharpless dihydroxylation and $NaIO_4$ oxidation. The Wittig reaction of the aldehyde (**169**) with the ylide of 2-iodopropane furnished shikonin derivative (**170**), which was converted to shikonin (**163**) by Braun's method.

4.2 ALKENONES AND CYCLOALKENONES

In this section, the annulation chemistry of the acceptors having structural motifs (**171**, **172**, and **173**) is presented in Fig. 4.5. Of the three types, examples of α,β-unsaturated aldehydes as acceptors for Hauser annulation are scarce.

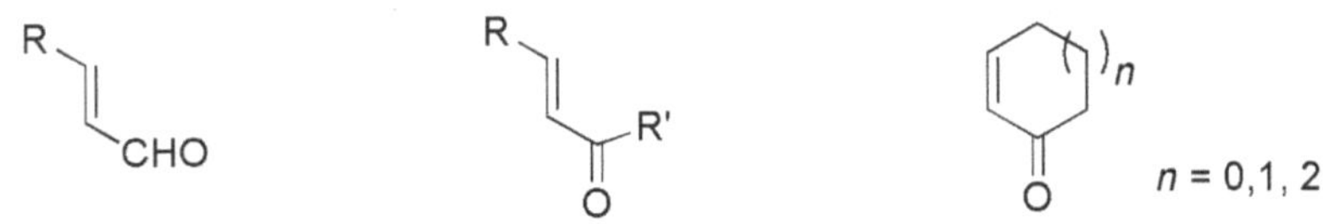

171 **172** **173**

FIG. 4.5 Typical structural motifs for unsaturated aldehydes and ketones.

In 1997, Katritzky et al. first studied annulation of crotonaldehyde and cinnamaldehyde with 3-(benzolriazol-1-yl)phthalide (**7**) (X=H). But, the results were not conclusive due to difficulties in the isolation of the annulation products.[14] However, Wege et al. demonstrated that an unsaturated aldehyde can be successfully with a cyanophthalide. The reaction of the cyanophthalide (**14**) with 3-furylacrylaldehyde (**174**) in the presence of LDA at −78°C, followed by *O*-methylation with MeI-K_2CO_3, afforded naphthalene carboxaldehyde (**175**) in 76% overall yield (Scheme 4.40).[28] Compound (**172**) was then oxidized with perselenous acid to give β-naphthol (**176**), a key intermediate for the synthesis of 3-hydroxymethylfuro[3,2-*b*]naphtho[2,3-*d*]furan-5,10-dione (**177**), a metabolite isolated from *Crescentia cujete*.

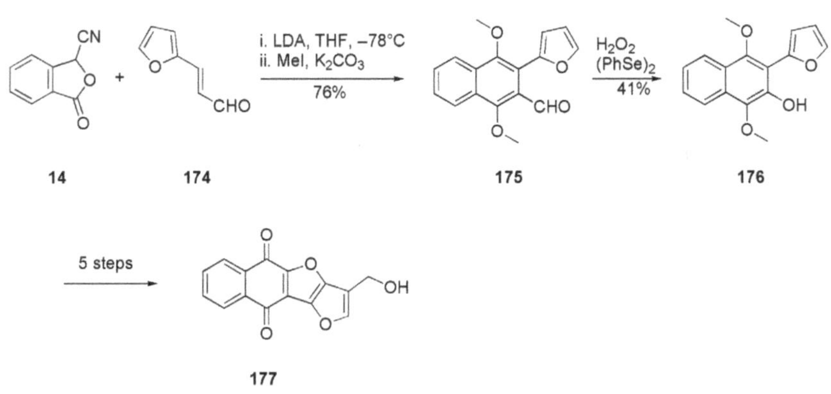

SCHEME 4.40 Hauser annulation with an unsaturated aldehyde [2009 THN 4569 Wege D].

4.2.1 Acyclic Conjugated Ketones

Since the discovery of Hauser annulation in 1978, α,β-unsaturated ketones (e.g., **172**) have received significant attention as the acceptors. Their annulations have been exercised to achieve total syntheses of many natural products. The yields of the annulations with α,β-unsaturated ketones are often high-yielding, except when the hydrogen at α position of the acceptors leads to side reactions. The yields tend to be lower when β-C does not bear a substituent (vide reaction 1, Scheme 4.41). The overall outcomes are dependent on the structures of both donors and acceptors. The reactions (**2**) and (**3**) indicate that both phthalide sulfone and cyanophthalide are equally efficient (Scheme 4.40).[5a,7]

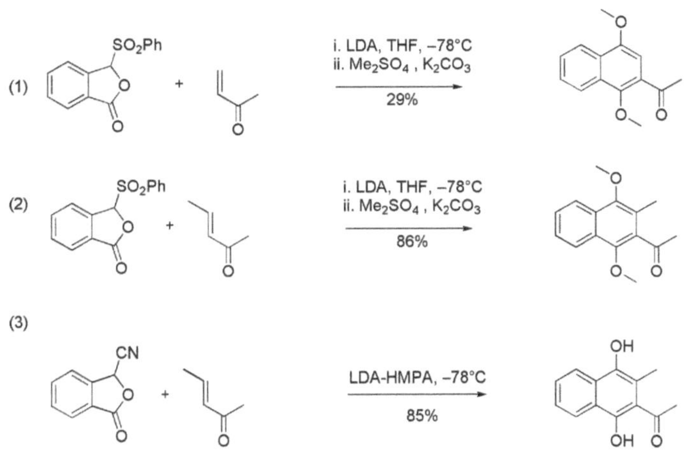

SCHEME 4.41 Hauser annulation with acyclic enones [1978 JOC 178 Hauser FM; 1983 JOC 3439 Kraus GA].

The annulation of an acyclic enone has been applied to the synthesis of kalafungin (**178**), a member of the pyranonaphthoquinone family. Naphthoquinone (**179**) was synthesized in two steps from phenylthiophthalide (**180**). The annulation of the phthalide (**180**) with methyl vinyl ketone (**181**) in the presence of LDA-HMPA complex gave naphthoquinol (**182**). Oxidation of (**182**) with Ag_2O gave naphthoquinone (**179**). Addition of *tert*-butoxyfuran to the quinone (**179**) in toluene gave quinol (**184**). In situ *O*-methylation of the quinol (**184**) followed by LAH reduction gave alcohol (**185**). Treatment of the alcohol (**185**) with TFA gave detertbutyl derivative (**186**), which was treated sequentially with DBU, Ag_2O, and BCl_3 to furnish kalafungin (**178**) (Scheme 4.42).[5a]

Couladouros synthesis of shikonin (**163**) utilized the Hauser annulation of vinyl ketone (**187**).[29] The reactivity of the ketone (**187**) to the Hauser phthalide (**164**) is markedly different from that with the Kraus phthalide (**188**). When the phthalide (**164**) was reacted with (**187**) in the presence of LTB at −78°C followed by reflux temperature, Michael adduct (**189**) was obtained as the major product. The desired annulation product (**190**) was obtained in 8% yield. In contrast, the

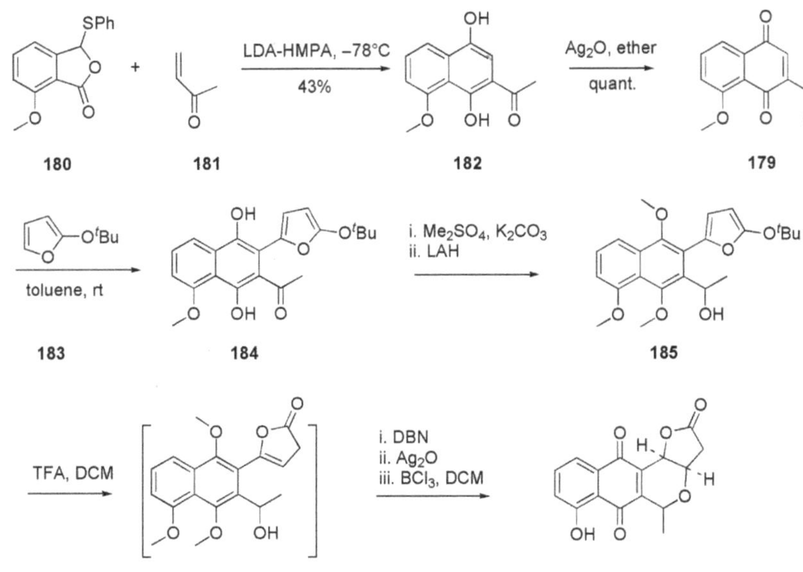

SCHEME 4.42 Synthesis of kalafungin (**178**) [1983 JOC 3439 Kraus GA].

reaction of the Kraus phthalide (**188**) with the same enone (**187**) in the presence of LDA afforded the desired product as the major product (**190**) and the corresponding Michael adduct (**191**) as the minor product (Scheme 4.43).[29]

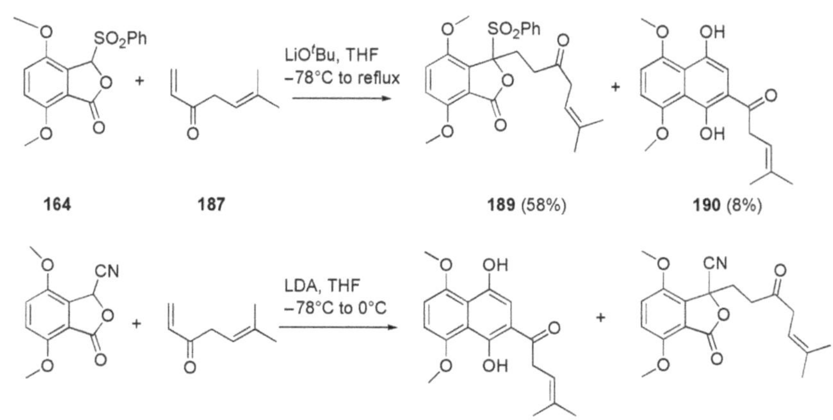

SCHEME 4.43 Couladouros annulation for the synthesis of shikonin (**163**) [2002 CEJ 1795 Couladouros EA].

It should be noted that no isomerization of β,γ-ketone moiety takes place during the annulation. The failure of the Hauser annulation of (**187**) with sulfone phthalide (**164**) was explained by a competing deprotonation of (**190**) to give more stable thermodynamic enolate (**192**) (Scheme 4.44).

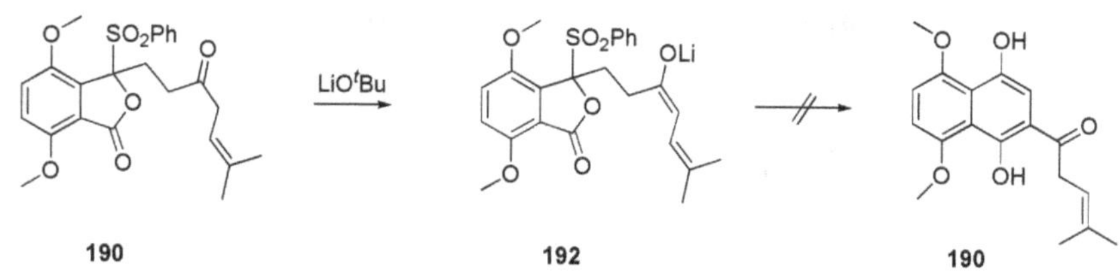

SCHEME 4.44 Couladouros explanation for the failure of the annulation with phthalide sulfone (**164**).

For the completion of the shikonin (**163**), the annulation adduct (**190**) was reduced with catechol borane in the presence of Corey catalyst following oxidative work-up to give naphthoquinone (**193**) (Scheme 4.45). Sodium dithionite reduction of quinone (**193**) followed by acetylation (Ac_2O, Py) afforded (**194**). Finally, oxidation of (**194**) with CAN to (**195**) followed by alkaline hydrolysis afforded shikonin (**163**) in >75% yield from the intermediate (**190**).

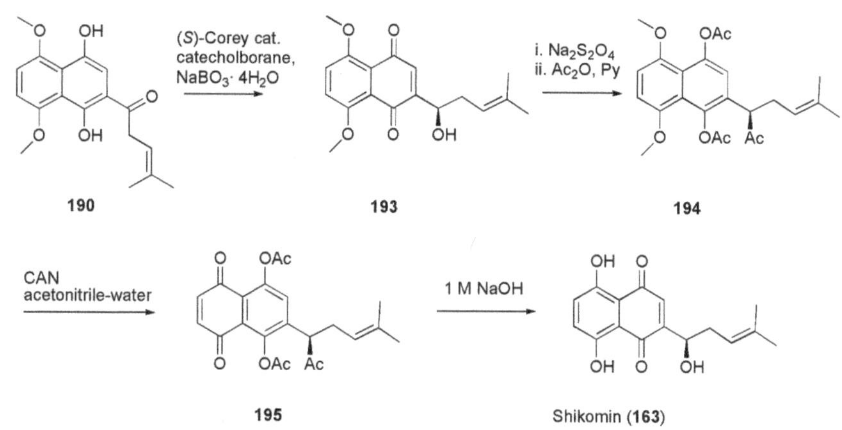

SCHEME 4.45 Final stage of Couladouros synthesis of shikonin (**163**) [2002 CEJ 1795 Couladouros EA].

For the β-substituted enones, the annulations with both Hauser phthalides and Kraus phthalides are efficient. The reaction of chalcone with 3-cyanophthalide in the presence of LDA gave the corresponding annulation product in 72% yield.[5a] For the synthesis of granaticin (**196**) (Fig. 4.6), Yoshii et al. utilized the Hauser annulation in joining two advanced intermediates, and they identified 4-furyl-2-butenones (**197**) as Hauser acceptors.

The Hauser annulation of the cyanophthalide (**14**) with 4-furyl-3-butene-2-ones (**197**) in the presence of dimsyllithium in DMSO-THF, followed by *O*-methylation of the resulting hydroquinone with dimethyl sulfate, provided the expected annulation products (**198**) in moderate yields (56%–72%) (Scheme 4.46).[30] This study has ultimately resulted in the total synthesis of granaticin (**196**).

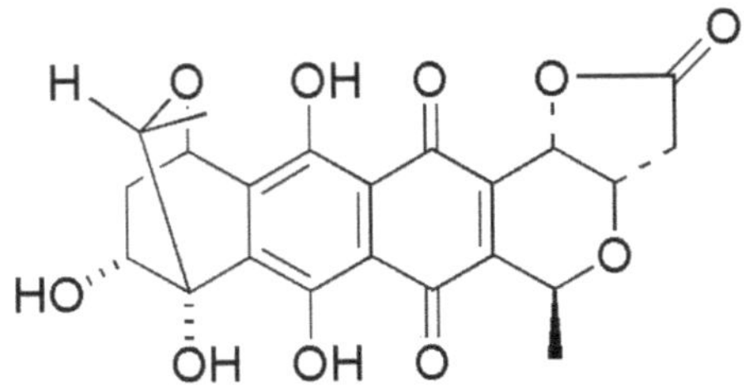

FIG. 4.6 Structure of granaticin (**196**).

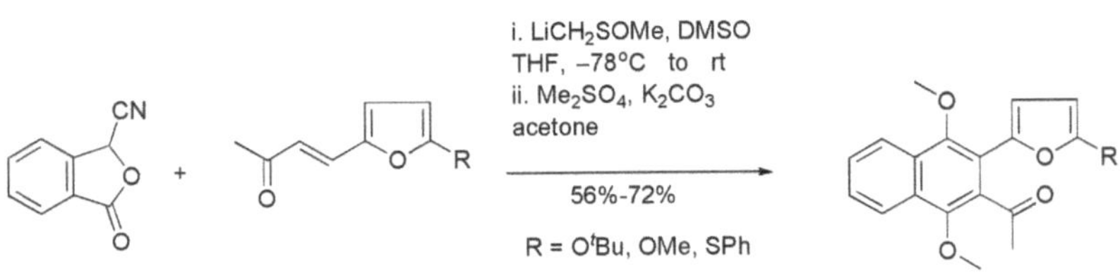

14 **197** **198**

SCHEME 4.46 Annulation of 4-furyl-3-butene-2-ones [1986 CPB 3175 Yoshii E].

Applications of the annulations with acyclic unsaturated enones have been furthered by Brimble et al. and Wege et al. in the synthesis of pyranonaphthoquinone natural products. Hydroxy-3-methylnaphtho[2,3-*c*]furan-4,9-dione (**199**) is a metabolite isolated from *Aloe ferox*, the healing properties of which have been the main driving force for the synthesis of the active principles. Wege synthesis of the quinone (**199**) relied upon the annulation of (**37**) with pentenone (**172**) (R, R′=Me) to produce quinol (**200**). Elaboration of (**200**) through oxidation, bromination, hydrolysis, and cyclodehydration provided quinone (**201**). Aluminum trichloride-mediated demethylation of (**201**) furnished (**199**). Oxidation of naphthoquinol (**200**) to quinone (**202**) was carried out with $FeCl_3$ in methanol. Displacement of bromine in (**203**) by acetate produced (**204**). Methanolysis of quinol acetate (**204**) produced naphthofuran (**201**) (Scheme 4.47).[31]

The annulation chemistry for the acyclic enones with γ-leaving groups is not straightforward. All such acceptors (**205**) studied for the annulation with the cyanophthalide (**37**) gave complex mixture of products. One of the explanations for the failure is the formation of reactive *o*-quinone methide intermediate (**206**) from the initial condensation products (**207**), and its susceptibility to an array of reactions including cyanide addition (Scheme 4.48).

With an alkynone acceptor, the desired annulation does not stop at the desired intermediate quinone, and proceeds further to give the cyanide addition product. When alkyne ketone (**208**) was submitted to annulation with cyanophthalide

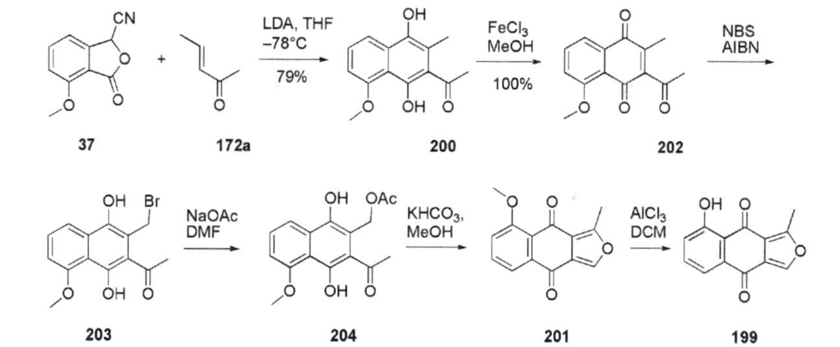

SCHEME 4.47 Wege synthesis of 5-hydroxy-3-methylnaphtho[2,3-*c*]furan-4,9-dione [2003 AJC 691 Wege D].

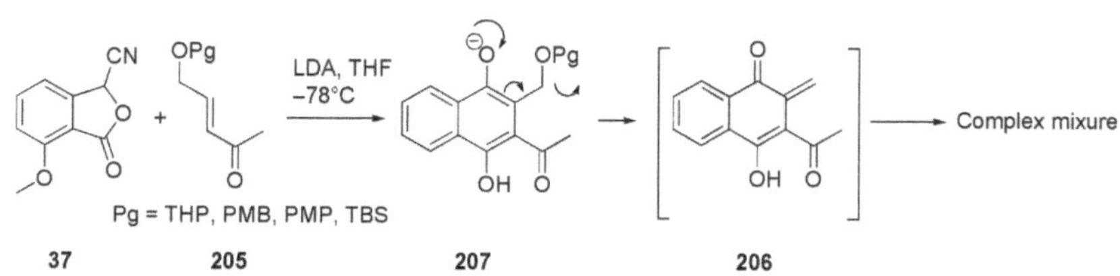

SCHEME 4.48 Annulation with acyclic enone γ-nucleofuges [2003 AJC 691 Wege D].

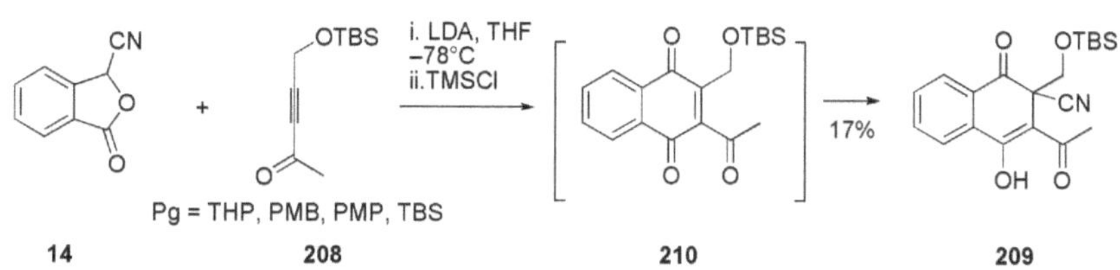

SCHEME 4.49 Annulation with an alkyne ketone [2003 AJC 691 Wege D].

(**14**), the product (**209**), derived by conjugate addition of cyanide ion to putative quinone (**210**), was isolated in 17% yield. The quinone intermediate (**210**) was not isolable (Scheme 4.49).

In a very similar looking annulation with substituted vinyl methyl ketone (**211**) that contains a γ-oxygen, unprecedented problems arose in the context of the synthesis of nanaomycin D (**212**).[32] For the reaction of phthalide (**37**) with (**211**) in the presence of KO^{t}Bu, LiO^{t}Bu, NaH and KH, the corresponding Michael adduct was isolated. When the annulation was carried out with LDA at −78°C and combined with reductive methylation ($Na_2S_2O_4$, Cs_2CO_3, MeI), the desired product (**213**) was obtained in only 35% yield. Neopentyl steric effect exerted the γ-substituent in the acceptor (**211**) may be suggested to inhibit the initial Michael addition. The key intermediate (**213**) was subjected to deprotection of the MOM group to give pentasubstituted naphthalene (**214**). Triethylsilane mediated reduction of (**214**) in the presence of BF_3 etherate gave

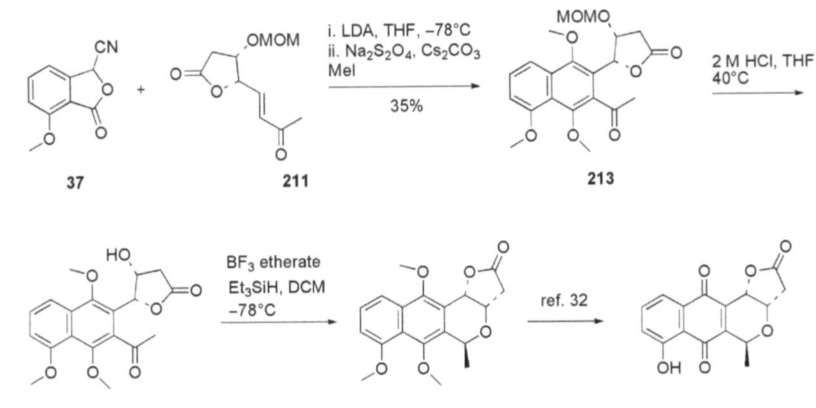

SCHEME 4.50 Brimble synthesis of nanaomycin D (**212**) [2015 THN 7137 Brimble MA].

Bruckner intermediate (**215**),[24] constituting a formal synthesis of nanaomycin D (**212**) (Scheme 4.50).

Crisamicin A (**216**) is a dimeric naphthoquinone natural product. It was isolated from *Micromonospora purpureochromogenes* and shown to display activity against B16 murine melanoma cells, and the herpes simplex and vesicular stomatitis viruses (Fig. 4.7). It is, therefore, an attractive target for the synthetic chemists.

Till date, only one total synthesis of crisamicin A (**216**) in racemic form

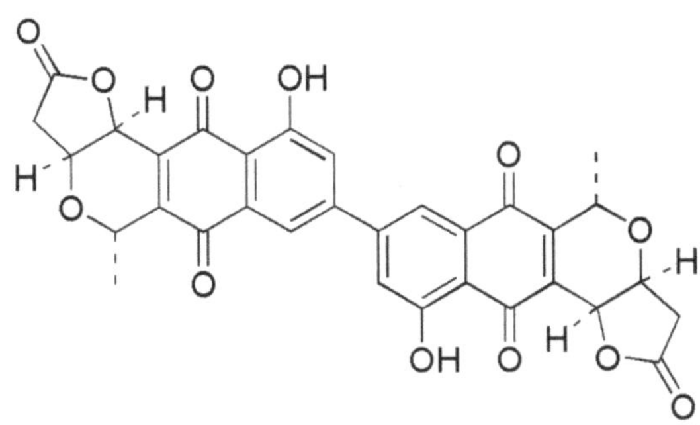

FIG. 4.7 Structure of crisamicin A (**216**).

has been reported, by Yang al. The 19-step synthesis consisted in three key reactions, namely, Pd-catalyzed biaryl formation, Diels-Alder reaction and Pd-catalyzed carbonylative lactonization.[33] Brimble et al., in their first approach, explored Suzuki coupling followed by double Hauser annulation with an advanced acyclic enone to assemble crisamicin A scaffold (Scheme 4.51).[34] The triflate (**217**) was dimerized by Suzuki-Miyaura homocoupling to give (**218**) in 100% yield. Directed lithiation of (**218**) with ^{t}BuLi followed by formylation with DMF gave amide (**219**). Yoshii method[35] (TMSCN, KCN, 18-c-6) was

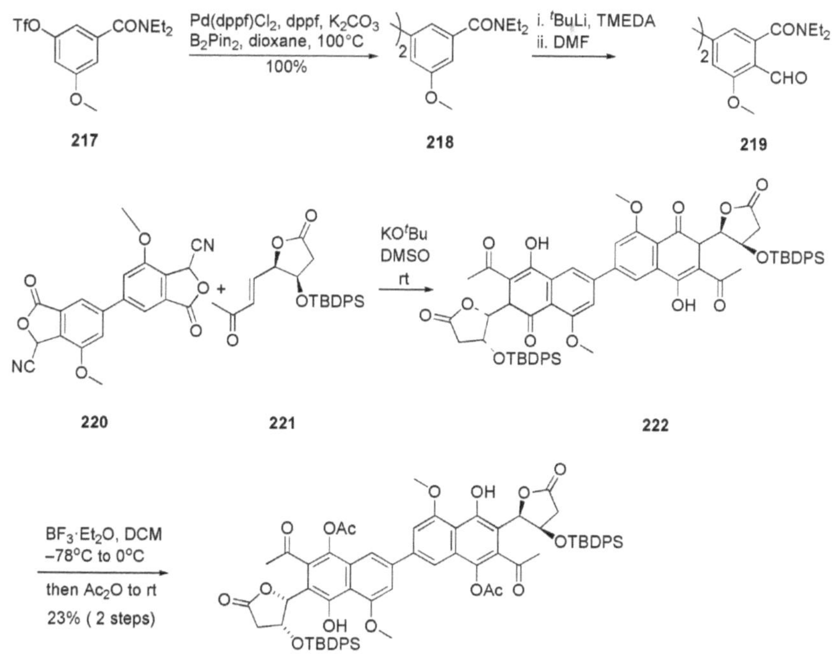

SCHEME 4.51 Synthesis of crisamicin A (**216**).

applied to the amide (**219**) to generate Hauser donor (**220**). The Hauser annulation of the dimeric cyanophthalide (**220**) with enone (**221**) gave annulation product (**222**). Derivatization of the product (**222**) with BF_3 etherate and acetic anhydride furnished the advanced intermediate (**223**). Since the annulation was not efficient, Brimble et al., in a later approach, adopted annulation at the early stage with modest improvement and achieved the total synthesis of dimethyl ether of crisamicin A (**216**). However, the total synthesis was incomplete due to the failure in the demethylation in the penultimate step.

The total synthesis of eleutherin (**228**) by Brimble et al. is another example of an anionic annulation with acyclic alkenone with oxygen functionality in the δ position. The planned annulation between the phthalide (**37**) and enone (**224**) proceeded smoothly in the presence of KO^{t}Bu in DMSO followed by *O*-methylation to afford (**225**) in 76% yield. Desilylation of the adduct (**225**) with TBAF to (**226**), followed by reductive cyclization gave tricyclic intermediate (**227**) in 56% yield. CAN oxidation of the dimethyl ether (**227**) in a mixture of acetonitrile and water provided (+)-eleutherin (**228**) (Scheme 4.52).[36]

One of the finest applications of the enone annulation is the Hauser-initiated domino reaction for assembling griseusin scaffold. The annulation of enone (**229**) with the cyanophthalide (**37**) proceeded in domino fashion give griseusin framework (**230**) in one pot. All four conceivable steps, namely Hauser annulation, *O*-methylation, double desilylation and spirocyclization took place under the reaction conditions (Scheme 4.53).[37]

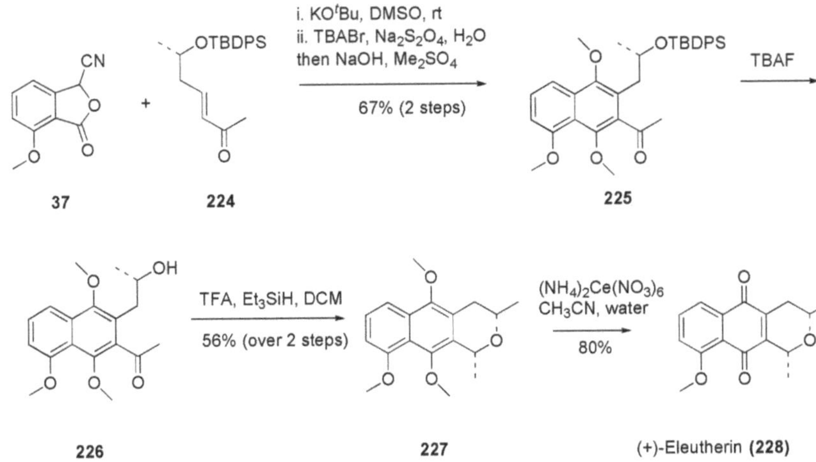

SCHEME 4.52 Brimble synthesis of eleutherin.

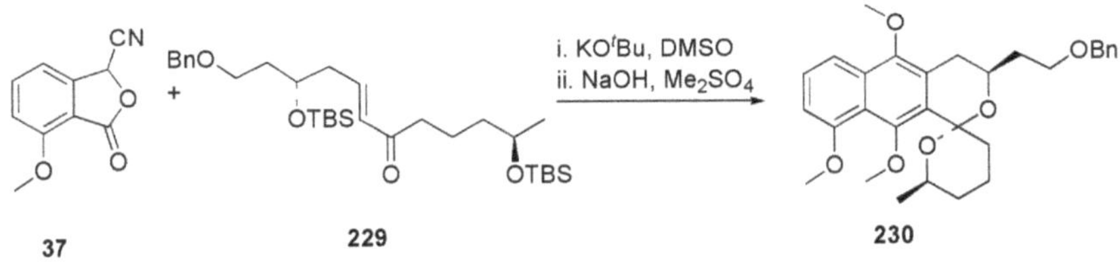

SCHEME 4.53 Synthesis of griseusin scaffold.

4.2.2 Cyclic Conjugated Ketones

Cyclopentenones have been studied as acceptors for the Hauser annulation in the context of the total synthesis of kinamycins (kinamycin D, **231**) and fredericamycin A (**232**) (Fig. 4.8), which feature benz[*f*]indanone scaffold.

To this end, the annulation of the parent cyclopentenone (**233**) with the phthalide sulfone (**11**) was studied in the presence of LTB, and the quinol

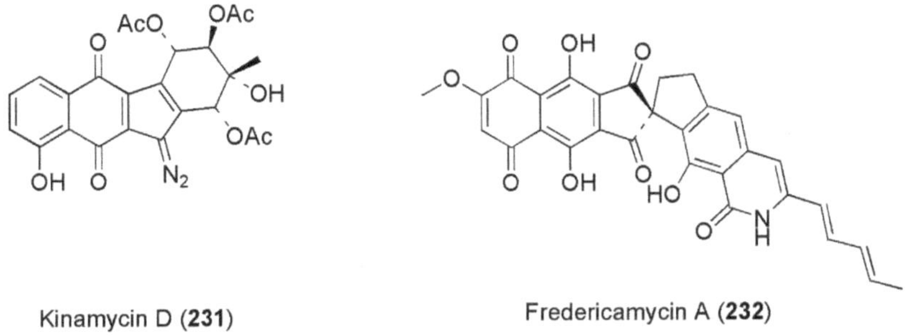

FIG. 4.8 Structures of kinamycin D and fredericamycin A.

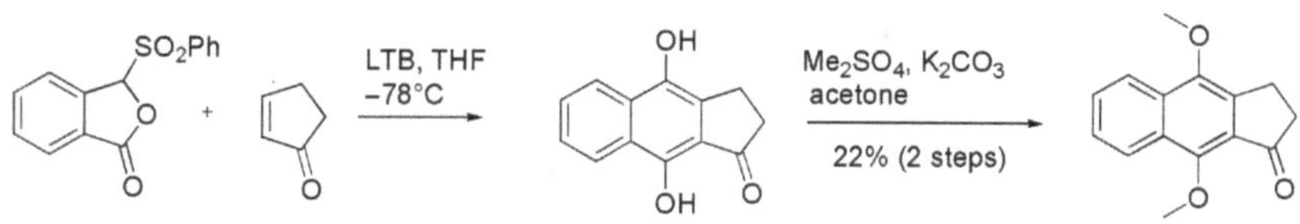

11 233 235 234

SCHEME 4.54 Cyclopentenone as Hauser acceptor.

derivative (**234**) was obtained in a markedly low yield (22% for two steps). One of the reasons for the low yield was base-catalyzed self-condensation of cyclopentenone (Scheme 4.54). On the other hand, similar annulation with phthalide phosphonate (**35**) gave the product (**235**) in 95% yield.[6]

In a model study for fredericamycin (**232**), Parker et al. successfully annulated 2,2-dimethylcyclopentendione in 45% yield.[38] Bach annulation between the cyanophthalide (**14**) and enedione (**236**), when carried out with ^{t}BuLi at −78°C, furnished annulation product (**237**) as a white crystalline material in 80% yield (Scheme 4.55).[39]

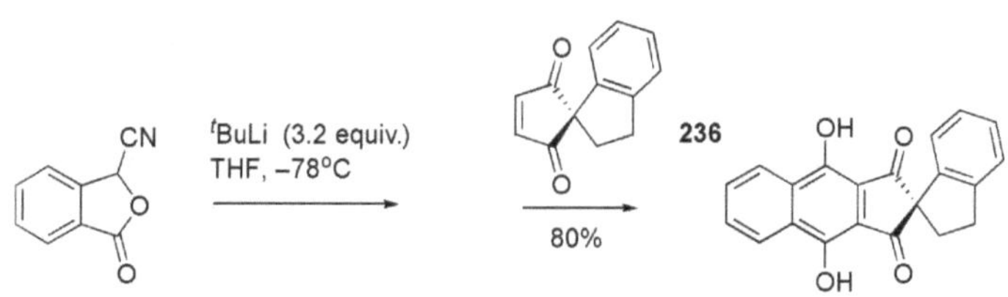

14 237

SCHEME 4.55 Parker model annulation for fredericamycin.

Although the above annulations were successful with the model acceptor (**236**), the required annulation of cyanophthalide (**238**) with acceptor (**239**) with all structural requirements for the total synthesis was impeded by initial failures. Carrying out the reaction in two stages, i.e., (i) treatment of the reactants (**238** and **239**) with LDA and (ii) treatment of the crude product (**240**) with NaOEt-ethanol-THF at −78°C afforded annulation product (**241**) in overall 78% yield. Boron tribromide promoted dimethylation of (**241**) followed by aerial oxidation provided fredericamycin A (**232**) (Scheme 4.56).[40]

When cyclopentenone ring is a part of polycyclic molecule like (**242**) which is stable to bases, the annulations pose no problem. The reaction of phthalide sulfone (**11**) with enone (**242**) in the presence of LTB followed by *O*-methylation with MeI-K_2CO_3 afforded the pentacycle (**243**) in 79% yield. Flash vacuum pyrolysis (450°C, 0.1 mm) gave corresponding benz[*f*]indenone (**244**) in quantitative yield (Scheme 4.57).[41] For the preparation of pentacycle (**243**), 3-phenylthiophthalide (**22**) can also used in the place of the sulfone (**11**) with equal or better efficacy.[42]

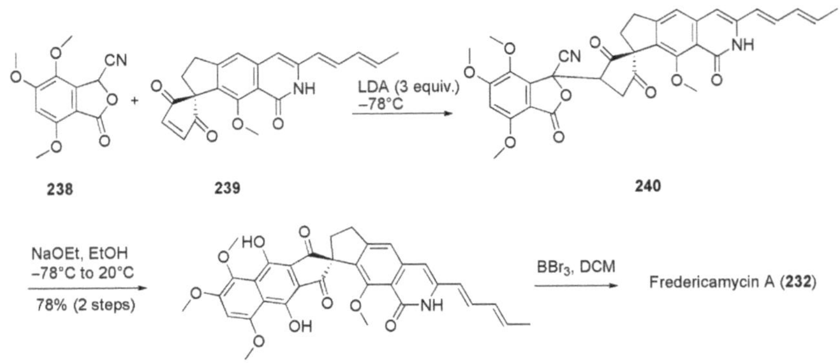

238 239 240

241

SCHEME 4.56 Bach fredericanycin synthesis.

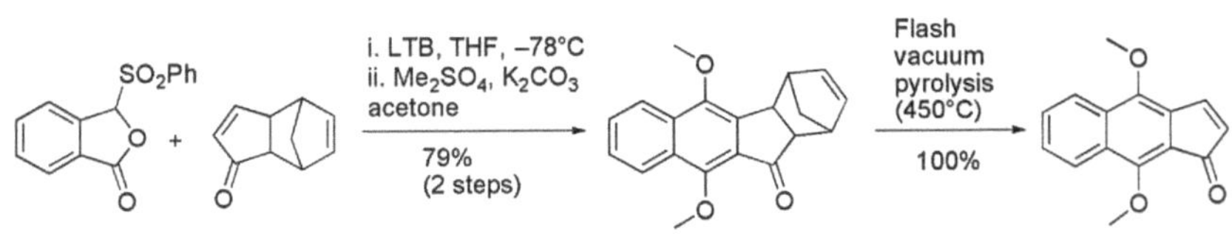

11 242 243 244

SCHEME 4.57 Annulation with a masked cyclopentadienone.

4.2.3 Indenones as Acceptors

Indenones, unsubstituted at C2 and C3, are sensitive to acids and heat, but appear to be stable to basic conditions. Therefore, indenone (**245**), obtained by FVP of (**246**) at 450°C, 0.5 mm pressure followed by quick chromatography, underwent smooth annulation with the anion of the sulfone (**11**) to produce benzo[*b*]fluorenone (**247**) in 73% yield (Scheme 4.58).[43]

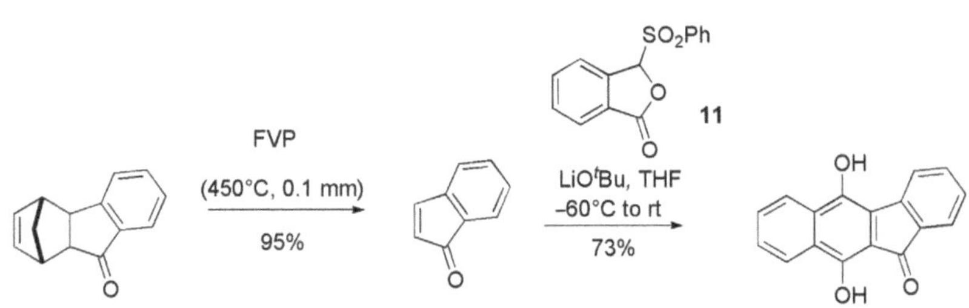

246 245 247

SCHEME 4.58 Annulation with indenone.

Based on the similar annulation, Hauser and Zhou reported the first total synthesis of prekinamycin (**254**), a diazobenzo[*b*]fluorenone using indenone (**248**).

Indanone (**249**) was obtained from a dihydroisocoumarin. It was converted to its enol silyl ether (**250**). Without isolation, the silyl ether (**250**) was submitted to Saegusa oxidation with $Pd(OAc)_2$ to produce the indenone (**248**). Hauser annulation of phthalide sulfone (**251**) with the indenone (**248**) gave benzafuranone (**252a**). Demethylation of (**251a**) with BBr_3 produced tetrahydroxy tetracycle (**252b**). Heating of (**252b**) with hydrazine gave hydrazone (**253**). Fetizon oxidation (Ag_2CO_3) of the hydrazone (**253**) gave prekinamycin (**254**) (Scheme 4.59).[44] This total synthesis of the synthesis was adopted by Feldman and Eastman in 2006 to study the biomechanism of kinamycins antibiotics. Likewise, the annulation of the cyanophthalide (**188**) with cyclopentenone fused tricycle (**256**) gave annulation product (**257**) in excellent yield in a model study on the synthesis of lomaiviticin (**258**) (Scheme 4.60).[45]

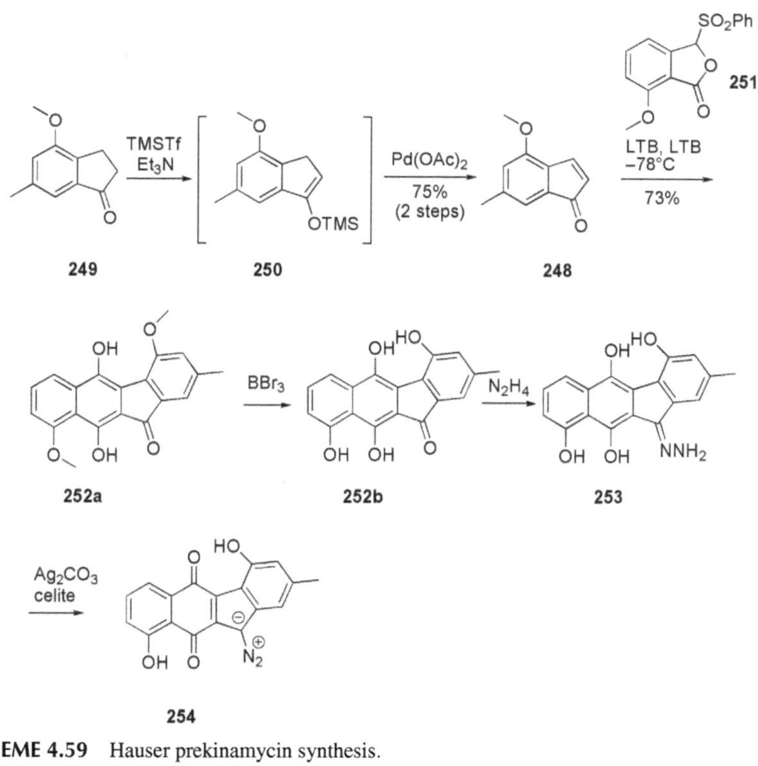

SCHEME 4.59 Hauser prekinamycin synthesis.

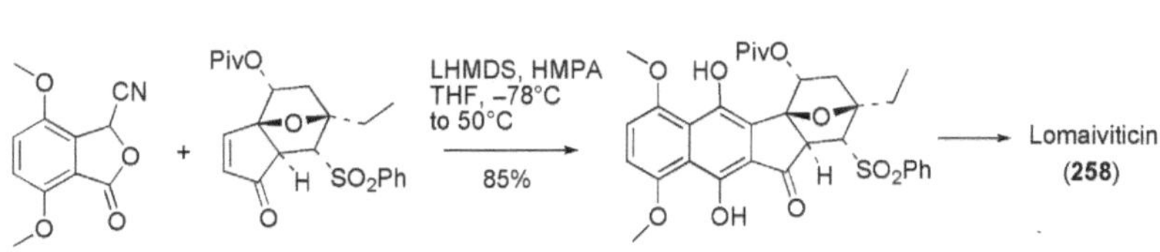

SCHEME 4.60 Annulation for lomaiviticin scaffold.

4.2.4 Cyclohexenones

In general, cyclohexenones (**258**) are well suited for Hauser annulations with various Hauser donors (**259**) to give 9,10-dihydroxy-3,4-dihydroanthracen-1(2*H*)-ones (**260**) (Scheme 4.61) via the intermediates (**261** and **262**). The yields of the products vary across a wide range.

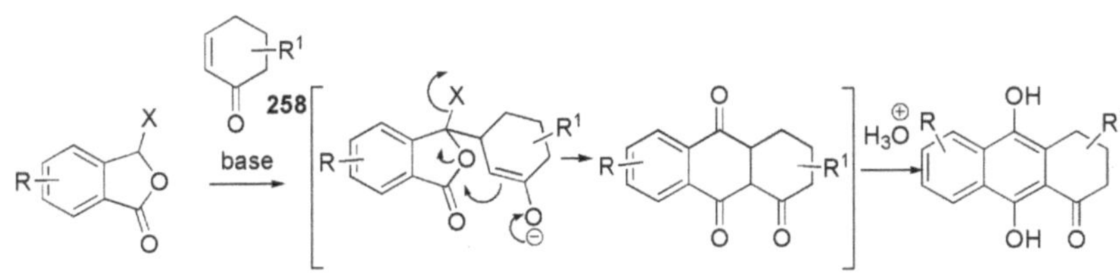

X = SPh, SO_2Ph, CN, NC, etc.

259 **261** **262** **260**

SCHEME 4.61 Annulation of cyclohexenones.

Under certain circumstances, the Michael adducts derived from the addition of phthalide anions to cyclohexenones are obtainable. In the presence of LDA at −78°C, the reaction of the donors (**22** and **11**) gave Michael adducts (**261a** and **261b**), respectively (Fig. 4.9). Under similar conditions, both 7-methoxy-3-cyanophthalide (**37**) and 7-methoxyphthalide sulfone (**251**) gave anthracenone (**263**) in moderate to good yields, the cyanophthalide (**37**) being more efficient. However, cyclohexenone (**264**) is successfully annulated with three types of donors (**2**, **3**, and **4**), if the reactions are carried out with LDA-HMPA complex (Kraus protocol) or LiO^{t}Bu (Hauser protocol).

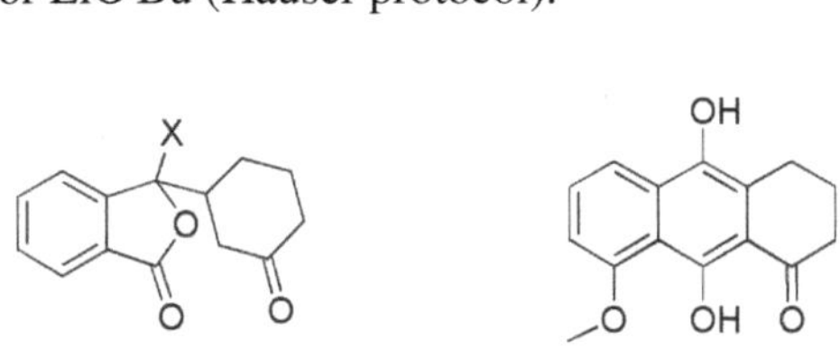

261a = SPh
261b = SO_2Ph

263

FIG. 4.9 Products from the attempted annulation with 2-cyclohexenone.

The use of LTB is crucial for the success of annulation with cyclohexenones. For example, the annulation between the phthalide sulfone (**24** and **264**) was best carried out with LTB, giving the product (**262**) in 67% yield (Scheme 4.62).[46]

The annulations with cyclohexenones are severely impeded for the donors having 5,6-dialkoxy substituted Hauser donors or Watanabe donors. As shown in Scheme 4.63, the annulation between phthalides (**266**) and cyclohexanone (**267**) fail under different conditions. The annulation was marginally successful with cyanophthalide (**266d**).[47]

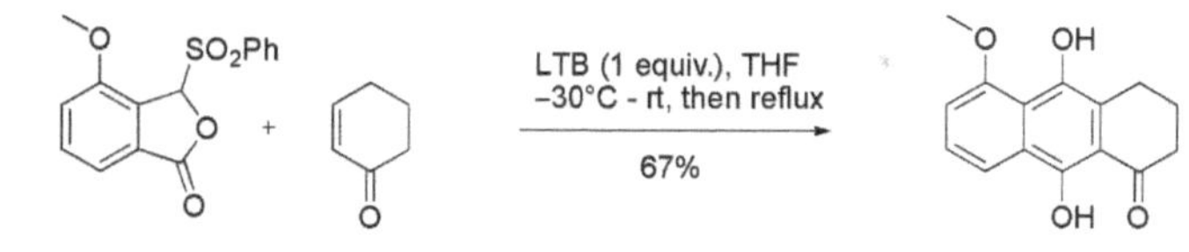

24 **264** **265**

SCHEME 4.62 Annulation of parent cyclohexenone.

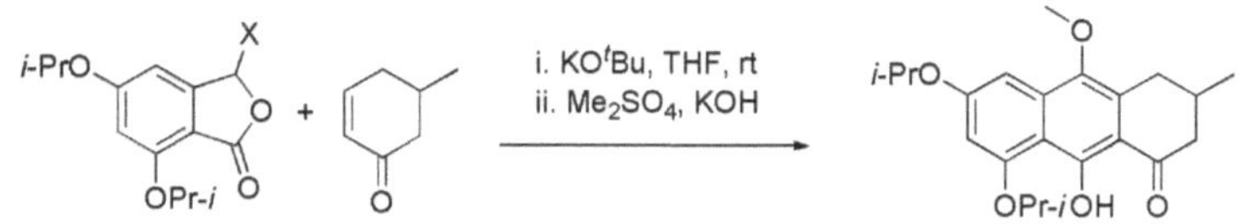

a : X = SO_2Ph; b : X = SPh; c = $PO(OEt)_2$
d : X = CN

266 **267** **268** (42% when X = CN)

(0% when X = SO_2Ph, SPh or $P(O)(OEt)_2$)

SCHEME 4.63 Substituent effect on cyclohexenone annulation.

Brimble et al., in their synthesis of 2-carbathysanone (**271**), found that NaH could be a suitable base for the Hauser annulation of 5,7-dialkoxy-3-cyanophthalide with cyclohexenones. The reaction of phthalides (**269**) with cyclohexanone (**267**), followed by methylation, provided the annulated product (**270**) in 79% yield. The dimethyl ether derivative (**270**) was converted into 2-carbathysanone (**271**) in three steps. DIBALH reduction of (**270**) gave alcohol (**272**), which was oxidized by CAN in acetonitrile-water to produce quinone (**273**). Bismuth triflate-catalyzed deprotection of MOM groups afforded carbathysanone (**271**) (Scheme 4.64).[47]

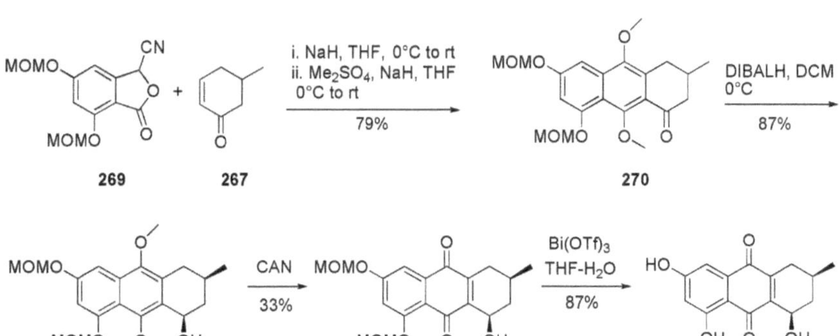

272 **273** **271**

SCHEME 4.64 Substituent effect on cyclohexenone annulation.

With fully nuclear substituted cyanophthalide (**274**), the efficiency of its annulation with 5-acetoxymethyl cyclohexenone (**275**) is modest. The annulation product dihydroxyanthracenone (**276**), on aerial oxidation, furnished anthraquinone (**277**), a potential intermediate for proisocrinin natural products.[48] The acetoxy group of the enone (**275**) remained intact during the process (Scheme 4.65).

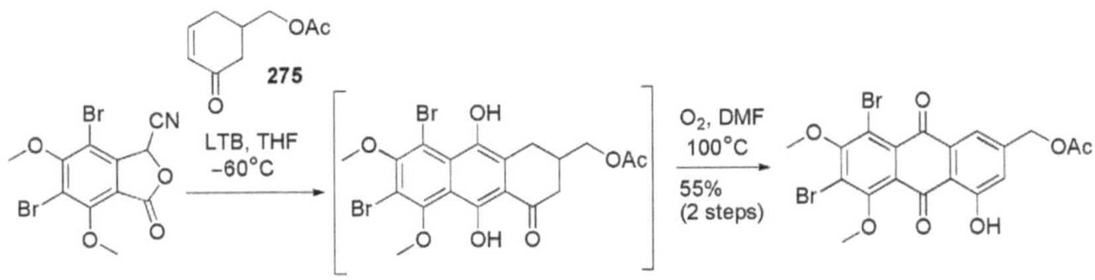

274 276 277

SCHEME 4.65 Annulation with a brominated Hauser donor.

The chemistry of the annulation with 5-substituted cyclohexenones was further applied to the annulation with a dimeric form of sulfonylphthalide to achieve a neat synthesis of (±)-biphyscion (**278**). It is noteworthy that the anionic annulation of a two-pronged phthalide donor can be performed in a one-pot operation. The annulation of (**279**) with cyclohexenone (**267**) in the presence of LTB gave bis-anthracenone (**280**), the oxidation of which with Ag_2CO_3, Et_3N furnished bis-anthraquinone (**281**) in 36% yield (two steps). MgI_2-assisted demethylation of (**281**) afforded (±)-biphyscion (**276**) (Scheme 4.66).[49]

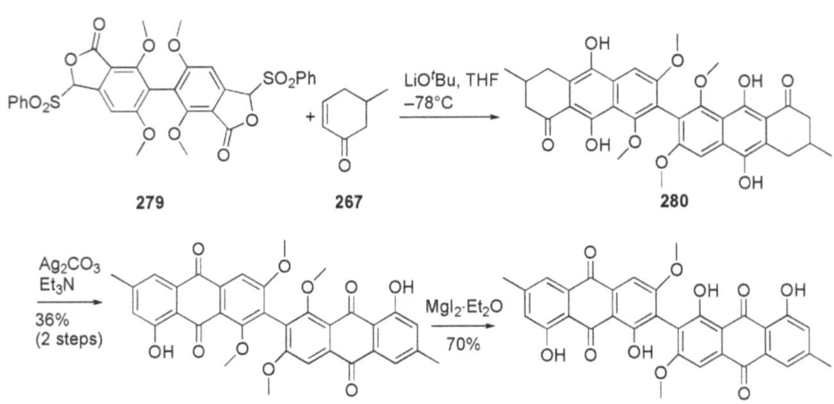

279 267 280

281 (±)-Biphyscion (**278**)

SCHEME 4.66 Double Hauser annulation.

With 6-methoxy-2-cyclohexenone (**282**), the annulation of the phthalide sulfone (**251**) in the presence of LiO^tBu furnished anthracenediol (**283**) in high yield. Upon oxidative aromatization with oxygen bubbling in DMF at 110°C, the quinol (**283**) was transformed to 1,2-dioxygenated anthraquinone (**284**) in >85% yields, which represent morindaparvin A derivatives (Scheme 4.67).[50]

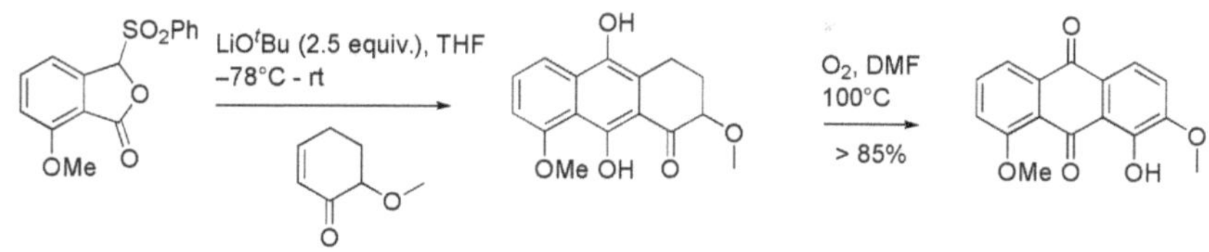

251 282 283 (> 85%) 284

SCHEME 4.67 Annulation of an oxy-substituted cyclohexenone.

The transformation of 9,10-dihydroxyanthracenones (e.g., **283**) to 1-hydroxy-9,10-anthraquinones (i.e., **283** to **284**), can best be performed by oxidation with Fetizon's reagent (Ag_2CO_3-celite), followed by addition of a catalytic amount of triethylamine.[51]

More strikingly, the Hauser annulations work well even with 5-oxysubstituted cyclohexenones, which are often susceptible to elimination of the oxygen functionality at C5. These annulations form the foundation of the work of Snider et al. and Nicolaou et al. on the chemistry of cytoskyrins and rugulosins.[52,53] Snider's work on annulation between the cyanophthalide (**285**) and enones (**286**) in the presence of KTB and DMSO or THF showed that the yields of the annulations vary with the nature of R group at C5 in the cyclohexenone acceptors (**286**). For 5-dimethylphenylsilyl derivative (**286a**), the yields of (**287a**) varied from 10% to 60%. In the case of *O*-benzyl derivative (**286b**), no annulation product (**287b**) was isolated, possibly due to destruction of cyclohexenone (**286**) through elimination of benzyl alcohol. In the case of 5-methyl- and 5-acetoxymethyl-2-cyclohexenones (**267** and **275**), the reaction proceeded smoothly to give the corresponding annulated products (**287c** and **287d**), respectively, in moderate yields (Scheme 4.68). These anthracenones (**287**) were finally converted to the rugulosin analogs by $Pb(OAc)_4$—pyridine-mediated dimerization.

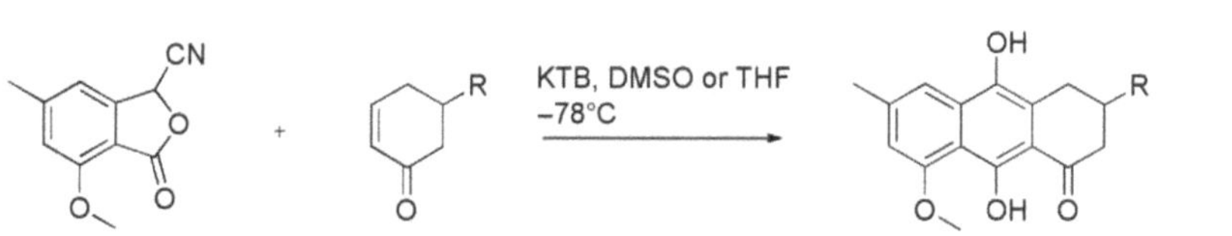

285

286a: R = $SiPhMe_2$
286b: R = OBn
267: R = Me
275: R = CH_2OAc

287a: R = $SiPhMe_2$ (10%-60%)
287b: R = OBn (0%)
287c: R = Me (>48%)
287d: R = CH_2OAc (>40%)

SCHEME 4.68 Annulation of 5-substituted cycloxenones.

This trend of reactivity of cyclohexenones was also observed with dimethoxyisobenzofuranones (e.g., **288**), indicating that nuclear methoxy substituents in isobenzofuranones have a critical role in the efficiency of the reactions (Scheme 4.69). As shown in Scheme 4.69, yields of the annulation products are moderate.[54]

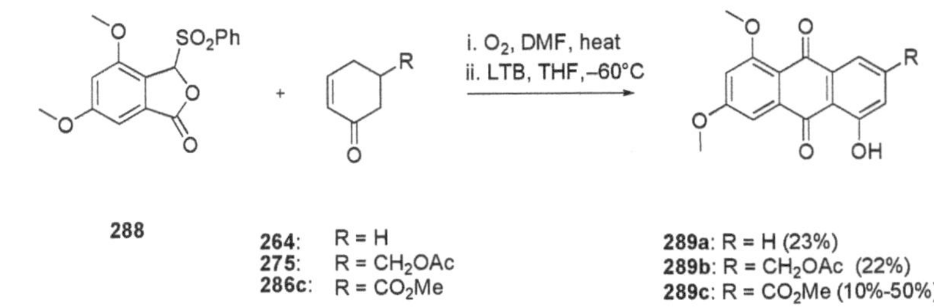

SCHEME 4.69 Substituent effect on cyclohexenone annulation.

A minute modification in the structure of the acceptor, as in the *O*-MOM substituted cyclohexenone (**290**), permitted smooth annulation with cyanophthalide (**291**) to give corresponding dihydroanthraquinone (**292**) in 91% yield. Similar annulations with cyclohexenones with OSEM, OTBS, OMe, and TMS at C5 work well in the presence of LHMDS to give the corresponding annulation products in good to excellent yields. Without any purification, anthracenone (**292**) was subjected to oxidative dimerization with MnO_2 via quinone (**293**). The initial dimerization product (**294**) was further isomerized with Et_3N to give the rugulosin derivative (**295**) through a cascade of Michael additions (Scheme 4.70).[53]

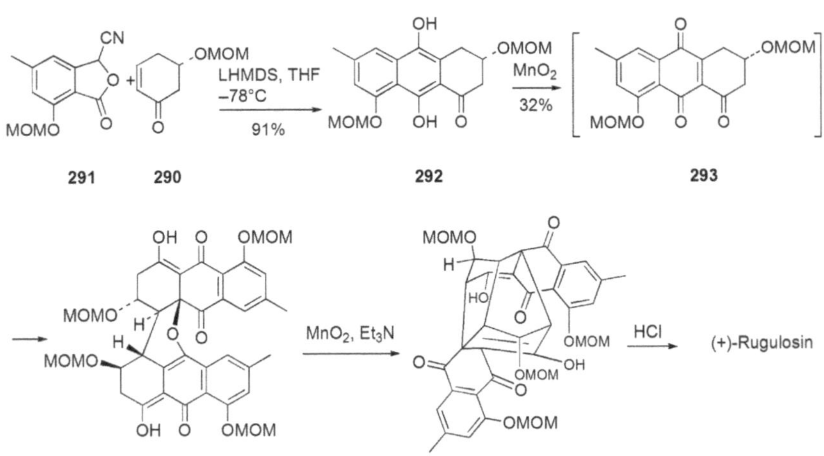

SCHEME 4.70 Rugulosin synthesis.

Suzuki synthesis[55] of aquayamycin (**296**) and Zhu synthesis[56] derhodinosylurdamycin (**297**) (Fig. 4.10) are unique examples of building up complexities in a single-step operation by carefully designed annulation of Hauser donor and its acceptor. Zhu synthesis of derhodinosylurdamycin A (**297**) closely parallels that of aquayamycin (**296**) by Suzuki et al. Both the syntheses share two key reactions: Hauser annulation and intramolecular Pedersen pinacol coupling. Zhu synthesis is presented in Scheme 4.71.

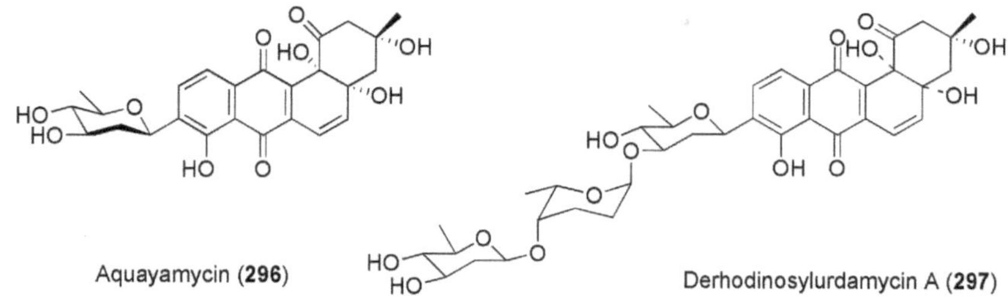

FIG. 4.10 Structures of aquayamycin (**296**) and derhodinosylurdamycin A (**297**).

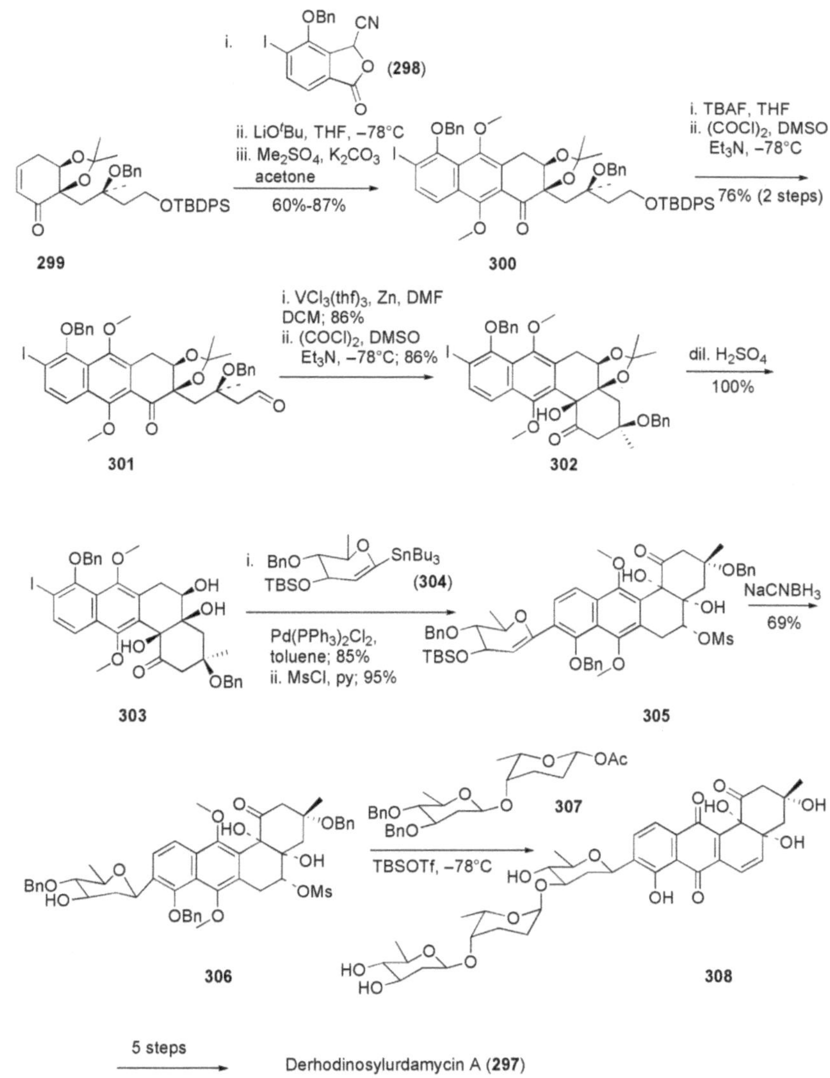

SCHEME 4.71 Zhu synthesis of derhodinosylurdamycin.

The phthalide donor (**298**) was prepared from 2-formyl-hydroxybenzamide in five steps. The acceptor cyclohexenone (**299**) was obtained from 2-bromo-2-cyclohexenol in 10 steps. The Hauser annulation between the phthalide (**298**) and the acceptor (**299**) in the presence of LTB followed by methylation gave, as expected, tricyclic hydroquinone dimethyl ether (**300**). Removal of the TBDPS ether of (**300**) followed by Swern oxidation of the resultant primary alcohol afforded aldehyde (**301**). Pedersen pinacol coupling of (**301**) and Swern oxidation of the resulting pinacol gave angular tetracycle (**302**). Deprotection of the acetonide of (**302**) with sulfuric acid afforded idodobenzanthracene (**303**) (Scheme 4.71). Stille coupling of the aryl iodide (**303**) with glycal stannane (**304**) in the presence of a Pd-catalyst followed by mesylation afforded product (**305**). Reduction of (**305**) using $Na(CN)BH_3$ and HCl afforded desired *C*-angucycline glycoside (**306**). TBS triflate-catalyzed glycosylation between disaccharide-derived donor (**307**) and complex acceptor (**306**) furnished derhodinosylurdamycin A skeleton (**308**). A sequence of five steps including several deprotection steps and CAN oxidation permitted the total synthesis of derhodinosylurdamycin A (**297**).

Trioxacarcin A (**309**) is a representative member of the family of naturally occurring trioxacarcins (Fig. 4.11), which inhibit the growth of cultured human cancer cells. Trioxacarcin DC-45-A2 (**310**), which is also naturally occurring, serves as a biosynthetic precursor to a variety of other members of the family. Besides being structurally complex, trioxacarcin A (**309**) exhibits subnanomolar growth inhibition values in a number of human cancer cell lines.

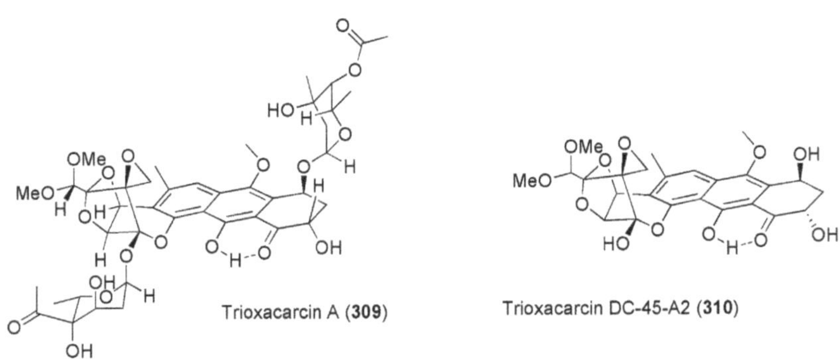

FIG. 4.11 Structures of trioxacarcins.

Total synthesis of (**309**) and the analogs has been reported by the Myers group and Nicolaou group.[57] Both groups resorted to applying Hauser annulation in constructing anthracenones motifs (**311** and **312**). Incidentally, both approaches utilized the same 4,6-dioxygenated cyclohexenone (**313**) (Fig. 4.12).

The Nicolaou synthesis was built upon hydroanthracenone (**314**), which is regiospecifically assembled by annulation of (**315**) with the cyclohexenone (**313**) in the presence of LTB, followed by partial *O*-methylation (Scheme 4.72). Deprotection of the MOM group in (**316**) by magnesium bromide etherate,

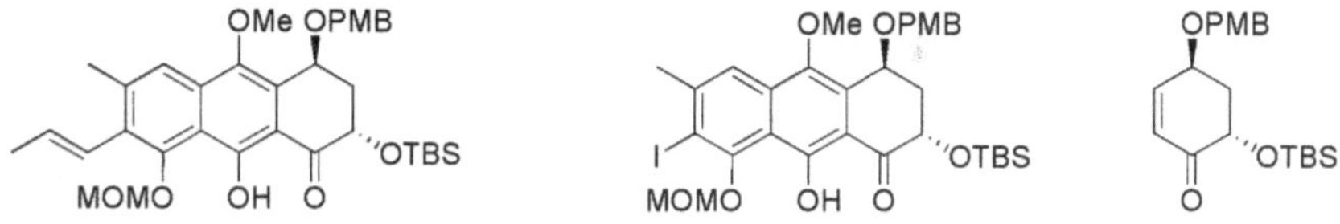

311 (Myers intermediate) 312 (Nicolaou intermediate) 313

FIG. 4.12 Synthetic precursors of trioxacarcin A (**307**).

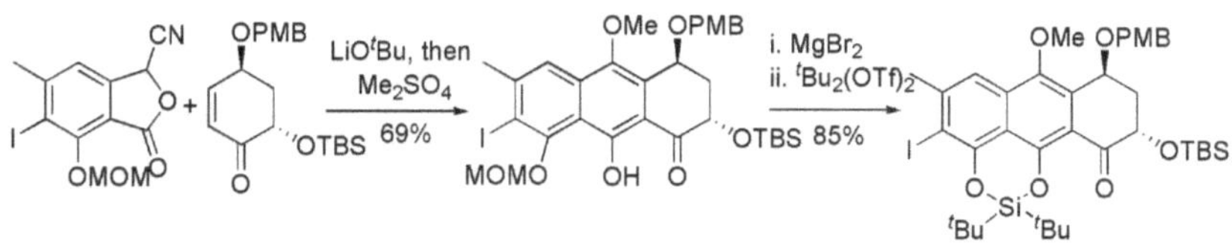

315 313 316 314

SCHEME 4.72 Synthesis of Nicolaou intermediate for trioxacarcin A [2015 ACIE 3074 Nicolaou KC].

followed by treatment with ^{t}Bu$_2$(OTf)$_2$ and lutidine, gave (**314**), which formed the key intermediate for further elaborations.[57c]

Annulation of 4,4-cyclohexen-2-ones has received much attention in various contexts. Generally, the respective annulations are successful, unless the acceptors are sterically crowded. The annulation of 4,4-dimethylcyclohex-2-enone (**317**) with the cyanophthalide (**14**) in the presence of LDA and $ZnCl_2$ gave a mixture of three products (**318**, **319**, and **320**) with the former (**318**) being the major product (Scheme 4.73).[58] The mixture, without separation, was oxidized with DDQ to give naturally occurring quinone (**321**) in an overall 84% yield, essentially constituting a single-step synthesis. The success of the annulation with the enone (**317**) is clear evidence that the *gem* dimethyl groups do not cause any neopentyl steric hindrance to the annulation, and synthesis of a quinonoid can be abbreviated by the use of Hauser annulation. The anthracenetrione (**321**) is found to possess potent in vitro lytic activity against *Trypanosoma cruzi*.

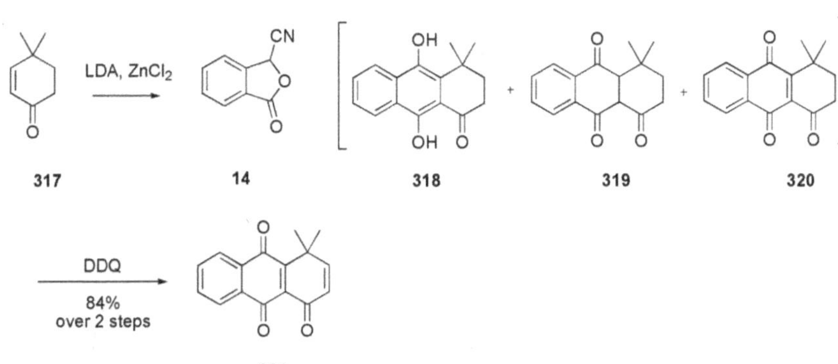

SCHEME 4.73 Annulation with a 4,4-disubstituted cyclohexanone [2003 JCRS 470 Markgraf JH].

A similar annulation is found to be useful in the total synthesis of tetarimycin (**322**), which is reported to possess potent activity against methicillin-resistant *Staphylococcus* infections. Structurally, it is a linearly fused tetracycle with a 1,4-quinone functionality and three methoxy substituents. Unlike common polyketides, it contains a *gem*-dimethyl group. The retrosynthesis of (**322**) identified the synthons phthalide (**323**) and 4,4-dimethylnaphthalenone (**324**) for the necessary Hauser annulation.[59] For the synthesis of phthalide (**323**), pyrone (**325**) was heated in neat DMAD (**326**) at 210°C to give Diels-Alder adduct (**327**) (73%). Alkaline hydrolysis of (**327**) afforded corresponding mono ester in 96% yield (Scheme 4.74). BH_3-mediated chemoselective reduction of the ester gave hydroxymethyl ester derivative (**328**), the acid-catalyzed lactonization of which provided phthalide (**329**) in 93% yield. The phthalide (**329**) was converted in two steps to the corresponding benzyloxyphthalide (**330**). Phthalide (**330**) was reacted with diethylamine under catalysis with $AlMe_3$ to form the corresponding amide alcohol, which was oxidized with MnO_2 to afford amide aldehyde (**331**) in 81% yield. Application of the Yoshii method (TMSCN-KCN cat. 18-c-6, AcOH) of cyanophthalide formation to compound (**331**) afforded cyanophthalide (**323**) in 81% yield.

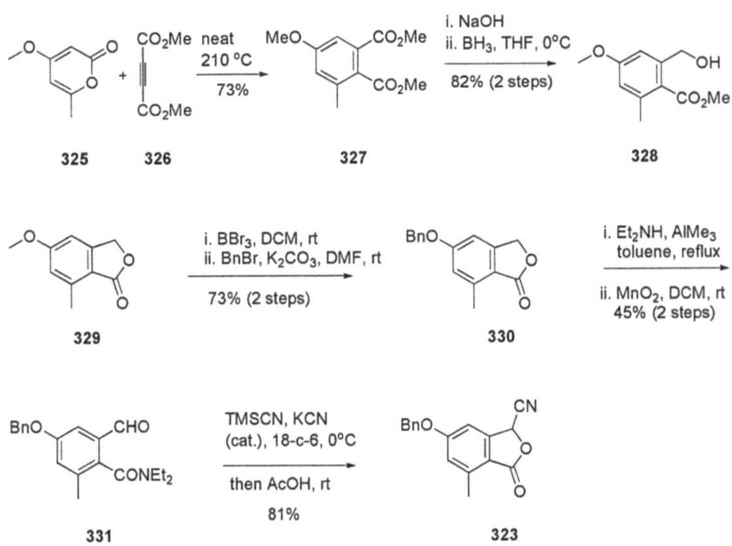

SCHEME 4.74 Synthesis of cyanophthalide (**323**) for the synthesis of tetarimycin [2015 ORS 4248 Shia K-S].

The synthesis of the enone (**324**) began from 1,3-dimethoxybenzene. It was converted into tetralin (**332**) in five steps. Oxidation of (**332**) with *tert*-butylhydroperoxide and CuI as a catalyst afforded tetralone (**333**) in 74% yield, which was converted to (**334**). Further oxidation of (**334**) with DDQ led to the desired enone (**324**) in 86% yield (Scheme 4.75). Hauser annulation of the phthalide (**323**) with the enone (**324**) using LHMDS and $ZnCl_2$ afforded a

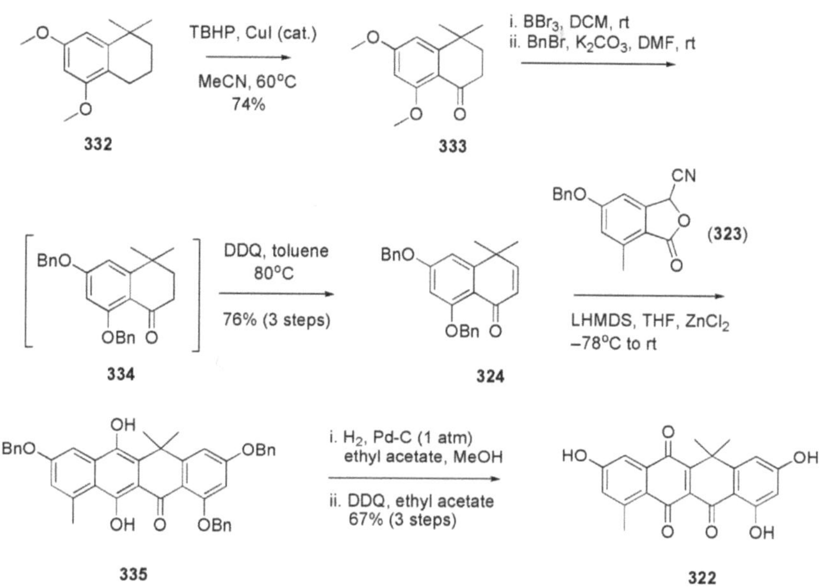

SCHEME 4.75 Preparation of acceptor tetralenone (**324**) en route to the synthesis of tetarimycin (**322**) [2015 ORS 4248 Shia K-S].

mixture of a annulation product (**335**) and its quinone. Treatment of the crude mixture with H_2/Pd-C followed by oxidation with DDQ resulted in the target (**322**) in 67% yield (three steps). This total synthesis led to a fruitful SAR study, revealing that C-10 hydroxyl group is essential for antibacterial activities.

In the above cases (Schemes 4.73 and 4.74), the substituents at C4 of the cyclohexenone moieties did not impede the Hauser annulations (Scheme 4.75). However, the annulation studies directed toward the model precursor (**336**) of halenaquinone and xestoquinone clearly indicate that annulations are not feasible if the substituents are in a rigid decalin system and exert a steric effect (Scheme 4.76). The annulation between the phthalide sulfone (**164**) and cyclohexenone derivative (**337**) under standard conditions, LTB, THF, −78°C afforded annulation product (**338**).[60] Under similar conditions, the attempted annulations between a decalenone and the phthalide sulfone (**164**) or the cyanophthalide (**188**) were not successful, as will be described in the next section.

4.2.5 Decalenones

The annulation chemistry of decalenones (**339–341**, Fig. 4.13) has thoroughly been studied in the context of anthracyclines. Their suitability as the acceptors in Hauser annulation is a significant advancement in the field of linearly fused *peri*-oxygenated tetracenes. It should be reiterated that decalenones have been the subject of Robinson annulation in the fabrication of angularly fused systems like hydrophenanthrenes. The early examples of decalenone annulations focused on the targets belonging to anthracyclines.

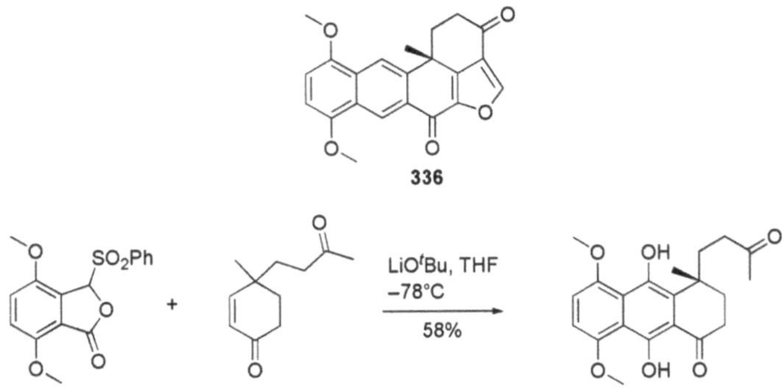

164 337 338

SCHEME 4.76 Model studies on halenaquinone and xestoquinone [Unpublished results Hauser FM].

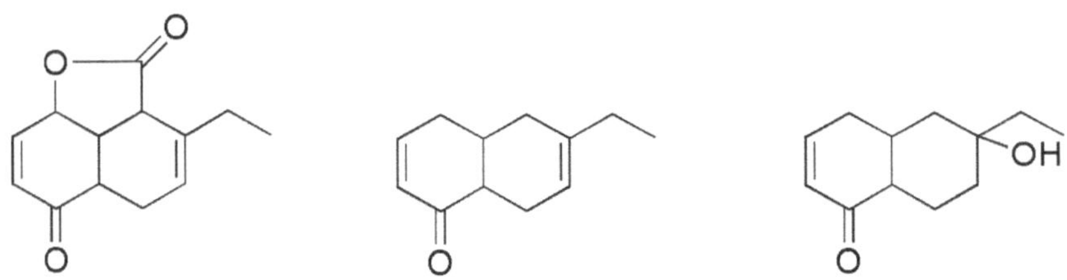

339 340 341

FIG. 4.13 Types of decalenones studied for [4+2] annulations.

The first use of a decalenone in the Hauser annulation is due to Li and Wu in 1981 in the context of the total synthesis of alkavinone, the aglycon of aclacinomycin A, a highly promising chemotherapeutic agent.[61] Both sulfone phthalides (Hauser donor) (**2**) and cyanophthalides (Kraus donor) (**3**) were used for assembling the tetracene cyclic framework of alkavinone (**342**). The synthesis started with *p*-benzoquinone (**343**). Its reaction with diene (**344**) in DCM at room temperature under 15 Kbar afforded enedione (**345**) in 60% yield. Meerwein-Ponndorf-Verley reduction of the enedione (**345**) followed by Jones oxidation gave tricyclic lactone (**346**) in good yield. The reaction of the cyanophthalide (**37**) with the acceptor (**346**) in the presence of LDA-HMPA at −78°C followed by aerial oxidation gave tetracyclic acid (**347**) via the intermediates (**348** and **349**). Air oxidation of (**347**) gave (**350**). Conversion of (**350**) into the methyl ester by diazomethane treatment and subsequent epoxidation of the ester (**351**) with *m*-CPBA gave epoxide (**352**) in good yield. Epoxide-opening of (**352**) with sodium bromide and *p*-toluenesulfonic acid at room temperature gave bromohydrin (**353**). Bromine of the bromohydrin (**353**) was reductively (ammonium formate, Pd-C) removed to give alcohol (**354**). Three more steps: *O*-demethylation, radical initiated bromination, and hydrolysis, applied to (**354**), resulted in the synthesis of racemic alkavinone (**342**) (Scheme 4.77).

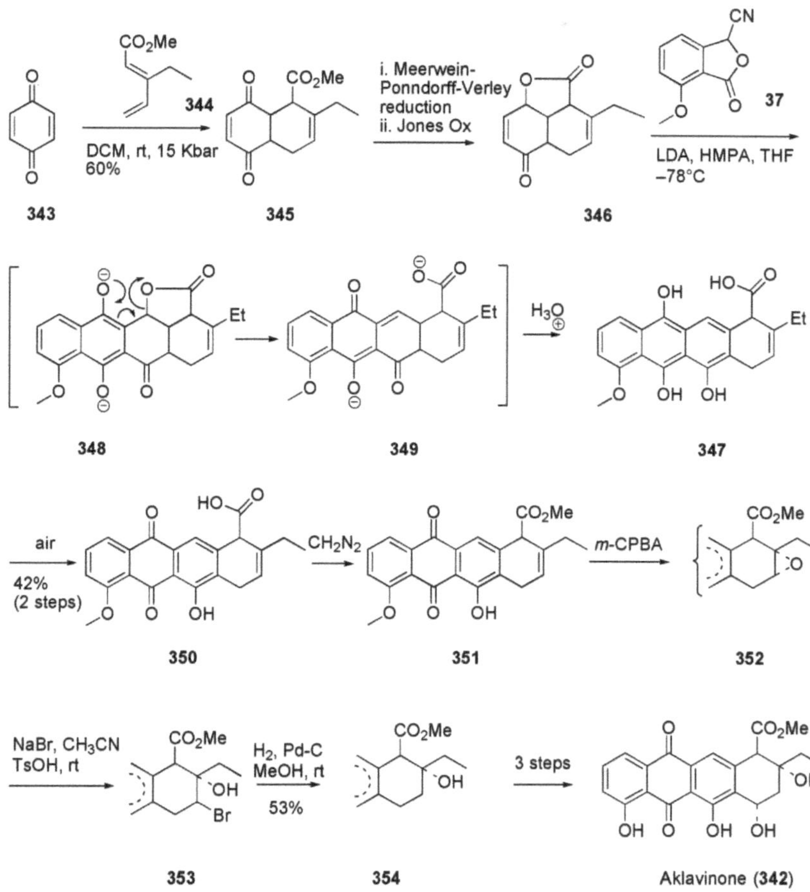

SCHEME 4.77 Li synthesis of racemic alkavinone (X) [1981 JACS 7007 Li T].

The Hauser total synthesis[62] of aklavinone (**342**) looks somewhat similar to the Li synthesis. Although both the syntheses utilize the Hauser annulation as the key step, the former is much more concise and efficient. It involves the ready accessibility of starting materials and high yields of the annulation of a decalenone. Furthermore, the Hauser total synthesis, which was completed in 12 steps, features homologation of a keto aldehyde to the corresponding keto anthraquinone acetic ester via the intermediary of a ketene thioacetal. Condensation of the anion of the phthalide sulfone (**251**), generated with lithium *tert*-butoxide in THF, with the naphthalenone (**355**) furnished hexahydronaphthacenone (**356**) in 83% yield as a mixture of *cis* and *trans* isomers. Conversion of (**356**) to quinone tetracenone (**357**) was accomplished by heating (**356**) in DMF under an oxygen atmosphere. Methylation of (**357**) with Me_2SO_4-K_2CO_3 gave methyl ether derivative (**358**). Ozonolysis of the tetracycle (**358**) in DCM-methanol gave keto aldehyde (**359**) in 96% yield. The Wittig reaction of the ylide of keto aldehyde (**360**) with (**359**) afforded ketene thioacetal (**361**) in 89% yield. Mercuric chloride catalyzed methanolysis of (**361**) gave the homologated ester (**362**) (80%). Intramolecular aldol cyclization of the keto ester (**362**) using magnesium methoxide afforded a diastereoisomeric mixture of hydroxy esters (**363**). Demethylation of the major isomer (**363**)–(**364**) was carried out with aluminum chloride in dichloromethane. Finally, the 7-hydroxy group of (**342**) was introduced through NBS bromination of (**364**) followed by hydrolysis (Scheme 4.78).

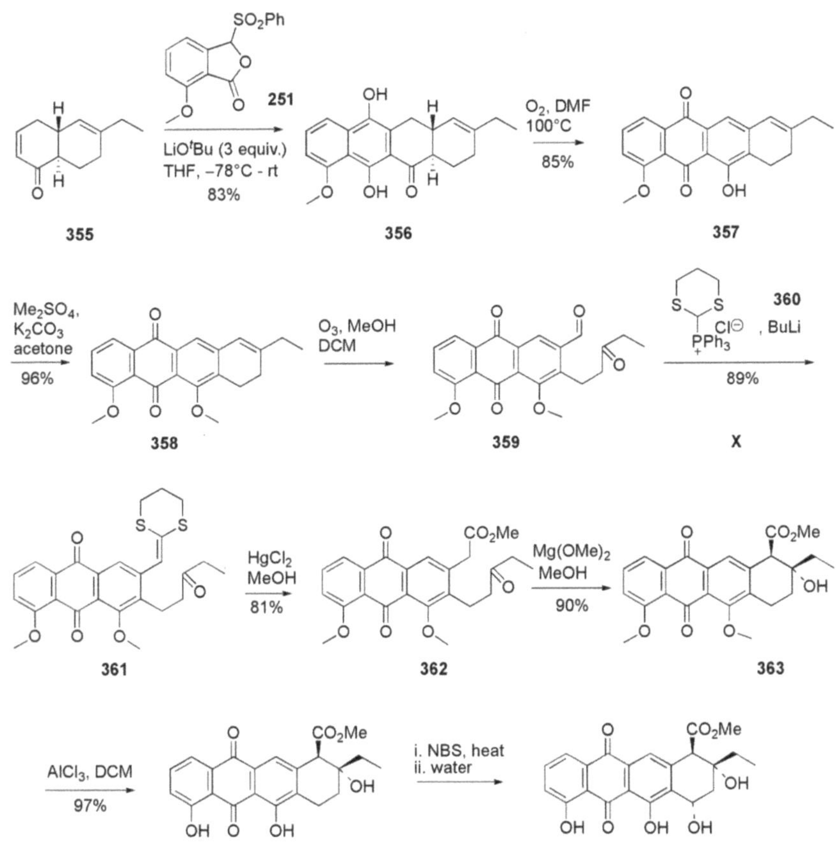

SCHEME 4.78 Hauser decalenone approach to alkavinone (**340**) [1989 JOC 5110 Hauser FM].

Another important feature of the Hauser total synthesis is the synthesis of the naphthalenone (**355**) forming AB ring system of aklavinone (**342**) (Scheme 4.79). The reaction of the Grignard reagent (**365**) with 2-ethylacrolein (**366**) gave the alcohol intermediate (**367**). Acetylation of (**367**) with acetic anhydride-pyridine furnished allylic acetate (**368**) in 63% yield. Treatment of

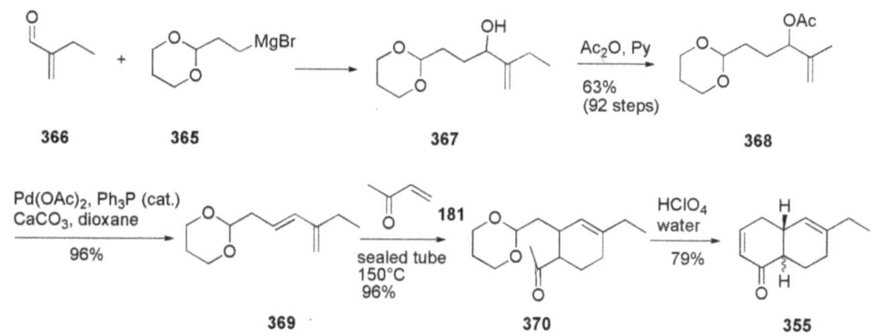

SCHEME 4.79 Synthesis of decalenone (**353**) for alkavinone (X) [1989 JOC 5110 Hauser FM].

(**368**) with a catalytic amount of palladium acetate and triphenylphosphine in refluxing dioxane resulted in *E*-diene (**369**) (purity >96%). The Diels-Alder cycloaddition of (**369**) with methyl vinyl ketone (**181**) (neat, sealed tube, 150°C) furnished cyclohexene (**370**) (96%) as a mixture of *cis* and *trans* isomers. Treatment of (**370**) with dilute perchloric acid in THF resulted in decalenone (**355**) through the intramolecular aldol condensation of the keto aldehyde obtainable by deketalization of (**370**).

The approach of Hauser and Mal to 11-deoxydaunomycinone, the characteristic structural feature of alkavin type anthracyclines, involved the annulation of the sulfone phthalide (**251**) with decalenone (**371**) in the presence of LiO^{t}Bu, to produce (**372**).[63] Aerial oxidation of (**372**) in hot DMF gave tetracene (**373**). The use of 3 equiv. of LiO^{t}Bu in the annulation gave the highest ever yield of a Hauser annulation. The decalenone (**371**) representing the AB ring of anthracyclines were prepared from bicyclic compound (**374**). The bicyclooctane (**374**), in turn, was synthesized in six steps from 1,3-cyclohexadiene and methyl methacrylate using Diels-Alder reaction as a key step. Further six-step elaboration of (**374**) through (i) RuO_4 oxidation of CC double bond, (ii) acid chloride formation, (iii) organocopper mediated methylation, (iv) deacetylation, (v) CrO_3-py oxidation of alcohol, and (vi) acid-catalyzed intramolecular aldol condensation, yielded decalenone (**371**). This sequence proceeded through the intermediates (**375–377**) (Scheme 4.80). This study was then extended to the synthesis 11-deoxydaunomycinone.

A modification of the above approach by Hauser and Tommasi culminated in the first total synthesis of (−)-(9*R*)-7,11-dideoxy-13-deoxydaunomycinone (**378**), which is not only enantioselective, but also regiospecific (Scheme 4.81).[64] It is founded on the combined applications of an enantioselective Diels-Alder reaction and a Hauser annulation. The naphthalenone (**379**), designed to serve as the AB rings of the target was assembled by Evan asymmetric Diels-Alder methodology. Condensation of the anion of the sulfone (**251**) with the naphthalenone (**379**) in the presence of LiO^{t}Bu, followed by aerial oxidation of the immediate annulation product (**380**), furnished the optically active anthracyclinone (**378**) in 85% yield.

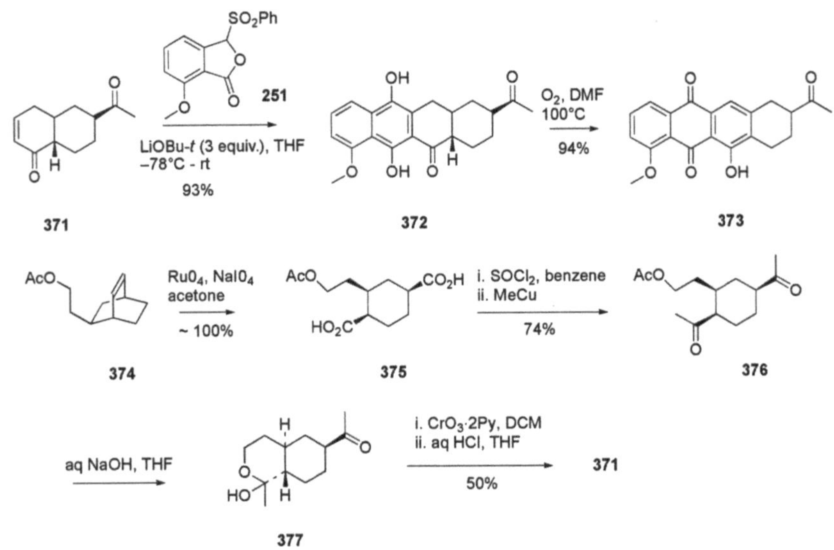

SCHEME 4.80 Total synthesis of 11-deoxydaunomycinone [1983 JACS 5688 Hauser FM].

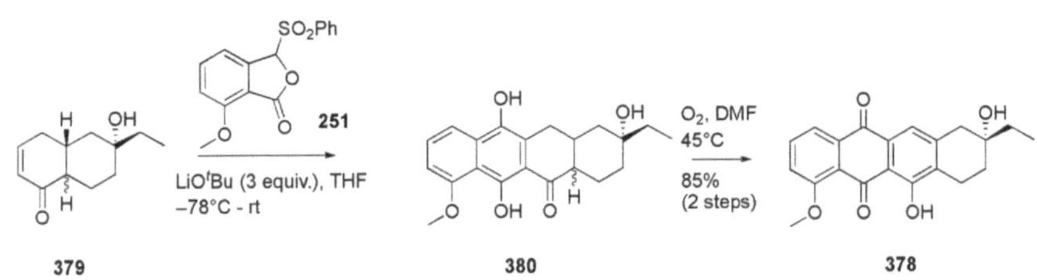

SCHEME 4.81 Asymmetric synthesis of (−)-(9*R*)-7,11-dideoxy-13-deoxodaunomycinone [1991 JOC 5758 Hauser FM].

The synthesis of naphthalenone (**379**) was started with 4-bromobutyl phenyl sulfide (Scheme 4.82). It was transformed to diene (**381**) in three steps involving Tsuji diene synthesis. Dimethylaluminum chloride catalyzed Diels-Alder reaction between diene (**381**) and oxazolidinone (**382**) gave adduct (**383**) (de > 100:1). Removal of the auxiliary group with LiOBn furnished benzyl ester (**384**). Treatment of the benzyl ester (**384**) with TMSCl-NaI in acetonitrile promoted debenzylative lactonization to give bicyclic compound (**385**). This process transferred the chirality of the ester function to the masked alcohol corresponding to C9 of anthracyclines. Oxidation of the sulfide group in (**385**) with *m*-CPBA led to formation of sulfoxide (**386**). Dieckmann-type cyclization of the sulfoxide (**386**) in the presence of LDA followed by thermal elimination of benzenesulfenic acid afforded decalenone (**379**).

Menogaril, also known as 7-con-*O*-methylnogarol (**387**), the semisynthetic derivative of nogalamycin, has been an attractive synthetic target because of its broad spectrum activity and lower cardiotoxicity than that observed for

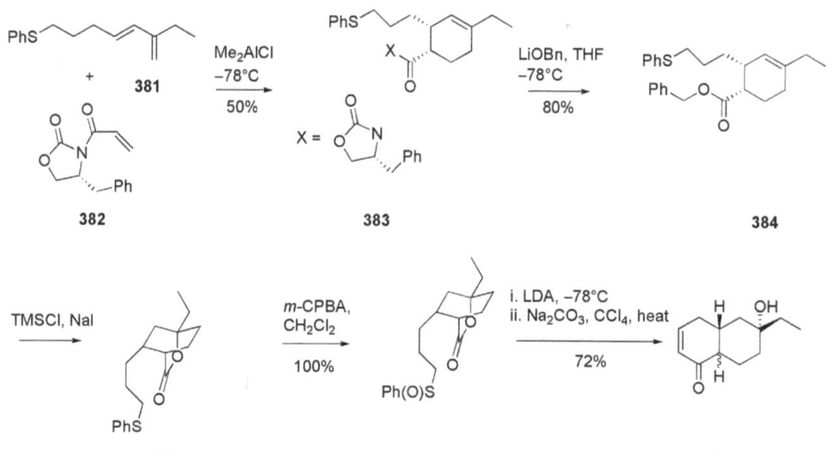

SCHEME 4.82 Asymmetric synthesis of naphthalenone X [1991 JOC 5758 Hauser FM].

adriamycin. The suitability of decalenones (cf. **340**) as the acceptors in the Hauser annulation has been exploited by Hauser group in the synthesis of (±)-7-*con*-*O*-methylnogarol (**387**).[65] In contrast to the previous uses of decalenones in the annulation, the strategy consisted in the synthesis of highly functionalized glycosyl Hauser donor (**388**). Other crucial stages were the synthesis of decalenone (**389**), and the elaboration of its annulated product. The installation of the 9-hydroxy group of the target was executed by epoxidation of the double bond at C9 followed by its reductive ring-opening of the resultant epoxide with hydrogenolysis with hydrogen in the presence of Pd-C. To prevent the oxidation of N at the stage of epoxidation, N of oxocin amine group was protected by a trimethylsilylethoxycarbonyl (Tec) group (Scheme 4.83).

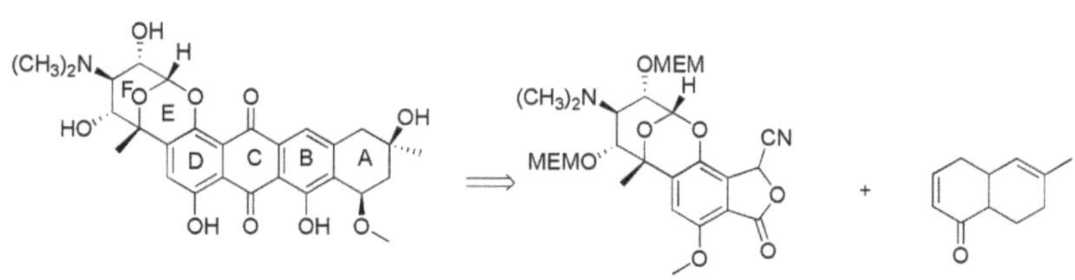

SCHEME 4.83 Retrosynthesis of 7-*con*-*O*-methylnogarol.

For the synthesis of the donor (**388**) representing DEF-ring system, an Achmatowicz rearrangement of (**390**) was executed to give (**391**). The overall route for the synthesis of donor (**388**) is shown in Scheme 4.84. Addition of 2-furyllithium to (**390**) in THF gave the adduct (**390**). Oxidation of the alcohol (**390**) with *m*-CPBA gave the hexenulose (**391**), which on methanolysis

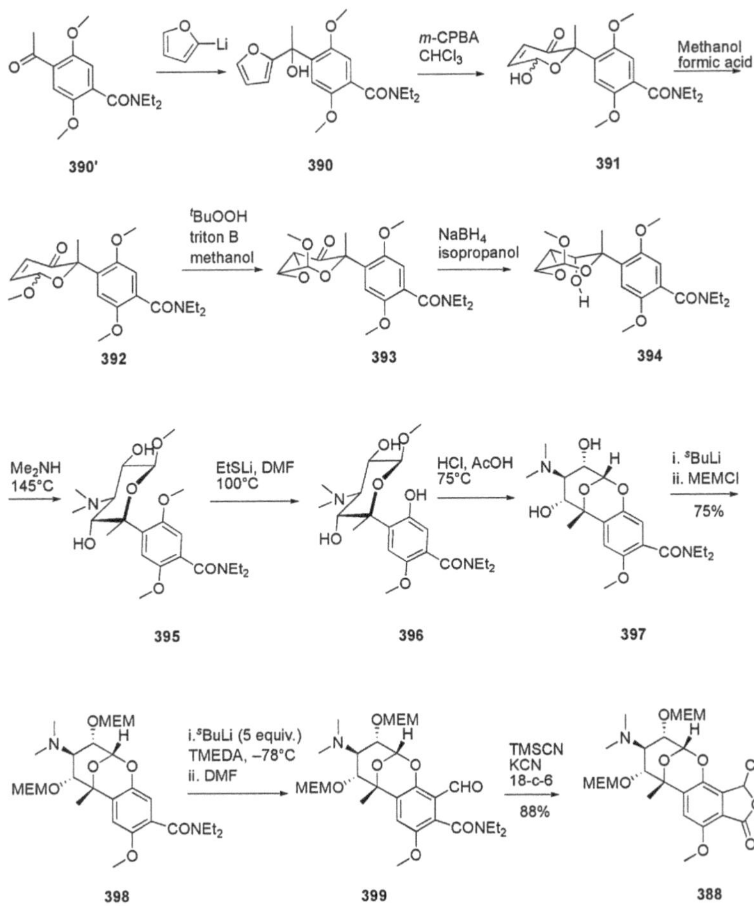

SCHEME 4.84 Synthesis of Hauser donor for 7-*con*-*O*-methylnogarol (**387**) [1991 JOC 5248 Hauser FM].

(HCO_2H, MeOH) afforded the methyl glycosides (**392**) in 90% combined yield. The mixture was submitted to epoxidation (^{t}BuOOH, CH_2Cl_2, triton B) to give epoxy ketone (**393**). $NaBH_4$ reduction of the ketone (**393**) stereospecifically furnished alcohol (**394**) in 95% yield. Opening of the oxirane in (**394**) with dimethylamine at 145°C yielded amino diol (**395**) in 95% yield according to the Fürst-Plattner rule. Regioselective demethylation (NaSEt, DMF, 100°C) of dimethyl ether (**395**) and cyclization (AcOH, 3 M HCl) of the resultant phenol (**396**) gave the epoxybenzoxocin (**397**). Free OH groups in (**397**) were protected as MEM groups. Lithiation of (**398**) with five equivalents of ^{s}BuLi-TMEDA in THF at −78°C, followed by quenching with DMF, gave the formyl derivative (**399**). Yoshii protocol (TMSCN, 18-c-6, KCN) was applied to the aldehyde (**399**) to furnish the donor (**388**).

Annulation of cyanophthalide donor (**388**) with the naphthalenone (**389**) in the presence of LDA −78°C followed oxidation (O_2, Co-salcomine, DMF, 50°C) of the resultant hydroquinone intermediate yielded hexacycle (**400**) in 73% yield (Scheme 4.85). Subsequent demethylations of (**400**) with LiI-pinacolone-$PhCO_2H$ and 1-chloroethylchloroformate followed by derivatization with TecCl afforded compound (**401**). For an effective epoxidation of the 9,10-olefinic moiety in (**401**), its nitrogen-protecting group was changed to the Tec group. Oxidation of (**401**) with m-CPBA gave epoxide (**402**). Reductive opening of the epoxide (**402**) with 10% Pd/C, H_2, EtOH-EtOAc, triethanolamine provided a separable mixture of alcohols (**403** and **404**). The major isomer (**403**) was carried forward to the final product (**387**) in eight steps.

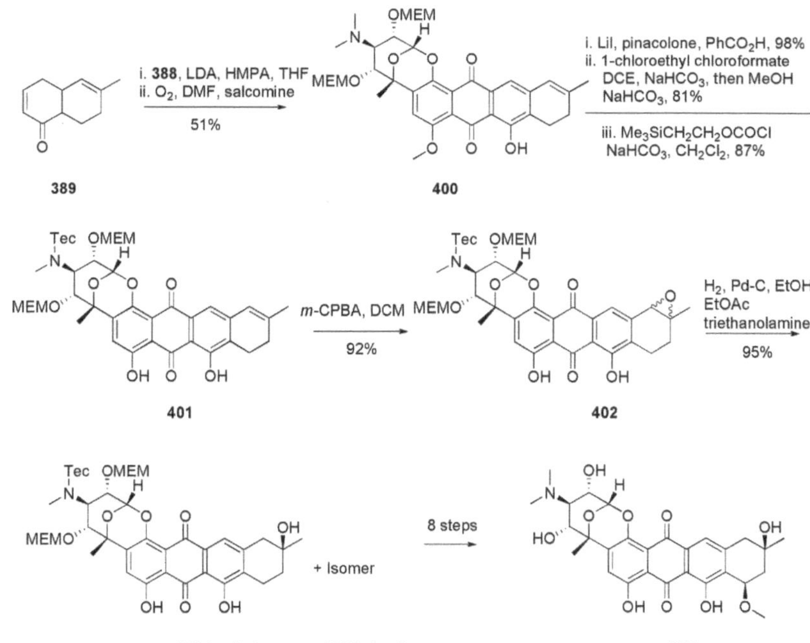

SCHEME 4.85 Final stages of the Hauser synthesis of 7-*con*-*O*-methylnogarol (**387**) [1991 JOC 5248 Hauser FM].

Hauser annulation has been successfully extended to an angular benzodecalenone en route to the total synthesis of trimethyl ether of G2N (**405**), a natural product isolated from an actinomycetes.[66] The decalenone (**406**) was prepared in a linear sequence from aldehyde (**407**). The aldehyde (**407**) was elaborated to precursor keto aldehyde (**408**). Intramolecular aldol condensation of the aldehyde (**408**) in the presence of $HClO_4$ afforded cyclohexanone (**409**). Titanium tetrachloride-mediated cyclization of the enone (**409**) furnished angularly fused decalenone (**410**). Treatment of a *cis*/*trans* mixture of decalenone (**410**) with

TBDMS-triflate in the presence of triethylamine afforded (**411**) as a mixture of *cis* and *trans* isomers in 95% yield triflate. The reaction of the mixture (**411**) with palladium acetate furnished the enone (**406**) in 53% yield. In the penultimate step, Hauser annulation between the sulfone (**164**) and the (**406**) was performed with lithium *tert*-butoxide in THF to assemble the G2N skeleton (**412**). Aerial oxidation of the quinol (**412**) allowed the synthesis of trimethyl ether (**405**) (Scheme 4.86).

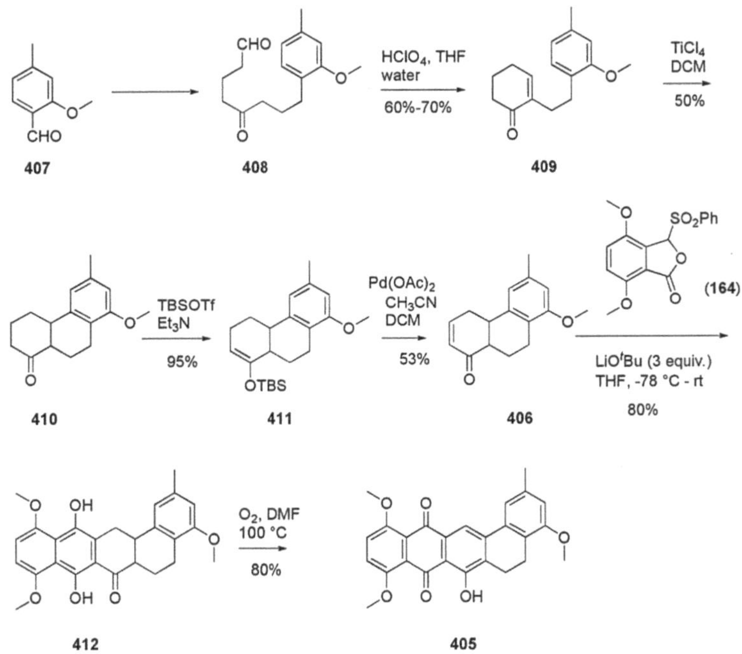

SCHEME 4.86 Synthesis of trimethyl ether of G2N, a benzo[*a*]naphthacene-8,13-dione natural product [1990 JOC 555 Hauser FM].

4.2.6 Tatsuta Total Synthesis of UCE6

The process of the Hauser annulation is utilized in the first total synthesis of UCE6 (**413**), an antitumor antibiotic that contains an aromatic A-ring. Tatsuta et al. accomplished the total synthesis in a regiospecific manner in 24 linear steps, as outlined in Scheme 4.87.[67] The decalenone acceptor (**414**) was assembled by the Diels-Alder reaction between cyclohexenone (**264**) and diene (**415**) as the key step.

The [4+2] annulation between (**414**) and (**416**) was carried out by treatment with LHMDS to yield the tetracyclic quinol (**417**), which was aromatized to (**418**) as a single product by reduction with benzyl methyl ether in the presence of Pd-C. It should also be noted that the *cis*-stereochemistry of the ring

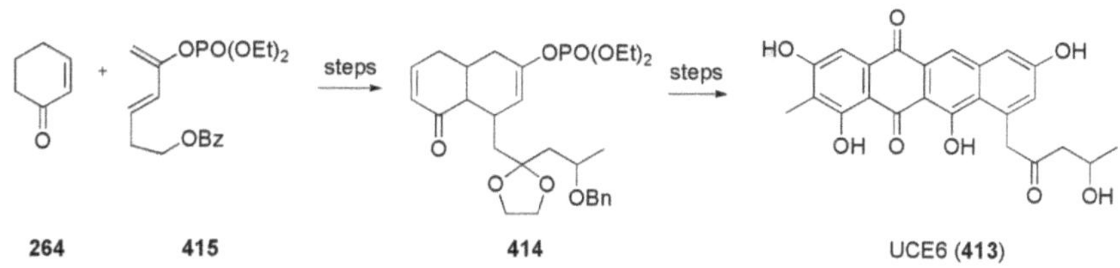

SCHEME 4.87 Outline of Tatsuta synthesis of UCE6 (413).

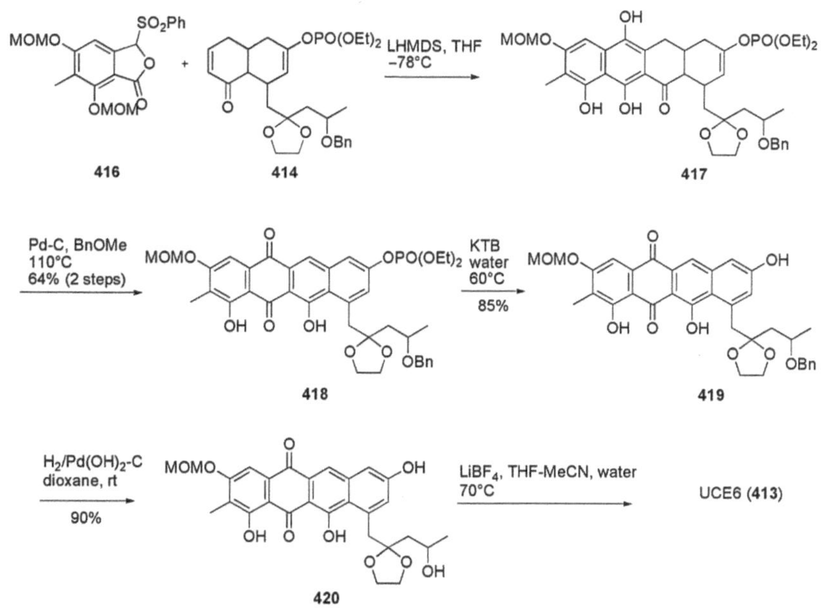

SCHEME 4.88 Tatsuta synthesis of UCE6 [2002 JABS 1076 Tatsuta K].

juncture and that of C-8 of the decalenone (**414**) did not affect the yield of the product (**417**) (Scheme 4.88). The phosphate group in (**418**) was hydrolyzed by Gassman protocol to give tetracenone (**419**) and the benzyl group in (**419**) was removed by hydrogenolysis to give compound (**420**). Lastly, the remaining protective groups of the compound (**420**) were removed by $LiBF_4$ to give the desired product, UCE6 (**413**).

The synthesis of the decalenone (**414**) was started from the diene (**415**), which, in turn, was prepared from 1,2,4-butanetriol (**421**) via ketone (**422**) via (**423**) in four steps, which were: (i) $NaIO_4$ oxidation, (ii) Wittig reaction, (iii) *O*-benzoylation, and (iv) LHMDS-mediated formation of enol phosphate (Scheme 4.89).

Boron trifluoride aided Diels-Alder reaction between cyclohexenone (**264**), and the diene (**415**) gave a mixture of diastereomers of (**422**). The mixture was converted to enone (**424**) by phenylselenylation and oxidative elimination of

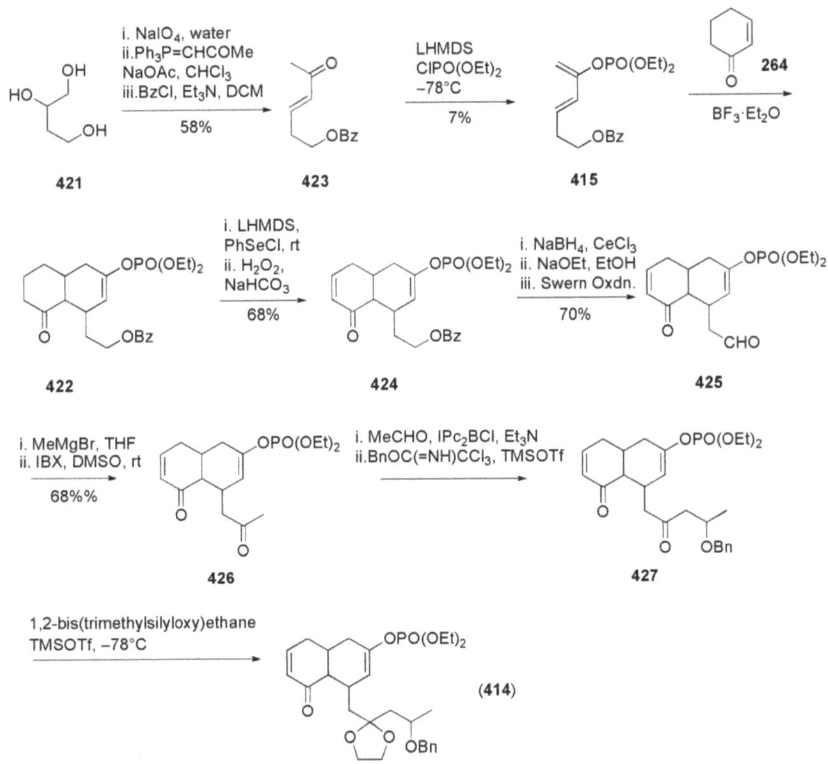

SCHEME 4.89 Synthesis of UCE6 decalenone synthon [2002 JABS 1076 Tatsuta K].

phenylselenic acid. $NaBH_4$ reduction of (**424**), followed by removal of the benzoyl group and Swern oxidation, afforded a 3:1 mixture of keto aldehyde (**425**). The major isomer was homologated to (**426**) by sequential methyl Grignard addition and IBX oxidation. The compound (**426**) was treated with acetaldehyde in the presence of optically active diisopinocamphenylchloroborane, followed by *O*-benzylation to give (**427**), which was converted into (**414**), by ethylene acetal formation.

4.2.7 Unsuccessful Annulation With a Decalenone

In connection with their studies on halenaquinone (**428**) and xestoquinone (**429**) (Fig. 4.12), the representative members of the xestoquinone family which are well known for their wide spectrum of biological activities, Hauser et al. attempted to condense the phthalide (**164**) or the cyanophthalide (**188**) with decalenone (**430**) to assemble the structure (**431**). Strikingly, all the attempts failed, in contrast to the successes described in the previous section (vide Scheme 4.76), to provide tetracycle (**431**) (Fig. 4.14 and Scheme 4.90).[68]

Halenaquinone (**428**) Xestoquinone (**429**)

FIG. 4.14 Structures of halenaquinone (**428**) and xestoquinone (**429**).

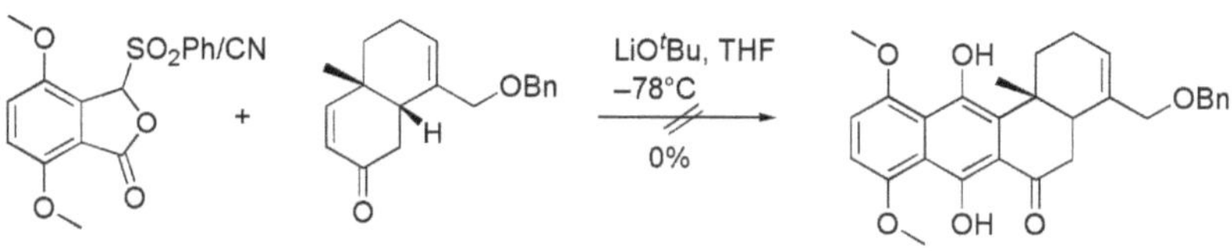

164 or **188** **430** **431**

SCHEME 4.90 Attempted Hauser annulation toward xestoquinone [Unpublished results Hauser FM].

4.2.8 Cycloheptenones

Although the simplest 2-cycloheptenone was never investigated, benzocycloheptenones (**432–434**) were shown to undergo Hauser annulation with phthalide the sulfone (**11**) to give respective benzo[4,5]cyclohepta[1,2-*b*]naphthalenes (Fig. 4.15).

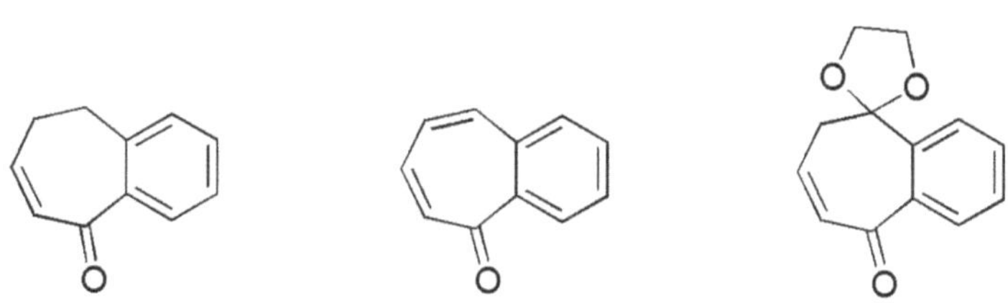

432 **433** **434**

FIG. 4.15 Structures of benzocyclohept-6-en-5-one acceptors.

For the total synthesis of radermachol (**435**), Hauser and Yin reported the annulation of the phthalide sulfone (**11**) with (**434**) to give the corresponding annulation product, which was methylated by an atypical method, i.e., TBAF-MeI-DMF, to give annulation product (**436**). PPTS-catalyzed hydrolysis of (**436**) gave the deketalized compound (**437**) in 97% yield. The acyclic version of Friedel-Crafts reaction of (**437**) with dimethylacrolyl chloride provided (**438**). TMSI-assisted demethylation of (**438**) followed by PTSA promoted cyclization gave radermachol (**435**) (Scheme 4.91).[69]

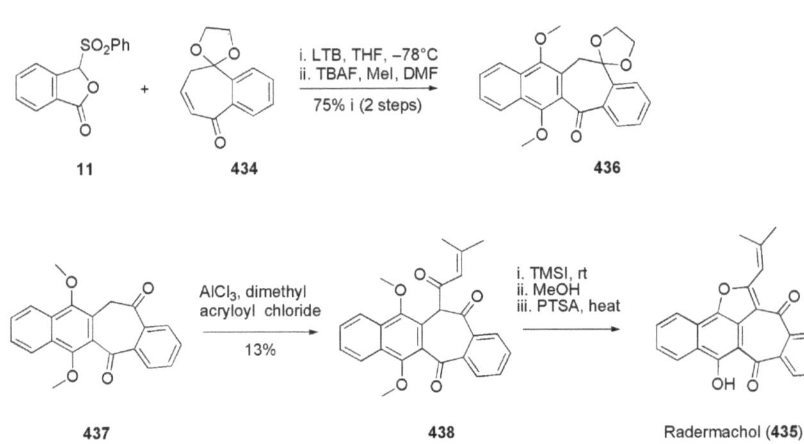

SCHEME 4.91 Total synthesis of radermachol [2000 ORS 1045 Hauser FM].

4.3 FURANONES, PYRANONES, AND BENZOPYRANONES

4.3.1 Unsaturated Lactones

The Hauser annulation of butenolides or furan-(5*H*)-2-ones has randomly been examined at various stages of the development of the annulation, although butenolides, benzobutenolides (phthalides), and naphthobutenolides (naphthalides) are prevalent in nature. The simplest butenolide (**439**) was first examined by the Kraus group in annulation with 3-cyanophthalide (**14**) in the presence of LDA giving dihydroxy naphthofuranone (**440**) in 50% yield. When the annulation was carried out with phthalide-3-phosphonate (**35**) using LTB as the base, the product (**440**) was obtained in 95% yield as reported by Watanabe group. The Danishefsky group reported 80% yield for the same reaction en route to the synthesis of lactonamycin scaffold (**441**) (Scheme 4.92).[70]

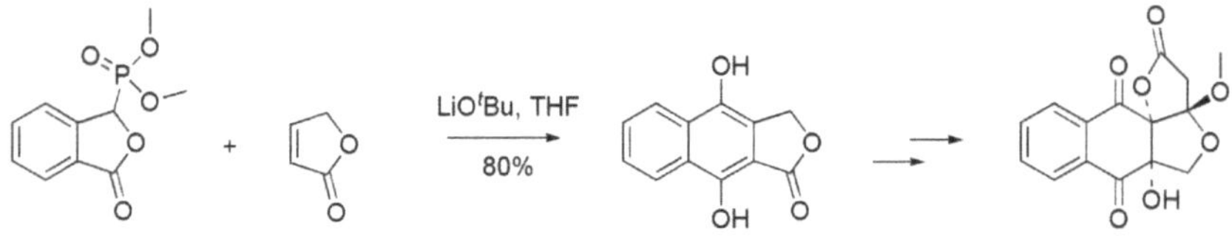

35 **439** **440** **441**

SCHEME 4.92 Annulation of phthalide-3-phosphonate with butenolide [2000 ORS 3493 Danishefsky SJ].

Very recently, the Ham group extended the annulation chemistry of the butenolide (**439**) to more substituted 3-cyanophthalides for the practical synthesis of various arylnaphthalene lactones featuring natural lignans. The primary reaction was the Hauser annulation of cyanophthalides (**442a** and **442b**) with

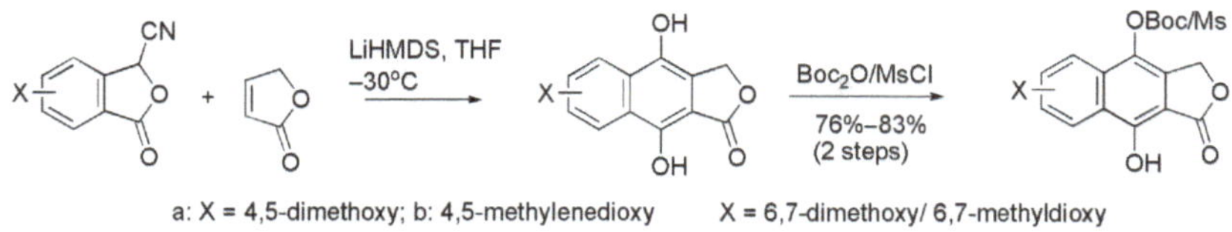

442 **439** **443** **444**

SCHEME 4.93 Synthesis of naphtho[*c*]furanones (**443**) [2017 EJOC 1704 Ham J].

γ-crotonolactone (**439**) in the presence of LHMDS (3 equiv.). The corresponding naphtho[*c*]furanones (**443**) were isolated as monoprotected derivatives (**444**) (Scheme 4.93).[71] The yields were sensitive to the reaction temperature, and the optimum temperature was −30°C.

The annulated products (**444**) were further utilized for the synthesis of a series of arylnaphthalene lactones such as justicidin A (**445**), justicidin B (**446**), justicidin C (**447**), cilinapthalide (**448**), taiwanin E (**449**), chinensinaphthol (**450**), etc. (Fig. 4.16). Six biologically important lignans synthesized by Ham et al. in this work are given below.

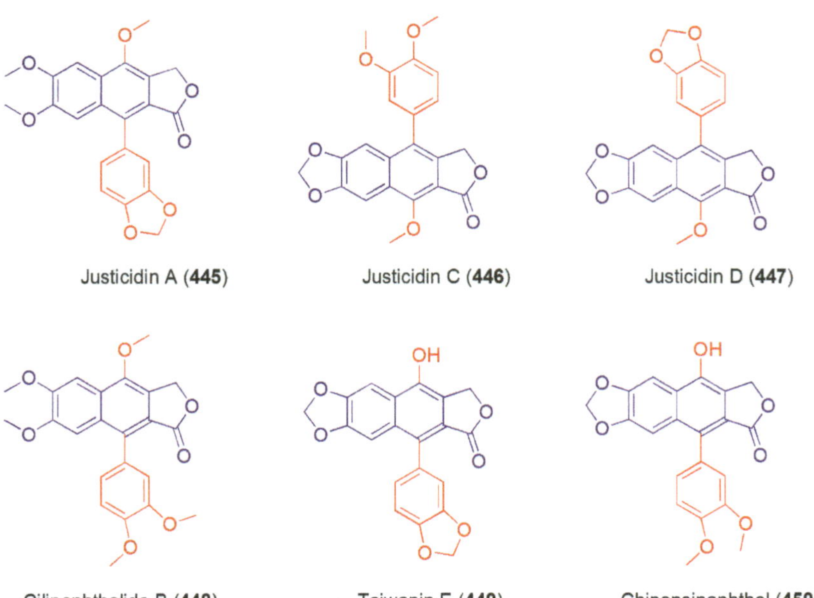

FIG. 4.16 Structures of lignans synthesized by Ham group.

Suzuki cross-coupling of protected naphtho[*c*]furanone (**451** and **452**) with potassium aryltrifluoroborates (**453**) produced isomeric arylnaphthalene lignans (**454** and **455**). This tactic allowed synthesis of nine natural lignans in excellent overall yields. As shown in Scheme 4.94, the free OH in (**444a**) was converted into triflate resulting in (**451**), which was coupled with aryltrifluoroborates

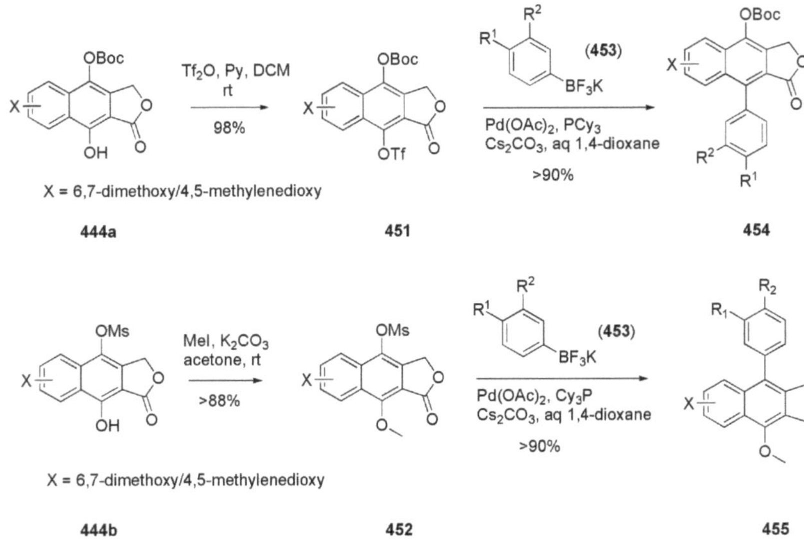

SCHEME 4.94 Synthetic protocols for arylnaphthalene lignans [2017 EJOC 1704 Ham J].

(**453**) in the presence of $Pd(OAc)_2$ and tricyclohexylphosphine to give lignans (**454**). In the second protocol, monoprotected naphtho[*c*]furanones (**444b**) were converted to their methyl ethers (**452**), which, in turn, were submitted to coupling with aryltrifluoroborates (**453**) under similar conditions to give (**455**) through replacement of OMs groups. A total of nine aryl naphthalene lactones have been synthesized starting from two cyano phthalides (**444a** and **444b**), using the above strategy.

Butenolides are compatible with annulation with phenylthiophthalides. The preformed anion of phenylthiophthalide (**456**), generated by the action by MeLi, on reaction with butyrolactone (**457**), afforded (+)-9-hydroxyeleutherol (**458**). The lactone was prepared in two steps from methyl *trans*-pent-3-enoate. Notably, the stereochemical integrity of the only chiral center remained unaltered during annulation (Scheme 4.95).[72]

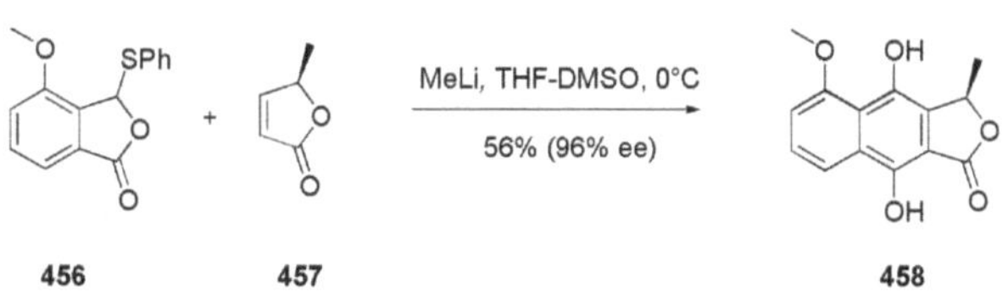

SCHEME 4.95 Synthesis of (+)-9-hydroxyeleutherol (**457**) [2006 EJOC 2110 Brueckner R].

Similarly, 5-ethoxy-2-furanone (**459**) is well-suited for annulation with the phthalide sulfones (**11**, **24**, and **251**). The anions of the sulfones, generated with LDA, underwent condensation with 5-ethoxy-2-furanone (**459**) and furnished, after *O*-methylation (Me_2SO_4, K_2CO_3, acetone), naphthalides (**460**) in

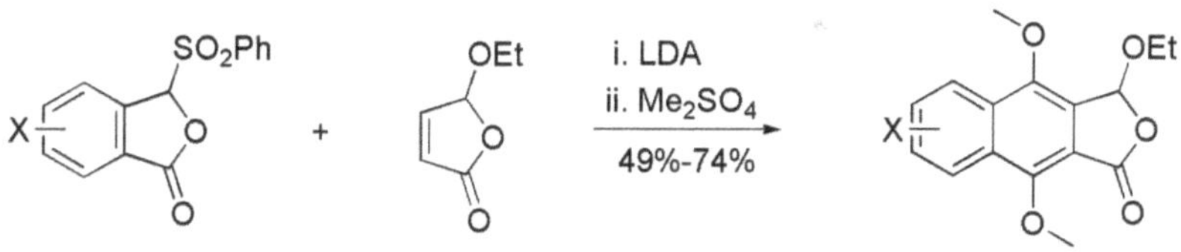

X = H, 4-OMe or 7-OMe

X = H, 5-OMe or 8-OMe

11, **24** and **251** **459** **460**

SCHEME 4.96 Annulation of phthalide sulfones with 5-ethoxy-2-furanone [1981 JACS 6378 Hauser FM].

49%–74%, the lowest yield being for 4-methoxy-3-phenylsulfonylphthalide.[10] It is to be noted that butenolide (**459**) has an acidic hydrogen alpha to the ethoxy group. Yet no complication was observed (Scheme 4.96).

4.3.2 Annulation With Pyranones

The unsaturated monocyclic pyranones often occur in four forms (**461–464**) (Fig. 4.17). Barring (**461**) and (**462**), pyrone-3-one (**463**), and pyrone-4-one (**464**) have found wide uses in the [4+2] anionic annulations.

461 **462** **463** **464**

FIG. 4.17 Structures of monocyclic pyranones.

The sole example of the use of an unsaturated δ-lactone in the Hauser annulation is due to Kimpe et al., who assemblef 3,4-dihydro-1*H*-benz[*g*]isochromene-5,10-diones. The yield of the reaction of the cyanophthalide (**14**) with the unsaturated lactone (**461**) afforded the corresponding 1,4-quinol in an exceedingly high yield of 97%. *O*-Methylation of the annulation product gave dimethyl ether (**465**). DIBALH reduction of the lactone (**465**) led to lactol (**466**). CAN oxidation of the lactol (**466**) delivered (**467**), the spectral data of which were different from those for the isomer of psychorubrin. Thus, the isolated compound was suggested to be psychorubrin (**468**).[73] This structure revision was possible due to robust regiochemical integrity of the Hauser annulation (Scheme 4.97).

The suitability of 3-pyrones (e.g., **463**) as the Hauser acceptors for a rapid entry to pyranonaphthoquinone natural product was disclosed by the Tatsuta group and Swenton groups simultaneously in 1985.[74] Pyranones featuring the structural motifs (**469–471**), displayed in Fig. 4.18, have thoroughly been studied in connection with the synthesis of naturally occurring pyranonaphthoquinones.

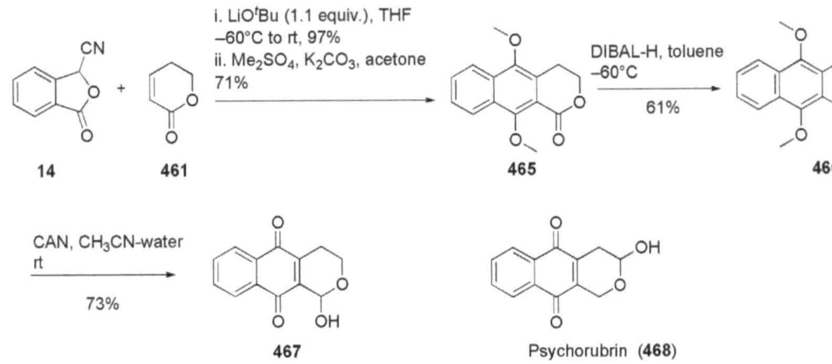

SCHEME 4.97 2-Pyrone studied for Hauser annulation [2008 THN 412 Kimpe ND].

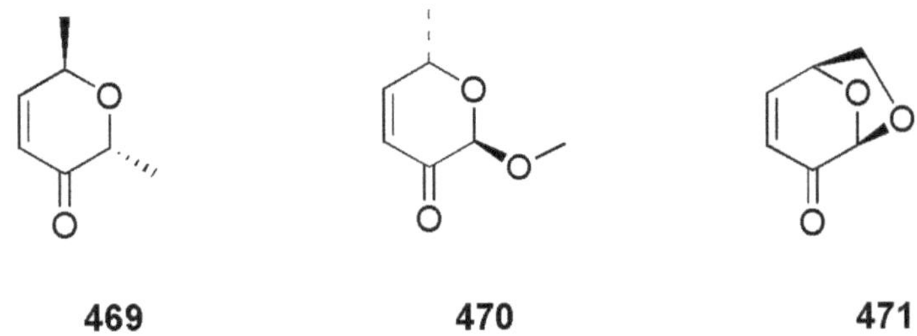

FIG. 4.18 3-Pyrones studied for the Hauser annulation.

Subsequently, many other groups including the Brimble group adopted the strategy for the convenient total syntheses of naphthopyranone natural products. The Tatsuta group and Lackner group completed the chiral synthesis of nanaomycin, kalafungin, and medermycin, Hauser annulation being the key common element of the strategies.[75] Also common to all the syntheses were the carbohydrate-based Hauser pyrone acceptors. Laevoglucenone (**471**) is readily available from pyrolysis of starch or waste paper.

The De Kimpe research group, a pioneer in the field of the synthesis of naphthopyranone antibiotics, found Hauser annulation to be superior to the existing methodologies and accomplished a short and efficient synthesis (Scheme 4.98) of the simplest pyranonaphthoquinone antibiotic pentalongin (**472**), from the pyrone (**463**) via naphthoquinol (**473**). The quinol (**473**) was readily assembled by the condensation of the pyrone (**463**) with the cyanophthalide (**14**). Following the protection of the quinol (**473**) as dimethyl ether (**474**), it was reduced to alcohol (**475**) with $NaBH_4$. CAN oxidation of the ether (**475**) gave quinone (**476**), which upon PTSA catalyzed dehydration furnished pentalongin (**472**).[76]

The application of the same approach to the synthesis of azapentalongin was futile due to the lack of a suitable transformation of the corresponding azaanthracene derivative (**477**). The annulation between (**14**) and (**478**) followed by *O*-methylation of the initial product with methyl sulfate and potassium carbonate gave dimethylether (**479**). Treatment of (**479**) with TFA gave azathracene (**480**) (Scheme 4.99).[77]

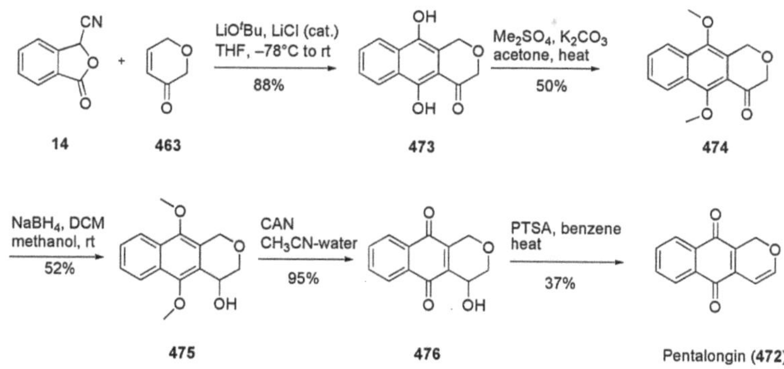

SCHEME 4.98 Synthesis of pentalongin (**472**) [2006 SYS 621 Kimpe ND].

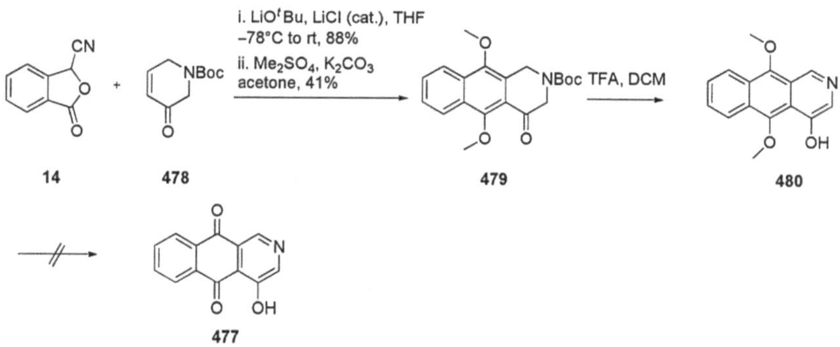

SCHEME 4.99 Hauser annulation with a 3-pyridone [2007 SYT 741 Kimpe ND].

When the annulation was switched to pyridones (**481**), their annulation with the cyanophthalide (**14**) in the presence of LTB provided a new synthesis of azaanthraquinones (**482**). But, a cyano group was incorporated in the anthraquinone products (**482**) through intermediates (**483–485**) via a conjugate addition. Interestingly, the annulation failed for the pyridone (**481**) when R=H (Scheme 4.100).

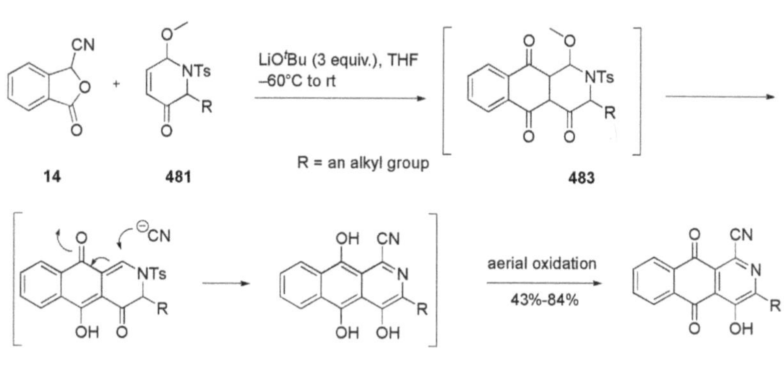

SCHEME 4.100 Hauser annulation with 6-methoxy-3-pyridones [2007 SYT 741 Kimpe ND].

Both Swenton group and Baker group applied the annulation strategy in achieving enantioselective total synthesis of hongconin (**486**), a pyranonaphthoquinone with antianginal activity. The Baker's synthesis is concise and more convergent than that of Swenton group.[78] It involves the Hauser reaction as the last step, and thus avoids the peripheral synthetic operations. No epimerization of the chiral centers were observed during the annulation. The required 3-pyranone (**487**) was prepared from 6-deoxy D-galactal (**488**) in three steps. The reaction of D-galactal (**488**) with trimethylaluminum in the presence of $TiCl_4$ resulted in C-1 alkylation with the desired stereochemistry to give monoacetate (**489**). Aminolysis of the acetate (**489**) followed by PDC oxidation afforded (**487**). Annulation of the pyrone (**487**) with (**490**) in the presence of LTB, produced quinol (**491**) in 11% yield along with 70% of the Michael adduct (**492**), which was cyclized with potassium *tert*-butoxide to give hongconin (**486**) in 96% yield (Scheme 4.101).

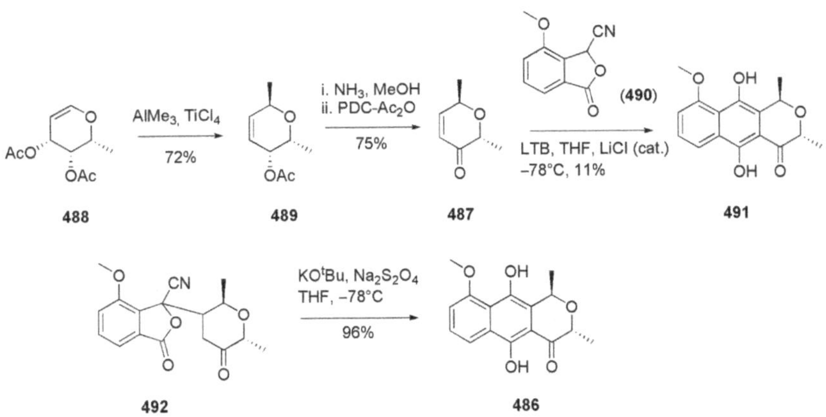

SCHEME 4.101 Baker Synthesis of (−)-hongconin [1996 JOC 455 Baker DC].

A similar approach was pursued by Reutrakul group to synthesize (**493**), a naturally occurring pyranonaphthoquinone (Scheme 4.102). The phthalide donor (**494**) was obtained through a new formylation and then condensed with pyranone (**495**), derived from L-rhamnose, and the resultant quinol was dimethylated to give (**496**). Sequential application of $Na(CN)BH_3$ reduction and CAN oxidation to (**496**) furnished (**493**).

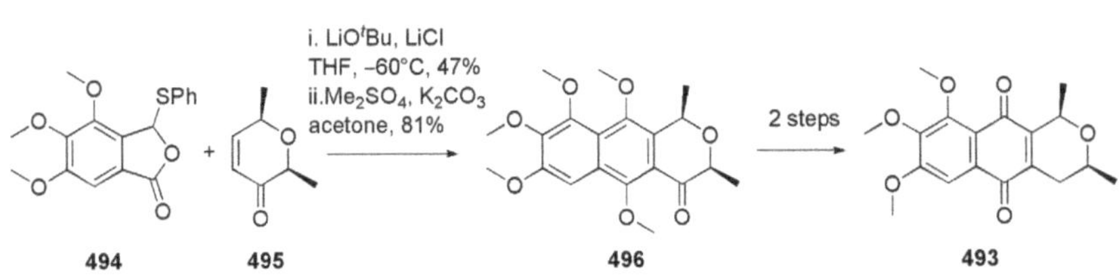

SCHEME 4.102 Synthesis of a natural pyranonaphthoquinone [2012 PAC 1435 Reutrakul V].

Perhaps, the Tatsuta group is the first to use a carbohydrate derived 3-pyrone in the Hauser annulation. In the enantiospecific total synthesis of nanaomycin D (**497**) and its enantiomer, pyrone (**470**) was prepared in four steps from methyl α-L-rhamnoside.[79] Lithium *tert*-butoxide mediated condensation of the phthalide sulfone (**24**) with pyranone (**470**) gave quinol (**498**) in 80% yield. *O*-Methylation of (**498**) with methyl sulfate followed by reduction of the resultant product with $NaBH_4$ afforded (**499**). Hydrolysis of (**499**) with HCl-AcOH gave pyran diol (**500**) quantitatively. The reaction of (**500**) with ethoxycarbonylmethylenetriphenylphosphorane in refluxing toluene gave products (**501**) and (**502**). The annulated product (**501**), formed through Wittig olefination and subsequent intramolecular Michael addition, was submitted to CAN oxidation to produce (**503**). De-*O*-methylation of (**503**) with aluminum chloride furnished (−)-nanaomycin D (**497**). Similarly, the intermediate (**502**) was elaborated to kalafungin (Scheme 4.103).

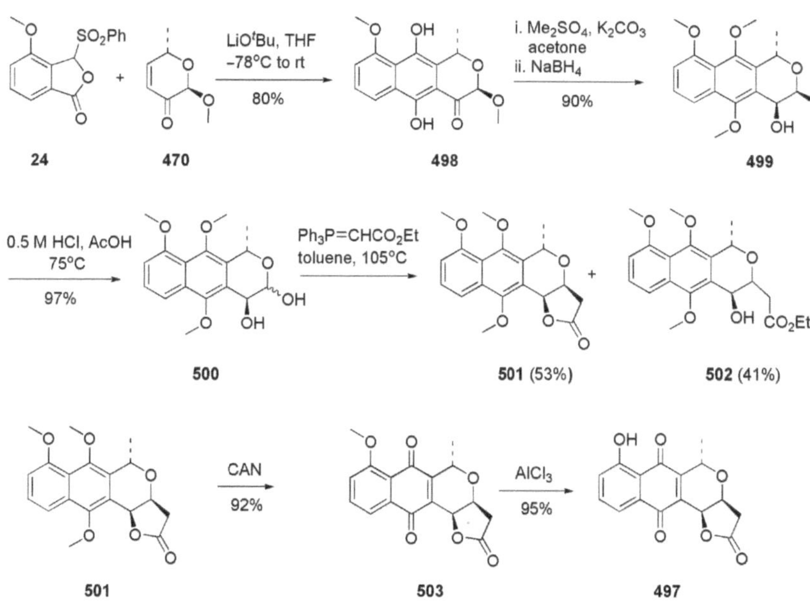

SCHEME 4.103 Tatsuta total synthesis of nanaomycin D [1985 BCSJ 1699 Tatsuta K].

Lackner group quickly followed the strategy to generate new analogs of nanaomycin D (**497**). For example, the annulation of phthalide (**504**) with pyranone (**470**) followed by *O*-methylation afforded (**505**) in 52% overall yield. This was then elaborated to 7-methylisonanaomycin D (**506**), which was found to be more potent than the parent nanaomycin, supporting the formation of a quinone methide intermediate as a putative intermediate (Scheme 4.104).[80]

Naquihexcin A (**508**) and (+)-BE-52440A (**507**) (Fig. 4.19) are two unusual naphthopyranones with S-bridges. Naquihexcin A (**508**), recently isolated from

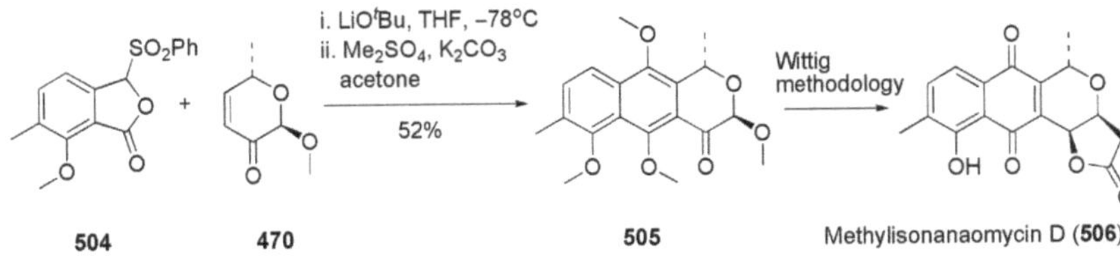

SCHEME 4.104 Lackner synthesis of nanaomycin analogs [1995 LAN 87 Lackner H].

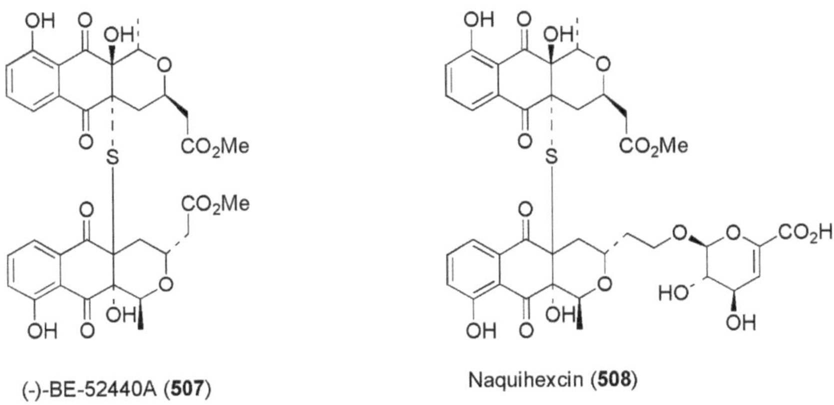

FIG. 4.19 Select structures of S-bridges dimeric pyranonaphthoquinones.

a *Streptomyces sp.*, showed inhibitory activity against the proliferation of the adriamycin resistant human breast cancer cell line MCF-7 ADM. Consequently, it arose interest among the synthetic chemists. Tatsuta accomplished a total synthesis of (+)-BE-52440A (**507**).[81]

The synthesis was started from pyranonaphthalene (**509**) (Scheme 4.105). It was oxidized to naphthoquinone (**510**) by treatment with AgO. Treatment of (**510**) with sulfuric acid promoted lactonization proceeded with epimerization at C1 and C4 positions to give lactone (**511**). Hydrogenolysis of (**511**) followed aerial oxidation and *O*-methylation with $TMSCHN_2$ afforded quinone (**512**). Quinone (**512**) was immediately treated first with TBHP-KO^{t}Bu and then BCl_3 to produce epoxide (**513**). Treatment of epoxide (**513**) with Na_2S gave (+)-BE-52440A (**507**) in 98% yield.

Tatsuta total synthesis of medermycin (514): The Hauser annulation of *C*-glycosyl phthalide sulfone (**515**) with enone (**516**) in the presence of LTB gave respective hydroquinone. This was *O*-methylated (Me_2SO_4/K_2CO_3) and then reduced with $NaBH_4$ to give alcohol (**517**). Acid hydrolysis (aq. HCl-AcOH) of (**517**) afforded the corresponding hemiacetal, which was subjected to the Wittig reaction with $Ph_3P{=}CHCO_2Et$ to give (**518**). Subsequently, functional group manipulation of the sugar part in (**518**) for installation of the dimethylamino group led to the total synthesis of medermycin (**514**) (Scheme 4.106).[82]

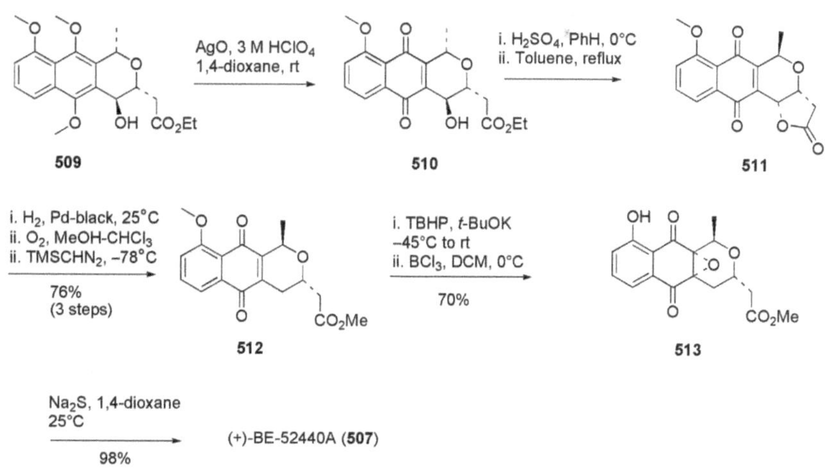

SCHEME 4.105 Tatsuta total synthesis of (+)-BE-52440A (**507**) [2007 THS 8018 Tatsuta K].

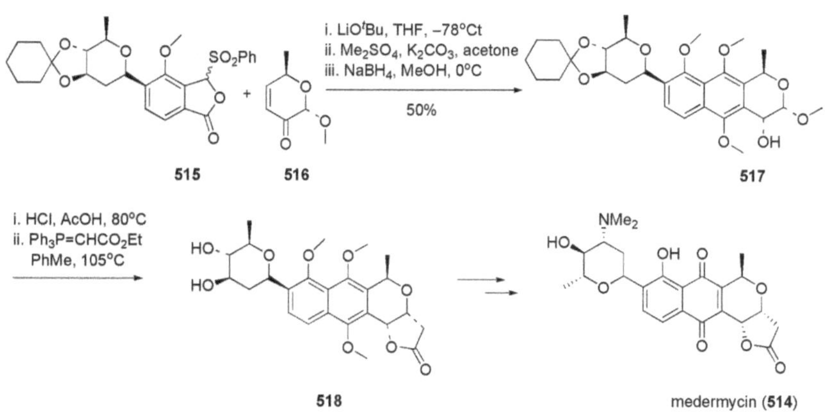

SCHEME 4.106 Tatsuta total synthesis of medermycin (**470**) [1990 THS 5495 Tatsuta K].

The total synthesis of more complex naphthopyranone antitumor agent, i.e., BE-54238B (**519**) was also achieved by the Tatsuta groups using the Hauser annulation of 3-pyranones of carbohydrate origin. As shown in Scheme 4.107, the reaction between 4-methoxy-3-phenylsulfonylphthalide (**24**) and pyranone (**520**) derived from carbohydrate (**521**), followed by *O*-methylation of the resultant quinol with dimethyl sulfate gave naphthopyranone (**522**) in 83% yield in an early stage of a 15-step sequence.[83] The sequence of bromination (NBS), $NaBH_4$ reduction and *O*-benzylation (NhBr, NaH, DMF) was applied to (**522**) to arrive at the intermediate (**523**). The synthon (**524**) was synthesized from L-pyroglutamic acid in four steps. *Tert*-butyllithium mediated coupling of (**523**) with (**524**) smoothly gave ketone (**525**). This was reduced to the corresponding alcohol, which was cyclized through *O*-mesylation to give a 3:1 diastereomeric

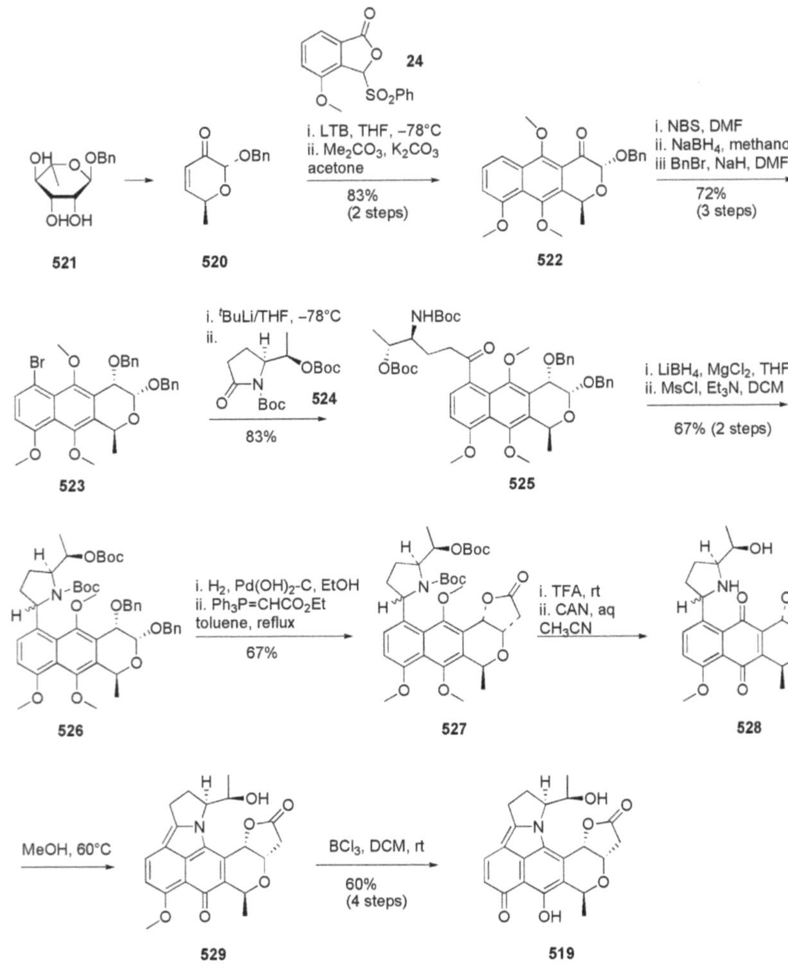

SCHEME 4.107 Tatsuta synthesis of pyranonaphthoquinone BE-54238B (**519**) [2004 *JABS* 291 Tatsuta K].

mixture of the pyrrolidine (**526**). Since the chiral at C8b is no of consequence in the final stages, hydrogenolysis of (**526**) was performed on the mixture to give corresponding hemiacetal, which was submitted to the Wittig reaction with ethoxycarbonylmethylene phosphorane in refluxing toluene to give product (**527**) as the major one. The lactone (**527**) results from a two-step sequence which includes Wittig reaction and intramolecular oxa-Michael addition.

Removal of the *O*-Boc group of (**527**) was followed by CAN oxidation to give the quinone (**528**). This was expectedly cyclized to (**529**) on heating. Without purification, (**529**) was de-*O*-methylated by BCl_3 to give the tautomerized compound (**519**), hydrochloride salt of which was identical in all respects with the hydrochloride salt of the natural BE-54238B (**519**). A similar approach

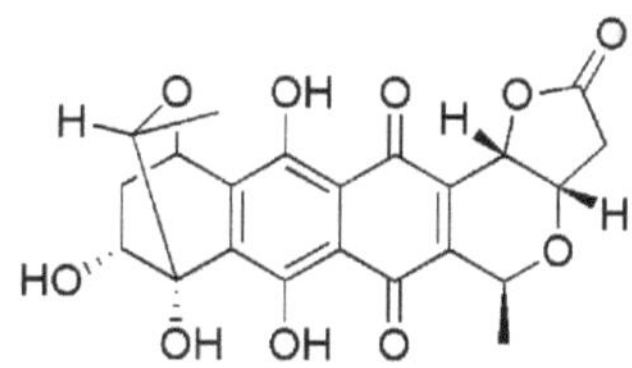

Granaticin (**530**)

FIG. 4.20 Structure of granaticin.

involving a C-glycosidic 3-pyranone and a cyanophthalide allowed a total synthesis of granaticin (**530**) (Fig. 4.19). The details of the synthesis are presented in Chapter 10 (Fig. 4.20).

Swenton synthesis of chiral hongconin parallels the naphthopyranone syntheses of Baker group and Tatsuta group in that all the syntheses used a Hauser annulation as the key step. However, Swenton synthesis differs from the other two in using levoglucosenone (**471**), a bicyclic 3-pyranone, readily available by pyrolysis of waste paper. Dimsyllithium promoted annulation between cyanophthalide (**490**) with levoglucosenone (**471**) at 0°C furnished quinol (**531**) in 74%. Reductive cleavage of (**531**) with the Zn-Cu couple, followed by conversion of the resulting alcohol to its iodide derivative, and reductive deiodination with Zn-Cu, furnished naphthopyranone (**532**). The pyranone (**532**) was transformed to (−)-hongconin (**486**) in three more steps (Scheme 4.108).[84]

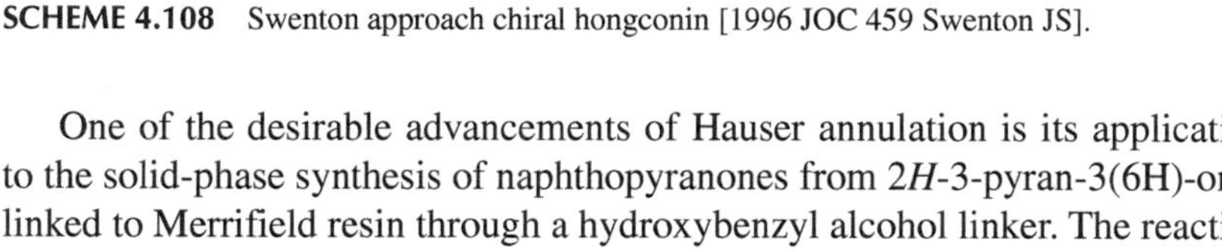

490 **471** **431** **532**

SCHEME 4.108 Swenton approach chiral hongconin [1996 JOC 459 Swenton JS].

One of the desirable advancements of Hauser annulation is its application to the solid-phase synthesis of naphthopyranones from 2*H*-3-pyran-3(6H)-ones linked to Merrifield resin through a hydroxybenzyl alcohol linker. The reaction of cyanophthalide (**188**) with polymer-supported pyranones (**533**) followed by TFA treatment resulted in naphthopyranone (**534**) in 80% yield, which is comparable to that of a solution phase synthesis (Scheme 4.109).[85]

Like 3-pyrones, unsaturated 4-pyrones are also well suited for Hauser annulations. As demonstrated by the Tatsuta group in their total synthesis of (±)-napyradiomycin A1 (**535**), the reaction between phthalide sulfone (**536**) and pyranone (**537**) followed by oxidation with MnO_2 afforded (**538**) in 64% overall yield (Scheme 4.110).[86] Chlorination of (**538**) with SO_2Cl_2 afforded chloro product (**539**). The geranyl side chain was incorporated into (**539**) using

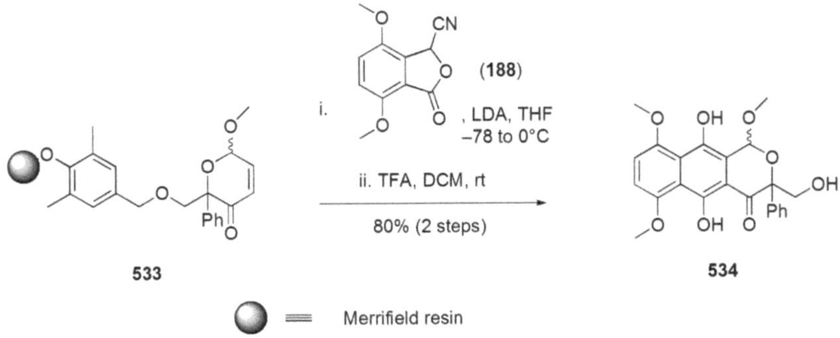

SCHEME 4.109 Annulation of a resin-bound 2*H*-3-pyran-3(6H)-one [2002 ACIE 3677 Couladouros EA].

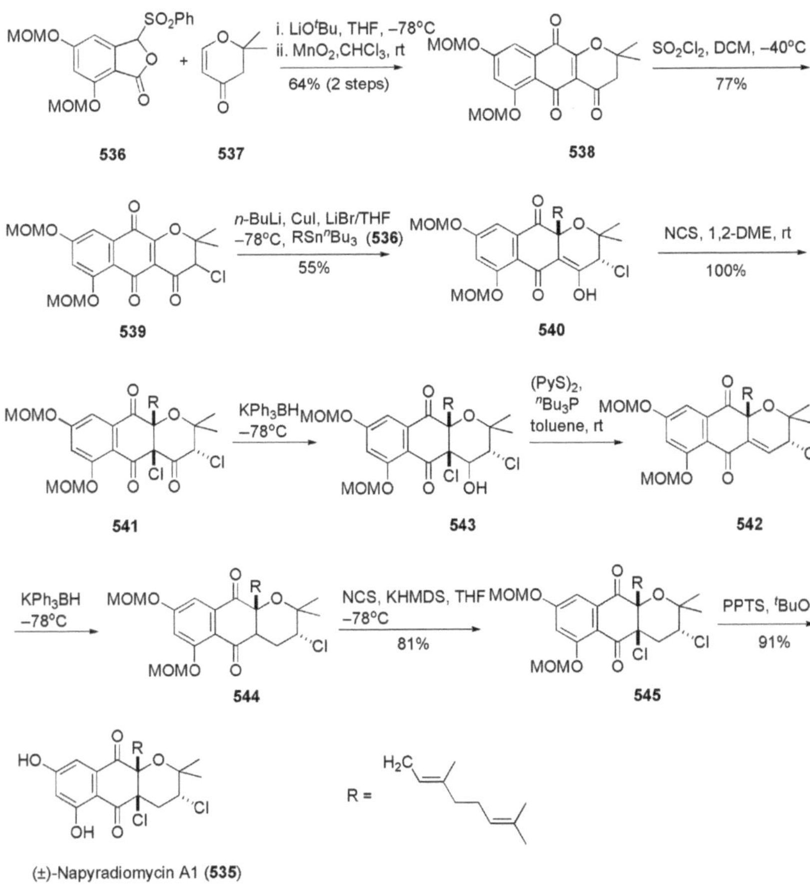

SCHEME 4.110 Tatsuta synthesis of napyradiomycin A1 (**535**) [2002 CLS 14 Tatsuta K].

stannyl derivative (**540**) through its cuprate derivative. Further chlorination of (**540**) proceeded smoothly to give dichloro derivative (**541**) in quantitative yield. Selective deoxygenation at C4 in (**541**) was achieved in four steps through the unsaturated ketone (**542**). The ketone (**541**) was reduced with KPh_3BH to give a diastereomeric mixture of the alcohol (**543**). This was treated with 2,2′-dipyridyl disulfide and tributylphosphine to afford the ketone (**542**). The selective 1,4-hydride reduction of (**542**) was carried out with KPh_3BH again to give exclusively the tetrahydropyran (**544**). The next crucial chlorination was accomplished by exposure of (**544**) to KHMDS followed by treatment with NCS to afford compound (**545**). Removal of MOM groups with PPTS-butanol furnished (±)-napyradiomycin A1 (**535**).

The possibility of ring-opening of the 4-pyranone ring in (**546**) during the annulation can be ruled out on the basis of an unfavorable stereoelectronic effect in the initial Michael adduct (Scheme 4.111). For the annulation of the simple phthalide sulfone (**11**) and 4-pyrone (**542**), it is expected that the anion (**11′**) of the sulfide (**11**) would add to (**546**) in Michael fashion to give the new carbanion (**547**). Since the oxygen atom of the pyran ring is β to the carbanion, its elimination to form ring-opened adduct (**554**) is justified at first glance. In a stereoelectronic analysis, the enolate (**549**) is likely to assume the conformation (**550**), which is conceivably the most stable of all possible isomers. In this conformation, the CO bond is not placed anti to the enolate orbital for possible fragmentation to (**548**). Such an analysis proved useful for the extension of the Hauser annulation to chromenones.

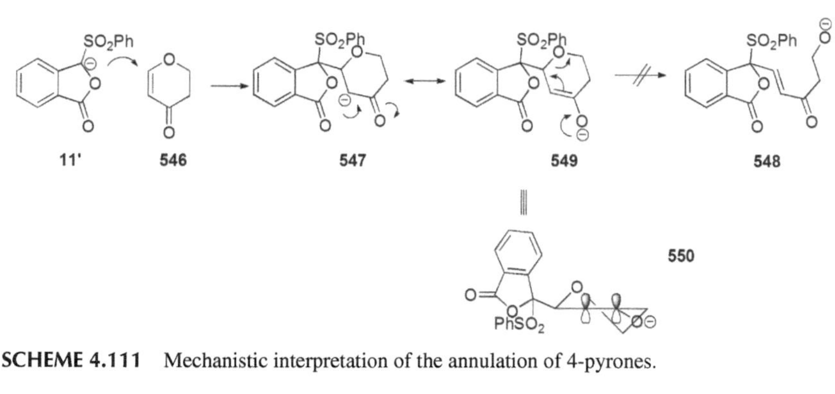

SCHEME 4.111 Mechanistic interpretation of the annulation of 4-pyrones.

Among benzopyranones, 2*H*-chromen-2-ones (e.g., **551**) and 4H-chromenones (e.g., **552**) have proved to be useful acceptors for Hauser annulations (Fig. 4.21).

The suitability of a 4*H*-chromen-4-one to Hauser annulation has been utilized in the total synthesis of bikaverin (**553**), a xanthone natural product with antiprotozoal activity (Scheme 4.112). Condensation of phthalide (**554**) with chromone (**555**) in the presence LiO^{t}Bu in THF at −78°C to room temperature followed by brief reflux regiospecifically furnished (**556**) in 27% yield. Oxidation of

2*H*-chromen-2-one 4*H*-chromen-4-one

551 552

FIG. 4.21 Structures of chromenones.

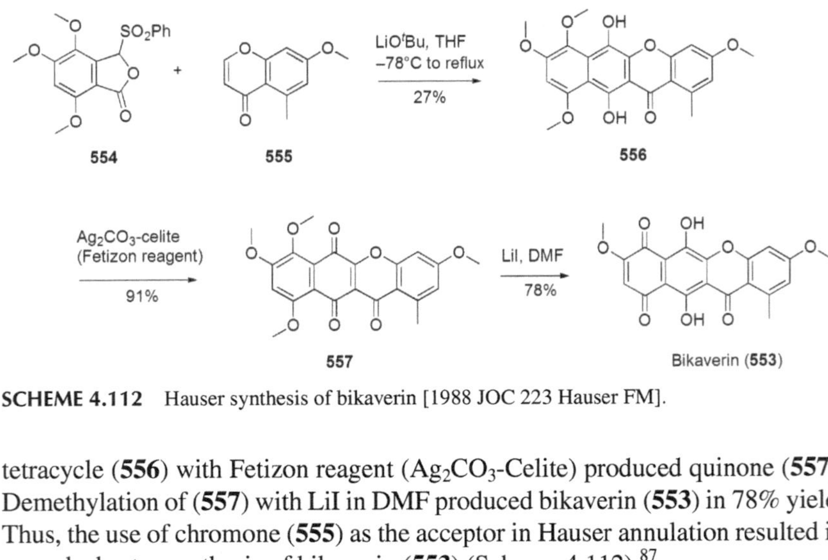

SCHEME 4.112 Hauser synthesis of bikaverin [1988 JOC 223 Hauser FM].

tetracycle (**556**) with Fetizon reagent (Ag_2CO_3-Celite) produced quinone (**557**). Demethylation of (**557**) with LiI in DMF produced bikaverin (**553**) in 78% yield. Thus, the use of chromone (**555**) as the acceptor in Hauser annulation resulted in a much shorter synthesis of bikaverin (**553**) (Scheme 4.112).[87]

Chromen-(2*H*)-2-ones (e.g., **551**), also known as coumarins, seem to be better suited for the anionic [4+2] annulations. They can be annulated with all three common types of Hauser donors, namely 3-phenylthiophthalides (**4**), 3-phenysulfonylphthalides (**2**), and 3-cyanophthalides (**3**), in good yields. For example, phthalide sulfide (**22**), on reaction with coumarin in the presence of LiO^{t}Bu at −60°C, gave (**558**) in 94% yield via intermediate (**559**) (Scheme 4.113).[5b]

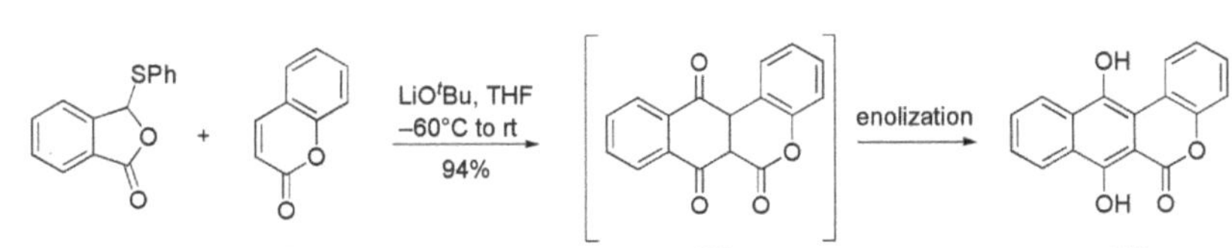

SCHEME 4.113 Annulation with a coumarin [1999 JCRS 30 Mal D].

The manifestations of the above coumarin annulation are found useful in the total synthesis of chartreusin (**560**), which consists of a disaccharide and an unusual benzonaphthopyranone aglycon, named chartarin, as shown in Fig. 4.21 Chartreusin (**560**) possesses potent antiproliferative activity against various tumor cell lines, such as murine P388 and L1210 leukemia, and B16 melanoma cells. IST-622 (**561**), a semisynthetic prodrug of chartreusin, has reached phase II clinical trials. Pharmacological studies have established that chartreusin (**560**) exerts its antitumor activities through binding to DNA, radical-mediated single strand breaks, and inhibition of topoisomerase II. Consequently, the aglycon chartarin (**562**) has been the subject of many syntheses (Fig. 4.22).

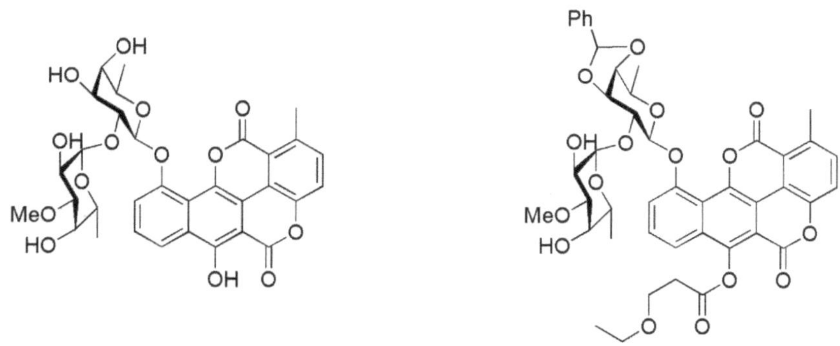

FIG. 4.22 Structures of chartreusin (**565**) and IST-662 (**566**).

Most of them rely on the employment of Hauser annulation. That a coumarin could be an acceptor for Hauser annulation was first established by Hauser and Combs in their synthesis of chartarin (**562**). It involved the annulation of the sulfone phthalide (**24**) with coumarin (**563**) to produce tetracycle (**564**) in 65% yield in the presence of LDA (Scheme 4.114).[88] Appropriate ring-opening of (**564**) followed by $COCl_2$-mediated intramolecular Friedel-Crafts formed the second pyrone ring of (**562**). A more convergent synthesis of chartarin (**562**) was later achieved by Mal and co-workers. It involved the tandem annulation between coumarin (**565**) and the cyanophthalide (**490**), assembling both benzene rings and both the lactone rings of (**562**) in a one-pot operation. The required coumarin (**565**) was prepared from benzoate (**566**) in two steps. Duff reaction of (**566**) gave benzaldehyde (**567**) in 30% yield. Treatment of (**567**) with (carbethoxymethylene)triphenylphosphorane in *N,N*-diethylaniline (DEtA) furnished coumarin (**565**) in 95% yield via a tandem Wittig olefination and lactonization. The Hauser reaction-initiated tandem reaction between the coumarin (**565**) and phthalide (**490**) provided pentacycle (**568**) in one pot in 86% yield via the intermediate (**569**). Demethylation of (**569**) with HBr in refluxing AcOH provided chartarin (**562**) in 81% yield (Scheme 4.115).[89]

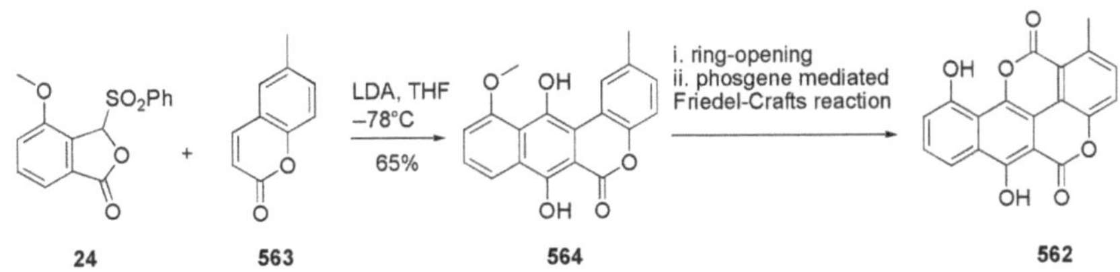

SCHEME 4.114 Hauser synthesis of chartarin [1980 JOC 4071 Hauser FM].

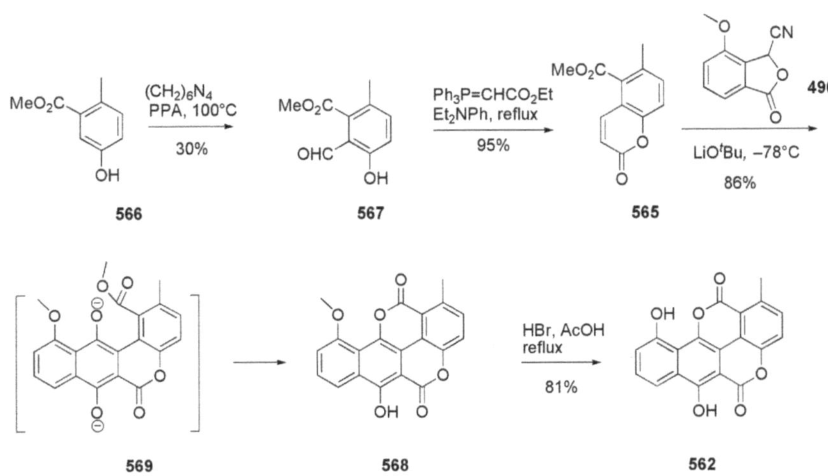

SCHEME 4.115 Mal synthesis of chartarin (**567**) [2004 THS 7895 Mal D].

The Mal synthesis of chartarin (**562**) has greatly facilitated generation of newer analogs of chartreusin (**560**). Hertweck et al. combined the chemical synthesis with biosynthetic engineering for the synthesis of 11 new chartreusin analogs.[90a] Their synthesis of norchartreusin (**570**) is presented in Scheme 4.116. Bromination of benzoate (**571**) with 2,4,4,6-tetrabromocyclohexa-2,5-dienone in acetic acid gave (**572**) in 30% yield. When it was subjected to Heck coupling with methyl acrylate, coumarin (**573**) was formed in 78% yield. The Hauser annulation of the coumarin (**573**) and phthalide (**574**) in the presence of KTB at −60°C furnished pentacycle (**575**) in 66% yield. Treatment of (**575**) with boron tribromide afforded norchartarin (**576**). Addition of a DMSO solution of synthetic norchartarin (**576**) to cultures of the ΔchaABC mutant (S. albus:pXU41/pXU343) formed norchartreusin (**570**).

As demonstrated in the preparation of chartarin analogs, Hauser annulation is general for coumarin acceptors. However, yields are sensitive to the substituents at C6 position of the coumarins. With the parent coumarin (**551**), the yield of the corresponding norchartarin was 92%, whereas the yields with substituted coumarins ranged from 59% to 71% under similar conditions. Most recently, the Chi group combined their de novo synthesis of C5 ester-appended coumarins with Mal chartarin synthesis for the rapid total synthesis of defucogilvocarcins.[90b]

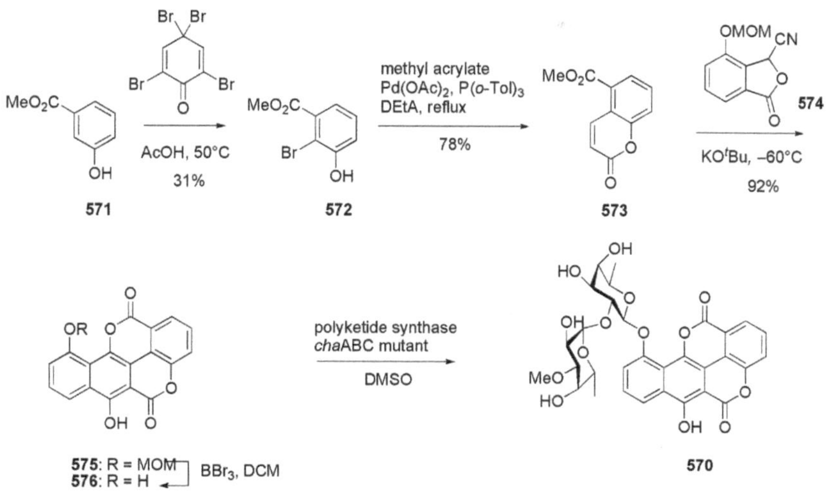

SCHEME 4.116 Hertweck total synthesis of norchartreusin (**570**) [2013 JACS 17408 Hertweck C].

4.4 QUINONES, QUINONE MONOKETALS, QUINONE IMINES, AND QUINONOIDS

1,4-Benzoquinones and 1,4-naphthoquinones are well-known to undergo nucleophilic addition to various carbon nucleophiles.[91a] They also undergo facile Diels-Alder reactions with a variety of dienes.[91b] However, their [4+2] annulation reactivities involving the vinyl group between the carbonyl groups were not known until 2007. The initial failure of Hauser annulation of benzoquinone was rationalized by competing aromatization of the initial Michael adduct of (**11**) and (**341**), i.e., (**577**), resulting in the formation of stable phenolate (**578**) (Scheme 4.117).

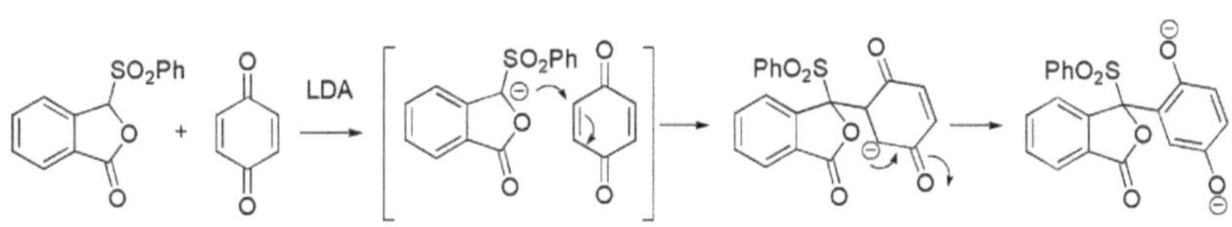

SCHEME 4.117 Mechanistic interpretation of failure of Hauser annulation with 1,4-quinones.

In the quest to unravel the mechanistic pathways, Mal et al. reinvestigated the annulations with 1,4-quinones, and found that all three common classes of Hauser donors, namely phenylthio-, phenylsulfonyl-, and cyano-phthalides (**2–4**), can be annulated with benzoquinones under appropriate reaction conditions (Scheme 4.118). The best results for the products (**579**) were obtained with phenylthiophthalide (**22**) as the donor and LiO^{t}Bu as the base for the quinones (**580**) (Scheme 4.118).[92]

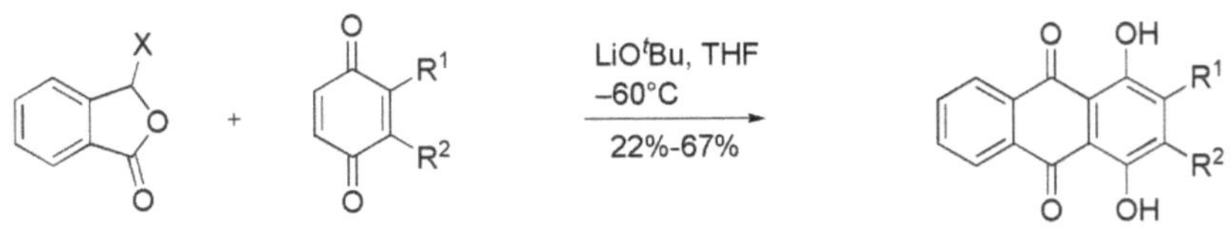

a: X = SPh (**22**)
b: X = SO_2Ph (**11**)
c: X = CN (**14**)

580

R^1, R^2 = H, Me, OMe
1-hydroxyethyl, etc.

579

SCHEME 4.118 Hauser annulation with 1,4-benzoquinones [2007 JOC 4981Mal D].

The clue to success was derived from the analysis of a mechanism based on cycloaddition of isobenzofuran oxyanion (**581**) with benzoquinones (**580**) giving cycloadducts (**582**). Under the influence of a strong base like LDA, they form dianions (**583**) and then tautomerize to aromatic anion (**584**). It was anticipated that a weaker base like LiO^{t}Bu is unable to form the dianion, thereby preventing the *retro*-Dieckmann step, i.e., (**583**) to (**584**) (Scheme 4.119).

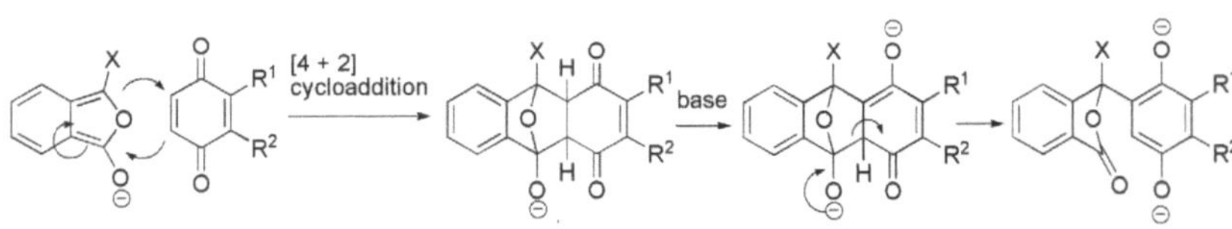

581 **580** **582** **583** **584**

SCHEME 4.119 [4+2] mechanism for the Hauser annulation with 1,4-benzoquinones.

Bicyclic *p*-quinones (**585–589**) were successfully annulated with phenylthiophthalide (**22**) to give products (**590–592**) in moderate yields, as shown in Fig. 4.23.

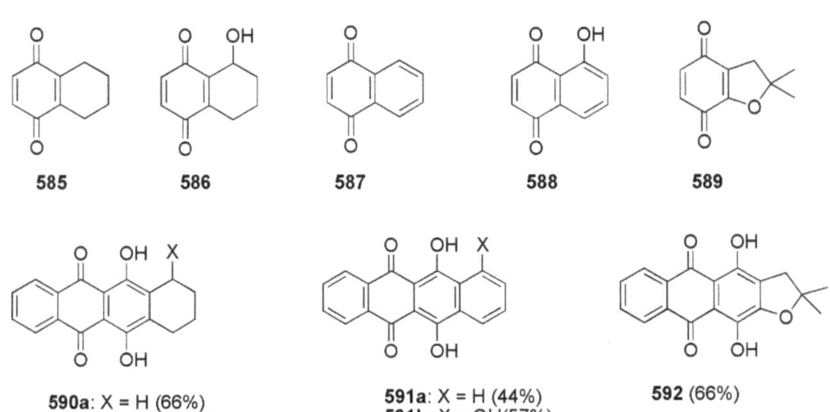

FIG. 4.23 Tetracyclic quinones from the annulations of quinones with 3-phenylthiophthalide (**22**) [2007 JOC 4981 Mal D].

When coupled with an intramolecular ene reaction, the annulation of benzoquinone (**593**) provided an easy access to the idarubicin skeleton. The annulation between the phthalide (**22**) and (**593**) in the presence of LTB provided dihydroxyanthraquinone (**594**), which did not require any protection of OH groups for subsequent transformation. Removal of THP ether of (**594**) with PPTS-promoted methanolysis furnished (**595**). Oxidation of alcohol (**595**) with PCC gave (**596**), setting the stage for an intramolecular ene reaction. Treatment of the aldehyde (**596**) with $SnCl_4{\cdot}5H_2O$ in DCM afforded tetracycle (**597**) with an exocyclic CC double bond in 89% yield (Scheme 4.120).[93]

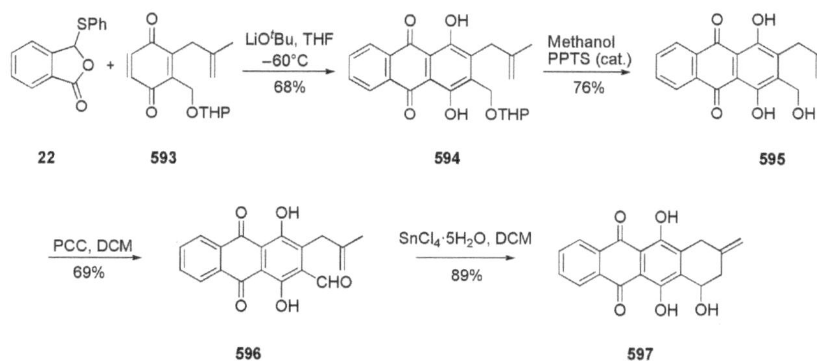

SCHEME 4.120 Synthesis of the tetracycle core of idarubicinone [2017 PhD Thesis, IIT Kharagpur, Basak S].

The quinone (**593**) was prepared from acetal (**598**) in five steps. *Ortho* lithiation of (**598**) with ^{n}BuLi followed by quenching with trimethyl borate and acidic hydrolysis afforded a mixture of boronic acids (**599a** and **599b**). This mixture was, without purification, treated with methallyl bromide in the presence of $PdCl_2(PPh_3)_2$-Na_2CO_3 in aq THF to provide methallyl benzaldehyde (**600**). $NaBH_4$ reduction of aldehyde (**600**) gave alcohol (**601**) in 86% yield. Oxidation of (**601**) with CAN in acetonitrile-water furnished benzoquinone (**602**). Protection of OH group in (**602**) as THP furnished (**593**) (Scheme 4.121).[94]

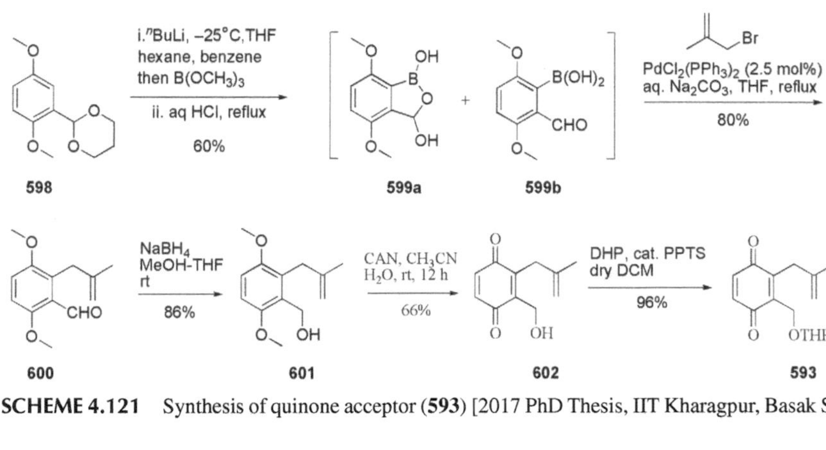

SCHEME 4.121 Synthesis of quinone acceptor (**593**) [2017 PhD Thesis, IIT Kharagpur, Basak S].

4.4.1 *p*-Quinone Monoketals

The first report on the use of 1,4-quinone monoketals as acceptors was concurrently published in 1981 by Russell et al. and Swenton et al.[95] The Hauser annulation of the ketals provides a direct synthesis of 1,4-dioxygenated anthraquinone substructures. Contrary to their poor reactivity of *p*-quinone monoketals to the Michael addition, for which very few examples are known, and which take place only under special reaction conditions and with special reagents, their Hauser annulation is well-known in the literature. The quinone monoacetals, accessible by hydrolysis of one acetal group of electrochemically derived quinone bisacetals, undergo the Hauser annulations, providing 1-hydroxy-4-methoxyanthraquinones in a single step. All the steps occur under basic conditions, but at different temperatures. The major by-products are the initial Michael adducts. Quinone ketals have been very popular acceptors in the Hauser reaction for the regiocontrolled target synthesis of quinonoids. The synthetic utility of the Hauser annulation was primarily and extensively established by the study on quinone monoketals that serve as the building blocks in the construction of anthracyclines and angucyclines. As shown in Scheme 4.122, a variety of 5,6-unsubstituted quinone monoketals (**603**) can be annulated with all major Hauser donors (**2–4**) to construct 1-hydroxy-4-methoxyanthraquinones (**604**) in good yields. Mechanistically, the reactions are believed to proceed through tetracyclic oxyanions (**605**) to form stable quinolates (**606**). On acidification, they transform to protonated quinols (**607**). In the presence of acid, methanol is lost from (**607**) to form methoxyanthraquinones (**604**).[95a]

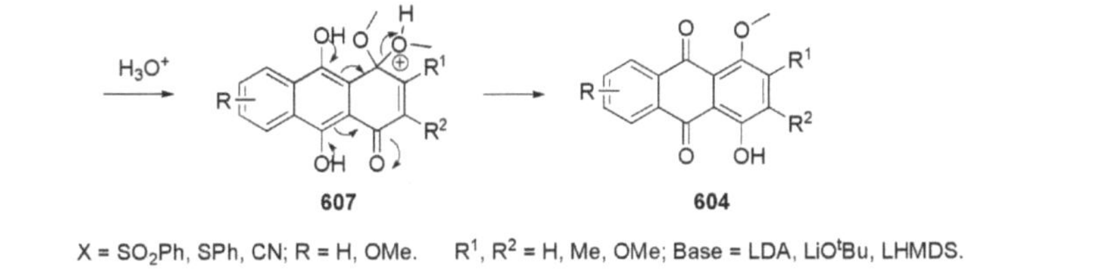

SCHEME 4.122 Hauser annulation of *p*-benzoquinone monoketals [1981 JCSCCS 108; 1997 THN 4387 Russell RA].

Obvious extension of this annulation is the study of the Schiff bases of quinone monoketals as acceptors. However, such a study is limited to Swenton condensation as depicted in (Scheme 4.123), which suggests the feasibility of the annulation. The reaction of the cyanophthalide (**14**) with quinone imine (**608**) afforded aminoanthraquinone (**609**) through the intermediate (**610**).[96]

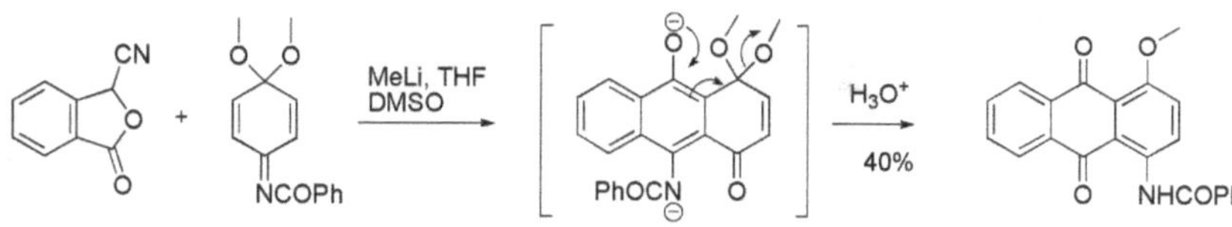

14 608 610 609

SCHEME 4.123 Swenton annulation with quinone Schiff base.

Recently, Mal et al. used bromine substituted benzoquinone monoketals as acceptors to synthesize brominated anthraquinones.[97] The reactions with quinone monoketals are selective to the unsubstituted double bonds and the disposition of the new phenolic hydroxyl group in the products is the result of the initial Michael addition. R^1 and R^2 in (**603**) have been alkyl substituents. It has been shown that two annulations are achievable in one-pot if an acceptor has two quinone monoketal units (Scheme 4.124).[98] The acceptor (**611**), prepared by electro-oxidation of (**612**) in methanol, gave alkane spacer linked bis-anthraquinone (**613**), a potential DNA bisintercalater in 70% yield on annulation with cyanophthalide (**14**). The efficiency of the annulations was not sensitive to the chain length of the spacer.

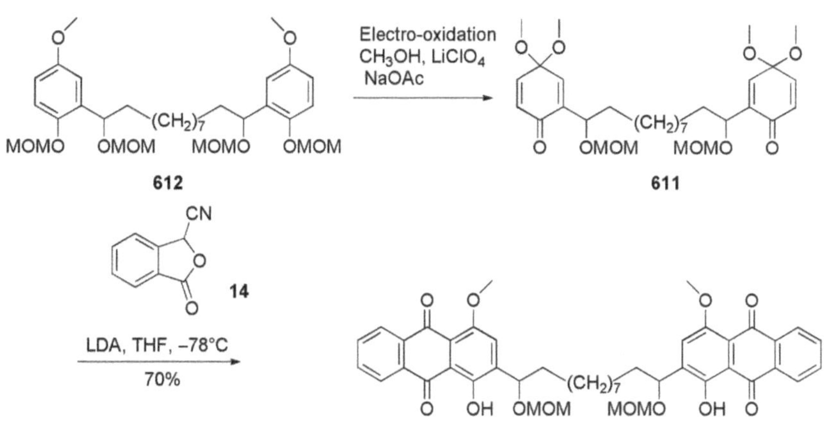

613

SCHEME 4.124 Double annulation [1997 THN 17477 Russell RA].

4.4.2 Nicolaou Annulation

Nicolaou et al. studied in depth the annulation of aminals and found that the reaction of the cyanophthalide (**14**) with quinone aminal (**614**) resulted in anthraquinone (**615**) in 85% yield via loss of methanol from the intermediate (**616**). Of the two leaving groups, i.e., OMe and N(Me)(alloc) in the intermediate (**616**), the methoxy group is eliminated. This reaction has been generalized with respect

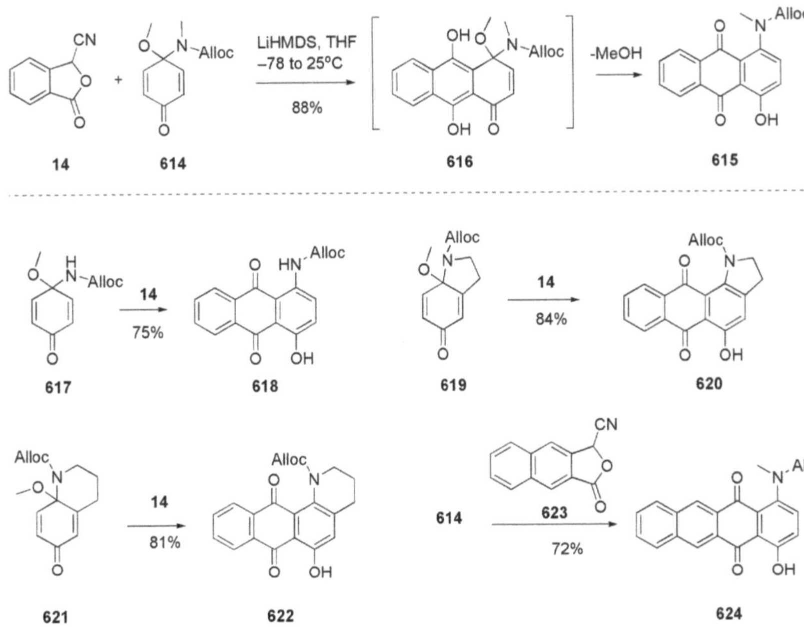

SCHEME 4.125 Nicolaou annulation [2015 ACIE 12687 Nicolaou KC].

to both aminals and cyanophthalides as displayed in Scheme 4.125.[99] Some examples of the annulation are the transformations: (**617** → **618**), (**619** → **620**), (**621** → **622**), and (**623** → **624**).

Placement of a methyl group at C3 of the aminals (e.g., **625**) diverts the course of the annulation. The reaction of aminal (**625**) with cyanophthalide (**14**) forms *p*-methoxyphenolic anthraquinones (**626**) through the expulsion of the C4 amino functionality (Scheme 4.126) from the intermediate (**627**). The general trend of the reaction is established by several examples displayed in the scheme. These are the conversions: (**628** → **629**), (**630** → **631**), (**632** → **633**), and (**634** → **635**), respectively. The reversal of the selectivity, i.e., synthesis of the *p*-aminophenolic anthraquinones can be achieved by replacing methoxy with trifluoroethoxy group in the acceptors.

The synthesis of acceptors (e.g., **614**) are typically carried out by oxidative aromatization of the phenols. For example, phenol (**636**) was converted to (**614**) by oxidation with $PhI(OAc)_2$ and (**637**) was transformed to (**619**) by $Pb(OAc)_4$ oxidation in methanol (Scheme 4.127).[99]

The [4 + 2] anionic annulation chemistry works well with bicyclic quinone ketals. It has largely been explored by groups of Swenton, Monneret, Achmatowicz and Russell in connection with their synthesis of anthracyclines.[95b,100] Swenton daunomycinone synthesis involved anodic oxidation of decalin derivative (**637**) to bisketal (**638**).[95b] Interestingly, acid-hydrolysis of the bisketal (**638**) selectively furnished monohydrolysis product (**639**). The Hauser annulation of

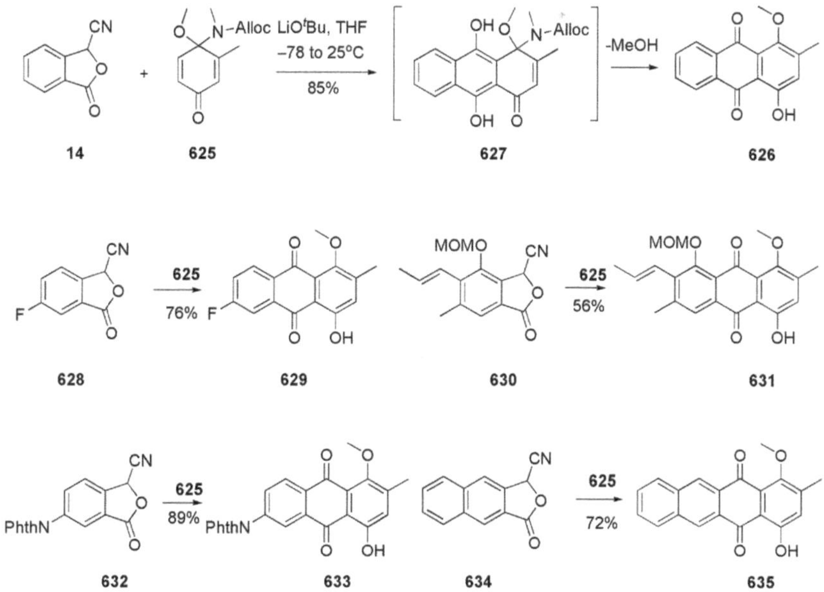

SCHEME 4.126 Nicolaou annulation for *p*-methoxyphenolic anthraquinones [2015 ACIE 12687 Nicolaou KC].

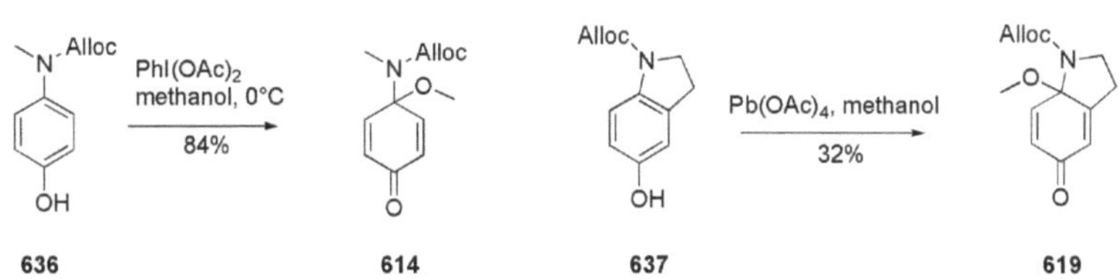

SCHEME 4.127 Representative synthesis of acceptors for amino/methoxy anthraquinones.

phthalide sulfone (**251**) with (**639**) in the presence of dimsyl anion gave the tetracyclic product (**640**). For completion of the synthesis, tetracenone (**640**) was demethylated (BCl_3, −70°C, >90%) and then desilylated product (90%) to furnish (**641**). The epi-7-methoxy-7-deoxydaunomycinone (**641**) was convertible to (±)-daunomycinone (**642**) by Wong protocol (CF_3CO_2H followed by NH_4OH) (Scheme 4.128).

The similar application of the Swenton strategy has been utilized by various groups. One example each from Russell group and Achmatowicz group are presented in Scheme 4.129. Both the similar looking annulations (i.e., **643**→**644** and **645**→**646**) could be carried out in excellent yields under different reaction conditions (Scheme 4.129).

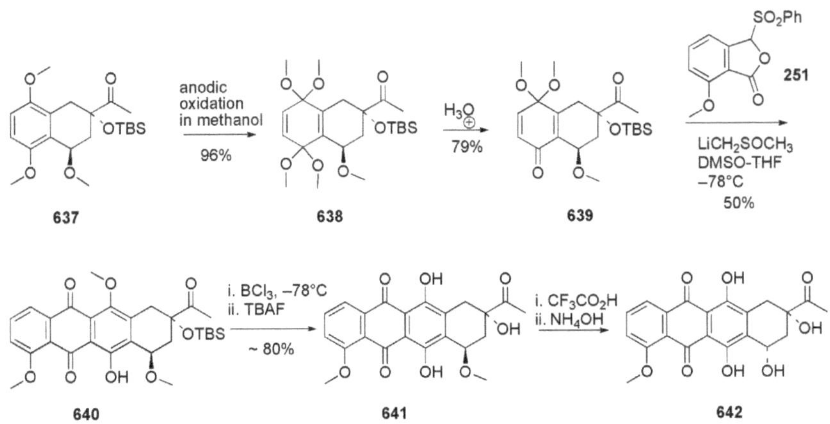

SCHEME 4.128 Synthesis of (±)-daunomycinone [1981 JACS 5263 Swenton JS].

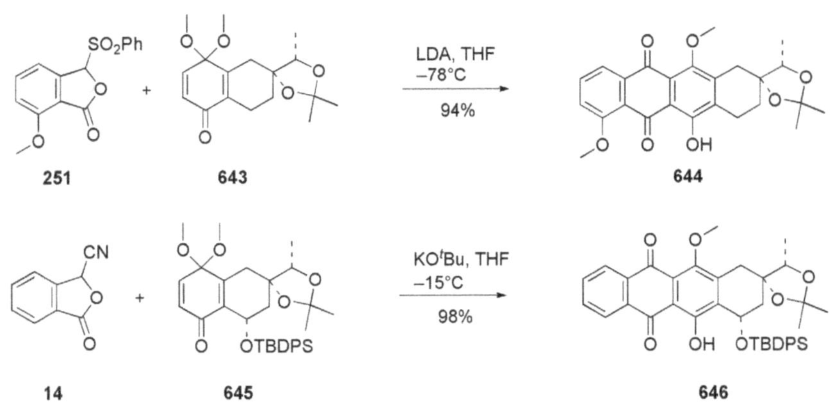

SCHEME 4.129 Illustrative examples of the annulation with decalin quinone monoketals.

Annulations of iminoquinone (e.g., **647**) and aminals (e.g., **648**) have been investigated by Danishefsky group, Myers group and Nicolaou group in order to achieve the total synthesis of dynemicin (**649**) and uncialamycin (**650**).[101] These natural products are classified as enediynes with aminoanthraquinone motifs. Historically, dynemicin (**649**) was the target of early synthetic studies (Fig. 4.24).

Through rigorous experimentations, it was possible to assemble polycyclic skeleton of dynemicin (**649**) as shown below in late stages of the total synthesis exemplifying the strength of the Hauser annulation. Myers et al. used quinone imine (**651**) as the Hauser acceptor in their synthesis of dideoxydynemicin. The reaction between the phthalide (**14**) and quinone imine (**651**) proceeded in 47% yield to produce (**652**) (Scheme 4.130).

The similar annulations of acceptor (**653**) with 4,7-dimethoxycyanophthalide (**188**) or 3-cyanophthalide-4,7-bisphosphonate (**654**) failed to give dynemicin skeleton. Instead, the addition products (**655** and **656**) were obtained, respectively (Scheme 4.131). The acceptor (**653**) was obtained from the corresponding

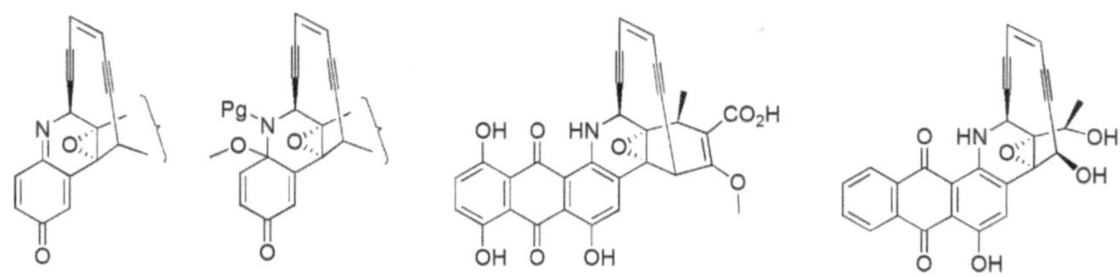

647 648 Dynemicin A (649) Uncialamycin (650)

FIG. 4.24 Structures of dynemicin (**649**) and uncialamycin (**650**), and their subunits.

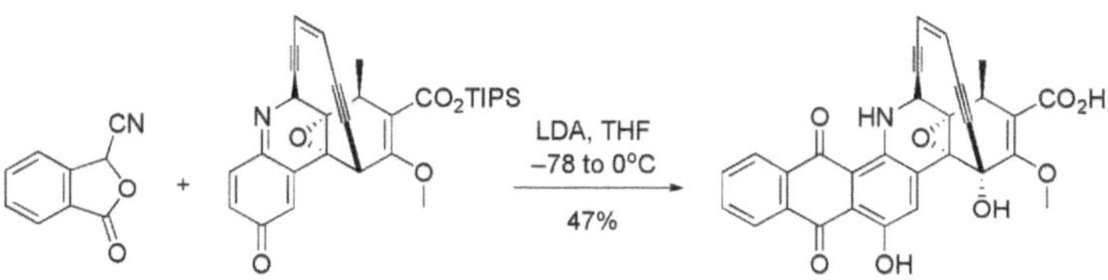

14 651 652

SCHEME 4.130 Myers successful annulation with a bicyclic quinone imine [1997 JACS 6072 Myers AG].

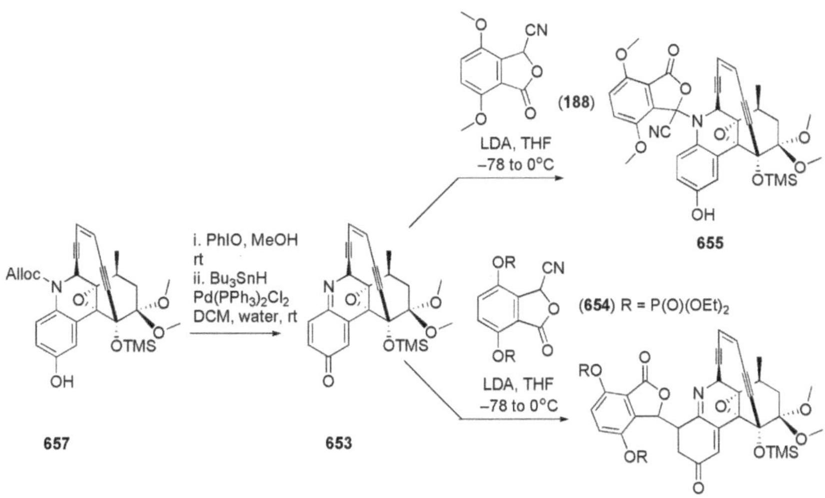

656

SCHEME 4.131 Myers annulation with a bicyclic quinone imine [1997 JACS 6072 Myers AG].

aminophenol (**657**) by oxidative dearomatization with iodosobenzene in methanol followed by palladium catalyzed reduction with tetrabutylation hydride of the intermediate aminal.

Compared to the quinone imines of the type (**651**) or (**653**), aminal (**658**) seems to be more reactive to the annulation. In the presence of LTB and LiCl, its reaction with the dimethoxyphthalide (**188**) gave hexacyclic intermediate (**659**) (Scheme 4.132).

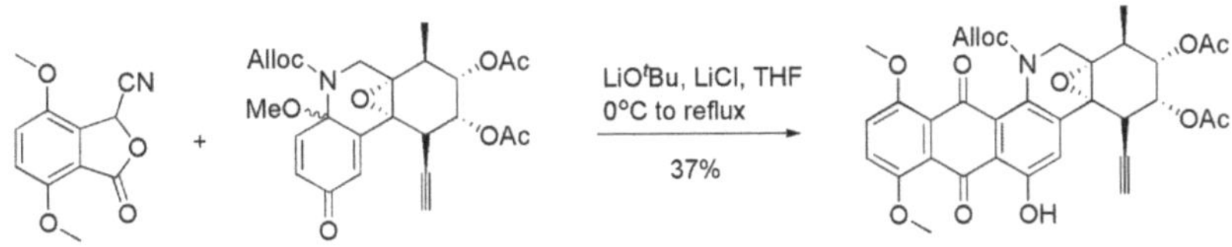

188 658 659

SCHEME 4.132 Model studies for dynemicin [1994 THS 6259 Danishefsky SJ].

But, the quinone aminal bearing cyclic enediyne moiety, i.e., (**660**) did not respond to the Hauser annulation with phthalide (**188**) (Scheme 4.133). The quinone aminal (**660**) was obtained from (**661**) by desilylation followed by oxidative dearomatization with $PhI(OAc)_2$. The reaction between the phthalide (**188**) and (**660**) provided only the Michael addition (**662**). All attempts to induce cyclization of (**662**) to hexacycle (**663**) failed.

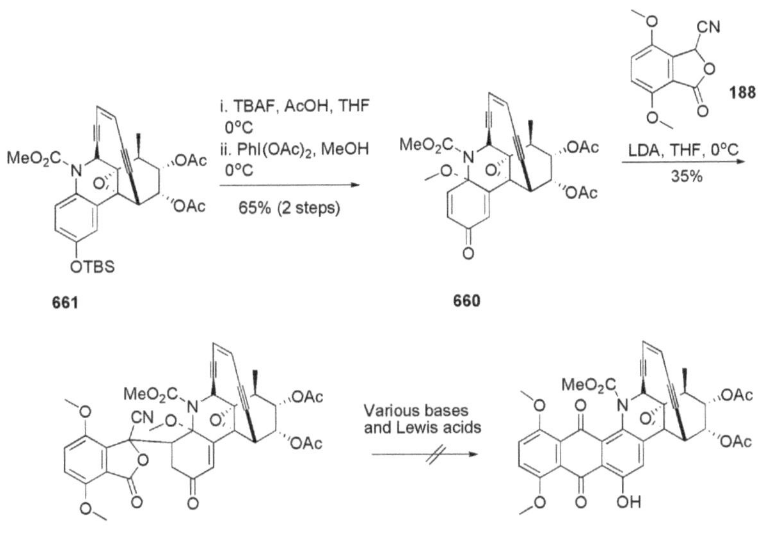

661 660 662 663

SCHEME 4.133 Failed annulation with an aminal [1996 JACS 9509 Danishefsky SJ].

However, the total synthesis of uncialamycin (**650**) could neatly be achieved by applying the Hauser annulation to the phthalide (**14**) and quinone imine (**664**) in the penultimate stage.[102] The annulation in the presence of LHMDS, followed by HF promoted desilylation gave the desired product (**665**) in 58% overall yield. It appears that the *peri*-methoxy groups in phthalide donors as in (**188**) inhibit the annulations (Scheme 4.134).

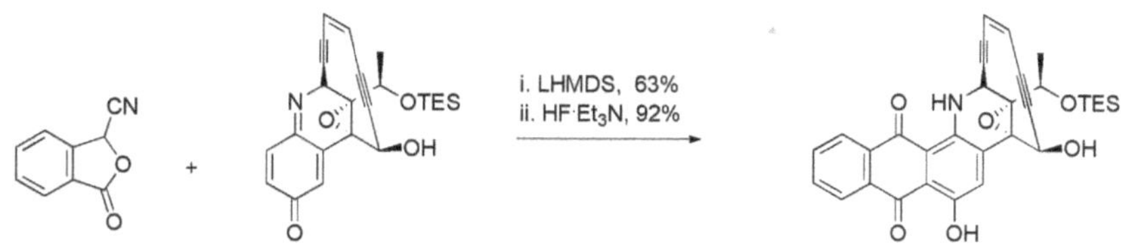

14 664 665

SCHEME 4.134 Nicolaou synthesis of uncialamycin from a quinone imine [2007 ACIE 4707 Nicolaou KC].

4.4.3 *o*-Quinone Monoketals

Like *para*-quinone monoketals, *ortho*-quinone monoketals have also been recognized as important acceptors for the Hauser annulation. In 1997, Mal group and Russell group concomitantly published their results on the annulations with *o*-quinone monoketals.[103] While Mal group focused on the bicyclic quinone monoketals derived from β-naphthols, Russell group dealt with monocyclic quinone monoketals, which are more unstable at room temperature. However, all the monoketals are sufficiently stable under basic conditions for their use in the anionic [4+2] annulations. Typical examples for *o*-benzoquinone monoketals are presented in Scheme 4.135. Phenols (**666**) are oxidized with PhI(OAc)$_2$ in methanol at room temperature. After flash chromatographic purifications, the ketals (**667**) are submitted to the reaction with anion of the phthalide (**14**) at −78°C. On quenching with HCl and purifications, the reactions afford alizarins (**668**) in very good yields. These reactions are believed to take place via the intermediates (**669**).

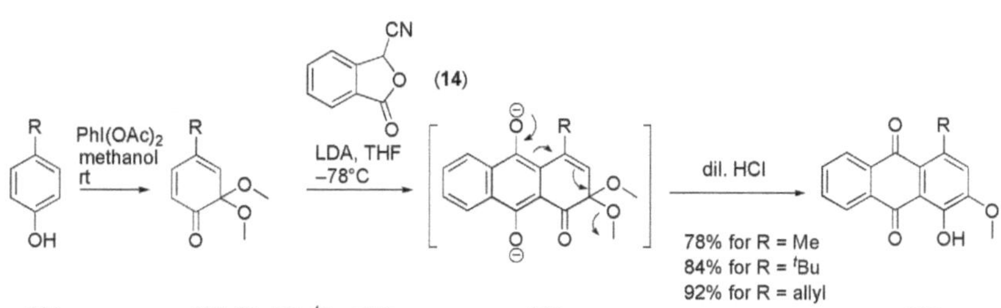

666 667 (R = Me, ^{t}Bu, allyl) 669 668

SCHEME 4.135 Annulation *o*-benzoquinone monoketals [1997 THN 4387 Mitchell AS].

The course of the Nicolaou reaction, as discussed in the context of *p*-benzoquinone monoketals, is dictated by the nature of the *N*-protecting group. For *ortho* quinone aminals (**670**), annulation with the phthalide (**14**), under the influence of LiO^{t}Bu, proceeds through elimination of *N*(Me)Alloc, giving methoxyanthraquinone (**671**). Mechanistically, it is assumed that intermediates (**672** and **673**) are formed during the reactions. These annulations are generalized with the transformations (**674**→**675** and **676**→**677**) (Scheme 4.136).

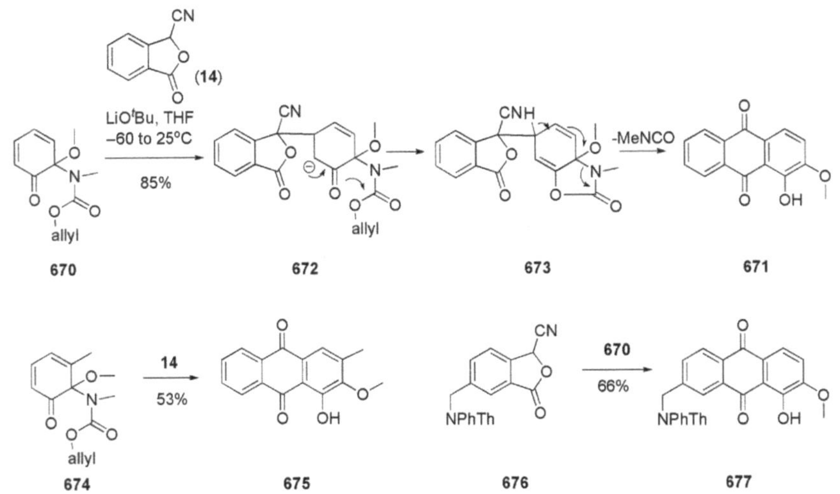

SCHEME 4.136 Nicolaou annulation with *o*-quinone aminals [2015 ACIE 12687 Nicolaou KC].

If the annulation is carried out with a trifluoroethoxy-substituted acceptor (**678**), 2-aminoanthraquinone (**679**) is formed via expulsion of CF_3CH_2OH from (**680**) because of the better leaving group property of the trifluoroethoxy group. This trend is general, as evident from the transformations (**681**→**682**) and (**683**→**684**) (Scheme 4.137).

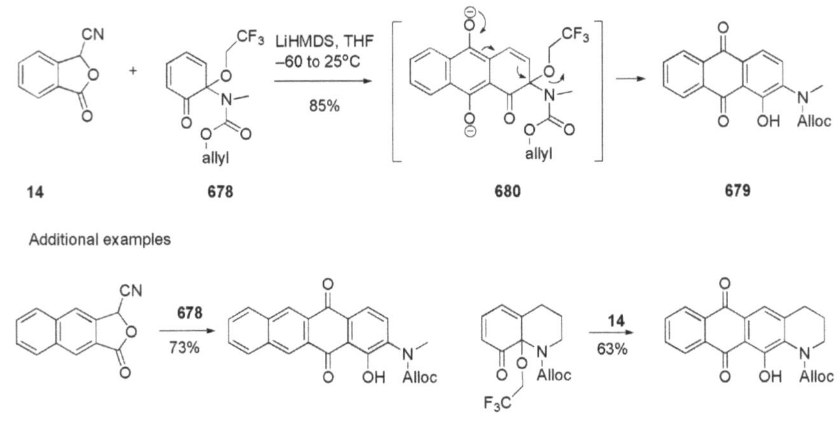

SCHEME 4.137 Nicolaou annulation with *o*-quinone aminals [2015 ACIE 12687 Nicolaou KC].

In 1997, Mal et al. reported ready accessibility of *ortho*-naphthoquinone monoketals (**685**) from β-naphthols (**686**). These acceptors could be readily be prepared in 43%–88% yields through oxidative dearomatization of β-naphthols (**686**) with $PhI(OAc)_2$ in methanol at 0°C to room temperature, and be engaged in the Hauser annulation with phenylsulfonylphthalides in the presence of LTB, resulting in rapid preparations of 5,6-dioxygenated benz[*a*]anthraquinones (**687**) (Scheme 4.138).[103b] The Hauser annulation cascade does not stop at the quinol (**688**) corresponding to the dioxide (**689**). When a strong acid, such as HCl, is used during the work-up, it further progresses to give an anthraquinone core (**687**) through expulsion of a methoxy group from the intermediate hydroquinone. Hauser et al. have shown that the elimination can be arrested using a weak acid, e.g., acetic acid during the work-up of the reaction, and the expected quinol form (e.g., **688**) can be exclusively isolated as the only product.

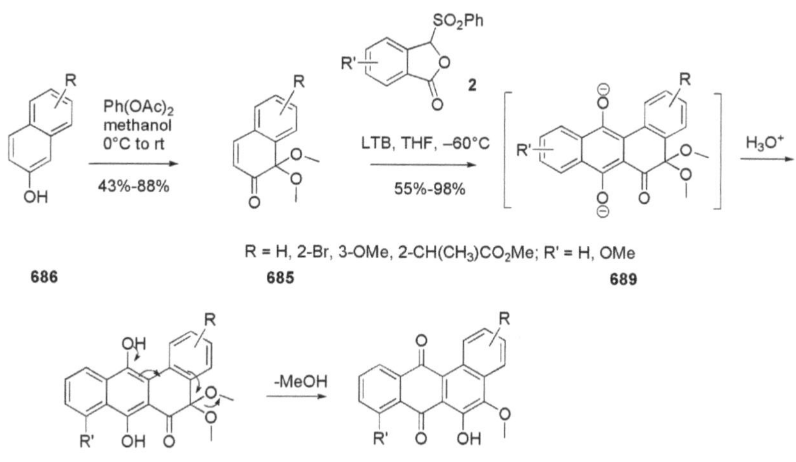

SCHEME 4.138 [4+2] Annulation with *o*-naphthoquinone monoketals [1997 THN 2177 Mal D].

Hauser et al. applied the annulation of *o*-naphthoquinone monoketals to the total synthesis of methyl ether (**690**) of PD116740, an angucycline antibiotic.[104] Condensation of the phthalide sulfide (**180**) with quinone monoketals (**691**) in the presence of LTB followed by acetic acid quench cleanly afforded annulation product (**692**) in 72% yield. It was dimethylated with K_2CO_3 and $(CH_3O)_2SO_2$, to produce (**693**) exclusively. NBS bromination of (**693**) with NBS afforded the corresponding bromomethyl intermediate, which was reacted with NaOAc in DMF to afford the acetoxymethyl compound (**694**). Deketalization of (**694**) was effected with TFA in $CHCl_3$-H_2O to give the corresponding *ortho*-quinone. The Harvey protocol for conversion of quinone to diol, i.e., $NaBH_4$ in the presence of air, was employed to convert the quinone to a 3:1 mixture of *trans* and *cis*-diols (**695**). The diastereomeric mixture (**695**) was acetylated with Ac_2O-py to its diacetate. The major isomer was oxidized with CAN to give the methyl ether acetate derivative (**690**) of PD 116740 (Scheme 4.139).

The above total synthesis of Hauser et al. additionally illustrates a complexity in the synthesis of an apparently simple-looking naphthalene derivative,

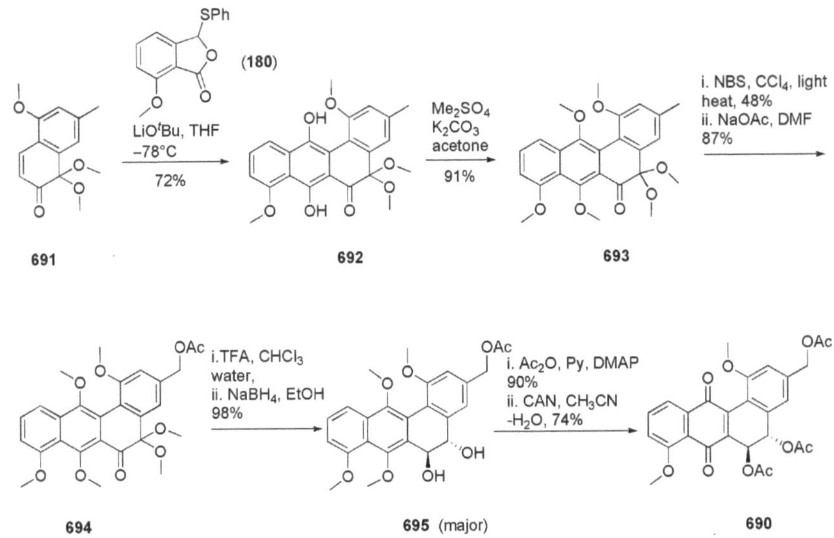

SCHEME 4.139 Hauser total synthesis of PD116740 derivative [2002 ORS 2237 Hauser FM].

which was accomplished via a lateral lithiation. The *ortho*-naphthoquinone monoketal (**691**) required for the total synthesis was prepared in a lengthy linear sequence of 11 steps from 3,4-dimethylanisole. It was converted to toluic acid (**690**) with two successive oxidations. Treatment of (**696**) with 2 equiv. of LDA afforded the dianion intermediate, which was quenched with dimethyl carbonate followed by hydrolysis of the resultant ester to furnish homophthalic acid (**697**). A sequence of three steps—(i) treatment of (**697**) with Ac_2O-py, (ii) alkaline hydrolysis with decarboxylation, and (iii) cyclization of the keto carboxylic acid intermediate with Ac_2O and catalytic $HClO_4$—afforded the benzopyranone (**698**) (70% overall yield). The pyranone (**698**) was then converted into benzyl ether (**699**) in two steps. Demethylation of (**698**) with BBr_3 followed by benzylation (BnBr and K_2CO_3) of the resultant phenol furnished the ether (**699**). The transformation of the ether (**699**) naphthalene to (**700**) was accomplished by the reaction of the anion of dimethyl methylphosphonate with the benzopyranone (**699**), involving the intermediate (**701**) followed by methylation with dimethyl sulfate and K_2CO_3. Hydrogenolysis of (**700**) followed by oxidation of the resultant phenol (**702**) with $PhI(OAc)_2$ in MeOH afforded the *ortho*-quinone ketal (**691**) (Scheme 4.140).

Mal and *co*-workers also applied Hauser annulation to the synthesis of pentacyclic benzonaphthopyranone cores of the glycosidic polyketide antibiotics chrymutasins and hayumicins.[105] With incorporation of a proximate ester group in *o*-quinone monoketal (e.g., **703**), two rings can be formed in tandem in the annulation. The *ortho*-quinone monoketal (**703**) was successfully condensed with the anion of the cyanophthalide (**490**) to afford the pentacycle (**704**) in good yield. Finally, acid-mediated ketal hydrolysis of (**704**) provided a hayumicinone

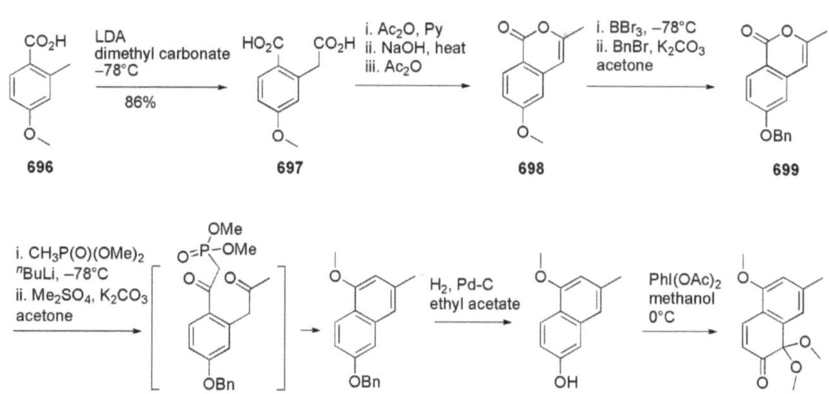

696 697 698 699 701 700 702 691

SCHEME 4.140 Synthesis of a substituted β-naphthol for PD116740 [2002 ORS 2237 Hauser FM].

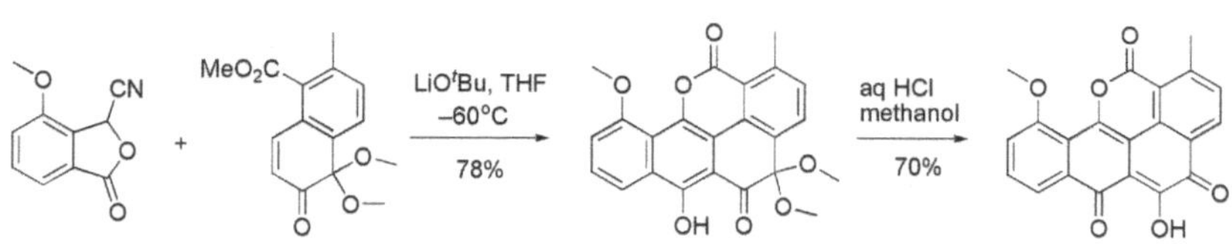

490 703 704 705

SCHEME 4.141 Model studies on chrymutasin/hayumicinone (**705**).

derivative (**705**), which features all of the important structural components of the hayumicin A aglycon (Scheme 4.141).

This naphthalenone annulation has been extended to 1-substituted naphthalene-2-ones (**706**), which were readily prepared by $PhI(OAc)_2$-MeOH promoted dearomatization of the corresponding β-naphthols (**707**). Their annulations with the cyanophthalide (**14**) in the presence of LTB were highly efficient, the yields of the products (**708**) ranging between 88% and 98% (Scheme 4.142).[106]

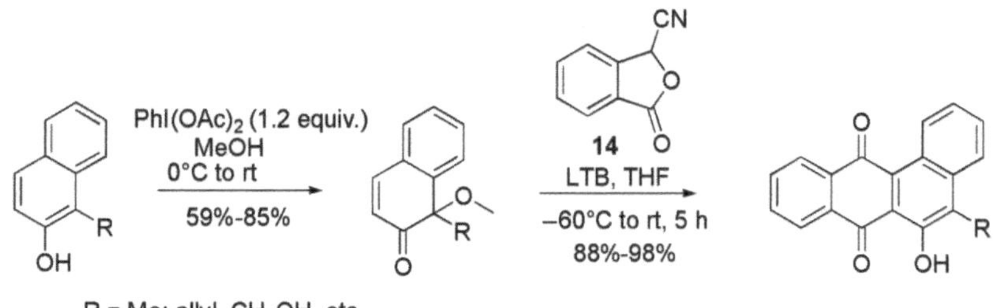

707 706 708

SCHEME 4.142 Annulation of 1-substituted naphthalene-2-ones [2013 JOC 9748 MaL D].

The above combination of the dearomatization of β-naphthols and Hauser annulation formed the basis of model synthetic studies of mayamycin (**709**). This mayamycin (**709**), isolated in 2010 from the cultures of marine *Streptomyces* sp. strain HB202, is a unique angucycline C-glycoside. It is the first example of a C-angucycline glycoside with carbohydrate residue at C5. Furthermore, its biological profile is also extraordinary. It exhibited potent in vitro cytotoxic activity against eight human cancer cell lines, and activity against several bacteria including antibiotic-resistant strains. Adaptation of the methodology described in Scheme 4.142 allowed fabrication of the mayamycin motif (**710**). Beta-naphthol was readily glycosylated with azido sugar (**711**) to give *C*-glycoside (**712**) (60%) along with its epimer. The glycoside (**712**) was dearomatized with $PhI(OAc)_2$ in methanol to furnish the Hauser acceptor (**713**). Its annulation with the cyanophthalide (**14**) in the presence of LTB provided the motif (**710**) resembling the structure of mayamycin (**709**) (Scheme 4.143).

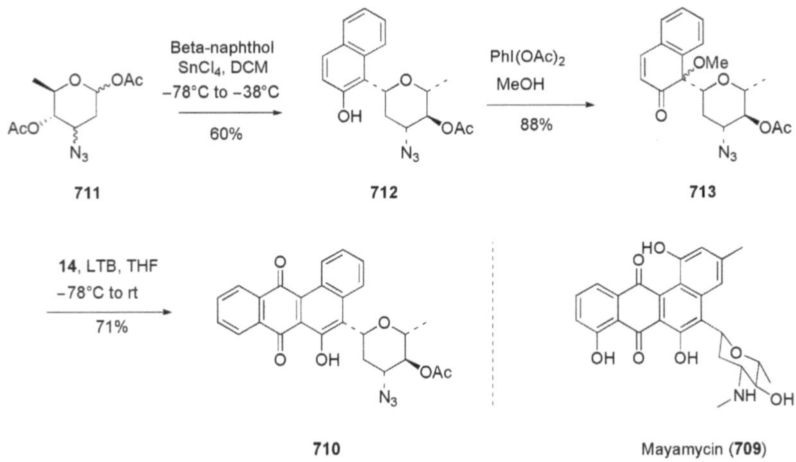

SCHEME 4.143 Annulation route to the mayamycin motif [2013 JOC 9748 MaL D].

4.4.4 Quinonoids/Masked Quinones

1,4-anthraquinones, pivotal building blocks in the synthesis of linear fused polycyclic systems such as anthracyclines, tetracenomycins, and polyacenes, can be conveniently prepared in three steps, the first step being an anionic [4+2] annulation. The annulation of the sulfones (**2**) with the masked quinone (**714**) in the presence of LTB followed by *O*-methylation with dimethyl sulfate and potassium carbonate gives pentacycles (**715**) in moderate yields. Flash vacuum pyrolysis of the pentacycle (**715**) at 500°C and 0.5 Torr furnish 1,4-anthraquinones (**716**) in excellent yields (Scheme 4.144).[107] Since *p*-benzoquinones often give complex mixtures of products in Hauser annulation, the masked quinones (**714**) can be used as substitutes for the quinones.

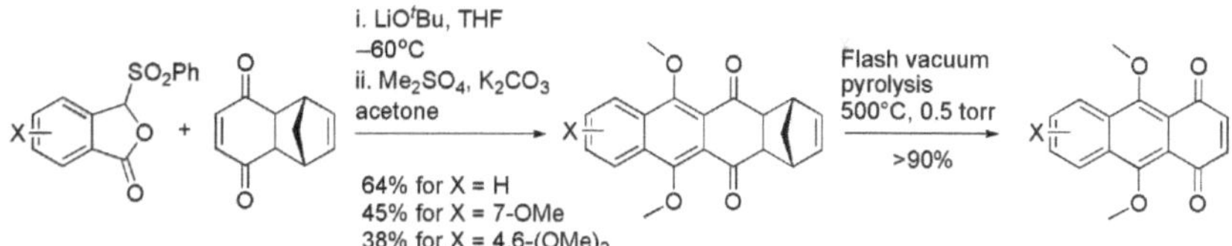

2 714 715 716

SCHEME 4.144 Synthesis of 1,4-anthraquinones via annulation with a masked quinone [1994 THS 6139 MaL D].

Cyclohex-2,5-dienones containing a heteroatom at C4 also exhibit quinone-like reactivity in the annulation with 3-nucleofugal phthalides. For example, cyclohexadienone (**717**) underwent annulation with the cyanaophthalide (**14**) in the presence of LDA to give anthraquinone (**718**) in 96% yield (Scheme 4.145). The reaction proceeded through the intermediate (**719**), similar to those proposed for *p*-quinone monoketals.

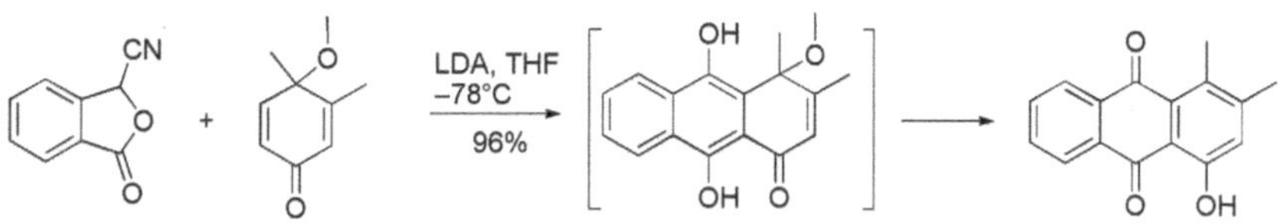

14 717 719 718

SCHEME 4.145 Annulation with a 2,5-cyclohexadienone with C4-methoxy group.

Similarly, dioxygenated cyclohexadienone (**720**), on annulation with the phthalide (**14**), resulted in a tetrasubstituted anthraquinone (**721**) in 82% yield via the intermediate (**722**) (Scheme 4.146).[108]

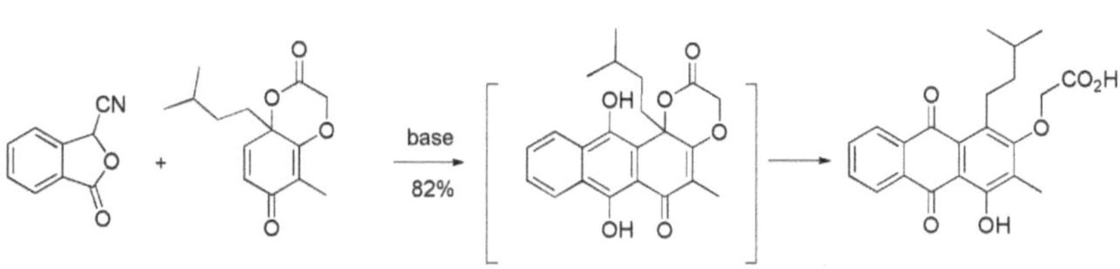

14 720 722 721

SCHEME 4.146 Annulation with a quinonoid [2004 THS 1793 Pettus TRR].

Brasiliquinones A–C (e.g., **723**) are a small group of three cytotoxic benz[*a*]anthraquinones with an ethyl group attached at C3 position. They were isolated from the actinomycete Nocardia brasiliensis IFM 0089. Besides being active against Gram-positive bacteria including Mycobacterium sp., they were also active against multiple drug-resistant P388/ADR tumor cells. Mal and Roy reported a synthetic entry to these targets based upon Hauser annulation of quinonoids (e.g., **724**) and Krohn photo-oxidation of tetracycle (**725**) (Scheme 4.147).[109]

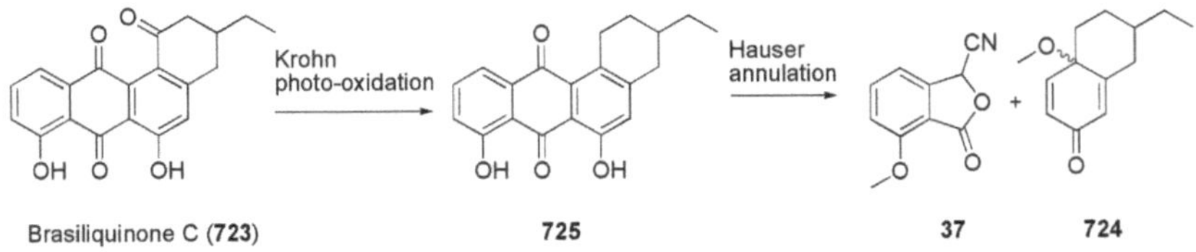

SCHEME 4.147 Anionic annulation-based retrosynthesis of brasiliquinone C (**723**).

Commercially available 7-methoxy-α-tetralone (**726**) was used for the preparation of decalenone (**724**) (Scheme 4.148). First, the tetralone (**726**) was converted into its methylene derivative (**727**), by the action of paraformaldehyde in the presence of *N*-methylanilinium trifluoroacetate. The derivative (**727**) was homologated to (**728**) through the use of cuprous iodide promoted conjugate addition of CH_3MgI. Reduction of the ketone (**728**) with hydrazine-KOH in ethylene glycol afforded (**729**). Demethylation of methyl ether (**729**) to its phenol derivative followed by $Ph(OAc)_2$ oxidation of the resultant naphthol in methanol afforded acceptor (**724**). The naphthalenone (**724**), obtained as a 1:1 mixture of diastereomers, was treated with the anion of 7-methoxycyanophthalide (**37**) to give the tetracycle (**730**) in 92% yield. Installation of the keto functionality at C1 of (**730**) was accomplished by its aerial sunlight promoting oxidation, thus completing the total synthesis of brasiliquinone B (**731**). $AlCl_3$-induced demethylation of (**731**) furnished brasiliquinone C (**723**). Similar reactions of the acceptors (**724**) with phenylsulfonylphthalides (**2**) did not occur, possibly due to the steric bulk of phenylsulfonyl group.

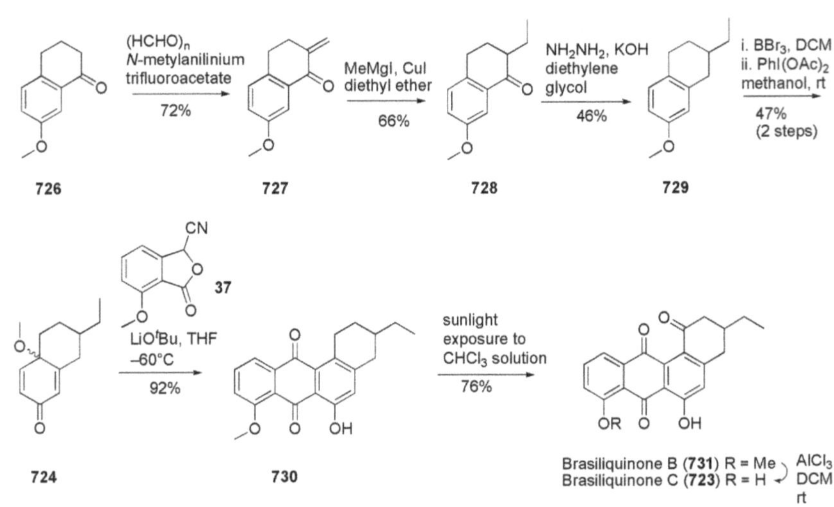

SCHEME 4.148 Mal total synthesis of brasiliquinone C (**723**) [1999 JCSPT 3167 Mal D].

In an attempt to conjugate a steroid to an anthraquinone motif, estrone (**732**) was deconjugated with a hypervalent iodine reagent, i.e., phenyliodonium ditrifluoroacetate along with HF-Py to give fluoro derivative (**733**). This was submitted to annulation with the parent Hauser donor (**11**) to furnish hexacyclic angucycline-like conjugate (**734**) (Scheme 4.149). This reaction exemplifies two aspects of Hauser annulation. It is selective to the less hindered double bond, and a suitably placed fluorine in a 2,5-cyclohexadienone unit can serve as a nucleofuse.[110]

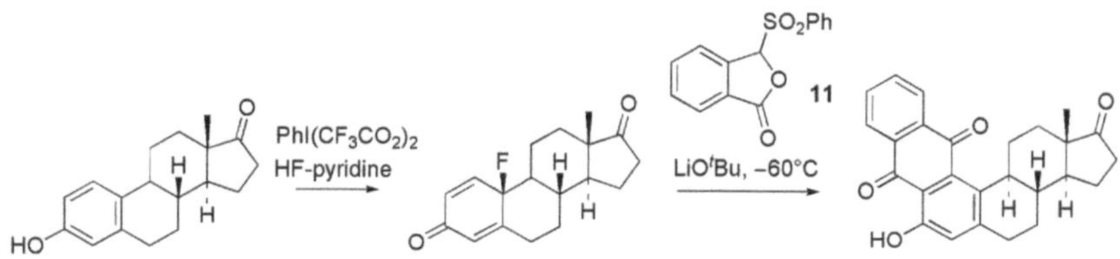

732 733 734

SCHEME 4.149 Synthesis of a steroid-angucycline conjugate [2007 THS 2635 Mal D].

4.5 ARYNE PRECURSORS

"Arynes" are referred to as derivatives of benzyne (1,2-dehydrobenzene) and their heterocyclic analogues (heteroarynes). Structurally, they are considered to have a CC triple bond within the aromatic ring. As a consequence, they are extremely reactive, and they undergo a variety of reactions even at low temperatures.[111] The typical reactions are: (i) nucleophilic additions, (ii) pericyclic reactions, and (iii) transition metal-catalyzed reactions. The commonest reactions, shown in Scheme 4.150, are additions of nucleophiles and [4+2] cycloadditions. Reaction 1 is a nucleophilic addition of benzyne (**735**), which is formed through the loss of triflic acid from (**736**), and the product is (**737**). The reaction of bromophenyl triflate (**738**) with diphenylisobenzofuran (**739**) in the presence of ^{n}BuLi gives tetracycle (**740**) in 90% yield via [4+2] cycloaddition the classic benzyne intermediate (**735**). Reaction 2 can also be classified as an annulation reaction.

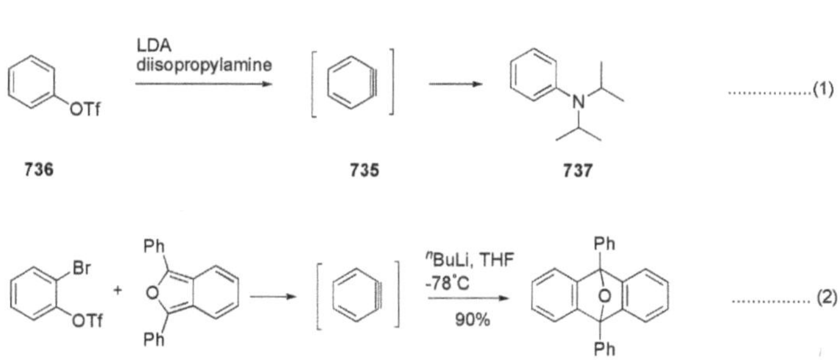

736 735 737

738 739 735 740

SCHEME 4.150 Typical reaction of arynes.

That the arynes (e.g., **735**) can be intercepted by a phthalide anion resulting in an annulation was first discovered in 1979 by Sammes and Dodsworth.[112] In 1981, Russell and Warrener reported that Hauser donors (**2**) can also be engaged with benzynes for direct and one-pot synthesis of anthraquinones. The sulfone substituent at C3 of Hauser donors (**2**) provides necessary oxidation level to give quinone products. As shown in Scheme 4.151, the reaction between phthalide (**24**) and (**741**) in the presence of LDA provides anthraquinone (**742**) in a one-pot operation. LDA serves two purposes: it removes HBr to produce benzyne (**743**) from (**742**), and promotes formation of carbanion (**744**) from the phthalide (**24**). The resulting carbanion (**744**) undergoes annulation to form intermediates (**745, 746**, and **747**). The intermediate (**747**) undergoes fragmentation to anthraquinone (**742**). The annulation can also be described as a two-step process via direct formation of (**746**).

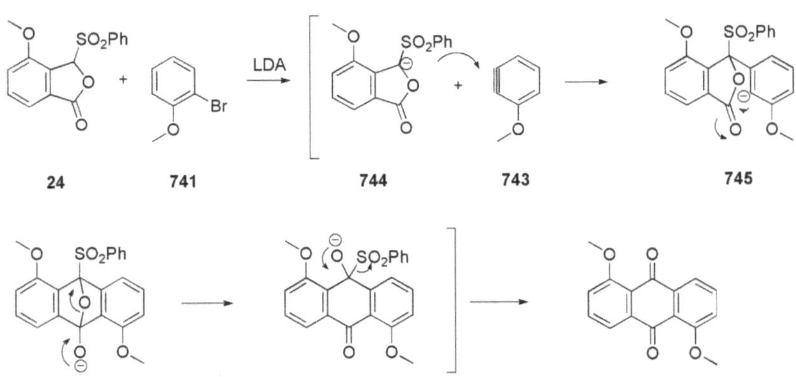

SCHEME 4.151 Russell annulation with benzyne precursors [1981 JCSCC 108 Russell RA].

The Biehl group advanced the strategy to the synthesis of various naturally occurring anthraquinones, and preferred to use 3-cyanophthalides (**3**) as the donors. For the synthesis of rubiadin (**748**) and damnacathol (**749**), anthraquinone (**750**) was prepared by the annulation of the cyanophthalide (**14**) with bromobenzene (**751**) in the presence of LDA (Scheme 4.152). Under the benzyne-forming conditions (LDA, THF, −78°C), the bromoarene (**751**) transformed to benzyne (**752**) and the cyanophthalide (**14**) to its anion. These two reactive species underwent annulation to form (**750**), which served as the common intermediate for the synthesis of both the natural products (**748** and **749**). Demethylation of (**750**) with boron tribromide gave rubiadin (**748**). CAN oxidation of the anthraquinone (**750**) in acetic acid followed by selective demethylation of the resulting alcohol (**753**) with HBr furnished damnacathol (**749**).[113]

Similarly, the cyanophthalide (**14**) reacted with the bromoarene (**754**) under the benzyne-forming conditions (LDA, THF, −78°C) via the reactive intermediates (**755** and **756**) to give morindaparvin A (**757**) in 65% yield. The oxolane oxygen in (**754**) did not promote any lithiations (Scheme 4.153).[114]

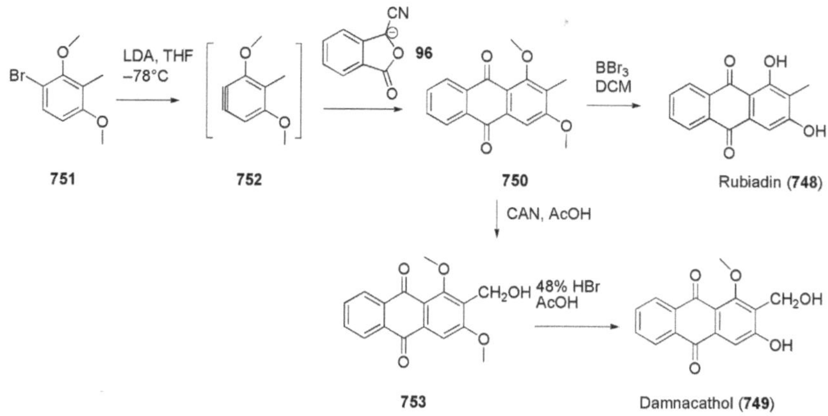

SCHEME 4.152 Synthesis of rubiadin (**748**) and damnacathol (**749**) [1995 JNP 1970 Biehl ER].

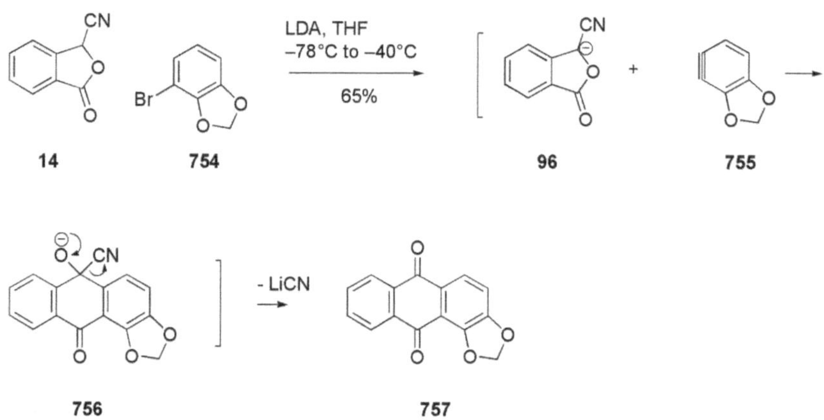

SCHEME 4.153 Biehl synthesis morindaparvin A (**757**) [1989 JNP 1357 Biehl ER].

Naphthalynes, generated in situ from the bromonaphthalenes by interaction with LDA at −78°C, react with the Kraus donor (**14**) to give benz[*a*]anthracenones. For example, the reaction of (**14**) with bromonaphthalene (**758**) gave benz[*a*]anthraquinone (**759**) in 47% yield via (**760**).[115] No linearly fused product was formed from the potential 2,3-naphthalyne intermediate (Scheme 4.154).

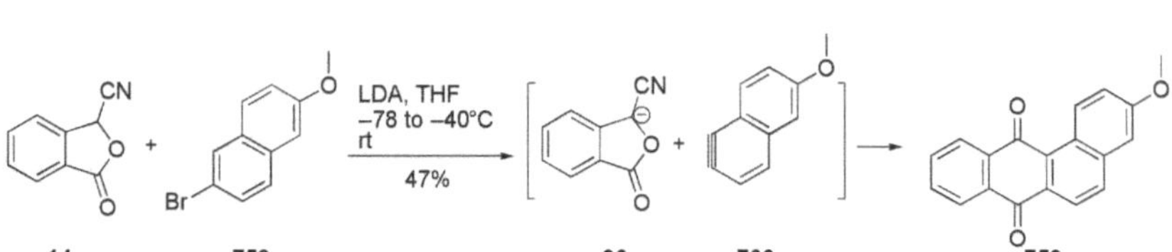

SCHEME 4.154 Biehl synthesis of benz[*a*]anthracene-7,12-dione [1993 SYS 885 Biehl ER].

Conjugation of a naphthoquinone to a pyrene can be achieved by using bromopyrenes as pyrenyne precursors. Thus, the reaction of (**761**) with the cyanophthalide (**762**) in the presence of LDA afforded dibenzo[*de,qr*]naphthacene-9,14-dione (**763**) via (**764**).[116] The *peri*-methyl groups present in the donor (**760**) do not seem to have any deleterious effect on the yield of the reaction (Scheme 4.155).

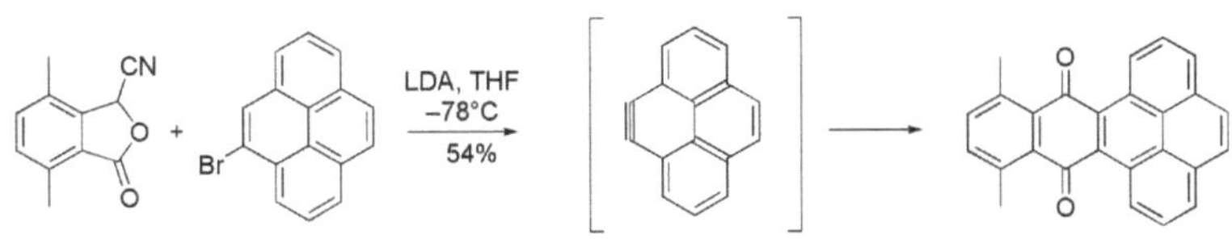

762 761 764 763

SCHEME 4.155 Annulation with a bromopyrene [1995 SYS 827 Biehl ER].

With the aryne-methodology, Kaiser et al. could achieve a regioselective synthesis of oxygenated anthracene-1-carboxaldehyde ketal (**765**) by two complementary routes (Scheme 4.156) in their synthetic study of mumbaistatin (**766**). These two routes differ in the donor-acceptor properties of the annulating agents. The annulation between bromoarene (**767**) and cyanophthalide (**768**) is more efficient than that between cyanophthalide (**769**) and bromoarene (**770**), in giving anthraquinone (**765**).[117]

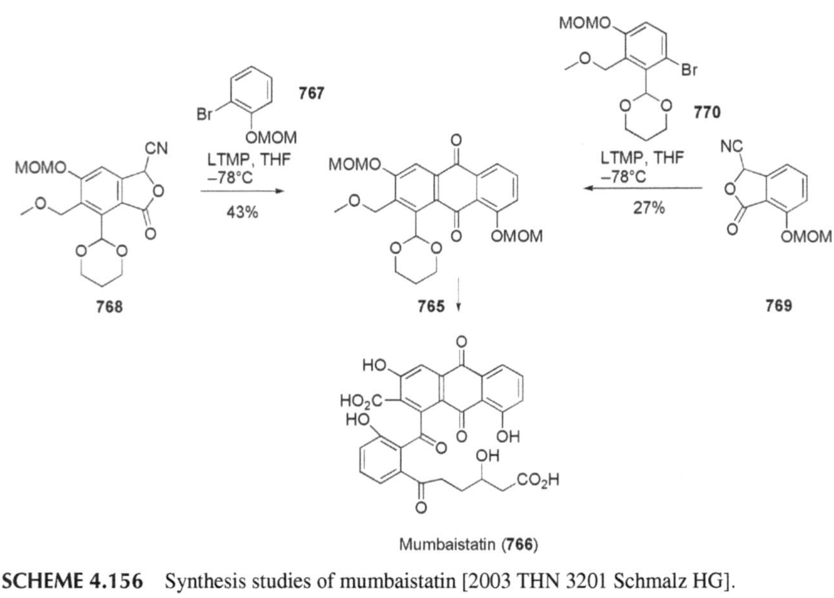

SCHEME 4.156 Synthesis studies of mumbaistatin [2003 THN 3201 Schmalz HG].

The annulation of cyanophthalides (**3**) with hetarynes has been a successful methodology for the entry to azaanthraquinones. The reaction of the phthalide (**14**) with pyridyne precursor (**771**) in the presence of LDA proceeded through

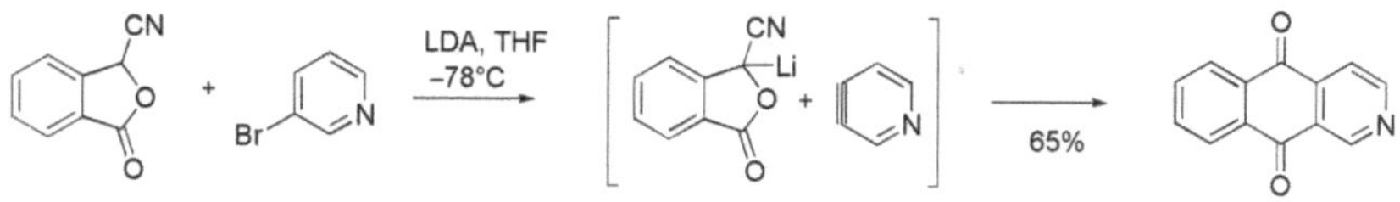

14 771 96 772 773

SCHEME 4.157 Annulation with a pyridyne [1988 HCS 2643 Biehl ER].

intermediates (**96**) and (**772**) to yield 2-azaanthraquinone (**773**) in 65% yield. Interestingly, 3-bromopyridine (**771**) provided only 3-pyridyne and not the 2-pyridyne (Scheme 4.157).[118]

As inherent with reactivities of the most arynes, the regioselectivity is not very pronounced. The cyanophthalide (**37**) provided mixtures of azaanthraquinones (**774** and **775**) through pyridyne (**772**). The major products are those arising out of preferential attack of the phthalide anion at C4 of the 3,4-hetaryne (Scheme 4.158).

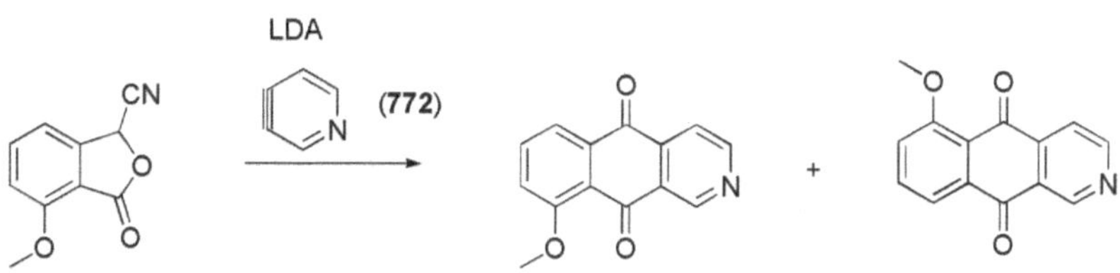

37 **775** (31%) **776** (15%)

SCHEME 4.158 Annulation of a pyridyne with an unsymmetrical cyanophthalide [1988 HCS 2643 Biehl ER].

4.6 ALKENYLSULFONES

Although the electron withdrawing effect of a sulfone group is well-established, the consequent intermolecular Michael acceptor property of conjugated alkenylsulfones is not.[119] The first report on the use of an alkenyl sulfone as an acceptor in the Hauser annulation is due to Tso and Chen in 1995.[120] Subsequently, Mal et al. examined alkenyl sulfones (**776** and **777**) toward the Hauser annulation in connect with their studies on gilvocarcin. The annulation of the sulfone (**776**) with the Hauser donor (**22**) in the presence of LTB at −60°C furnished the desired product (**778**) through enolization of the intermediate diketo form (**779**). The similar reaction of the substituted sulfone (**777**) with the cyanophthalide (**14**) under similar conditions provided sulfone-free naphthoquinone (**780**) via loss of phenylsulfinate anion from (**781**) (Scheme 4.159).[121]

The reaction of phthalide sulfides (**3**) with more complex α,β-unsaturated sulfone (**782**) in the presence of NaHMDS gave sulfone-free naphthoquinones (**783**) through the intermediates (**784**) (Scheme 4.160), unlike normal cases. The activating sulfone group was eliminated as phenylsulfinate ion after the initial annulation under the reaction conditions. The most appropriate donor

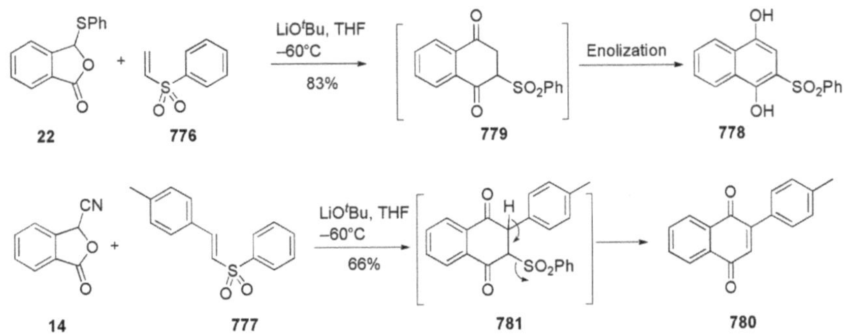

SCHEME 4.159 Annulation of simple alkenylsulfones [2005 PhD Thesis, IIT Kharagpur, Patra A].

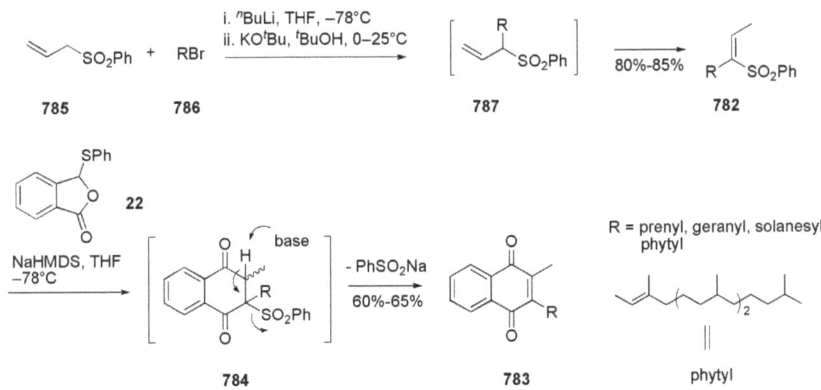

SCHEME 4.160 Tso synthesis of vitamin K by annulation a key step [1995 JCRS 104 Tso H-H].

for such annulations is 3-phenylthiophthalide (**22**). This study was extended to the synthesis of difficult-to-obtain K-vitamins in moderate to good yields. The reactions were performed using NaHMDS as the base. The methodology has been extended to the synthesis of cyclopentanaphthoquinone and tetrahydroanthraquinone. All the sulfones acceptors (**782**) were synthesized in two steps from allyl phenyl sulfone (**785**) by ^{n}BuLi-promoted alkylation with appropriate bromides (**786**) to give (**787**), followed by CC double bond isomerization of (**787**) with KO^{t}Bu in *tert*-butanol in the same flask.

The similar reaction of the styryl sulfone (**788**) with the cyanophthalide (**37**) was not straight forward. Three different annulation products (**789**, **790**, and **791**) were obtained in compliance with the initial Hauser annulation, and all of them were free from the phenylsulfonyl group (Scheme 4.161).[122]

In striking contrast to the reactivity of the sulfone (**788**), the more congested styryl sulfone (**792**) provided only one product (**793**) in excellent yield. The sulfone (**792**) was prepared by the Heck reaction of the diazonium salt of (**794**) with phenyl vinyl sulfone. Desulfonylation of (**793**) with Bu_3SnCl and sodium cyanoborohydride gave the tricycle (**795**) that features the structure of gilvocarcin antibiotics (Scheme 4.162).

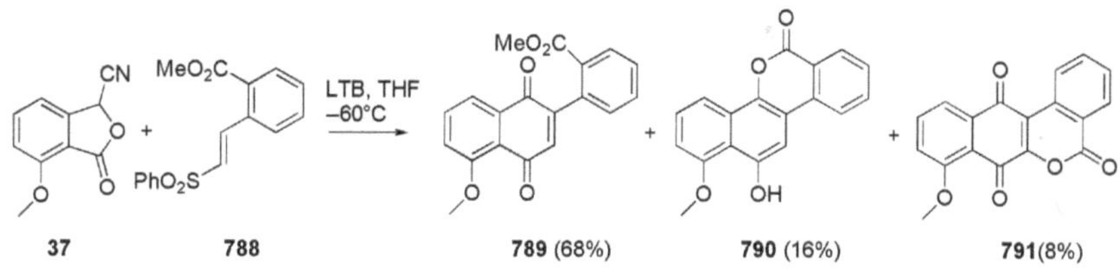

SCHEME 4.161 Annulation with a styryl sulfone [2005 JOC 9017 MaL D].

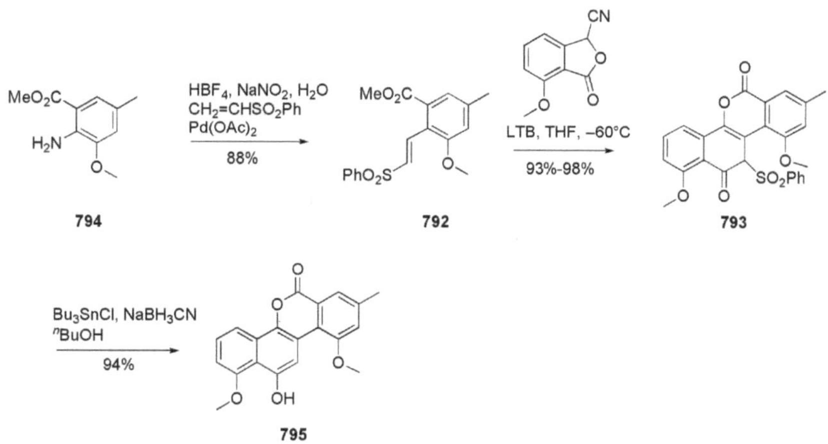

SCHEME 4.162 Model annulation study with a styryl sulfone for gilvocarcin [2005 JOC 9017 MaL D].

4.7 NITROALKENES

Nitroalkenes are established Michael acceptors.[123] But, their reactivity toward the Hauser annulation has not systematically been investigated. In the presence of a quinine base, the reaction between the phthalide (**796**) and nitrostyrenes (**797**) at room temperature results in the Michael adducts (**798**) (Scheme 4.163).[124]

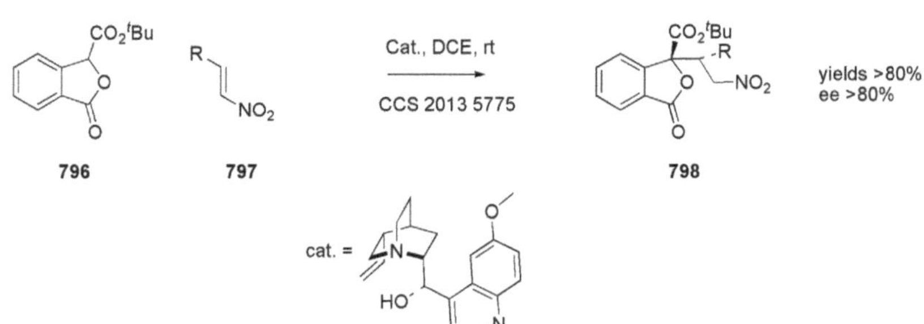

SCHEME 4.163 Michael addition of a phthalide to nitroalkenes [2013 CCS 5775 Lu Y].

The first report on the annulation of a nitroalkene appeared in the work of Vasella et al. For the total synthesis of (−)-cryptosporin (**799**), representing a naphtho[2,3-*b*]pyrandione nucleus, Brade and Vasella explored the Hauser annulation with 1-nitroglycals. The reaction of the sulfonyl phthalide (**11**) with glycal (**800**) in the presence of LDA at −70 to 40°C gave the expected product (**801**) in 89% yield. The same reaction, when quenched at room temperature yielded Michael adduct as the major product, i.e., the protonated forms of (**802**) and the annulation product (**801**) in 13% yield (Scheme 4.164).[125]

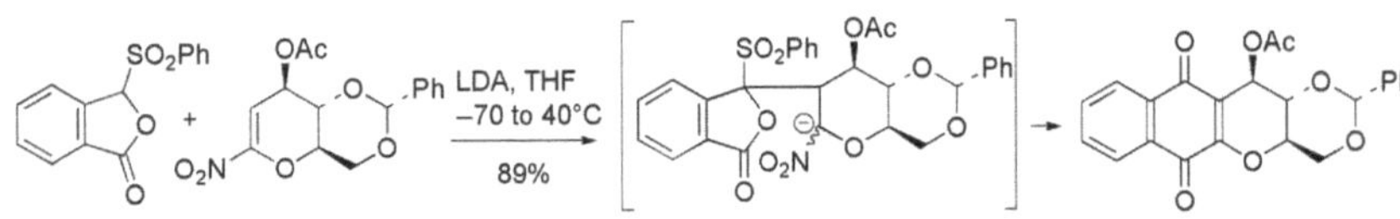

SCHEME 4.164 Hauser annulation with a 1-nitroglycal [1989 HCA 1649 Vasella A].

The similar annulation between (**803**) and (**804**) led to the synthesis of (−)-cryptosporin (**799**), the enantiomer of the fungal metabolite (+)-cryptosporin. The required nitroglycal (**804**) was prepared from L-fucose. MEM (methoxyethoxymethoxy) sulfonylphthalide (**803**) was prepared from the corresponding benzamide via *ortho*-lithiation followed by formylation, phenylthiolation and peracid oxidation. The annulation product (**805**) was obtained in 65% yield. Deketalization of the product (**805**) resulted in the formation of (−)-cryptosporin (**799**) in 62% yield (Scheme 4.165).

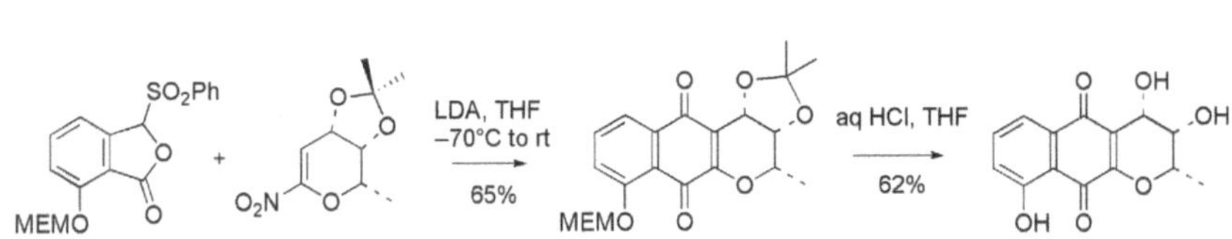

SCHEME 4.165 Synthesis of (−)-cryptosporin (**799**) [1989 HCA 1649 Vasella A].

With nitrostyrene (**806**), the annulation of the cyanophthalide (**37**) proceeded in moderate yield (45%) to give quinone (**807**). Although the corresponding quinol product (**808**) was expected, the product (**807**) was formed by a rapid aerial oxidation (Scheme 4.166).[126]

Namboothiri et al. advanced the annulation chemistry of nitrostyrenes to a useful general synthesis of 2-substituted naphthoquinones (**809**) under very mild conditions. In the presence of K_2CO_3 in THF at room temperature, 2-arylnaphthoquinones (**810**) are obtained at 40–52% yields along the Michael adducts (**811**) as minor products in the reaction between the phthalide sulfone (**11**) and nitrostyrenes (**809**). Unlike Gould's work (Scheme 4.166), the Namboothiri annulation involved the loss of HNO_2 (Scheme 4.167).[127]

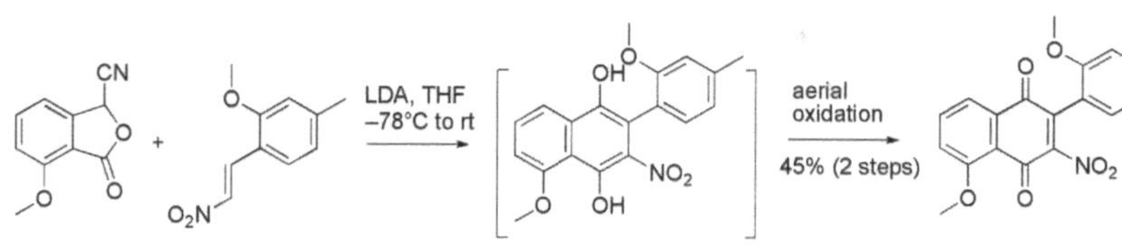

37 806 808 807

SCHEME 4.166 Hauser annulation with a nitrostyrene [1991 JOC 2289 Gould SJ].

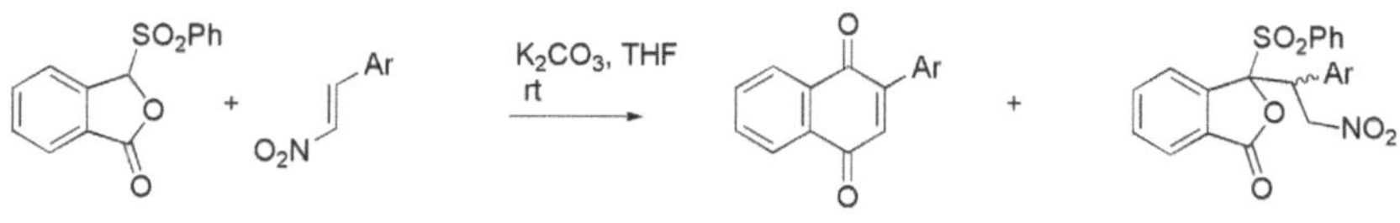

11 809 810 811

SCHEME 4.167 Hauser annulation with nitroalkenes [2016 EJOC 3316 Namboothiri INN].

When extended to *ortho* hydroxynitrostyrenes (**812**), the Namboothiri annulation proceeded through a cascade of anionic reactions to furnish indenofurans (**813**) in good yields. The reaction is proposed to involve formal [4+4] annulation forming intermediates (**814**). Further reaction of the intermediates (**814**) with phenylsulfinate anion involving an intramolecular cyclization forms the products (**813**) (Scheme 4.168).

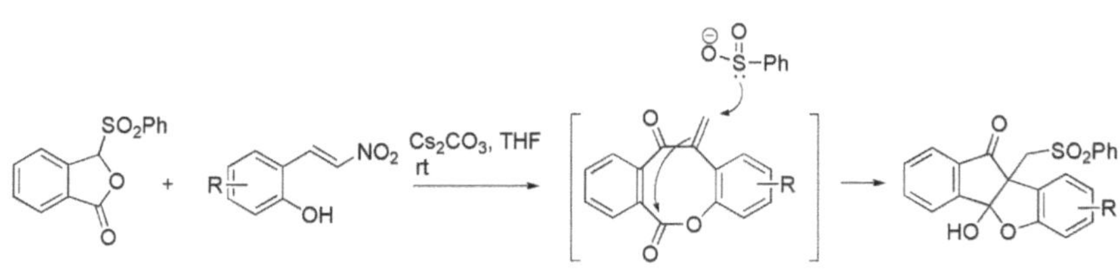

11 812 814 813

SCHEME 4.168 Hauser annulation with *ortho* hydroxynitrostyrenes.

More interestingly, the indenofuran sulfones (**813**) are sensitive to bases like K_2CO_3 or Cs_2CO_3 in acetone or THF. Under these conditions, they undergo ring-opening followed by cyclization to give sulfone-free spirocyclic phthalides. For example, the reaction of the sulfone with nitrostyrene (**815**) gave phthalide (**816**) (Scheme 4.169).

When the nitrostyrene annulation chemistry was applied to the nitrostyrenes (**817**), substituted at the alpha position, 2,3-disubstituted naphthoquinones (**818**) with a butanone appendage were obtained through intermediates (**819**) (Scheme 4.170).[128]

The quinone products (**818**) serve as versatile building blocks. They undergo cyclization with primary and secondary amines in the presence of PTSA to form aminophenanthrenes (**820**). On the other hand, they provide benzoquinolines (**821**) on reaction with ammonium acetate in refluxing acetic acid (Scheme 4.171).

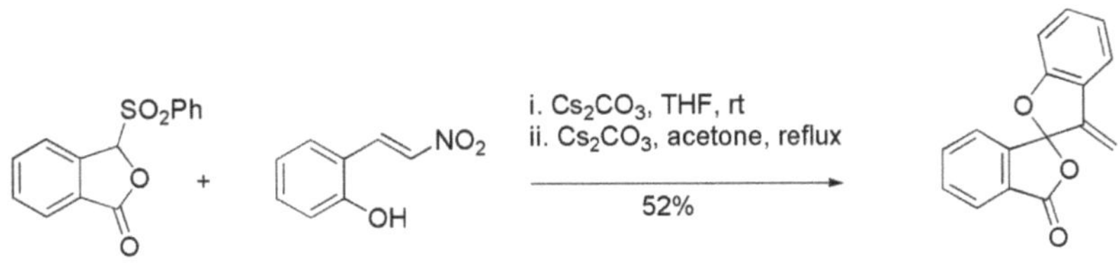

11 815 816

SCHEME 4.169 Base catalyzed rearrangement of indenofurans to spirocyclic phthalides.

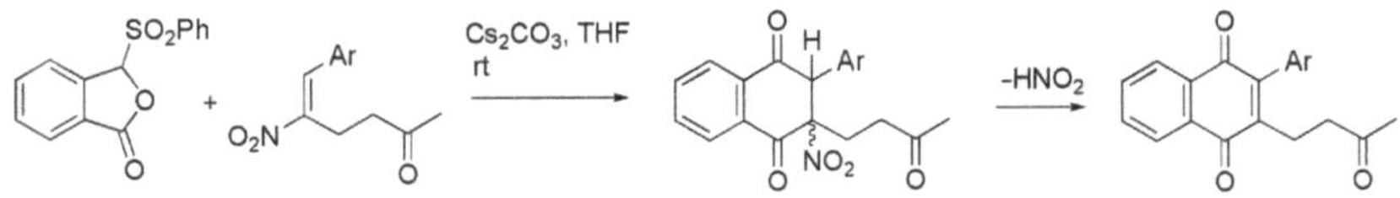

11 817 819 818

SCHEME 4.170 Annulation with substituted nitrostyrenes [2017 ORS 4283 Namboothiri INN].

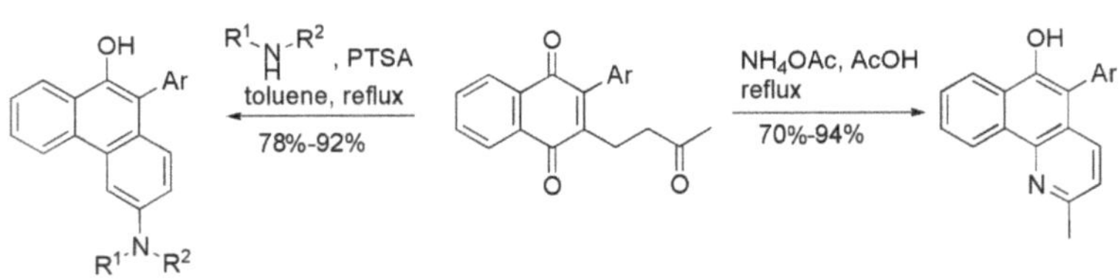

820 818 821

SCHEME 4.171 Synthesis of aminophenanthrenes and benzoquinolines [2017 ORS 4283 Namboothiri INN].

4.8 MISCELLANEOUS HAUSER DONORS AND ACCEPTORS

This section deals with (i) the uncommon/miscellaneous Hauser donors, such as naphthalides, thiophthalides, 3-isocyanophthalides, xanthones, isooxazoles, and furoindolones and (ii) less frequently used Hauser acceptors: alkynylphosphonates and cyclopropenes. The most obvious yet uncommon donors are benzene ring homologated phthalides, i.e., naphthalides, which can be viewed as benzophthalides. They can be broadly classified into three categories (**822–824**), as shown in Fig. 4.25.

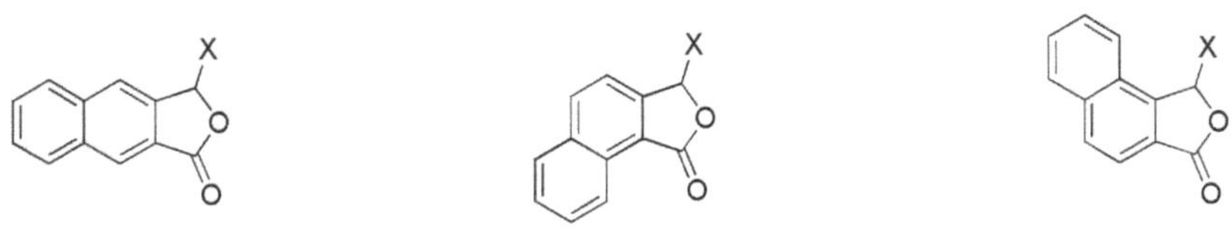

X = SPh, SO_2Ph, CN, etc.

822 823 824

FIG. 4.25 Types of naphthalides with a nucleofuge at the benzylic positions.

The annulation chemistry of the linearly naphthalides (**822**) was studied by the groups of Hauser, Angle, Rho, Nicolaou, and Mal. The work of the first three groups was in connection with the chemistry of anthracyclines. In a study directed toward the understanding of the mechanism of action of the anthracyclines, Angle et al. resorted to the Hauser annulation of the naphthalide (**825**) to produce the quinone methide (**826**). The reaction of (**825**) with cyclohexenone (**285d**) in the presence of LDA furnished quinol (**827**), which was converted into its pivaloate derivative (**828**). $NaBH_4$ reduction of (**828**) gave the deoxygenated product (**829**). The formation of the aimed quinone methide (**826**) was accomplished by oxidation using silver(I) oxide. Characterization of the quinone methide (**826**) was carried out by analysis of ^{1}H NMR data as well as its reaction with various nucleophiles. The reaction of (**826**) with thiophenol gave the conjugate addition product (**830**) as the single diastereomer (Scheme 4.172).[129]

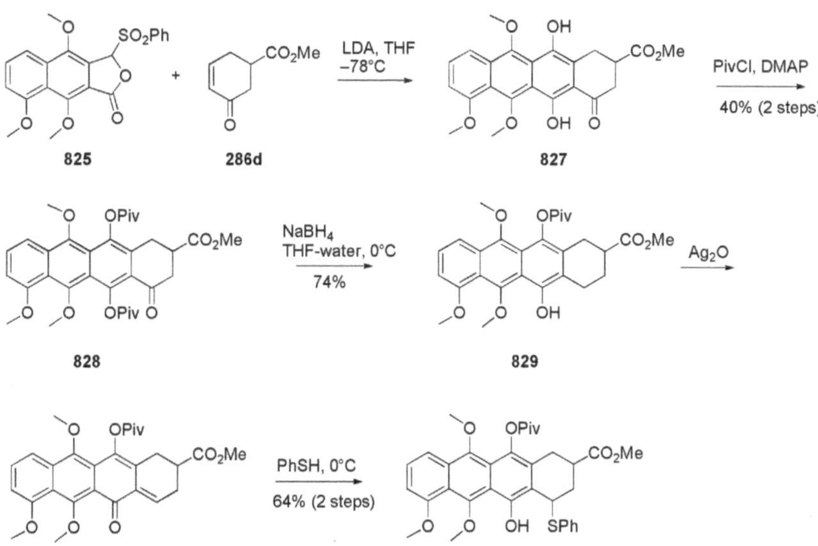

SCHEME 4.172 Annulation of a Hauser naphthalide donor [1997 JOC 5884 Angle SR].

In the context of the total synthesis of galtamycin (**831**), Hauser and Zheng explored the [4+2] anionic annulation reactivity of the naphthalide (**832**) with the *p*-benzoquinone monoketal-containing C-glycoside (**833**) (Scheme 4.173). Their reaction in the presence of LTB expectedly furnished tetracene glycoside (**834**) in 84% yield. Selective deoxygenation of (**834**) through its triflation (LTB, Tf_2O) into (**835**), and subsequent hydrogenation of (**835**), yielded optically active galtamycin (**831**).[130]

In connection with the work on the synthesis of analogues of uncialamycin (**650**), an exceptionally potent antibiotic, the cyanonaphthalide (**836**) was submitted to annulation with aminal (**837**). The annulation between them followed by deblocking of the alloc group with $Pd(Ph_3P)_4$, and desilylation of the annulation product provided uncialamycin analog (**838**) (Scheme 4.174).[102]

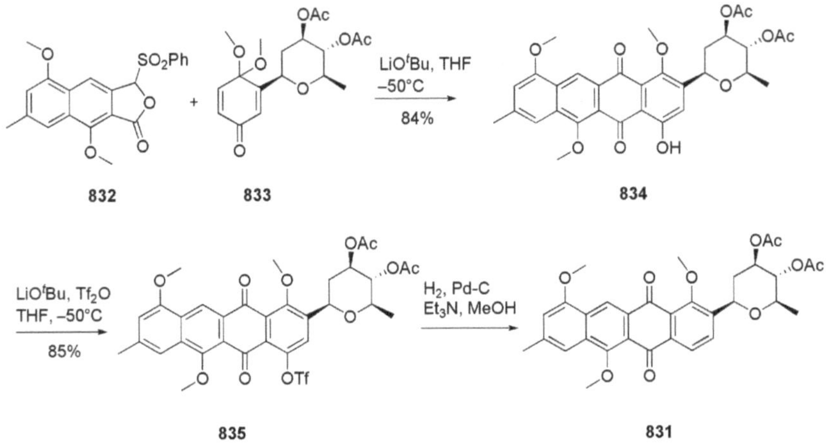

SCHEME 4.173 Naphthalide annulation for the synthesis galtamycin [2010 PhD Thesis, SUNY Zheng R].

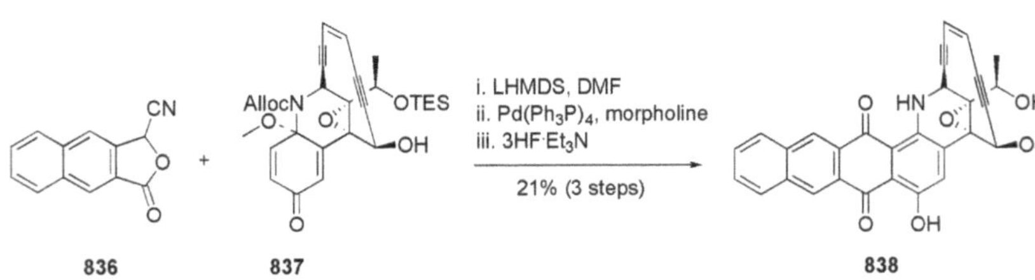

SCHEME 4.174 Annulation of a naphthalide cyanide donor [2016 JACS 8235 Nicolaou KC].

When extended to the allene carboxylate (**93b**), annulation of the naphthalide (**839**) proceeded to give products (**840** and **841**) in accordance with results obtainable from simple phthalides (Scheme 4.175). The required naphthalide (**839**) was prepared from simple naphthalide (**842**). This was prepared by annulation of the parent phthalide (**843**) with the parent allene carboxylate (**93a**) in the presence of LDA, followed by *O*-methylation with MeI-K_2CO_3. Incorporation of the cyano group at C3 of (**842**) required three steps: (i) $AlCl_3$ assisted lactone-opening with diethylamine, giving (**844**), (ii) PPC oxidation of alcohol (**844**) to aldehyde (**845**), and (iii) TMS-promoted cyanolactone formation to (**839**).[21]

A tetracycle Hauser donor can also be annulated to construct tetracycles like 5,12-tetracenediones. LTB promoted annulation of the linearly fused phthalide sulfone (**846**) with methyl vinyl ketone (**181**) yielded tetracenone derivative (**847**) in 85% yield (Scheme 4.176).[131]

The reactivity of angular naphthalides of the type II (Fig. 4.25) has recently been investigated, in brief, by the Mal group in connection with their work on salviamines. Cyanophthalide (**848**) was prepared from naphthalene-1-carboxylic acid in four steps. Yu lactonization of (**849**) with dibromomethane in the presence of $Pd(OAc)_2$ gave angular naphthalide (**850**) in 70% yield. Benzylic bromination of (**850**) with NBS followed by hydrolysis of the bromo

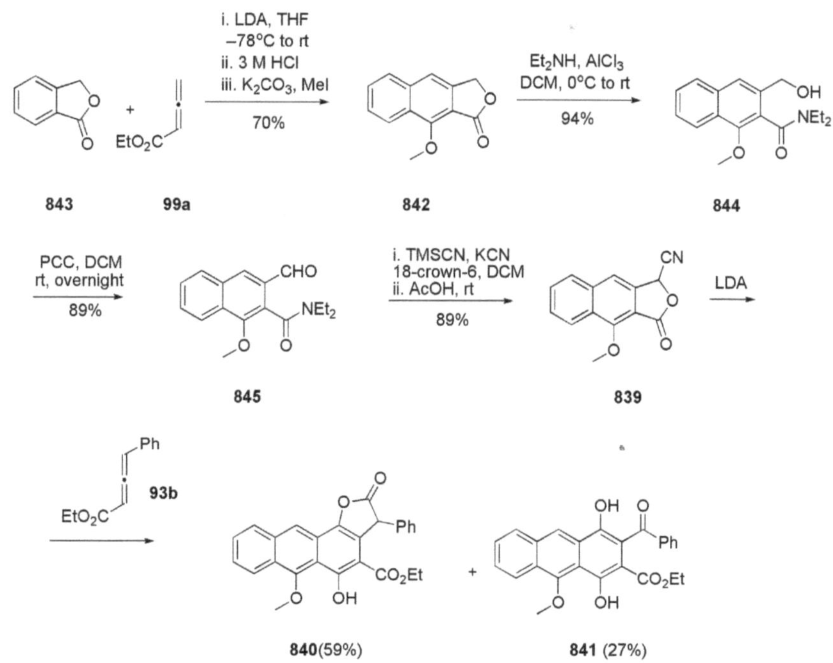

SCHEME 4.175 Annulation of a linear napthalide [2018 JOC 0000 Mal D].

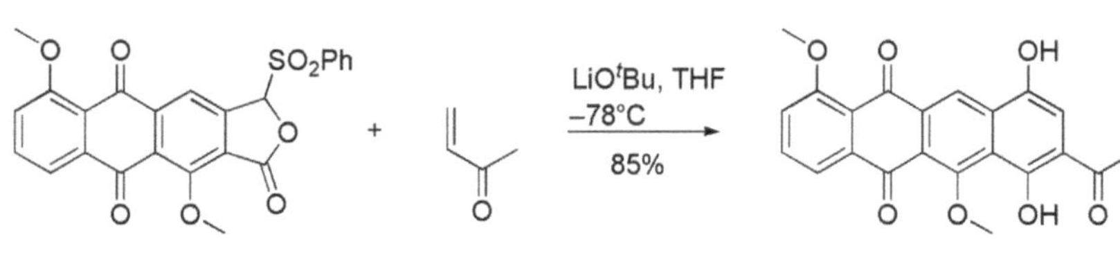

SCHEME 4.176 Annulation with a tetracyclic donor [1993 JKCS 141 Rho YS].

derivative (**851**) afforded naphthofuranone (**852**) in 54% overall yield. It was readily converted to (**848**) via formation of cyanohydrins with HCN followed by lactonization with Vilsmeier reagent. Annulation of the donor (**848**) with acrylate (**853**) afforded the annulation product (**854**) and its oxidized form (**855**) in a combined yield of 81% (Scheme 4.177). The sole example reflects comparable donor reactivity of phthalides (**2–4**) and angular naphthalides (e.g., **848**).[23]

In their study on the synthesis of cervinomycin A_2, the Hauser group explored annulation reactivity of the more complex angular phthalide sulfone (**856**). Only modest yield of the annulation product was observed for the annulation of napthalide (**856**) with methyl crotonate in the presence of LTB. The annulation product (**857**) was oxidized with Fetizon reagent to give quinone (**858**) in 41% overall yield (Scheme 4.178).[132]

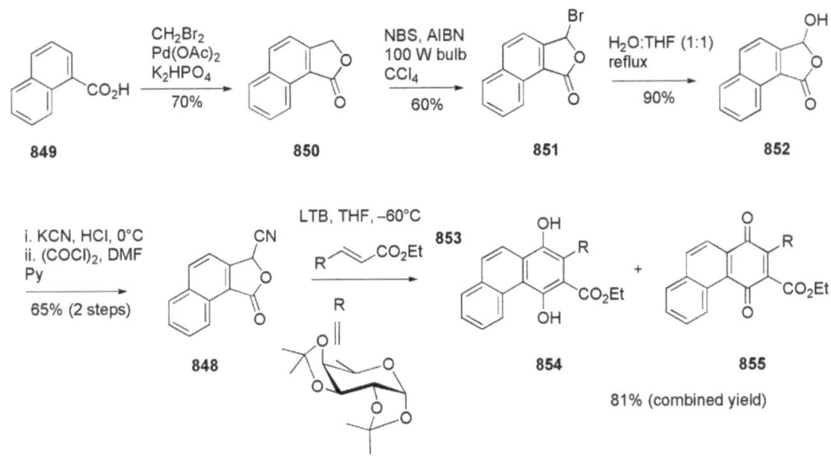

SCHEME 4.177 Annulation of an angular napthalide [2016 SYS 1235 Mal D].

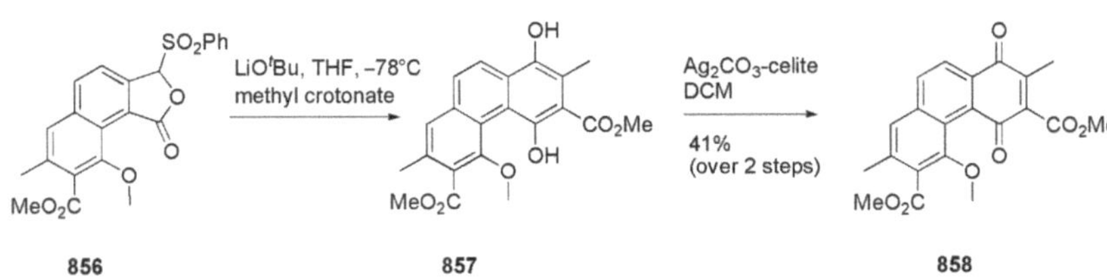

SCHEME 4.178 Annulation of an angular naphthalide [1997 unpublished results Hauser FM].

While the expected results on the [4+2] annulations were obtained with naphthalides of types I and II, the reactivity of naphthalides of type III (Fig. 4.25) is distinctly different. In context of the total synthesis of chlorocyclinone A, a model study was carried out with naphthalide (**859**) of type III (Scheme 4.179). Its reaction with 2-cyclohexenone under a variety of conditions provided only the Michael adduct (**860**), the ring closure of which failed to yield (**861**) (Scheme 4.179).[133]

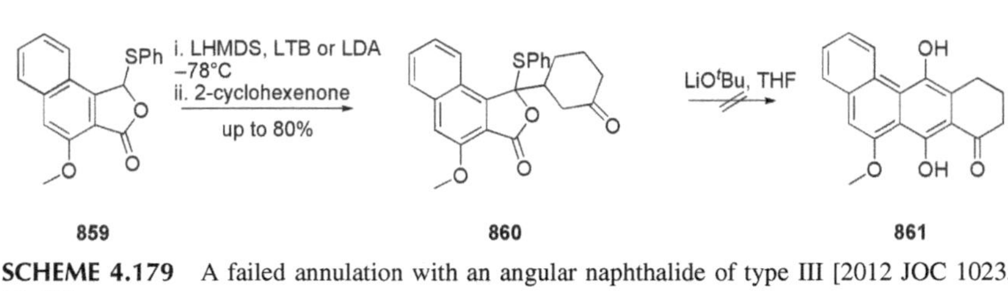

SCHEME 4.179 A failed annulation with an angular naphthalide of type III [2012 JOC 10235 Mal D].

4.8.1 Chemistry of Isocyanophthalides

The concept of using 3-isocyanophthalides (**862**) as the Hauser donors was recently evaluated to replace 3-cyanophthalides on the ground of toxicity of KCN in their preparations. The isocyanides (**862**) were accessible by the route shown in Scheme 4.180. Treatment of phthalaldehydic acids (**863**) with formamide (neat) at 90°C afforded 3-formamidophthalides (**864**) in good yields. Treatment of phthalides (**864**) with $POCl_3$ in the presence of triethylamine provided isocyanophthalides (**862**), which were fairly stable for the study of the anionic annulations.[8]

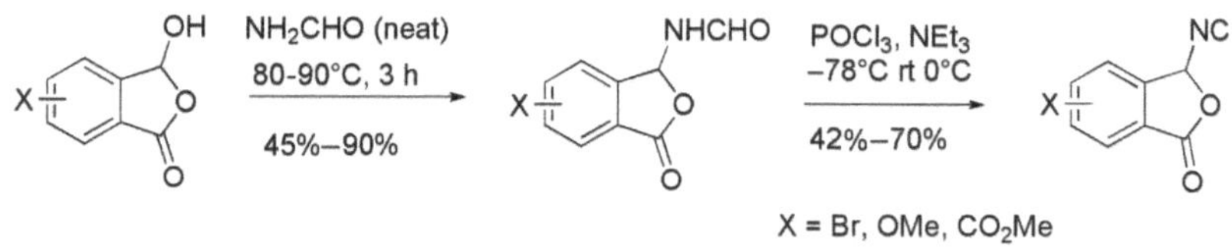

863 **864** **862**

SCHEME 4.180 Synthesis of 3-isocyanophthalides [2015 SYS 2473 Mal D].

As is evident from Table 4.1, the annulations with isocyanides are, in general, inferior to those with the corresponding cyanophthalides.

TABLE 4.1 Annulation of Isocyanophthalides (862)

Donor	Acceptor	Product	%Yield (%Yield)[a] From the Corresponding Cyanophthalide)
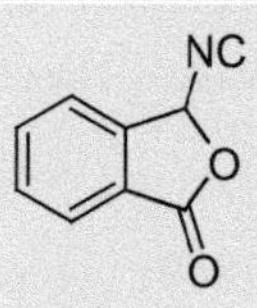	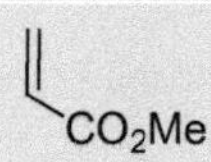	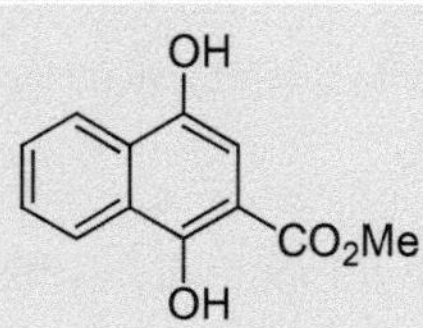	31 (68)
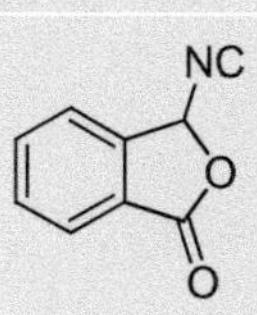	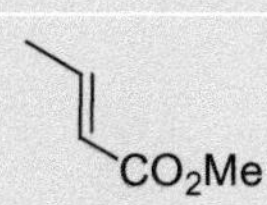	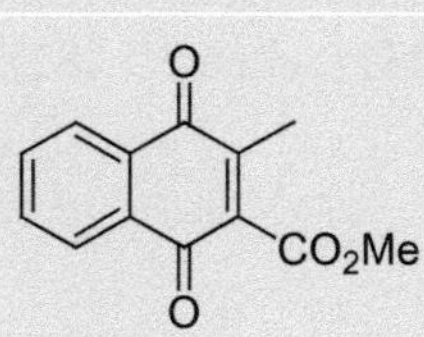	36 (73)
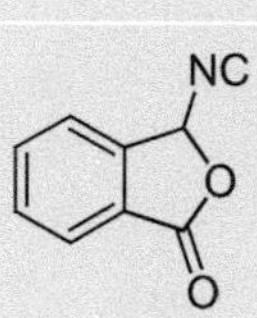	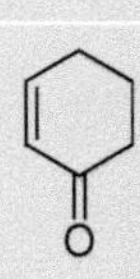	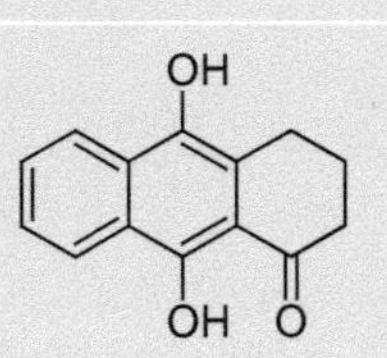	46 (75)

[a] *The yields in parentheses correspond to those obtained with the corresponding cyanophthalide (**14**).*

In 1993, the use of phthalide-3-phosphonates (**5**) in Hauser annulation was developed by Watanabe et al., and the yields of the annulations were, in most cases, excellent when carried out with LiO^{t}Bu as the base. Some of the known poor Michael acceptors like cyclopentenone (**233**) and methyl vinyl ketone (**181**) showed best compatibility with the Watanabe phosphonate donors. Unfortunately, the difficulties in preparation of the phthalide phosphonates rendered them less attractive. Brimble et al. indicated that the phthalides-3-phosphonates (**5**) may be comparable to the Hauser sulfone donors (**2**) in their reactivity.

Chemistry of thiophthalides: In 1994, Mal et al. introduced the sulfur analogs of the Hauser phthalide donors (**865** and **866**).[134] Their annulations with typical acrylate and enone acceptors are complicated with additional Michael addition of the initial annulation products (cf. **867**) or the oxidative dimerizations. However, the desired products (**868**) were obtained when the annulations were quenched with MeI. As shown in Scheme 4.181, the reactions of both the reagents (**865** and **866**) with cyclohexenones furnished sulfurated anthracenones (**868**) via the intermediates (**867**) in moderate to good yields.

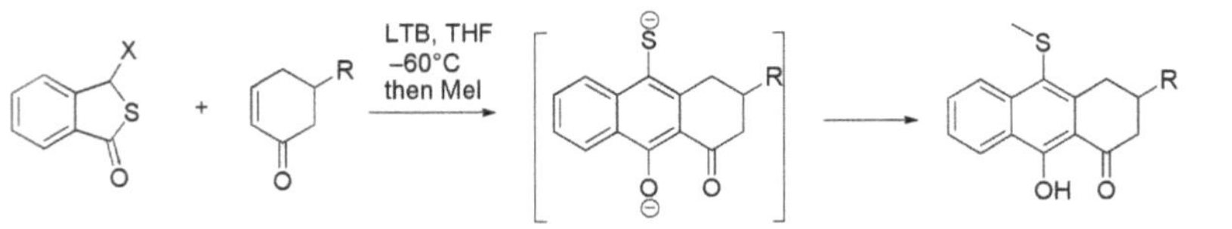

865: X = SPh
866: X = SO_2Ph
X
867
868a: R = H (70% from sulfide) (72% from sulfone)
868b: R = CO_2Me (56% from sulfide)
868c: R = CH_2OAc (38% from sulfide) (38% from sulfone)

SCHEME 4.181 Annulation reactivity 3-(phenylthio)thiophthalide and 3-(phenylsulfonyl)thiophthalide [1994 JCSPT 309 Mal D].

Similar annulation of the thiophthalide (**865**) with enedione (**869**), which seems to be stable to base-catalyzed destruction, gave excellent yield of the desired product (**870**) via (**871**) (Scheme 4.182).[135]

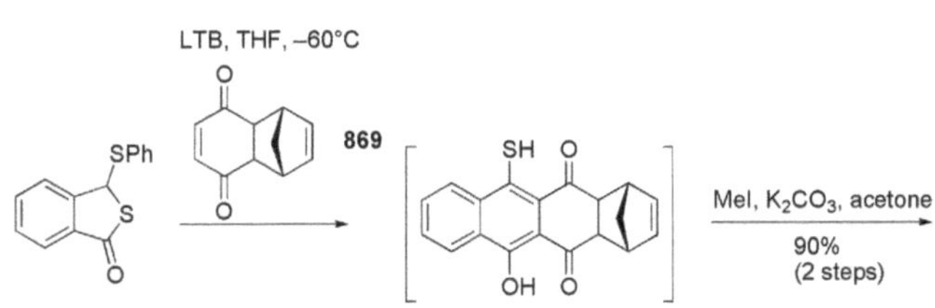

865 **871** **870**

SCHEME 4.182 Annulation of thiophthalide with a bicyclic enedione [1994 PhD Thesis, IIT Kharagpur, Gita Majumder].

To conclude, the annulation chemistry of thiophthalides (**6**) has been found useful as the Hauser donors in the Tatsuta total synthesis of hibarimicin discussed in Chapter 10.

4.8.2 Isoxazole Analog

Systems formed by the fusion of two rings via two neighboring atoms of the rings are known as fused systems. Thus, the phthalides, the basic requirements of Hauser annulation, are fused systems derived from benzenes and butenolides. While the extension of the Hauser reaction to the heterocyclic analogs of isobenzofuranones by replacing the benzene ring of the phthalide is quite conceivable, it was not reported until 1995 by Alguacil et al. In their pursuit to generate heterocyclic analogues of anthracyclines, they investigated the reactivity of the isoxazole-fused furanone sulfone (**872**), with functionalized quinone monoketals, and the corresponding annulated products could be prepared in moderate yields.[136] The reaction of the isoxazole sulfone (**872**) with benzoquinone monoketal (**873**) in the presence of LHMDS at −15°C for 15 days resulted in a mixture of naphthoisoxazole (**874**) in 40% and adduct (**875**) in 17% yield (Scheme 4.183). The atypical formation of the Michael adduct (**875**) through "O" is explained in terms of the nucleophilic attack of the resonance-stabilized anion through the oxygen at the 4-position of the donor.

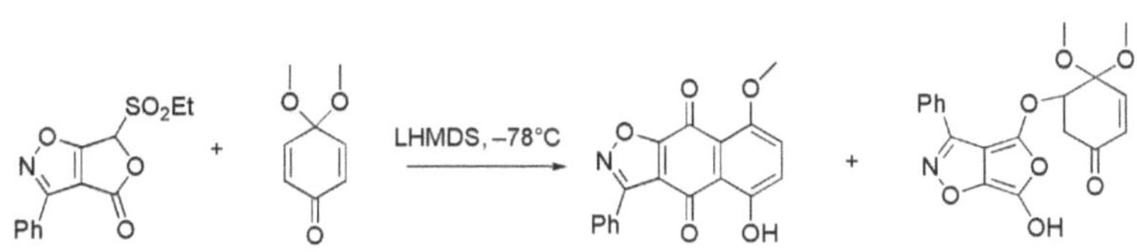

872 **873** **874** (40%) **875** (17%)

SCHEME 4.183 Annulation of a furo-oxazole donor [2000 HCS 1029 Paredes MC].

A similar reaction of the donor (**872**) with naphthoquinone monoketals (**876**) in the presence of LHMDS for 90 min provided a mixture of products (**877–879**) (Scheme 4.184).

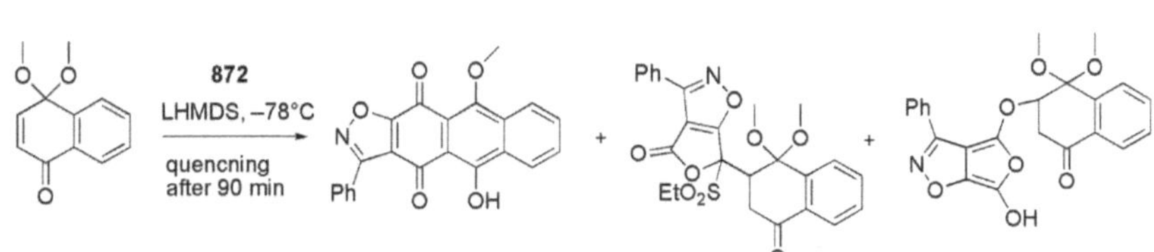

876 **877** (3%) **878** (14%) **879** (14%)

SCHEME 4.184 Annulation with a naphthoquinone acetal under a shorter duration [2000 HCS 1029 Paredes MC].

The same reaction, if continued for 5 days at −2°C, affords product (**877**) in 40% yield. Under the similar optimized reaction conditions, the isoxazole donor (**872**) was successfully annulated with quinone ketal (**880**) to give anthracycline analog (**881**) via (**882**). Notably, quinone ketal (**880**) needs no protection of its alcoholic hydroxy group (Scheme 4.185).

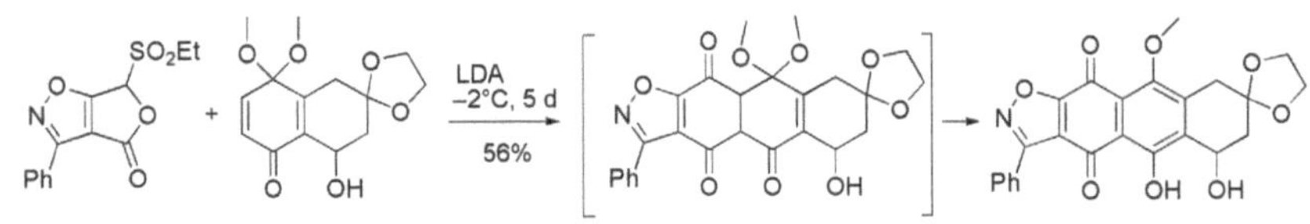

872 **880** **882** **881**

SCHEME 4.185 Synthesis of an anthracyclinone analog [2000 HCS 1029 Paredes MC].

4.8.3 Xanthone Donors

In assembling the xanthone skeleton of bikaverin (**553**), Hauser et al. employed condensation of the phthalide sulfone (**554**) with benzopyranone (**555**) (Scheme 4.112). Alternatively, they considered transposing of the reactivity of the reactants, and extended their work to the development of a general synthesis of xanthones by using benzopyranofuranone donors. Benzopyranofuranone (**883**) reacts smoothly with methyl crotonate (**21**) to give the corresponding xanthone (**884**) in 88% yield, after acetic anhydride quench. Likewise, acceptors (**264**, **551**, and **686**) were converted to xanthone derivatives (**885–887**, respectively), in good yields. It is interesting to note that the SPh group in (**883**) is sufficiently activating to assist the formation of incipient carbanion at its alpha position (Scheme 4.186).[137]

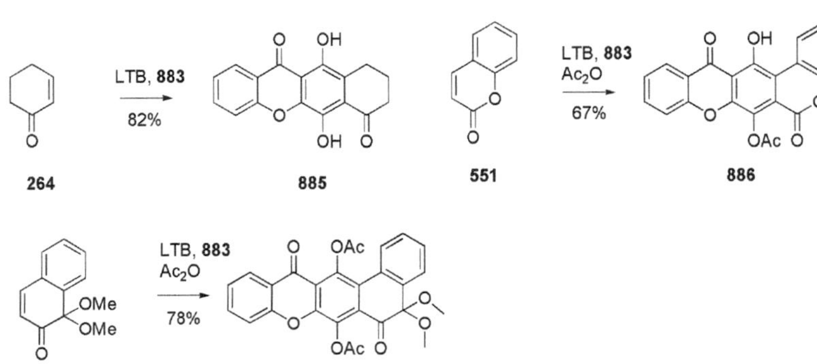

883 **21** **884**

264 **885** **551** **886**

686 (R = H) **887**

SCHEME 4.186 Annulation reactivity of xanthone analogs [2003 ORS 3753 Hauser FM].

4.8.4 Furoindolone Donors

The idea of extending the Hauser chemistry to an indole analog of phthalide (**888**) en route to heterocyclic quinones deemed trivial (Scheme 4.187) at early stages of investigations. Mechanistically, the annulation of furoindolones (**888**) with acceptors (**889**) would give quinols (**890**), which on oxidation are expected to give quinones (**891**). However, this chemistry was not straightforward. The initial investigation on the topic by Mal et al. was of limited success. The problems were twofold: (i) donors were somewhat difficult to be prepared and (ii) their annulation reactivity was sensitive to the nature of *N*-protecting groups in furoindolones (**888**).

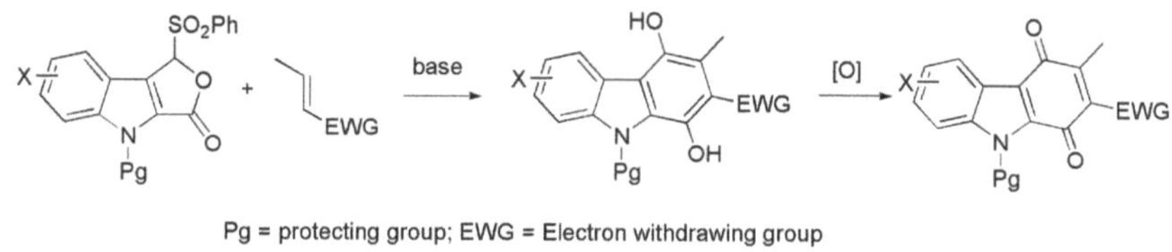

888 **889** **890** **891**

SCHEME 4.187 Proposed annulation route for the synthesis of indoloquinones.

Three 1-nucleofugal furoindolones (**888a–888c**) were prepared from 3-acetyl-dihydrofuran-2(3H)-one (**892**). Japp-Klingemann reaction of (**892**) with anilines followed by Fischer indolization of the resulting hydrazones furnished furoindolones (**893**), which were appropriately protected to give furoindolones (**894**). The bromo derivatives (**895**) were prepared by NBS bromination of (**894**). They were converted into their sulfones (**888**) by direct displacement of bromine by with sodium benzenesulfinate (Scheme 4.188). All the three donors (**888**), differing in the nature of the protecting, were examined for their annulation reactions. Ester protected donor (**888a**) readily underwent annulation with 2,5-cyclohexadienones as the acceptors in the presence of LDA to give the respective products. Under similar conditions, their annulation failed with simple acceptors like crotonates. These failures were attributed to the

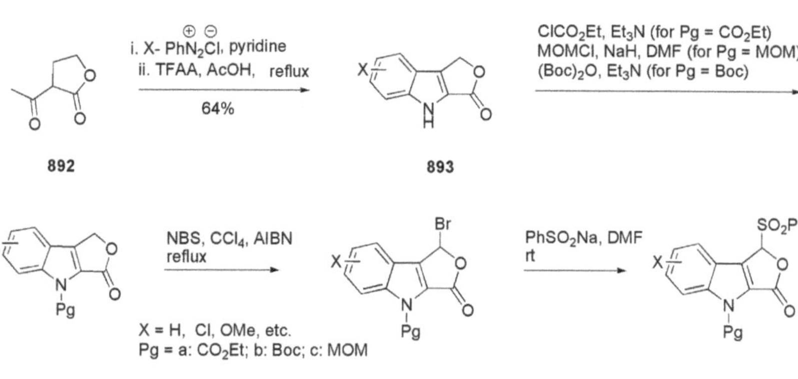

892 **893** **894** **895** **888**

SCHEME 4.188 Synthesis of furoindolone donors [2007 THN 3768 Mal D].

instability of the *N*-CO_2Et under the reaction conditions. The results with the Boc-protected donors (**888b**) were similar. In the annulation with *N*-MOM furoindolones (**888c**), the desired annulation products with intact MOM group were obtained in good yields.[138]

As shown in Scheme 4.189, the reaction of (**888c**) with methyl crotonate in the presence of LDA in THF afforded (**896**). During the work-up of the reaction mixture, the annulation product readily underwent aerial oxidation to give the quinone (**897**) (reaction 1). Similarly, furoindolone (**888c**) in reaction with cyclohexenone (**264**) gave tetracyclic quinone (**898**) in good yield via (**899**) (reaction 2). Benzo[*b*]carbazole-6,11-dione core is also accessible from benzyne precursor, i.e., bromobenzene. The reaction of (**888c**) with bromobenzene (**900**) with excess LDA furnished (**901**) via (**735**) in one pot without involving any oxidation reaction (reaction 3).[139]

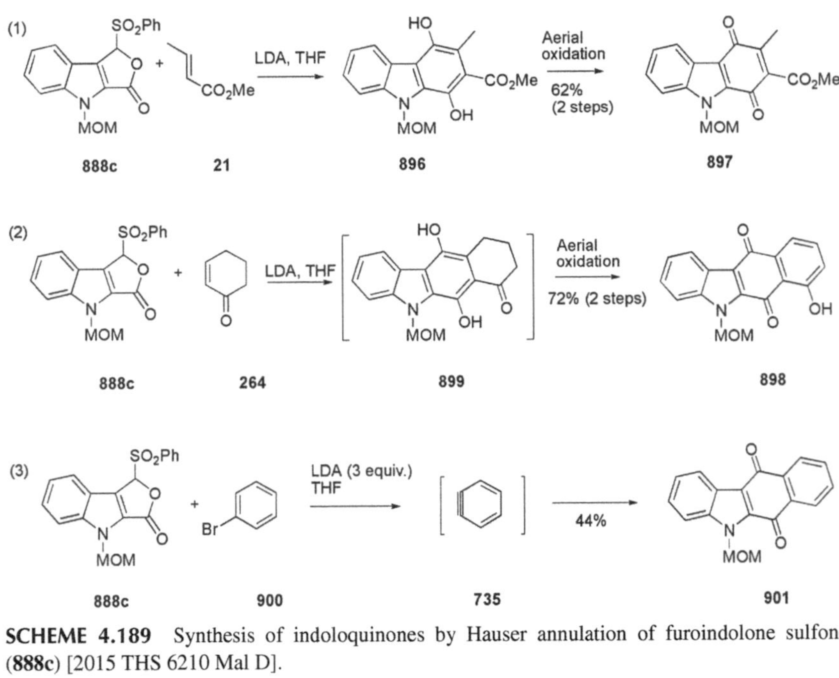

SCHEME 4.189 Synthesis of indoloquinones by Hauser annulation of furoindolone sulfone (**888c**) [2015 THS 6210 Mal D].

4.8.5 Miscellaneous Acceptors

This section deals with few atypical acceptors, namely, acrylamide, conjugated phosphonates, cyclopropane-fused enedione, and cyclopropene embedded in a bicyclic system. Acrylamides are not often encountered as Michael acceptors. The same is true of Hauser acceptors. The sole example executed by the Watanabe group is presented in Scheme 4.190. The reaction between the phthalide phosphonate (**35**) and acrylamide (**902**) yielded the annulation product (**903**) in 42% yield.[6]

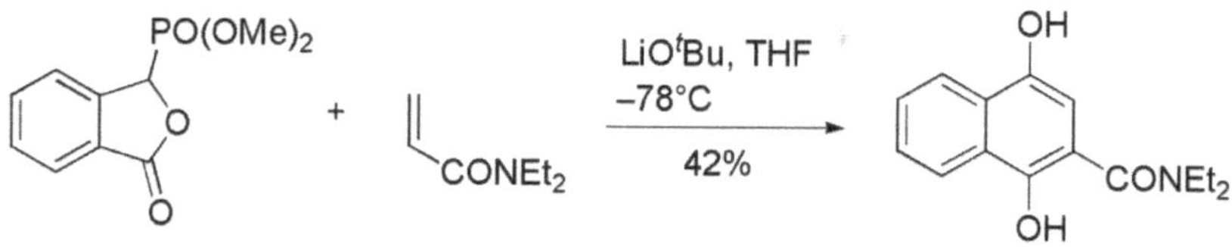

35 902 903

SCHEME 4.190 Annulation of an acrylamide [1993 CPB 968 Watanabe M].

Although it was indicated by Mal et al. that the phthalide sulfone (**11**) undergoes anionic annulation with dimethyl vinylphosphonate to yield the desired product, the detailed study on the reactivity of styryl phosphonates as Hauser acceptors was made by the Rastogi group. They explored the chemistry with less-known Katritzky benzotriazolylphthalides (**8**). As expected, the reaction between (**8**) and aryl substituted vinyl phosphonates (**904**) in the presence of LHMDS furnished annulated products (**905**) in good yields. NBS-water promoted oxidation of the quinol (**905**) gave the naphthoquinones (**906**).[140] The annulation is general with regard to both types of the reactants (Scheme 4.191).

More interesting results were obtained in the with alkynylphosphonates (**907**) under the similar reaction conditions. The same products (**905**) as those obtained from alkenylphosphonates (**904**) were formed (Scheme 4.192).

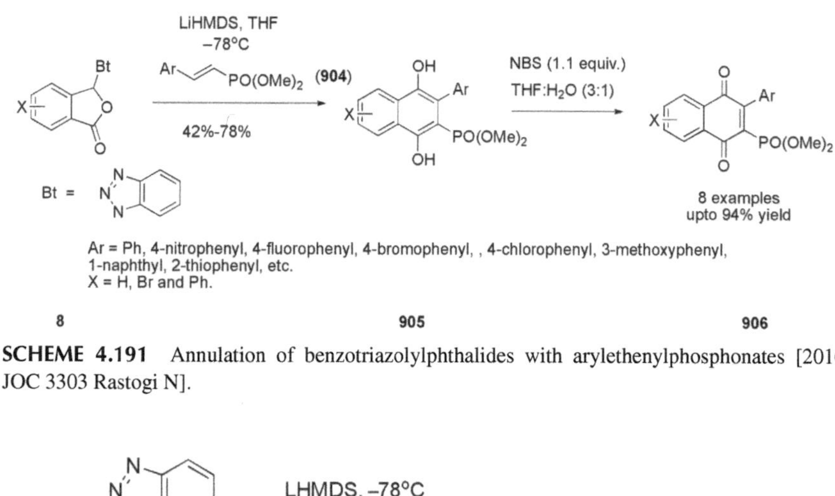

8 905 906

SCHEME 4.191 Annulation of benzotriazolylphthalides with arylethenylphosphonates [2016 JOC 3303 Rastogi N].

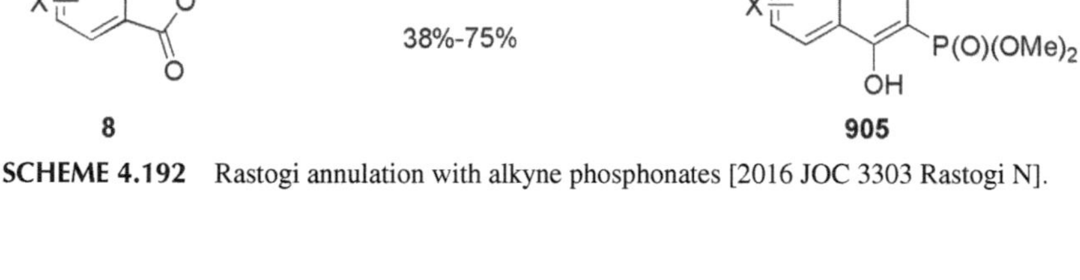

8 905

SCHEME 4.192 Rastogi annulation with alkyne phosphonates [2016 JOC 3303 Rastogi N].

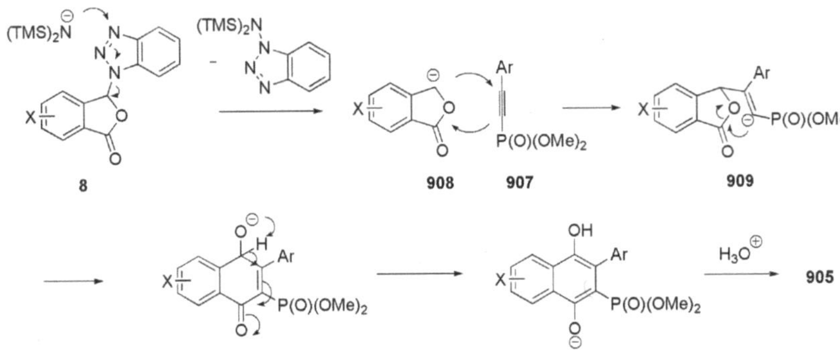

8 908 907 909

910 911 905

SCHEME 4.193 Mechanism of the Rastogi annulation [2016 JOC 3303 Rastogi N].

The mechanistic interpretation is displayed in Scheme 4.193. It is suggested that HMDS anion adds to azo nitrogen of the benzotriazolylphthalides (**8**) to produce the corresponding phthalide anions (**908**) which, in turn, add to alkyne phosphonates (**907**) to furnish anions (**909**). Cyclization of (**909**) results in the formation of intermediates (**910** and **911**), and finally the 1,4-naphthoquinol-phosphonates (**905**).

Cyclopropanated enedione (**912**) (Scheme 4.194) represents an uncommon acceptor in that it leads to the formation cycloheptanaphthalenone (**913**). The typical annulation product (**914**), obtained from the phenylthiophthalide (**22**) and (**912**), furnished (**913**) on reductions with Zn-AcOH in 66% yield.[141]

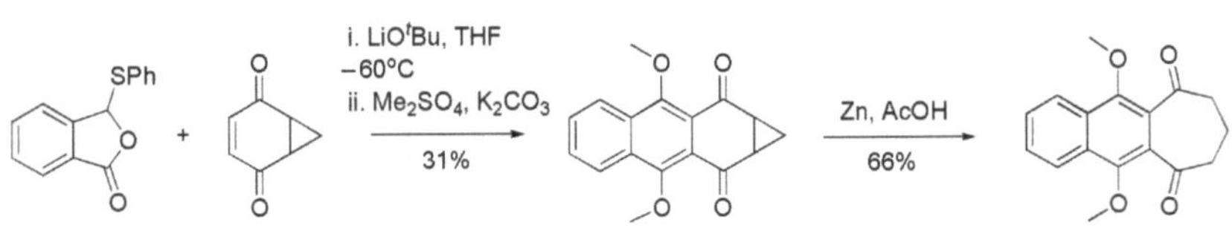

22 912 914 913

SCHEME 4.194 Synthesis of a cucloheptanaphthalenone [2008 EJOC 3014 Mal D].

With highly activated alkenes such as bicyclo[4.1.0]heptenes, the Hauser annulation proceeds smoothly to give cyclopropanaphthoquinones. Such an annulation has led to the total synthesis of naturally occurring tetracycle favelanone (**915**) (Scheme 4.195). The required cyanophthalide (**916**) was prepared from bromobenzene ketal (**917**) in four steps. BuLi promoted lithiation of (**917**) followed by carbon dioxide quench gave phthalaldehydic acid (**918**). Cyanohydrin formation of (**918**) followed by Vilsmeier lactonization gave (**916**) in 64% overall yield. The cyclopropane precursor (**919**) was prepared from a bromocyclohexanone. The reactive intermediates (**920** and **921**) were then generated under anion generating conditions. The former, i.e., (**920**) was prepared by deprotonation of (**921**) with LDA followed by trapping with TMSCl. The cyclopropene

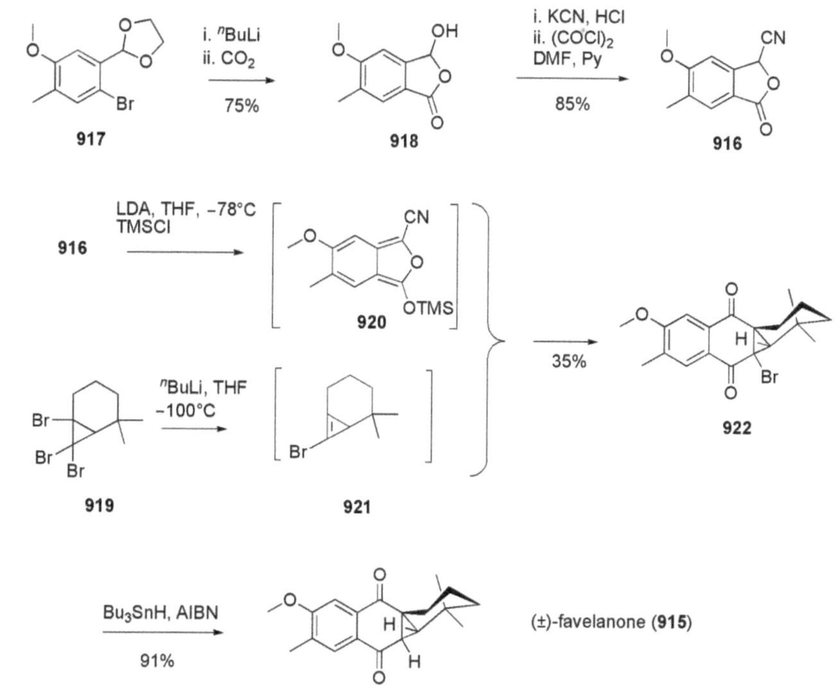

SCHEME 4.195 Wege total synthesis of favelanone (**915**) [1996 THS 6797 Wege D].

intermediate (**918**) was obtained by treating the tribromo compound with ^{n}BuLi at −100°C. Without isolation, these intermediates were allowed to react together at low temperature to produce (**922**) as the single isomer. Debromination of (**922**) by Bu_3SnH afforded favelanone (**915**) in overall 15% yield.[142]

In concluding this chapter, it can be said that [4+2] anionic annulations of 3-nucleofugal phthalides (Hauser annulations) are perhaps the best strategies to synthesize 1.4-naphthoquinols and 1,4-naphthoquinones. In most cases, the regiochemical dispositions of the new functionalities are undisputed. These strategies have made major impacts in the synthesis of *peri*-oxygenated fused aromatics and hydroaromatics, more specifically oxygenated anthraquinones and quinonoids. Furthermore, the required synthons for the annulations are accessible via a variety of synthetic avenues. However, much remains to be done in the field of heterocyclic and alicyclic versions of the annulations.

REFERENCES

1. (a) Mal, D.; Pahari, P. *Chem. Rev.* **2007**, *107*, 1892; (b) Rathwell, K.; Brimble, M. A. *Synthesis* **2007**, 643; (c) Mitchell, A. S.; Russell, R. A. *Tetrahedron* **1995**, *51*, 5207.
2. Hauser, F. M.; Mal, D. *J. Am. Chem. Soc.* **1983**, *105*, 5688.
3. Tatsuta, K.; Tokishita, S.; Fukuda, T.; Kano, T.; Komiya, T.; Hosokawa, S. *Tetrahedron Lett.* **2011**, *52*, 983.

4. Kumar, T.; Satam, N.; Namboothiri, I. N. N. *Eur. J. Org. Chem.* **2016**, *2016*, 3316.
5. (a) Kraus, G. A.; Cho, H.; Crowley, S.; Roth, B.; Sugimoto, H.; Prugh, S. *J. Org. Chem.* **1983**, *48*, 3439; (b) Ghorai, S. K.; Roy, H. N.; Bandopadhyay, M.; Mal, D. *J. Chem. Res. (S)* **1999**, 30.
6. Watanabe, M.; Morimoto, H.; Nogami, K.; Ijichi, S.; Furukawa, S. *Chem. Pharm. Bull.* **1993**, *41*, 968.
7. Hauser, F. M.; Rhee, R. P. *J. Org. Chem.* **1978**, *43*, 178.
8. Ghosh, K.; Chakraborty, S.; Mal, D. *Synthesis* **2015**, *47*, 2473.
9. Brimble, M. A.; Houghton, S. I.; Woodgate, P. D. *Tetrahedron* **2007**, *63*, 880.
10. Hauser, F. M.; Prasanna, S. *J. Am. Chem. Soc.* **1981**, *103*, 6378.
11. Kraus, G. A.; Dong, P. *Nat. Prod. Commun.* **2015**, *10*, 1025.
12. Cho, I. H.; Rho, Y. S.; Lee, J. Y.; Soh, S.; Kim, S.; Sin, H.; Yoo, D.; Kim, S. I. *J. Korean Chem. Soc.* **1991**, *35*, 756.
13. Gore, M. P.; Gould, S. J.; Weller, D. D. *J. Org. Chem.* **1992**, *57*, 2774.
14. Katritzky, A. R.; Zhang, G.; Xie, L. *Synth. Commun.* **1997**, *27*, 3951.
15. Mal, D.; Ray, S.; Patra, A. *Tetrahedron* **2008**, *64*, 3253.
16. Mal, D.; Jana, A.; Ray, S.; Bhattacharya, S.; Patra, A.; De, S. R. *Synth. Commun.* **2008**, *38*, 3937.
17. Chakraborty, S.; Das, G.; Ghosh, S.; Mal, D. *Org. Biomol. Chem.* **2016**, *14*, 10636.
18. Basak, S.; Mal, D. *J. Org. Chem.* **2017**, *82*, 11035.
19. Lokesh, K.; Kesavan, V. *Eur. J. Org. Chem.* **2017**, 5689.
20. Sivasankara, C.; Satham, L.; Namboothiri, I. N. N. *J. Org. Chem.* **2017**, *82*.
21. Jana, S.; Mal, D. *J. Org. Chem.* **2018**, *83*, 4537.
22. Mal, D.; Ghosh, K.; Jana, S. *Org. Lett.* **2015**, *17*, 5800.
23. Dinda, B.; Basak, S.; Ghosh, B.; Mal, D. *Synthesis* **2016**, *48*, 1235.
24. Fernandes, R. A.; Brueckner, R. *Synlett* **2005**, 1281.
25. Mal, D.; Ghosh, B., IIT Kharagpur, Unpublished Results, 2018.
26. Kraus, G. A.; Sugimoto, H. *Synth. Commun.* **1977**, *7*, 505.
27. Couladouros, E. A.; Plyta, Z. F.; Strongilos, A. T.; Papageorgiou, V. P. *Tetrahedron Lett.* **1997**, *38*, 7263.
28. Nielsen, L. B.; Slamet, R.; Wege, D. *Tetrahedron* **2009**, *65*, 4569.
29. Couladouros, E. A.; Strongilos, A. T.; Papageorgiou, V. P.; Plyta, Z. F. *Chem. Eur. J.* **2002**, *8*, 1795.
30. Nomura, K.; Okazaki, K.; Hori, K.; Yoshii, E. *Chem. Pharm. Bull.* **1986**, *34*, 3175.
31. Piggott, M. J.; Wege, D. *Aus. J. Chem.* **2003**, *56*, 691.
32. Hassan, N. P. S.; Naysmith, B. J.; Sperry, J.; Brimble, M. A. *Tetrahedron* **2015**, *71*, 7137.
33. Li, Z.; Gao, Y.; Tang, Y.; Dai, M.-j.; Wang, G.; Wang, Z.; Yang, Z. *Org. Lett.* **2008**, *10*, 3017.
34. Andrey, O.; Sperry, J.; Larsen, U. S.; Brimble, M. A. *Tetrahedron* **2008**, *64*, 3912.
35. Okazaki, K.; Nomura, K.; Yoshii, E. *Synth. Commun.* **1987**, *17*, 1021.
36. Gibson, J. S.; Andrey, O.; Brimble, M. A. *Synthesis* **2007**, 2611.
37. Naysmith, B. J.; Brimble, M. A. *Org. Lett.* **2013**, *15*, 2006.
38. Parker, K. A.; Koziski, K. A.; Breault, G. *Tetrahedron Lett.* **1985**, *26*, 2181.
39. Evans, J. C.; Klix, R. C.; Bach, R. D. *J. Org. Chem.* **1988**, *53*, 5519.
40. Wendt, J. A.; Gauvreau, P. J.; Bach, R. D. *J. Am. Chem. Soc.* **1994**, *116*, 9921.
41. Mal, D.; Hazra, N. K.; Murty, K.V.S.N.; Majumdar, G. *Synlett* **1995**, 1239.
42. De, S. R.; Ghorai, S. K.; Mal, D. *J. Org. Chem.* **2009**, *74*, 1598.
43. Mal, D.; Hazra, N. K. *Tetrahedron Lett.* **1996**, *37*, 2641.
44. Hauser, F. M.; Zhou, M. *J. Org. Chem.* **1996**, *61*, 5722.

45. Lee, H. G.; Ahn, J. Y.; Lee, A. S.; Shair, M. D. *Chem. Eur. J.* **2010**, *16*, 13058.
46. Bates, M. A.; Sammes, P. G.; Thomson, G. A. *J. Chem. Soc., Perkin Trans. 1* **1988**, 3037.
47. Schuenemann, K.; Furkert, D. P.; Choi, E. C.; Connelly, S.; Fraser, J. D.; Sperry, J.; Brimble, M. A. *Org. Biomol. Chem.* **2014**, *12*, 905.
48. Roy, J.; Mal, T.; Jana, S.; Mal, D. *Beilstein J. Org. Chem.* **2016**, *12*, 531.
49. Hauser, F. M.; Gauuan, P. J. F. *Org. Lett.* **1999**, *1*, 671.
50. Rho, Y. S.; Park, S. H.; Kwon, Y. J.; Kang, H. S.; Cho, I. H. *J. Korean Chem. Soc.* **1995**, *39*, 218.
51. Hauser, F. M.; Tacheuchi, C.; Yin, H.; Corlett, S. A. *J. Org. Chem.* **1994**, *59*, 258.
52. Snider, B. B.; Gao, X. *J. Org. Chem.* **2005**, *70*, 6863.
53. Nicolaou, K. C.; Lim, Y. H.; Papageorgiou, C. D.; Piper, J. L. *Angew. Chem. Int. Ed.* **2005**, *44*, 7917.
54. Murty, K.V.S.N.; Hazra, N. K.; Datta, K.; Mal, D. *Ind. J. Chem.* **1997**, *36B*, 126.
55. Matsumoto, T.; Yamaguchi, H.; Tanabe, M.; Yasui, Y.; Suzuki, K. *Tetrahedron Lett.* **2000**, *41*, 8393.
56. Khatri, H. R.; Nguyen, H.; Dunaway, J. K.; Zhu, J. *Chem. Eur. J.* **2015**, *21*, 13553.
57. (a) Magauer, T.; Smaltz, D. J.; Myers, A. G. *Nat. Chem.* **2013**, *5*, 886; (b) Nicolaou, K. C.; Chen, P.; Zhu, S.; Cai, Q.; Erande, R. D.; Li, R.; Sun, H.; Pulukuri, K. K.; Rigol, S.; Aujay, M.; Sandoval, J.; Gavrilyuk, J. *J. Am. Chem. Soc.* **2017**, *139*, 15467; (c) Nicolaou, K. C.; Cai, Q.; Qin, B.; Petersen, M. T.; Mikkelsen, R. J. T.; Heretsch, P. *Angew. Chem. Int. Ed.* **2015**, *54*, 3074.
58. Crawley, M. L.; Hein, K. J.; Markgraf, J. H. *J. Chem. Res. (S)* **2003**, 470.
59. Huang, J.-K.; Yang Lauderdale, T.-L.; Shia, K.-S. *Org. Lett.* **2015**, *17*, 4248.
60. Hauser F.M., Unpublished results.
61. Li, T.; Wu, Y. L. *J. Am. Chem. Soc.* **1981**, *103*, 7007.
62. Hauser, F. M.; Hewawasam, P.; Rho, Y. S. *J. Org. Chem.* **1989**, *54*, 5110.
63. Hauser, F. M.; Mal, D. *J. Am. Chem. Soc.* **1983**, *105*, 5688.
64. Hauser, F. M.; Tommasi, R. A. *J. Org. Chem.* **1991**, *56*, 5758.
65. Hauser, F. M.; Chakrapani, S.; Ellenberger, W. P. *J. Org. Chem.* **1991**, *56*, 5248.
66. Hauser, F. M.; Caringal, Y. *J. Org. Chem.* **1990**, *55*, 555.
67. Tatsuta, K.; Inukai, T.; Itoh, S.; Kawarasaki, M.; Nakano, Y. *J. Antibiot.* **2002**, *55*, 1076.
68. Hauser, F.M., Unpublished results.
69. Hauser, F. M.; Yin, H. *Org. Lett.* **2000**, *2*, 1045.
70. Cox, C.; Danishefsky, S. *J. Org. Lett.* **2000**, *2*, 3493.
71. Kim, T.; Jeong, K. H.; Kang, K. S.; Nakata, M.; Ham, J. *Eur. J. Org. Chem.* **2017**, *2017*, 1704.
72. Braukmueller, S.; Brueckner, R. *Eur. J. Org. Chem.* **2006**, 2110.
73. Jacobs, J.; Claessens, S.; De Kimpe, N. *Tetrahedron* **2008**, *64*, 412.
74. (a) Tatsuta, K.; Akimoto, K.; Annaka, M.; Ohno, Y.; Kinoshita, M. *J. Antibiot.* **1985**, *38*, 680; (b) Freskos, J. N.; Swenton, J. S. *J. Chem. Soc. Chem. Commun.* **1985**, 658.
75. Hoffmann, B.; Schoenebaum, A.; Lackner, H. *Liebigs Ann. Chem.* **1993**, 333.
76. Claessens, S.; Naidoo, D.; Mulholland, D.; Verschaeve, L.; van Staden, J.; De Kimpe, N. *Synlett* **2006**, *621*.
77. Claessens, S.; Jacobs, J.; De Kimpe, N. *Synlett* **2007**, 741.
78. Deshpande, P. P.; Price, K. N.; Baker, D. C. *J. Org. Chem.* **1996**, *61*, 455.
79. Tatsuta, K.; Akimoto, K.; Annaka, M.; Ohno, Y.; Kinoshita, M. *Bull. Chem. Soc. Jpn.* **1985**, *58*, 1699.
80. Hoffmann, B.; Lackner, H. *Liebigs Ann. Chem.* **1995**, 87.
81. Tatsuta, K.; Suzuki, Y.; Toriumi, T.; Furuya, Y.; Hosokawa, S. *Tetrahedron Lett.* **2007**, *48*, 8018.

82. Tatsuta, K.; Ozeki, H.; Yamaguchi, M.; Tanaka, M.; Okui, T. *Tetrahedron Lett.* **1990**, *31*, 5495.
83. Tatsuta, K.; Hirabayashi, T.; Kojima, M.; Suzuki, Y.; Ogura, T. *J. Antibiot.* **2004**, *57*, 291.
84. Swenton, J. S.; Freskos, J. N.; Dalidowicz, P.; Kerns, M. L. *J. Org. Chem.* **1996**, *61*, 459.
85. Couladouros, E. A.; Strongilos, A. T. *Angew. Chem. Int. Ed.* **2002**, *41*, 3677.
86. Tatsuta, K.; Tanaka, Y.; Kojima, M.; Ikegami, H. *Chem. Lett.* **2002**, 14.
87. Hauser, F. M.; Hewawasam, P.; Baghdanov, V. M. *J. Org. Chem.* **1988**, *53*, 223.
88. Hauser, F. M.; Combs, D. W. *J. Org. Chem.* **1980**, *45*, 4071.
89. Mal, D.; Patra, A.; Roy, H. *Tetrahedron Lett.* **2004**, *45*, 7895.
90. (a) Ueberschaar, N.; Xu, Z.; Scherlach, K.; Metsae-Ketelae, M.; Bretschneider, T.; Dahse, H.-M.; Goerls, H.; Hertweck, C. *J. Am. Chem. Soc.* **2013**, *135*, 17408; (b) Huang, X.; Zhu, T.; Huang, Z.; Zhang, Y.; Jin, Z.; Zanoni, G.; Chi, Y. R. *Org. Lett.* **2017**, *19*, .
91. (a) Kutyrev, A. A. *Tetrahedron* **1991**, *47*, 8043; (b) Nawrat, C. C.; Moody, C. J. *Angew. Chem. Int. Ed.* **2014**, *53*, 2056.
92. Mal, D.; Ray, S.; Sharma, I. *J. Org. Chem.* **2007**, *72*, 4981.
93. Basak, S.; Ray, S.; Mal, D. *Arkivoc Part III* **2018**, 257.
94. Basak, S.; Mal, D. *Tetrahedron* **2016**, *72*, 1758.
95. (a) Russell, R. A.; Warrener, R. N. *J. Chem. Soc. Chem. Commun.* **1981**, 108; (b) Dolson, M. G.; Chenard, B. L.; Swenton, J. S. *J. Am. Chem. Soc.* **1981**, *103*, 5263.
96. Swenton, J. S.; Bonke, B. R.; Clark, W. M.; Chen, C. P.; Martin, K. V. *J. Org. Chem.* **1990**, *55*, 2027.
97. Md Raja, S. K.; Chakraborty, S.; Mal, D. *Synth. Commun.* **2018**, *48*, 309.
98. Ge, P.; Russell, R. A. *Tetrahedron* **1997**, *53*, 17477.
99. Nicolaou, K. C.; Lu, M.; Chen, P.; Shah, A. A. *Angew. Chem. Int. Ed.* **2015**, *54*, 12687.
100. (a) Deguin, B.; Florent, J.-C.; Monneret, C. *J. Org. Chem.* **1991**, *56*, 405; (b) Achmatowicz, O.; Szechner, B. *J. Org. Chem.* **2003**, *68*, 2398; (c) Irvine, R. W.; Kinloch, S. A.; McCormick, A. S.; Russell, R. A.; Warrener, R. N. *Tetrahedron* **1988**, *44*, 4591.
101. (a) Yoon, T.; Shair, M. D.; Danishefsky, S. J. *Tetrahedron Lett.* **1994**, *35*, 6259; (b) Myers, A. G.; Tom, N. J.; Fraley, M. E.; Cohen, S. B.; Madar, D. J. *J. Am. Chem. Soc.* **1997**, *119*, 6072; (c) Nicolaou, K. C.; Zhang, H.; Chen, J. S.; Crawford, J. J.; Pasunoori, L. *Angew. Chem. Int. Ed.* **2007**, *46*, 4704.
102. Nicolaou, K. C.; Wang, Y.; Lu, M.; Mandal, D.; Pattanayak, M. R.; Yu, R.; Shah, A. A.; Chen, J. S.; Zhang, H.; Crawford, J. J.; Pasunoori, L.; Poudel, Y. B.; Chowdari, N. S.; Pan, C.; Nazeer, A.; Gangwar, S.; Vite, G.; Pitsinos, E. N. *J. Am. Chem. Soc.* **2016**, *138*, 8235.
103. (a) Mal, D.; Hazra, N. K.; Roy, H. N.; Adhikary, S. *Tetrahedron* **1997**, *53*, 2177. (b) Ge, P.; Russell, R. A. *Tetrahedron* **1997**, *53*, 17469.
104. Hauser, F. M.; Dorsch, W. A.; Mal, D. *Org. Lett.* **2002**, *4*, 2237.
105. Mal, D.; Patra, A.; Roy, H. *Tetrahedron Lett.* **2004**, *45*, 7895.
106. Mal, D.; Mitra, P.; Behera, B.; Maiti, T. K. *J. Org. Chem.* **2013**, *78*, 9748.
107. Majumdar, G.; Murty, K.V.S.N.; Mal, D. *Tetrahedron Lett.* **1994**, *35*, 6139.
108. Wang, J.; Pettus, L. H.; Thomas, R. R. *Tetrahedron Lett.* **2004**, *45*, 1793.
109. Mal, D.; Roy, H. N. *J. Chem. Soc. Perkin Trans.* **1999**, *1*, 3167.
110. Pahari, P.; Senapati, B.; Mal, D. *Tetrahedron Lett.* **2007**, *48*, 2635.
111. Tadross, P. M.; Stoltz, B. M. *Chem. Rev.* **2012**, *112*, 3550.
112. Sammes, P. G.; Dodsworth, D. J. *J. Chem. Soc. Chem. Commun.* **1979**, 33.
113. Zhao, H.; Biehl, E. R. *J. Nat. Prod.* **1995**, *58*, 1970.
114. Khanapure, S. P.; Biehl, E. R. *J. Nat. Prod.* **1989**, *52*, 1357.
115. Biehl, E. R.; Deshmukh, A. R.; Dutt, M. *Synthesis* **1993**, 885.

116. Han, W.; Tran, J.; Zhang, H.; Jeffrey, S.; Swartling, D.; Ford, G. P.; Biehl, E. *Synthesis* **1995**, 827.
117. Kaiser, F.; Schwink, L.; Velder, J.; Schmalz, H.-G. *Tetrahedron* **2003**, *59*, 3201.
118. Khanapure, S. P.; Biehl, E. R. *Heterocycles* **1988**, *27*, 2643.
119. El-Awa, A.; Noshi, M. N.; Mollat du Jourdin, X.; Fuchs, P. L. *Chem. Rev.* **2009**, *109*, 6920.
120. Tso, H.-H.; Chen, Y.-J. *J. Chem. Res. (S)* **1995**, 104.
121. Patra A., 2005, IIT Kharagpur.
122. Mal, D.; Patra, A.; Pahari, P.; Ray, S. *J. Org. Chem.* **2005**, *70*, 9017.
123. Halimehjani, A. Z.; Namboothiri, I. N. N.; Hooshmand, S. E. *RSC Adv.* **2014**, *4*, 31261.
124. Luo, J.; Wang, H.; Zhong, F.; Kwiatkowski, J.; Xu, L.-W.; Lu, Y. *Chem. Commun.* **2013**, *49*, 5775.
125. Vasella, A.; Brade, W. *Helv. Chim. Acta* **1989**, *72*, 1649.
126. Gore, M. P.; Gould, S. J.; Weller, D. D. *J. Org. Chem.* **1991**, *56*, 2289.
127. Kumar, T.; Satam, N.; Namboothiri, I. N. N. *Eur. J. Org. Chem.* **2016**, *2016*, 3316.
128. Kumar, T.; Mane, V.; Namboothiri, I. N. N. *Org. Lett.* **2017**, *19*, 4283.
129. Angle, S. R.; Rainier, J. D.; Woytowicz, C. *J. Org. Chem.* **1997**, *62*, 5884.
130. Zheng R. 2010, PhD Thesis, SUNY Albany.
131. Cho, I. H.; Rho, Y. S.; Park, S. H.; Ahn, K. H.; Sin, H. S.; Han, B. K. *J. Korean Chem. Soc.* **1993**, *37*, 141.
132. Hauser, F. M. SUNY Albany, Unpublished results, 1997.
133. Karmakar, R.; Mal, D. *J. Org. Chem.* **2012**, *77*, 10235.
134. Majumdar, G.; Pal, R.; Murty, K.V.S.N.; Mal, D. *J. Chem. Soc. Perkin. Trans.* **1994**, *1*, 309.
135. Majumdar, G., PhD Thesis, IIT Kharagpur, 1994.
136. Alguacil, R.; Martin, M. V.; Paredes, M. C. *Heterocycles* **2000**, *53*, 1029.
137. Hauser, F. M.; Dorsch, W. A. *Org. Lett.* **2003**, *5*, 3753.
138. Mal, D.; Senapati, B. K.; Pahari, P. *Tetrahedron* **2007**, *63*, 3768.
139. Roy, J.; Pal, R.; Mal, D. *Tetrahedron Lett.* **2015**, *56*, 6210.
140. Chaturvedi, A. K.; Rastogi, N. *J. Org. Chem.* **2016**, *81*, 3303.
141. Mal, D.; Ray, S. *Eur. J. Org. Chem.* **2008**, 3014.
142. Ng, W.; Wege, D. *Tetrahedron Lett.* **1996**, *37*, 6797.

Chapter 5

[4+2] Annulation With Nonactivated Phthalides (Sammes Annulation)

Sammes annulation, i.e., [4+2] annulation chemistry of simple phthalides (phthalides without a nucleofuge at C3), was reported in 1978, concomitantly with the Hauser's classic publication.[1,2] It can be described as the condensation between a phthalide (e.g., **1**) with an acceptor (e.g., **2**) in the presence of a strong base like LDA to give 4-hydroxytetralone (**3**) or its dehydration product (**4**). It is usually carried out in two steps to obtain the α-naphthols (**4**) (Scheme 5.1).

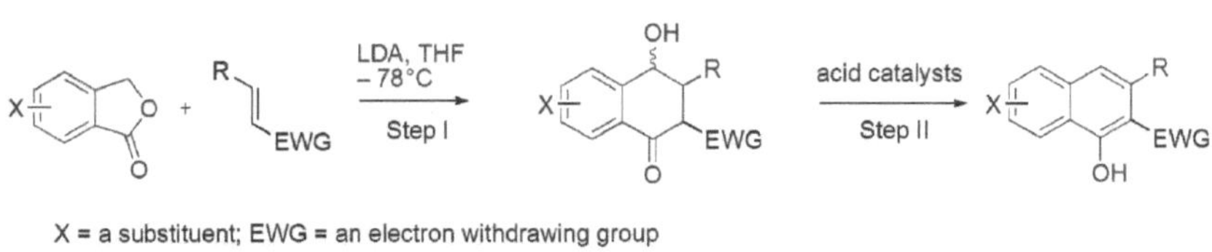

1 2 3 4

SCHEME 5.1 General scheme for the Sammes annulation [1978 JCSCCS 163 Sammes PG].

The underlying principles of this annulation are largely similar to those of the Hauser phthalide annulation discussed in Chapter 4. Treatment of phthalide (**5**) with LDA at −78°C generates orange/yellow colored phthalide anion (**6**) through deprotonation (Scheme 5.2). The anion (**6**) then undergoes conjugate addition to the acceptor (**2**) to produce anion (**7**), which then undergoes cyclization through the tricyclic intermediate (**8**). The newly generated alkoxide (**9**), on acid quenching gives hydroxy tetralone (**10**). Treatment of (**10**) with BF_3-etherate or trifluoroacetic acid furnishes α-naphthol (**11**).

Thus, the Sammes annulation involves phthalides without a nucleofuge at C3 unlike the Hauser annulation. Consequently, it provides a 1-naphthol-2-carboxylate in the reaction with ethyl acrylate, whereas the Hauser annulation gives a 1,4-naphthalenediol-2-carboxylate. Due to the lack of the nucleofuge,

Anionic Annulations in Organic Synthesis. https://doi.org/10.1016/B978-0-12-813800-7.00005-1

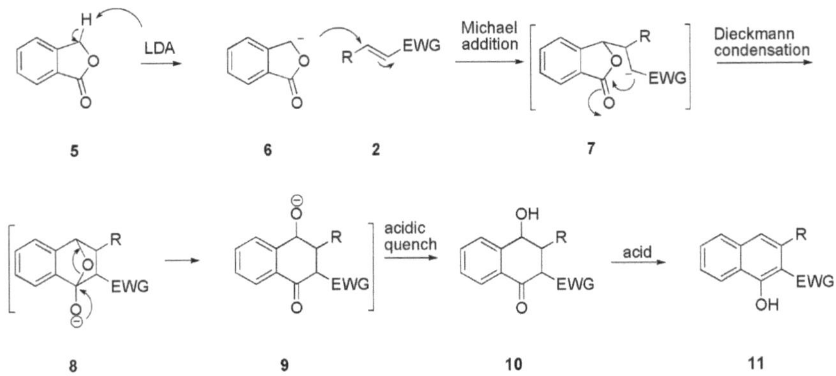

SCHEME 5.2 Mechanism of the Sammes annulation.

the oxidation level[3] of the products of the Sammes annulation is one less than that of the Hauser annulation. Compared to the Hauser annulation, the Sammes annulation has been investigated with fewer types of acceptors. These are acrylates, conjugated lactones, allenecarboxylates, conjugated ketones, aryne precursors, etc.

5.1 CONJUGATED ESTERS, NITRILES, LACTONES, ALLENECARBOXYLATES

Conjugated unsaturated esters and lactones are appropriate acceptors for the annulation with phthalides leading, in two steps, to α-naphthols. The scope and limitations are evident from the results in Table 5.1. Methyl acrylate (**12**), dimethyl fumarate (**13**), methyl 3-methylcrotonate (**14**), benzyl acrylate (**15**), methyl crotonate (**16**) and acrylonitrile (**17**) react with phthalides (**5**), (**18**), (**19**), and (**20**) to give respective hydroxytetralones (**21–27**). These tetralones, on reaction with BF_3-etherate or trifluoroacetic acid furnish α-naphthols (**28–34**). For methyl 3-methylcrotonate (**14**), the tetralones (**23** and **24**) undergo rearrangement to trisubstituted naphthols (**30** and **31**, respectively) during the acid-catalyzed dehydrations.

For the acceptors like alkylidene malonate (**35**) and mesityl oxide (**36**), the reaction of the parent phthalide (**5**) gives only the Michael adducts (**37** and **38**) in poor yields (Scheme 5.3).

The extension of the Sammes annulation of phthalide (**20**) to β-methoxyacrylate (**39**) resulted in the elimination of methoxy group during the process giving annulation product (**40**) in only 17% yield along with addition-elimination product (**41**) in 18% yield. The formation of (**40**) can be explained by elimination of methoxy group from intermediate (**42**) to form (**43**), which undergoes aromatization to naphthalene dioxide (**44**) (Scheme 5.4).[7]

Sammes annulation has found applications in the total synthesis of many natural products. The simplest example is the synthesis of *O*-methylasparvenone (**45**). It was synthesized in four steps from benzyl alcohol (**46**). Nuclear lithiation

TABLE 5.1 Examples of the Sammes Annulation [1981 JCSPT 465 Sammes PG]

Entry	Phthalide Donor	Michael Acceptor	Product (Step I)	Product (Step II)
1[4]	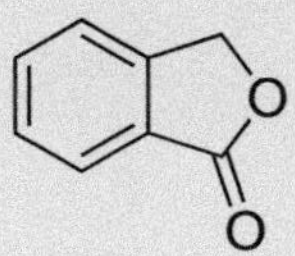 **5**	 **12**	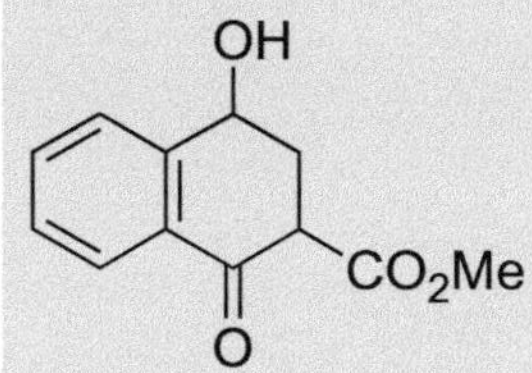 **21** (37%)	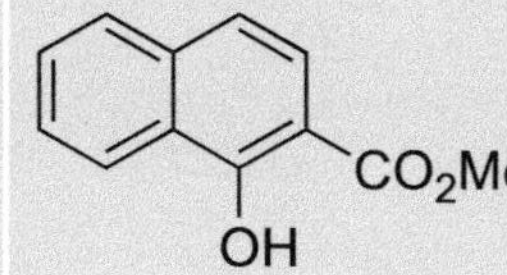 **28** (72%)
2[4]	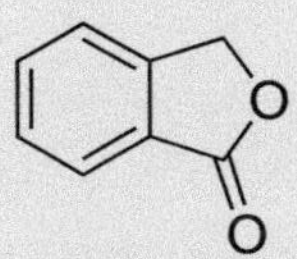 **5**	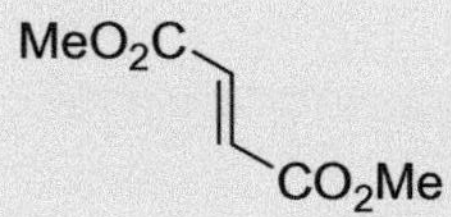 **13**	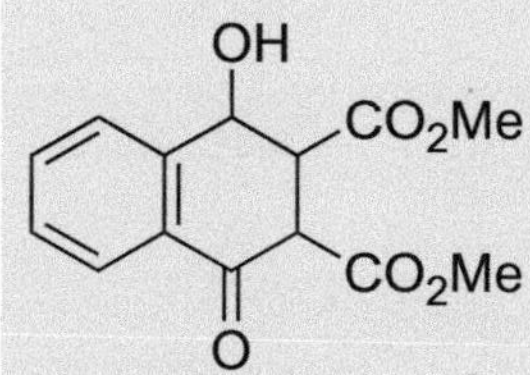 **22** (44%)	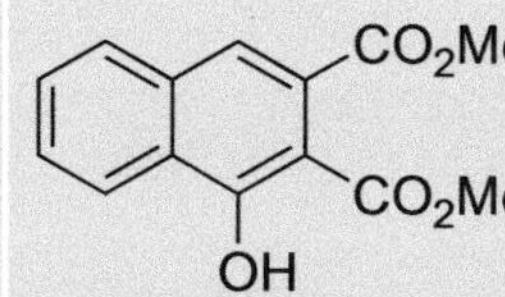 **29** (82%)

Continued

TABLE 5.1 Examples of the Sammes Annulation [1981 JCSPT 465 Sammes PG]—cont'd

Entry	Phthalide Donor	Michael Acceptor	Product (Step I)	Product (Step II)
3[4]	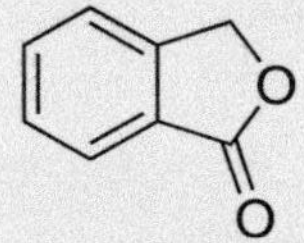 **5**	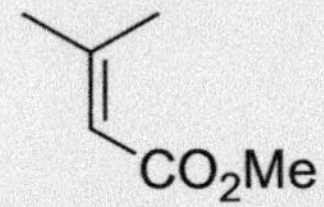 **14**	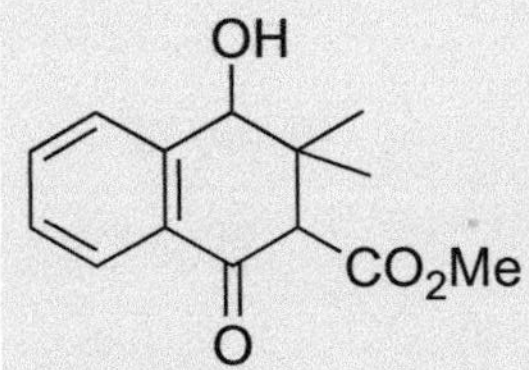 **23** (37%)	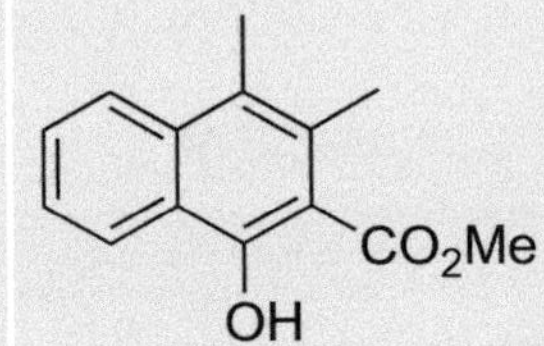 **30** (67%)
4[4]	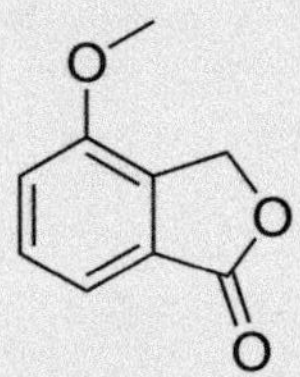 **18**	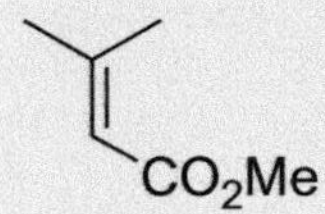 **14**	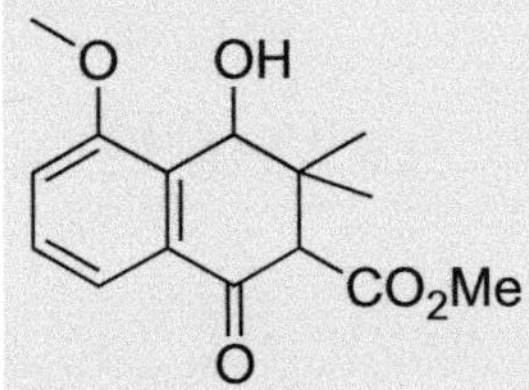 **24** (31%)	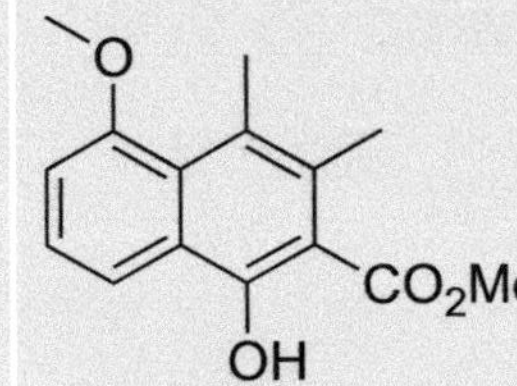 **31** (32%)
5[5]	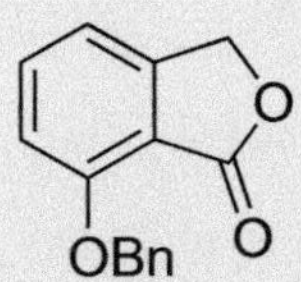 **19**	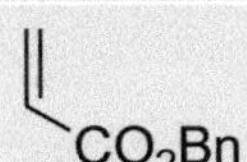 **15**	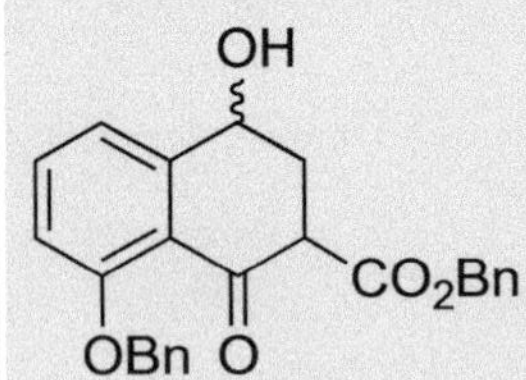 **25** (41%)	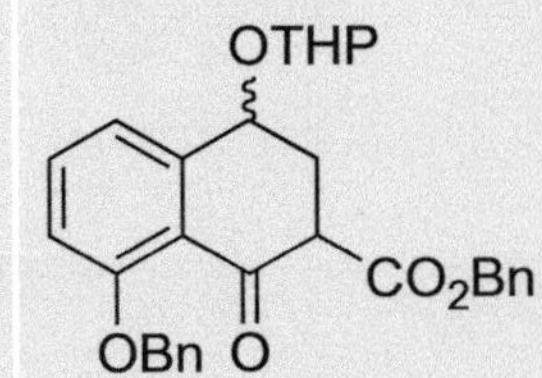 **32** (73%)

6[6]	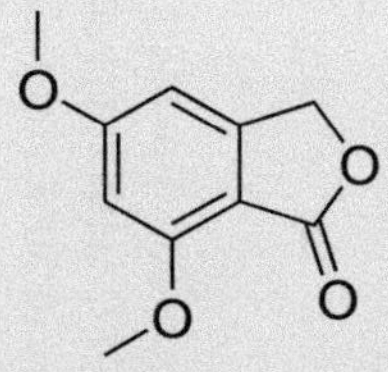 **20**	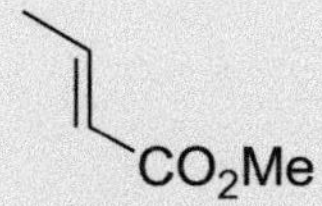 **16**	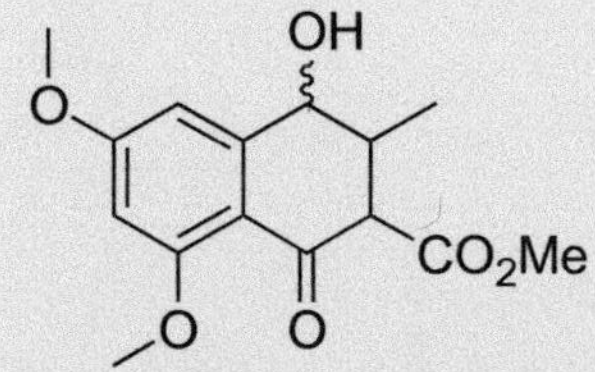 **26**	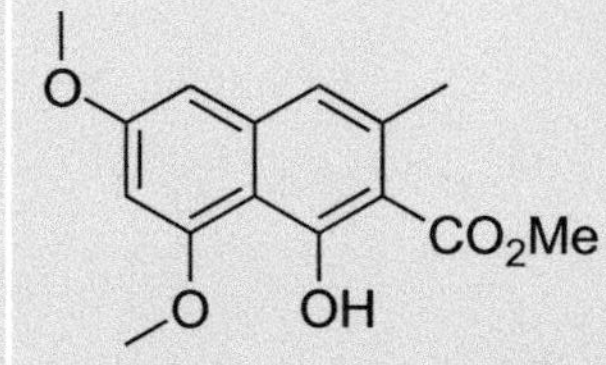 **33** (52% overall yield)
7[4]	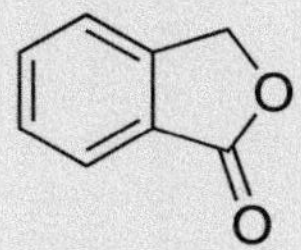 **5**	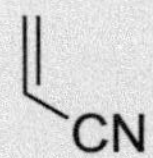 **17**	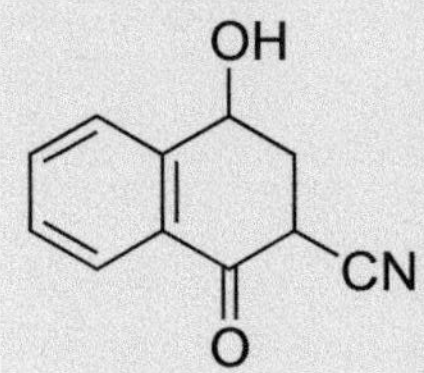 **27** (39%)	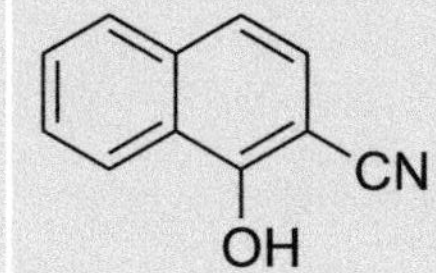 **34** (74%)

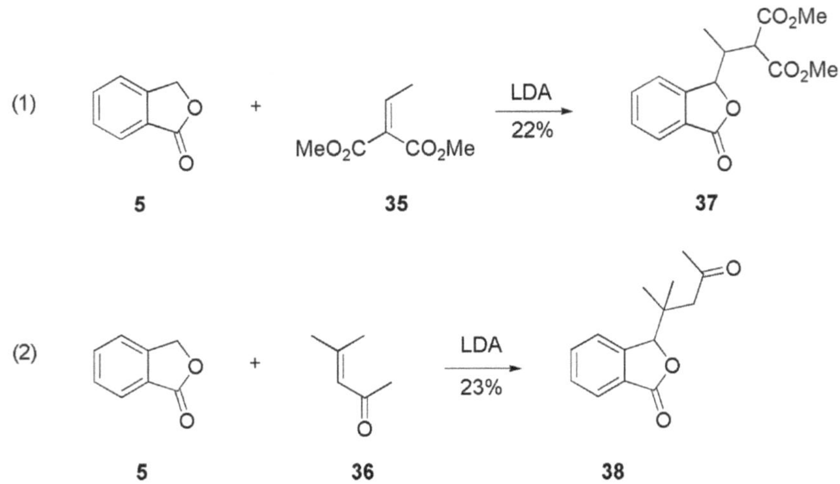

SCHEME 5.3 Failed Sammes annulations with unsaturated esters and ketones.

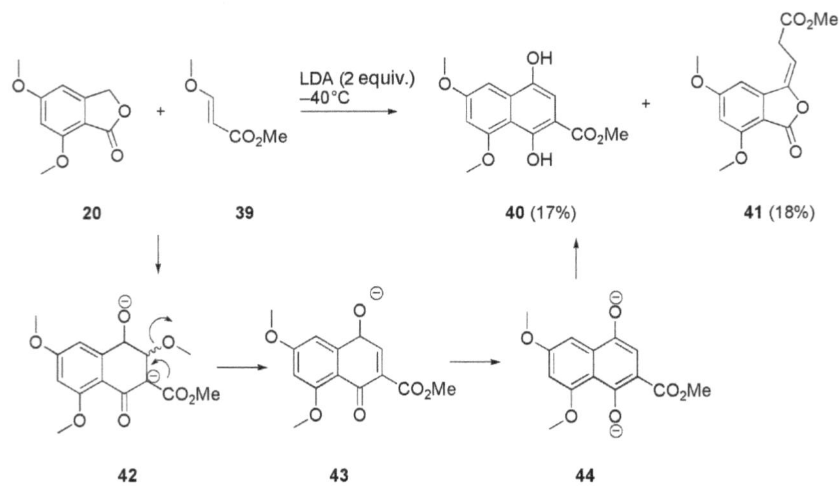

SCHEME 5.4 Sammes annulation with a β-methoxyacrylate [2009 HCA 903 Couche E].

of the alcohol (**46**) with ^{n}BuLi followed by quenching with carbon dioxide gave phthalide (**47**) (Scheme 5.5).[8] Annulation of the phthalide (**47**) with butyl acrylate (**48**) in the presence of LDA gave hydroxy tetralone (**49**) in 50% yield along with substantial amount of the conjugate adduct (**47**) and (**48**). Alkaline hydrolysis of (**49**) followed by HCl-promoted decarboxylation gave tetralone (**50**) in 71% yield. Boron tribromide mediated selective demethylation furnished *rac-O*-methylasparvenone (**45**) in 57% yield. The salient feature is selective debutyloxycarbonylation in the presence of an acid.

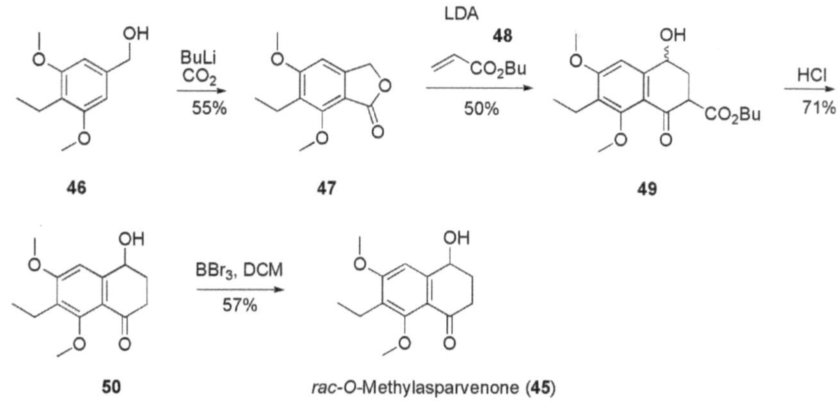

SCHEME 5.5 Synthesis of *O*-methylasparvenone (**45**) [1997 BMC 2165 Bos M].

Mal and co-workers combined Sammes annulation with *retro*-Kolbe-Schmitt reaction to develop a shorter synthesis of α-naphthol (**51**), a key intermediate in the synthesis of PD 116740, an angucycline antibiotic. Benzyloxy phthalide (**52**), on reaction with methyl acrylate in the presence of LDA gave annulation product (**53**).[9] Treatment of (**53**) with aq. KOH at 80°C furnished naphthol (**51**) (Scheme 5.6). Decarboxylation, by the way of *retro*-Kolbe-Schmitt reaction, proceeded during the hydrolysis. No complication arises from possible lateral lithiations during the Sammes annulation due to the presence of benzyloxy group in (**52**).

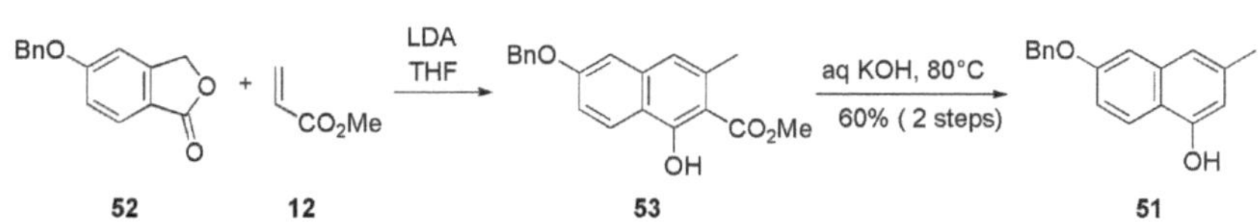

SCHEME 5.6 Combined application of Sammes annulation and *retro*-Kolbe-Schmitt reaction [2015 THN 5610 Mal D].

C3 alkyl substituted phthalides also take part in the Sammes annulation. Allyl phthalide (**54**), obtained through In-mediated allylation of phthalaldehydic and (**55**), underwent smooth annulation with methyl crotonate (**16**) to give tetrasubstituted naphthalene (**56**) (Scheme 5.7).[10]

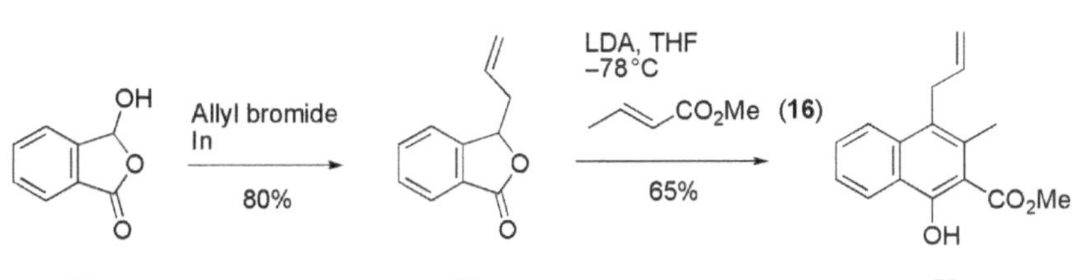

SCHEME 5.7 Sammes annulation of a 3-allylphthalide [2012 SYT 1769 Mal D].

Roush et al. adopted the Sammes annulation for the total synthesis of (+)-olivin (**57**), the naturally occurring enantiomer of the aglycon of olivomycin **A**. For the model study, the annulation of phthalide (**58**) with methyl crotonate (**16**) was carried out to obtain naphthol (**59**) in 65% overall yield. Dehydration of the hydroxytetralone intermediate of the annulation was carried with mesyl chloride and triethylamine. A similar reaction of (**58**) with chiral acrylate (**60**) gave annulation product (**61**) in only 30%–35% yields. For both the annulations, the initial annulation products were dehydrated with MsCl-Et_3N. The intermediate (**61**) was submitted to a sequence of 14 more steps to achieve a total synthesis of natural product olivin (**57**) (Scheme 5.8).[11]

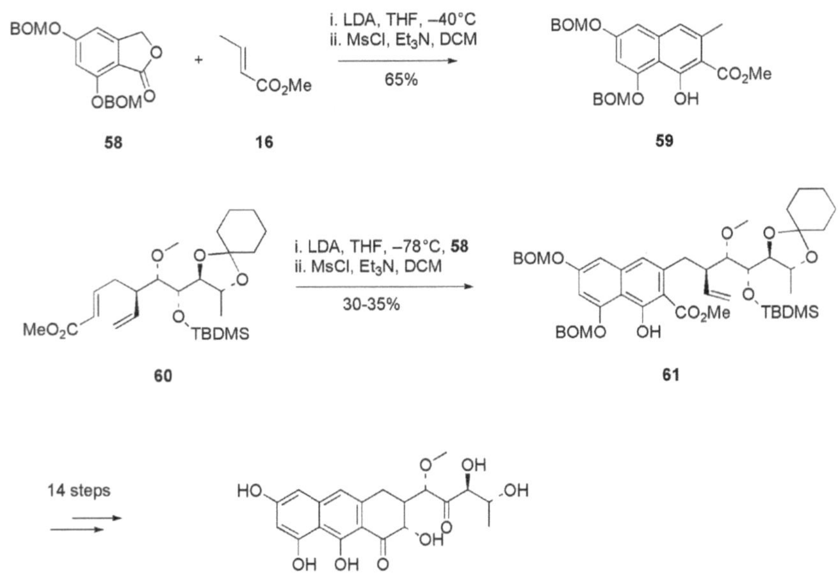

SCHEME 5.8 Roush synthesis of olivin naphthoate [1989 JACS 2984 Roush WR].

Although typical cinnamate acceptors are not studied for the Sammes annulation, related β-pyridylacrylate (**62**) is found to act as an excellent acceptor (Scheme 5.9).[12] The reaction between (**62**) and (**63**) in the presence of LDA gave annulation product (**64**) in 96% yield. Aromatization of the adduct (**64**) with PTSA in $CHCl_3$ furnished corresponding pyridyl naphthoate, which was converted into pivaloate (**65**) by using pivaloyl chloride in 87% overall yield in two steps. Treatment of (**65**) with LDA induced lateral lithiation mediated cyclization to give 1-azabenz[*a*]anthracene (**66**) via the intermediate (**67**). Oxidation of (**66**) with ceric ammonium nitrate (CAN) in acetonitrile-water produced the anthraquinone (**68**) (74%) as an orange solid, a model compound for the study of dynemicin A cytotoxicity.

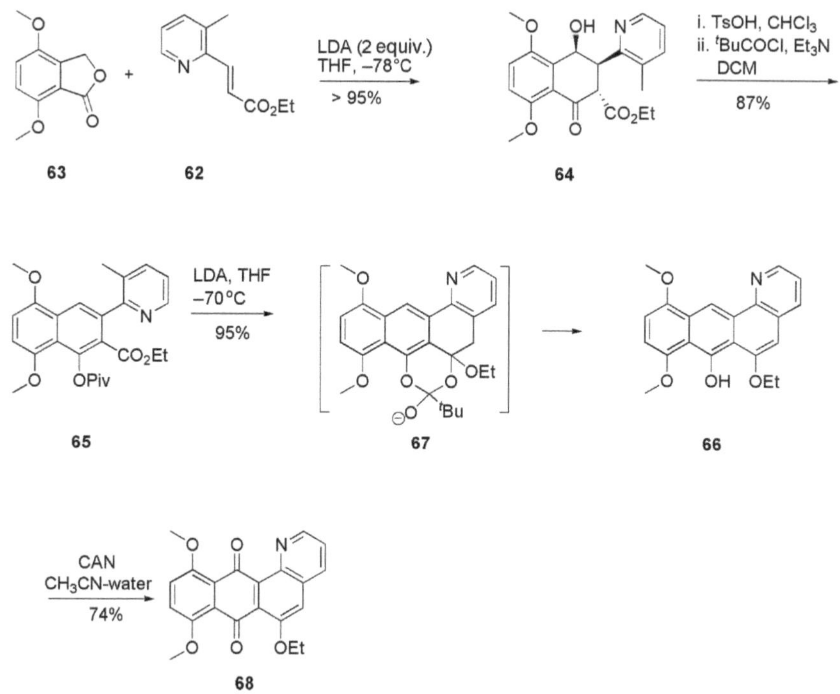

SCHEME 5.9 Annulation with a β-pyridyl acrylate [1997 JACS 5591 Magnus P].

The annulation of bromophthalides (**69**) with methyl crotonate (**16**) have been carried out on gram-scales to give bromonaphthalenes (**70**) (Scheme 5.10).[13] No complication arises due to the possible formation of benzyne from the bromophthalides, although the generation of benzyne by the action of LDA on bromobenzene is well established. The above Sammes annulation is regiospecific as well as chemoselective. The only demerit of the reactions is moderate yields.

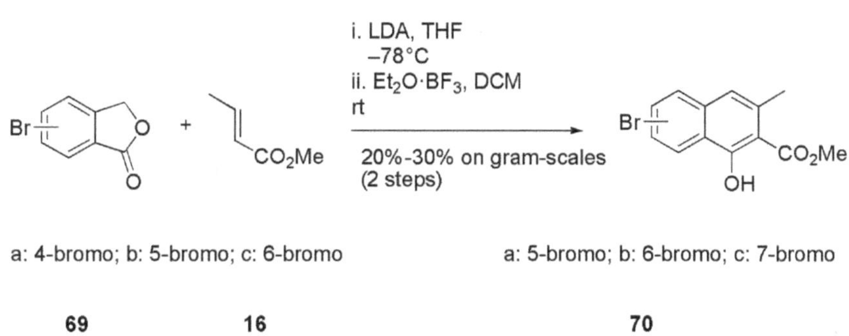

SCHEME 5.10 Sammes annulation of bromophthalides [2011 JMC 3704 Xiao X-Y].

The study of furan-2(5H)-ones like (**71**) as the Sammes acceptor is scanty. The annulation of the furanones is somewhat sluggish compared to acrylates. The reaction of 4-methoxyphthalide (**18**) with furanone (**71**) in the presence of LDA gave the corresponding hydroxytetralone derivative in 30% yield, which was aromatized with boron trifluoride etherate to produce 9-hydroxy-5-methoxy-naphtho[2,3-*c*]furan-1(3*H*)-one in 52% yield.[4] This reaction was extended to 3-arylphthalides (**72**) for the synthesis of lignan analogs (**73**). The reaction of the anions of (**72**) reacted with furanone (**71**) in the presence of LDA to afford naphthalides (**74**). These naphthalides were aromatized with iodine or thionyl chloride to give lignan analogs (**75**) (Scheme 5.11).[14] For characterization, phenolic OH groups were methylated by diazomethane or CH_3I-K_2CO_3.

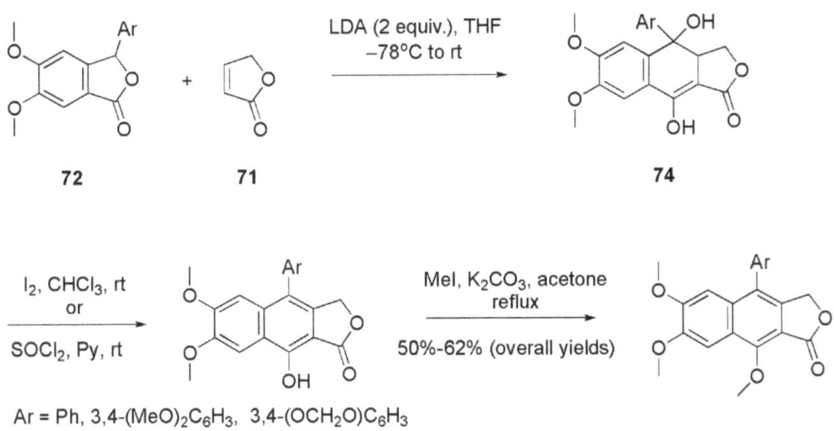

SCHEME 5.11 Sammes annulation with 2-furanone [1997 JCSPT1 443 Kobayashi K].

5,6-Dihydropyran-2-ones: The Sammes annulation, though mechanistically applicable to 5,6-dihydropyran-2-ones, has been studied with only one example, i.e., 5,6-dihydropyran-2-one, i.e., (**76**). Its annulation with phthalide (**77**) constitutes one of the key steps in the total synthesis of antibiotics K1115 B_1 (**78**) (Scheme 5.12).[15] The annulation was carried by a modification of the Sammes procedure. In the place of LDA, LHMDS was used as the base. The differentially protected phthalide (**77**) was prepared in two steps from phthalide (**79**) via selective demethylation with $MgBr_2$ followed by benzylation. Its annulation with dihydropyranone (**76**), derived from D-rhamnal, in the presence of LHMDS furnished tricycle (**80**). The crude product (**80**) was treated MsCl and triethylamine and then the resultant product (**81**) was benzylated to yield (**82**) in 89% yield over three steps. This was then reduced with DIBAL to give lactol (**83**). TMS triflate promoted Sakuari allylation of (**83**) followed by Wilkinson hydrogenation

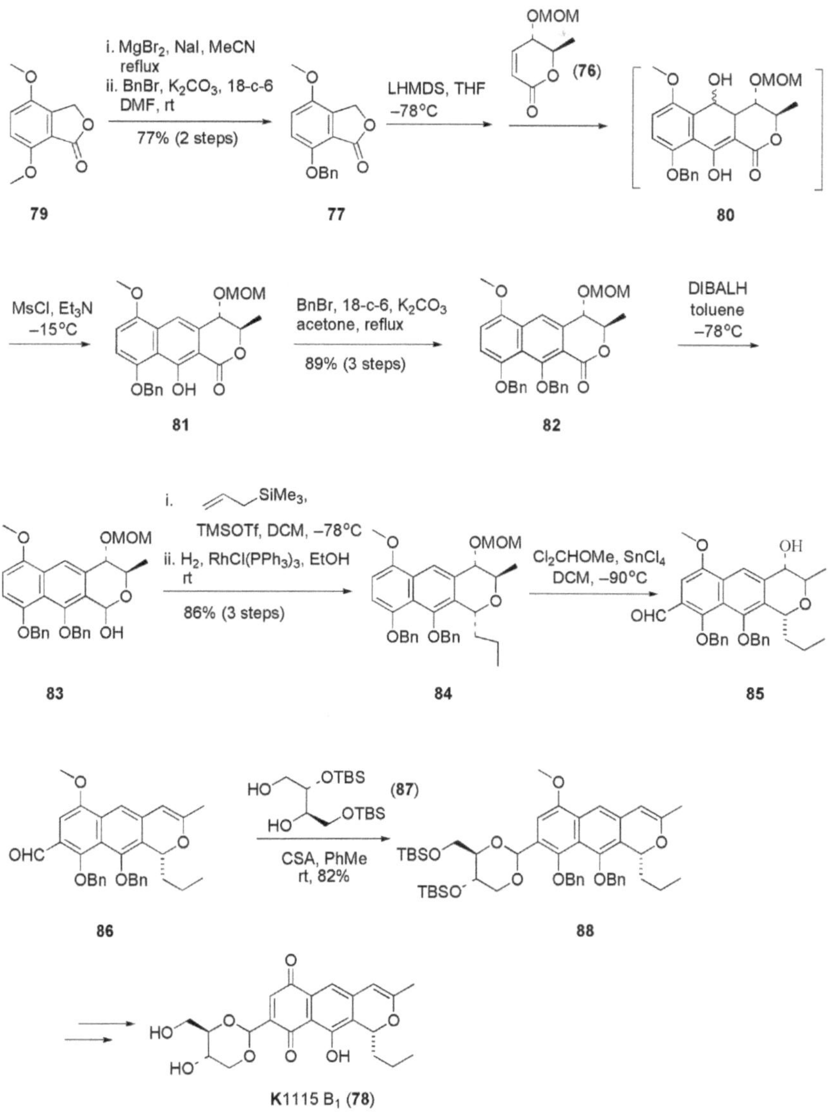

SCHEME 5.12 Total synthesis of K1115, a naphthopyranomycin [2011 THS 983 Tatsuta K].

afforded naphthopyran (**84**). Treatment of (**84**) with dichloromethyl methyl ether and stannic chloride promoted selective formylation as well as demethoxymethylation yielding (**85**), which upon dehydration with Burgess reagent afforded naphthopyran (**86**). Coupling between (**86**) and (**87**) was effected in the presence of camphor sulfonic acid to give (**88**). A few more functional group manipulations including a hypervalent iodine assisted dearomatization afforded the target molecule, i.e., (**78**).

Allenecarboxylates (2,3-allenoates): The study of annulation reactivity of 2,3-allenoates has recently been carried out by Mal group, although the original Sammes annulation was discovered 40 years ago in 1978, and the extension of the annulation to allenoates was obvious.[16] The annulation between phthalide (**5**) with allenecarboxylates (**89**) was expected to give tetralone (**90**) like an acrylate or its CC double bond isomerized aromatic product (**91**) (Scheme 5.13). This annulation study of 2,3-allenoates was conceptualized to avoid the use of cyanophthalides (Chapter 4) in the Hauser annulation for the synthesis of quinol like compound (**91**), because of toxicity of cyanide in their preparations.

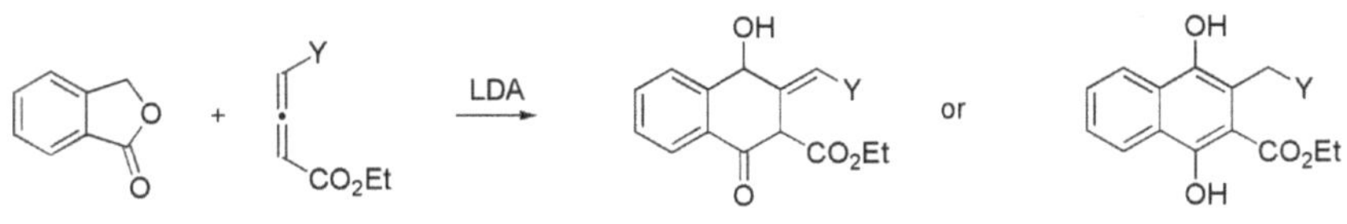

5 **89** **90** **91**

SCHEME 5.13 Proposed Sammes annulation with allenoates.

The difference between the phthalide donors of the Hauser annulation and the Sammes annulation is their level of oxidation, which is higher in the former. The increased oxidation level is due to a nucleofuge at C3 of phthalides. As a result, the Hauser annulations give 1,4-naphthoquinols as the products, whereas the Sammes annulations give 1-naphthols. The proposed employment of allenoates in the place of acrylates as shown in Scheme 5.13 increases oxidation level in the acceptors. Therefore, the net oxidation levels of the two annulations are same. Hence, the product from the Hauser annulation with an acrylate is same as that of the Sammes annulation of an equivalent allenoate.

Mal and Jana, for the first time, showed that nonactivated phthalides (**92**) can undergo smooth annulation with allenecarboxylates (**89**) in the presence of LDA in THF at −78°C, and give naphthalides (**93**) in good yields (Scheme 5.13).[16] The expected products (**94**) were not obtained.

The reaction is quite general with regard to the substituents: methoxy, bromo, cyano, methyl and ethyl in the phthalides. The selected examples of naphthalides (**95–100**) synthesized by the route are presented in Fig. 5.1.

This transformation is interpreted by stepwise Michael addition followed by Dieckmann condensation in the initial stages. In the second phase, the newly generated hydroxyl group is transposed leading to the formation of the products (**101**). The annulation cascade is initiated by LDA promoted generation of 3-lithiophthalides (**102**), which undergo Michael addition to the central carbon of the allene carboxylate (**89**) to give enolate (**103**). The Dieckmann condensation of new carbanionic species (**103**), and produces (**104**) and then (**105**) through deprotonation. During work-up of the mixture, intermediates (**105**) combine with H^+ to give (**106**). Attack of H_2O at the electron deficient double bond of (**106**) results in the formation of naphthols (**101**). Further cyclization of the naphthols (**101**) with mild bases gives desired naphtho[*c*]furanones (**93**) (Scheme 5.15).

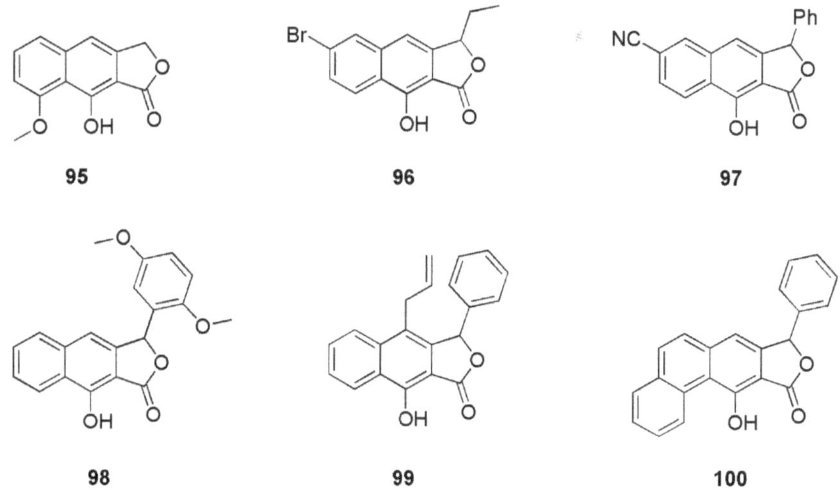

FIG. 5.1 Naphthalides synthesized by annulations with allenecarboxylates.

The most striking feature of the annulation is the exclusive formation of single regioisomers, i.e., (**93**), refuting the above proposed mechanism, which leaves room for another possible set of products, i.e., (**107**). As shown in Scheme 5.16, the carbanions (**108**) obtainable from (**103**) are expected to undergo annulation to give tetralones (**109**), which, in term, are likely to undergo isomerization to yield naphthol (**107**). In effect, this mode of annulation was not realized suggesting a concerted cycloaddition as the initial step for the formation of (**104**).

The annulation described in Scheme 5.14 has been employed as a key step in the synthesis of justicidin B (**110**) as shown in Scheme 5.17. The reaction of phthalide (**111**) with the allenecarboxylate (**112**) in the presence of LDA followed by treatment with K_2CO_3 in acetone provided (**113**) in 59% yield. Two-step conversion of naphtho[*c*]furanone (**113**) to justicidin B (**110**) was reported by Shin et al. The naphthalide (**113**) was converted to its triflate derivative by phenyl triflimide. The triflate derivative was then coupled with boronic acid (**114**) in the presence of tetrakis(triphenylphosphine)palladium(0) to furnish justicidin B (**110**).

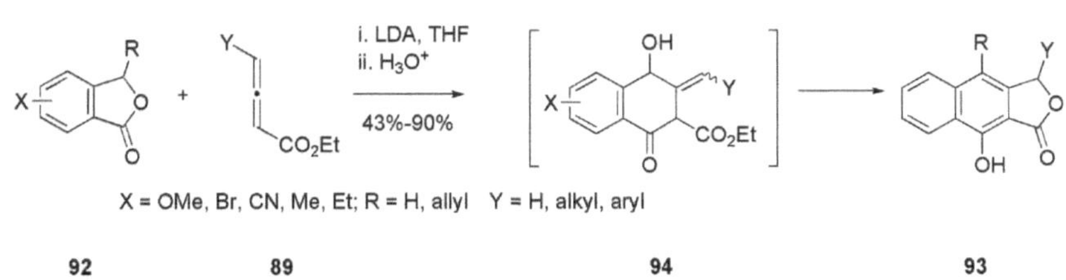

SCHEME 5.14 Annulation of phthalides with allenecarboxylates [2016 JOC 11857 Mal D].

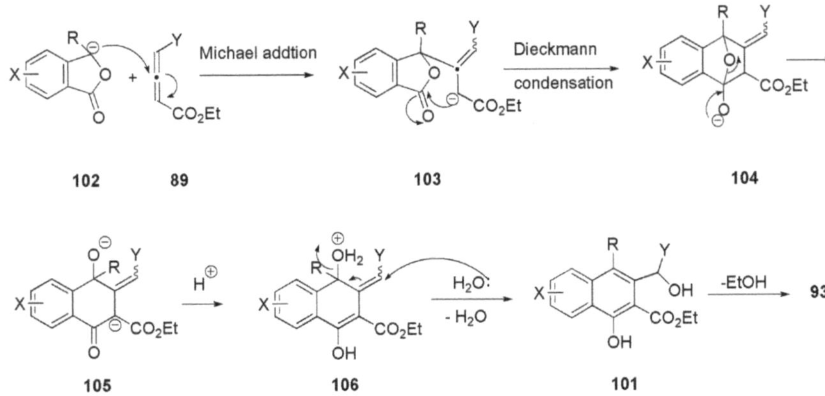

SCHEME 5.15 Mechanism for the synthesis of naphthofuranones (**93**).

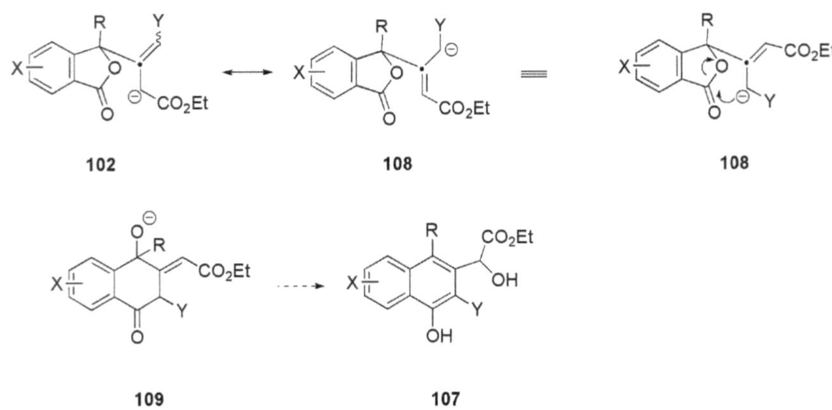

SCHEME 5.16 Annulations that are not realized.

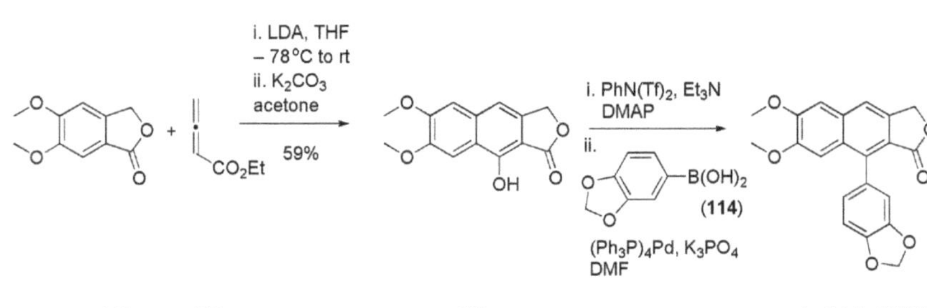

SCHEME 5.17 Synthesis of justicidin B (**110**) [2016 JOC 11857 Mal D; 2015 THN 2945 Shin D].

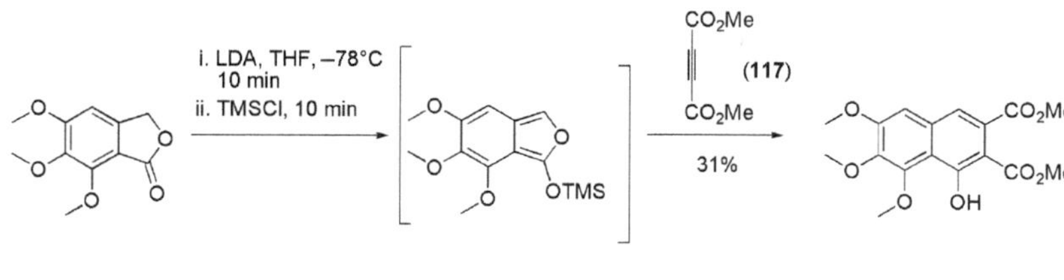

115 116 118

SCHEME 5.18 Annulation of an alkyne ester with a phthalide [1993 OPPI 633 Mali RS].

Propiolates: Activated alkynes have not been systematically examined as acceptors in an annulation with a phthalide. The reactivity of an alkyne ester to the anion of a phthalide was first examined by Mali and Jagtap in 1993. It is proposed to consist of a sequence of trapping of initial phthalide anion as silyl ether, Diels-Alder reaction followed by a fragmentation and an isomerization. In the presence of LDA, phthalide anion derived from (**115**) was trapped as isobenzofuran (**116**), which in situ reacted with dimethyl acetylenedicarboxylate (**117**) to produce annulation product (**118**) in 31% yield. This strategy has been generalized with three more analogous naphthalene carboxylates, highest yield being 41% (Scheme 5.18).[17] There is no mention of the fate of the reaction without the mediation of TMSCl.

5.2 ALPHA-SUBSTITUTED ACCEPTORS (MAL DECARBOXYLATIVE ANNULATION)

Although the Sammes annulation was discovered in 1978, for more than 30 years no α-substituted acrylates except (**35**) were tested for their reactivity to phthalide anions. In 2011, Mal et al. systematically investigated the Sammes annulation with α-substituted acrylates.[18] The outcomes of the studies are unique. The reaction of phthalides with α-substituted acrylates in the presence of bases like LDA or LHMDS provides α-naphthols, devoid of the ester group of the acrylates, in one pot and good yields. The gap of 33 years between the two investigations[1,18] is perhaps due to the apprehension that the initial annulation product (**119**) would be susceptible to *retro*-Dieckmann ring-opening giving (**120**). In a normal Dieckmann cyclization, the equilibrium is shifted to the forward direction by enolate formation (cf. **121**) of the initial Dieckmann product with an α-hydrogen (Scheme 5.19).

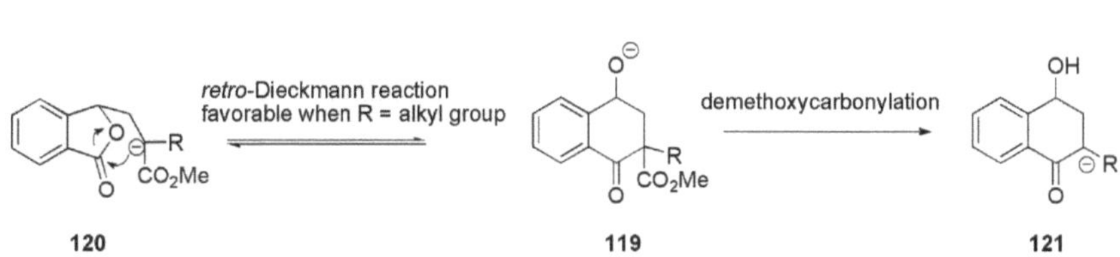

120 119 121

SCHEME 5.19 Epitome of the Mal dealkoxycarbonylative annulation.

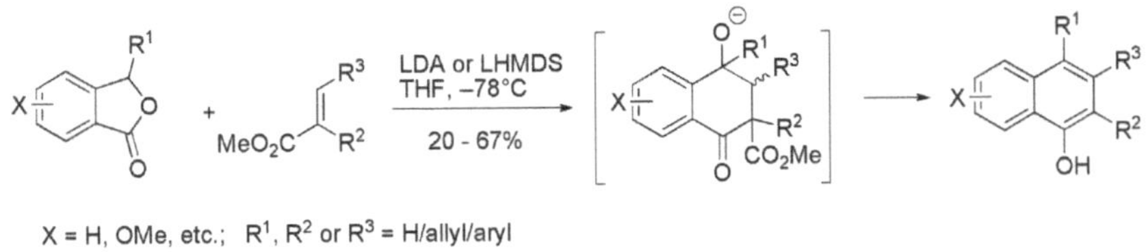

122 123 125 124

SCHEME 5.20 Mal decarboxylative benzannulation [2011 JOC 3392 Mal D].

The reaction of phthalides (**122**) with α-substituted acrylates (**123**) in the presence of LHMDS or LDA furnished naphthols (**124**) in 20%–67% yields through loss of the ester group of (**125**) (Scheme 5.20).[18]

The Mal annulation (Scheme 5.20) differs from the Sammes annulation (5.1) in that the former accommodates the α-substituent of the acrylates at C2 of the α-naphthol products and the electron-withdrawing group, often the ester groups, of the acceptor is eliminated during the annulation.

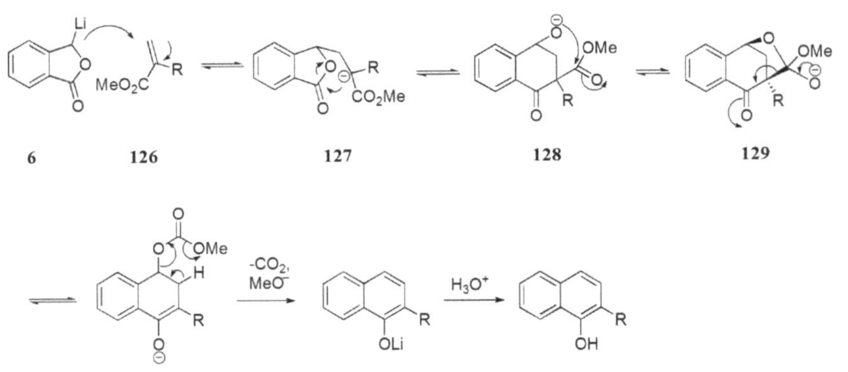

6 126 127 128 129

130 131 132

SCHEME 5.21 A plausible mechanism of the Mal dealkoxycarbonylative annulation.

As shown in Scheme 5.21, the annulation cascade is initiated by generation of 3-lithiophthalide (**6**) from the phthalide (**5**), which undergoes Michael addition with acrylate (**126**) to form carbanion (**127**). This anion then undergoes Dieckmann cyclization, leading to the tetralone (**128**). Intramolecular nucleophilic attack of the alkoxy anion in oxytetralone (**128**) gives (**129**), which fragments to carbonate (**130**). Loss of CO_2 and MeO^- from (**130**) then generates (**131**), eventually forming (**132**) on acidic work-up. The mechanism is, in part, supported by a fortuitous isolation of a hydroxytetralone.

The above decarboxylative annulation has been found effective for a concise synthesis of arnottin I (**133**), a natural product surmised to possess antibiotic properties. As shown in Scheme 5.22, LHMDS-promoted annulation of phthalide (**134**) with acrylate (**135**) provided arnottin (**133**) in one step with

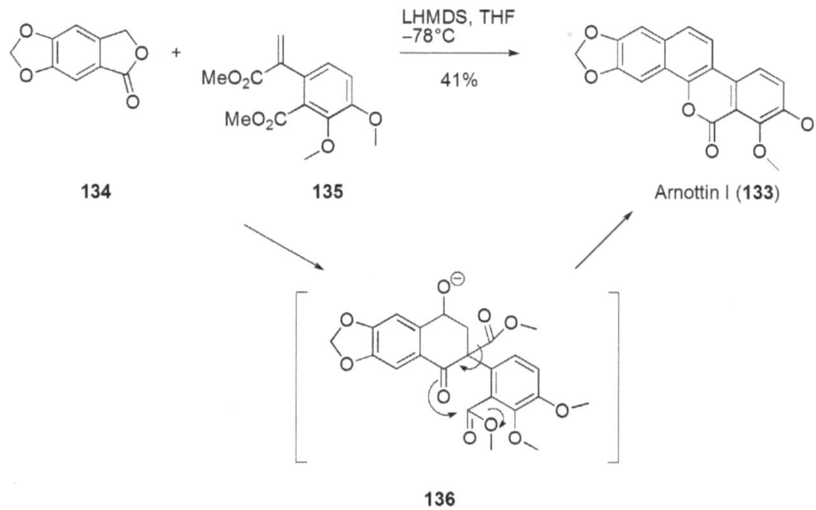

SCHEME 5.22 Synthesis of arnottin I by decarboxylative annulation [2011 JOC 3392 Mal D].

concomitant formation of two new rings. The reaction is proposed to take place via the intermediate (**136**).

In similar fashion, neotanshinlactone (**137**), an attractive synthetic target because of its bioactivity profile, has been synthesized by Mal decarboxylative annulation.[19] Neo-tanshinlactone exhibits anti-breast cancer activity and is more potent and selective than the estrogen receptor modulator tamoxifen citrate. Hydroxy benzoate (**138**) was converted in three steps into phthalide (**139**). Treatment of the phthalide (**139**) with LHMDS at −78°C followed by addition of furyl acrylate (**140**) afforded (**141**) as the single product. Furonaphthopyranone (**141**), was then converted into triflate derivative (**142**). The removal of the TBDMS group of (**141**) with tetra-*n*-butylammonium fluoride (TBAF) followed by treatment of the resultant phenol with triflic anhydride and 2,6-lutidine afforded (**142**) in 40% yield (over two steps). The Fürstner cross-coupling reaction of the triflate (**142**) with MeMgI in the presence of $Fe(acac)_3$ and *N*-methylpyrrolidone gave neotanshinlactone (**137**) (Scheme 5.23).

The decarboxylative annulation, when extended to phthalide (**5**) and 2-acetaminoacrylate (**143**), provided desired annulation product (**144**) was obtained. But, the yield of the annulation was only 20%. The poor yield may be due to the presence of active NH functionality in the acceptor (Scheme 5.24).[20]

An extension of the above decarboxylative annulation was considered to synthesize 2-azidonaphthols (**145**) on the way to 2-aminonaphthols (**146**) from phthalide (**5**) and azidoacrylates (**147**) as depicted in Scheme 5.25. Quite surprisingly, such an annulation was not realized. The reaction between phthalide (**5**) and azidoacrylate (**148**) in the presence of LHMDS at −78°C furnished benzazepinone (**149**) in 90% yield (Scheme 5.26).[21] The unusual formation of the azepinone (**149**) in the above reaction can is explained by the mechanism given

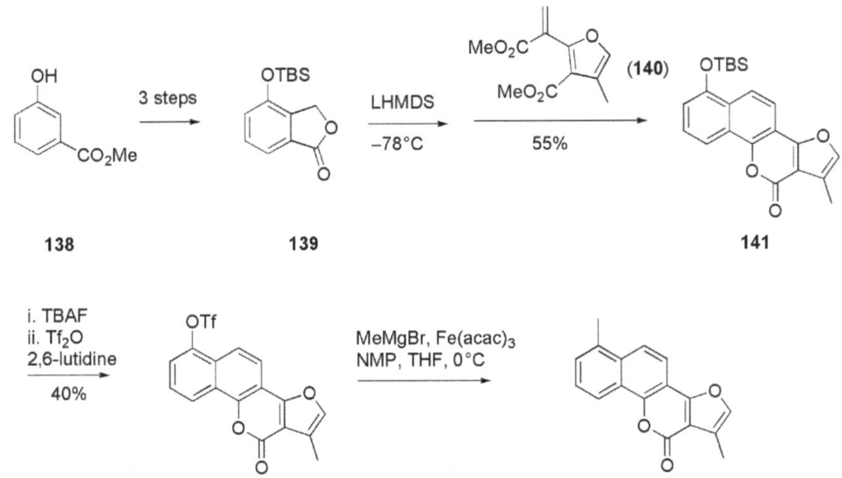

138 139 141

142 137

SCHEME 5.23 Synthesis of neotanshinlactone [2013 EJOC 4037 Mal D].

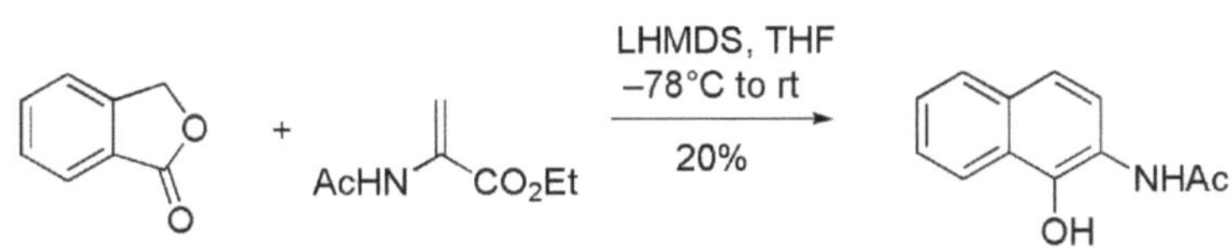

5 143 144

SCHEME 5.24 Annulation with acetaminoacrylate (**143**).

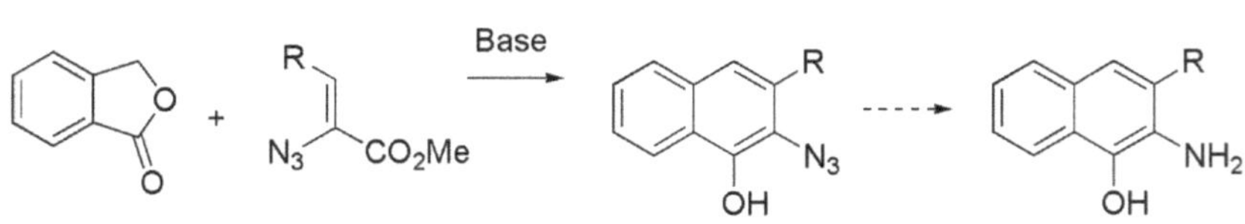

5 147 145 146

SCHEME 5.25 Nonrealized annulation with 2-azidoacrylates.

in Scheme 5.27. At low temperature, the incipient yellow phthalide anion (**6**) is proposed to add to azidoacrylate (**148**) in Michael fashion to form anion (**150**). The enolate ion (**150**) expels N_2 to form imine anion (**151**) in a manner similar to 2-azido esters. This, in turn, undergoes lactamization to form seven-membered lactam (**152**). Finally, isomerization of (**152**) followed by protonation of the anion leads to the azepinone (**149**). This reaction has been applied to the synthesis of a series of benzazepinones (**147–151**) shown in Fig. 5.2.

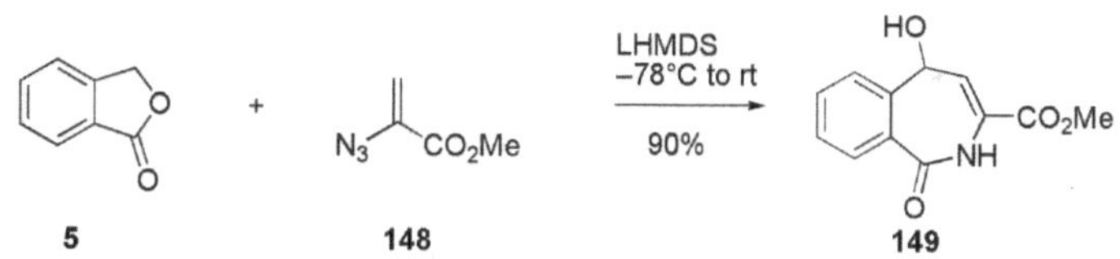

The yields vary with bases. 45% (with LDA); 13% (with NaHMDS); 0% (with KHMDS); 0% (with LiO^{t}Bu)

SCHEME 5.26 Anionic [4+3] heteroannulation with *α*-azidoacrylates.

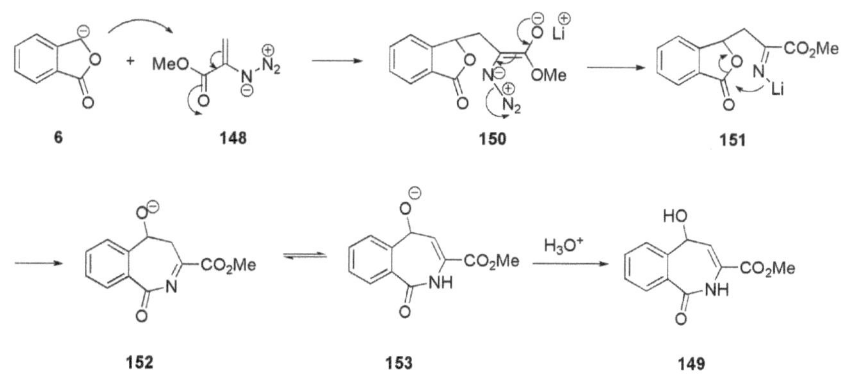

SCHEME 5.27 Probable mechanism of the formation of benzazepinone (**149**).

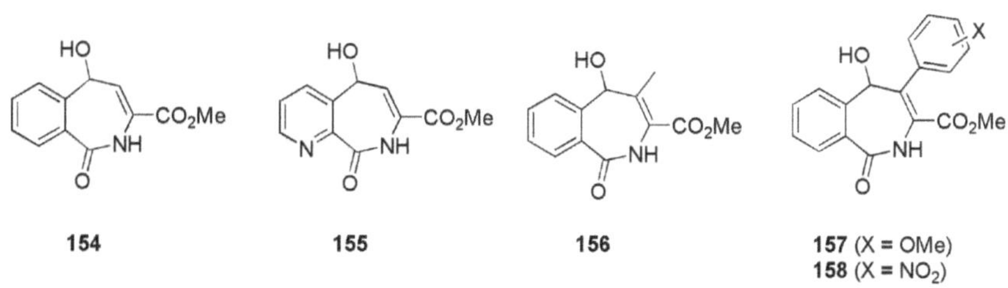

FIG. 5.2 Some examples of benzazepinones synthesized by the Mal [4+3] annulation.

5.3 CONJUGATED KETONES

The annulations of phthalides with simple unsaturated ketones are very uncommon. For simple ketonic acceptor, the reactivity of potassium phthalide-3-carboxylate (**159**) has been studied. It is shown to undergo facile decarboxylation in DMSO at 145°C to generate phthalide anion (**6**), which reacts with α,β-unsaturated ketones to give two kinds of products. The reaction with benzalacetophenone gave Michael adduct (**160**) in 44% yield and the rearranged annulation product (**161**) in 30% yield.[22] Cesium salts of phthalide-3-carboxylic acids seem to be favorable for annulation products (Scheme 5.28).

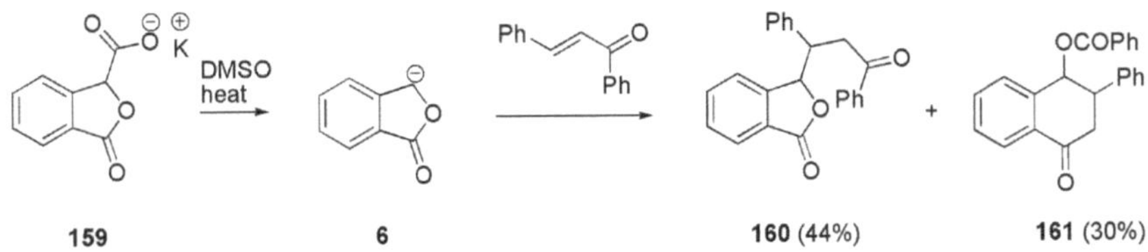

SCHEME 5.28 [1985 AJC 921 Prager RH].

Tetracyclines and Sammes annulation: The Sammes annulation studies with cyclohexenones were started with an aim to fabricate the anthracenone moieties of auerolic acids, so to speak, olivins. It has been found that cyclohexenones are notorious acceptors for phthalides without a nucleofuge at C3 in contrast to their annulation with Hauser sulfone donors (**162**) and Kraus cyanide donors (**163**) (Fig. 5.3).[23] However, 3-methylphthalide was successfully annulated with a decalenone to establish an asymmetric synthesis of a tetracycline.[24]

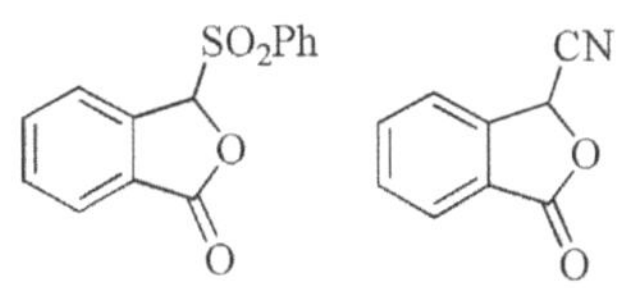

162 **163**

FIG. 5.3 Structures of Hauser donors and Kraus donors.

Tetracyclines **164–167** are naturally occurring or semi-synthetic antimicrobial agents. They are characterized by a unique highly oxygenated hydronaphthacene structure. The first tetracycline antibiotic, chlorotetracycline (Fig. 5.4) was discovered in 1948, by Benjamin Duggar of Lederle Laboratories from the culture broth of a novel species of *Streptomyces*. It was found effective against a broad spectrum of microorganisms including Gram-positive and negative bacteria as well as eukaryotic protozoan parasites. Since then Pfizer, Lederle and Wyeth scientists have led many discoveries of new tetracycline derivatives which are clinically approved.

All tetracyclines approved for the human or veterinary use are fermentation products. The use amounts to about 5000 metric tons. The third generation semisynthetic tetracyclines **168–171** shown in Fig. 5.5 show improved activity.

Although the tetracycline antibiotics have been the mainstay of antibacterial chemotherapy for decades, they had eluded efficient total synthesis. The synthetic chemistry of tetracycline families began with Woodward's synthesis of 6-demethyl-6-deoxytetracycline in 1962 and Muxfeldt's terramycin synthesis in 1968. These were followed by Stork's 12a-deoxytetracycline synthesis in 1996. However, all these syntheses ended in racemic forms. In a landmark accomplishment, Tatsuta extended the concept of anionic [4+2] benzannulation to the synthesis of tetracyclines. In 2000, Tatsuta group reported the first total

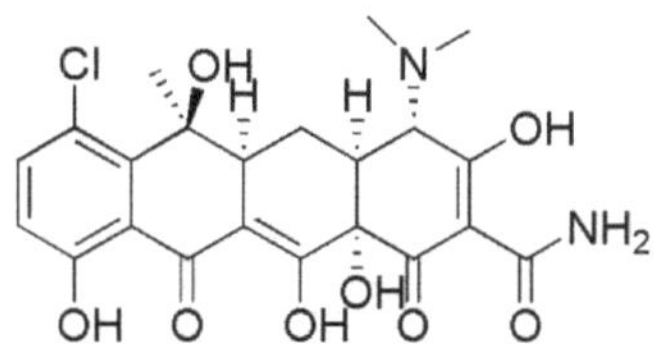

Chlorotetracycline (**164**)
(Aureomycin)

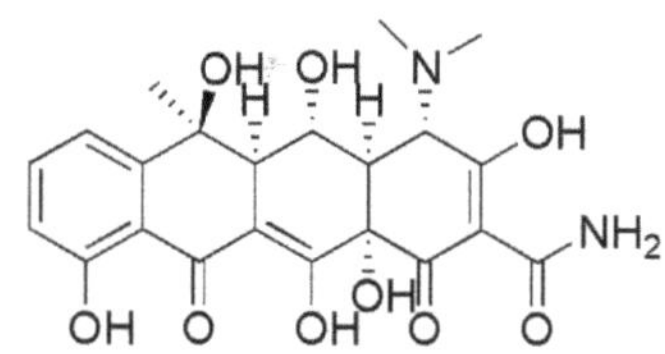

Oxytetracycline (**165**)
(Terramycin)

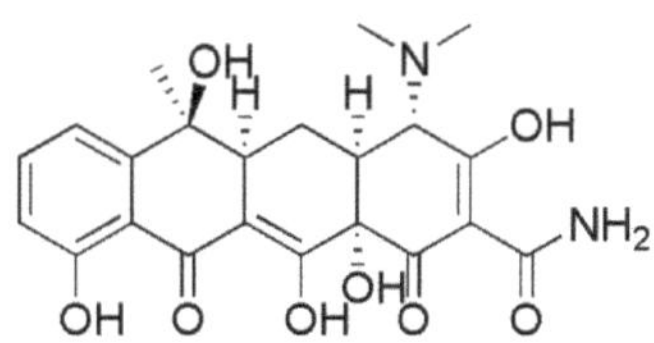

Tetracycline (**166**)

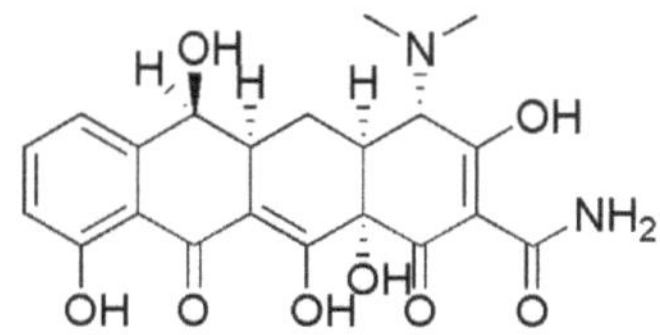

6-Demethyltetracycline (**167**)

FIG. 5.4 Natural tetracyclines.

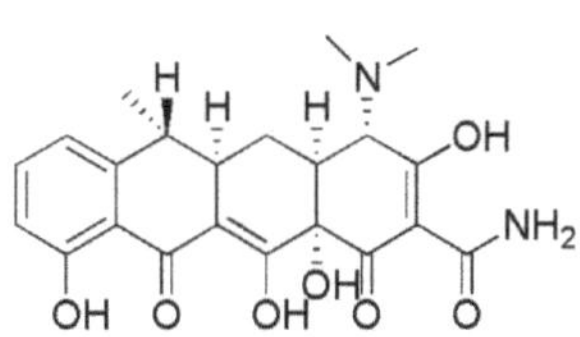

6-Deoxytetracycline (**168**)

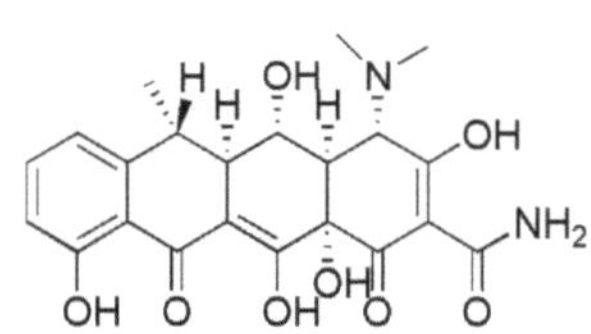

Doxycycline (**169**)

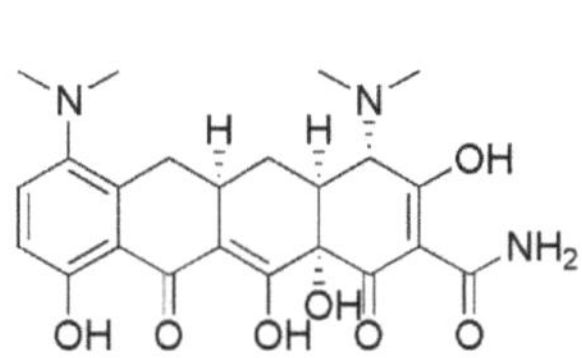

Minocycline (**170**)

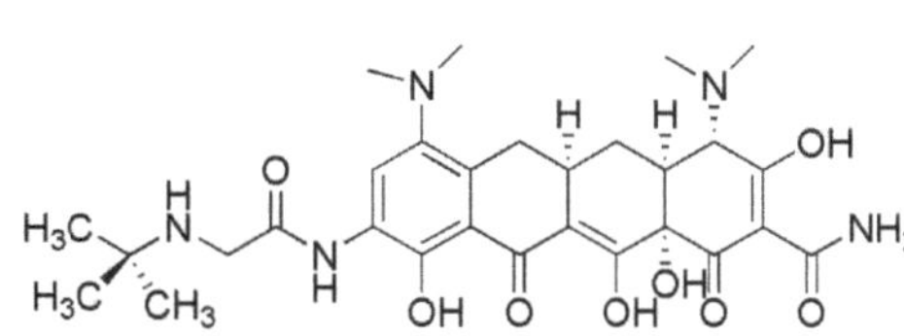

Tigecycline (**171**)

FIG. 5.5 Semi-synthetic tetracyclines.

synthesis of natural (−)-tetracycline (**166**), the key step being a Sammes annulation (Scheme 5.29).[24] Monosaccharide (**172**), providing the structural features of the A ring, was converted in 11 steps to enone (**173**) in 26% overall yield. Diels-Alder of (**173**) with 1-trimethylsiloxybutadiene followed by oxidation of the resultant adduct with CrO_3 gave decalenone (**174**). Annulation of (**175**) with (**174**) proceeding through (**176**) afforded tetracycle (**177**). Thionyl chloride promoted aromatization of (**177**) gave (**178**). Twelve more steps applied to (**178**)

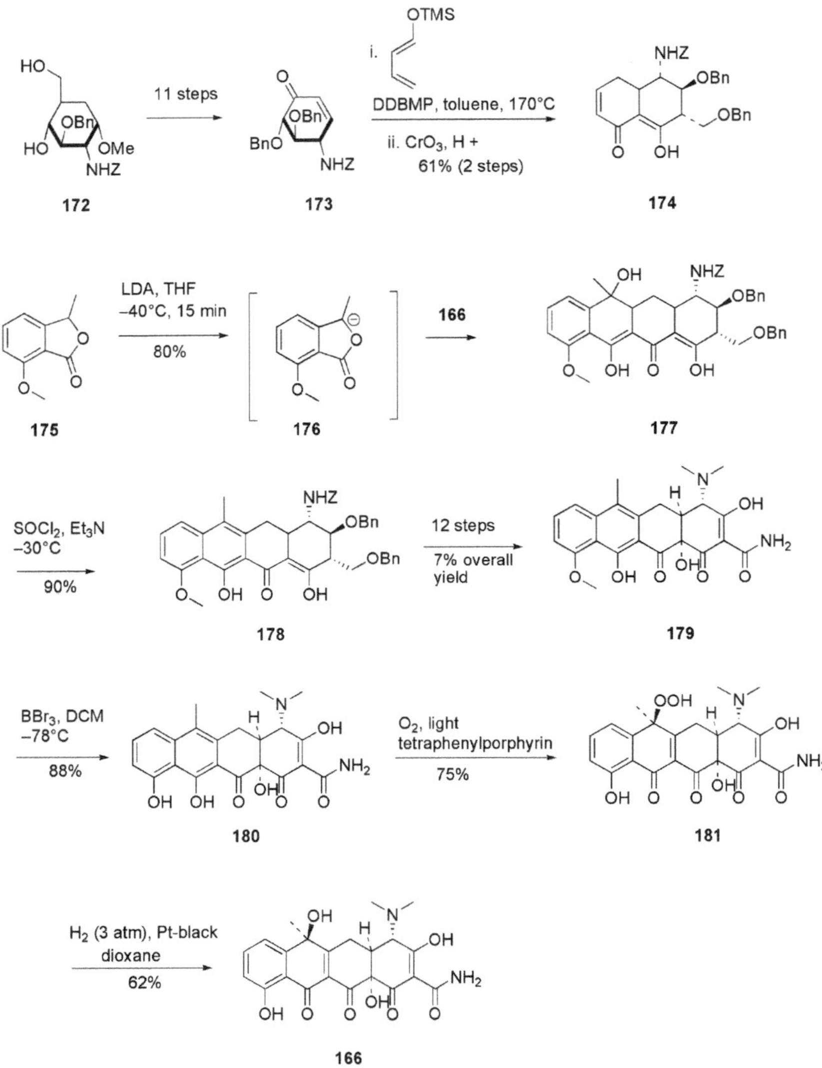

SCHEME 5.29 Tatsuta asymmetric synthesis of a tetracycline [2000 CLS 646 Tatsuta K].

produced angularly oxygenated tetracycle (**179**). BBr_3 promoted demethylation gave tetracycle (**180**). Singlet oxygen oxidation of (**180**) followed by catalytic reduction of (**181**) furnished parent tetracycline (**166**).

In contrast to the behavior of 2-cyclohexenone, 2-azidocyclohexenone enters into a new cascade of reactions with phthalides, which can be described as a [4+3] annulation. A representative example of this annulation is presented in Scheme 5.30. The reaction of phthalide (**5**) with 2-azido-α,β-unsaturated ketone (**182**) in the presence of LHMDS yielded benazepinone (**183**) (Scheme 5.30).[21] A probable mechanistic pathway is presented in Scheme 5.31. The conjugate

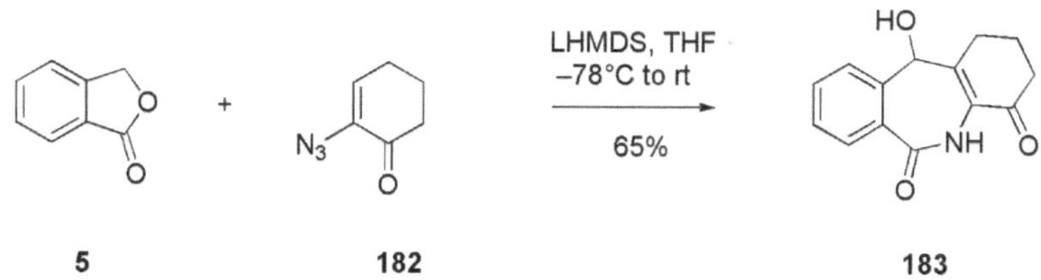

SCHEME 5.30 A new [4+3] annulation to benzazepinones [2012 CCS 3999 Mal D].

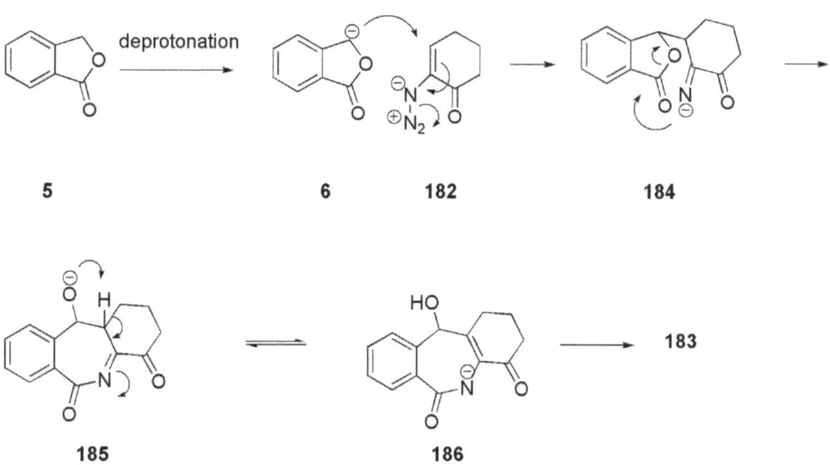

SCHEME 5.31 Mechanistic proposal for the formation of 2-benzazepinone (**183**).

base (**6**), derived from (**5**), adds to azidoenone (**182**) to form imide anion (**184**). The imide ion (**184**) then undergoes lactamization to produce intermediate (**185**), which then transforms to (**186**) and finally to (**183**).

Quinone imines, unlike quinones, succumb to the Sammes annulation. In a modified version, Myers et al. utilized the Sammes annulation in their first total synthesis of dynemicin (**187**). The anion of phthalide (**188**) was converted into its silyl ether (**189**). Diels-Alder cycloaddition of the quinone imine (**190**) with the isobenzofuran (**189**) gave adduct (**191**), oxidative work-up of which provided (+)-dynemicin (**187**) in 40% yield (Scheme 5.32).[25] This annulation is flexible in terms of the quinone imine and isobenzofuran components to access dynemicin analogs.

The reaction of a similar phthalide, i.e., (**192**) with quinone imine (**193**) in the presence of LHMDS without the mediation of TMSCl resulted in the formation of a 1.3:1 diastereomeric mixture of Michael adducts (**194**). The subsequent Dieckmann condensation step seems to be a thermodynamically unfavorable (Scheme 5.33).

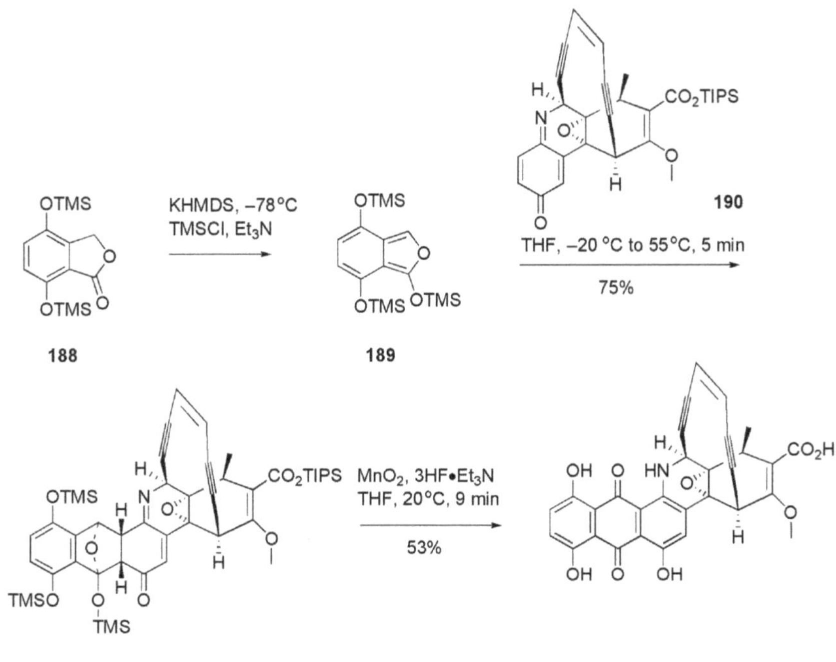

SCHEME 5.32 Annulation of a quinone imine [1997 JACS 6072 Myers AG].

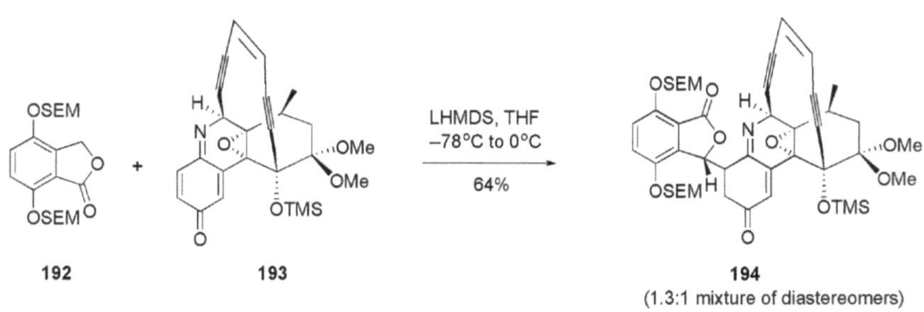

SCHEME 5.33 Attempted direct Sammes annulation with a quinone imine.

5.4 SCHIFF BASES

The reaction between phthalides (**195**) with Schiff base (**196**) in the presence of LDA yields isoquinolone products (**197**) and (**198**) in 2:1 ratio (Scheme 5.34).[26] The preponderance of the *cis* products (**197**) is explained by the formation of lithium coordinated complex (**199**) derived from (**200**). This result may be compared with the work of Magnus et al. on 2-pyridyl acrylate, which yielded only the *cis* product (vide 5.1.7). The *cis*-alcohols (**197**) undergo elimination to give 2,3-disubstituted isoquinolones, on treatment with mesyl chloride in pyridine.

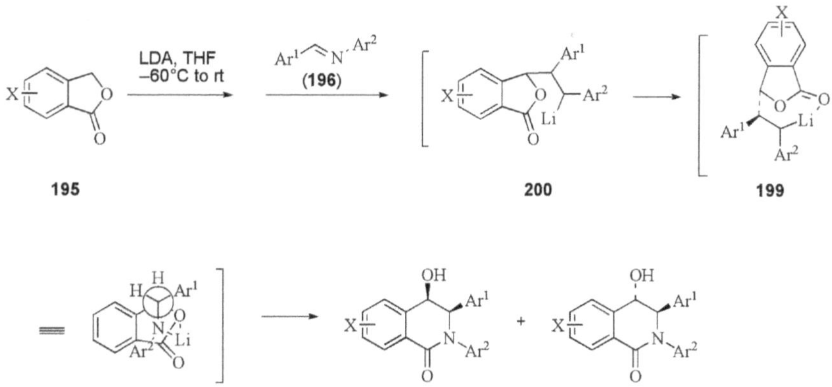

195 200 199

199 197 198

SCHEME 5.34 Reaction of phthalides with Schiff bases [1983 JCSPTI 1455 Sammes PG].

Attempted annulations with imines of aliphatic amines failed to deliver annulated products, except in a few cases.

On the other hand, the phthalide anions (**201**), prepared by the reaction of the corresponding phthalides with LDA in THF, react with iminium salts (**202**) gives phthalide isoquinolines (**203a** and **203b**) (Scheme 5.35).[27]

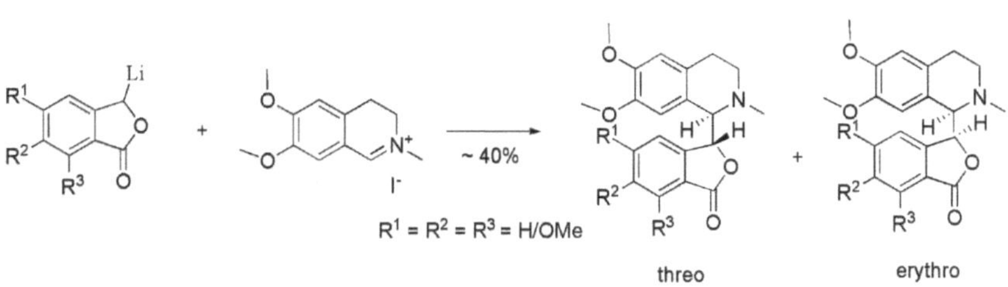

201 202 203a 203b

SCHEME 5.35 Synthesis of phthalide isoquinolines [1984 CJC 306 MacLean DB].

Like Schiff bases, 3,4-dihydroisoquinolines react with phthalide anions in THF to give 13-hydroxyprotoberberine derivatives. For example, the reaction of (**204**) with (**205**) in the presence of LDA gave the annulation product (**206**) as a single isomer (Scheme 5.36).[28]

The annulation of cyclic imine has been used in the total synthesis of protoberberine alkaloid, cavidine. Treatment of the cyclic imine (**207**) with the anion, generated by the action of LDA at −78°C on the phthalide (**208**), selectively produced single diastereoisomers, i.e., (**209**) in 80% yield. Dehydration of (**209**) with PTSA yielded tetracycle (**210**). A sequence of LAH reduction and $NaBH_4$ reduction of (**210**) is reported to furnish cavidine (**211**) in racemic form (Scheme 5.37).

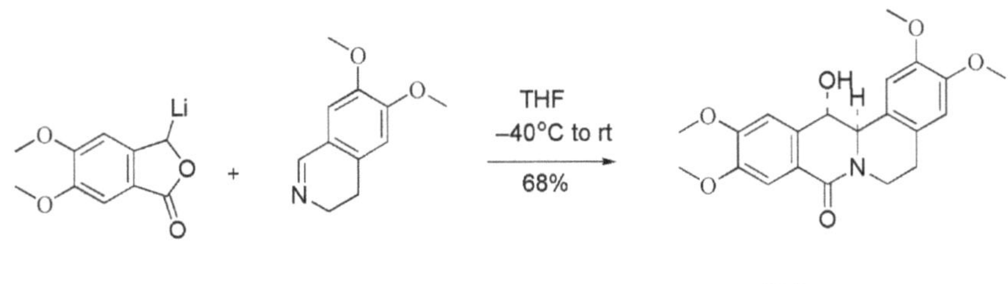

204 205 206

SCHEME 5.36 Annulation with a cyclic imine en route to protoberberine [1984 CJC 1392 MacLean DB].

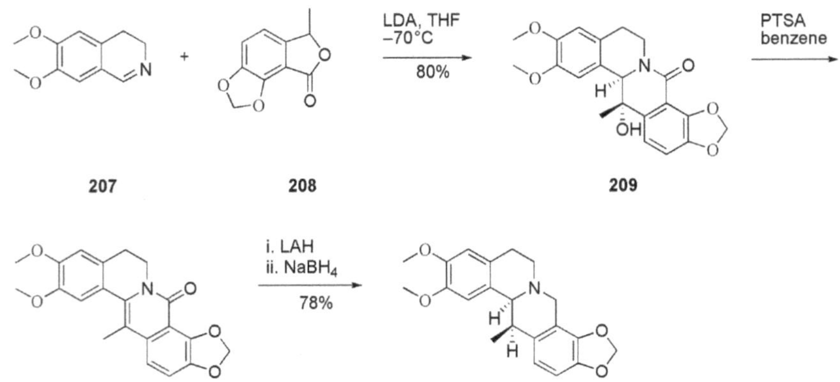

207 208 209

210 Cavidine (211)

SCHEME 5.37 Synthesis of cavidine [1984 CJC 1392 MacLean DB].

5.5 ARYNE PRECURSORS

The annulation by the way of interception of arynes with a phthalide anion was first discovered in 1979 by Sammes and Dodsworth.[29] It provides highly functionalized anthraquinones in one step from aryne precursors and phthalides. Unlike the phthalides with C3 nucleofuges discussed in details in Section 5.5, Sammes one-pot synthesis of anthraquinones require phthalides without any nucleofuge at C3. The necessary oxidation level to give quinone products is compensated by aerial oxidation of the initial hydroxyanthrone or 9,10-anthracenediol products. As shown in Scheme 5.38, the reaction between phthalide (**5**) and bromobenzene (**212**) in the presence of LDA provides anthraquinone (**213**) in one-pot operation up to 74.5% yield. LDA serves two purposes: it removes HBr to produce benzyne (**214**) from (**212**), as well as promotes formation of carbanion (**6**) from phthalide (**5**). The resulting carbanion (**6**) undergoes annulation to form intermediates (**215**), (**216**), and (**217**). The intermediate (**216**) undergoes fragmentation to anthrone (**217**). The annulation can also be described as a two-step process via direct formation of (**216**) from the oxyanion structure of (**6**).

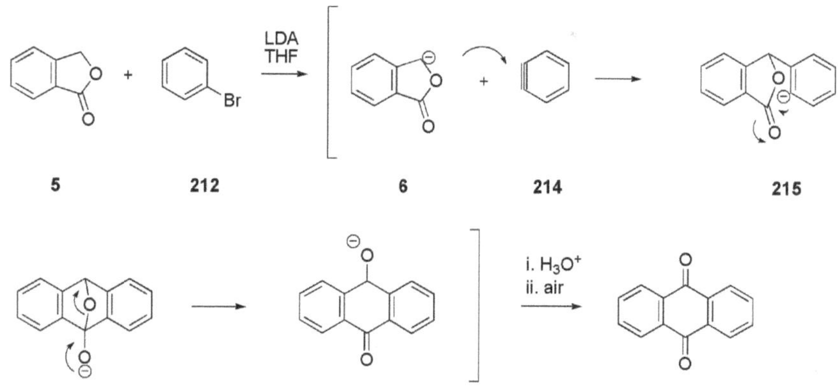

5 212 6 214 215

216 217 213

SCHEME 5.38 Sammes annulation [1979 CCS 33 Sammes PG].

The substituents of the anthraquinones are usually alkyl, alkoxy and benzo groups. Most often, bromoarenes are used as aryne precursors. Thus, the acetal (**218**) reacted with phthalide (**5**) under the influence of LDA to give quinizarin (**219**) in 41% yield.[30] The corresponding intermediates are (**220–222**) (Scheme 5.39).

Sammes annulation can be used for the synthesis of angularly fused anthraquinones as well. The aryne produced from 1-bromonaphthalene reacted with 3-lithiophthalide to give benz[*a*]anthraquinone in 62% yield. Extension of the Sammes annulation to 7-substituted phthalides suffers from the regioselectivity. The reaction between phthalide (**214**) and bromoarene (**224**) gave a mixture of (**225**) and (**226**) (Scheme 5.40).[30]

However, a methoxy group in the benzyne precursor can induce high regioselectivity in the Sammes annulation with a formally unsymmetrical phthalide. For example, the reaction of phthalide (**227**) with bromoarene (**228**) in the presence of

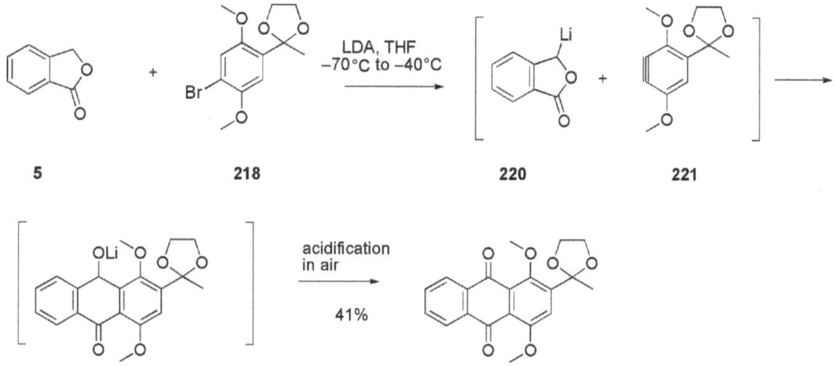

5 218 220 221

222 219

SCHEME 5.39 [1988 JCSPT1 3037 Sammes PG].

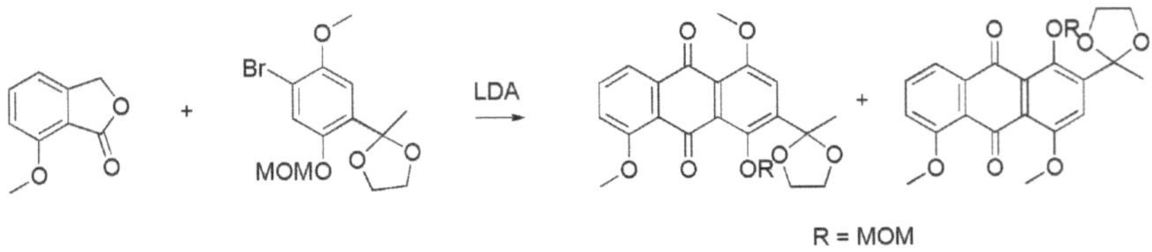

SCHEME 5.40 Nonregioselective Sammes annulation [1988 JCSPT1 3037 Sammes PG].

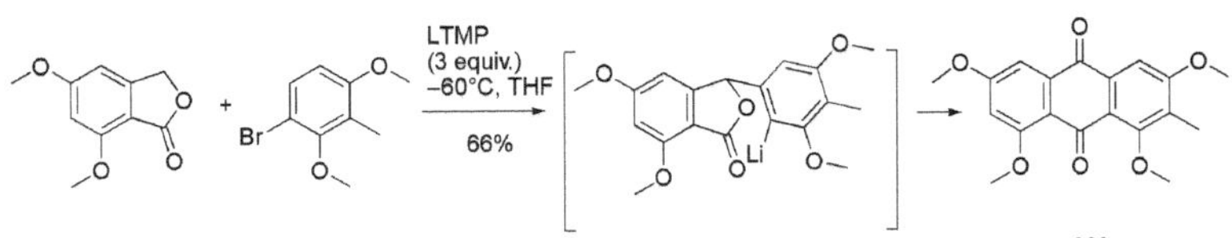

SCHEME 5.41 Methoxy-directed Sammes annulation [1981 JACS 6885 Townsend CA].

LTMP at −60°C to −40°C afforded anthraquinone (**229**) in 66% yield as the sole product (Scheme 5.41).[31] The regioselectivity is explained by the additional stability of the organolithium intermediate (**230**) due to inductive effect and chelation effect of the adjacent methoxy group. Similar regioselectivity exerted by a MOM group is manifested in the Townsend total synthesis of averufin (Chapter 10).

Schmalz work towards the synthesis of mumbaistatin involved a direct comparison of the efficacies of Sammes benzyne annulation and Biehl benzyne annulation. Both (**231**) and (**232**), on the reaction with bromoarene (**233**), furnished the same annulated product, i.e., anthraquinone (**234**) (Scheme 5.42).[32] The other possible regioisomer was not identified. The regioselectivity

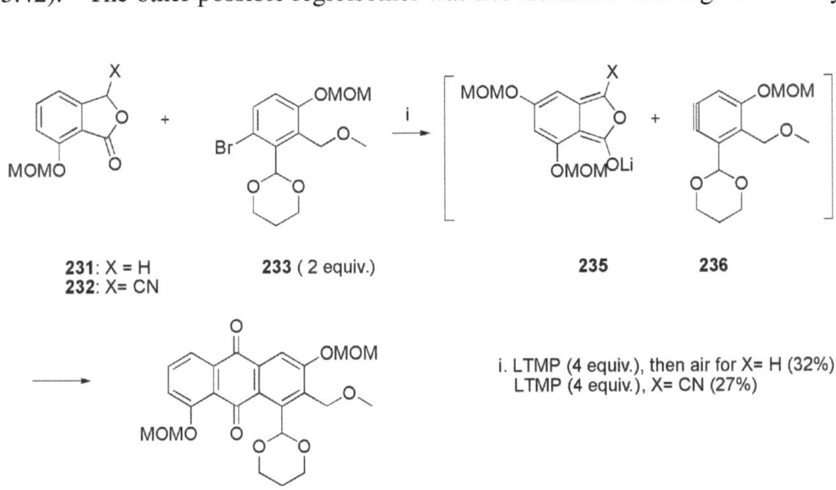

SCHEME 5.42 Comparison of Sammes annulation with Biehl annulation [2003 THS 3201 Schmalz H-G].

is attributed to the attack of the phthalide anion (**235**) at the less electron-rich position of the aryne (**236**). The Biehl annulation is marginally better than the Sammes annulation with regard to efficacy.

High regioselectivity of phthalide-aryne annulation is observed for a peri-MOM substituted aryne precursor. The annulation between phthalide (**237**) and bromoarene (**238**) gave highly oxygenated anthraquinone (**239**) as the single isomer. The regioselectivity is due to coordination of the initial lithioadduct with the MOM group (Scheme 5.43).[33]

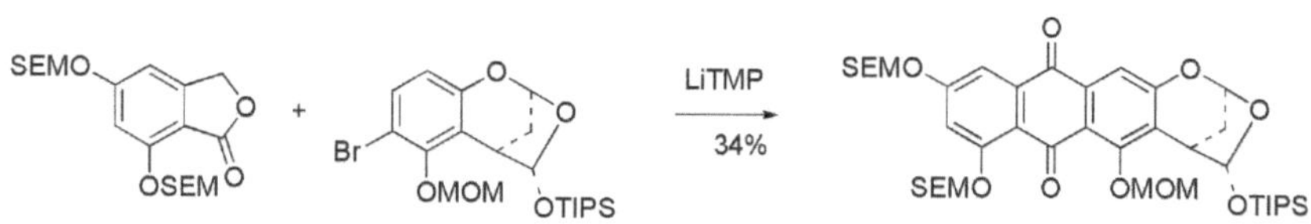

237 **238** **239**

SCHEME 5.43 MOM-directed Sammes annulation [1999 JACS 7730 Townsend CA].

Although the yield was low, the cyclophaneanthraquinone (**240**) could be synthesized by Miki et al. through Sammes annulation of phthalide (**241**), obtained by LAH reduction of (**242**), with bromobenzene (**212**) in the presence of LDA (Scheme 5.44).[34]

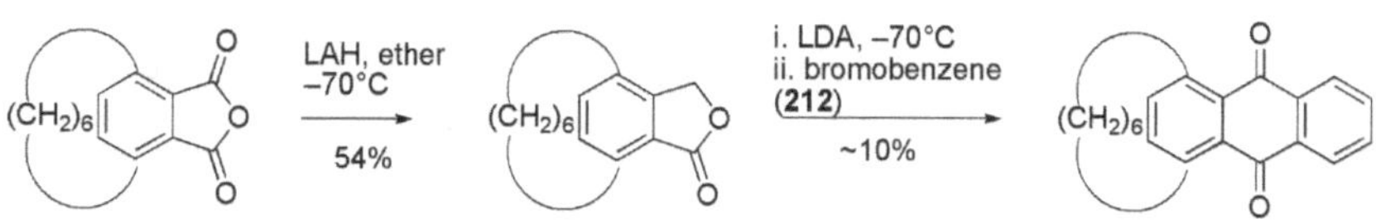

242 **241** **240**

SCHEME 5.44 Synthesis of [6]-1,4-cyclophaneanthraquinone [1992 THS 953 Miki S].

The reactivity of hetaryne precursors towards phthalide anions is attended with duality of reaction mechanisms.[35] The results with four examples displayed in Table 5.2 suggest that both hetaryne and addition-cyclization-elimination mechanisms are probable with the reaction with phthalide anions with 4-bromoquinoline (**243**). In contrast, hetaryne mechanism prevail in the reaction with 3-bromoquinoline (**244**) as the aryne precursor. The reaction of the phthalide (**5**) with 4-bromoquinoline (**243**) produced two products (**245** and **246**). The major product (**245**) is believed to be formed by addition-elimination mechanism. Formation of the minor product (**246**) is interpreted by annulation of 3,4-quinolyne intermediate. The same trend of reactivity is found with 3-methylphthalide (**247**) forming products (**248** and **249**). However, both the phthalides (**5**) and (**247**) gave typical annulation products (**250**), (**251**), and (**252**) on reaction with 3-bromoquinoline (**244**) under aryne-forming conditions.

TABLE 5.2 Reaction of Phthalide Anions With Hetaryne Precursors

Entry	Donor	Acceptor	Product (% Yield)
1	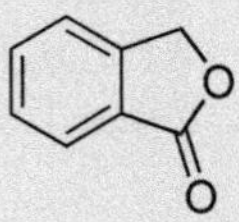 **5**	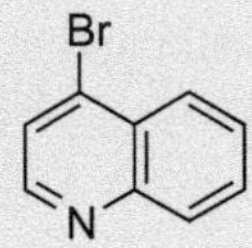 **243**	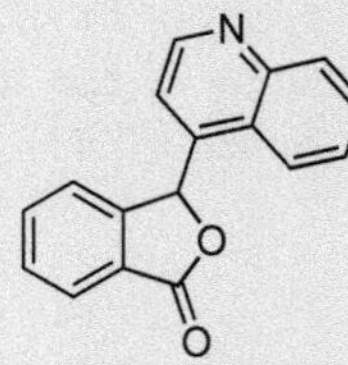 **245** (42%) 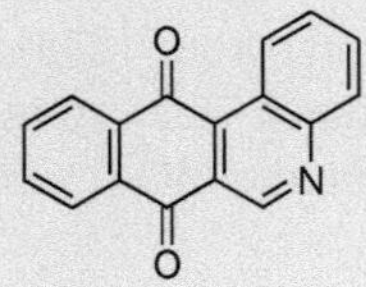 **246** (15%)
2	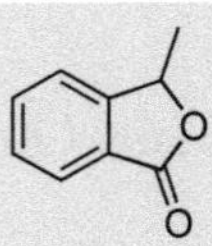 **247**	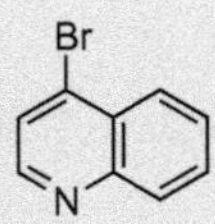 **243**	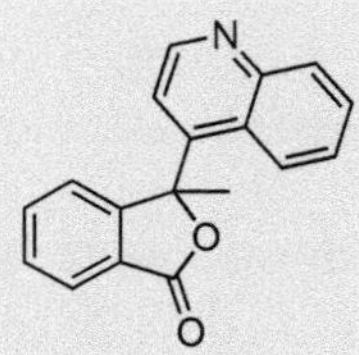 **248** (38%) 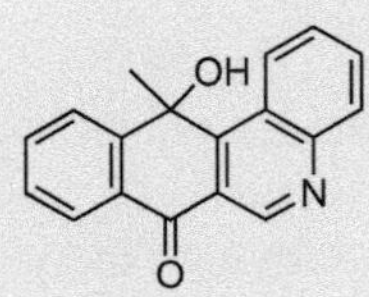 **249** (9%)
2	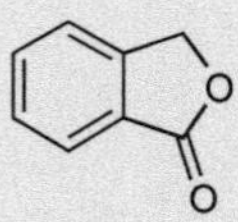	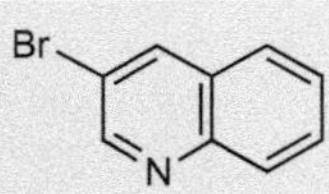 **244**	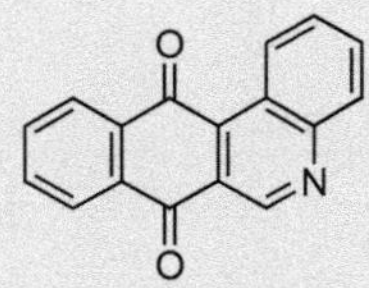 **250** (43%)
4	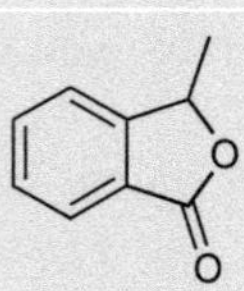 **247**	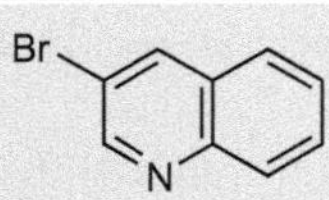 **244**	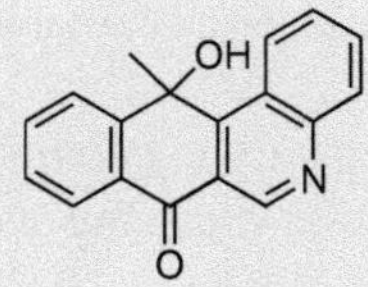 + 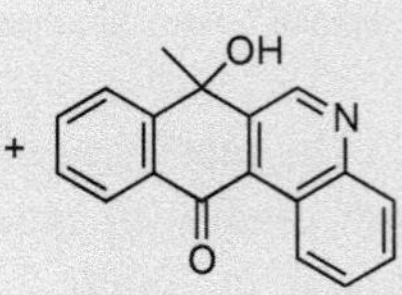 **251** (28%) **252** (31%)

5.6 MISCELLANEOUS SAMMES DONORS AND ACCEPTORS

Thiophthalide donors: The annulation chemistry of thiophthalides was developed in analogy with that of phthalides with a view to developing more efficient annulations. More specifically, it was designed to construct 3,4-dihydroanthracenone moieties of olivin and pillaromycinone for which the Sammes annulation fails.

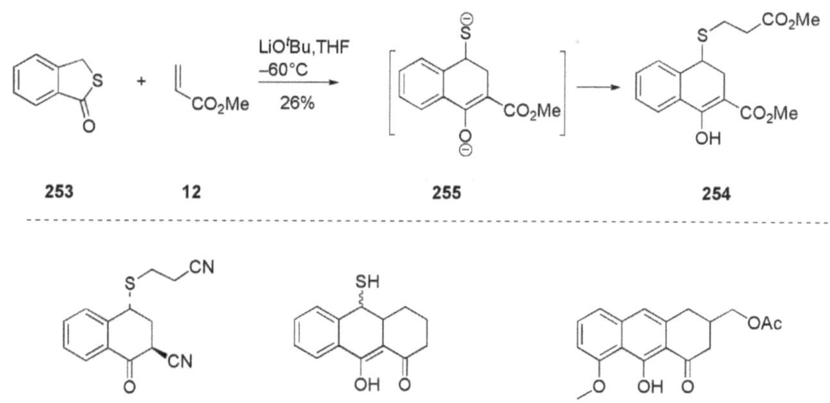

253 12 255 254

2567 (43%) 257 (27%) 258 (23%)

SCHEME 5.45 Annulation of thiophthalide [1994 JCSPT1 309 Mal D].

The respective reactions of phthalides without a leaving at C3 with the cyclohexenones are not effective although the annulations of 3-sulfonylphthalides and cyanophthalides with 2-cycloxenones in the presence of LDA/LiO^{t}Bu are facile. This problem was partly alleviated by the use of thiophthalide analogs (cf. **253**). Contrary to the expectation, the reaction of thiophthalide (**253**) with methyl acrylate (**12**) in the presence of LDA afforded an intractable mixture of products. The change of base to LTB from LDA in the above reaction resulted in formation of the annulated product (**254**) in 26% yield (Scheme 5.45).[23a] The formation of the product (**254**) is explicable in terms of Sammes annulation followed by selective Michael addition of the resultant dianion (**255**) to methyl acrylate. Similar reaction of the thiophthalide (**253**) with acrylonitrile and 2-cyclohexenone gave annulation products (**256** and **257**, respectively). On the other hand, the reaction of 7-methoxythiophthalide with 5-acteoxymethyl-2-cyclohexenone furnished sulfur-free anthracenone (**258**).

The trend of low to moderate yields of the annulation was also observed with other substituted thiophthalides. The annulation of thiophthalide (**253**) with enedione (**259**) underwent in fairly good yield to give sulfur-free annulation product which was characterized as (**260**) after *O*-methylation (Scheme 5.46).[36]

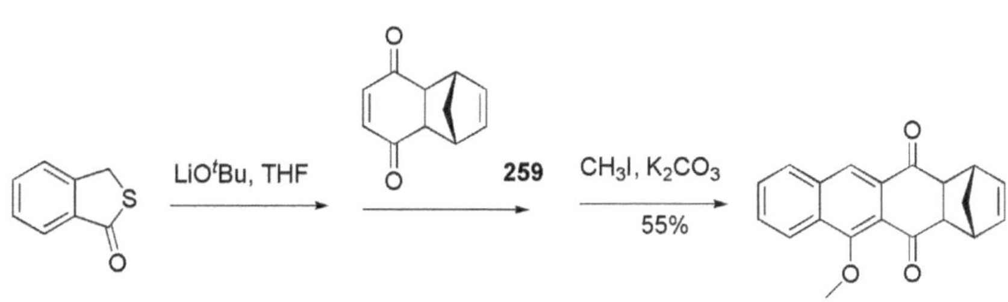

253 260

SCHEME 5.46 Annulation of thiophthalide [1994 THS 6139 Mal D].

The superiority of thiophthalides over phthalides is found useful by Tatsuta et al. in constructing structurally complex molecule like hibarimicin (vide Chapter 10), for which many established annulations were not seemingly suited.

Pyridine analogs: In general, very little work has been done on the reactivity of heterocyclic analogs of Sammes phthalide donors. In an attempt to develop a single step version of Sammes annulation, Sarkar et al. attempted to couple it with Peterson olefination. Such a concept was expected to result in direct aromatization of the original Sammes annulation products. In the corresponding exploratory study, they used pyridine analogs (**261** and **262**). As apparent from Scheme 5.47, Sammes annulation of (**261**) and (**262**) with methyl acrylate gives Sammes products (**263** and **264**, respectively) in modest yields.[37] The planned annulation with silyl substituted acrylate (**265**) was reasonably successful with (**262**) giving hydroxy quinoline (**266**) in 52% yield. The reaction between (**261**) and (**265**) gave two products namely, Sammes product (**267**) and Sammes-Peterson product (**268**). Apparently, the *cis* isomer undergoes Peterson olefination in situ. On the other hand, the *trans* isomer (**267**) remains resistant to the olefination.

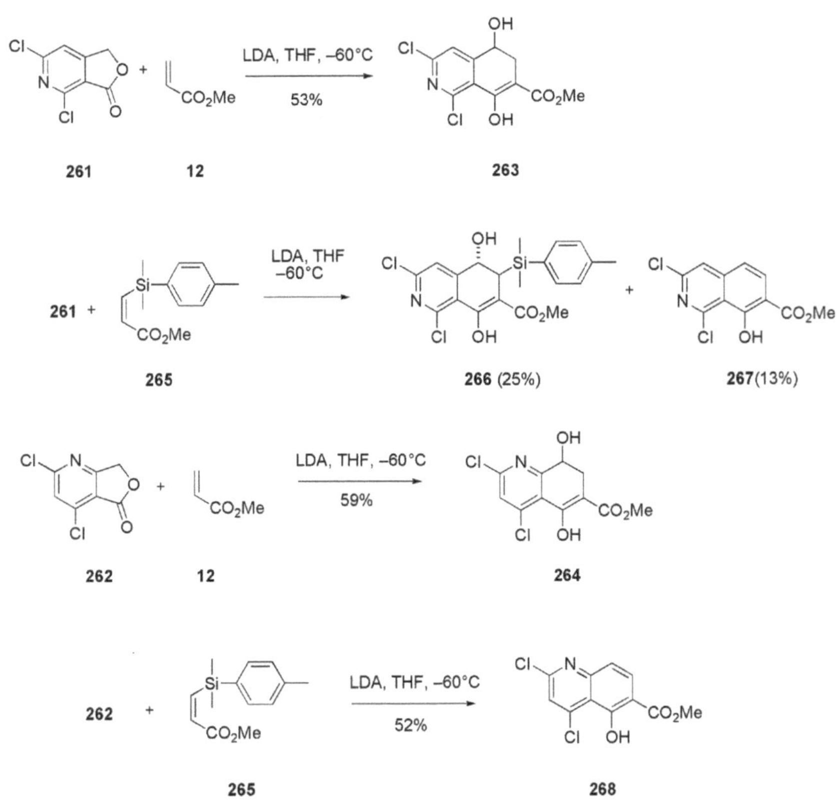

SCHEME 5.47 Sammes annulation with pyridine donors [2010 JCR 109 Sarkar TK].

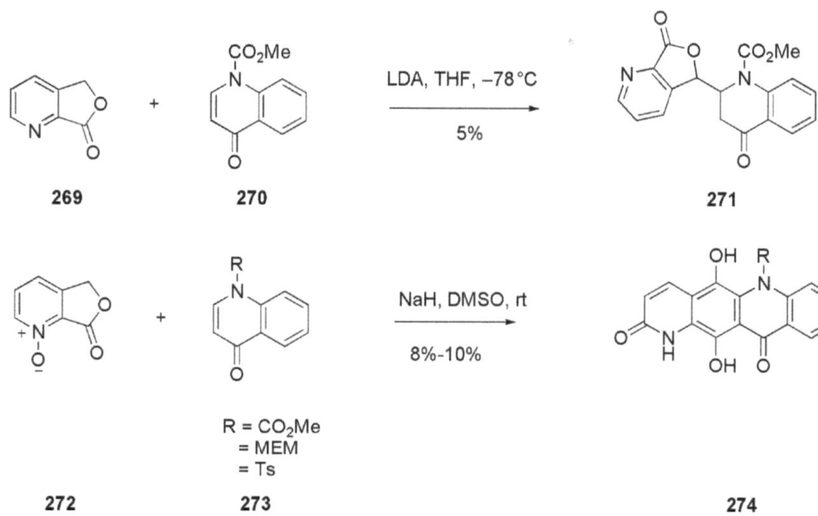

SCHEME 5.48 Annulation of a pyridine analog of phthalides [1997 THN 341 J JA].

The Joule group may have been the first one to report an annulation of pyridine analogs with 4-quinolone derivatives. As shown in Scheme 5.41, the annulation of (**269**) with (**270**) in the presence of LDA gave only Michael adduct (**271**) (Scheme 5.48).[38] The sign of the expected annulation was evident with the corresponding *N*-oxide pyridine donor (**272**) with quinolones (**273**). The yields of the annulation products (**274**) are only marginal.

Furo[3,4-b]indol-3-one donors: Furoindolone donors have been studied in depth to understand their reactivity towards anionic [4+2] cycloadditions. The primary objective was to access hydroxy carbazoles alkaloids which abound literature. Carbazoles nuclei are ubiquitous in materials as well as medicines. For example, polyvinyl carbazoles behave as conducting polymers. They are used in the photoreceptors of photocopiers. Carvedilol is a carbazole based medicine used for the management of congestive heart failure and hypertension. For the synthesis of simple carbazoles and naturally occurring carbazoles, there are numerous methods and methodologies. Mal et al. adopted [4+2] anionic annulation of furoindolones (**275**), which, in turn, are accessible in few steps from commercially available starting materials (vide Chapter 4). Typically, one of two benzene rings of the carbazole nucleus of the naturally occurring carbazoles is more heavily substituted than the other. It is conveniently constructed by the application of the [4+2] anionic annulation of furoindolones (**275b** and **275d**). These annulations permit not only the construction of the benzene ring but also disposition of the substituents. The survey displayed in Fig. 5.6 clearly reveals that the [4+2] anionic annulation of furoindolones is versatile. All of the carbazole natural products, namely, ekeberginine (**276**), clausamine D (**277**), clausine F (**278**), clausamine E (**279**), clausevatine D (**280**), clausamine

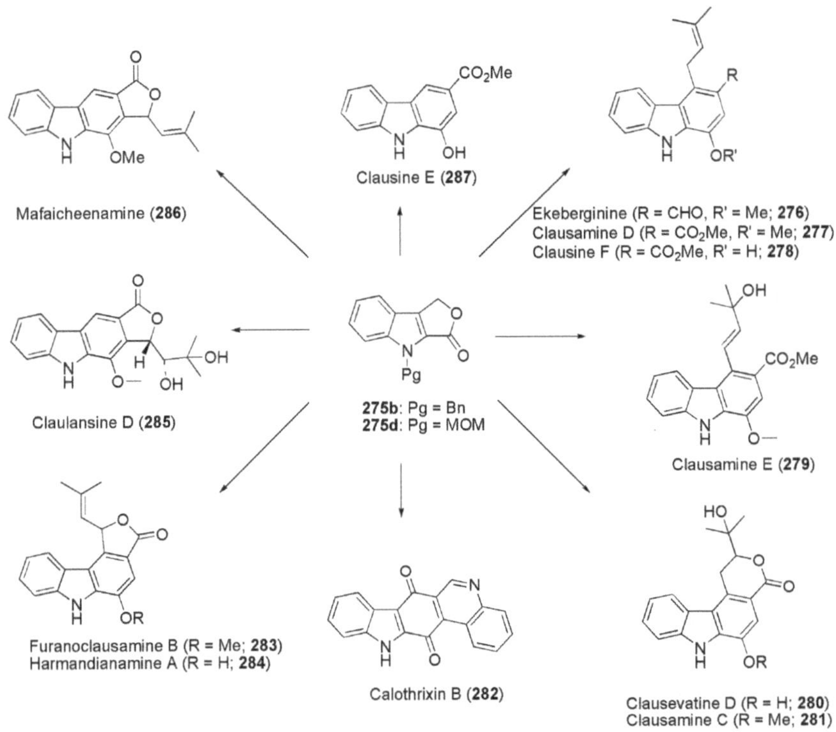

FIG. 5.6 Naturally occurring carbazoles synthesized by anionic [4+2] annulations.

C (**281**), calothrixin B (**282**), furanoclausamine B (**283**), harmandianamine A (**284**), claulansine D (**285**), mafaicheenamine (**286**) and clausine E (**287**) have been synthesized by the common concept, i.e., anionic annulation.

As shown in Scheme 5.49, the annulation reactivity 3-furoindolones is dependent on the protecting groups of indole "N." For the annulation of furoindolones (**275**) with dimethyl maleate, *N*-Boc (**275b**) is not suitable. Base stable protecting groups such as MOM, PMB and Bn survive the reaction conditions,

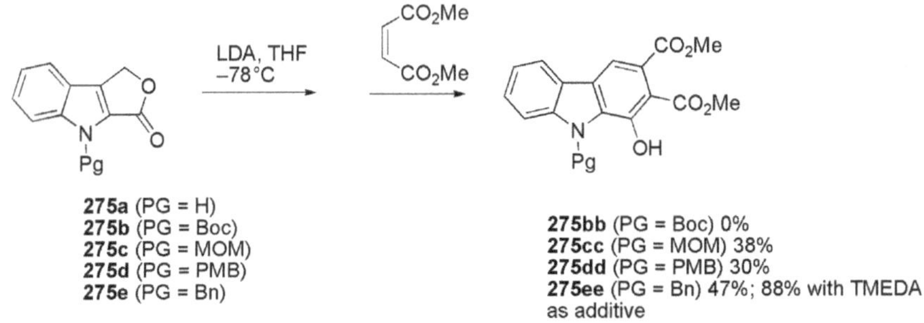

SCHEME 5.49 Annulation of furoindolones with dimethyl maleate [2012 SYT 1769 Mal D].

and the respective furoindolones (**275c–275e**) give carbazoles (**275cc–275ee**) in good yields.[39]

The 1-oxygenated carbazoles with 3,4-disubstitution pattern as in clausine F (**278**), clausamine E (**279**), clausevatine D (**280**) and furoclausamine B (**283**) are readily accessible by the combined applications of the annulations and *retro*-Kolbe-Schmitt reaction. However, the success of the total synthesis of the above carbazole natural products is critically dependent on the nature of the *N*-protecting groups in (**275**). Both clausamine E (**279**) and furanoclausamine B (**283**) have been synthesized from the common intermediate (**288**), which bears Boc protective group. While the former was obtained by a Heck reaction, the latter was obtained by an appropriate lithiation followed by a 1,2-addition of a butenal (Scheme 5.50).[40]

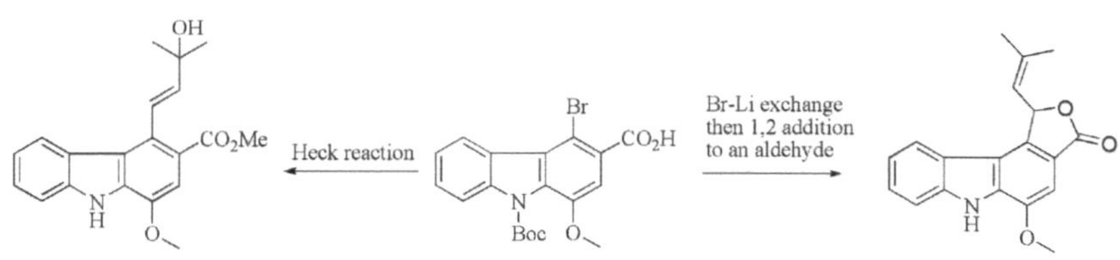

Clausamine E (**279**)

288

Furanoclausamine B (**283**)

SCHEME 5.50 Outline of the approaches to clausamine E (**279**) and furanoclausamine B (**283**).

The furoindolone (**275b**) was subjected to annulation with methyl crotonate and the resultant annulation product was *O*-methylated with MeI and K_2CO_3 to give carbazole trisubstituted carbazole (**289**) (Scheme 5.51). The carbazole ester (**289**) was hydrolyzed with NaOH in MeOH to give the corresponding acid, which was, without purification, subjected to decarboxylation using Cu_2O, TMEDA in NMP to yield natural product murrayafoline A (**290**). In order to introduce a bromine at C4, compound (**290**) was converted to its N-Boc derivative (**291**) using Boc_2O, Et_3N in ethyl acetate. Bromination of 1-methoxycarbazole (**291**) using NBS gave carbazole (**292**) in 68% yield. The methyl group in (**292**) was oxidized to carboxaldehyde group using DDQ in MeOH-THF. The aldehyde (**293**), without any purification, was subjected to Pinnick oxidation ($NaClO_2$, NaH_2PO_4, ^{t}BuOH, 2-methyl-2-butene and water) to give acid (**288**) and then the acid was treated with DBU-MeI in acetone to give its methyl ester. Heck coupling of the methyl ester with 3-methyl-but-2-en-3-ol using $Pd(OAc)_2$, TBAB, triphenylphosphine and potassium carbonate in DMF furnished clausamine E (**279**) in 76% yield. Thus, the first synthesis of clausamine E (**279**) was achieved in 11 steps from readily accessible furoindolone (**275b**). The facile access to the bromocarbazole acid (**288**) allowed Br-Li exchange reaction en route to the synthesis of furanoclausamine B (**293**). The application of the Kato's protocol of lithiation worked out well with the bromocarbazole acid (**288**). Treatment of (**288**) with $^n Bu_2Mg$ and ^{n}BuLi, followed by treatment of the intermediate (**294**)

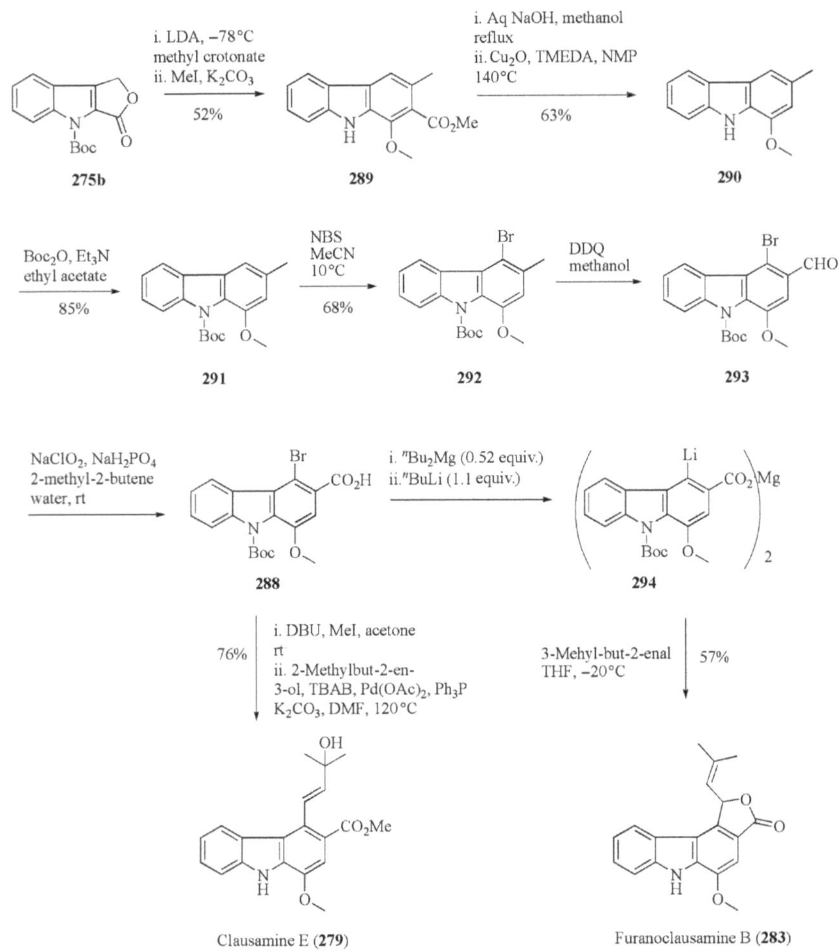

SCHEME 5.51 Total synthesis of clausamine E (**279**) and furanoclausamine B (**283**) [2015 THN 1247 Mal D].

with 3-methyl-but-2-enal in THF at −20°C gave furanoclausamine B (**293**) in 57% yield. Furanoclausamine B (**293**) was alternatively synthesized from a carbazole aldehyde. But, both the syntheses entail an anionic annulation.

The total synthesis of ekeberginine (**276**) by Mal et al. exemplifies both alkylation reactivity and annulation reactivity of furoindolone (**275c**). The furoindolone (**275c**), on treatment with LHMDS followed by prenyl bromide, furnished monoprenylated indolone (**295**) in 80% yield. It may be mentioned that the alkylations of *N*-protected furoindolones with alkyl halides are often accompanied by dialkylated products. The synthesis of monoalkylated furoindolones requires rigorous optimizations. Prenylfuroindolone (**295**), on annulation with dimethyl maleate in the presence of LDA, furnished the nonaromatized

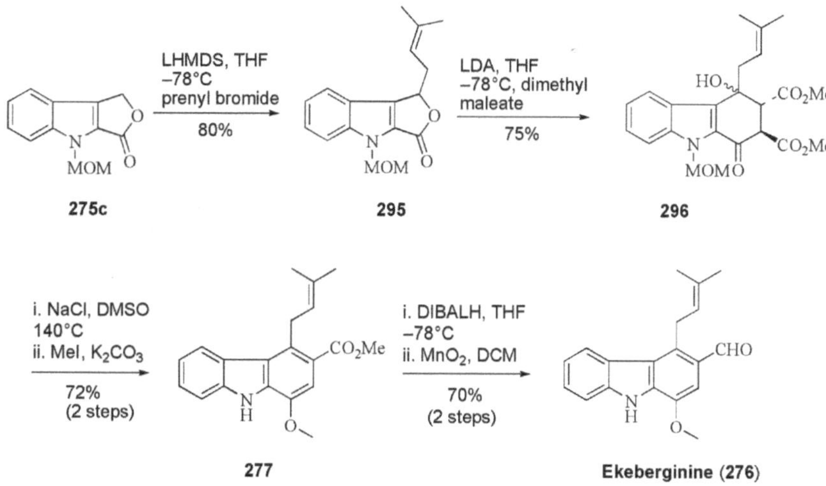

SCHEME 5.52 Total synthesis of ekeberginine (**276**) [2014 EJOC 1873 Mal D].

adduct (**296**). The reaction of (**296**) with sodium chloride in dimethyl sulfoxide (DMSO) at reflux resulted in decarboxylated carbazole, which was used in the next step without further purification. Notably, the MOM group was cleaved during this step. On selective methylation with MeI, the decarboxylated carbazole formed clausamine D (**277**) in 72% yield over two steps. Reduction of (**277**) with DIBALH followed by oxidation of the resulting mixture with MnO_2 furnished ekeberginine (**276**) in 70% yield, overall yield from commercially available 2-acetoxy-γ-butyrolactone and aniline being 31.5% (Scheme 5.52).[41]

For the access to the carbazole alkaloids with 1-oxy-2,3-disubstitution as in claulansine D (**285**) and mafaicheenamine (**286**), two more key reactions other than the [4+2] anionic annulation, were employed.[42] These are i) selective ester reduction of 1-hydroxycarbazole-2,3-dicarboxylates and ii) acid-catalyzed carbon-carbon double bond isomerization. The carbazole diester (**275cc**) was treated with LAH (2 equiv) in THF at room temperature and then the resultant product, without purification, was methylated CH_3I-K_2CO_3 to give furocarbazolone (**297**) in 64% yield. The selective ester reduction is due to preferential intramolecular reaction of C1 oxygen coordinated metal hydride. For the elaboration of furocarbazolone (**297**) to the target molecule (**286**), it was converted to the lactol (**299**) in two steps. Bromination of the furocarbazolone (**297**) with NBS-AIBN in refluxing CCl_4 afforded (**299**), which was, without purification, hydrolyzed with dioxane-water to afford lactol (**298**) via concomitant deprotection of the MOM group with in situ release of HBr (Scheme 5.52).

The reaction of the lactol (**298**) with methallylindium bromide generated in situ from methallyl bromide and indium metal furnished furocarbazole (**300**) in 85% yield. The isomerization of the methallyl group in (**300**) to propenyl group was effected in 95% yield using *p*-toluenesulfonic acid in refluxing toluene leading to the completion of the total synthesis of the target alkaloid i.e., mafaicheenamine

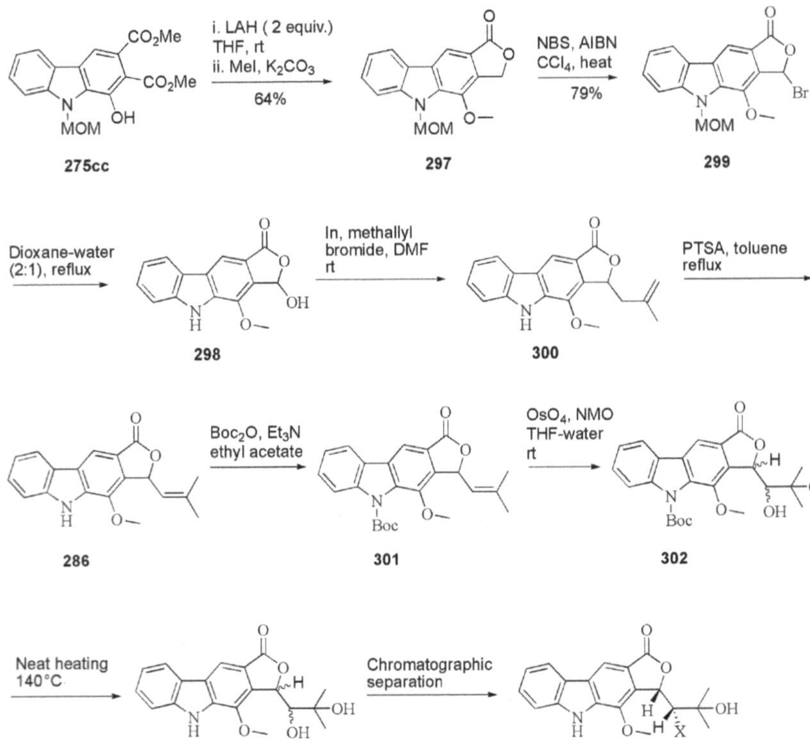

SCHEME 5.53 Total synthesis of claulansine D and mafaicheenamine [2015 OBC 6344 Mal D].

E (**286**) (Scheme 5.52). Since the attempted dihydroxylation of carbazole (**286**) with OsO_4, NMO in THF: water failed, the NH in (**286**) was protected as its Boc derivative with boc-anhydride, triethylamine in ethyl acetate. The resulting furocarbazolone (**301**) was then subjected to dihydroxylation with OsO_4, NMO to provide a diastereomeric mixture of (**302**). Neat heating of the sample (**302**) at 140°C culminated in a 2:1 mixture of (**303**), which were separated to furnish (±) claulansine D (**285**) and their nonnatural isomer (Scheme 5.53).

The annulation reactivity of furo[3,4-*b*]indol-3-one donors (**275**) towards aryne precursors widely varies with their structural features. With benzyne (**214**), generated from bromobenzene by LDA, the reaction of the donor (**304**) expectedly gave benzcarbazoloquinone (**305**) in 64% yield, and the corresponding NCO_2Et containing annulation product was not obtained, although the respective intermediate were probably (**214**) and (**306**) (Scheme 5.54).

Similarly, 3-bromopyridine (**307**) underwent annulation with the anion of (**275b**) via 3-pyridyne (**308**) under the influence of LDA to give ellipticine quinone (**309**) and isoellipticine quinone (**310**) in equal amounts.[43] This reaction constitutes a formal synthesis of ellipticine (**311**).

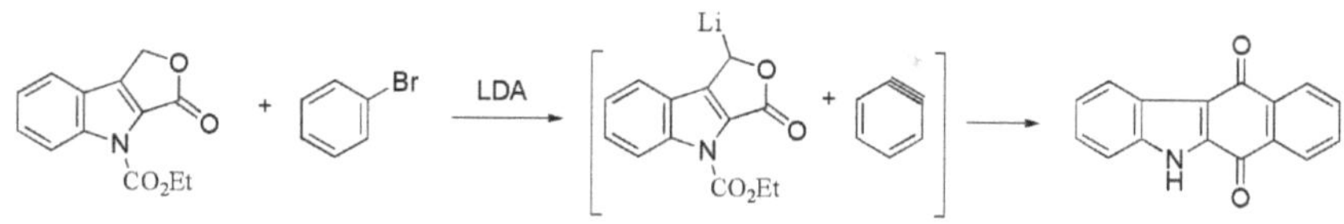

304 212 306 214 305

SCHEME 5.54 Annulation of furoindolones with benzyne [2005 SYT 994 Mal D].

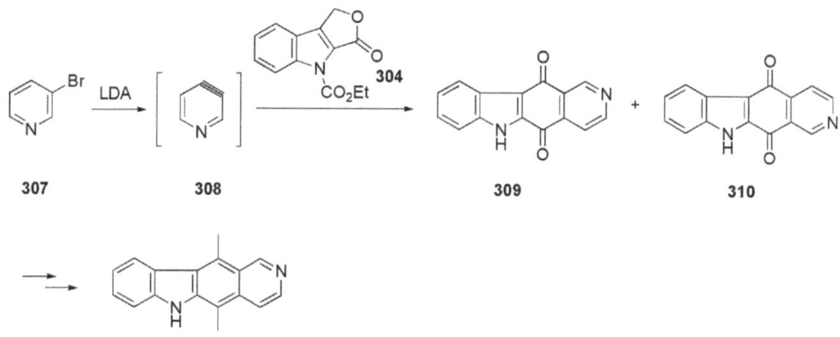

307 308 309 310

Ellipticine (311)

SCHEME 5.55 Annulation of furoindolones with pyridyne [2005 SYT 994 Mal D].

Extension of the hetaryne annulation to 4-bromoquinoline (**243**) and 3-bromoquinoline (**244**) has led to an extraordinarily facile synthesis of calothrixin B (**312**) (Scheme 5.55). The reaction of the bromoquinolines (**243** and **244**) with furoindolone (**275b**) gave a pair of regioisomeric quinones (**313** and **314**, and **315** and **316**), respectively (Scheme 5.55).[35]

At first glance, the mechanism of the formation of **313–316** appears to involve the common quinolyne (**317**) (Scheme 5.56). But, the careful scrutiny of the ratios of the annulation products led to a nonhetaryme mechanism, i.e., addition-cyclization-elimination-oxidation mechanism for the reaction with 4-bromoquinoline (**243**).

The mechanism proposed by Mal et al. for 4-bromoquinoline (**243**) as the precursor is presented in Scheme 5.57. The anion (**318**), generated by deprotonation of C3 of furoindolone (**275b**) adds to C3 of quinoline (**243**), forming carbanion (**319**). The driving force for the formation of (**319**) may be twofold: (i) electronegativity of bromine and (ii) stabilization of the anion by an adjacent benzene ring. In the next step, cyclization of (**319**) occurs to form pentacycle (**320**). Elimination of hydrogen bromide from (**320**) gives pentacycle (**321**), aerial oxidation of which furnishes the desired pentacyclic quinone (**313**), acidic work-up of which yields calothrixin (**312**) (Scheme 5.58).

A similar mechanism applied to 3-bromoquinoline (**244**) explains formation of isocalothrixin B quinone (**316**).

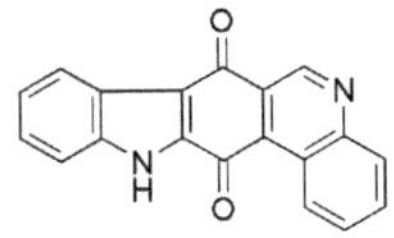

Calothrixin B (**312**)

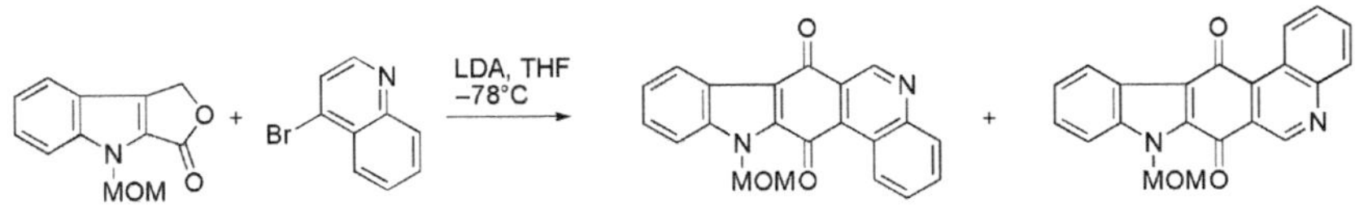

275b **243** **313** (42%) **314** (6%)

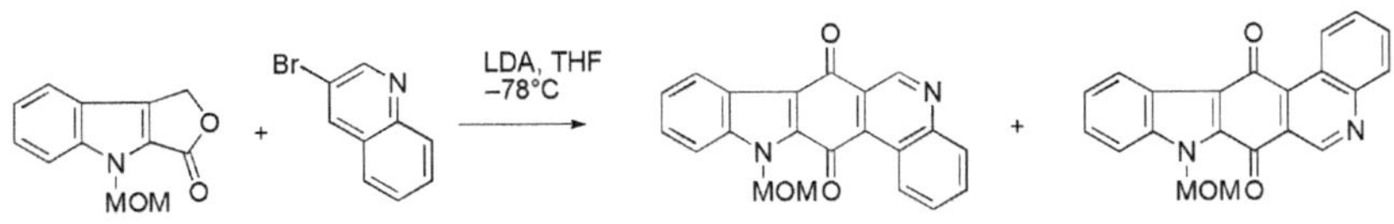

275b **244** **315** (6%) **316** (36%)

SCHEME 5.56 Annulation with bromoquinolines [2014 OBC 8196 Mal D].

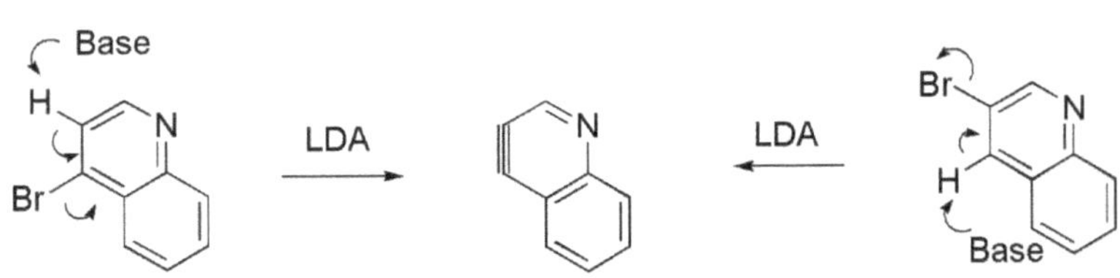

243 **317** **244**

SCHEME 5.57 Proposed dehydrobromination of bromoquinolines.

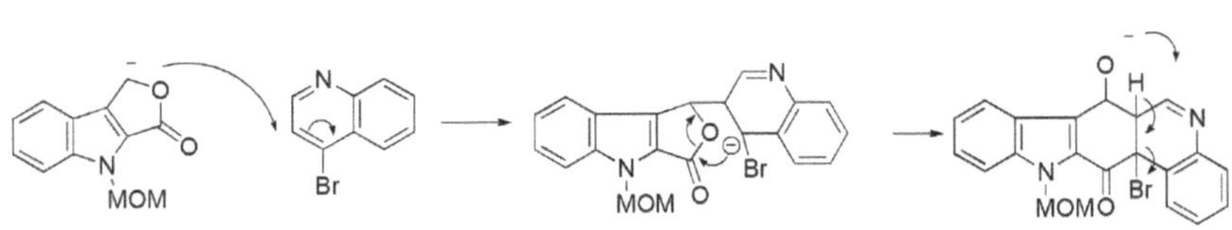

318 **243** **319** **320**

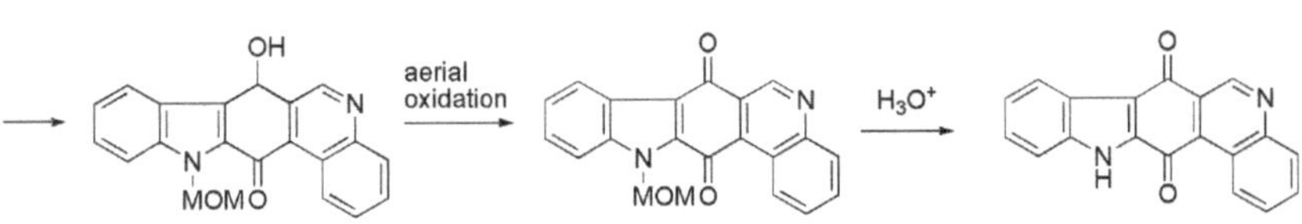

321 **313** Calothrixin (**312**)

SCHEME 5.58 Annulation with bromoquinolines [2014 OBC 8196 Mal D].

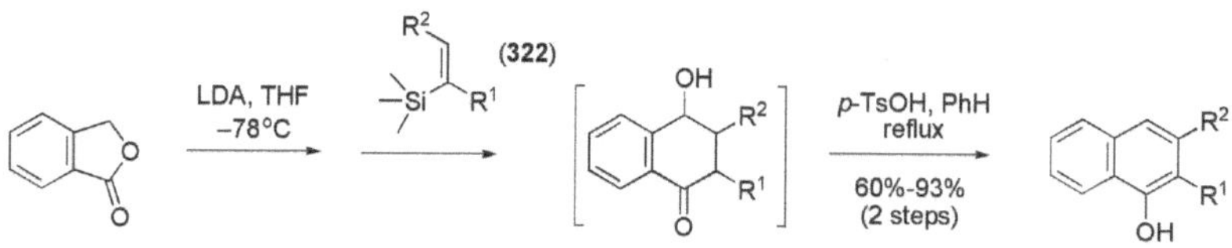

a: R^1 = SPh, R^2 = H: b: R^1 = SO_2Ph, R^2 = H; c: R^1 = SO_2Ph, R^2 = 2-MeO-C_6H_4; d: R^1 = SO_2Ph, R^2 = 4-MeO-C_6H_4.

5 **323** **324**

SCHEME 5.59 Annulation of α-trimethylsilyl vinyl sulfides and vinyl sulfones [1990 CPB 902 Watanabe M].

Miscellaneous acceptors: Very rarely, a vinyl sulfide is used as a Michael acceptor. In the sole example of Sammes annulation, vinyl sulfide (**322a**) was cyclocondensed with the parent phthalide (**5**) to yield (**323a**), which, without purification, was heated with *p*-toluenesulfonic acid in refluxing to give naphthol (**324a**) in 75% yield.[44] Likewise, vinylsulfones (**322b–322d**) furnished (**324b–324d**) via (**323b–323d**) (Scheme 5.59).

In conclusion, the Sammes annulation is a useful reaction for the synthesis of α-naphthols. It encompasses a good variety of acceptors. Most often, α,β-unsaturated esters/lactones and aryne precursors are found useful. Studies on the heterocyclic donors remain underrepresented. Further systematic studies are required to unravel the structure-reactivity relationships.

REFERENCES

1. Broom, N. J. P.; Sammes, P.G.J.C.S. *Chem. Comm.* **1978**, 162.
2. Hauser, F. M.; Rhee, R. P. *J. Org. Chem.* **1978**, *43*, 178.
3. Kjonaas, R. A. *J. Chem. Educ.* **1986**, *63*, 311.
4. Broom, N. J. P.; Sammes, P. G. *J.C.S. Perkin Trans.* **1981**, *1*, 465.
5. Couche, E.; Fkyerat, A.; Tabacchi, T. *Helv. Chim. Acta* **2003**, *86*, 210.
6. Franck, R. W.; Bhat, V.; Subramaniam, C. S. *J. Am. Chem. Soc.* **1986**, *108*, 2455.
7. Couche, E.; Fkyerat, A.; Tabacchi, T. *Helv. Chim. Acta* **2009**, *92*, 903.
8. Bos, M.; Canesso, R.; Inoue-Ohga, N.; Nakano, A.; Takehana, Y.; Sleight, A. *Bioorg. Med. Chem.* **1997**, *12*, 2165.
9. Mitra, P.; Mandal, S.; Chakraborty, S.; Mal, D. *Tetrahedron* **2015**, *71*, 5610.
10. Jana, A. K.; Pahari, P.; Mal, D. *Synlett.* **2012**, *23*, 1769.
11. Roush, W. R.; Michaelides, M. R.; Tai, D. F.; Lesur, B. M.; Chong, W. K. M.; Harris, D. J. *J. Am. Chem. Soc.* **1989**, *111*, 2984.
12. Magnus, P.; Eisenbeis, S. A.; Fairhurst, R. A.; Iliadis, T.; Magnus, N. A.; Parry, D. *J. Am. Chem. Soc.* **1997**, *119*, 5591.
13. Sun, C.; Hunt, D. K.; Clark, R. B.; Lofland, D.; O'Brien, W. J.; Plamondon, L.; Xiao, X.-Y. *J. Med. Chem.* **2011**, *54*, 3704.
14. Kobayashi, K.; Maeda, K.; Uneda, T.; Morikawa, O.; Konishi, H. *J. Chem. Soc. Perkin Trans.* **1997**, *1*, 443.
15. Tatsuta, K.; Tokishita, S.; Fukuda, T.; Kano, T.; Komiya, T.; Hosokawa, S. *Tetrahedron Lett.* **2011**, *52*, 983.

16. Mal, D.; Jana, S. *J. Org. Chem.* **2016**, *81*, 11857.
17. Mali, R. S.; Jagtap, P. G. *Org. Prep. Proc. Int.* **1993**, *25*, 633.
18. Mal, D.; Jana, A.; Mitra, P.; Ghosh, K. *J. Org. Chem.* **2011**, *76*, 3392.
19. Ghosh, K.; Karmakar, R.; Mal, D. *Eur. J. Org. Chem.* **2013**, 4037.
20. Dinda, B.; Basak, S.; Ghosh, B.; Mal, D. *Synthesis* **2016**, *48*, 1235.
21. Dinda, B. K.; Jana, A. K.; Mal, D. *Chem. Commun.* **2012**, *48*, 3999.
22. Janowski, W. K.; Prager, R. H. *Aus. J. Chem.* **1985**, *38*, 921.
23. (a) Majumdar, G.; Pal, R.; Murty, K.V.S.N.; Mal, D. *J. Chem. Soc. Perkin. Trans.* **1994**, *1*, 309. (b) Hill, B.; Rodrigo, R. *Org. Lett.* **2005**, *7*, 5223.(c)Li, J.; Mootoo, D. R. *Synthesis* **2013**, *45*, 2287.
24. Tatsuta, K.; Yoshimoto, T.; Gunji, H.; Okado, Y.; Takahashi, M. *Chem. Lett.* **2000**, 646.
25. Myers, A. G.; Tom, N. J.; Fraley, M. E.; Cohen, S. B.; Madar, D. J. *J. Am. Chem. Soc.* **1997**, *119*, 6072.
26. Dodsworth, D. J.; Pia-Calcagno, M.; Ehrmann, E. U.; Quesada, A. M.; Oswaldo, N. S.; Sammes, P. G. *J. Chem. Soc. Perkin Trans.* **1983**, *1*, 1453.
27. Marsden, R.; MacLean, D. B. *Can. J. Chem.* **1984**, *62*, 306.
28. Marsden, R.; MacLean, D. B. *Can. J. Chem.* **1984**, *62*, 1392.
29. Sammes, P. G.; Dodsworth, D. J. *J. C. S. Chem. Comm.* **1979**, 33.
30. Bates, M. A.; Sammes, P. G.; Thomson, G. A. *J. Chem. Soc. Perkin Trans.* **1988**, *1*, 3037.
31. Townsend, C. A.; Davis, S. G.; Christensen, S. B.; Link, J. C.; Lewis, C. P. *J. Am. Chem. Soc.* **1981**, *103*, 6885.
32. Kaiser, F.; Schwink, L.; Velder, J.; Schmalz, H.-G. *Tetrahedron* **2003**, *59*, 3201.
33. Graybill, T. L.; Casillas, E. G.; Pal, K.; Townsend, C. A. *J. Am. Chem. Soc.* **1999**, *121*, 7729.
34. Miki, S.; Shimzu, R.; Nakatsuji, H. *Tetrahedron Lett.* **1992**, *7*, 953.
35. Mal, D.; Roy, J.; Biradha, K. *Org. Biomol. Chem.* **2014**, *12*, 8196.
36. Majumdar, G.; Murty, K.V.S.N.; Mal, D. *Tetrahedron Lett.* **1994**, *35*, 6139.
37. Panda, B.; Basak, S.; Hazra, A.; Sarkar, T. K. *J. Chem. Res. (S)* **2010**, *34*, 109.
38. Ajana, W.; Lopez-Calahorra, F.; Joule, J. A.; Alvarez, M. *Tetrahedron* **1997**, *53*, 341.
39. Jana, A. K.; Pahari, P.; Mal, D. *Synlett* **2012**, *23*, 1769.
40. Mal, D.; Roy, J. *Tetrahedron* **2015**, *71*, 1247.
41. Roy, J.; Mal, D. *Eur. J. Org. Chem.* **2014**, 1873.
42. Roy, J.; Mal, D. *Org. Biomol. Chem.* **2015**, *13*, 6344.
43. Mal, D.; Senapati, B.; Pahari, P. *Synlett* **2005**, 994.
44. Date, M.; Watanabe, M.; Furukawa, S. *Chem. Pharm. Bull.* **1990**, *38*, 902.

Chapter 6

[4+2] Annulation of Homophthalic Anhydrides

Cyclic anhydrides, besides being acylating agents, can act nucleophiles if they are enolizable. More specifically, they undergo annulation in presence of a base with a variety of acceptors with multiple bonds, such as CO, CN, CC double bonds.[1] Activated alkynes are also suitable for the annulations. The typical anhydrides are succinic anhydrides, glutaric anhydrides, and homophthalic anhydrides. In this chapter we focus on the anionic chemistry of the homophthalic anhydrides, which behave as a 1,4-dipolar synthon in a benzannulation. The typical acceptors for such annulations are carbonyl compounds, propiolates, bicyclic 1,4-quinones, imines, and nitroethenes. The annulation chemistry of homophthalic anhydrides is of two kinds: Tamura annulation and Cushman-Haimova annulation. In the subchapters below, the reactivities of various kinds of acceptors to the annulation with homophthalic anhydrides are described.

6.1 ACTIVATED ALKENES AND ALKYNES (TAMURA ANNULATION)

Tamura annulation is a [4+2] anionic annulation that was developed during the 1980s with an aim to facilitate the total synthesis of anthracyclines. It is the cyclocondensation of a homophthalic anhydride with an activated alkene or an alkyne to provide *peri*-oxygenated bicyclic or polycyclic compounds in a regiospecific manner. It provides a complementary route to the Hauser annulation (vide Chapter 4) in the synthesis of fused aromatics and hydroaromatics. In most cases, the new ring annulated is of a phenol. For example, homophthalic anhydride (**1**) reacts with diethyl acetylenedicarboxylate (**2**) in the presence of NaH at 0°C–20°C to give 1-hydroxynaphthalene diester (**3**) (Scheme 6.1).[2] This benzannulation is viewed as a Michael addition of the conjugate base of (**1**), i.e., (**4**) to alkyne (**2**) to generate carbanion (**5**), which undergoes Dieckmann type cyclization to give bicyclic intermediate (**6**). Decarboxylative fragmentation of (**6**) takes place to give (**7**), the conjugate base of the product (**3**). Likewise, the anhydride (**1**) reacts with propiolate esters to afford the corresponding 1-hydroxynaphthalene carboxylates. The Tamura annulation is also suited for an activated alkene. However, the alkene must have activating groups at both of its termini. Foe example, the annulation between anhydride (**1**) with dimethyl

Anionic Annulations in Organic Synthesis. https://doi.org/10.1016/B978-0-12-813800-7.00006-3

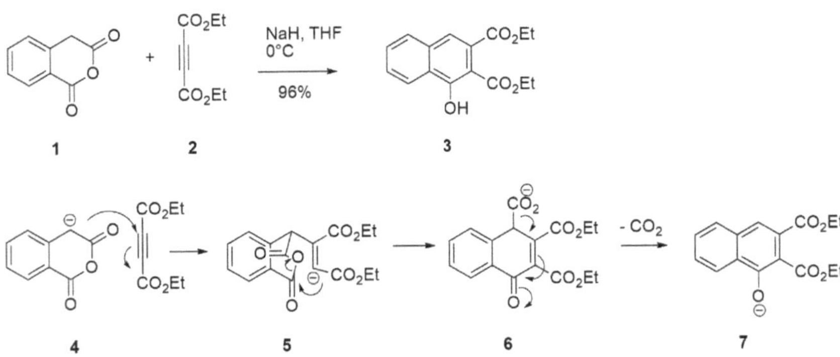

SCHEME 6.1 Tamura annulation with an acetylene dicarboxylate [1984 JOC 473 Tamura Y].

maleate in the presence of NaH in THF at 80°C gave a tetralone tricarboxylate in 46% yield. Similarly, the reaction between homophthalic anhydride and dibenzalacetone produced the respective nonaromatized annulation product.

The most successful class of acceptors for the Tamura annulation is 1,4-naphthoquinones. The first report on the cyclocondensation of homophthalic anhydrides with 1,4-naphthoquinones was published in 1981 by Tamura et al. In the original thermal version, a homophthalic anhydride was reacted with a variety of 1,4-naphthoquinones. As shown in Scheme 6.2, the parent homophthalic anhydride (**1**) and the parent naphthoquinone (**8**) undergo annulation in refluxing dichlorobenzene to give naphthacenedione (**9**) in 44% yield. The reaction presumably proceed through adducts (**10** and **11**) and the decaboxylated intermediate (**12**).[3]

Under the anionic conditions (NaH, THF), the reaction between the same reactants (**1** and **8**) readily take place below 25°C to give the same product, i.e., (**9**) (Scheme 6.3) via intermediate (**10′**).

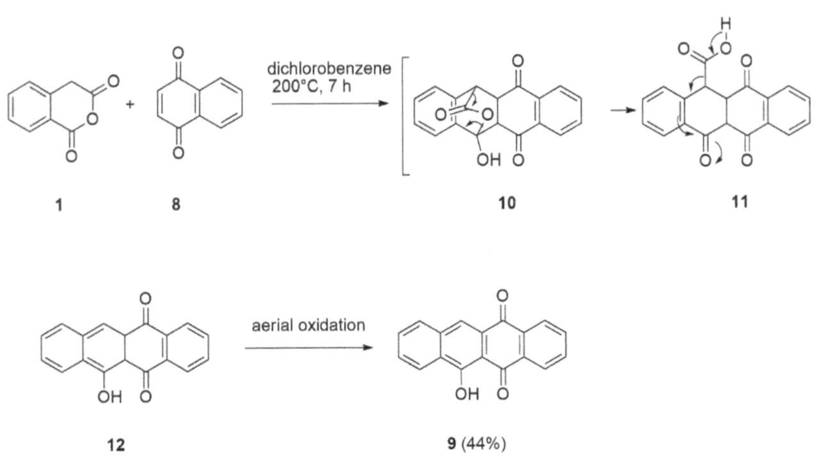

SCHEME 6.2 Thermal cyclocondensation of homophthalic anhydrides [1981 THS 4283; 1983 CPB 2691 Tamura Y].

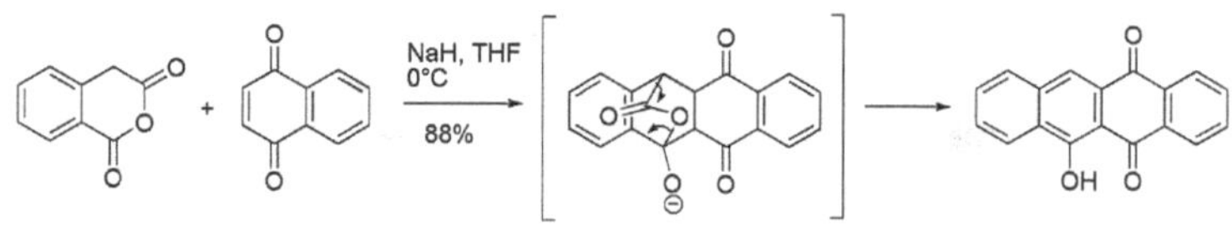

SCHEME 6.3 Anionic cyclocondensation of homophthalic anhydrides [1984 JOC 473 Tamura Y].

Acetylene ester (**13**), on annulation with anhydride (**1**), gave naphthalene carboxylate (**14**) (Table 6.1). However, dimethyl maleate (**15**) afforded tetralone (**16**), for which the stereochemistry was unknown (vide entry 2). For unsymmetrical substrates, a good degree of the regioselectivity can be attained by introducing an electronegative group in the acceptors. For naphthoquinone acceptors (**17** and **18**), bromine at the alpha position of the quinone carbonyl controls the regiochemistry and lead to single regioisomers (**19** and **20**, respectively) (vide entries 3, 4). Entry 5 exemplifies nonaromatic quinone (**21**) giving Tamura annulation product (**22**). Entry 6 is an example where the oxidation level of the product (**23**) has been increased by placing a methoxy group in the anhydride donor (**24**) for the reaction with quinone (**25**). Similarly, entry 7 indicates the feasibility of a Tamura annulation of a nonaromatic analog of homophthalic anhydride, i.e., (**26**) with heterocyclic quinone (**27**) producing annulation product (**28**).[5]

Over the years, the Tamura annulation has proved a very useful methodology for synthesis of polycyclic aromatic compounds. The merit of the annulation lies in operational simplicity, short reaction times, and its applicability to 1,4-quinone acceptors. Furthermore, the regioselectivity of the reaction is most often very high. This attribute has been very useful in determination of the absolute configuration of natural fredericamycin A. Many classes of complex natural products such as galtamycinone, dynemicin A, fredericamycins, lactonamycins, and viridicatumtoxins have synthesized by adaptation of this methodology. In the following paragraphs, few total syntheses are presented to illustrate the prowess of the Tamura reaction.

The antibiotic SS-228R (**29**) is produced from another antibiotic SS-228Y by thermal as well as photochemical reaction. Originally, it was given a regioisomeric structure. Through its synthesis by the Tamura annulation in two steps, it was reassigned the present structure. The reaction of homophthalic anhydride (**30**) with 2-bromonaphthoquinone (**18**) in the presence of NaH afforded naphthacenedione (**31**) in 87% yield. Demethylation of (**31**) with boron tribromide furnished SS-228R (**29**) in 51% yield (Scheme 6.4).[6]

The regiochemical outcome of the Tamura annulation is conveniently utilized in a short synthesis of tetracenomycin D (**32**) (Scheme 6.5).[7] Tetracenomycin D (**32**) is a naturally occurring quinone and closely related to cytotoxic quinone saintopin. For its synthesis, Hagemann's ester was elaborated to *ortho* toluate (**33**). It was then homologated to homophthalate (**34**) through lateral lithiation with LDA followed by diethyl carbonate quench. Alkaline hydrolysis

TABLE 6.1 Tamura Annulation With Different Acceptors [2008 TCC 299 Kita Y]

Entry		Acceptors	Products (%Yield)
1	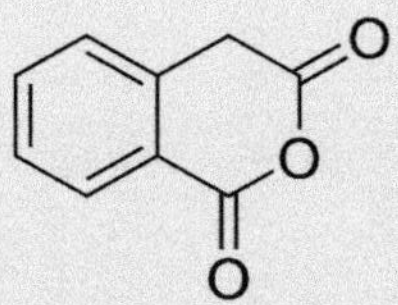 **1**	 **13**	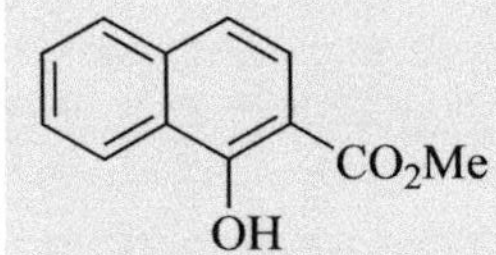 **14** (49)
2	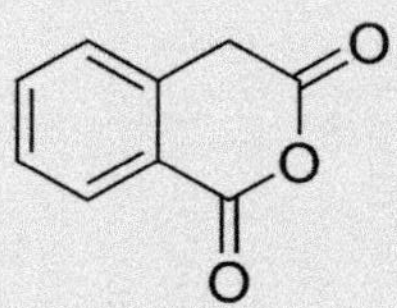 **1**	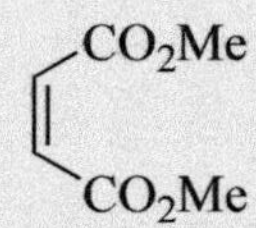 **15**	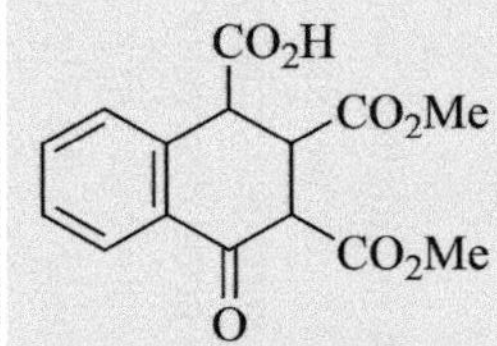 **16** (46%)
3	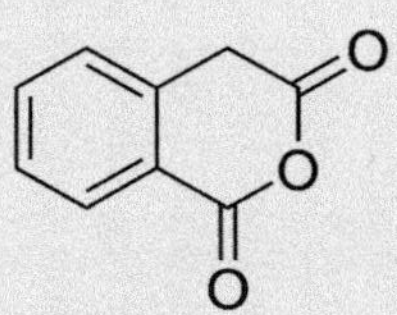 **1**	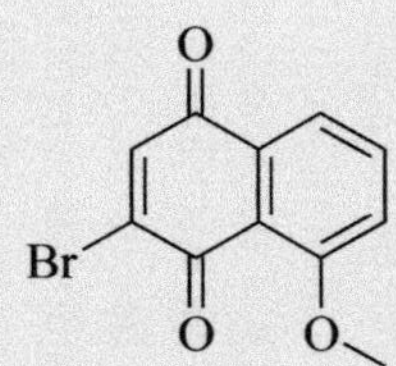 **17**	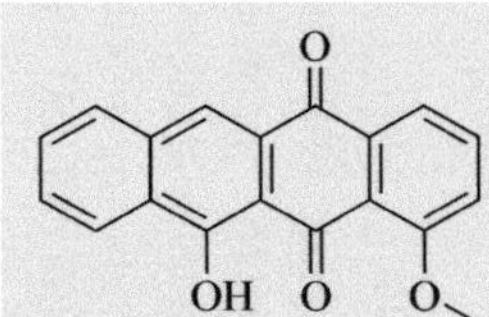 **19** (90)

4	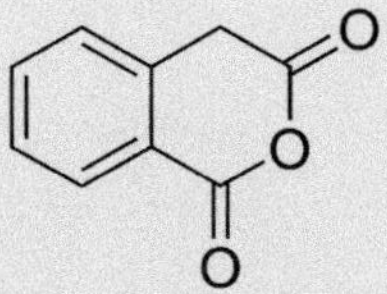 **1**	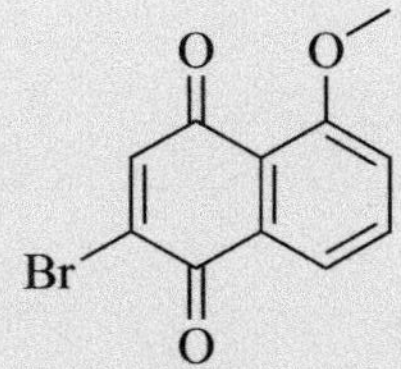 **18**	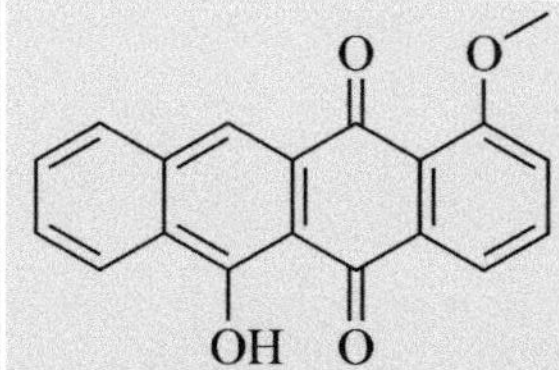 **20** (73)
5	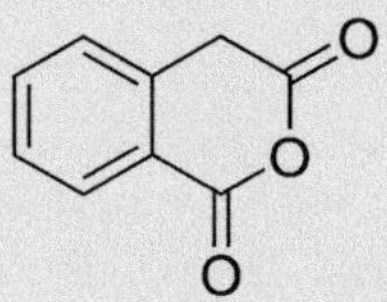 **1**	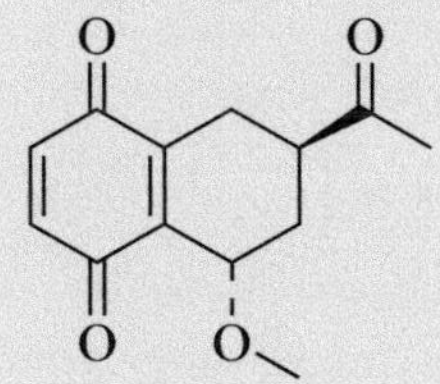 **21**	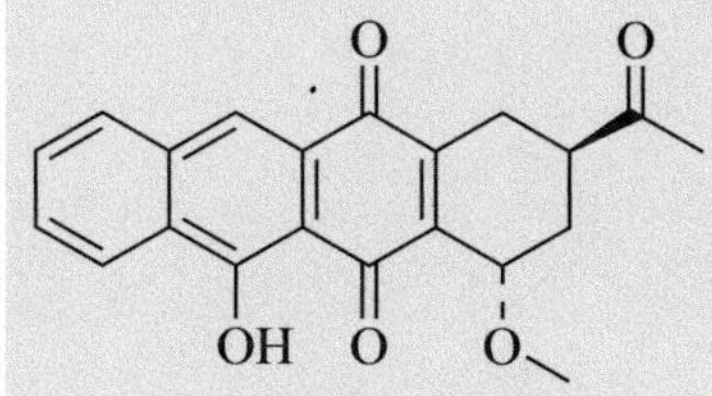 **22** (58)
6	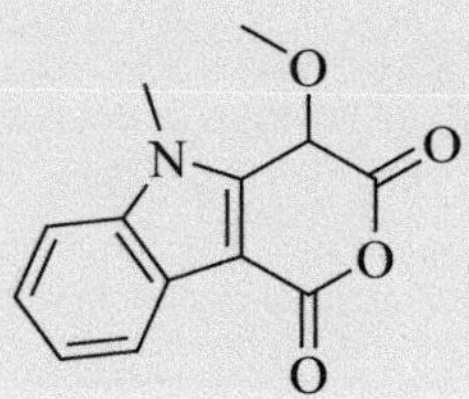 **24**	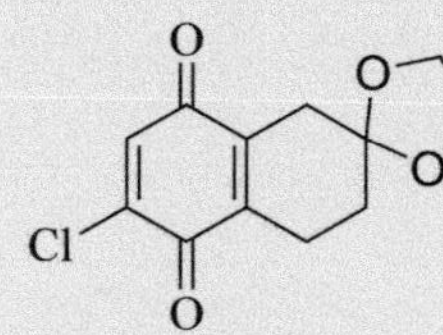 **25**	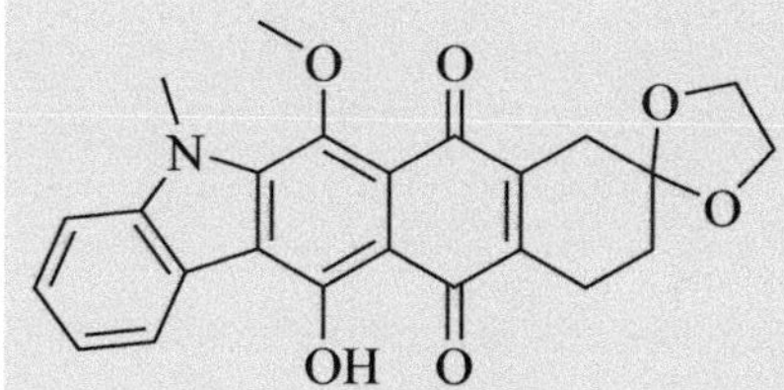 **23** (57)

Continued

Table 6.1 Tamura Annulation With Different Acceptors [2008 TCC 299 Kita Y]—cont'd

Entry		Acceptors	Products (%Yield)
7		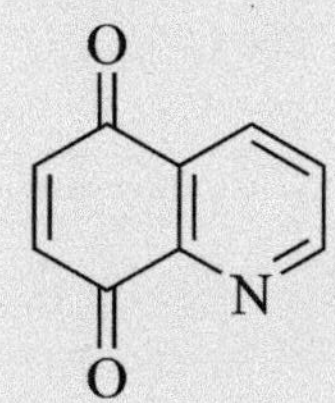 **27**	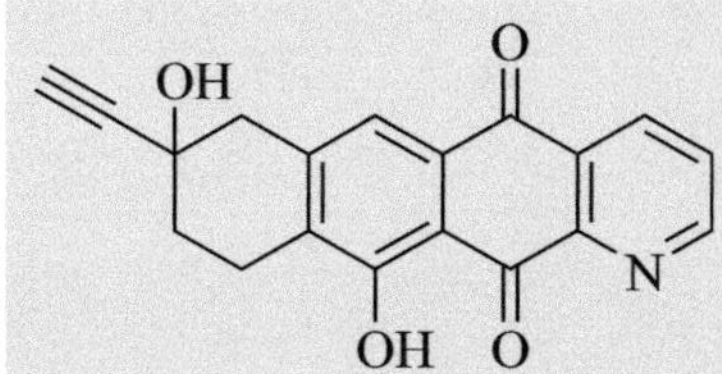 **28** (77%)

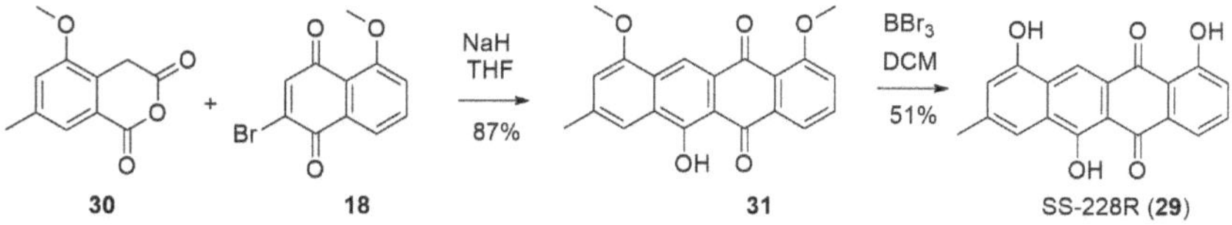

SCHEME 6.4 Synthesis of antibiotic SS-228R (**29**) [1985 JOC 2273 Tamura Y].

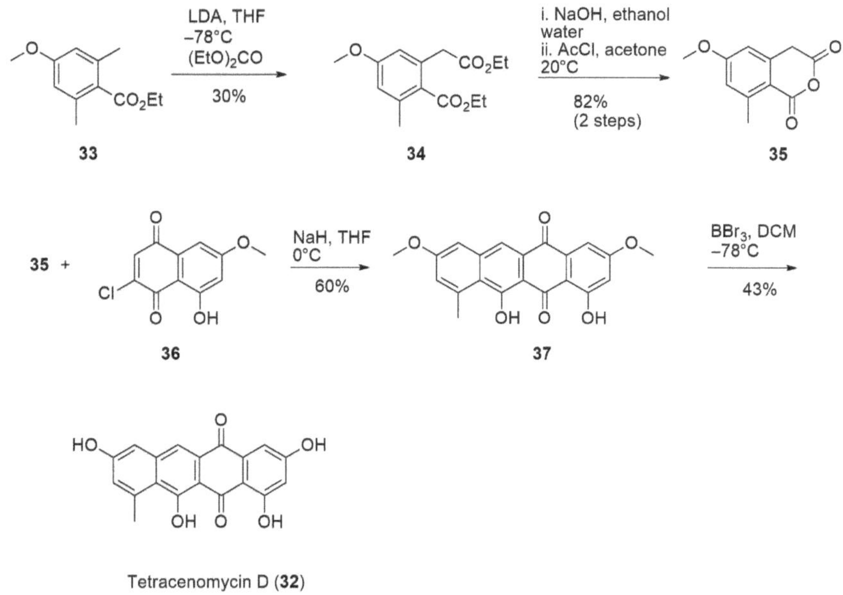

SCHEME 6.5 Synthesis of tetracenomycin D (**X**) [2002 BMC 253 Gesson J-P].

of (**34**) followed by dehydrative cyclization of the resultant diacid afforded homophthalic anhydride (**35**). The Tamura annulation of the anhydride (**35**) with hydroxynaphthoquinone (**36**) in the presence of NaH at 0°C furnished naphthacenedione (**37**). Didemethylation of (**37**) with boron tribromide at −78°C produced tetracenomycin D (**32**) as a dark red solid.

Extension of the Tamura annulation to 2-chlorobenzoquinone (**38**) has led to a short synthesis of hydroxyviocristin (**39**), an advanced intermediate for the total synthesis of variecolortide A, which contains a diketopiperazine joined to an anthrone moiety. Between the two CC double bondsof the benzoquinone (**38**), one having chlorine underwent regioselective Tamura annulation with homophthalic anhydride (**40**) in the presence of NaH to give 1,4-anthraquinone (**41**) in 52% yield. Demethylation of (**41**) with $AlCl_3$ furnished hydroxyviocristin (**39**) in 75% yield (Scheme 6.6).[4]

The Tamura annulation has been applied to *C*-glycosyl juglone for the first total synthesis of galtamycinone (**42**) (Scheme 6.7). *C*-Glycosyl juglone (**43**) was synthesized in two steps from triflate (**44**). Treatment of the triflate (**44**) with

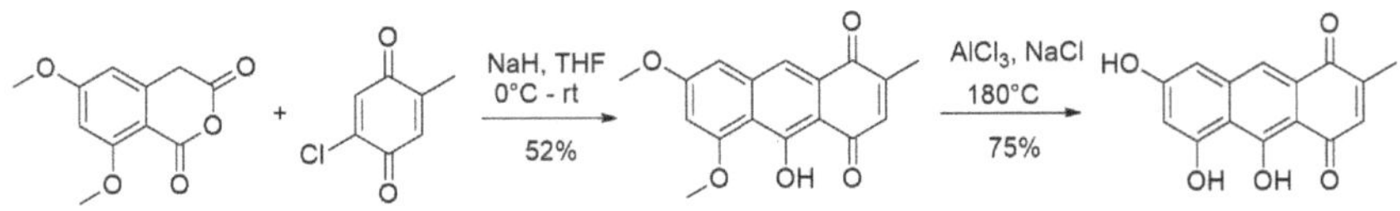

40 38 41 39

SCHEME 6.6 Synthesis of hydroxyviocristin (**39**) [2011 ACIE 1402 Trauner D].

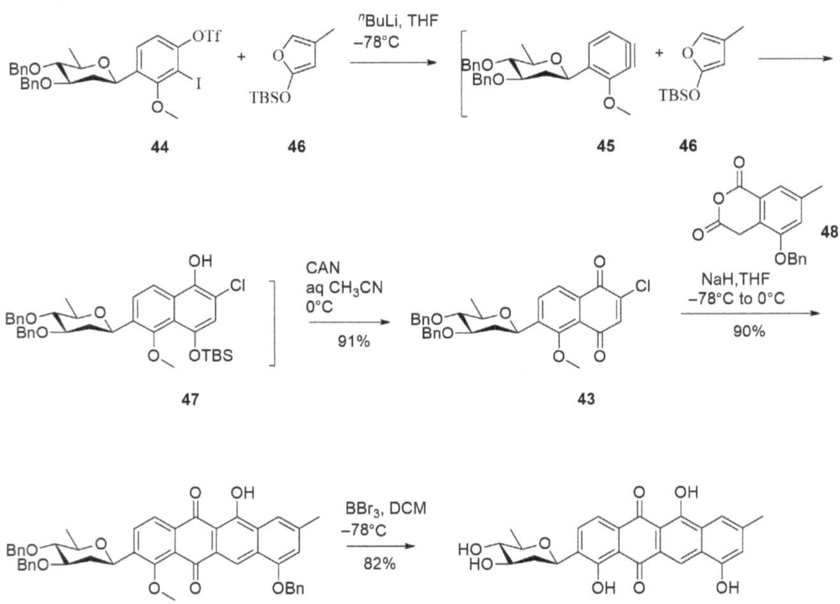

44 46 45 46 48 47 43 49 42

SCHEME 6.7 Total synthesis of galtamycinone (**42**) [1997 THN 16533 Suzuki K].

^{n}BuLi generated benzyne (**45**), which reacted with furan (**46**) in situ to give (**47**). The crude product (**47**) was oxidized with CAN to furnish (**43**), the key intermediate for the Tamura annulation. NaH promoted annulation of the juglone (**43**) with homophthalic anhydride (**48**) at 0°C provided protected galtamycinone (**49**). BBr_3 promoted global removal of the protecting groups, yielded galtamycinone (**42**).[8]

Like halogenated 1,4-naphthoquinones, naphthoquinones without a substituent at C2 and C3 are equally suited for the Tamura annulations. The high regioselectivity of the annulations are achieved if a quinone carbonyl forms an intramolecular hydrogen bond. Such a manifestation is found in the studies on nogalamycin analogs by the Terashima group. Nogalamycin and its congeners are a group of notable anthracycline antibiotics. They are characterized by a bicyclic epoxyoxocin moiety. The reaction of homophthalic anhydride (**50**) with naphthoquinone derivative (**51**) in the presence of NaH quantitatively afforded nogarene analog (**52**) as the single regioisomer in one operation (Scheme 6.8).[9]

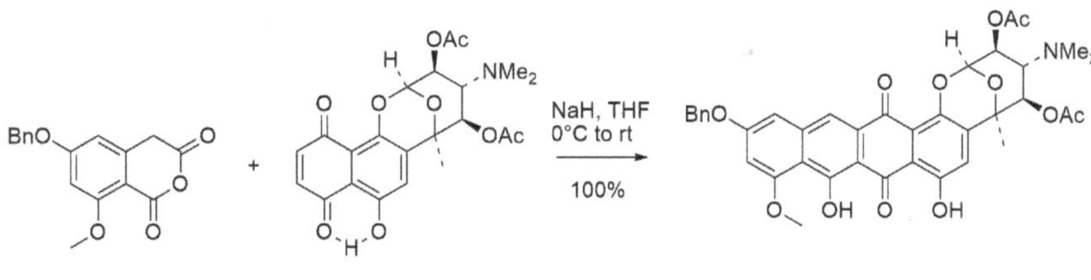

50 51 52

SCHEME 6.8 Synthetic studies on nogalamycin analogs [1988 THN 5745 Terashima S].

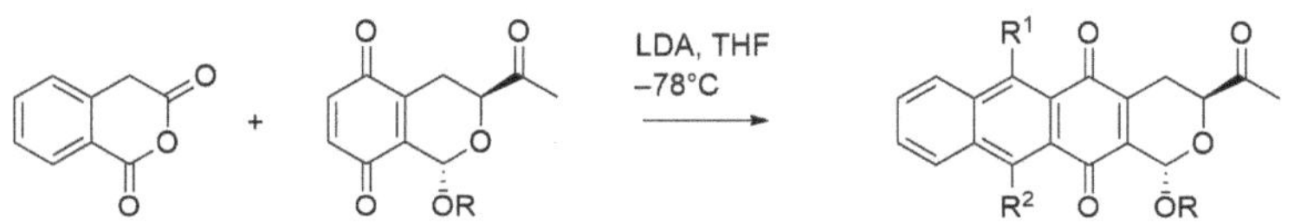

1

53: R = H
55: R = Me

54a: R = H, R^1 = OH, R^2 = H
54b: R = H, R^1 = H, R^2 = OH
in 2:1 ratio
56 (55%): R = Me, R^1 = OH, R^2 = H

SCHEME 6.9 Synthetic studies on idarubicin analog [1993 THS 5745 Lavallee JF].

For the quinones without a directing group like pyranoquinone (**53**), the regiochemistry of the annulations is unpredictable. The LDA promoted annulation of (**1**) with lactol (**53**) afforded two idarubicin analogs (**54a**) and (**54b**) in 2:1 ratio, whereas the annulation with the protected acetal (**55**) provided single product, i.e., (**56**) (Scheme 6.9).[10]

The Tamura annulation is the seemingly only [4+2] anionic annulation, which has witnessed an intramolecular version. Two new carbocyclic rings are concomitantly produced and the resulting products are angularly fused hydroaromatic systems. Successes have so far been found with activated alkyne acceptors activated by ester, ketone and aldehyde groups. The length of the tethers between the anhydride and the alkyne has typically been three or four carbons. The precursors are prepared from the corresponding homophthalates by alkylation with iodo compounds with trimethylsilyl acetylene or diethoxyacetylene moieties. The anhydride (**57**), on reaction with NaH, gave phenanthrene derivative (**58**). Anhydride with shorter chain (**59**) accordingly furnished (**60**). Tricyclic anhydride (**61**) was no exception. The action of NaH on (**61**) afforded tetracycle (**62**) (Scheme 6.10).[11] Further studies on the intramolecular version of the annulation are warranted for an application.

6.1.1 Variants

Both Hauser annulation and Tamura annulation give similar products but differ in the level of oxidation states of their products. With regard to oxidation states of the products, Tamura annulation is akin to Sammes annulation. However, the

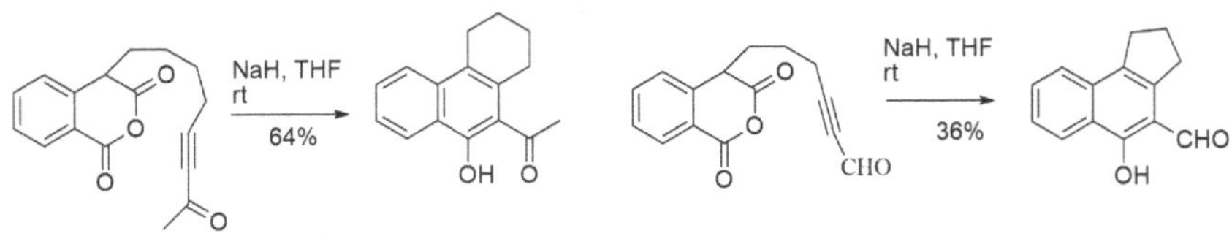

57 58 59 60

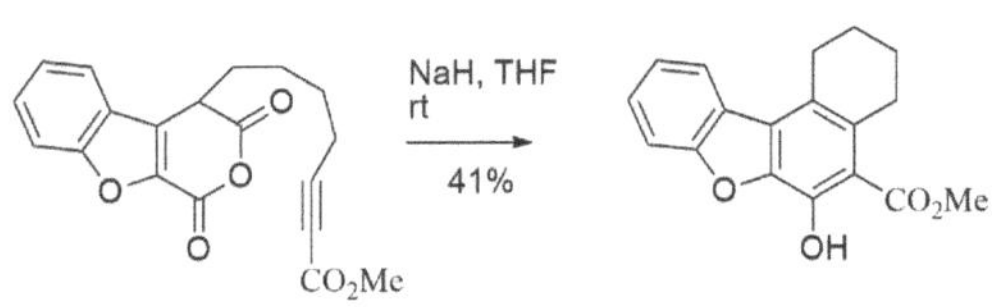

61 62

SCHEME 6.10 Intramolecular Tamura annulation [1991 JOC 119 Kita Y].

oxidation states of the products of the former can formally be increased by employing a starting material with increased oxidation state like 3-alkoxyhomophthalic anhydride (vide entry 6, Table 6.1). Similarly, 4-acetoxyhomophthalic anhydride can be used in the place of homophthalic anhydride to increase the oxidation level of the products. This strategy has been utilized in the regioselective synthesis of daunomycinone (**63**) (Scheme 6.11). It relied upon the annulation between 4-acetoxyphthalic anhydride (**64**) and chloronaphthoquinone (**65**) in the presence of NaH to yield tetracycle (**66**) in 62% yield. The acetal group in (**66**) was hydrolyzed to produce (**67**). The compound (**67**) was treated with trimethylsilylethynyl ceric chloride (**68**) to furnish side chain elaborated tetracycle (**69**). Mercuric sulfate catalyzed hydrolysis of (**69**) gave (**70**). Finally, the C7 hydroxy group of (**63**) was introduced by benzylic bromination of (**70**) followed by hydrolysis to complete the synthesis of daunomycinone (**63**).[12]

6.1.2 Anthracycline Antibiotics

Anthracycline antibiotics are one of most useful classes of anticancer drugs. Several of them have been synthesized by Tamura annulation. Kita synthesis of oxaunomycin (**71**) illustrates an elegant convergent approach to anthracycline antibiotics. The secondary benzylic hydroxy group of the acceptor appears to control the regiochemistry of the annulation (Scheme 6.12). The Tamura annulation of (**72**) with (**73**) in the presence of NaH produced tetracycle (**74**) in 55% yield as a single regioisomer. The acetate group of (**74**) was hydrolyzed with aq trifluoroacetic acid to give (**75**). This annulation is the first example of regio-controlled annulation of a 2,3 unsubstituted naphthoquinone derivative. The regioselectivity is rationalized by invoking hydrogen bonding or chelation through H or Na as shown in the intermediate (**76**).[13]

The Tamura annulation has tactically been utilized in the synthesis of 4-demethoxyadriamycinone (**77**). A ruthenium-catalyzed oxidation of allyl

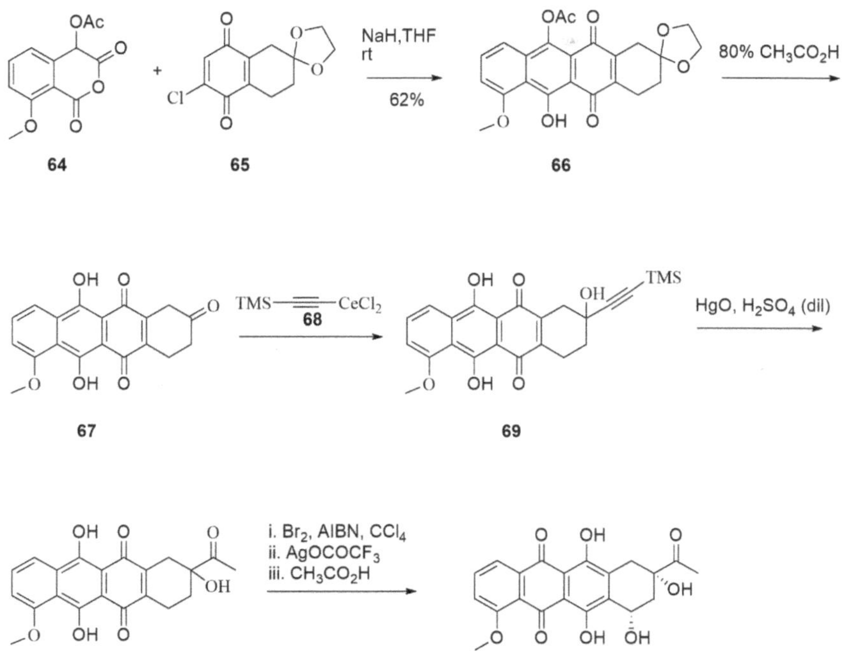

SCHEME 6.11 Tamura total synthesis of daunomycinone (**63**) [1987 CPB 1405 Tamura Y].

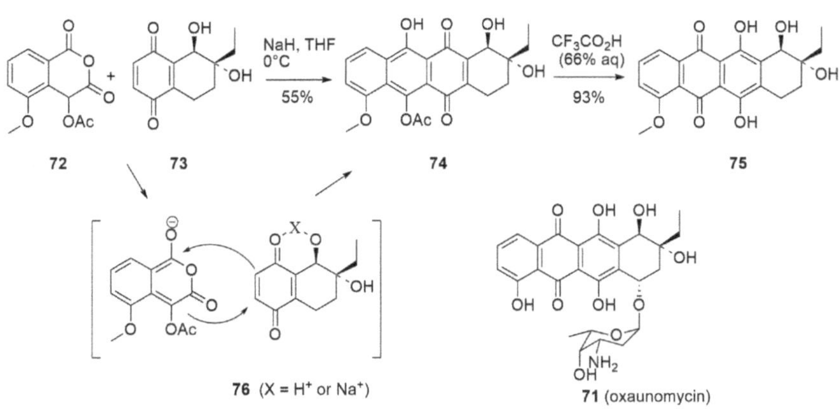

SCHEME 6.12 Synthesis of oxaunomycin precursor [1992 CPB 32 Kita Y].

acetates to the corresponding hydroxyketone has also been employed. The regiochemical outcome of the annulation is of no consequence with regard to the total synthesis. However, the annulation between (**78**) and (**79**) with sodium hydride at −40°C regioselectively gave the tetracycle (**80**) in 70% yield as a single product via migration of the acyl group in (**81**) (Scheme 6.13).

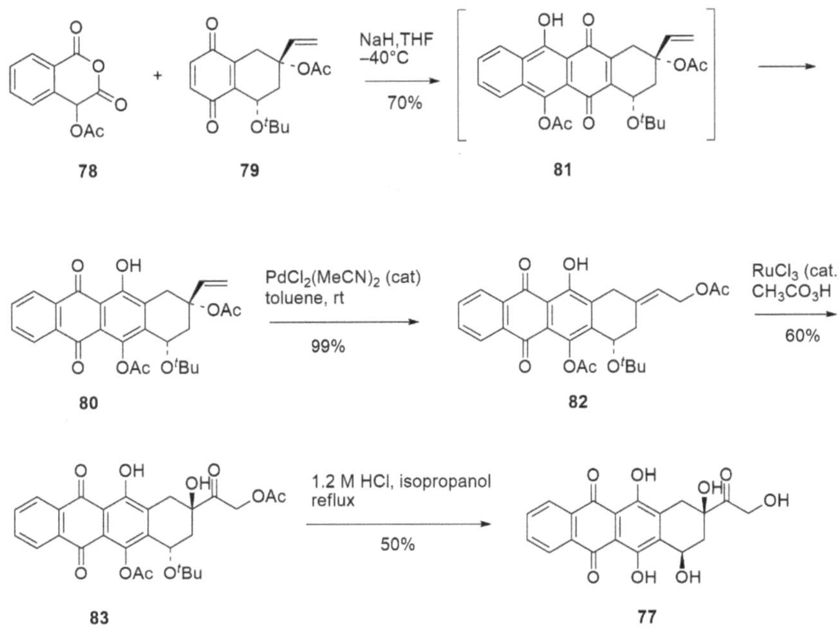

SCHEME 6.13 Synthesis of 4-demethoxyadriamycinone [2001 THS 3343 Murahashi S.-I].

The palladium-catalyzed allylic rearrangement of (**80**) gave the allyl acetate (**82**) in 99%yield. Oxidation of (**82**) with CH_3CO_3H in the presence of $RuCl_3$ catalyst in a mixture of water, acetonitrile, and dichloromethane (1:1:1) afforded (**83**) in 60% yield. The presence of an acetoxy group at C-5 in (**82**) seems to affect the stereochemistry of the oxidation product (**83**). Deprotection and epimerization atC-7 were readily accomplished by heating (**83**) with diluted hydrochloric acid to furnish 4-demethoxyadriamycinone (**77**).[14]

6.1.3 Kita Annulation

The parent 2-cyclohexenone and 3-chlorocyclohexenone do not react with lithium salt of homophthalic anhydride to give Tamura product. On the contrary, enones (**84**), equipped with an α-phenylsulfinyl group, can undergo annulation with homophthalic anhydrides in the presence of NaH, and give the benzannulated products (e.g., **85**) (Scheme 6.14).[15]

The mechanism of this reaction has been proposed to involve an initial [4+2] cycloaddition (Diels-Alder type) of the enolate (**86**) or a base-catalyzed Michael addition-Dieckmann type reaction of the intermediates (**86**) and (**84b**) to give cycloadduct (**87**) (Scheme 6.15). This then undergoes decarboxylation to give (**88**). Elimination of phenylsulfinate from (**88**) forms the product (**85c**).

The Kita protocol has been very fruitful in solving some of the long standing problems in the field of aromatic polyketides. It has been utilized in the

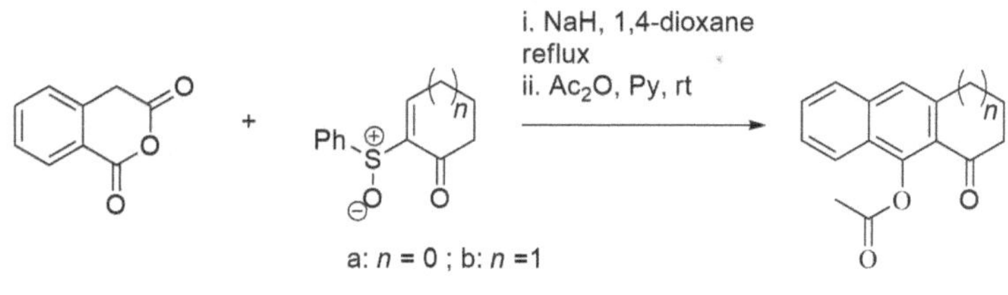

1 84 85a (62%) 85b (70%)

SCHEME 6.14 Kita nnulation of enolizable ketones [2000 JOC 89 Kita Y].

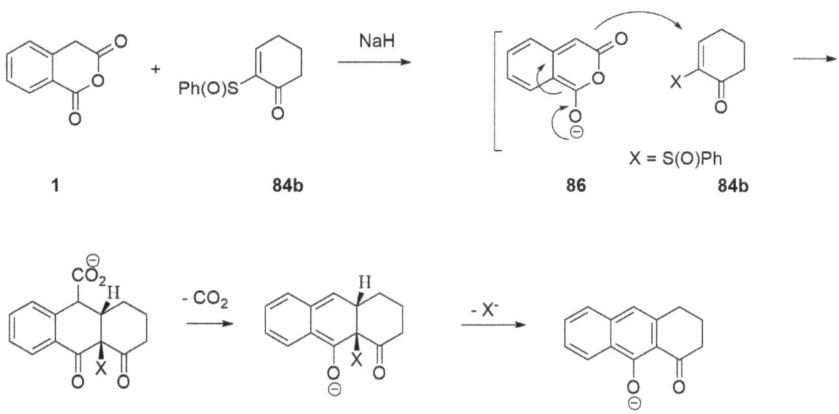

1 84b 86 84b

87 88 85c

SCHEME 6.15 Mechanism of the Kita annulation.

stereoselective synthesis of 2-deoxyolivin, which was elusive for quite some time. Kita annulation between a homophthalicanhydride and a chiral cyclohexenone derivative is the key step. Note that the conditions for the annulation are a unique one. Although the structural requirement of a α-phenylsulfinyl group in the cyclohexenone acceptor is in place, the annulation between (**89**) and (**90**) required thorough experimentation to obtain the annulation product (**91**) in good yield. Contrary to the uses of common alkali metal bases like NaH, LDA, etc., this annulation was successful with proton sponge, an organic base. Thus, the annulation between (**89**) and (**90**) in the presence of the proton sponge, MS 4Å and $ZnBr_2$ afforded (**91**) in 71% yield (Scheme 6.13). The allyl protecting group in **91** was effectively removed using Chandrasekhar method ($Pd(PPh_3)_4$, PHMS, $ZnCl_2$). The resultant alcohol was oxidized by Dess-Martin periodinane to furnish (**92**). Hydrogenolysis of (**92**) with $H_2/Pd(OH)_2$ followed by treatment of the resultant product with Dowex gave 2-deoxyolivin (**93**) (Scheme 6.16).[16]

The strength of the Kita annulation is best illustrated by the asymmetric total synthesis of fredericamycin A (**94**) (Scheme 6.17). It was isolated 1981 and was found to have strong in vitro and in vivo cytotoxicities against several tumor

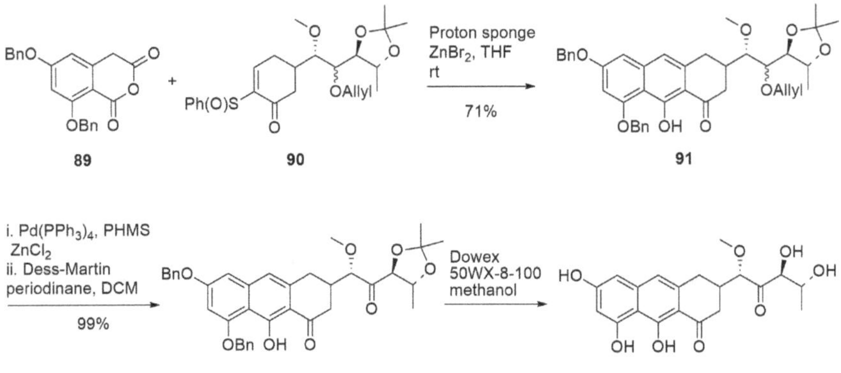

89 90 91

92 93

SCHEME 6.16 Synthesis of 2-deoxyolivin (**93**) [2008 THN 7211 Sasaki S].

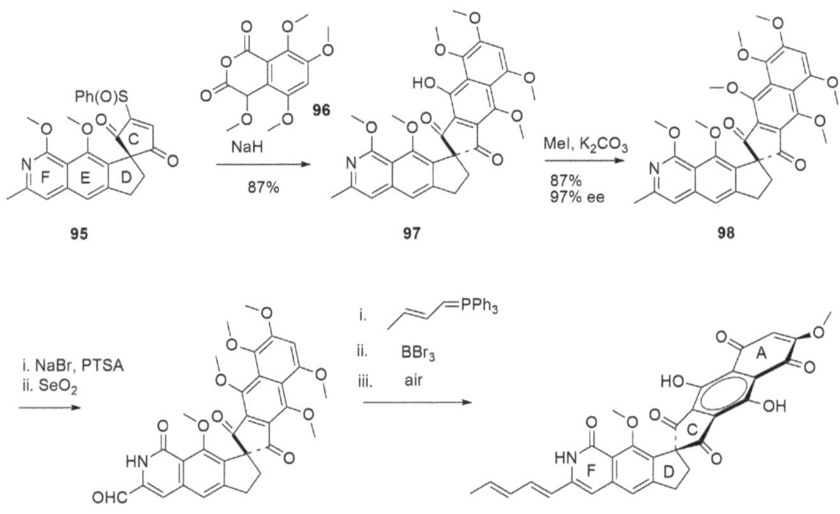

95 97 98

99 Fredericamycin A (**94**)

SCHEME 6.17 Total synthesis of fredericamycin A (**94**) [2001 JACS 3214 Kita Y].

models, such as P388 leukemia, B16 melanoma, and CD8F mammary carcinoma. Although it has only one asymmetric center, the spirocyclic and hexacyclic structures have been of interest to the numerous chemists. The Kita group adopted two Tamura annulations, one for the isoquinolone part and the other for the benzindenone part, i.e., D ring and B ring. To construct the spiro center of fredericamycin A (**94**) with appropriate absolute configuration, the stereospecific rearrangement of an optically active epoxy acylate has been executed. Two advanced molecular fragments (**95** and **96**) were united through Tamura annulation in the presence of NaH to produce (**97**). *O*-Methylation of (**97**) afforded the hexacyclic product (**98**) in 87% yield and 97% ee (Scheme 6.17). Selective demethylation of (**98**) with NaBr and PTSA gave the corresponding pyridone, which was oxidized with SeO_2 to give aldehyde (**99**). The Wittig homologation of (**99**) with *trans*-2-butenyl triphenylphosphoniumbromide gave a 5:1 mixture of (*E,E*)- and (*E,Z*)-side chain isomers in 46% yield. Deprotection of the mixture with BBr_3, followed by aerial oxidation afforded a 5:1 mixture of (**94**) and its (*E,Z*)-side chain isomer in a total yield of 74%. The pure fredericamycin (**94**) was obtained in 40% yield after separation.[17]

The CDEF fragment (**95**) was built also by the application of a Tamura annulation (Scheme 6.18). The reaction between homophthalic anhydride (**100**) and sulfinylenone (**101**) yielded (**102**). Methylation of the phenolic hydroxy group (**102**), and deketalization with TFA gave (**103**). Intramolecular pinacol coupling of (**103**) with SmI_2 gave cyclic diol, which was oxidized with Swern method to give hydroxy ketone (**104**) in 62% yield from (**103**). The reaction of (**104**) with the Burgess reagent in refluxing THF proceeded smoothly to give enone (**105**) in quantitative yield. Very interestingly, the same reaction in refluxing benzene did not give the dehydration product. CBS asymmetric reduction of (**105**) using the chiral borane gave a quantitative yield of the (*R*)-alcohol (**106**) with 74% ee in 100% yield. Several more steps provided the required enone, i.e., (**95**).[17]

One of the finest applications of the Tamura annulation is found in the Danishefsky synthesis of dynemicin A (**107**) (Scheme 6.19). The crucial annulation between anhydride (**108**) and quinone imine (**109**) was sufficiently mild to accommodate acid-sensitive functional groups like epoxide, enol ether and

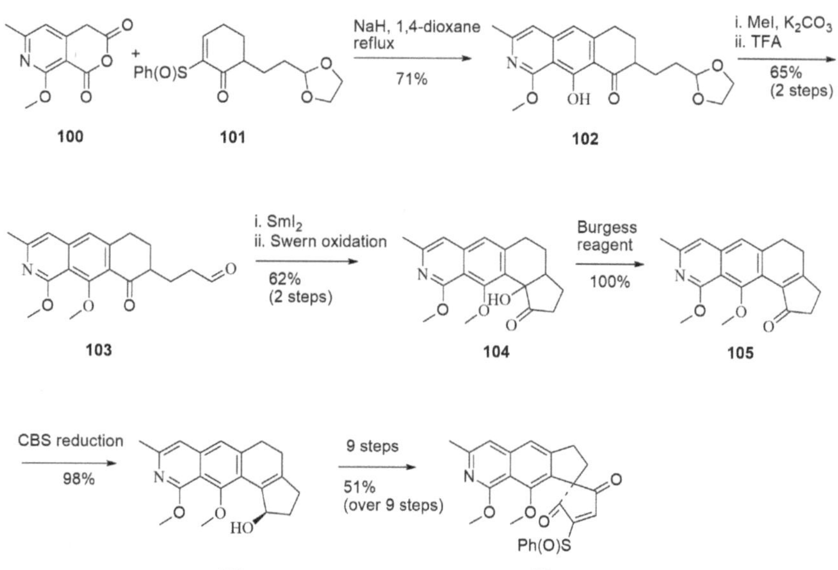

SCHEME 6.18 Synthesis of CDEF fragment of fredericamycin A (**95**) [2001 JACS 3214 Kita Y].

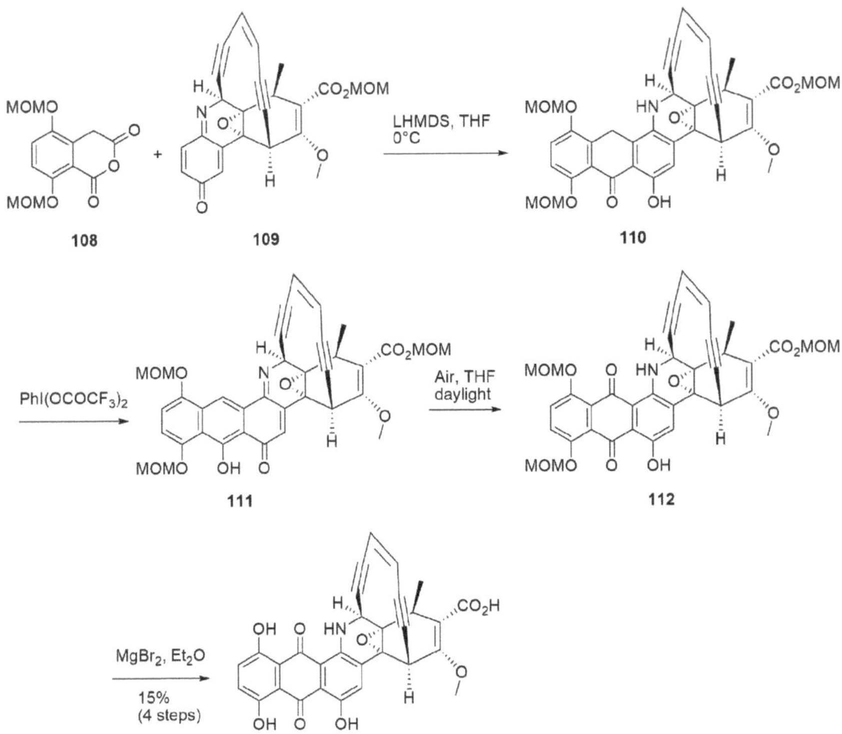

SCHEME 6.19 Synthesis of dynemicin A [1996 JACS 9509 Danishefsky SJ].

imine at the penultimate stage of the total synthesis. It also exemplifies reactivity of the first yet complex quinone imine (**109**). The annulation product (**110**) was oxidized with $PhI(OCOCF_3)_2$ to give (**111**). Further oxidation of (**111**) in air gave (**112**). Finally, removal of the MOM protecting groups of (**112**) furnished (**107**).[18]

Lactonamycin, possessing many uncommon structural attributes, has antimicrobial activity against Gram-positive bacteria, including methicillin- and vancomycin-resistant strains. It is also cytotoxic against various tumor cell lines. Consequently, it has been the target of many synthetic groups. Both the Danishefsky group and Nakata group relied upon the Tamura annulation for its synthesis as depicted in Schemes 6.20 and 6.21.[19] When homophthalic (**113**) was condensed with functionalized benzoquinone derivative (**114**) in the presence of NaH, the product (**115**) was formed as a single isomer in 42% yield. The free OH group in the acceptor (**114**) directed the Tamura annulation in a regioselective manner. Several more steps led to lactonamycinone (**116**).

In their total synthesis of lactonamycin, Nakata et al. utilized bicyclic anhydride (**117**) and chlorobenzoquinone derivative (**118**) for the LDA-mediated

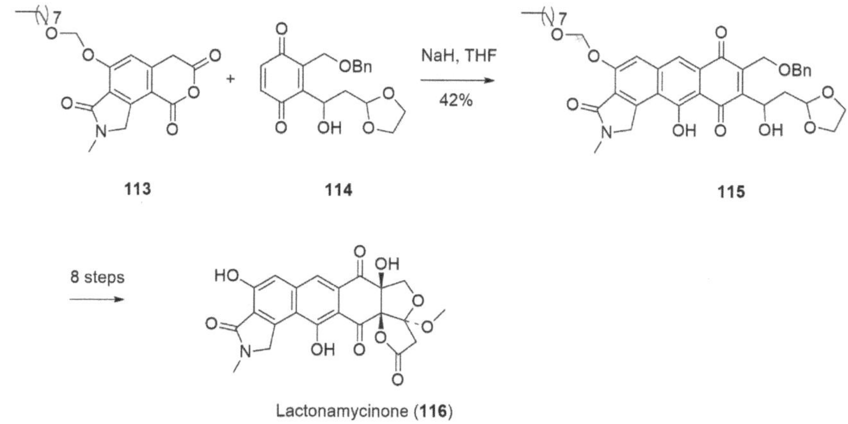

SCHEME 6.20 Hydroxy-directed Tamura annulation approach to the synthesis of lactonamycinone (**116**) [2003 ACIE 5629 Danishefsky SJ].

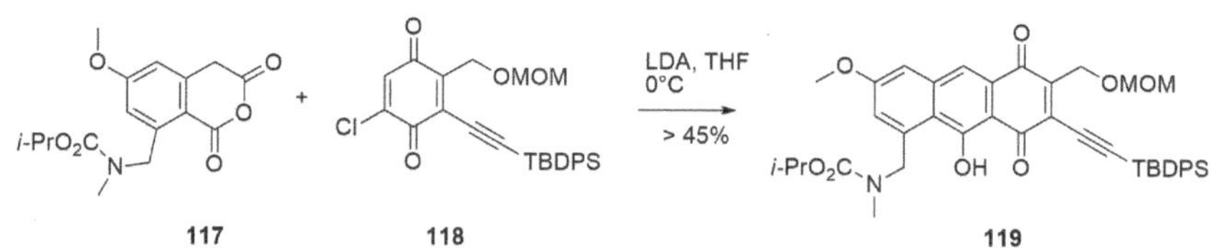

SCHEME 6.21 Chlorine-directed Tamura annulation in the synthesis of lactonamycinone (**X**) [2013 ACIE 2087 Nakata M].

Tamura annulation to give anthraquinone (**119**). As expected, the chlorine-directed annulation gave (**119**) as a single isomer. Nine more steps including a Pd-catalyzed methoxycarbonylation were applied to the intermediate (**119**) to complete synthesis of lactonamycinone (**116**). Three more steps were needed to synthesize the lactonamycin.

While Tamura annulation was primarily developed as a means for the benzannulations, it can be maneuvered for cyclohexenone annulation. Products without aromatization can be isolated. In one of the most recent developments in the use of Tamura annulation, such results were obtained with benzoquinone monoketals. After the initial annulation, decarboxylation is effected with DBU in a separate step. This modification is found useful in the total synthesis of viridicatumtoxin B (**120**) by Nicolaou et al. Viridicatumtoxin B (**120**) is a member of tetracycline antibiotics. In contrast to the most of the members of tetracyclines, it has a terpene unit embedded in its structure. Biologically, it exhibits potent activities against Gram-positive bacteria, including methicillin-resistant *Staphylococcus aureus*. Like most of the preceding examples, the donor is a homophthalic anhydride, but the acceptor is a benzoquinone monoacetal for the Tamura annulation. The annulation between (**121**) and (**122**) was effected by a sequential use of NaH and

DBU to give dihydroanthrone (**123**) via the intermediate (**124**). Treatment of the anthrone (**123**) with CSA gave (**125**). C-Alkylation of (**125**) with bromide (**126**) furnished (**127**), which upon treatment with BF_3 generated the crucial B-C-D-E-F ring subunit (**128**). $PhI(OAc)_2$ promoted dearomatization of the B ring of (**128**) provided (**129**). Using another annulation (vide Chapter 8), it was elaborated to the target, i.e., (**120**) (Scheme 6.22).[20]

Although methyl acrylate, acrylonitrile and 2-cyclohexenone are not reactive substrates for the Tamura annulation, alkylidene oxindole (**130**) is suitable for the annulation with cyclic anhydride (**131**) to give cyclohexenone annulated product (**132**), which was obtained with high degree of enantioselectivity in the presence of quinine-squaramide asymmetric catalyst (**133**) (Scheme 6.23).[21] More interestingly, the initial Michael addition of homophthalic anhydride anion takes place at the less substituted terminus. With nonaromatic anhydrides, annulation products undergo decarboxylation under the above reaction conditions. Furthermore, the temperature dependent reversal of the enantioselectvity was found in such annulations. At −50°C, the diastereomer (**134**) was formed. There was no explanation for such an extraordinary result.

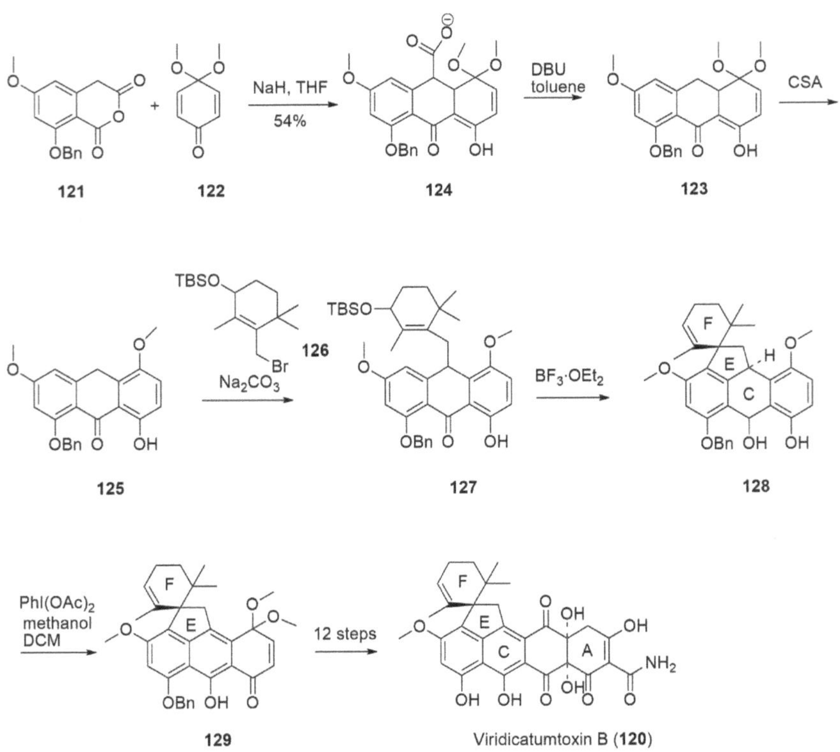

SCHEME 6.22 Tamura annulation for the synthesis of viridicatumtoxin B (**120**)[2014 JACS 12137 Nicolaou KC].

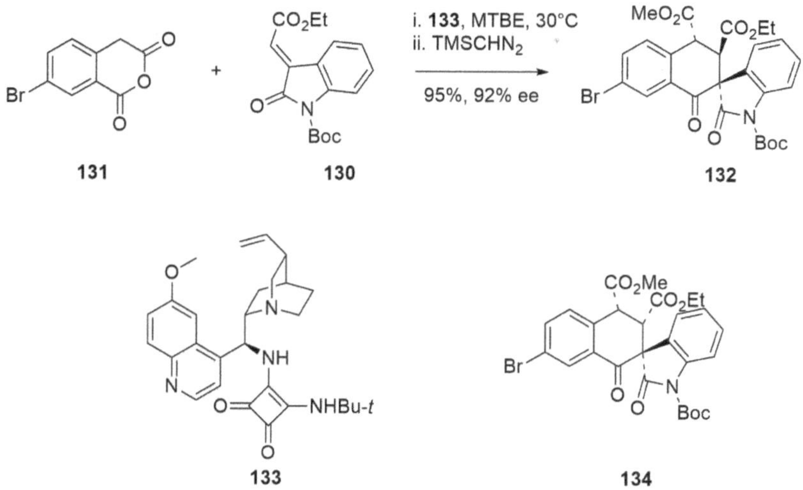

SCHEME 6.23 Asymmetric Tamura annulation [2014 ACIE 2628 Connon SJ].

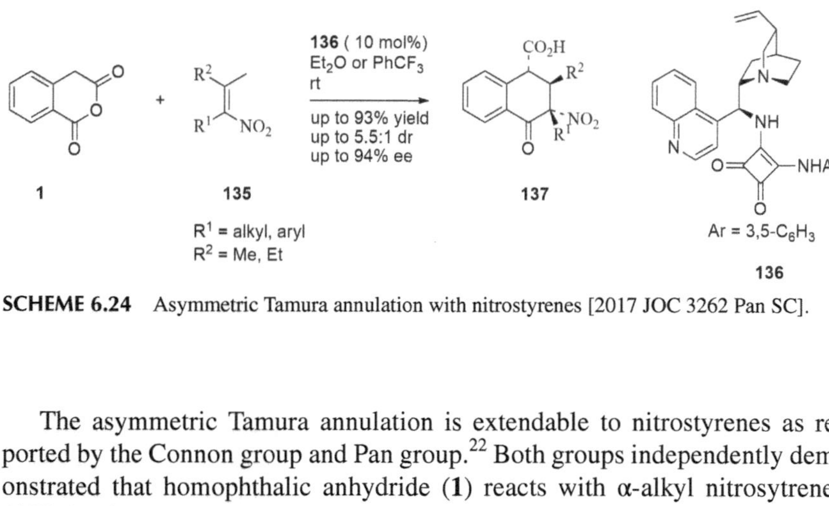

SCHEME 6.24 Asymmetric Tamura annulation with nitrostyrenes [2017 JOC 3262 Pan SC].

The asymmetric Tamura annulation is extendable to nitrostyrenes as reported by the Connon group and Pan group.[22] Both groups independently demonstrated that homophthalic anhydride (**1**) reacts with α-alkyl nitrosytrenes (**135**) in the presence of the cinchonidine-derived squaramide catalyst (**136**) to give α-tetralone derivatives (**137**) with three new chiral centres with high enantioselectivity and good diastereoselectivity (Scheme 6.24). The reaction between (**1**) and (**135**) in the presence of the catalyst (**136**) provided tetralones (**137**). The absolute configuration was determined by X-ray analysis. No additional bases were required for the completion of the reactions. The asymmetric induction in the above annulations was explained by activation of both the substrates by the squaramide catalyst through hydrogen bonding (Fig. 6.1), as in structure (**138**).

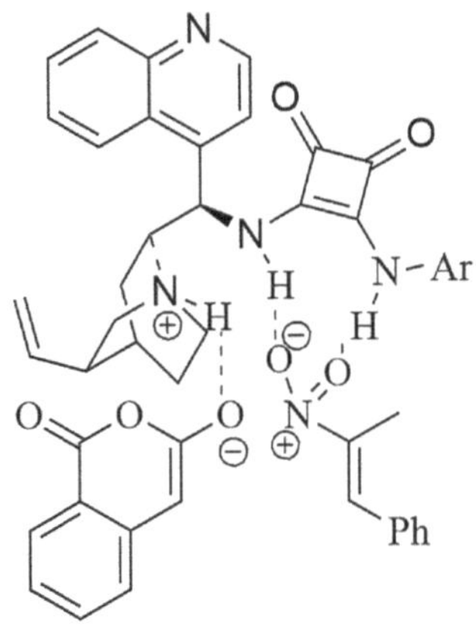

138

FIG. 6.1 Proposed transition state for the nitrostyrene annulation.

6.2 IMINES (CUSHMAN-HAIMOVA ANNULATION)

The Cushman-Haimova annulation reaction is a [4+2] thermal annulation, and does not belong to the anionic annulations. Since it strongly resembles the Tamura annulation being an aza-version, it was felt to highlight this annulation in this section. More importantly, it gives a historic perspective in that it was discovered in 1977, few years before the Tamura annulation. It consists in the cyclocondensation of homophthalic anhydrides with acceptors with CN double bond to give 3,4-hydroquinolone derivatives. This reaction may be called Cushman-Haimova annulation. The reaction of the homophthalic anhydride (**1**) with imine (**139**) in chloroform at room temperature furnishes isoquinolone acid (**140**) in 96% along with its *trans* isomer (Scheme 6.25).[23] It may be noted that the oxa-version of the Tamura annulation was discovered 25 years ago in 1958 by Pinder et al.[24]

The reaction is conjectured to proceed via one of two possible mechanisms (Scheme 6.26). The first one involves intermolecular proton transfer to form (**141**) and (**142**), followed by their Mannich-type addition giving (**143**). Intramolecular *N*-acylation of (**143**) gives the product (**140**). The second one is believed to proceed through *N*-acylation of the imine (**139**) with the anhydride (**1**) followed by intramolecular trapping of *N*-acyliminium species X. It is difficult to discern which of the above mechanisms is actually operative.

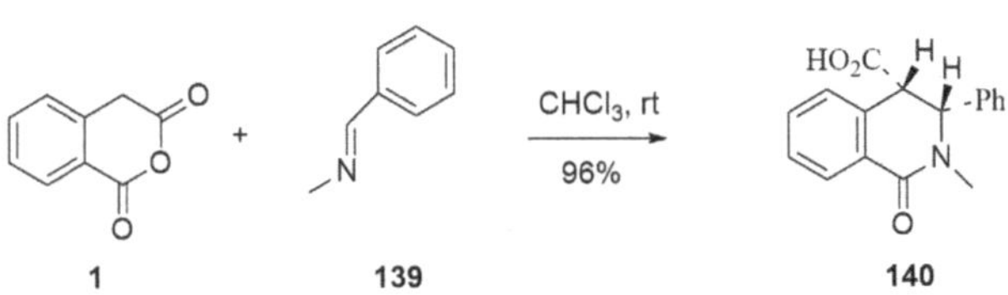

SCHEME 6.25 Example of a Cushman-Haimova annulation [1977 JOC 1111 Cushman M].

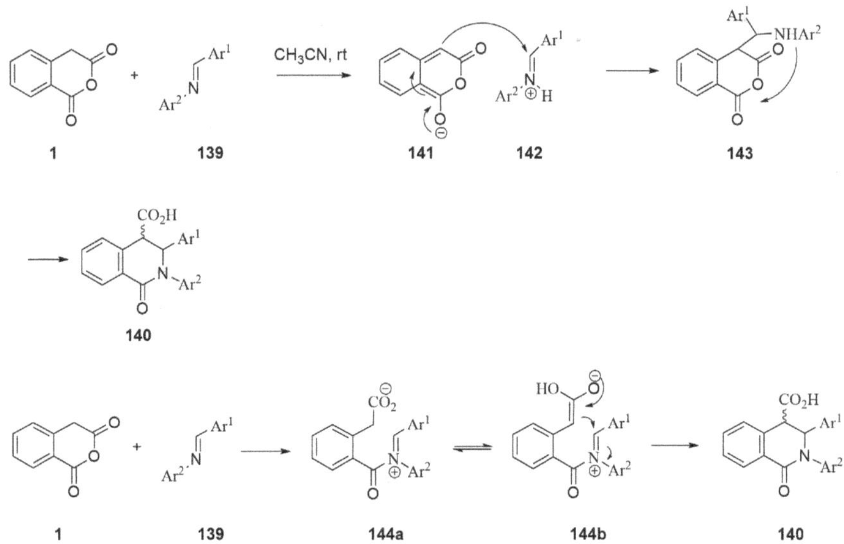

SCHEME 6.26 Mechanism of the Cushman-Haimova annulation.

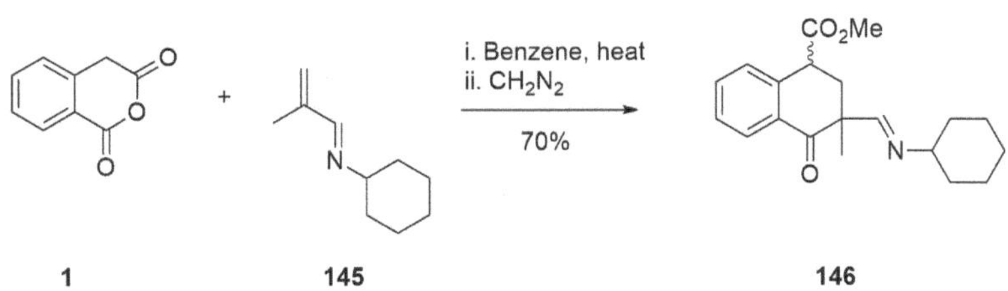

SCHEME 6.27 Cyclocondensation of homophthalic anhydride with 1-aza-1,3-dienes [1995 THN 6099 Haimova M].

The reaction of homophthalic anhydrides with conjugated imines gives diverse products. In most cases, conjugated CC double bonds take part in the reaction. As shown in Scheme 6.27, cyclocondensation of the anhydride (**1**) with imine (**145**) in hot benzene gives a new imine, which is derivatized with CH_2N_2 to furnish its methyl ester (**146**).[25]

The Cushman-Haimova annulation has been found immensely relevant for the synthesis of protoberberine alkaloids. For the sake of simplicity, the Xu synthesis of documbenine B (**147**), exemplifying a substituted isoquinoline is presented in Scheme 6.28. The annulation of anhydride (**148**) with imine (**149**) provided isoquinolone acid (**150**) in 61% yield. Decarboxylation of (**150**) in hot DMSO gave isoquinolone (**151**). LAH reduction of (**151**) converted the lactam into (**152**). Finally, DDQ oxidation of (**152**) yielded documbenine B (**147**).[26]

The sole example of the effect of a base in the Cushman-Haimova annulation, that can be classified as an anionic annulation, was reported by Knapp et al. The reaction between the anhydride (**1**) and (**153**) in the presence of two

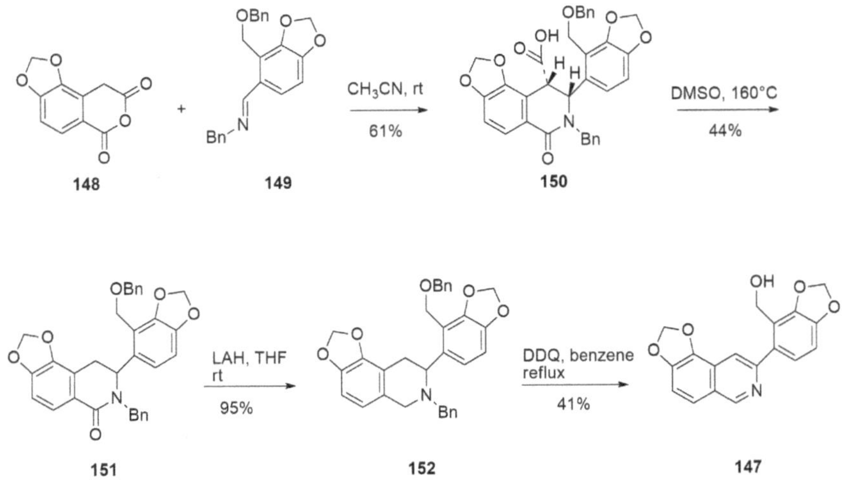

SCHEME 6.28 Synthesis of documbenine B (**147**) [1998 THN 14179 Xu X-Y].

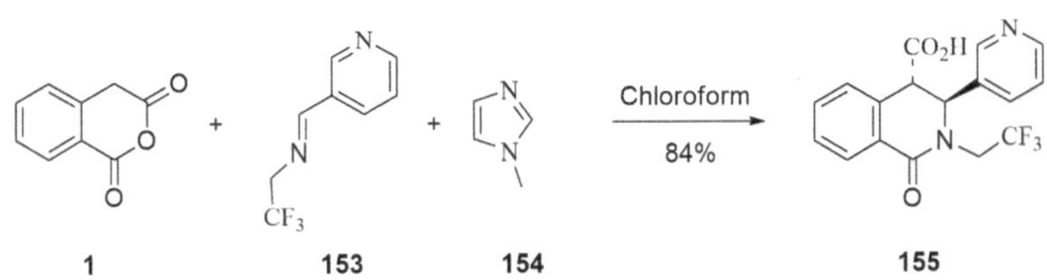

SCHEME 6.29 *N*-Methylimidazole promoted Cushman-Haimova annulation [2014 JOC 7593 Knapp S].

equivalents of *N*-methylimidazole (**154**) resulted in minimization of the formation of byproducts and improvement of the yield of the annulation product (**155**) (Scheme 6.29).[27]

Akin to the catalytic asymmetric Tamura annulation, Cushman-Haimova annulation has also witnessed an asymmetric version. As reported by Seidel et al., the annulation of the anhydride (**1**) with Schiff base (**156**) in the presence of catalyst (**157**) afforded the corresponding quinolone carboxylic acid, which was methylated to give ester (**158**) in 90% ee (Scheme 6.30).[28]

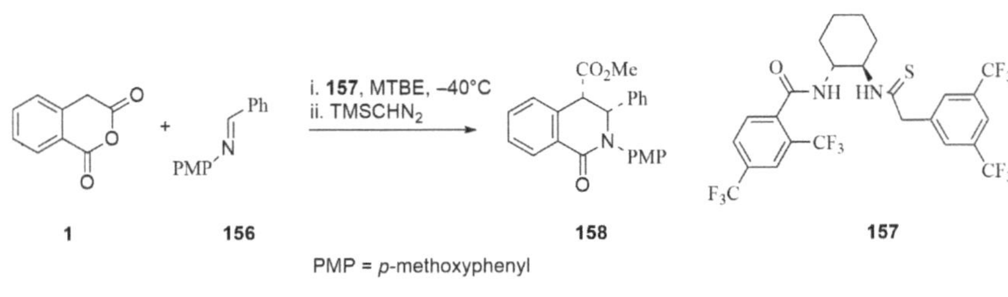

SCHEME 6.30 Organocatalytic asymmetric Cushman-Haimova annulation [2017 ACIE 2670 Seidel D].

In conclusion, the Tamura annulation works well with enolizable homophthalic anhydrides and their homologs. Among the acceptors, quinones and quinone monoacetals are well suited. Monoactivated acceptors like nitrostyrenes and diactivated acceptors like alpha-sulfenated unsaturated ketones are also compatible. The products with the nitrostyrenes are hydroaromatics. Among the bases, NaH is the most useful. However, LDA and LHMDS are occasionally used. It is to be seen if an anionic version of Cushman-Haimova annulation proceeds faster at subzero temperature.

REFERENCES

1. Gonzalez-Lopez, M.; Shaw, J. T. *Chem. Rev.* **2009**, *109*, 164.
2. Tamura, Y.; Sasho, M.; Nakagawa, K.; Tsugoshi, T.; Kita, Y. *J. Org. Chem.* **1984**, *49*, 473.
3. (a) Tamura, Y.; Wada, A.; Sasho, M.; Kita, Y. *Tetrahedron Lett.* **1981**, *22*, 4283; (b) Tamura, Y.; Wada, A.; Sasho, M.; Kita, Y. *Chem. Pharm. Bull.* **1983**, *31*, 2691.
4. Kuttruff, C. A.; Zipse, H.; Trauner, D. *Angew. Chem. Int. Ed.* **2011**, *50*, 1402.
5. Kita, Y.; Fujioka, H. *Top. Curr. Chem.* **2008**, *282*, 299.
6. Tamura, Y.; Fukata, F.; Sasho, M.; Tsugoshi, T.; Kita, Y. *J. Org. Chem.* **1985**, *50*, 2273.
7. Martin, P.; Ridier, S.; Mondon, M.; Renoux, B.; Pfeiffer, B.; Renard, P.; Pierre, A.; Gesson, J.-P. *Biorg. Med. Chem.* **2002**, *10*, 253.
8. Matsumoto, T.; Yamaguchi, Y.; Suzuki, K. *Tetrahedron* **1997**, *53*, 6533.
9. Matsuda, F.; Kawasaki, M.; Ohsaki, M.; Yamada, K.; Terashima, S. *Tetrahedron* **1998**, *44*, 5745.
10. Lavallee, J.-F.; Rej, R.; Courchesne, M.; Nguyen, D.; Attardo, G. *Tetrahedron Lett.* **1993**, *34*, 3519.
11. Kita, Y.; Okunaka, R.; Honda, T.; Shindo, M.; Taniguchi, M.; Kondo, M.; Sasho, M. *J. Org. Chem.* **1991**, *56*, 119.
12. Tamura, Y.; Sasho, M.; Akai, S.; Kishimoto, H.; Sekihachi, J.-I.; Kita, Y. *Chem. Pharm. Bull.* **1987**, *35*, 1405.
13. Kita, Y.; Maeda, H.; Kirihara, M.; Fujii, Y.; Nakajima, T.; Yamamoto, H.; Tamura, Y.; Fujioka, H. *Chem. Pharm. Bull.* **1992**, *40*, 61.
14. Hottop, T.; Gutke, H.; Murahashi, S.-I. *Tetrahedron Lett.* **2001**, *42*, 3343.
15. Iio, K.; Ramesh, N. G.; Okajima, A.; Higuchi, K.; Fujioka, H.; Akai, S.; Kita, Y. *J. Org. Chem.* **2000**, *65*, 89.
16. Haruta, Y.; Onizuka, K.; Watanabe, K.; Kono, K.; Nohara, A.; Kubota, K.; Imoto, S.; Sasaki, S. *Tetrahedron* **2008**, *64*, 7211.
17. Kita, Y.; Higuchi, K.; Yoshida, Y.; Iio, K.; Kitagaki, S.; Ueda, S.; Akai, K.; Fujioka, H. *J. Am. Chem. Soc.* **2001**, *123*, 3214.
18. Shair, M. D.; Yoon, T. Y.; Mosny, K. K.; Chou, T. C.; Danishefsky, S. J. *J. Am. Chem. Soc.* **1996**, *118*, 9509.
19. (a) Siu, T.; Cox, C. D.; Danishefsky, S. J. *Angew. Chem. Int. Ed.* **2003**, *42*, 5629; (b) Adachi, S.; Watanabe, K.; Iwata, Y.; Kameda, S.; Miyaoka, Y.; Onozuka, M.; Mitsui, R.; Saikawa, Y.; Nakata, M. *Angew. Chem. Int. Ed.* **2013**, *52*, 2087.
20. Nicolaou, K. C.; Hale, C. R. H.; Nilewski, C.; Ioannidou, H. A.; ElMarrouni, A.; Nilewski, L. G.; Beabout, K.; Wang, T. T.; Shamoo, Y. *J. Am. Chem. Soc.* **2014**, *136*, 12137.
21. Manoni, F.; Connon, S. J. *Angew. Chem. Int. Ed.* **2014**, *53*, 2628.
22. (a) Nath, U.; Pan, S. C. *J. Org. Chem.* **2017**, *82*, 3262; (b) Manoni, F.; Farid, U.; Trujillo, C.; Connon, S. *J. Org. Biomol. Chem.* **2017**, *15*, 1463.

23. (a) Cushman, M.; Gentry, J.; Dekow, F. W. *J. Org. Chem.* **1977**, *42*, 1111; (b) Haimova, M. A.; Mollov, N. M.; Ivanova, S. C.; Dimitrova, A. I.; Ognyanov, V. I. *Tetrahedron* **1977**, *33*, 331.
24. Jones, J. B.; Pinder, A. R. *J. Chem. Soc.* **1958**, 2612.
25. Georgieva, A.; Stanoeva, E.; Spassov, S.; Haimova, M.; De Kimpe, N.; Boelens, M.; Keppens, M.; Kemme, A.; Mishnev, A. *Tetrahedron* **1995**, *51*, 6099.
26. Xu, X.-Y.; Qin, G.-W.; Xu, R.-S.; Zhu, X.-Z. *Tetrahedron* **1998**, *54*, 14179.
27. Liu, J.; Wang, Z.; Levin, A.; Emge, T. J.; Rablen, P. R.; Floyd, D. M.; Knapp, S. *J. Org. Chem.* **2014**, *79*, 7593.
28. Jarvis, C. L.; Hirschi, J. S.; Vetticatt, M. J.; Seidel, D. *Angew. Chem. Int. Ed.* **2017**, *56*, 2670.

Chapter 7

[4+2] Annulation of Activated *o*-Toluate/*o*-Tolualdehyde Donors (Hauser-Leusen Annulation)

This chapter deals with yet another class of [4+2] anionic annulation. This class of the annulation involves activated derivatives of *ortho* toluate and *ortho* tolualdehyde as the 1,4-dipole equivalents. The corresponding dipole exits in the lateral position of the toluates/tolualdehydes. The activating groups are an anion-stabilizing group. These are esters, cyanide, nitro, sulfones, sulfoxides, thioacetals, etc. The respective annulations fall into two major groups: one in which the products retain the activating groups and the other in which activating groups are lost during the annulations. The first may be named as the "Schmid annulation" and the second as the "Hauser-Leusen annulation." Overall, this chapter covers the annulation reactivity of the reagents of the types (**1–10**) (Fig. 7.1).[1–8]

7.1 SCHMID HOMOPHTHALATE DONORS

The concept of [4+2] lateral metalation-initiated annulation with an *ortho*-toluate was first introduced by Schmid et al. in 1965.[1] The reactions of homophthalate (**11**) with methyl crotonate (**12b**), maleate (**12c**), and cinnamate (**12d**) in the presence of NaOMe in methanol gave cyclocondensed products (**13**) in 50%–56% (Scheme 7.1). These annulations are formulated to proceed through Michael addition of the lateral carbanions (**14**) with acceptors followed by Dieckmann reaction of the resultant carbanions (**15**). The marked acidity of the benzylic methylene group in (**11**) is the key to the annulations. Such a condensation was also effected with LDA in THF at −78°C.[9] Similarly, coumarin (**16**) gave product (**17**) in 39% yield, on reaction with homophthalate (**11**) in the presence of NaOMe via the intermediate (**18**) (Scheme 7.2).[10]

When combined with the Krapcho dealkoxycarbonylation, the Schmid annulation leads to a general route for the synthesis of α-tetralones. As exemplified in Scheme 7.3, Schmid annulation of (**19**) and (**12**) give dihydronaphthalene derivatives (**20**). Two esters in (**20**) can be removed by treatment with NaCl in refluxing DMSO to produce (**21**). Similar treatment of (**20**) with NaCl in

Anionic Annulations in Organic Synthesis. https://doi.org/10.1016/B978-0-12-813800-7.00007-5

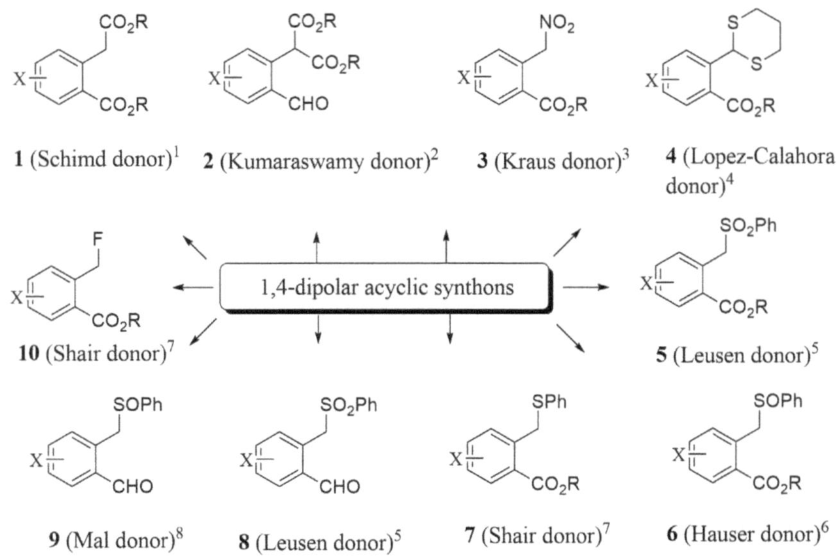

FIG. 7.1 Structures of toluate-based 1,4-dipolar reagents (**1–10**).

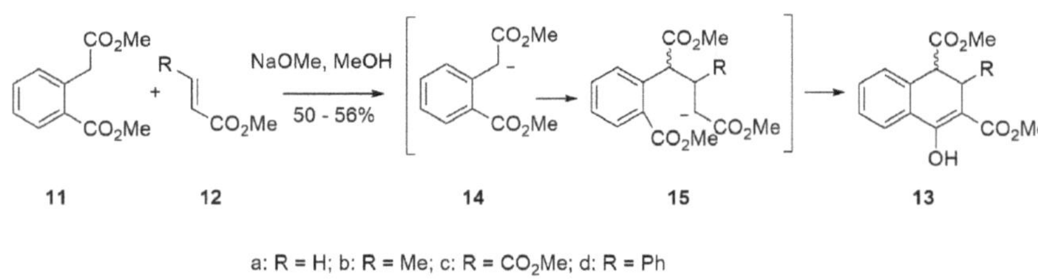

SCHEME 7.1 Annulation of homophthalate with acrylates [1965 HCA 375 Schmid H].

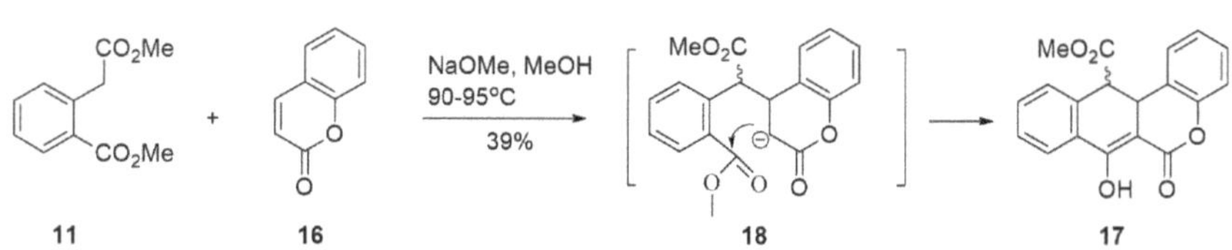

SCHEME 7.2 Schmid annulation with a coumarin [1965 HCA 375 Schmid H].

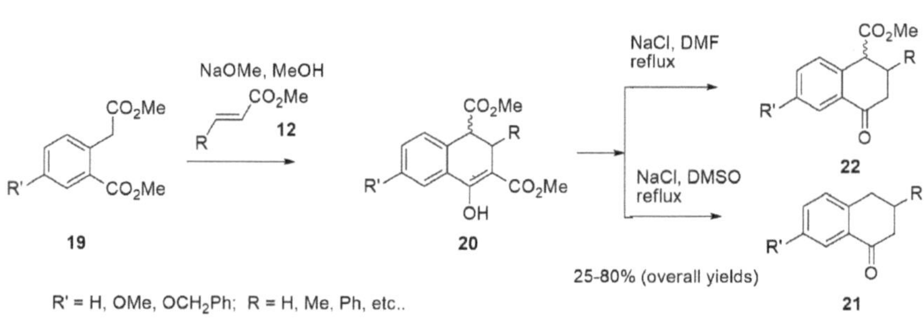

SCHEME 7.3 Schmid annulation for the synthesis of α-tetralones.

refluxing DMF results in selective demethoxycarbonylation of the ester group at C2, the yielding (**22**).[11]

The annulation reactivity of quinone monoketals toward a Schmid homophthalate donor was first reported by Swenton et al.[10] The reaction of the homophthalate (**11**) with monoketals (**23**) in the presence of NaH in THF at room temperature initially afforded anthracenones (**24**), which were treated with PTSA to give anthrones (**25**) (Scheme 7.4).[12] It may be noted that very few carbon nucleophiles are compatible with quinone monoketals for simple Michael additions.

Schmid annulation was successfully applied to indole donors (**26** and **27**) for the synthesis of 4-hydroxycarbazoles (**28** and **29**, respectively). For example, homophthalate analog **26** reacted with propiolate (**30**) in the presence of KTB at 0°C to give trisubstituted carbazole (**28**) in 45% yield (Scheme 7.5). Similarly, cyanoactivated ketone donor **27** reacted with acetylene carboxylate **31** under similar conditions to furnish tetrasubstituted carbazole (**29**), in which one of the two benzene rings was fully substituted.[13]

The annulation of *gem*-diester activated *o*-tolualdehydes (**32**) with allenecarboxylates takes place in two directions depending on the substituents at C2 of the acceptors. When the annulation was carried out with 2-methyl substituted

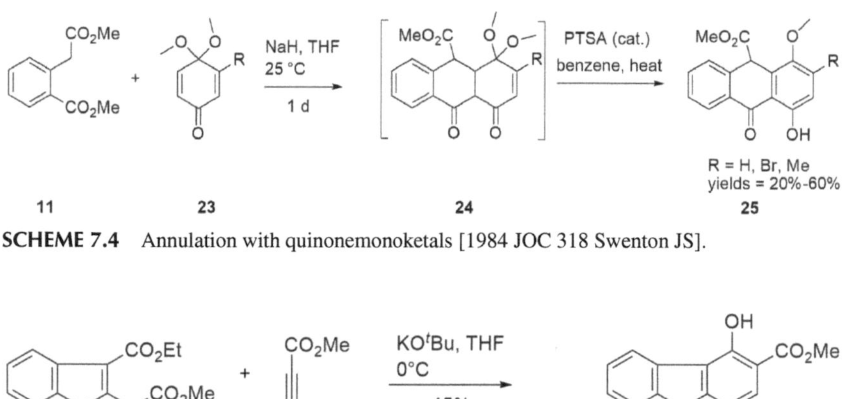

11 **23** **24** **25**

SCHEME 7.4 Annulation with quinonemonoketals [1984 JOC 318 Swenton JS].

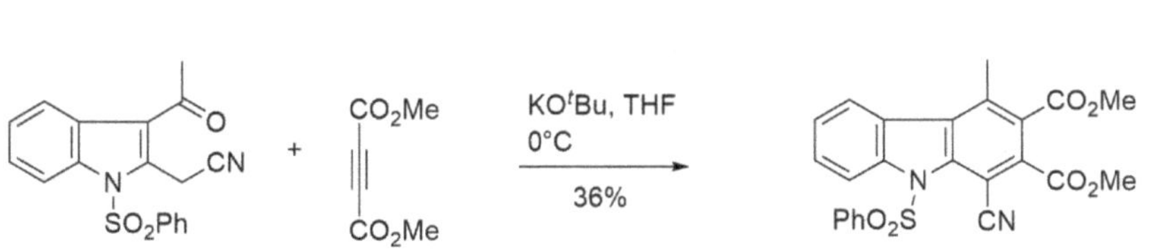

26 **30** **28**

27 **31** **29**

SCHEME 7.5 Synthesis of polysubstituted carbazoles [2009 THN 3592 Mohanakrishnan AK].

allenecarboxylates (**33**) in the presence of K_2CO_3 in DMSO at 80°C, the trisubstituted naphthalenes (**34**) were formed in 71%–85% yields, resulting from the fusion of β,γ bond of the acceptors (Scheme 7.6).[2] The reaction begins with abstraction of proton from the active methine carbon of compound (**32**) under the basic condition. The resulting carbanion (**35**) then attacks at the β carbon of the allene carboxylate (**33**), and forms the carbanion (**36**). Subsequent attack of the carbanion at the carbonyl group of *o*-tolualdehyde (**37**) forms the cyclic intermediate (**37** and **38**). Liberated water drives the reaction further to (**39**). Finally, decarboxylation and aromatization gives rise to (**34**).

Interestingly, the reactivity pattern of the allenecarboxylates without a C2 substituent markedly differs from that of (**33**). Both possible regioisomers of the annulation products were formed. The reaction of (**32**) with acceptors (**40**) in the presence of K_2CO_3 in DMSO at 80°C provided mixtures of products (**41** and **42**) in combined yields of 81%–94%. The products arising out of the fusion of β,γ bond of the acceptors, i.e., (**41**) are the major products (Scheme 7.7).

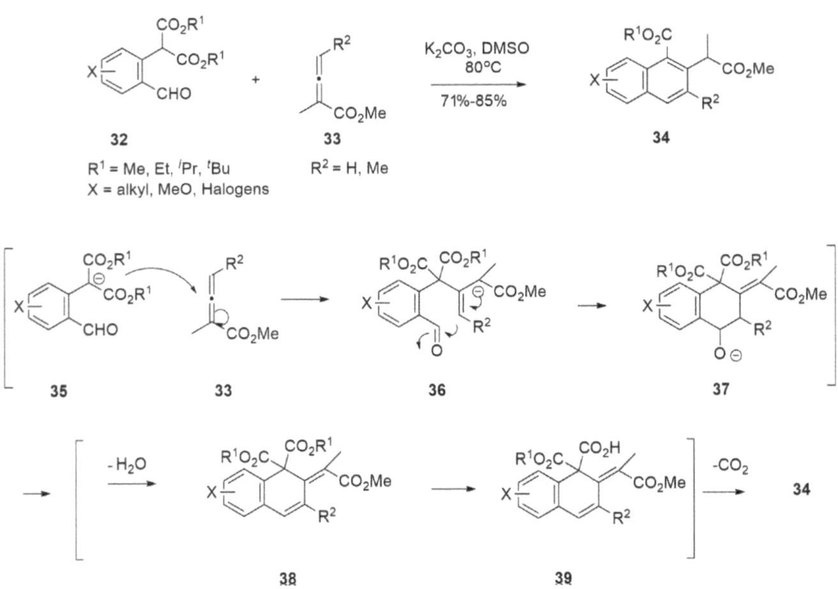

SCHEME 7.6 Kumara Swamy annulation of activated tolualdehydes with allenecarboxylates [2016 OBC 5079 Kumara Swamy KC].

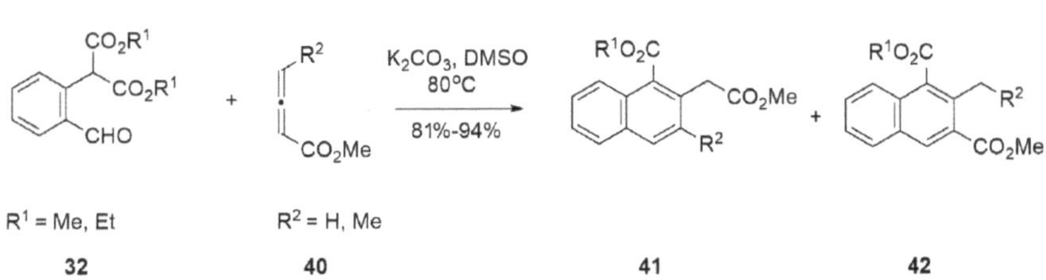

SCHEME 7.7 Kumara Swamy annulation of activated tolualdehydes with allenecarboxylates without α-substituents [2016 OBC 5079 Kumara Swamy KC].

Analogous annulations of *o*-tolualdehyde (**32**) have also been realized with allenephosphonates and allenylphosphine oxides, giving the corresponding naphthalene carboxylates in good yields.

Indole analogs of the activated tolualdehyde donors and their reactivities toward conjugated nitroalkenes have been reported by Samanta et al.[14] The most striking feature of their work for the synthesis of carbazoles is that free NH group of the indole donors does not interfere the annulations. The carboxy activated indole carboxaldehydes (**43**) have been shown to undergo annulations with the nitroalkene acceptors (**44**) in the presence of DABCO in water giving carbazole products (**45**) in one step via their dihydro derivatives (**46**) (Scheme 7.8).

2-Cyanomethylbenzaldehyde (**47**) reacted with cycloalkenones (**48**) in the presence of K_2CO_3 and Aliquot 336, a phase transfer catalyst, to give [4+2] annulation products (**49**) in good yields (Scheme 7.9). The initial products (**50**) underwent facile aerial oxidation under the reaction conditions to result in the products.[15]

Cycloalkenones, like cyclopentenone or cyclohexanone, are poor acceptors for the [4 + 2] annulations. However, if the reactions are out in two stages, the annulation with 2-nitromethyl benzoate (**51**) becomes moderately successful. The reaction between (**51**) and (**48a**) first gives Michael adduct (**52**) in the presence of tetrabutylammonium hydroxide. The adduct (**52**), on reaction with NaOMe, undergoes Claisen reaction to produce the corresponding tricyclic product, which is air-oxidized to furnish benz[f]indanone (**53**) (Scheme 7.10).[3] This protocol worked well indenone, dicyclopentadienone, tetrahydroindenone, and phenyl vinyl ketone to give the products (**54–57**), in 64%, 47%, 27%, and 22% yields, respectively (Fig. 7.2).

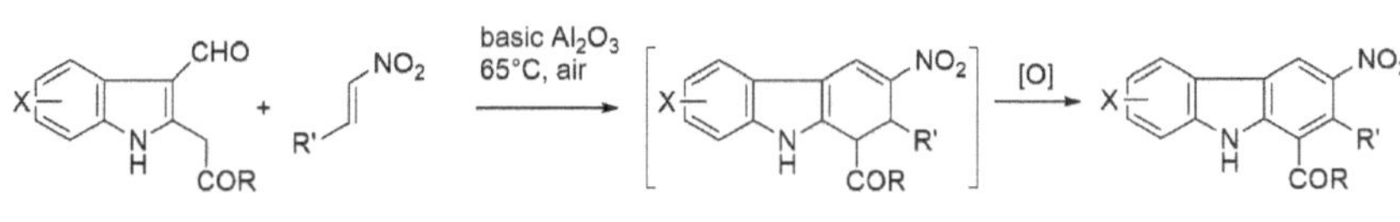

43 **44** **46** **45**

SCHEME 7.8 Annulation of activated carbazole aldehydes with nitroalkenes [2016 CST 6362 Samanta S].

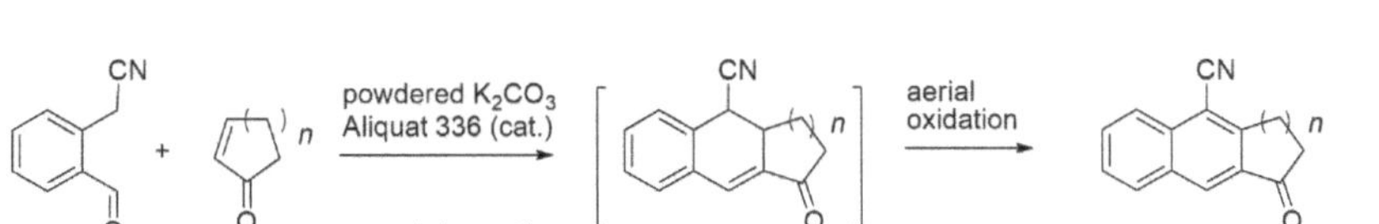

47 **48** **50** **49** $n = 1$ 55% $n = 2$ 80%

SCHEME 7.9 Annulation with a cyano-activated tolualdehyde [2009 Arkivoc 98 Jonczyk A].

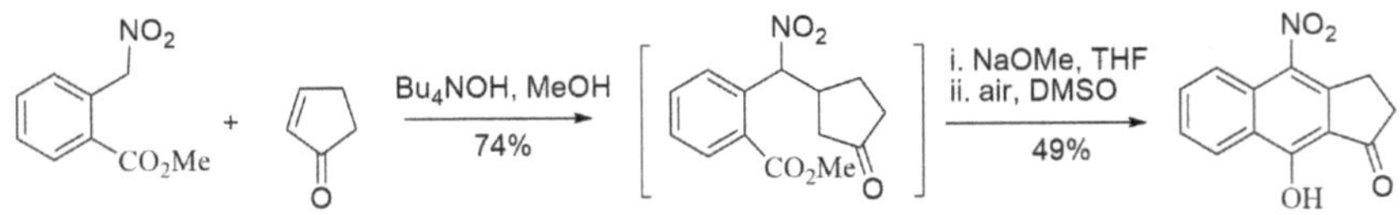

51 48a 52 53

SCHEME 7.10 Annulation with 2-nitromethyl benzoate [2012 THS 4444 Kraus GA].

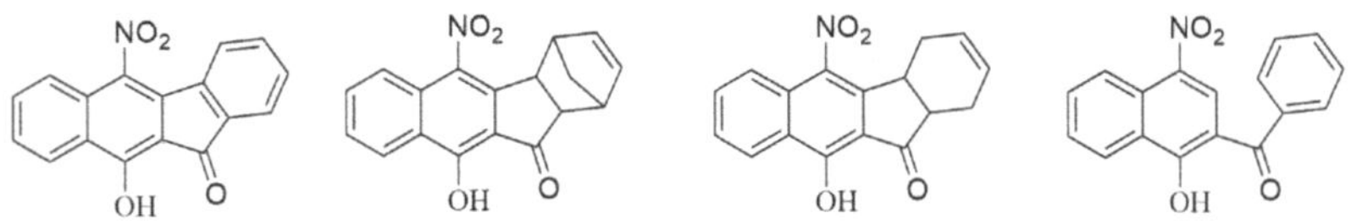

54 55 56 57

FIG. 7.2 Annulation products obtained from 2-nitromethyl benzoate.

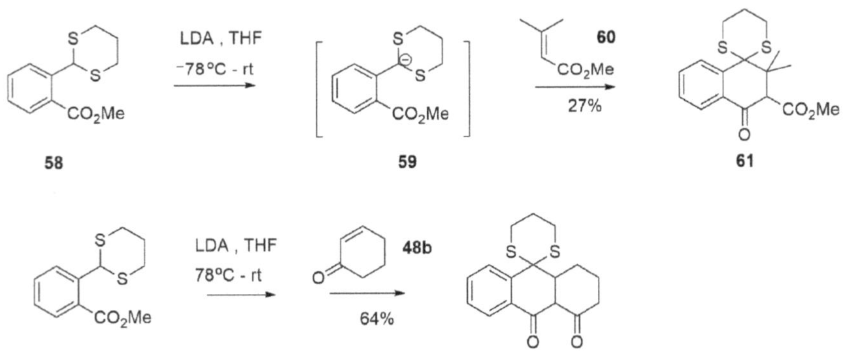

58 59 61

58 62

SCHEME 7.11 Annulation with 2-(1,3-dithian-2-yl)benzoate (1992 THS 3679 Lopez-Calahorra F).

Methyl 2-(1,3-dithian-2-yl)benzoate (**58**) was introduced in 1992 by Lopez-Calahorra as a 1,4-dipolar reagent in an annulation.[4] Its annulation reactivity arising from the activation with dithiane moiety is illustrated with two examples in Scheme 7.11. The reaction of (**58**) with LDA generated (**59**), which reacted with methyl butenoate (**60**) to give tetralone (**61**) in 27% yield. Similar annulation of (**58**) was feasible with cyclohexenone (**48b**). The reaction between (**58**) and (**48b**) gave anthracenone (**62**) in 64% yield.

The principle of annulation of the 2-dithianyl benzoate (**58**) has been used to construct complex spiroannulated structures. For example, dithianylbenzoate (**63**), prepared in one step from formylbenzoate (**64**), underwent annulation with spirocyclicenone (**65**) diastereoselectively to give (**66**) in 84% yield (Scheme 7.12).[16]

The annulation of a dithiane-activated aryl ketone donor was utilized by Harrowven in the synthesis of the cytotoxic arylnaphthalene lignin taiwanin E

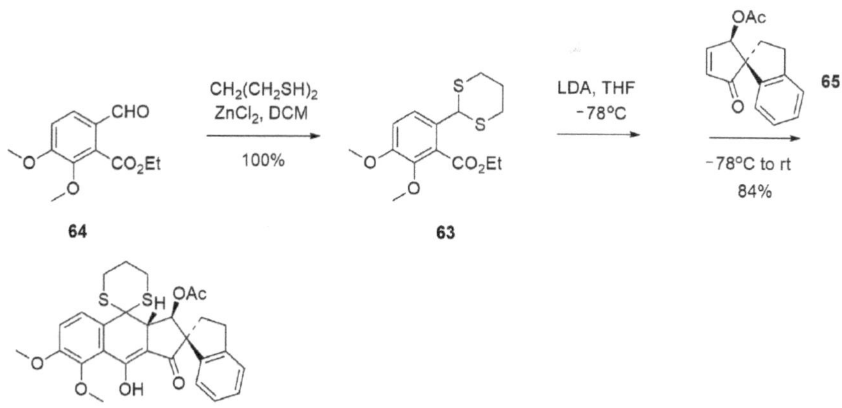

SCHEME 7.12 Annulation with a spiroannulatedcyclopentenone [2009 THS 7021 Burnell DJ].

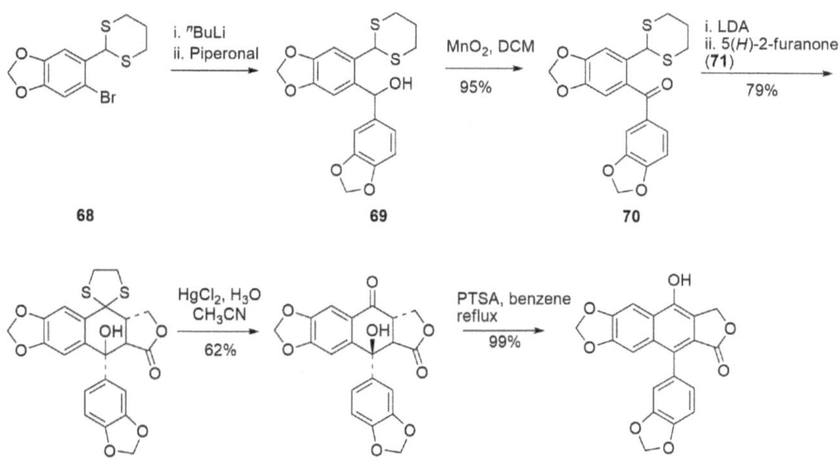

SCHEME 7.13 Synthesis of taiwanin E by an annulation [1993 THN Harrowven DC].

(**67**). Bromoarene (**68**) was lithiated with nBuLi and then treated with piperonal to obtain (**69**). This was then oxidized with MnO2 to give 1,4-dipolar reagent (**70**). Lithiation of (**70**) with LDA followed by addition of 5(H)-furan-2-one (**71**) yielded lignan precursor (**72**). This intermediate was subjected to standard dithiane hydrolysis to give (**73**). PTSA promoted dehydration of (**73**) afforded taiwanin E (**67**) (Scheme 7.13).[17] Kamal et al. adopted the dithiane annulation route and coupled it with a reductive aromatization step, which was achieved by a Ni-containing reducing agent. Thus, the reaction between (**74**) and (**71**) gave annulation product (**75**). Treatment of (**75**) with nickel-containing complex reducing agent caused desulfuration as well as aromatization to furnish justicidin B (**76**) (Scheme 7.14).[18]

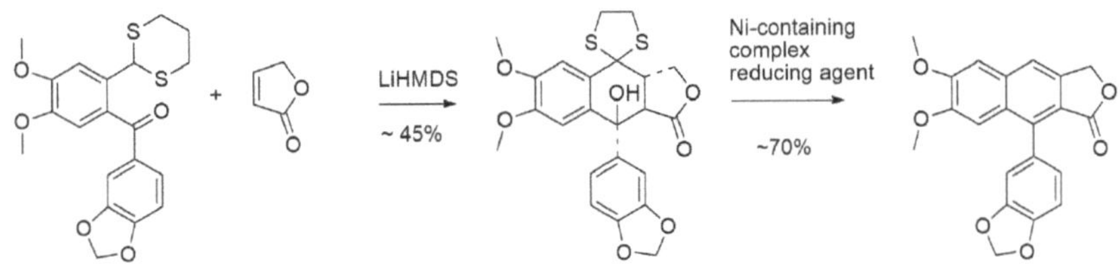

74 71 75 76 (Justicidin B)

SCHEME 7.14 Kamal justicidin B synthesis [1994 THS 3879 Kamal A].

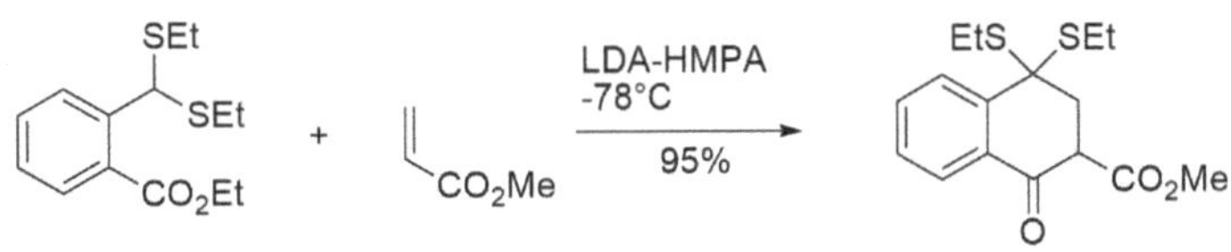

77 12a 79

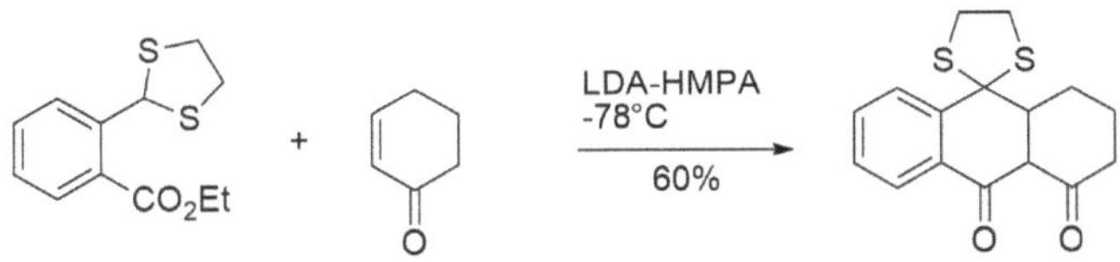

78 48b 80

SCHEME 7.15 Examples of annulations with bis(ethylthiomethyl)benzoate (**77**) and 1,3-dithilanylbenzoate (**78**) [1996 CPB 1785 Ozaki Y].

Similar to the dithianyl reagent (**58**), bis(ethylthiomethyl)benzoate (**77**) and 1,3-dithiolanylbenzoate (**78**) undergo successful annulations with acrylates and enones. Two examples of such annulations are displayed in Scheme 7.15. The reaction of (**77**) and (**12a**) in the presence of LDA-HMPA gave tetralone (**79**) in an excellent yield. Similarly, the reaction of (**78**) with (**48b**) furnished anthracenone (**80**) in 60% yield.[19]

The annulation reactivity of ethyl *o*-tosylmethylbenzoate (**81**) and *o*-tosylmethylbenzonitrile (**82**) as 1,4-dipolar reagents, which we term Leusen donors is quite varied.[5] The reaction of these donors with activated alkynes like ethyl propiolate, ethyl tetrolate, and diethyl acetylene dicarboxylate results in the formation of tri- or tetra-substituted naphthalenes in good to very good yields. The tosyl group of the donors is retained in the products. However, depending on the nature of the acceptors, the tosyl may be eliminated during the process (vide next section). As shown in Scheme 7.16, the annulation of sulfone (**81**) with ethyl propiolate (**83**) and ethyl tetrolate (**84**) give naphthalenes (**85** and **86**, respectively). Likewise, the reaction of (**82**) with the above two acceptors provide carbonitriles (**87** and **88**, respectively). The mechanism of these reactions is illustrated by the reaction of (**81**) with ethyl tetrolate. Lateral

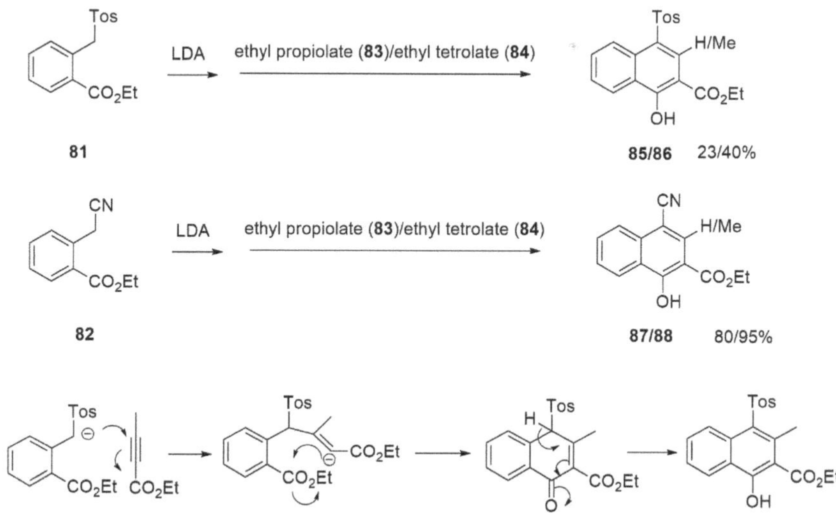

SCHEME 7.16 Annulation of Leusen donors [1978 THS 2213 Leusen AMV].

carbanion (**89**) undergoes Michael addition to form (**90**), which then internally displaces ethoxy group to provide tetralone (**91**). Tautomerization of (**91**) furnishes product (**92**).

7.2 HAUSER SULFOXIDE AND LEUSEN SULFONE TOLUATE DONORS

7.2.1 Hauser Sulfoxide Donors

In their original publication[6] in 1978, Hauser and Rhee reported two benzannulations. The one based upon the activated phthalides is already discussed in Section X. The other one based upon the annulation (Scheme 7.17) with phenylsulfoxide activated toluates forms the subject of this section. Unlike the phthalide annulation for 1,4-naphthalenediols, the sulfoxide annulations give the regiospecific synthesis of α-naphthols. The reaction of sulfoxide (**93**) with crotonate (**12b**) in the presence of LDA at −78°C afforded disubstituted naphthol (**94**). These reactions are distinguished from the Hauser phthalide annulation by the level of oxidation and the resulting pattern of hydroxy substituents in the products. The other notable difference is that the latter entails a sulfoxide stabilized toluate donor whereas the former involves a stabilized phthalide. This section also highlights also the reactivity of Leusen donors (e.g., **81**) involving the loss of the activating sulfone groups.

As to the mechanism, the sulfoxide (**93**) undergoes deprotonation in the presence of LDA to form carbanion (**93**). Michael addition of (**93**) to

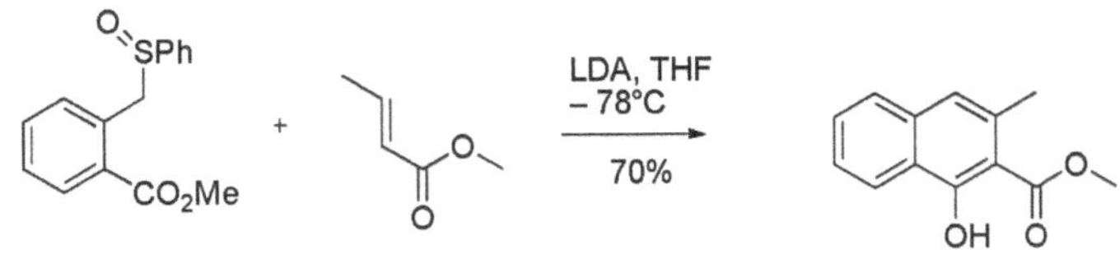

SCHEME 7.17 Hauser original toluate sulfoxide annulation [1978 JOC 178 Hauser FM].

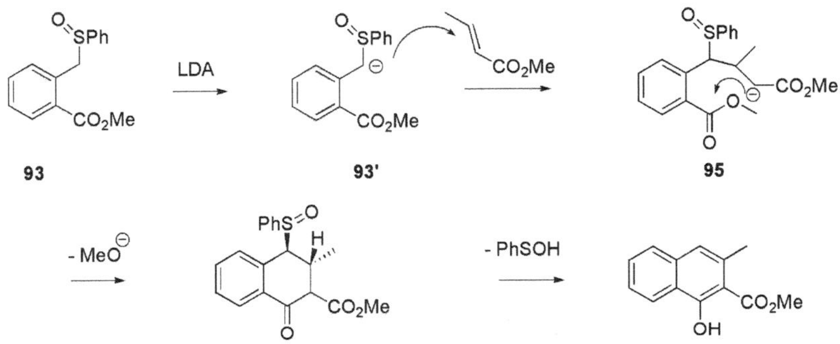

SCHEME 7.18 Mechanism of Hauser sulfoxide annulation.

methyl crotonate gives new carbanion (**95**), which participates in Dieckmann condensation to form tetralone (**96**). Through cis-elimination of phenylsulfinic acid, it transforms to naphthol (**93**). For aromatization of the terminal ring, the reaction mixture is warmed to room temperature and then heated to reflux to effect elimination of phenylsulfinic acid from the initial products (Scheme 7.18).

The Hauser sulfoxide annulation (Scheme 7.17) is quite general with regard to both donors and acceptors. Select examples are presented in Scheme 7.19 to illustrate the scope of the annulation. The donors could be heavily substituted toluates or heterocyclic analogs. The reaction between (**97**) and (**98**) took place in the presence of LiO^tBu in THF-DMSO to give (**99**). The annulation (reactions 2 and 3) of the donors (**100** and **101**) with acceptors (**102** and **103**) occurred at 20°C to give (**104** and **105**), respectively. The reaction (reaction 4) of furan donor (**106**) with acceptor (**98**), similar to (**97**), gave pentacycle (**107**), a key precursor to coumarins.

Recently, the Hauser sulfoxide annulation has been utilized in construction of naphthopyranones. The fusion of (**108**) with (**109**) in the presence of LDA furnished (**110**) in 29% yield (Scheme 7.20).[20]

One of the advanced applications is found in the total synthesis of *O*-methylkidamycinone (**111**). The required polyoxygenated anthracenone (**112**) was obtained in 71% by condensation of naphthoate sulfoxide (**113**) with pentenone (**114**) in the presence of LDA.[21] Unlike common annulations, this annulation needed refluxing of the reaction mixture for aromatization (Scheme 7.21).

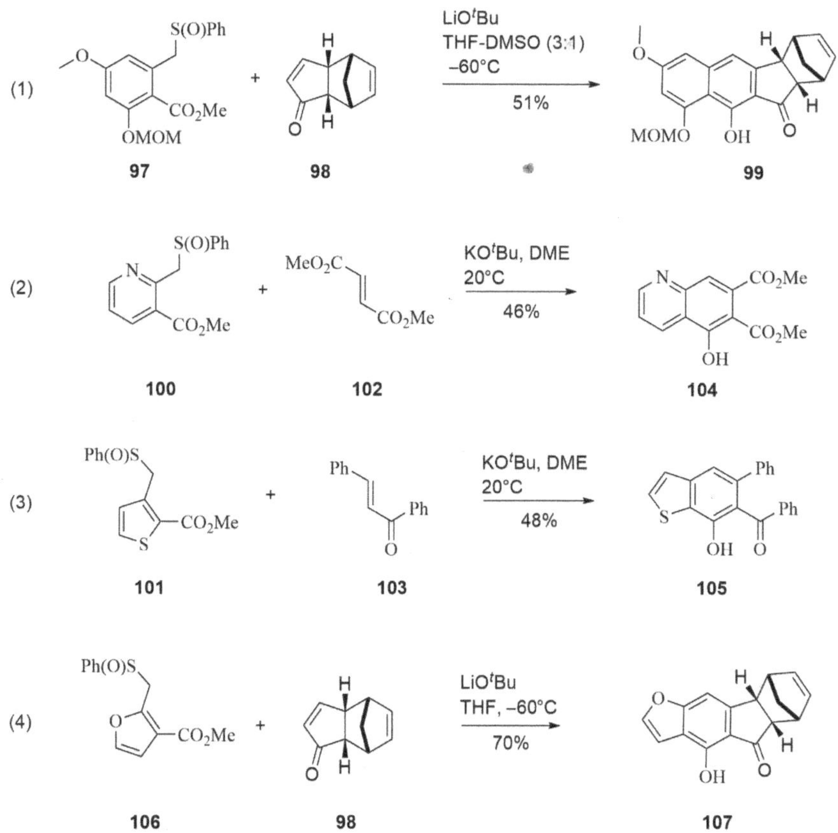

SCHEME 7.19 Examples of the Hauser sulfoxide annulation.

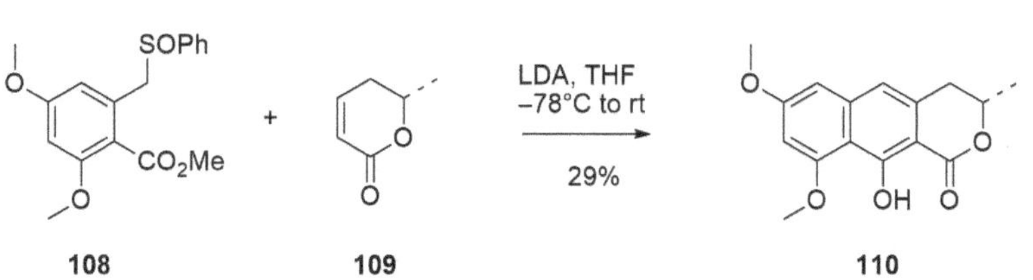

SCHEME 7.20 Annulation with 4,6-dimethoxytoluate [2009 SYS 2561 Brimble M].

The application of the Hauser sulfoxide annulation was first extended to an indole analog by Mohanakrishnan and Srinivasan in 1993. Sulfoxide activated indole ester (**115**), on reaction with acceptors (**116**) in the presence of NaH at 0°C, regiospecifically afforded 4-hydroxycarbazoles (**117**) in 50%–72% yields (Scheme 7.22).[22]

Further extensions of the annulation to indole donors (**118** and **119**) have shown broad applicability of the method in assembling a variety of oxysubstituent-free

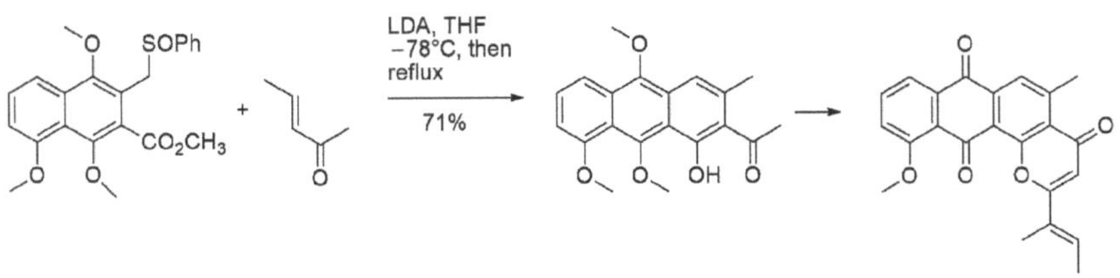

113 114 112 111 (O-Methylkidamycinone)

SCHEME 7.21 Synthesis of *O*-methylkidamycinone precursor [1980 JOC 3061 Hauser FM].

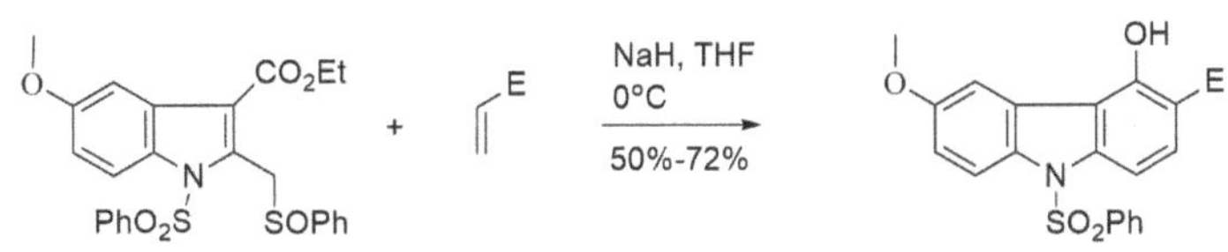

115 116 117

SCHEME 7.22 Annulation with indole sulfoxide donor [1993 THS 1343 Srinivasan PC].

carbazole derivatives such as (**120**) and (**121**). These annulations were carried with KO^{t}Bu in THF at 0°C, yields being in the range of 45%–50% (Scheme 7.23).[13]

7.2.2 Annulation With Phenylsulfinylmethyl Pyrrole (122)

For the direct synthesis of indole derivatives, pyrrole sulfoxide analog (**122**) was studied toward annulation with a variety of acceptors. As shown in Scheme 7.24, the process was expected to lead to first the sulfoxide-containing indoles (**123**) and then indoles (**124**) through *syn* thermal elimination of phenylsulfenic acid. The reaction of (**122**) (entry 1) with simple Michael acceptor like methyl crotonate (**12b**) in the presence of LHMDS in THF at −78°C-rt readily afforded (**125**). Entries 2 (**12d** → **126**) and 3 (**127** → **128**) exemplify reactivity of cinnamates. The expected annulation products (**126** and **128**) were obtained in excellent yields (82%–94%). Entries 4 (**48a** → **129**) and 5 (**98** → **130**) represent the reactivity of 2-cyclopentenone (**48a**) and bridged indenone (**98**).[23]

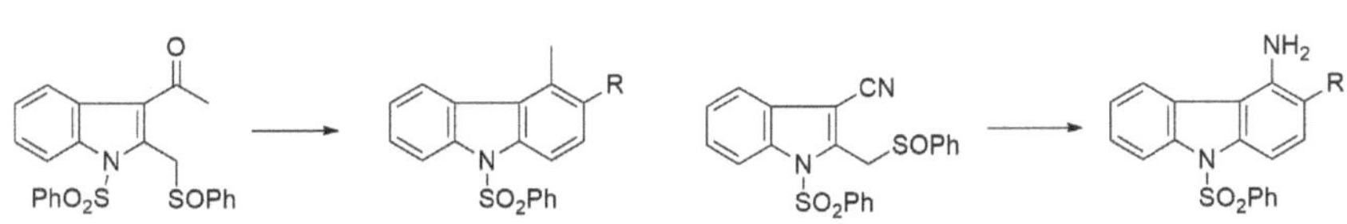

118 120 119 121

R = CO_2Me, COMe, CN

SCHEME 7.23 Mohanakrishnan carbazole synthesis [2009 THN 3592 Mohanakrishnan AK].

Entry	Acceptor	Product (% yield)
1	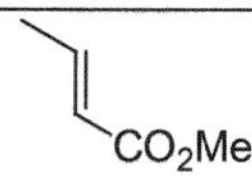 **12b**	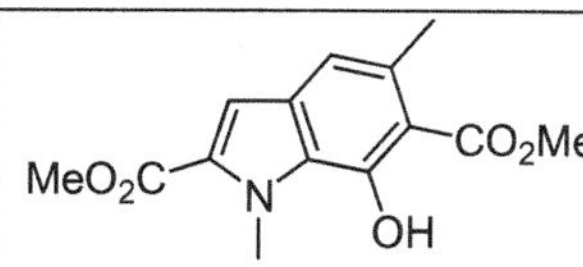 **125** (90)
2	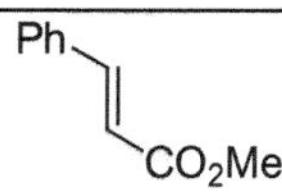 **12d**	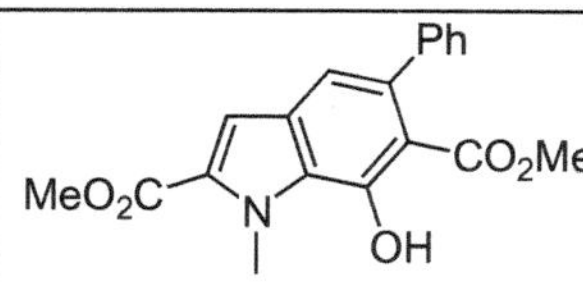 **126** (82)
3	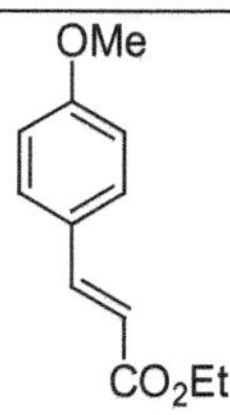 **127**	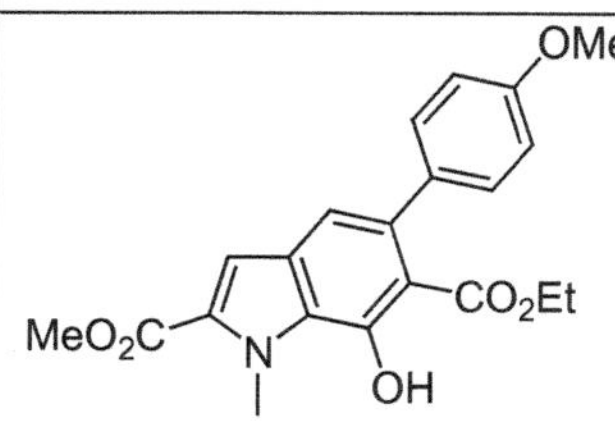 **128** (94)
4	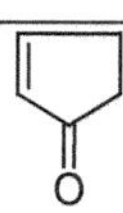 **48a**	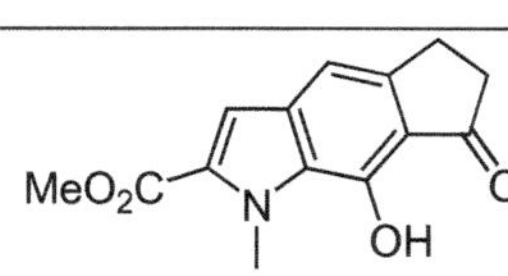 **129** (20)
5	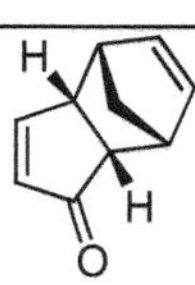 **98**	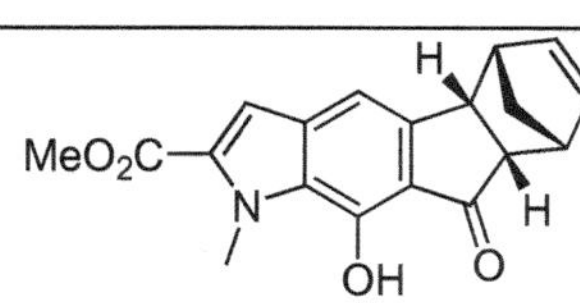 **130** (70)

In striking contrast, the entry to 1-hydroxycarbazoles (e.g., **131**→**132**) by the Hauser sulfoxide annulation was not possible due to formation of polymeric materials. Formation of the highly reactive ketene (**133**) has been suggested for the failure (Fig. 7.3).[13]

The use of the Hauser sulfoxide annulation in the synthesis of natural products, such as, pradimicinone analogs, kidamycinone, coriandrin, psoralen, and

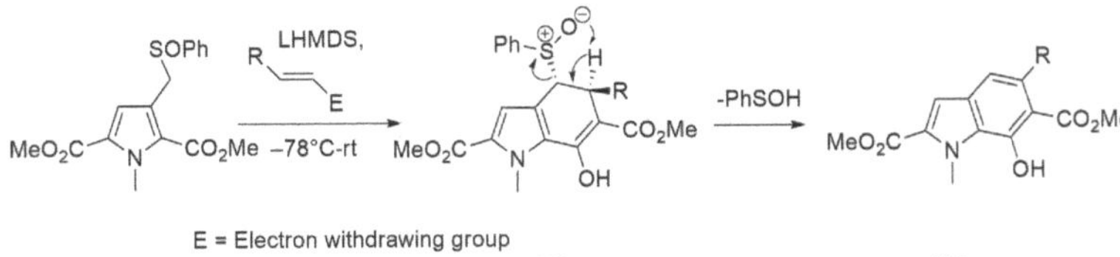

122 **123** **124**

SCHEME 7.24 Synthesis of indoles from pyrrole sulfoxide donors [2014 EJOC 5521 Mal D].

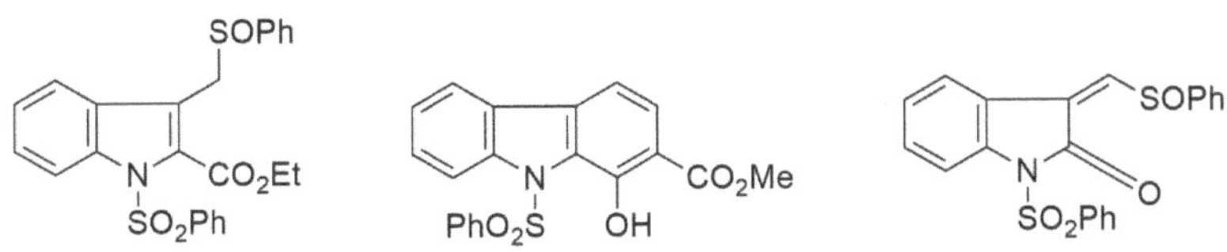

131 **132** **133**

FIG. 7.3 Indole that did not undergo anionic annulations.

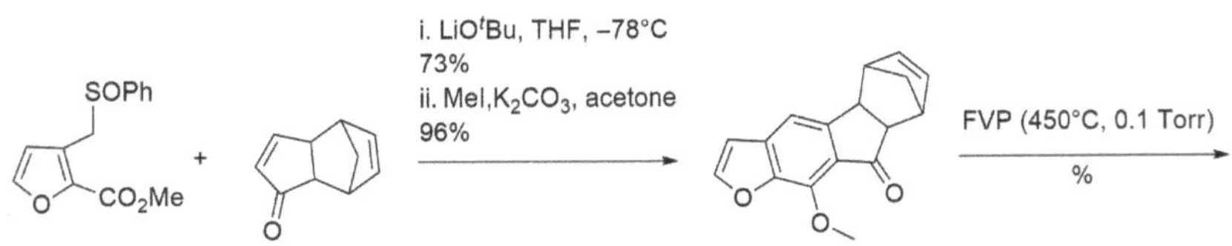

136 **98** **135**

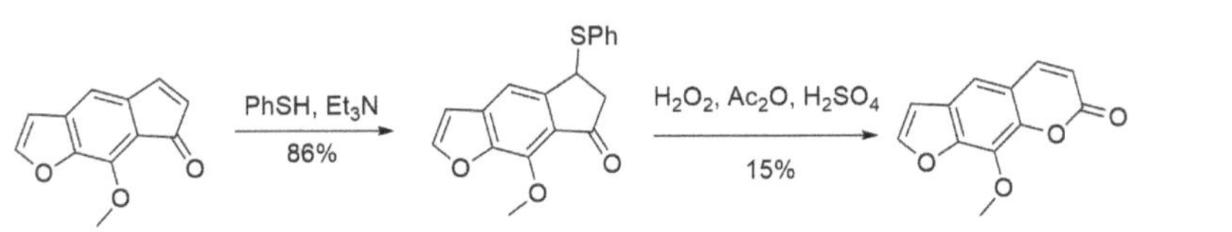

137 **138** **134** (8-methoxypsoralen)

SCHEME 7.25 Synthesis of 8-methoxypsoralen by Hauser sulfoxide annulation [1994 THS 9617 Mal D].

euplectin is well documented in the literature. Presented below are the highlights of the selected total syntheses.

The synthesis of 8-methoxypsoralen (**134**) is depicted in Scheme 7.25.[24] The annulation product (**135**) was obtained in 73% yield from the reaction between furan sulfoxide (**136**) and bicyclic enone (**98**) in the presence of LHMDS, followed by *O*-methylation of the resultant product. Flash vacuum pyrolysis of (**135**) at 450°C furnished furoindenone (**137**) in almost quantitative yield. Et_3N-catalyzed thiophenol addition to (**137**) gave adduct (**138**). Peracetic acid-promoted Baeyer-Villiger oxidation of (**138**) afforded 8-methoxypsoralen (**134**).

The Hauser sulfoxide annulation is also applicable to furan sulfoxide (**139**), an isomer the sulfoxide (**136**). When combined with rearrangement of cyclopentene epoxide rearrangement, it led to a regiospecific synthesis of coriandrin (**140**). Coriandrin is a natural product isolated from coriander leaves, also known as cilantro or Chinese parsley and found to exhibit UVA-dependent antiviral activity against a variety of enveloped viruses. The synthetic strategy involving annulation of furan sulfoxide carboxylate (**139**) provides a new efficient route to furoisocoumarins (Scheme 7.26) and eventually leads to the total synthesis of coriandrin (**140**).[25] The anion prepared from furosulfoxide (**139**) by treatment with LiO^{t}Bu was reacted with cyclopentadienone precursor (**98**). Their reaction resulted in the tandem process consisting of Michael addition, Dieckmann condensation and spontaneous loss of PhSOH from (**141**) by a *syn* elimination to give the benzannulated structure (**142**) in 73% yield. A two-step synthetic sequence consisting of *O*-methylation and *C*-methylation furnished (**143**), *retro*-Diels-Alder reaction of which gave tricyclic intermediate (**144**). Epoxidation of (**144**) with hydrogen peroxide afforded epoxide (**145**) in 40% yield. Epoxycyclopentenone rearrangement converted the compound (**145**) to coriandrin (**140**) in 88% yield.

That the complex structures can be assembled by the Hauser sulfoxide annulation is amply revealed by the Shair synthesis of lomaiviticin scaffold.

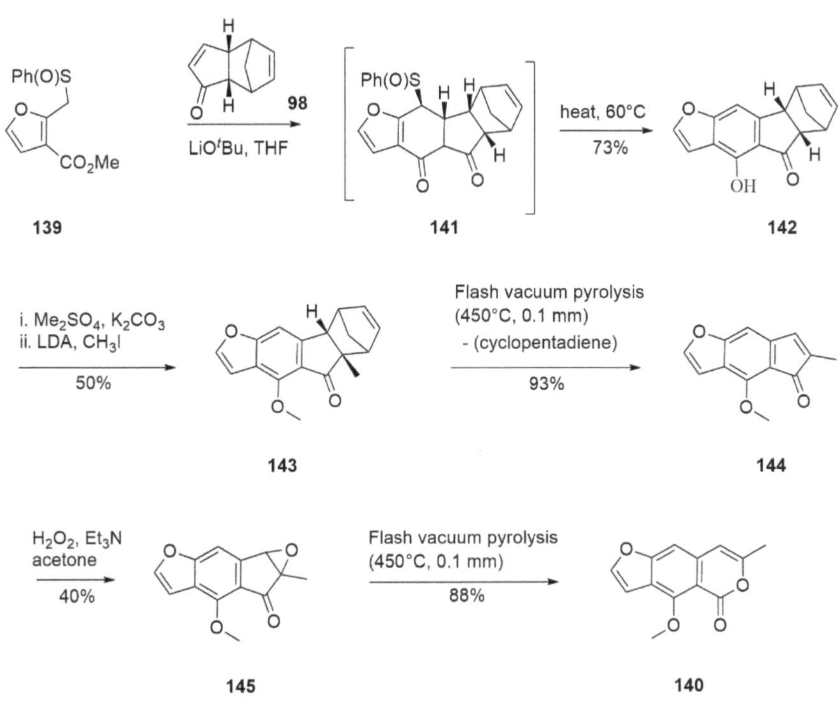

SCHEME 7.26 Mal synthesis of coriandrin, a component of coriander [2000 THS 3677 Mal D].

Lomaiviticin is a marine natural product with C2 symmetric diazobenzofluorene glycoside structure. It strongly inhibits the growth of cultured cancer cell lines at nanomolar scale. It also exhibits impressive activity against Gram-positive bacteria. The Shair synthesis entailed two intermediates (**146**) and (**147**) (Scheme 7.27).[26] Their fusion by Hauser annulation with LHMDS at −60°C delivered phenol (**148**) in 79% yield. The selection of phenyl ester (**146**) over commonly used methyl ester is useful for higher yields of such annulations. Ketalization followed by *O*-allylation of the phenol (**148**) afforded (**149**). Cleavage of the C1-pivaloyl group in (**149**) and subsequent oxidation of the resultant alcohol by TPAP-NMO, furnished ketone (**150**). The crucial oxidative dimerization of the ketone (**150**) was performed with LHMDS, HMPA, [Cp_2FePF_6], −60°C in THF and then the resultant product was deallylated to give (**151**) in 80% yield as a single diastereomer.

Hauser and co-workers extended their strategy to the synthesis of an analogue of the potent antifungal metabolites, pradimicins/benanomicins (Scheme 7.28).[27] The key sulfoxide intermediate (**152**) was obtained in three steps from naphthoate (**153**). Bromination of (**153**) with 2 equiv. of NBS cleanly afforded dibrominated product in 86% yield, the reaction of which with thiophenoxide afforded the thiophenyl methyl compound (**154**) in 93% yield,

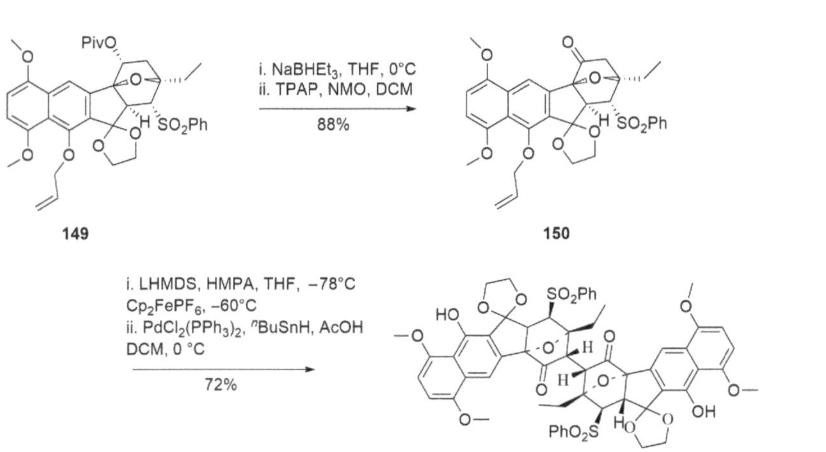

SCHEME 7.27 Hauser sulfoxide annulation route for lomaiviticin scaffold [2010 CEJ 13058 Shair MD].

and this was then oxidized with peroxyboric acid to sulfoxide (**152**) in 86% yield. Condensation of the anion of the sulfoxide (**152**) with *ortho*-quinone monoketal (**155**) was performed in a tetrahydrofuran/dimethylsulfoxide solvent mixture to give pentacycle (**156**), after *O*-methylation, in 31% overall yield (Scheme 7.28). Further 7-step transformations including a hydrogenative debromination led to the synthesis of pentacyclic analogue (**157**) resembling pradimicins (**158**) (Scheme 7.28).

Phenylsulfoxide activated tolualdehydelike (**159**) has scarcely been explored toward an annulation. In connection with the synthesis of parent benz[*f*]indenone, the sulfoxide (**159**), prepared by $NaIO_4$ oxidation of (**160**), was reacted with enone (**98**) to give (**161**) in 65% (Scheme 7.29).[8]

An alicylic version of the above example was studied by Reusch et al. The reactions of (**162**) with acceptors was carried out in two stages (Scheme 7.30).[28] The reaction of (**162**) with methyl acrylate smoothly proceeded through (**163a**) to give tetralin (**164a**). With methyl crotonate, the yield of the reaction product (**164b**) was only 24%. Methyl cinnamate failed to give the corresponding annulation product.

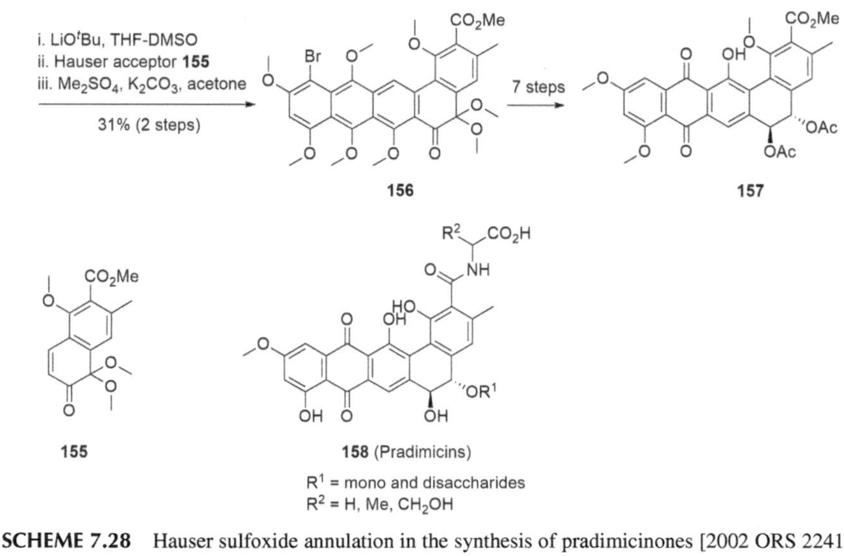

SCHEME 7.28 Hauser sulfoxide annulation in the synthesis of pradimicinones [2002 ORS 2241 Hauser FM].

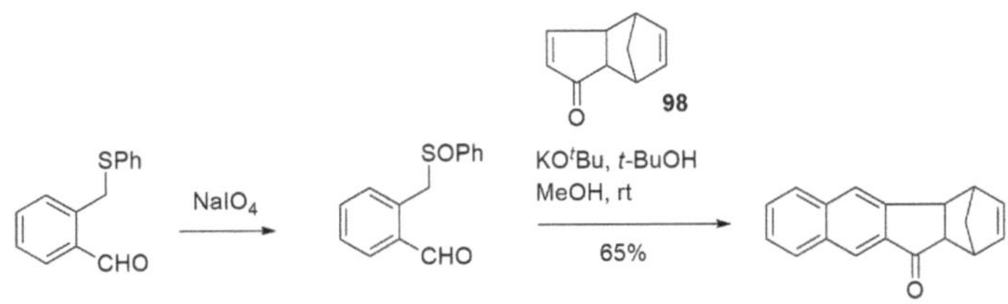

160 159 161

SCHEME 7.29 Annulation 2-phenylsulfinylmethylbenzaldehyde [1998 THS 689 Mal D].

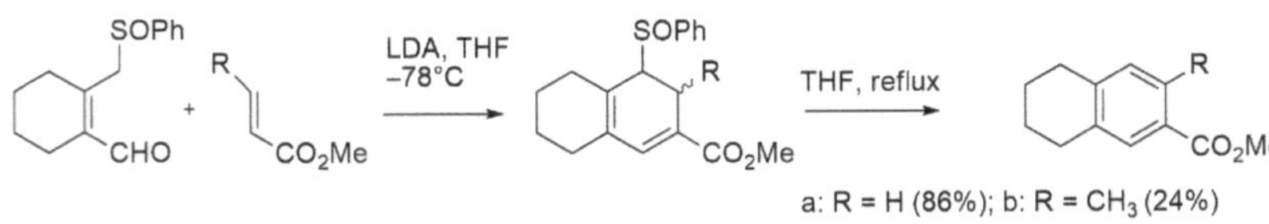

162 12 163 164

SCHEME 7.30 Nonaromatic Hauser sulfoxide annulation: [1983 SCS 303 Reusch W].

7.2.3 Leusen Sulfone Donors

The 1,4-dipolar reagents (**165–167**) for the [4+2] annulation, displayed in Fig. 7.4 were largely developed by Leusen group, beginning in 1978.[5] The chemistry of furan and pyrrole analogs (**168** and **169**) were studied by Mal et al.[24] The product profiles of all the reagents are very similar. However, the bases used and reaction conditions are quite different.

The example cited in Scheme 7.31 represents a one-step synthesis of 2,3-disubtituted naphthalenes. The reaction of (**170**) with chalcone (**171**) in the presence of NaH in DME at rt. to reflux furnished (**172**). The Michael addition of α-sulfonyl carbanion (**173**) to chalcone followed by sequential aldol reaction, elimination of water and tosic acid is the key step for the formation of the product (**172**). This reaction has been successfully applied to cinnamonitriles, cinnamates, crotonates, maleates, and pentenones.

This reaction has been extended to *o*-tosylmethylacetophenone for the synthesis of 1,2,3-trisubstituted naphthalenes. More condensed aromatics can be prepared by the metholodogy. As shown in Scheme 7.32, sulfone (**174**) was cyclocondensed with benz[*f*]indenone (**175**) in the presence of NaO^{t}Bu in THF to give dibenzofluorenone (**176**).[29]

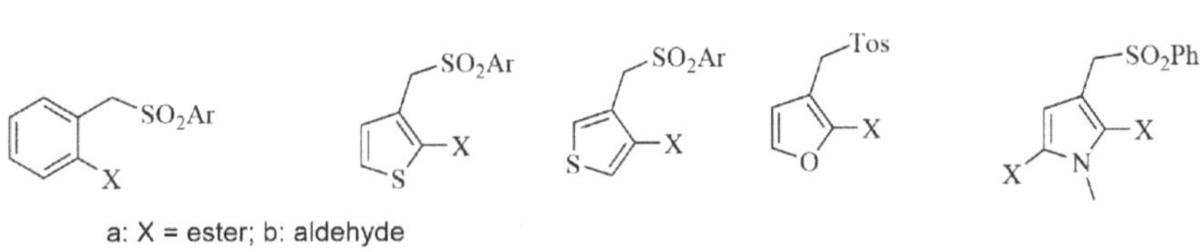

165 166 167 168 169

FIG. 7.4 Structures of common Leusen donors.

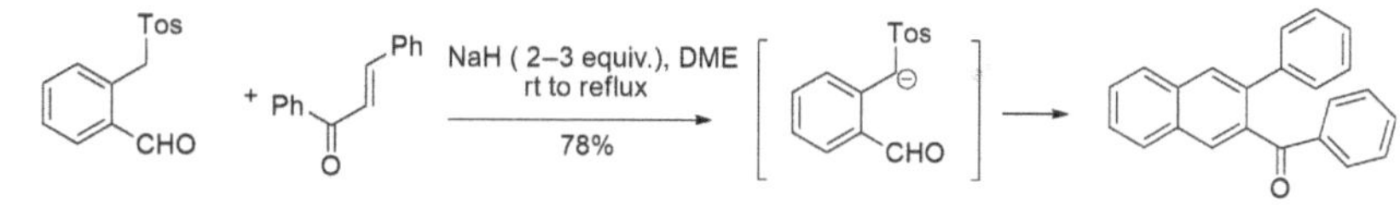

170 171 173 172

SCHEME 7.31 Leusen original annulation with tosyl-activated tolualdehyde [1978 THS 2213 Leusen AMV].

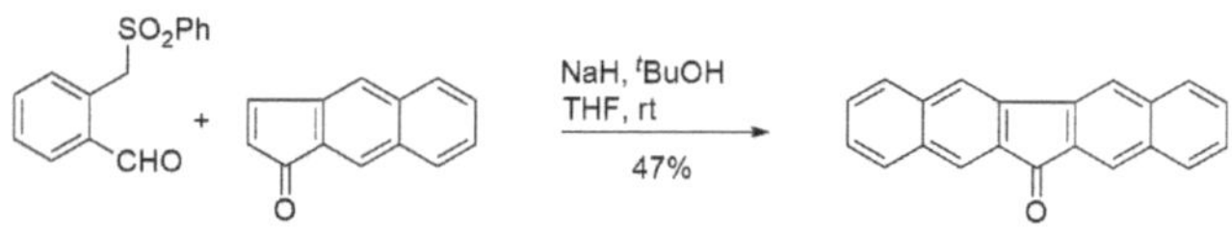

174 175 176

SCHEME 7.32 Synthesis of dibenzofluorenone by Leusen annulation [2006 SYS 2556 Mal D].

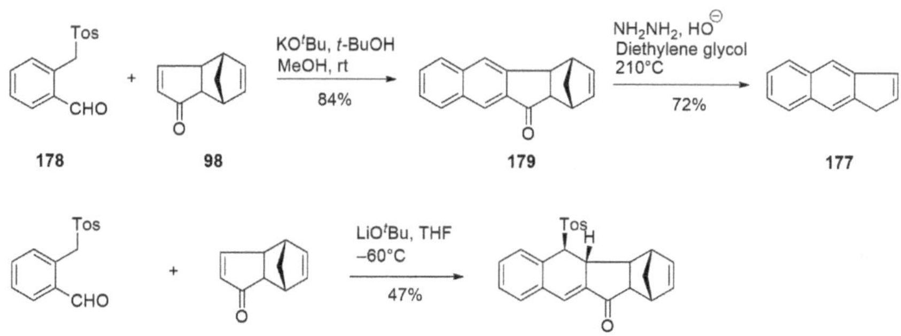

178 98 179 177

178 98 180

SCHEME 7.33 Synthesis of benz[*f*]indene [1998 THS 689 Mal D].

Similar principle was extended to enone (**98**) by Mal et al. to provide a convenient synthesis of benz[*f*]indene (**177**) in two steps.[8] The reaction of Leusen aldehyde (**178**) with tricyclic enone (**98**) in the presence of KO^{t}Bu gave annulation product (**179**) in 84% yield. Treatment of (**179**) with hydrazine in alkaline diethylene glycol afforded benz[*f*]indene (**177**) via Huang-Minlon reduction and retro-Diels Alder reaction. In the presence of LiO^{t}Bu, the reaction between (**178**) and (**98**) stopped to give (**180**), probably due to lesser basicity of the base used and *cis* orientation of the Tos group and the indicated hydrogen (Scheme 7.33).

The reaction of Leusen reagents with enone (**98**) is quite general. The furan analogs (**181** and **182**) similarly gave annulation products (**183** and **184**, respectively) in fair yields (Scheme 7.34).[30]

Evidently as noted in Scheme 7.35, the reactivity of thiophene analog (**185**) parallels the Leusen donor (**178**). Thus, the reaction of (**185**) with acceptors (**186**) in the presence of KO^{t}Bu at rt. gave disubstituted benzothiophenes (**187**).[31]

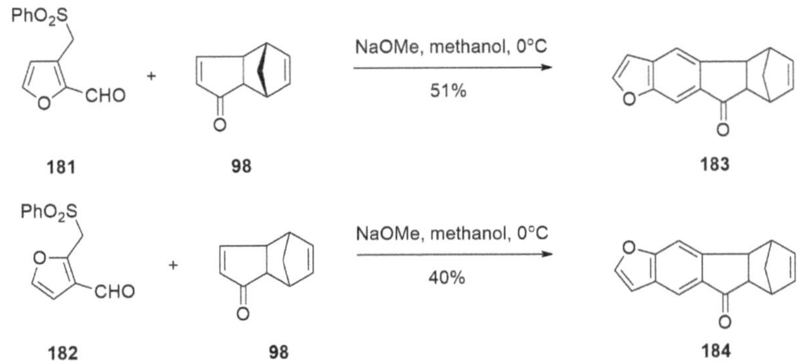

SCHEME 7.34 Annulation of furan analogs of Leusen donors [1999 JICS 551 Mal D].

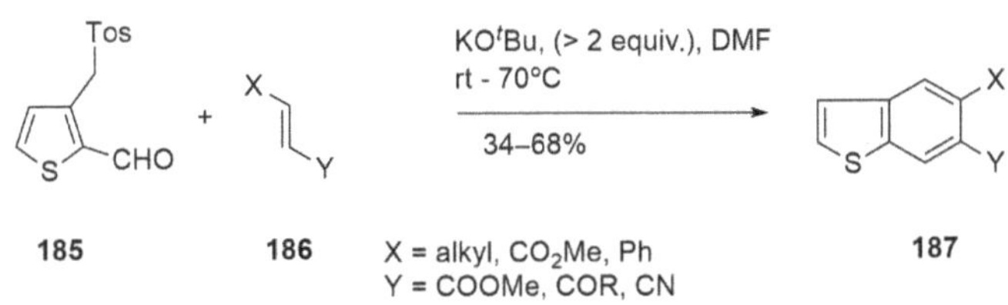

SCHEME 7.35 Annulation of thiophene analog of Leusen sulfone tolualdehydes [1986 JOC 20 Van Leusen AM].

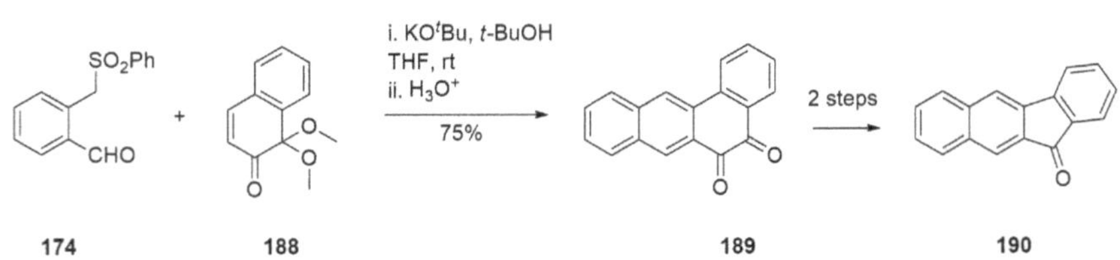

SCHEME 7.36 Annulation with sulfonyl tolualdehyde (**174**) en route to benz[*b*]fluorenones [2006 SYS 2556 Mal D].

The reactivity of (**174**) is well-suited for the quinone monoketal (**188**), which is otherwise not a good Michael acceptor. The annulation of (**174**) with (**188**) in the presence of KO^{t}Bu, followed by acid hydrolysis gave benz[*a*]anthracenedione (**189**) leading to a convenient synthesis of benzo[*b*]fluorenone (**190**) through a combined application of benzil-benzilic acid rearrangement and CrO_3 oxidation (Scheme 7.36).[29]

The chemistry of sulfonylmethyltoluates (**165b**) also allows one-step synthesis of substituted naphthalenes, more specifically, α-naphthol or their derivatives. The annulation of (**191**) with acceptors (**192**) in the presence of NaH followed by *O*-methylation afforded (**193**) (Scheme 7.37).[5]

In the similar fashion as above, annulation of thiophene analog (**194**) gives hydroxylated benzothiophenes (**195**) in moderate yields in reaction with

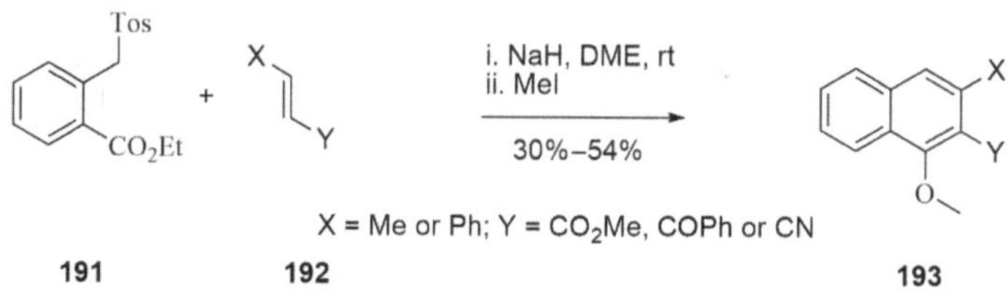

SCHEME 7.37 Annulation of Leusen sulfone toluate (**191**) [1978 THS 2213 Leusen AMV].

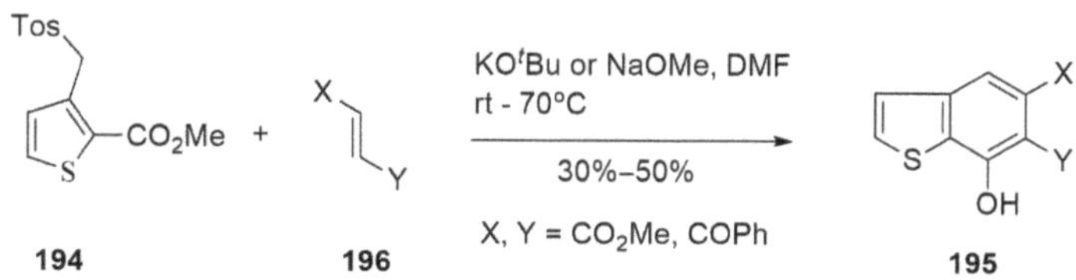

SCHEME 7.38 Annulation of thiophene analog of Leusen sulfone toluates [1986 JOC 230 Leusen AMV].

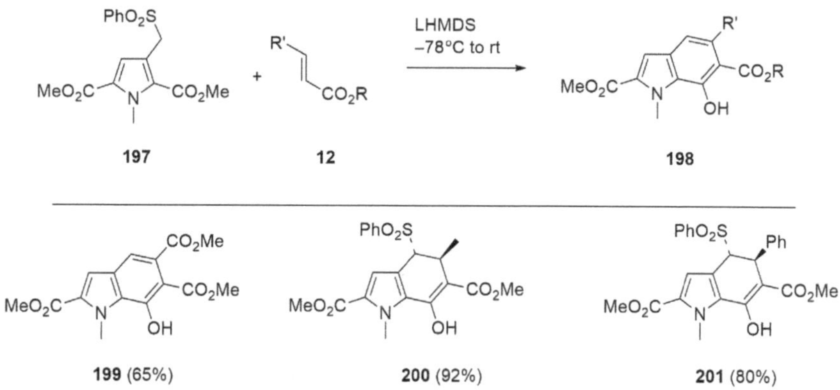

SCHEME 7.39 Annulations with pyrrolyl-3-methyl sulfones [2014 EJOC 5521 Mal D].

acceptors (**196**) with both termini occupied by an electron withdrawing group (Scheme 7.38).[31]

The annulation reactivity of the pyrrole analogs are also quite similar to the parent Leusen donor (**191**). The results obtained with (**197**) are presented in Scheme 7.39. In analogy with the above results with thiophene donor (**194**), it was expected that the pyrrole donor (**197**) would produce indole derivatives (**198**). In reality, the product varied with the nature of the acceptors. With dimethyl maleate, desired product (**199**) was obtained. With methyl crotonate and methyl cinnamate, dihydroindole derivatives (**200** and **201**, respectively) were produced.[23]

There are few instances where Leusen annulation fails to give desired annulation products. For example, sulfonyl toluate (**202**), on reaction with pyranone (**203**) in the presence of LDA did not yield naphthopyranone (**204**), a key intermediate for thysanone (Scheme 7.40).[20]

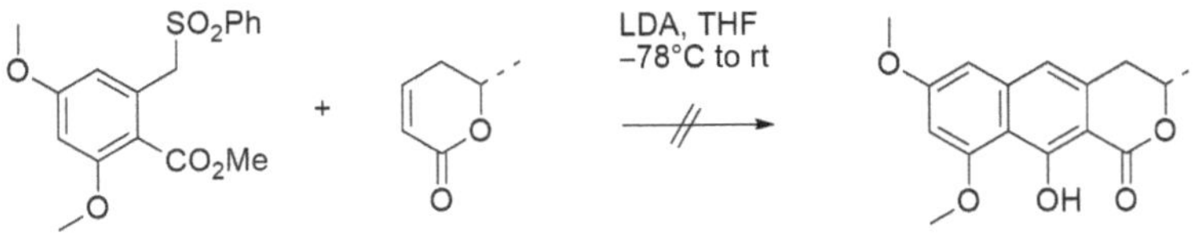

202 203 204

SCHEME 7.40 Leusen annulation with pyranone [2009 SYS 2561 Brimble MA].

Similarly, cyclohexenone (**48b**) was found unsuitable for the base-catalyzed reaction of bistoluate Leusen donor (**205**) for bisanthracenone (**206**), in a study directed to angelimicins (Scheme 7.41).[32]

7.2.4 Total Synthesis of Lactonamycin (207) With Leusen Annulation

The Tatsuta total synthesis of lactonamycin (**207**), a hexacyclic antibiotic is the classic example of an application of the Leusen annulation.[33] It relied upon the use of a modified Leusen donor (**208**), which is a thioester in place of the commonly used carboxylates. The intermediates (**208–210**), outlined in Scheme 7.42 represent a doubly convergent approach to the target. The Leusen annulation was used as a penultimate step.

Thioester (**208**) corresponding to the DEF unit was synthesized in 13 steps from 4-bromo-3,5-dimethylphenol. ABC ring unit (**209**) was synthesized in 11 steps from phenolic ester (**211**). Dearomatization of hydroxy alcohol (**212**)

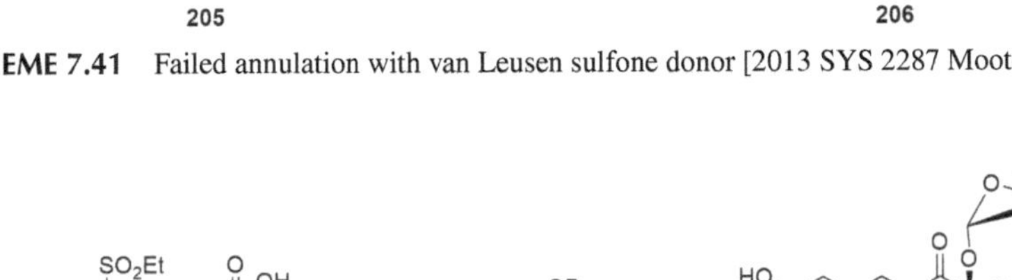

205 206

SCHEME 7.41 Failed annulation with van Leusen sulfone donor [2013 SYS 2287 Mootoo DR].

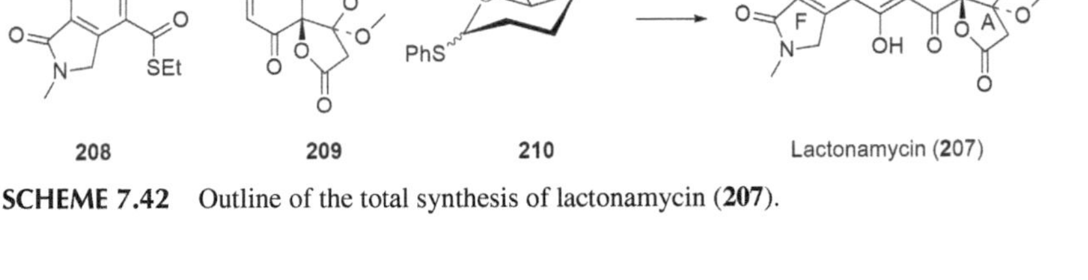

208 209 210 Lactonamycin (207)

SCHEME 7.42 Outline of the total synthesis of lactonamycin (**207**).

by $PhI(OCOCF_3)_2$ was used as one of the key steps to synthesize the acceptor synthon of ABC segment. Selective dihydroxylation of (**213**) provided the triol (**214**), which was elaborated to γ-lactone (**215**), followed by glycosylation to the 1,4-diketone (**216**). The glycosylation step provided a diastereomeric mixture of only α-glycosides from which the required acceptor, i.e., (**216**) was separated by silica gel column chromatography. Leusen annulation of (**208**) with (**216**) gave protected lactonamycin (**217**) in 37% yield. Removal of benzyl groups from (**217**) gave the final molecule. Although the yield of the crucial annulation was not very high, the regioselectivity was remarkable. The authors have not discussed the origin of the directing effect of the carbonyl groups of (**216**). Neither have they noted formation of the other possible regioisomer (Scheme 7.43). The stereoselective glycosylation of the tertiary alcohol in (**215**) with the 2,3-deoxysugar (**210**) served as a resolving agent for racemic enone (**216**).

In closing this chapter, presented below are two benzannulations developed by Shair et al. in connection with their studies on hibarimicinone (**218**), HMP-Y1 and HMP-P1. They utilized a phenylthio group and a fluorine for activating the lateral positions of the 1,4-dipolar toluates (Fig. 7.5).[34]

The resulting toluates serve as the substitutes for the Hauser sulfoxide and the Leusen sulfone donors. Although lees studied, sulfide-activated donors are found to useful in the generation of complex structures from large fragments

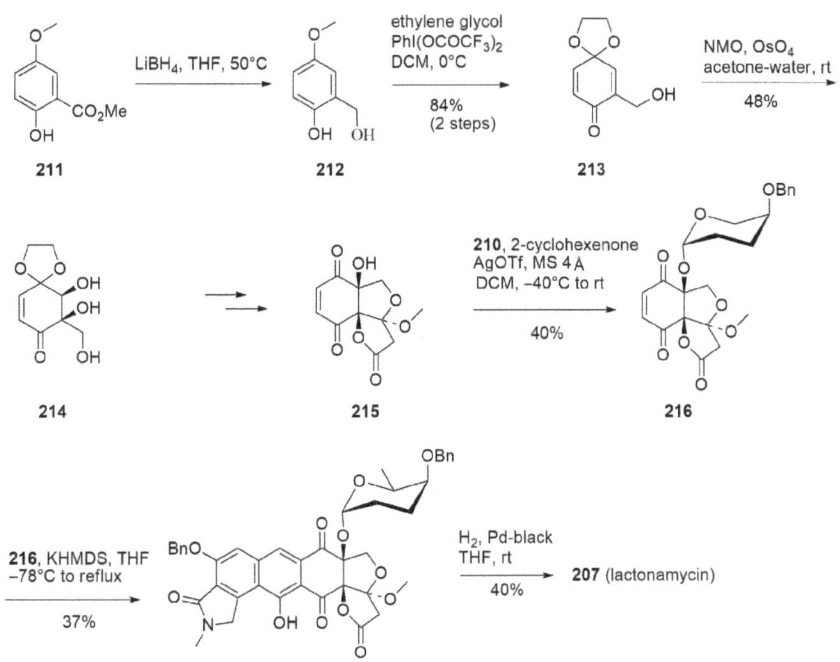

SCHEME 7.43 Tatsuta total synthesis of lactonamycin (**207**) [2010 THS 5546 Tatsuta K lactonamycin].

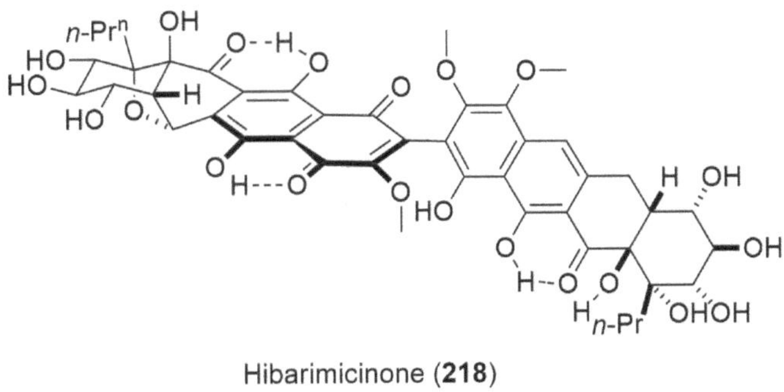

FIG. 7.5 Structure of hibarimicinone (**218**).

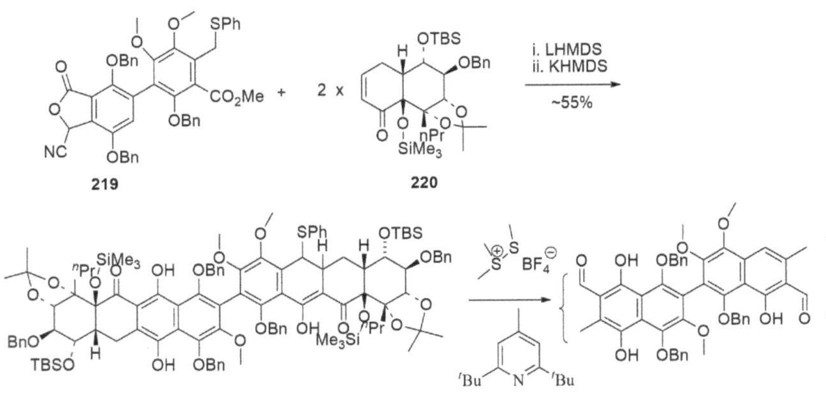

SCHEME 7.44 Annulation with a phenylthio-substituted toluate [2012 JACS 16765 Shair MD].

(**219** and **220**). Treatment of the mixture of (**219**) and (**220**) with LHMDS followed by KHMDS yielded double annulation products as a 1.3:1 mixture of atropisomers (**221**) (Scheme 7.44). Elimination of SPh group to form naphthol moiety in (**222**) was achieved by treatment with dimethyl(methylthio)sulfonium tetrafluoroborate and 2,6-di-*tert*-butyl-4-methylpyridine. It is worth to note that both Hauser sulfoxide and Leusen sulfone analogs of (**219**) failed to deliver the product (**221**) under similar conditions.

Shair annulation with a benzyl fluoride has proven to be yet another superior method for the double naphthol annulation. As shown in Scheme 7.45, Shair et al. eventually succeeded to assemble the challenging core structure of hibarimicinone (**218**) by fluorotoluate (**223**). Its anionic [4+2] annulation with (**224**) was developed to generate complex 1-hydroxy naphthalene moieties present in (**225**). The success of the annulation was based on the considerations of (i) the stability and the nucleophilicity of the carbanion, (ii) the stability of the acceptor to the base used, and (iii) the steric bulk of the both donors and

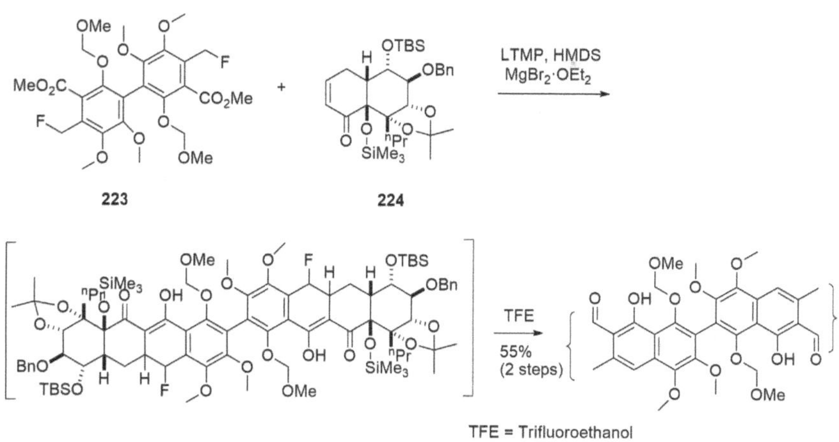

223 224 226 225

SCHEME 7.45 Double naphthol annulation with a benzyl fluoride [2012 JACS 16765 Shair MD].

acceptors. The annulation between (**223**) and (**224**) was carried out in the presence of LTMP and magnesium bromide etherate, followed by heating of the unpurified product (**226**) in 2,2,2-trifluoroethanol.

The majority of the [4+2] anionic annulations discussed above produce aromatic compounds, i.e., a naphthalene moieties resulting from the formation of a new benzene ring. The formation of the new ring takes place through a 2,3-dihydro derivatives of the naphthalene. It is often difficult to isolate the dihydro intermediate, because the activating group required at the benzylic for activation of the 1,4-dipolar synthon is lost during the annulation under the basic conditions. There are scant examples where an oxygen-containing activating group at the lateral benzylic position of the donors promotes the annulation. One such example is that of Kobayashi et al. displayed in Scheme 7.46.[35] The reaction between (**227**) and butenolide (**71**) in the presence of LDA proceeded through the anionic intermediates (**228** and **229**) to furnish (**230**) in 56% yield. Dehydration of (**230**) by the treatment with thionyl chloride in pyridine at room temperature afforded (**231**) dimethyl(methylthio)sulfonium tetrafluoroborate in 78% yield. Palladium-catalyzed dehydrogenation of (**231**) provided arylnaphthalide (**232**) in 75% yield.

In conclusion, this chapter has dealt with the anionic chemistry of toluates and tolualdehydes that bear an activated group (e.g., CO_2R, CN, NO_2, S(O)Ar, SO_2Ar, $(SR)_2$, F, etc.) at the benzylic positions. Their reactions with acceptors like acrylates, allenecarboxylates, vinyl ketones, nitroalkenes, etc. provide an access to substituted naphthalenes with defined patterns of the substituents. Among them, Hauser-Leusen annulations proved to be useful in the total synthesis of natural products. Hence, the chapter has been titled "Hauser-Leusen annulation."

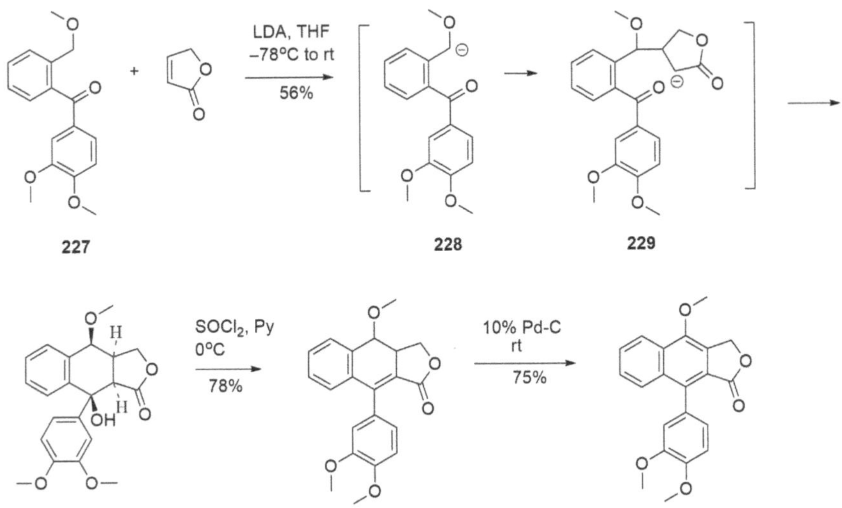

227 228 229

230 231 232

SCHEME 7.46 Annulation of a methoxy-activated methyl acetophenone [1997 HCS 1593 Kobayashi K].

REFERENCES

1. Eisenmuth, W.; Renfroe, H. B.; Schmid, H. *Helv. Chim. Acta* **1965**, *48*, 375.
2. Koppanathi, N.; KumaraSwamy, K. C. *Org. Biomol. Chem.* **2016**, *14*, 5079.
3. Kraus, G. A.; Chaudhary, D.; Yuan, Y.; Schuster, A. *Tetrahedron Lett.* **2012**, *53*, 4444.
4. Wadi, A.; Lopez-Calahora, F.; Alvarez, M.; Joule, J. A. *Tetrahedron Lett.* **1992**, *33*, 3679.
5. Wildeman, J.; Borgen, P. C.; Pluim, H.; Rouwette, P.H.F.M.; Leusen, A. M. V. *Tetrahedron Lett.* **1978**, 2213.
6. Hauser, F. M.; Rhee, R. P. *J. Org. Chem.* **1978**, *43*, 178.
7. Liau, B. B.; Milgram, B.; Shair, M. D. *J. Am. Chem. Soc.* **2012**, *134*, 16765.
8. Mal, D.; Ghorai, S. K.; Hazra, N. K. *Tetrahedron Lett.* **1998**, *39*, 689.
9. Kozikowski, A. P.; Schmiesing, R. *Synth. Commun.* **1978**, *8*, 363.
10. Chenard, B. L.; Anderson, D. K.; Swenton, J. S. *J. Chem. Soc. Chem. Commun.* **1980**, 932.
11. Roy, H. N.; Rahman, A.F.M.M.; Islam, M. A. *J. Chem. Res. (S)* **2003**, 594.
12. Chenard, B. L.; Dolson, M. G.; Sercel, A. D.; Swenton, J. S. *J. Org. Chem.* **1984**, *49*, 318.
13. Ramesh, N.; Rajeshwaran, G. G.; Mohanakrishnan, A. K. *Tetrahedron* **2009**, *65*, 3592.
14. Dagar, A.; Biswas, S.; Mobin, S. M.; Samanta, S. *ChemistrySelect* **2016**, *1*, 6362.
15. Panasiewicz, M.; Zdrojewski, T.; Chrulski, K.; Wojtasiewicz, A.; Jonczyk, A. *Arkivoc* **2009**, *part vii*, 98.
16. Morrison, C. F.; Stamp, C. T. M.; Burnell, D. J. *Tetrahedron Lett.* **2009**, *50*, 7021.
17. Harrowven, D. C. *Tetrahedron* **1993**, *49*, 9039.
18. Kamal, A.; Daneshtalab, M. *Tetrahedron Lett.* **1994**, *35*, 3879.
19. Ozaki, Y.; Imaizumi, K.; Okamura, K.; Morozumi, M.; Hosoya, A.; Kim, S.-W. *Chem. Pharm. Bull.* **1996**, *44*, 1785.
20. Sperry, J.; Yuen, T. Y.; Brimble, M. A. *Synthesis* **2009**, 2561.
21. Hauser, F. M.; Rhee, R. P. *J. Org. Chem.* **1980**, *45*, 3061.

22. Mohonakrishnan, A.; Srinivasan, P. C. *Tetrahedron Lett.* **1993**, *34*, 1343.
23. Dinda, B.; Basak, S.; Mal, D. *Eur. J. Org. Chem.* **2014**, 5521.
24. Mal, D.; Murty, K.V.S.N.; Datta, K. *Tetrahedron Lett.* **1994**, *35*, 9617.
25. Mal, D.; Bandyopadhay, M.; Ghorai, S. K.; Datta, K. *Tetrahedron Lett.* **2000**, *41*, 3677.
26. Lee, H. G.; Ahn, J. Y.; Lee, A. S.; Shair, M. D. *Chem. Eur. J.* **2010**, *16*, 13058.
27. Hauser, F. M.; Liao, H.; Sun, Y. *Org. Lett.* **2002**, *4*, 2241.
28. Dickenson, J. B.; Reusch, W. *Synth. Commun.* **1983**, *13*, 303.
29. Patra, A.; Ghorai, S. K.; De, S. R.; Mal, D. *Synthesis* **2006**, 2556.
30. Bandyopadhyay, M.; Datta, K.; Mal, D. *J. Ind. Chem. Soc.* **1999**, *76*, 551.
31. Terpstra, J. W.; Leusen, A. M. V. *J. Org. Chem.* **1986**, *51*, 230.
32. Li, J.; Mootoo, D. R. *Synthesis* **2013**, *45*, 2287.
33. Tatsuta, K.; Tanaka, H.; Tsukagoshi, H.; Kashima, T.; Hosokawa, S. *Tetrahedron Lett.* **2010**, *51*, 5546.
34. Liau, B. B.; Milgram, B.; Shair, M. D. *J. Am. Chem. Soc.* **2012**, *134*, 16765.
35. Kobayashi, K.; Kajimura, Y.; Maeda, K.; Uneda, T.; Morikawa, O.; Konishi, H. *Heterocycles* **1997**, *45*, 1593.

Chapter 8

[4+2] Annulation With Nonactivated *o*-Toluate Donors (Staunton-Weinreb-Myers Annulation)

This chapter covers annulation chemistry of simple *o*-toluates and their derivatives. It differs from the earlier chapter in that it describes chemistry of *o*-toluates without an electron withdrawing group at the benzylic position. The annulation of the simple *o*-toluates can be divided into two subclasses: Staunton-Weinreb annulation and Myers annulation. While the Staunton-Weinreb annulation involves annulation of an *ortho*-toluate with α,β-methoxyacrylate or β-methoxyenone, Myers annulation entails cycloalkenones as the acceptors.[1] The Staunton-Weinreb annulation can be viewed as a benzannulation. In a single operation, the reaction of toluates (**1**) with pyrones (**2**) in the presence of LDA affords tricycles (**3**) (Scheme 8.1). In the Myers annulation, phenyl toluates (**4**) are reacted with cycloalkenones (**5**) to give linearly fused systems (**6**), where the newly formed central ring is not an aromatic one (Scheme 8.2).

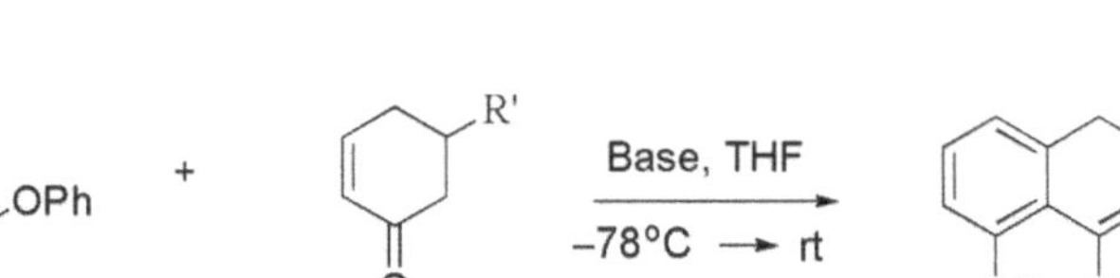

SCHEME 8.1 General scheme for Staunton-Weinreb annulation.

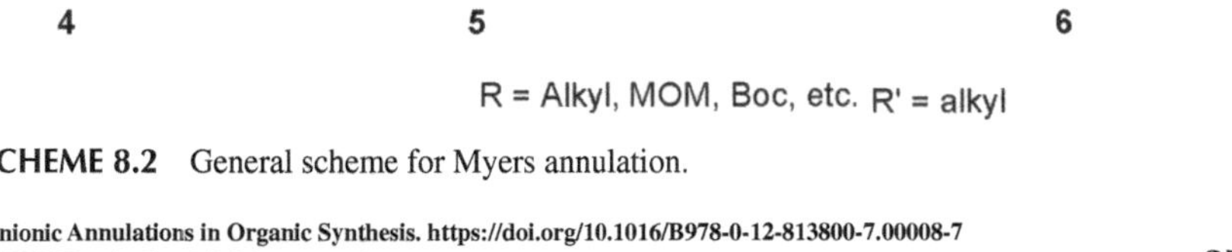

SCHEME 8.2 General scheme for Myers annulation.

Anionic Annulations in Organic Synthesis. https://doi.org/10.1016/B978-0-12-813800-7.00008-7

8.1 ANNULATION WITH β-METHOXY UNSATURATED ESTERS AND KETONES (STAUNTON-WEINREB ANNULATION)

This section provides mechanism, scope, limitations, and applications of the Staunton-Weinreb annulation. One of the original Staunton reactions is presented in Scheme 8.3. In one single operation, toluate (**7**) reacted with pyrone (**8**) in the presence of LDA to give naphthopyrone (**9**), which was derivatized with diazomethane to furnish naphthopyrone (**10**).[2]

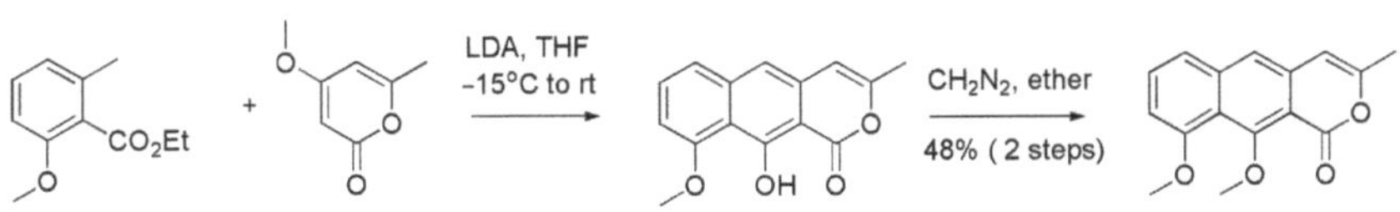

SCHEME 8.3 Staunton original annulation work [1979 CCS 205 Staunton J].

Weinreb's original work, displayed in Scheme 8.4, serves as the foundation of the annulation and its scope. All of the annulations involve orsellinate (**11**) as the donor.[3] The acceptors are β-methoxyenones (**12–14**). The reaction of the enones (**12–14**) afforded annulation products (**15–17**). Strikingly, the annulation of β-methoxy acrylate (**18**) was not successful to give (**19**). The failure may be attributed to electron donating resonance effect of the ester methoxy group.

The probable mechanism is given in Scheme 8.5 for the formation of (**9**) from (**7**) and (**8**). The very first step of this annulation is lateral deprotonation of orsellinate (**11**) to form anion (**20**). This anion then adds to (**8**) in Michael fashion to give enolate (**21**). Expulsion of the methoxy group from (**21**) forms the intermediate (**22**). Excess LDA further promotes lateral deprotonation to give (**23**), which subsequently ring-closes to afford tricycle (**24**). Tautomerization of (**24**) gives the final product (**25**). The only support for the mechanism is the isolation of the intermediate (**22**) under controlled conditions and it conversion into the product (**25**), upon further treatment with LDA.

The substrate scope for the Staunton-Weinreb annulation is fairly significant with regard to the type of the acceptors. The following three examples in Scheme 8.6 illustrate that pyrones are good acceptors for the orsellinate donor (**11**). Unsymmetrical 4-pyrone (**26**) gave product (**27**) in 24% yield. Mechanistically, orsellinate anion (**20**) undergoes 1,4-conjugate addition to the more electron deficient CC double bond of (**26**) resulting in the formation of the intermediate (**28**), which subsequently cyclizes to the product. For symmetrically substituted pyrone (**29**), the reaction proceeded through the intermediate (**30**) to give (**31**) in 79% yield. The product (**32**), corresponding to the acceptor (**33**), was formed through intermediate (**34**). Dihydropyrones are also widely used in the Staunton-Weinreb annulation, and such annulations have been useful in the total synthesis of many naphthopyranone natural products.

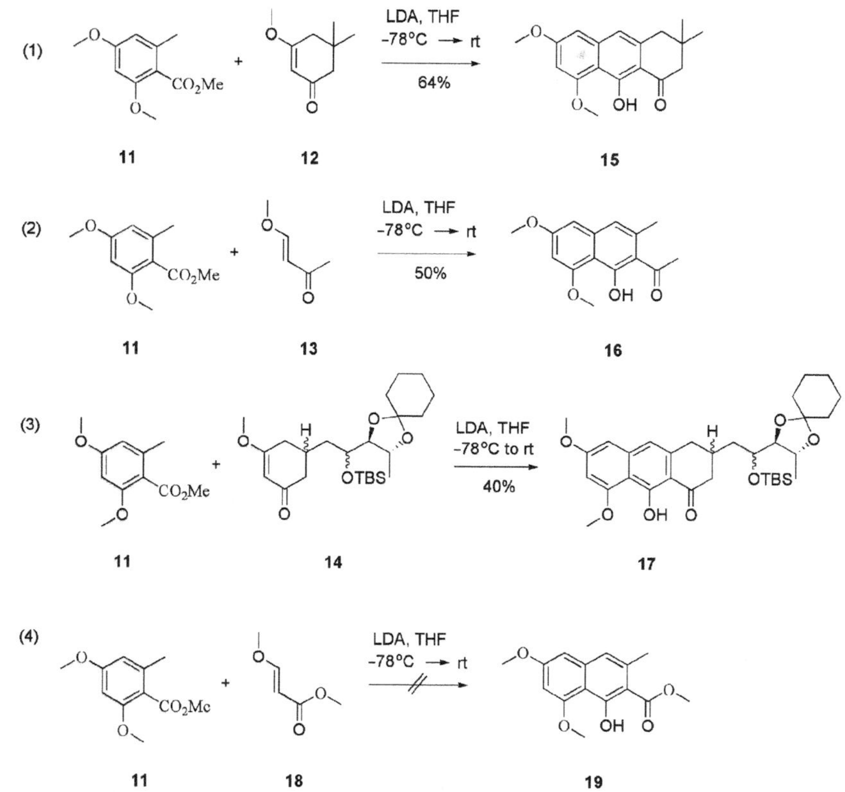

SCHEME 8.4 Weinreb original annulation work [1982 JOC 4045 Weinreb SM].

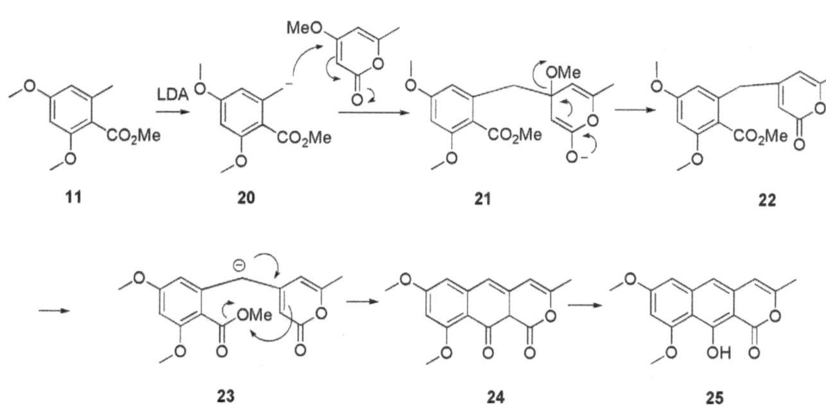

SCHEME 8.5 Mechanism of the Staunton-Weinreb annulation.

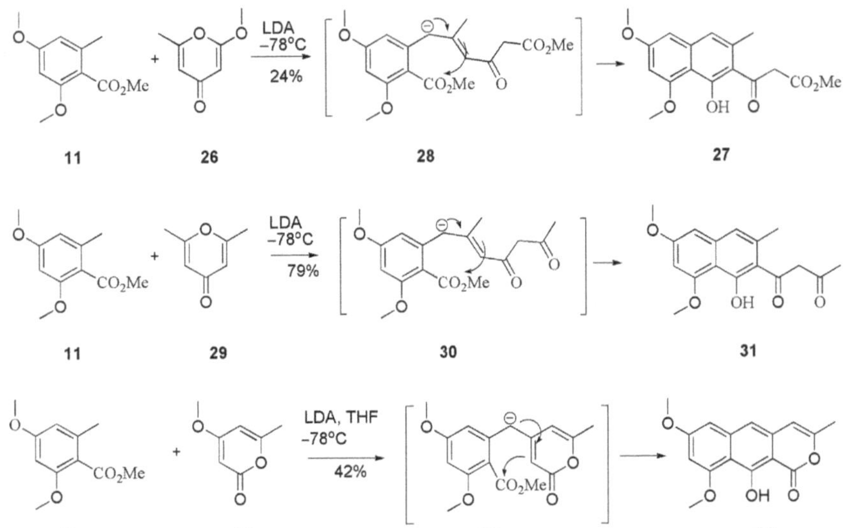

SCHEME 8.6 Examples of the Staunton-Weinreb annulation with different pyrone acceptors.

The effect of nuclear substituents of the toluate donors (**35**) on the annulation efficiencies has been systematically been examined by their reaction with pyrone (**36**) in producing annulation products (**37**).[4] These annulations were carried out under the same conditions (LDA, −65°C), keeping the Michael acceptor same for all but varying the oxygen substituents of toluate donors (**35**). The yield varied between 0% and 49%. The results in terms of the yields of the products with respect to the toluate donors (**11**), (**38–47**) are displayed in Scheme 8.7.

What may be concluded is that orsellinate/salicylate substitution is well suited for the Staunton-Weinreb annulation. The annulations with the substrates (**11**, **41**, **43**, **44**, and **47**) are moderately successful. TBS protected orsellinate (**40**), however, did not afford any annulation product. This failure may be due to the steric effect of the TBS in second step, i.e., Dieckmann condensation. It is well known that the anion (**48**) of the parent toluate (**38**) has a strong propensity for dimerization under the reaction conditions through 1,2-addition with the parent toluate to give vinylogous β-keto ester anion (**49**), which cyclizes to isocoumarin (**50**) (Scheme 8.8).[5] Thus, the required annulation of (**38**) was a failure. An oxy-functionality *para* to the lateral methyl group seems to affect the annulation due to destabilizing effect on the carbanion intermediates.

Like monomeric orsellinate (**11**), dimeric orsellinate (**51**) can be engaged in the Staunton-Weinreb annulation with (**52**), yielding a flavomannin analog (**53**) in 27% yield (Scheme 8.9).[6]

Applications in the total synthesis of natural products: Although the Staunton-Weinreb annulation is limited to resorcylate and salicyclate donors,

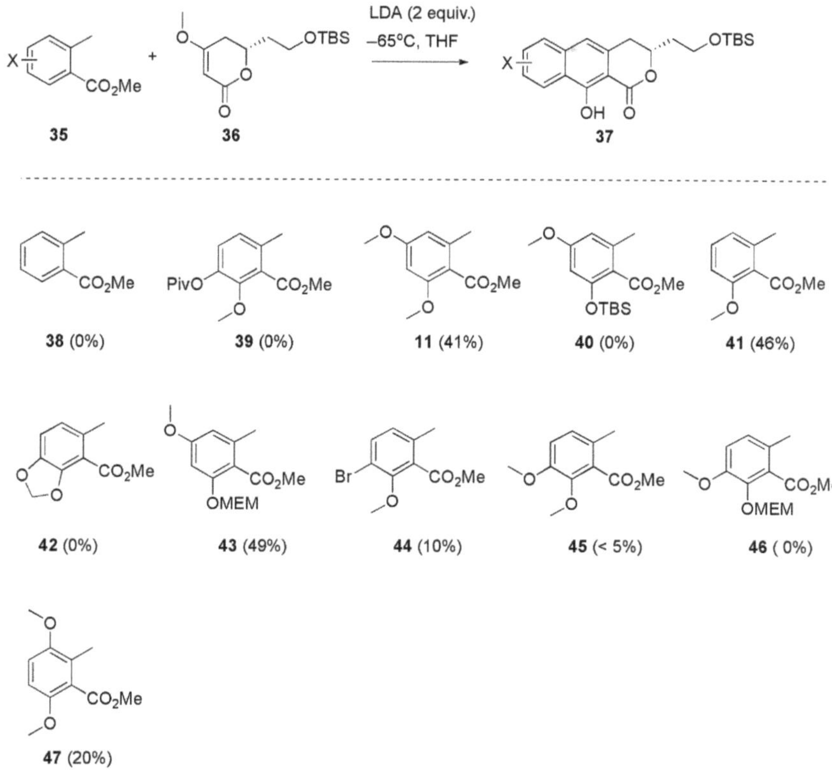

SCHEME 8.7 Effect of substituents of the toluate donors on their annulations [2008 THS 4160 Donner CD].

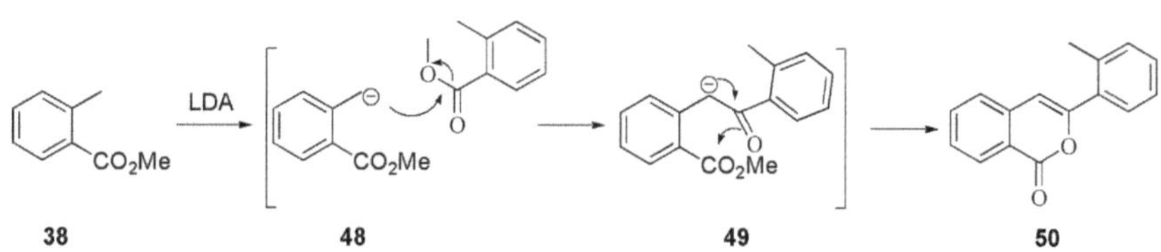

SCHEME 8.8 Anionic dimerization of methyl *o*-toluate [1980 SYS 72 Hauser FM].

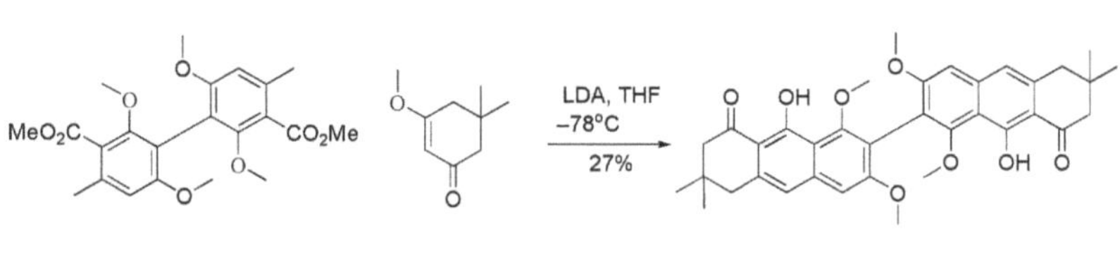

SCHEME 8.9 Staunton-Weinreb annulation with dimeric orsellinate (**51**) [2007 EJOC 1749 Steglich W].

its application to the total synthesis of natural products, more specifically naphthopyranones, is significant. For example, (−)-thysanone (**54**), a 3C-protease inhibitor, has been synthesized by using a Staunton-Weinreb annulation as the key step staring from orsellenic acid. As shown in Scheme 8.10, isopropyl protected orsellinate (**55**) reacted with β-methoxy lactone (**56**) in the presence of LDA at −78°C to give regiospecifically naphthopyranone (**57**) in 47% yield.[7] The lactone carbonyl in (**57**) was reduced with borane-dimethyl sulfide complex to produce (**58**). Salcomine-catalyzed oxygen-mediated oxidation of (**58**) gave pyranonaphthoquinone (**59**). Deisopropylation of (**59**) with aluminum chloride afforded (**60**). Finally, stereoselective introduction of the benzylic hydroxy group in (**60**) was achieved through radical bromination with bromine followed by hydrolysis of the resultant bromo derivative to furnish (**54**). It may be noted the similar sequence of reactions applied to the dimethyl ether derivative precluded the total synthesis of (**54**) due to the problems in demethylation of the analog of (**59**). The Hauser sulfoxide annulation for the preparation of the dimethyl analog of (**57**) was less efficient. The lactone (**56**) was accessed in three steps from commercially available (*S*)-ethyl 3-hydroxybutyrate. Thus, the total synthesis confirms the absolute stereochemistry of the natural product (**54**).

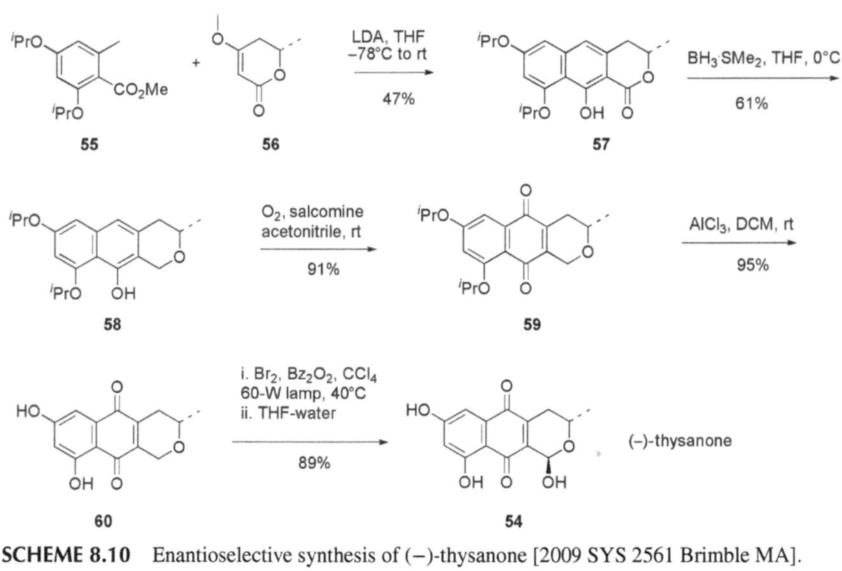

SCHEME 8.10 Enantioselective synthesis of (−)-thysanone [2009 SYS 2561 Brimble MA].

Donner total synthesis of semiviriditoxin (**61**) represents a straightforward approach to a naphthopyranone. The synthesis started from (*S*)-aspartic acid (**62**). It was converted into the epoxide (**63**) in four steps as shown in Scheme 8.11.[8] The epoxide (**63**) was then treated with the anion of methyl propiolate to give acetylene (**64**). Exposure of the acetylene ester (**64**) to sodium methoxide in methanol led to conjugate addition of methanol and then lactonization to give pyranone (**65**). Annulation of (**11**) with (**65**) in the presence of LDA at −60°C afforded naphthopyranone (**66**) in 36% yield. Removal of TBS ether in (**66**) followed oxidative transformations of the resultant alcohol (**67**) yielded corresponding pyranone ester, BCl_3-mediated selective demethylation of which furnished semiviriditoxin (**61**) (Scheme 8.11).

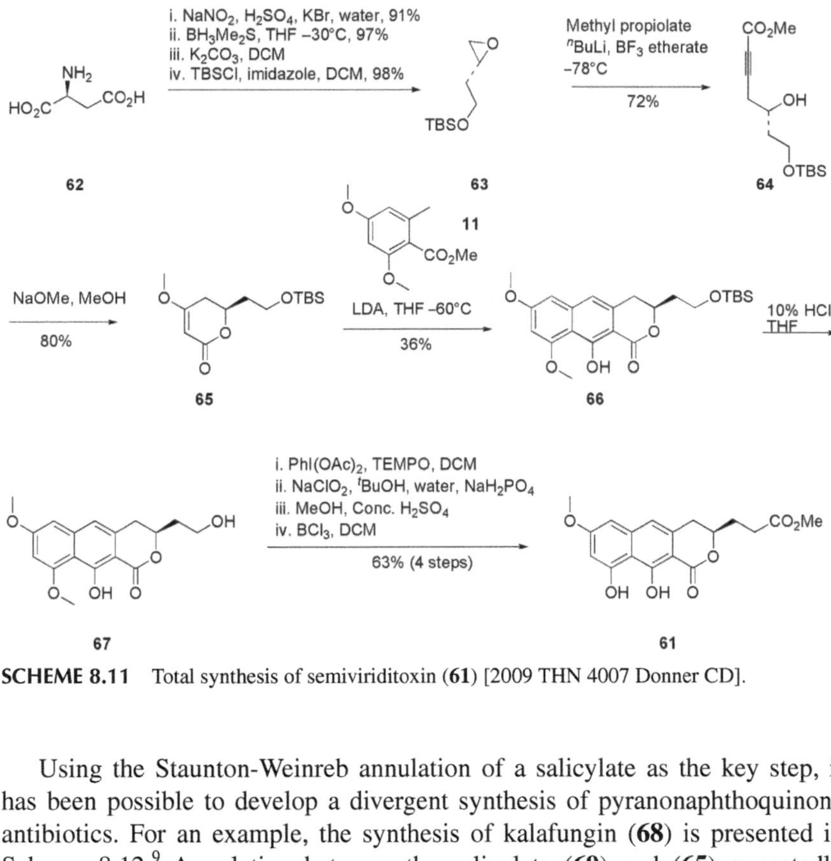

SCHEME 8.11 Total synthesis of semiviriditoxin (**61**) [2009 THN 4007 Donner CD].

Using the Staunton-Weinreb annulation of a salicylate as the key step, it has been possible to develop a divergent synthesis of pyranonaphthoquinone antibiotics. For an example, the synthesis of kalafungin (**68**) is presented in Scheme 8.12.[9] Annulation between the salicylate (**69**) and (**65**) expectedly furnished naphthopyranone (**70**) in 52% yield. Treatment of (**70**) with methylmagnesium bromide yielded the corresponding lactol, which on reduction with triethylsilane-TFA followed by deprotection of the TBS group, afforded *cis*-pyran (**71**). NBS promoted bromination of (**71**), followed CAN oxidation gave naphthoquinone (**72**). Treatment of (**72**) with $AlCl_3$ gave the corresponding phenol, which was submitted to two oxidations to produce phenolic acid (**73**). Quinone methide-mediated lactonization of (**73**) by oxygen followed by reductive work-up gave (**74**). Finally, acid-catalyzed isomerization of (**74**) provided kalafungin (**68**).

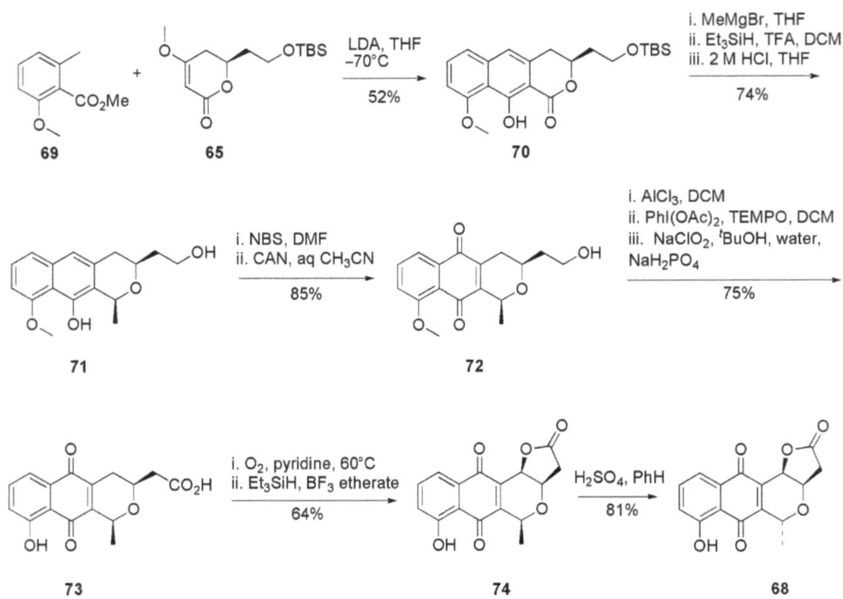

SCHEME 8.12 Synthesis of kalafungin (**68**) [2013 THN 377 Donner CD].

Pillaromycinone (**75**), the aglycon of pillaromycin A, is a challenging synthetic target. Till date, there is no synthesis reported for this target. It was with Staunton-Weinreb annulation, White et al. could assemble completely functionalized tetracyclic core (**76**) of pillaromycinone (**75**). The glycal (**77**), available in three steps from L-rhamnose, was elaborated to tetrahydrophthalide derivative (**78**), which was further developed to decalenone derivative (**79**), suitable for the Staunton-Weinreb annulation. Its condensation with salicylate (**80**) in the presence of LDA and $CeCl_3$ afforded tetracenone derivative (**81**) (Scheme 8.13).[10] Three steps—*O*-methylation, desilylation, and PCC oxidation—furnished (**76**), a core strongly resembling pillaromycinone (**75**).

The Staunton-Weinreb annulation is also suited for the β-alkylthio-substituted enones. The reaction between 2-(cyanomethyl)benzonitrile (**82**) and 3,3-bis(methylthio)-prop-2-enones (**83**) in the presence of KOH in DMSO gave 4-amino-3-aroyl/acetyl-2-(methylsulfanyl)naphthalene-1-carbonitriles (**84**) in very good yields (Scheme 8.14).[11] These reactions are believed to proceed through the addition-elimination intermediates (**85**).

8.1.1 Heterocyclic Donors

The application of the concept of Staunton-Weinreb annulation has not been applied to a heterocyclic donor. However, Mali et al., in their synthesis of carbazole derivatives, conceived of indole based donors (**86**) and alkyne acceptors, equivalent to β-alkoxy unsaturated esters. The reaction of these donors with dimethyl acetylene dicarboxylate in the presence of LDA regioselectively gave carbazoles (**87**) via (**88**) (Scheme 8.15).[12]

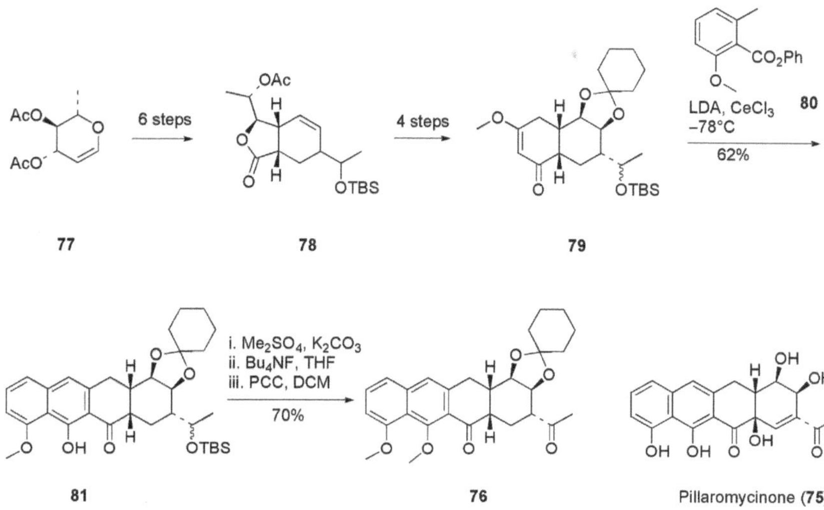

SCHEME 8.13 Synthesis of pillaromycinone core [1986 JOC 1150 White JD].

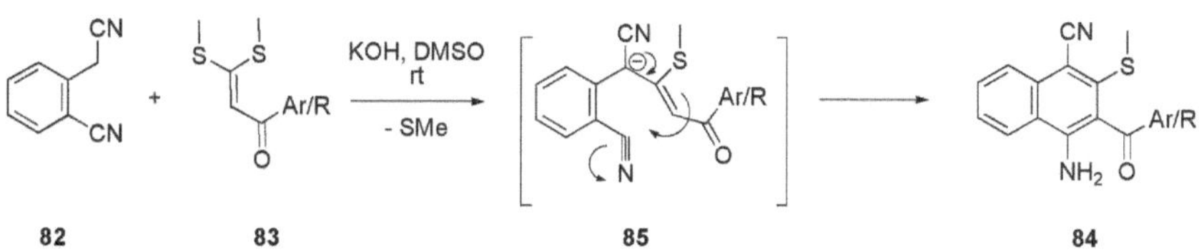

SCHEME 8.14 Annulation with β-alkylthioenones [2014 THN 8879 Pratap R].

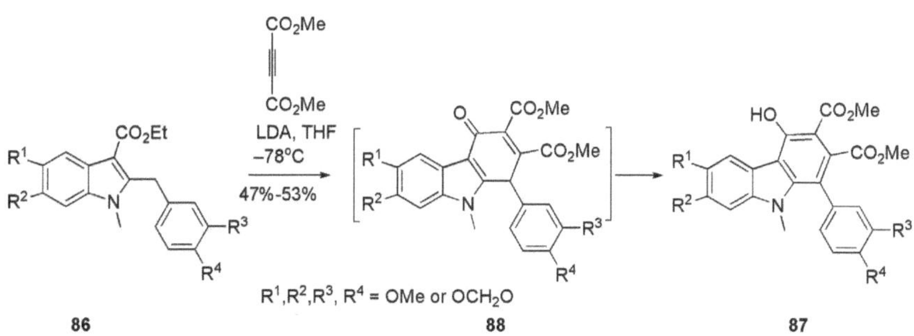

SCHEME 8.15 Indole analog of Staunton-Weinreb donors [1992 THS 1655 Mali RS].

As reported by Wada et al. (Scheme 8.16), the reaction of methyl pyrimidinecarboxylate (**89**) with dimethyl acetylenedicarboxylate and methyl propiolate in the presence of LDA afforded quinazolines (**90** and **9**)**1** in 38% and 24% yield, respectively, through the intermediates (**92** and **93**).[13]

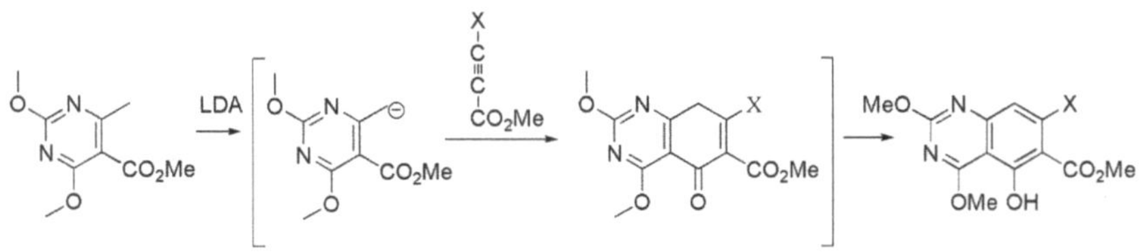

89 92 93 90 (38%, X = CO_2Me) 91 (24%, X = H)

SCHEME 8.16 Annulation methyl 2.4-dimethoxy-6-methyl-5-pyrimidinecarboxylate [1988 HCS 1345 Wada A].

8.2 ANNULATION WITH ACRYLATES AND CONJUGATED ALKENONES (MYERS ANNULATION)

The first reports on the successful annulation of *o*-toluates with acrylates/conjugated lactones are perhaps due to Thebtaranonth et al. and Franck et al.[14] Based on the annulation, Thebtaranonth et al. developed a convenient method for the synthesis of 8-methoxy-α-tetralones. The reaction of the toluates (**94**) with the acrylates (**95**) in the presence of LDA gave keto esters (**96**), which, without purification, was subjected to acid-catalyzed hydrolysis-decarboxylation to give tetralones (**97**). These annulations differ from the Staunton-Weinreb annulation in the use of acceptors without the β-alkoxy groups (Scheme 8.17).

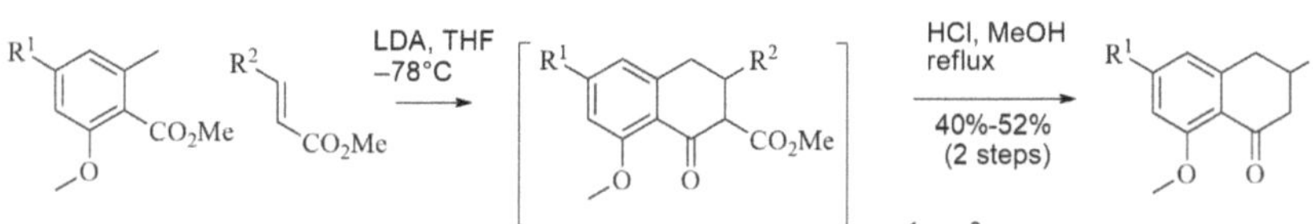

$R^1 = R^2$ = H or Me

94 95 96 97

SCHEME 8.17 Annulation of toluates with acrylates [1986 SYS 785 Thebtaranonth Y].

The Thebtaranonth annulation has successfully been employed in the total synthesis of diospyrol (**98**). The annulation of (**99**) with (**100**) in the presence of LDA gave annulation product (**101**). DDQ oxidation of (**101**) gave (**102**). Alkaline hydrolysis of (**102**) followed by decarboxylation and *O*-methylation afforded (**103**). Retro Diels-Alder reaction of (**103**) afforded acceptor (**104**) for second annulation. Annulation of (**99**) with (**104**) gave (**105**), which was subjected to four reactions: hydrolytic decarboxylation (aqueous NaOH), methyl ether formation ($(MeO)_3CH$, TsOH), dehydrogenation (DDQ) and exhaustive demethylation (BCl_3) to provide diospyrol (**98**) (Scheme 8.18).

Although the lateral anions of parent toluates are too reactive to undergo [4+2] annulation with acrylates, phenyl substituted toluate (**106**) underwent annulation with furan-2(5*H*)-one (**107**) to give (**108**), which was derivatized to (**109**) by dehydrogenation with 10% Pd-C, overall yield being 56% (Scheme 8.19).[15]

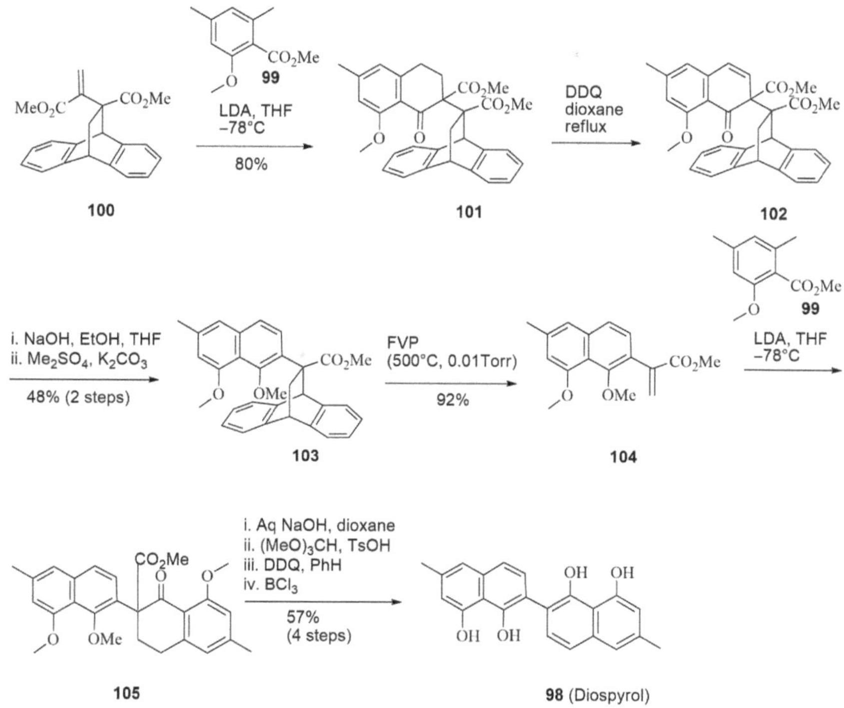

SCHEME 8.18 Synthesis of diospyrol [1989 THS 3861 Thebtaranonth Y].

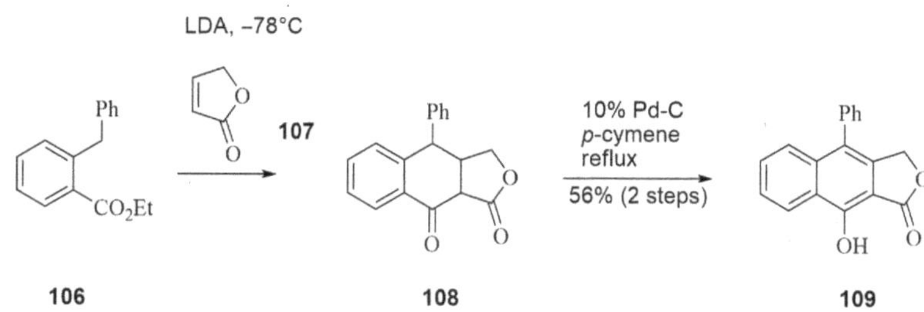

SCHEME 8.19 Annulation of 2-(phenylmethyl)benzoate with furan-2(5*H*)-one (**107**) [1997 JCSPT1 443 Kobayashi K].

The furanone annulation with a toluate was found applicable in the total synthesis of MS-444 (**110**) by the Tatsuta group (Scheme 8.20).[16] In the presence of LDA, the annulation between (**47**) and (**107**) proceeded smoothly to give (**111**) and its enol form in 70% yield. This mixture was treated with trimethylorthoformate and camphorsulfonic acid to give (**112**). Reaction of MeLi with (**112**) followed by CSA-promoted cyclization gave naphtho[2,3-*c*]furanone (**113**). Demethylation of (**113**) with BBr_3 afforded (**114**). Pd-catalyzed dehydrogenation of (**114**) yielded MS-444 (**110**) in 59% yield.

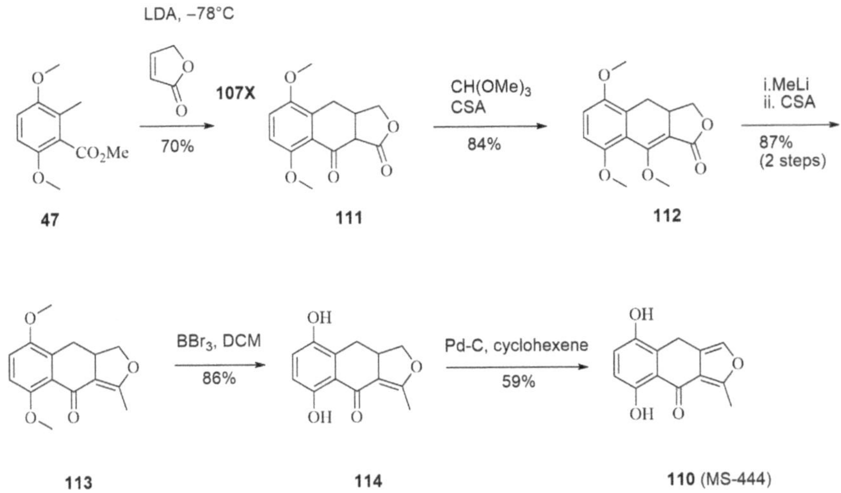

SCHEME 8.20 Total synthesis of MS-444 (**110**) [1997 JABTS 289 Tatsuta K].

A more sophisticated example of the above annulation is found in the asymmetric synthesis of olivinetrimethyl ether (**115**) by Franck et al. in 1986 (Scheme 8.21).[14b] It used naphthoate (**116**) as the donor and D-fucose derived δ-lactone (**117**) as the acceptor for the annulation. The annulation of (**116**) with the unsaturated lactone (**117**) (LDA, THF-DMPU, −78°C) gave tetracycle (**118**)

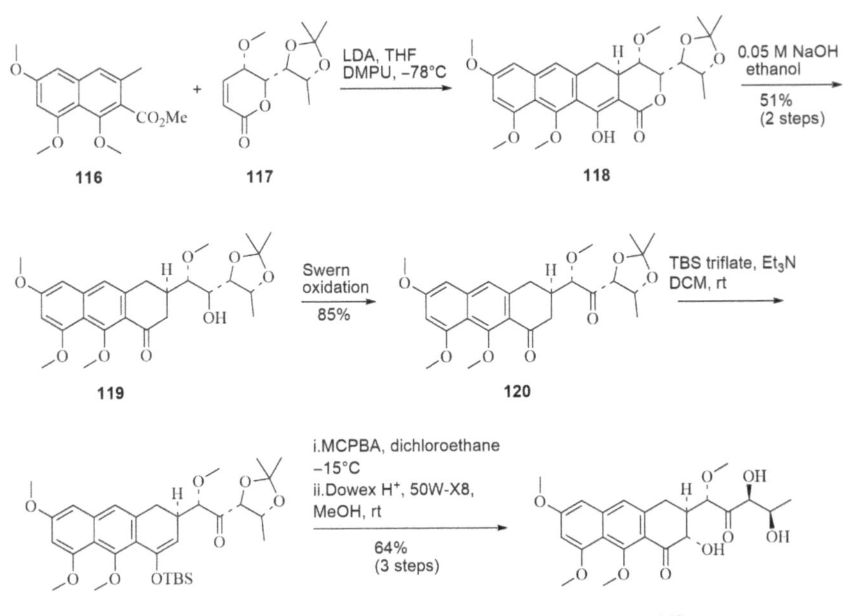

SCHEME 8.21 Total synthesis of natural form of olivintrimethyl ether (**115**) [1986 JACS 2455 Franck RW].

which, without purification, was hydrolyzed and decarboxylated (0.05 M aqueous NaOH, EtOH,75°C) to give hydroxyanthracenone (**119**). Swern oxidation of (**119**) furnished diketone (**120**). It was chemoselectively converted to enolsilyl ether (**121**), employing TBDMS triflate and Et_3N in DCM. Without purification, (**121**) was subjected to MCPBA oxidation followed by deprotection of the silyl group and ketal group in (**121**) to give (**115**). A remarkable feature of the annulation between (**116**) and (**117**) is its stereoselectivity. Of the two possible diastereomers due to generation of a new chiral center, only (**118**) was formed. This selectivity may be due to steric effect exerted by the allylic methoxy group during the initial Michael addition of the anion of (**116**).

Shaw et al. utilized a toluate-lactone annulation and a vanadium catalyzed phenol coupling for the synthesis of viriditoxin (**122**) Annulation between (**123**) and (**124**) in the presence of LDA provided the corresponding naphthopyranone, which was dehydrogenated with DDQ to give (**125**) (Scheme 8.22).[17] *O*-methylation of (**125**) followed by deprotection of ethoxymethyl ether yielded dimerization precursor (**126**). Oxidative dimerization of (**126**) with $VO(acac)_2$ afforded (**127**). *O*-methylation and subsequent desilylation of (**127**) gave (**128**). Three more steps as indicated in Scheme 8.22 furnished viriditoxin (**122**). It is noteworthy that the two-step synthesis of naphthopyranone (**125**) is more efficient than the one-step Staunton-Weinreb annulation (**65** → **66**). The application

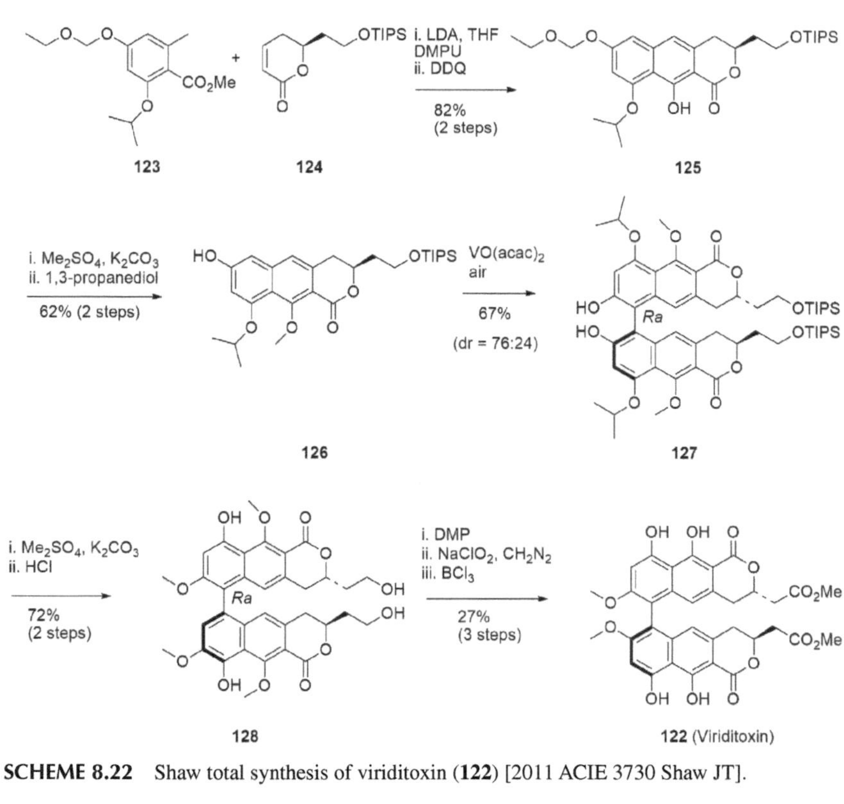

SCHEME 8.22 Shaw total synthesis of viriditoxin (**122**) [2011 ACIE 3730 Shaw JT].

of this strategy has culminated in an asymmetric synthesis of pigmentosin A via a vanadium catalyzed phenol coupling reaction.

The total synthesis of bioxanthracene (−)-ES-242-4 (**129**) was carried out in linear steps from two advanced intermediates (**11** and **130**) (Scheme 8.23).[18] The α,β-unsaturated lactone (**130**) was obtained from di-*O*-acetyl-L-rhamnal. The orsellinate (**11**) was prepared from 3,5-dihydroxytoluene. Lateral lithiation of (**11**) with LDA followed by treatment of the lactone (**130**) resulted in the formation of (**131**), which was dehydrogenated with DDQ to give naphthopyranone (**132**). Following *O*-benzylation of (**132**), the resultant compound was subjected to two reductions to obtain (**133**). Transfer hydrogenolysis of (**133**) yielded phenol (**134**). Oxidative dimerization of (**134**) with oxygen in the presence of CuCl(OH). TMEDA gave dimer (**135**). Base catalyzed tautomerization of (**135**) and subsequent deprotection of the MOM groups of the resultant product furnished (−)-ES-242-4 (**129**).

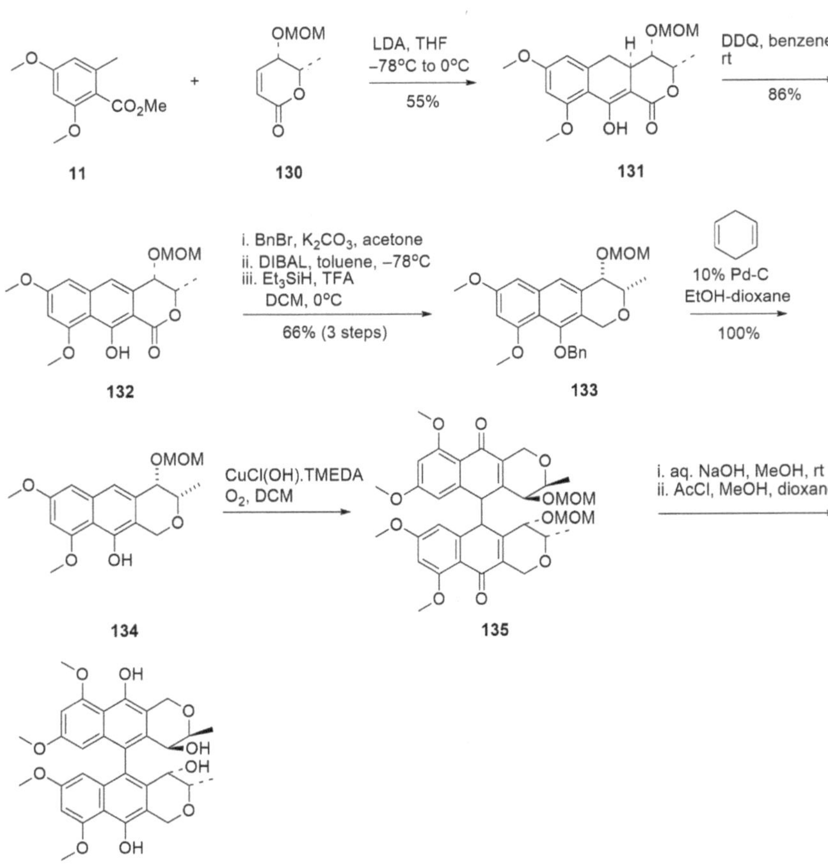

SCHEME 8.23 Total synthesis of bioxanthracene (−)-ES-242-4 (**129**) [1998 THS 1771 Tatsuta K].

In a synthetic approach towards K-259-2, a unique calmodulin antagonist, annulation of the skipped dienoate (**136**) with naphthoate (**137**) was examined. Interestingly, Hauser sulfoxide annulation did not work for the benzannulation. However, lithiated naphthoate (**138**), generated from bromide (**137**) with ^{s}BuLi, underwent annulation with dienoate (**136**) to give anthracenone (**139**) in 61% yield (Scheme 8.24).[19] Annulation of the intermediate (**138**), under the conditions of deprotonation of its methyl precursor, was not successful.

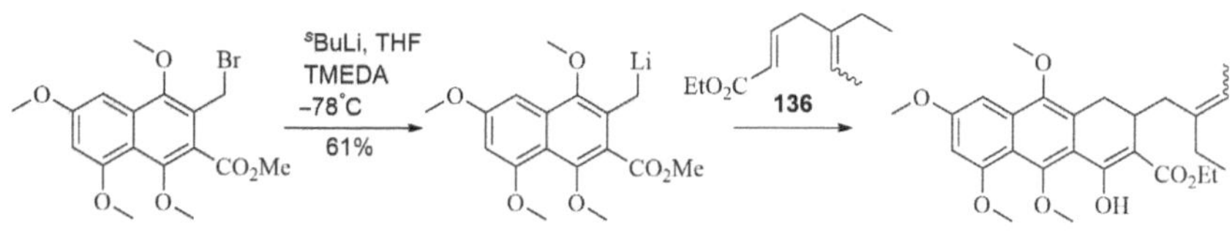

137 **138** **139**

SCHEME 8.24 Mal Synthetic approach towards K-259-2, a unique calmodulin antagonist.

Compound (**140**) is the only heterocycle that has been studied for the annulation with activated alkenes (**141**). Their reaction in the presence of LDA gave tetralones (**142**) in modest yields via (**143**) (Scheme 8.25).[13]

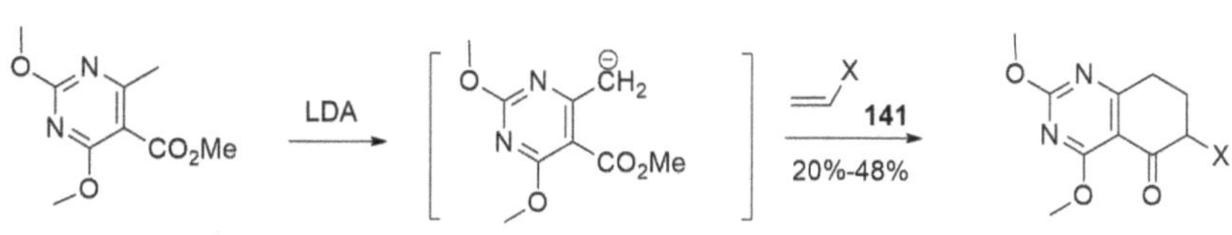

X = CN, COMe, CO_2Me, etc.

140 **143** **142**

SCHEME 8.25 Annulation of methyl 2.4-dimethoxy-6-methyl-5-pyrimidinecarboxylate with activated alkenes [1988 HCS 1345 Wada A].

Myers annulation: Annulations of *o*-toluates with 2-cycloalkenones or decalenones to construct 3,4-dihydro-9-hydroxyanthracenone/naphthacene cores have been a serious problem since the inceptions of the annulations, and consequently it became topics of detailed investigations. A good solution to this problem prohibited many total syntheses such as pillaromycinone, olivin, etc. The use of cycloalkenones in the annulation with an *o*-toluate was indicated by the Clive group in their two-step synthesis of a model compound for fredericamycin A, a naturally occurring antitumor antibiotic. As shown in Scheme 8.26, pyridine carboxylate (**144**), on reaction LDA followed by cyclopentenone, gave Michael adduct (**145**) in 62%–82% yield. In the next step, cyclization of (**145**) was effected by the action of sodium hydride in the presence of ethanol (cat.), giving product (**146**) in 75% yield. The central ring of (**146**) was aromatized with DDQ oxidation to furnish (**147**).[20]

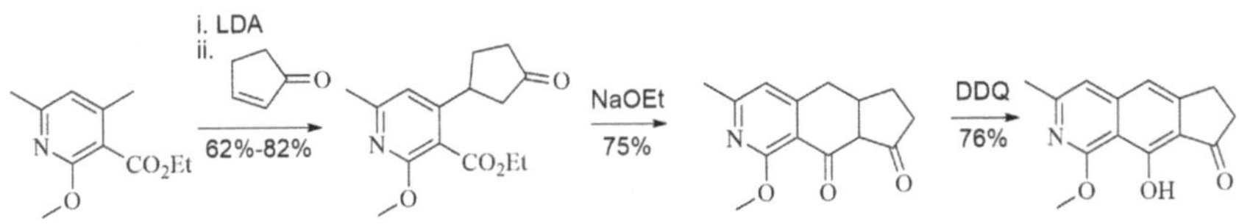

144 145 146 147

SCHEME 8.26 Two-step annulation of a toluate analog with cyclopentenone [1987 JHC 509 Clive DLJ].

In a similar instance (Scheme 8.27), pyridine derivative (**148**) was cyclocondensed with cyclopentenone in one step under controlled conditions to produce (**149**). The addition of ethanol in second phase significantly improved the yield of the annulation to 85%. The compound (**149**) was aromatized to (**150**) by DDQ dehydrogenation.[21]

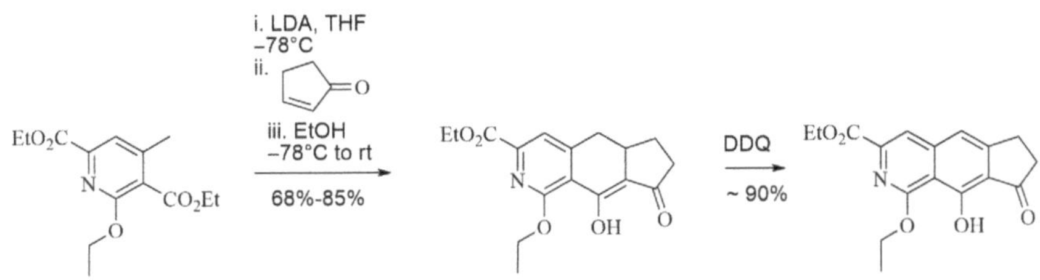

148 149 150

SCHEME 8.27 One-step annulation of pyridine analog with cyclopentenone [1995 JACS 11839 Boger DL].

The annulation of methyl toluates with cycloalkenones is usually unsuccessful, barring few instances like one in Scheme 8.28. The reaction of (**151**) with LDA followed by addition of cyclohexenone (**152**) gave anthracenone (**153**), which was characterized as centrally aromatized anthracenone (**154**).[22]

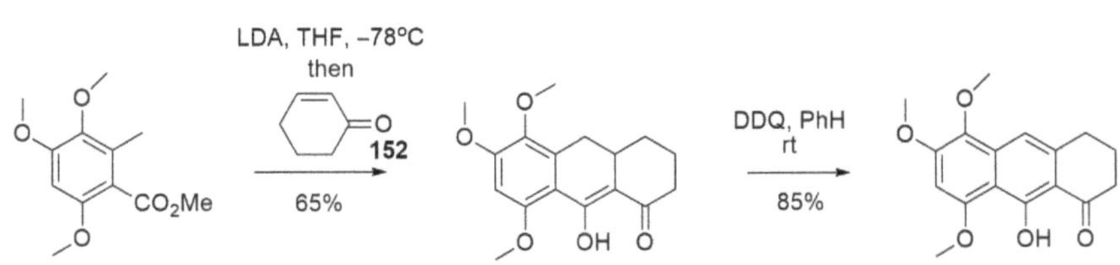

151 153 154

SCHEME 8.28 Successful Staunton-Weinreb annulation [2013 SYS 2287 Mootoo DR].

The pioneering work on annulation of cycloalkenones with toluates is due to Myers et al. They could accomplish the annulations between the above reacting partners in one step and utilize them to generate more than 3000 tetracycline molecules. Consequently, these annulations are identified as Myers annulation. Based upon the annulations, Myers et al. laid the foundation for the first asymmetric synthesis of 6-deoxytetracycline (**155**) in 2005. They followed White's clue of using phenyl ester and established that the anionic [4+2] annulation is the cornerstone of the synthetic chemistry of modern-day tetracyclines. They employed the anionic [4+2] annulation for the synthesis of 6-deoxytetracycline (**155**). Treatment of *tert*-butoxycarbonyl-protected phenyl ester (**156**) with LDA-TMEDA followed by addition of enone (**157**) and slow warming of the reaction mixture afforded annulation product (**158**) (Scheme 8.29). Desilylation of pentacycle (**158**) gave (**159**). The final step of demasking hydrogenolysis of (**159**) gave 6-deoxytetracycline (**155**).[23] The same strategy has been applied to the synthesis of many other tetracycline analogs.

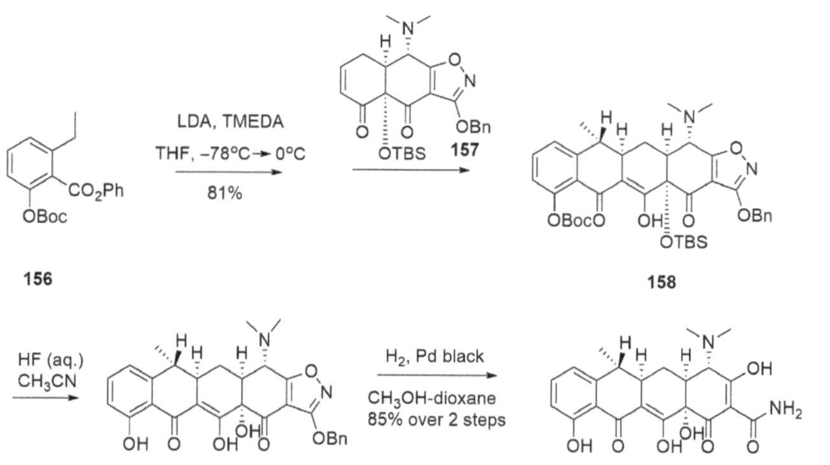

156 **157** **158** **159** **155** (6-deoxytetracycline)

SCHEME 8.29 Myers synthesis of deoxytetracycline [2005 SCE *308*, 395 Myers AG].

The condensation between (**156**) and (**157**) seems to proceed in stepwise fashion. On quenching at −78°C, the reaction affords the Michael adduct (**160**) (Scheme 8.30).[24]

The White group, in their study directed to the total synthesis of pillaromycinone (**75**), developed a two-step protocol for the benzannulation with a decalenone. As shown in Scheme 8.31, the annulation of phenyl *o*-toluate (**161**) with decalenone (**162**), obtained in eight steps from (**163**), gave the annulated product (**164**) in almost quantitative yield.[25] However, it led to another problem, i.e., selective dehydrogenation to obtain the naphthalene core, which was eventually feasible by treatment of (**164**) with *p*-chloranil in refluxing toluene. Further elaboration on installing the annular hydroxy group of pillaromycinone (**75**) was futile.

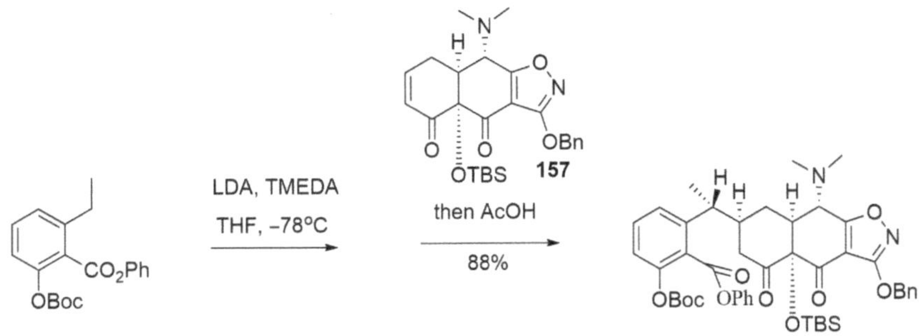

156 160

SCHEME 8.30 Isolation of the Michael adduct of (**156**) and (**157**).

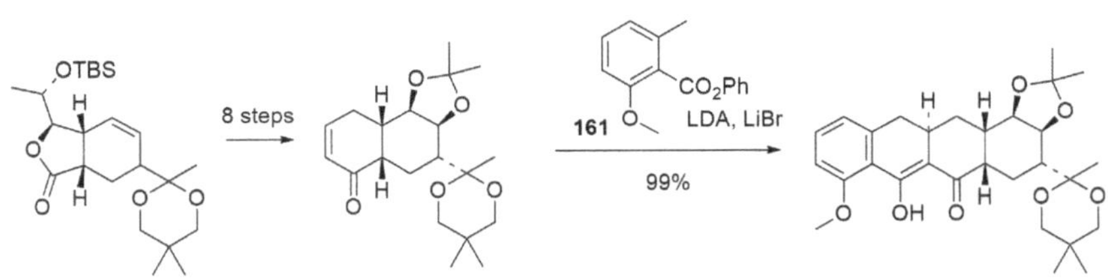

163 162 164

SCHEME 8.31 Synthetic studies on pillaromycinone [2008 ORS 2833 White JD].

Myers et al. have developed a new method of toluate annulations. It is based on toluate anion generation by bromine-lithium exchange under conditions involving a variety of reagents, additives, and conditions. These anions are suitable for annulation with enone acceptors. Although the simplest methyl *o*-toluate anion fails to undergo annulation, the anion (**165**), generated by Br-Li exchange of phenyl ester (**166**), smoothly underwent annulation with (**157**) to give (**167**) in 81% yield. This mode of enone annulation was applicable to bromo derivatives (**168–171**) (Scheme 8.32).[24]

The Myers annulation is so powerful that the hexacyclic analog (**172**) could be assembled with ease, starting from (**157**) and (**173**) (Scheme 8.33). The precursor (**173**) was obtained from (**174**) through its lithiation-mediated formylation followed by intramolecular cycloaddition of azomethine ylide from *N*-allylglycine and aldehyde (**175**). Annulation between (**157**) and (**173**) gave (**172**) in 98% yield. Deallylation ($Pd(PPh)_4$, dimethylbarbituric acid), reductive amination (CH_3CHO, $NaBH(OAc)_3$), and removal of the protecting groups, applied to (**172**), gave a tetracyclic analog (**176**).[26]

The Myers annulation has also witnessed significant success for isoxazole analogs of *o*-toluate. The synthesis of tetracycline AB synthon (**157**) relied upon an annulation. The experimental trick of annulating Stork-Hagedorn isoxazole

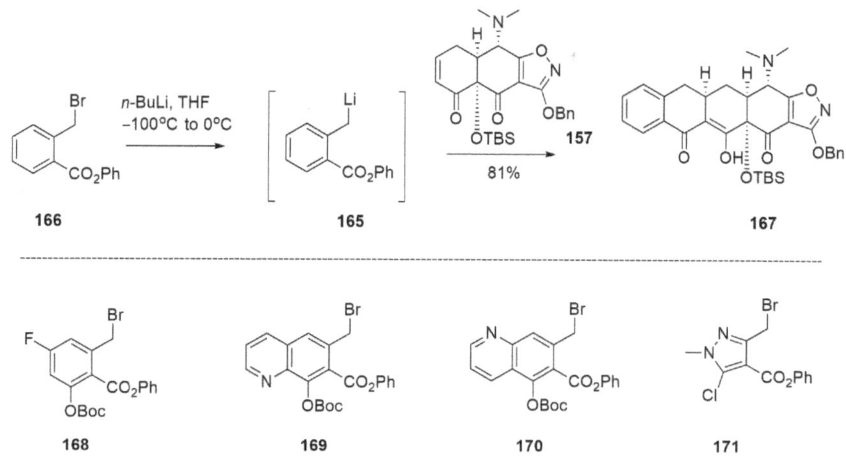

SCHEME 8.32 Bromine-lithium exchange-mediated annulation [2008 JACS 17913 Myers AG].

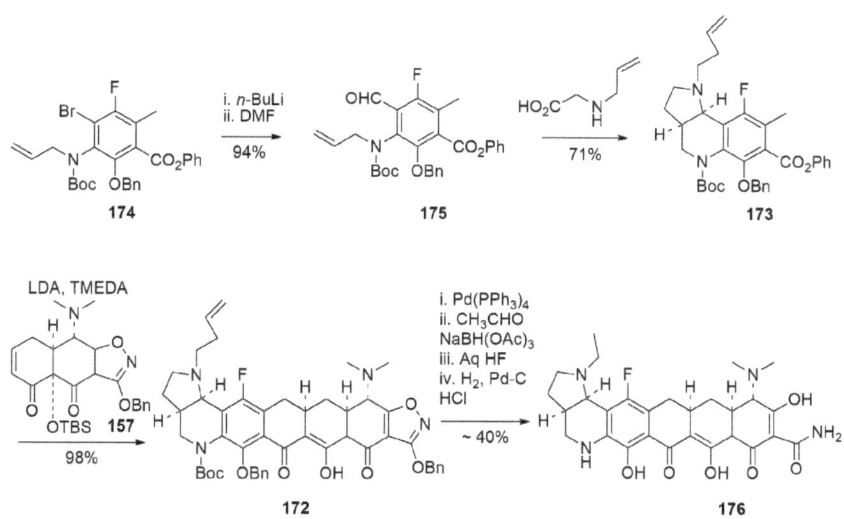

SCHEME 8.33 Myers annulation in the synthesis of the hexacyclic analog of tetracyclines [2015 JMC 4703 Xiao X-Y].

(**177**) with decalenone (**178**) in two stages with two different bases, sodium hexamethyldisilazide and potassium hexamethyldisilazide, successfully gave pentacycle (**179**) via Michael adduct (**181**) in 80% yield. *Retro*-Diels-Alder reaction of (**180**) provided tricycle (**181**). Oxidation of (**181**) with Davies reagent (**182**) led to angularly hydroxylated tricycle (**183**). Isomerization of (**183**) via lateral anion formation resulted in precursor (**184**). It was *O*-silylated with TBSOTf to furnish (**157**) (Scheme 8.34).[27]

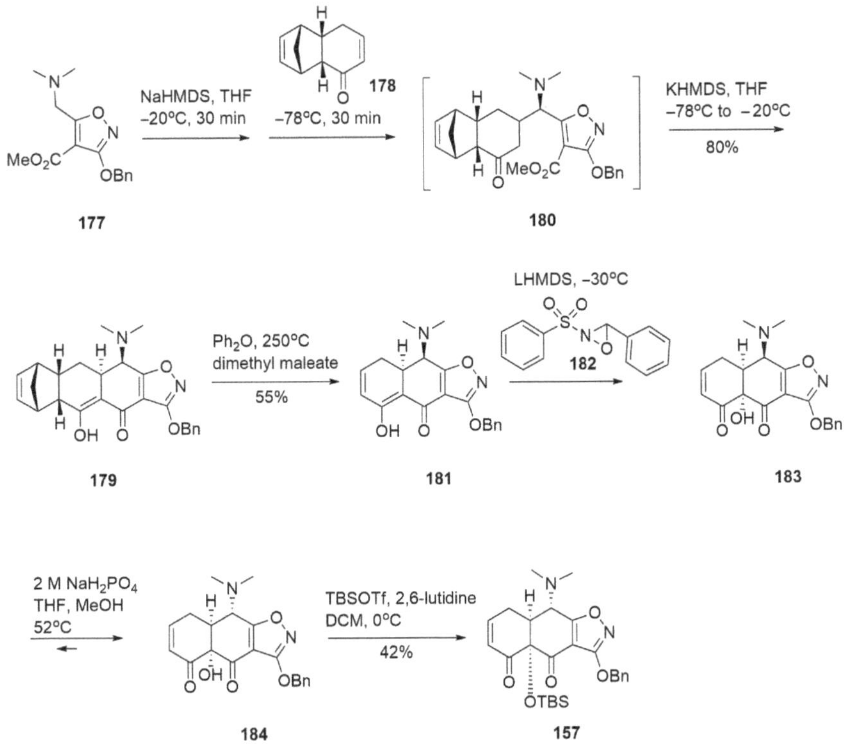

SCHEME 8.34 Synthesis of tetracycline AB synthon by [4+2] annulation [2011 CSC 1710 Myers AG].

In a study directed towards the synthesis of oxatetracyclines, Myers et al. explored annulation reactivity of the isoxazole (**178**) to arabinose-derived dihydropyrone (**185**) (Scheme 8.35).[28] The reaction between (**178**) and (**185**) in the presence of NaHMDS gave annulation product (**186**) in 40% yield with good diastereoselectivity. Strikingly, no Staunton-Weinreb product was formed, although there was a β-ether functionality in the enone moiety of (**185**). Furthermore, methyl ester analog of (**187**) afforded only the Michael adduct (**188**) due to the lesser reactivity of the methyl ester than a phenyl ester (cf. **178**). The annulation of isoxazole analogs is crucially dependent on the acidity of lateral CH bond. A much more efficient annulation occurred with *N*-Boc phenyl ester analog of (**178**), the yield being 92% yield.

Although the phenyl esters are more suited to the Myers annulations, Rodrigo et al. showed that generation of *ortho* toluate anion by transmetalation of a stannane precursor without a phenyl ester group with ^{n}BuLi is efficient for the annulation. The resulting toluate carbanions reacted with 2-cyclohexenones to give hydroanthracenone derivatives in good yields (47%–83%) which were readily dehydrogenated to the corresponding 1,2,3,4-tetrahydroanthracenone

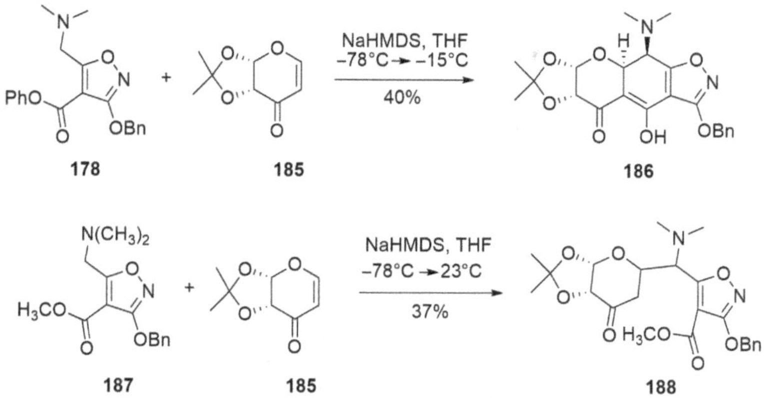

SCHEME 8.35 Annulation with a 4-pyrone [2017 ORS 206 Myers AG].

derivatives (Scheme 8.36). Toluate (**189**) was converted into stannane (**190**) by the action of LDA and Bu_3SnCl. In the presence of ^nBuLi, the stannane (**190**) was transformed to (**191**), which smoothly reacted with cyclohexenone (**152**) to give (**192**). It was dehydrogenated with DDQ to give (**193**).[29]

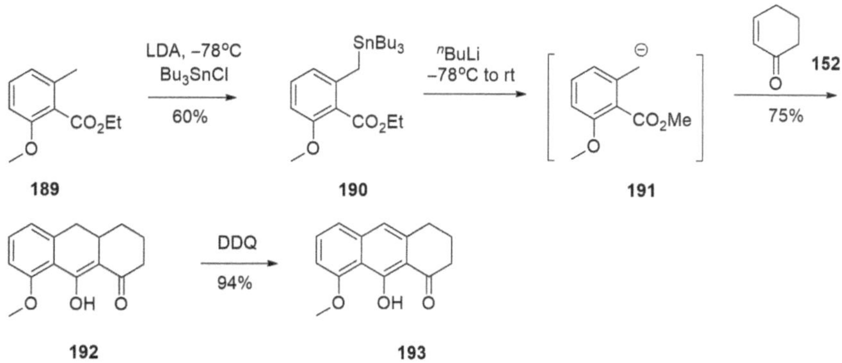

SCHEME 8.36 Rodrigo annulation with cycloalkenones [2005 ORS 5223 Rodrigo R].

The Rodrigo method worked well with decalenone (**194**) and cyclohexenone (**195**) to afford (**196**) and (**197**), respectively (Scheme 8.37). The o-toluate anion generated from the corresponding trimethylsilylmethyl precursor by the action of CsF gave poorer results with cycloalkenone acceptors.

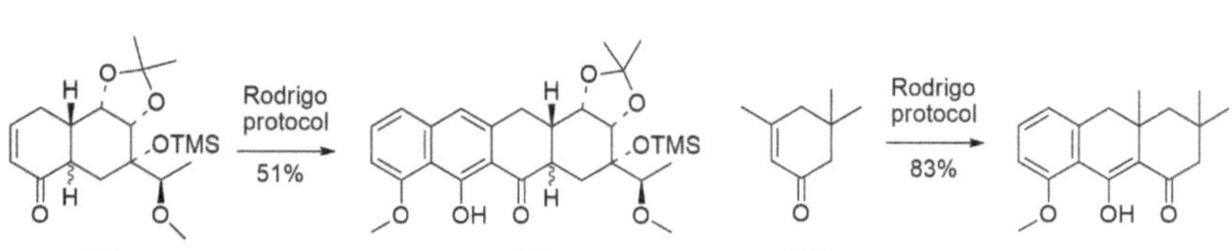

SCHEME 8.37 Examples of Rodrigo annulation.

8.3 MISCELLANEOUS TOLUATE DONORS AND THEIR ANNULATIONS

Although *ortho* tolunitriles (**198**) were expected to undergo [4+2] annulation with acrylates and acrylonitriles, their annulations require activation by ZnI_2 under strongly basic conditions, and provide 1-amino-3,4-dihydro-2-naphthalene carboxylates and nitriles. The mechanism of the annulations is based on the tandem Michael addition of (**199**) and enolate-nitrile coupling reaction of (**200**). The addition of ZnI_2 results in zinc enolate and bivalent zinc ion accelerate the enolate-nitrile coupling. The resultant products (**201**) were dehydrogenated with Pd-C in refluxing *p*-cymene to furnish α-napthylamines (**202**) (Scheme 8.38).[30]

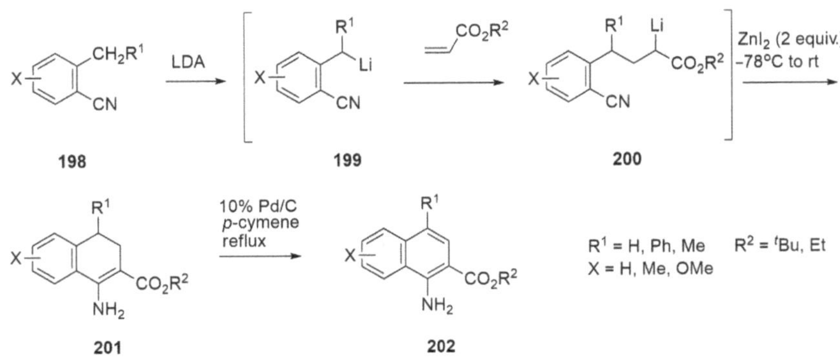

SCHEME 8.38 Annulation of *o*-tolunitriles [1997 JOC 664 Kobayashi K].

In a recent development, the Myers group developed a versatile method for rapid construction of highly substituted isoquinolines as depicted in Scheme 8.39 The method relies upon lateral lithiation of *o*-tolualdehyde *tert*-butylimine (**203**) with ^{n}BuLi and a catalytic amount of 2,2,6,6-tetramethylpiperidine in THF. The lithiated species (**204**) first undergoes 1,2-addition to alkylnitrile to give (**205**), which cyclizes to amino isoquinoline motifs (**206**). Intramolecular lithium-hydrogen exchange in (**206**) provides quinolines (**207**). Reaction of (**207**) with electrophilic reagents (EX) followed by loss of $^{t}BuNH_2$ leads to 3,4-disubstituted isoquinolines (**208**).[31] This heteroannulation, a three-component reaction, has been generalized with 13 examples.

In order to develop an organocatalytic atropselective synthesis of BINAM derivatives, the Kurti group examined the [4+2] anionic annulation reactivity of *o*-tolualdehyde *tert*-butylimine towards β-nitrostyrenes. When the Schiff base (**203**) was lithiated with ^{n}BuLi and catalytic TMP in THF at 0°C and treated with nitrostyrenes (**209**) at −78°C followed by HCl quench, 2-nitrodihydronaphtalenes (**210**) were obtained in good yields (Scheme 8.40).[32] Most of these annulation products were aromatized by dehydrogenation with NBS-benzoyl peroxide in DCE to the corresponding 3-alkyl/aryl-2-nitronaphthalenes in excellent yields.

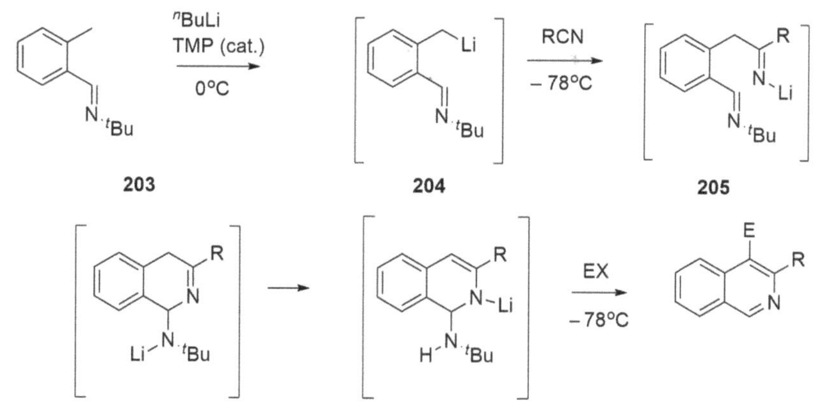

R = alkyl, aryl, alkynyl, N,N-dimethyl, piperinyl, etc.; E = alkyl, CO_2Me, F, Cl, SMe, etc.

SCHEME 8.39 Myers isoquinoline synthesis from 2-tolualdimines [2011 ACIE 10409 Myers AG].

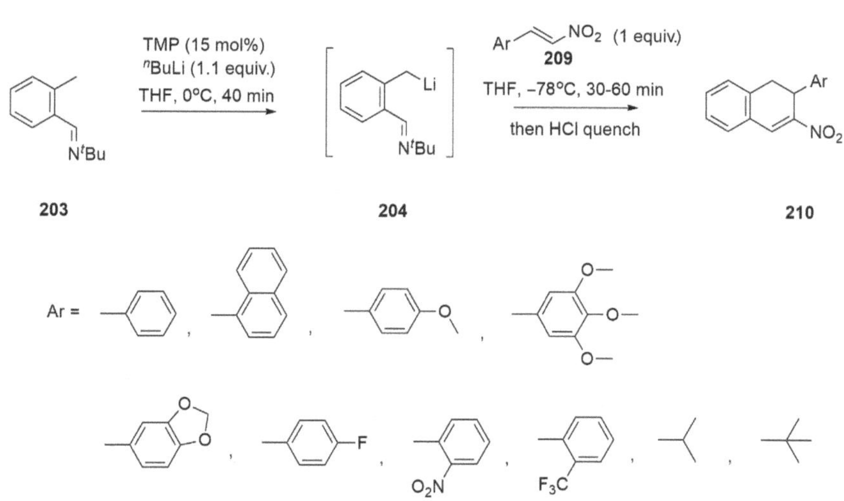

SCHEME 8.40 Kurti annulation of tolualdehyde Schiff base with nitrostyrenes [2013 SYS 1719 Kurti L].

The synthetic potential of indole analogs of the tolualdehyde imines (cf. **203**) have been rigorously studied by Andrade et al. for the synthesis of aspidosperma alkaloids. *N*-sulfinylimine (**211**), the 1,4-dipole donor, was readily prepared from 2-methyl-indole-3-carboxaldehyde (**212**) by first condensing with *N*-sulfinamide (**213**) to give (**214**) and then by *N*-sulfonylation of (**214**). LHMDS promoted annulation of (**211**) with acceptors (**215**) furnished tetrahydrocarbazoles (**216**) in one pot (Scheme 8.41).[33]

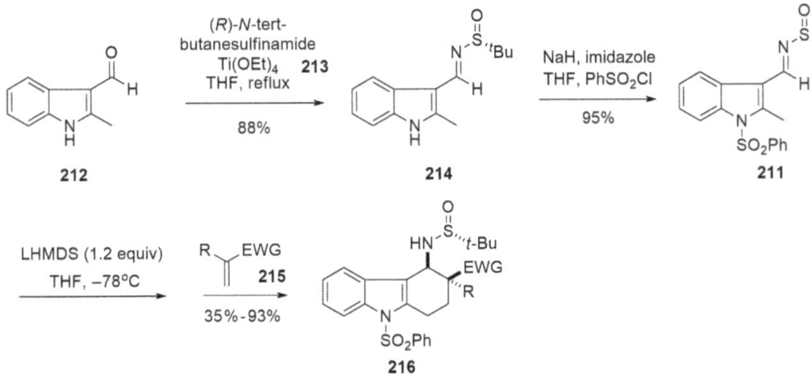

SCHEME 8.41 Andrade annulation [2017 JOC 521 Andrade RB].

This annulation is vastly general with regard to the nature of the acceptors. Details are given in Fig. 8.1. The ratios denote the diastereomeric ratios. Methacrolein (**217**), methyl vinyl ketone (**218**), methyl acrylate (**219**), and methyl ethacrylate (**220**), on reaction with (**211**), gave (**221**), (**222**), (**223**), and **224**), respectively. Likewise, unsaturated lactone (**225**) and Weinreb amide (**226**) provided (**227**) and (**228**), respectively.

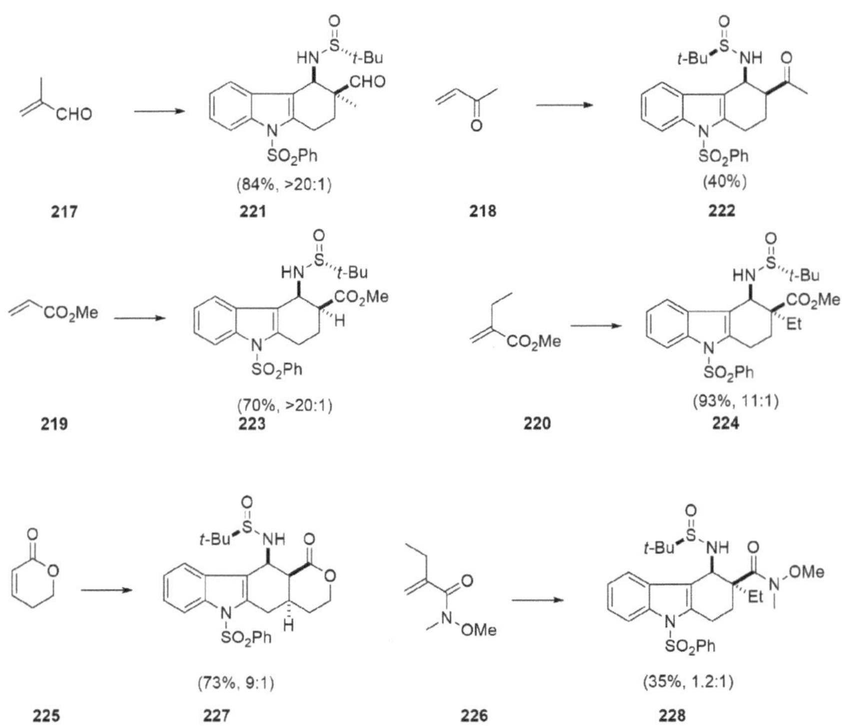

FIG. 8.1 Acceptors and their products from the Andrade donor (**211**).

The mechanism of the Andrade annulation is proposed to involve cycloaddition between lithio *o*-quinoned imethide analog (**229**) and ethacrylate (**220**) to give the lithium coordinated transition state (**230**) en route to (**218**) (Scheme 8.42).

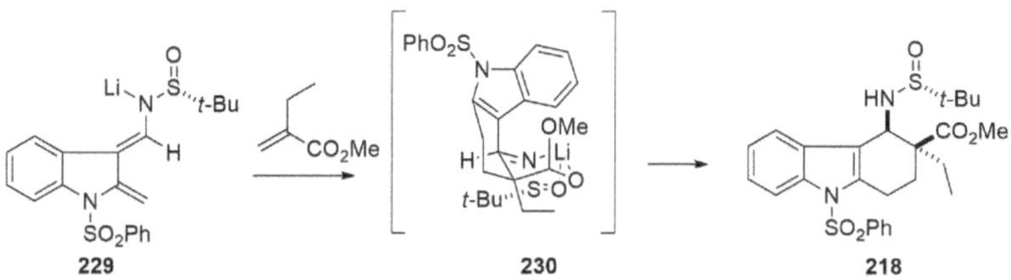

SCHEME 8.42 Mechanism of the Andrade annulation.

This section is concluded with elegant application of the Andrade annulation to the synthesis of aspidospermidine alkaloids, more specifically (−)-vincadifformine (**231**). It began with the imine (**211**). It was cyclocondensed with ethacrolein (**232**) to give tetrahydrocarbazole (**233**). *N*-allylation of (**233**), followed by Wittig methylenation, gave (**234**). Ring closing metathesis of (**234**) with the aid of Hoveyda-Grubbs second generation catalyst (HG-II) provided (**235**). The *N*-sulfinyl group in (**235**) was removed by methanolic HCl. Subsequent *N*-alkylation of the resultant intermediate with 2-bromoethanol and Na_2CO_3 delivered alcohol (**236**) in 80% overall yield. Thus, the Bosch-Rubiralta spiro cyclization was accomplished by the addition of KO^{t}Bu in THF at 0°C to afford indolenine (**237**) in 60% yield (Scheme 8.43). Lithiation of indolenine (**237**)

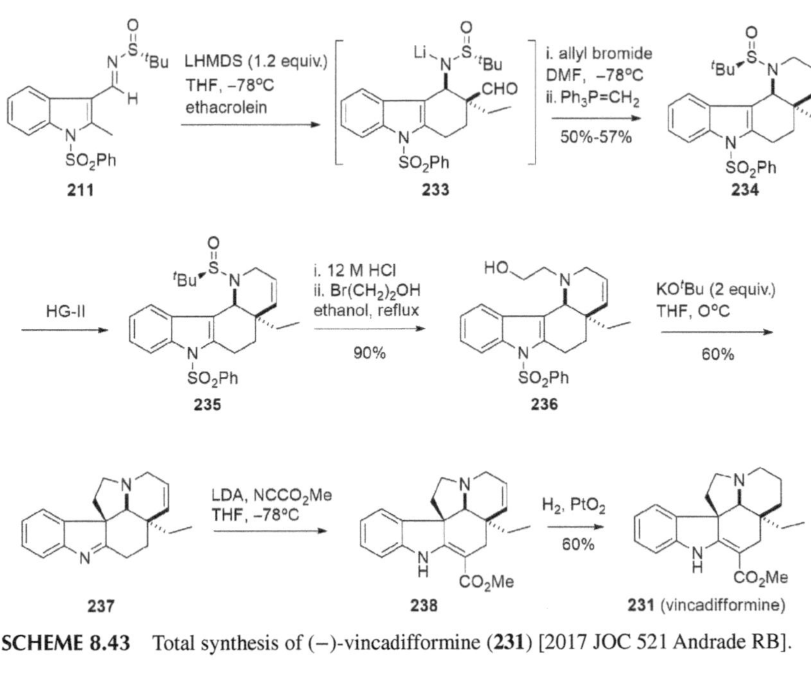

SCHEME 8.43 Total synthesis of (−)-vincadifformine (**231**) [2017 JOC 521 Andrade RB].

with LDA followed by addition of Mander's reagent furnished (−)-tabersonine (**238**) in 73% yield. Hydrogenation of (**238**) with Adams's catalyst and H_2 in EtOAc afforded (−)-vincadifformine (**231**) in 81% yield.[33]

In conclusion, this chapter has described the anionic chemistry of toluates and tolualdehydes without an activated group at the benzylic positions. Their [4+2] reactions with acceptors like β-alkoxyacrylates and β-alkoxy-α,β-unsaturated ketones under basic conditions are shown to provide an access to substituted naphthalenes with defined patterns of the substituents. These reactions are termed "Staunton-Weinreb annulations." These annulations have been useful in the total synthesis of various polycyclic natural products, embodying functionalized naphthalenes and anthracenes. Similarly, unactivated toluates participate in the reaction with α,β-unsaturated esters and α,β-unsaturated ketones. Those reactions with α,β-unsaturated ketones are called Myers annulation. It has been vitally important in the total synthesis of tetracyclines as well as development of their new analogs. Also included in this chapter is the anionic chemistry of *o*-tolualdehyde imines resulting in the synthesis of substituted naphthalenes, isoquinolines, and carbazoles.

REFERENCES

1. Donner, C. D. *Tetrahedron* **2013**, *69*, 3747.
2. Evans, G. E.; Leeper, F. J.; Murphy, J. A.; Staunton, J.J.C.S. *Chem. Comm.* **1979**, 205.
3. Dodd, J. H.; Garigipati, R. S.; Weinreb, S. M. *J. Org. Chem.* **1982**, *47*, 4045.
4. Tan, N. P. H.; Donner, C. D. *Tetrahedron Lett.* **2008**, *49*, 4160.
5. Hauser, F. M.; Rhee, R. P.; Prasanna, S.; Weinreb, S. M.; Dodd, J. H. *Synthesis* **1980**, 72.
6. Drochner, D.; Huttel, W.; Bode, S. E.; Muller, M.; Karl, U.; Nirger, M.; Steglich, W. *Eur. J. Org. Chem.* **2007**, 1749.
7. Sperry, J.; Yuen, T. Y.; Brimble, M. A. *Synthesis* **2009**, 2561.
8. Tan, N. P. H.; Donner, C. D. *Tetrahedron* **2009**, *65*, 4007.
9. Donner, C. D. *Tetrahedron* **2013**, *69*, 377.
10. White, J. D.; Nolen, E. G., Jr.; Miller, C. H. *J. Org. Chem.* **1986**, *51*, 1150.
11. Singh, S.; Althagafi, I.; Yadav, P.; Panwar, R.; Kumar, A.; Pratap, R. *Tetrahedron* **2014**, *70*, 8879.
12. Mali, R. S.; Jagtap, P. G. *Tetrahedron Lett.* **1992**, *33*, 1655.
13. Wada, A.; Yamamoto, H.; Kanatomo, S. *Heterocycles* **1988**, *27*, 1345.
14. (a) Tarnchompoo, B.; Thebtaranonth, C.; Thebtaranonth, Y. *Synthesis* **1986**, 785. (b) Franck, R. W.; Bhat, V.; Subramaniam, C. S. *J. Am. Chem. Soc.* **1986**, *108*, 2455.
15. Kobayashi, K.; Maeda, K.; Uneda, T.; Morikawa, O.; Konishi, H. *J. Chem. Soc. Perkin Trans. 1* **1997**, 443.
16. Tatsuta, K.; Yoshimoto, T.; Gunji, H. *J. Antibiotics* **1997**, *58*, 289.
17. (a) Park, Y. S.; Grove, C. I.; Gonzalez-Lopez, M.; Urgaonkar, S.; Fettinger, J. C.; Shaw, J. T. *Angew. Chem. Int. Ed.* **2011**, *50*, 3730. (b) Grove, C. I.; Maso, M. J. D.; Jaipuri, F. A.; Kim, M. B.; Shaw, J. T. *Org. Lett.* **2012**, *14*, 4338.
18. Tatsuta, K.; Yamazaki, T.; Mase, T.; Yoshimoto, T. *Tetrahedron Lett.* **1998**, *39*, 1771.
19. Basak, S.; Mandal, S.; Mal, D. *Tetrahedron* **2018**, *74*, 96.
20. Clive, D. L. J.; Sedgeworth, J. *J. Heterocyclic Chem.* **1987**, *24*, 509.

21. Boger, D. L.; Huter, O.; Mbiya, K.; Zhang, M. *J. Am. Chem. Soc.* **1995**, *117*, 11839.
22. Li, J.; Mootoo, D. R. *Synthesis* **2013**, *45*, 2287.
23. Charest, M. G.; Lerner, C. D.; Brubaker, J. D.; Siegel, D. R.; Myers, A. G. *Science* **2005**, *308*, 395.
24. Sun, C.; Wang, Q.; Brubaker, J. D.; Wright, P. M.; Lerner, C. D.; Noson, K.; Charest, M.; Siegel, D. R.; Wang, Y.-M.; Myers, A. G. *J. Am. Chem. Soc.* **2008**, *130*, 17913.
25. White, J. D.; Demnitz, F. W. J.; Xu, Q.; Martin, W. H. C. *Org. Lett.* **2008**, *10*, 2833.
26. Sun, C.; Hunt, D. K.; Chen, C.-L.; Deng, Y.; He, M.; Clark, R. B.; Fyfe, C.; Grossman, T. H.; Sutcliffe, J. A.; Xiao, X.-Y. *J. Med. Chem.* **2015**, *58*, 4703.
27. Kummer, D. A.; Li, D.; Dion, A.; Myers, A. G. *Chem. Sci.* **2011**, *2*, 1710.
28. Liu, F.; Wright, P. M.; Myers, A. G. *Org. Lett.* **2017**, *19*, 206.
29. Hill, B.; Rodrigo, R. *Org. Lett.* **2005**, *7*, 5223.
30. Kobayashi, K.; Uneda, T.; Takada, K.; Tanaka, H.; Kitamura, T.; Morikawa, O.; Konishi, H. *J. Org. Chem.* **1997**, *62*, 664.
31. Si, C.; Myers, A. G. *Angew. Chem. Int. Ed.* **2011**, *50*, 10409.
32. Keene, C.; Kurti, L. *Synthesis* **2013**, *45*, 1719.
33. Zhao, S.; Andrade, R. B. *J. Org. Chem.* **2017**, *82*, 521.

Chapter 9

Comparative Annulation Chemistry of Phthalides, Homophthalic Anhydrides and *o*-Toluates

The preceding chapters have exclusively dealt with annulation of phthalides, homophthalic anhydrides, and *o*-toluates. The prerequisite of the success of the above chemistry is ready accessibility of the starting 1,4-dipolar reagents and choice of reaction conditions. In this chapter, the synthesis of phthalides, homophthalic anhydrides, and *o*-toluates is presented. In the latter part, the factors that influence the successes and setbacks of the anionic annulation chemistry are highlighted.

9.1 SYNTHESIS OF PHTHALIDES, HOMOPHTHALIC ANHYDRIDES, AND *O*-TOLUATES

There are a plethora of synthetic methods for the synthesis of phthalides, homophthalic anhydrides, and *o*-toulates. In the sections below, the methods that have been chosen for large-scale preparations or those which are found to be useful in the context of a total synthesis are discussed.

9.1.1 Synthesis of 3-Phenylthiophthalides and 3-Phenylsulfonylphthalides

Structurally, 3-phenylthiophthalides and 3-phenylsulfonylphthalides differ from each other in the level of oxidation at the sulfur center. In terms of Hauser annulation chemistry, they are equivalent to each other, because the annulation products from both kinds of donors are the same. A survey of their synthetic methods is graphically presented in Fig. 9.1. The central precursors to the 3-phenylthiophthalides and 3-phenylsulfonylphthalides are phthalaldehydic acids (**1**), which are ring isomers of the corresponding 2-formylbenzoic acids. Phthalaldehydic acids, in turn, can be obtained from a wide variety of monosubstituted and disubstituted benzene derivatives (**2–6**).

Anionic Annulations in Organic Synthesis. https://doi.org/10.1016/B978-0-12-813800-7.00009-9

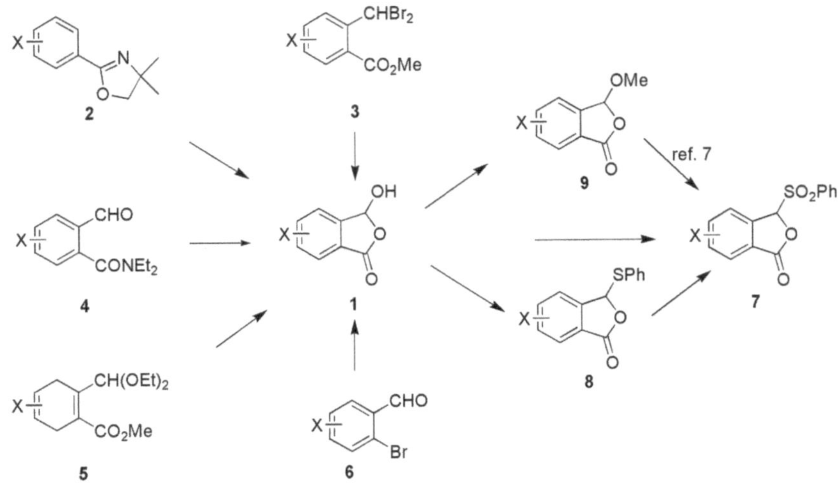

FIG. 9.1 Outline of the synthetic methods for 3-phenylthiophthalides and 3-phenylsulfonylphthalides.

The synthesis of 3-phenylthiophthalides (**8**) and 3-phenylsulfonylphthalides (**7**) are commonly achieved from readily accessible phthalaldehydic acids (**1**) and *ortho*-formylbenzamide derivatives (**4**), which are obtainable in a regiodefined manner via *ortho*-metalation methodologies.[1] Phthalaldehydic acids (**1**) can be directly converted to sulfone phthalide (**7**). Alternatively, they can be prepared through (**8**) or (**9**). In Scheme 9.1, representative examples of different synthetic routes are displayed. Reaction 1 is a direct condensation of phthalaldehydic acid (**10**) with arenesulfinic acids (**11**) in the presence BF_3 etherate at room temperature.[2] It is proposed to take place via attack of arenesulfinic acids (**11**) to the oxocarbonium ion (**12**). Arylsulfonylphthalides (**13**) are obtained in good yields. Reaction 2 is a two-step process for preparing benzenesulfonylphthalide (**14**) from phthalaldehydic acid (**10**) or its methyl derivative (**15**). Both (**10**) and (**15**) undergo reaction with thiophenol in the presence of PTSA in refluxing benzene, giving phenylthiophthalide (**16**) in good overall yield. MCPBA oxidation of (**16**) gives sulfone (**14**).[3] Reaction 3 is a direct conversion of 2-formylbenzamide (**17**) to its sulfone derivative (**18**). Treatment of (**1**) with sodium benzenesufinate in hot acetic acid afforded sulfone phthalide (**18**) in 80% yield.[4] The conversion (reaction 4) of *o*-toluate (**19**) to 3-phenylthiophthalide (**20**) is a unique one. It is limited to toluates with a methoxy group at C6. LDA-mediated lateral lithiation of (**19**) followed by diphenyl disulfide quenching gave dithioketal (**20**). Treatment of (**20**) with trifluoroacetic acid afforded (**21**) in excellent yield.[5] Direct preparation of phenylthiophthalides from 3-bromophthalides is not a common process, partly because of limited accessibility of the starting bromides. However, simple bromophthalide (**22**) was converted to phenylthiophthalide (**16**) in 74% by direct displacement of bromine by thiophenol. This method has been applied to 3-bromothiophthalides for the preparation of 3-(phenylthio)thiophthalides.[6] Similarly, indole analog of 3-bromophthalide was reacted with sodium benzenesulfinate in DMF to give its 3-phenylsulfonyl analog.[7]

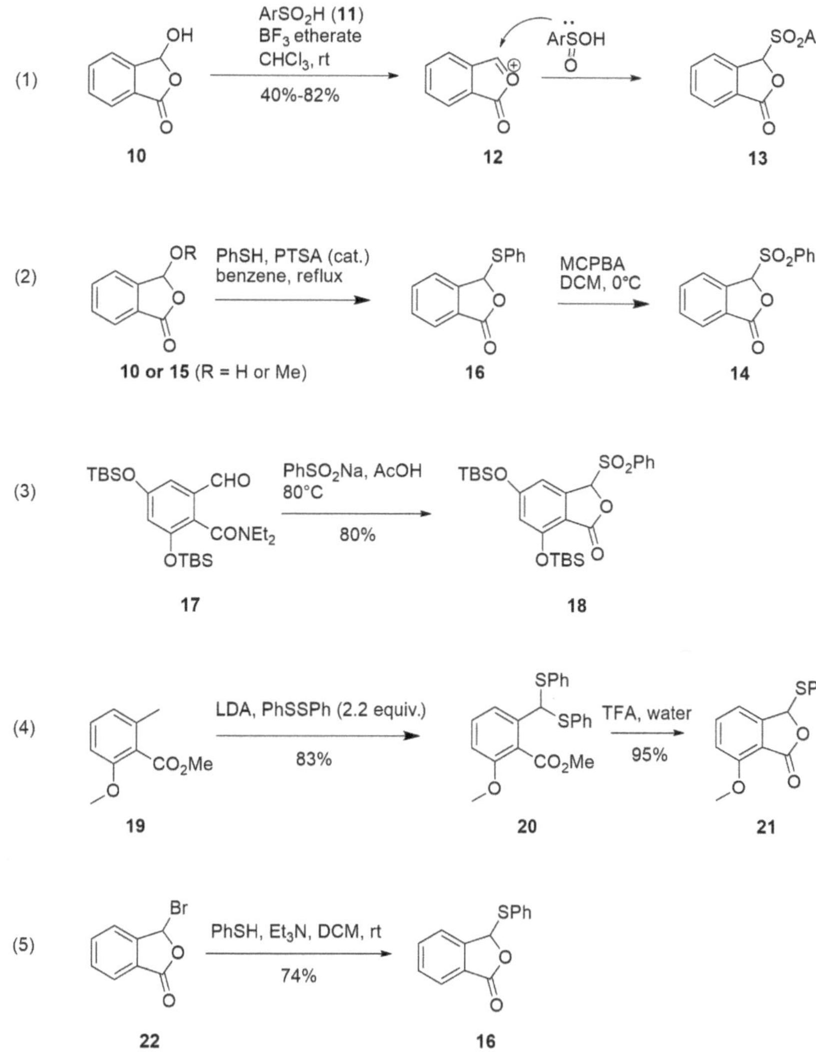

SCHEME 9.1 Synthesis of 3-phenylthiophthalides and 3-phenylsulfonylphthalides.

Regioselective aromatic formylation is an important way to the preparation of 3-methoxyphthalides. The reaction of benzoate (**23**) with dichloromethyl methyl ether in the presence of $TiCl_4$ followed by quenching with methanol afforded phthalide (**24**), which was transformed to sulfone phthalide (**25**) in two steps: phenylthiolation and oxidation (Scheme 9.2).[8]

Reutrakul synthesis of phenylthiophthalide (**26**) represents a distinguished example. The synthesis of precursor formyl ester (**27**) was a difficult task. Trimethoxybenzene carboxylate (**28**) resisted many standard formylations, which were based upon Friedel-Crafts reactions or *ortho*-metalations. The problem was solved by the use of phenylsulfanyldifluromethane as the electrophilic

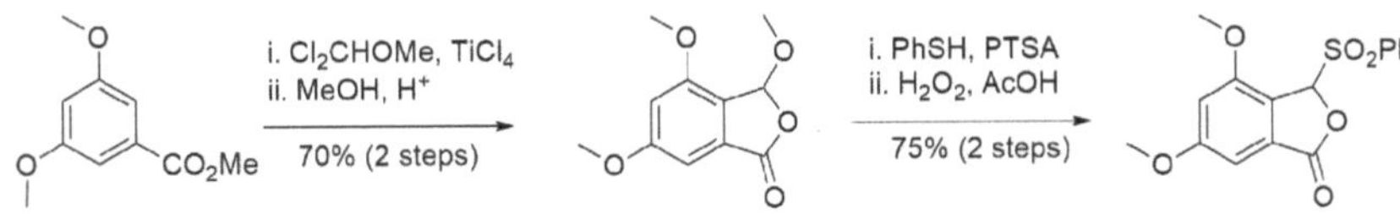

SCHEME 9.2 Synthesis of phthalide sulfone via formylation [1997 IJC 126 MaL D].

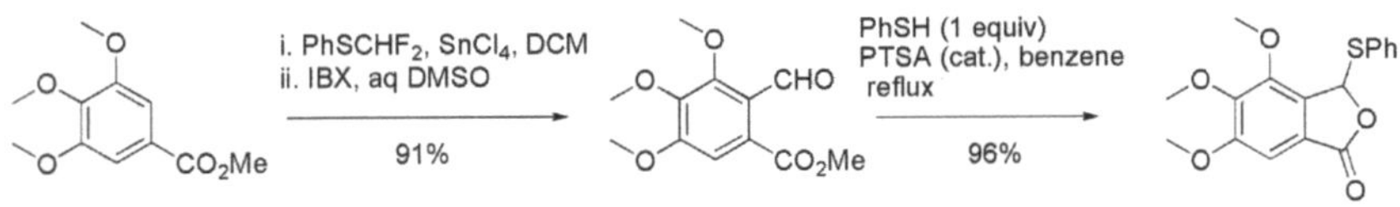

SCHEME 9.3 Reutrakul synthesis of a phenylthiophthalide via a new formylation [2012 PAC 1435 Reutrakul V].

agent followed by oxidative work-up with IBX. The overall yield of (**27**) was 91% (Scheme 9.3).[9] Condensation of (**27**) with thiophenol gave phthalide (**26**).

The Pummerer rearrangement has been found suitable for the synthesis of chromenone analog (**29**). Radical bromination of (**30**) with NBS in CCl_4 afforded bromo derivative (**31**). Its reaction with thiophenolate anion gave the corresponding thioether (**32**), which was hydrolyzed to give acid (**33**). MCPBA oxidation of acid (**33**) gave sulfoxide (**34**). This was then subjected to an acetic anhydride-promoted Pummerer rearrangement to furnish (**29**) via Pummerer intermediate (**35**) (Scheme 9.4).[10]

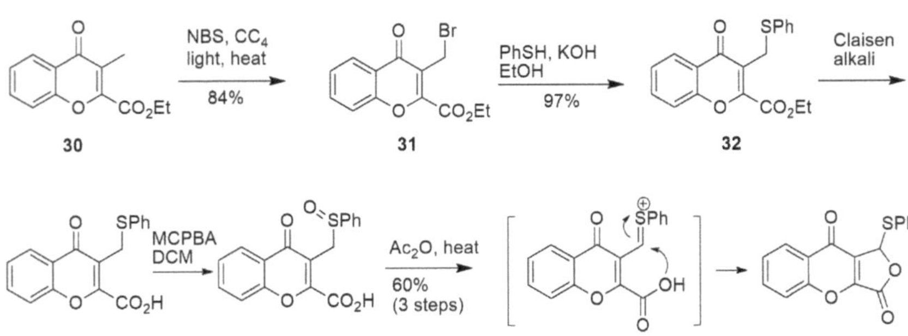

SCHEME 9.4 Synthesis of (phenylthio)chromenone by intramolecular Pummerer rearrangement [2003 ORS 3753 Hauser FM].

The synthesis of the only known 5-5 fused analog of the Hauser phthalide sulfone (**36**) relied upon an unconventional approach. Intermolecular cycloaddition of furanone (**37**) with benzonitrile oxide was the key step for the synthesis of isooxazoloisobenzofuranone (**38**). The loss of the resulting angular sulfone from (**38**) under the influence of silica gel gave the Hauser donor (**36**) (Scheme 9.5).[11]

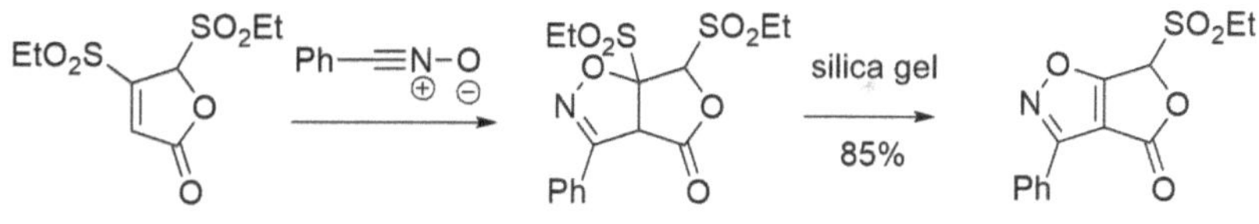

37 38 36

SCHEME 9.5 Synthesis of furoisoxazolone sulfone analogs [2000 HCS 1029 Paredes MC].

Tatsuta synthesis of UCE6 sulfone (**39**) represents a direct conversion of a 2-formylbenzamide to a phthalide sulfone. Resorcinol (**40**) was protected as its di-MOM ether and then carboxamidated via directed lithiation to give amide (**41**). Anionic formylation of (**41**) with DMF provided formyl derivative (**42**). Treatment of (**42**) with sodium benzenesulfinate in hot acetic acid provided sulfone phthalide (**39**). Under the conditions, one MOM group was partly removed, giving (**43**). This was protected with MOMCl chloride and Hunig base to furnish (**39**) (Scheme 9.6).[4]

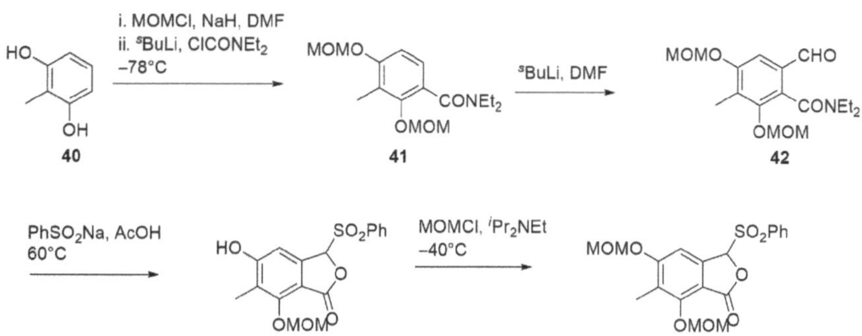

SCHEME 9.6 Tatsuta synthesis of UCE6 sulfone [2002 JABS 1076 Tatsuta K].

The synthesis of sulfone naphthalide (**44**) presented in Scheme 9.7 is a lengthy synthesis of a simple-looking molecule used in the total synthesis of galtamycinone. The key steps of the synthesis are selective bromination (**45** → **46**) and carboxylation (**46** → **47**). Aldehyde (**48**) was submitted to Stobbe condensation with dimethyl succinate (**49**) to form (**50**), which, on reaction with acetic anhydride, underwent an intramolecular Friedel-Crafts reaction to furnish naphthoate (**51**). LAH reduction of (**51**) gave carbinol (**52**). Swern-type oxidation of (**52**) afforded aldehyde (**45**). Selective bromination of (**45**) with NBS followed by *O*-methylation of the resultant bromo compound gave (**46**). Sinhababu carboxylation (BuLi, morpholine, CO_2) provided naphthaldehydic acid (**47**). The reaction of (**47**) with thiophenol in the presence of PTSA gave precursor sulfide, which was oxidized with MCPBA to furnish sulfone (**44**).

The synthesis of Suzuki's phthalide sulfone (**53**), used in the total synthesis of aquayamycin, is distinctly different from those discussed above. It does not involve commonly used phthalaldehydic acid precursors. It proceeded

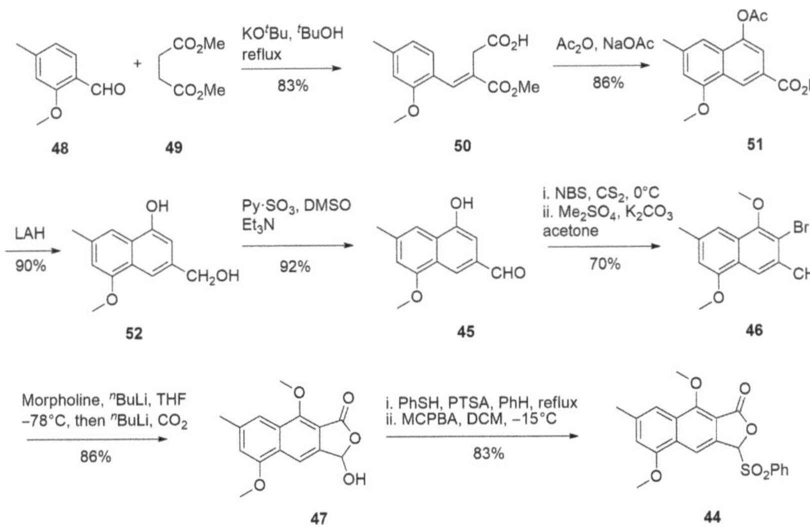

SCHEME 9.7 Synthesis of a 3-phenylsulfonylnaphthalide [Unpublished results Hauser FM].

through benzocyclobutane derivatives. The sequence features a regioselective [2 + 2] cycloaddition of benzyne with ketene silyl acetal, and an unusual Baeyer-Villiger reaction. Thus, the mixture of triflate (**54**) and acetal (**55**), when treated with ^{n}BuLi gave (**56**) as the single product. Desilylation of (**56**) generated ketone (**57**). $NaBH_4$ reduction of (**57**) followed by hydrolysis generated the corresponding hydroxy ketone, which was acetylated to give (**58**). Regioselective Baeyer-Villiger oxidation of (**58**) with MCPBA resulted in the formation of phthalide (**59**). Condensation of (**59**) with thiophenol in the presence of PTSA gave sulfide (**60**). MCPBA oxidation of (**60**) led to the preparation of the phthalide sulfone (**53**) in 93% yield (Scheme 9.8).[12]

The Hauser group, in their total synthesis of biphyscion, utilized a biphthalidesufone for double Hauser annulation. Its synthesis was based upon the lateral phenylsulfenation of bis-*o*-toluate (**61**), which was accessed through Cu-bronze-mediated Ullman coupling of monomeric toluate (**62**). Sulfenation (LDA, PhSSPh) of (**61**) gave tetrathiphenylated product (**63**). Treatment of (**63**) by TFA quantitatively afforded (**64**). MCPBA oxidation of (**64**) furnished bis(sulfonylisobenzofuranone) (**65**) in 100% yield (Scheme 9.9).[13]

9.1.2 Synthesis of Cyanophthalides

The applications and utilities of Hauser annulations crucially depend on the accessibility of cyanophthalide donors. They are logically derived from a 1,2-disubtitutedbenzene derivative. Functional group interconversions of the 1,2-substituents are interplayed to build the cyano δ-lactone motif of the

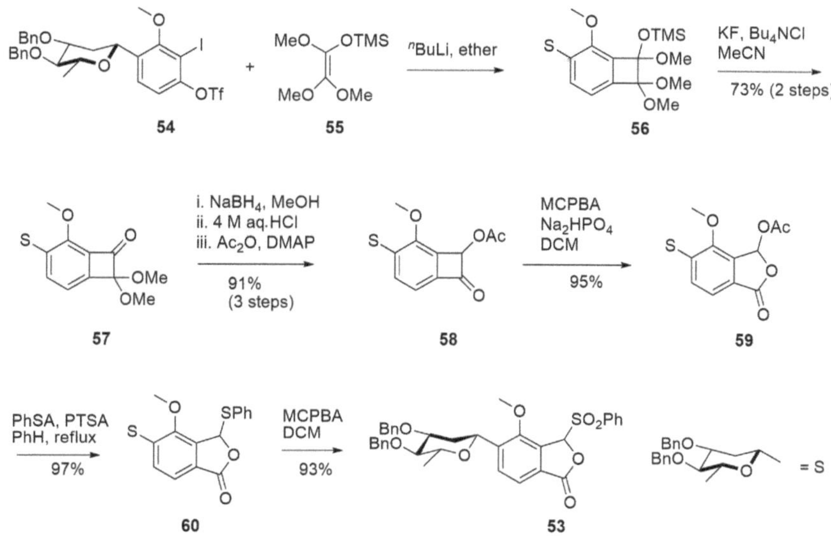

SCHEME 9.8 Suzuki synthesis of *C*-glycosidicphthalide sulfone [2000 THS 8383 Suzuki K].

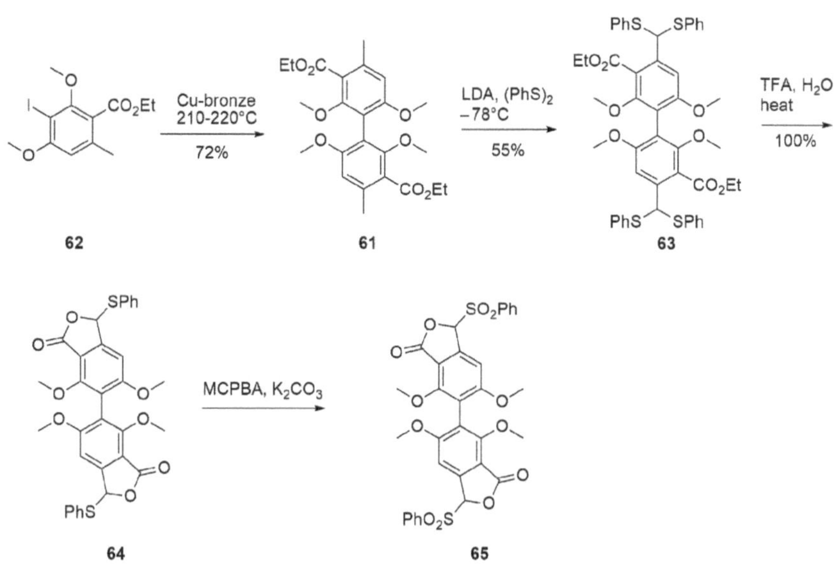

SCHEME 9.9 Synthesis of dimeric phthalide sulfone [1999 ORS 671 Hauser FM].

cyanophthalides (**66**) (Fig. 9.2). The starting compounds belong to the classes of *o*-toluates, phthalaldehydic acids, 2-formylbenzoates and *N,N*-dialkyl 2-formylbenzamides. Among the above classes, 2-formylbenzamides (**4**) are most frequently used, because they are easily prepared from benzamides in regiospecific manner by directed *ortho*-lithiation followed by formylation.

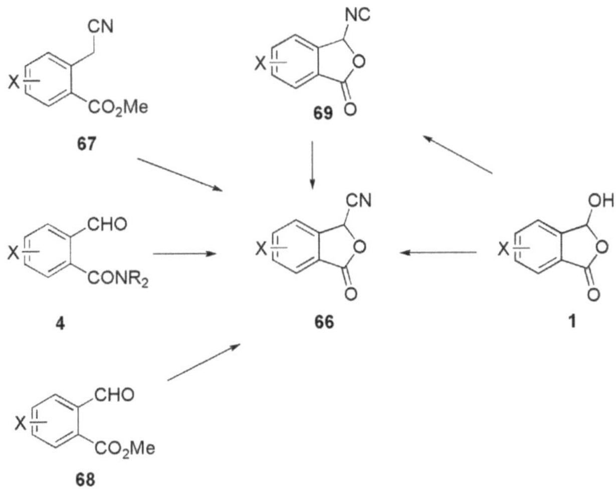

FIG. 9.2 Established precursors for the 3-cyanophthalides.

The methods existing in the literature for the preparation of 3-cyanophthalides are outlined in Fig. 9.2. There are essentially two kind of starting materials. They are 2-formylbenzoic acid derivatives (**1**), (**4**), and (**68**) and 2-cyanomethylbenzoates (**67**). Isocyanophthalides (**69**) were prepared from the phthalaldehydic acids **1**.

From 2-cyanomethylbenzoates: When the method of thermal lactonization was applied to cyanomethyltoluate (**70**), it resulted in the synthesis of cyanophthalide (**71**) in 49% overall yield (Scheme 9.10). Bromo derivative (**72**), prepared from (**70**) by benzylic bromination, on thermolysis gave phthalide (**71**) through intermediate (**73**).[14]

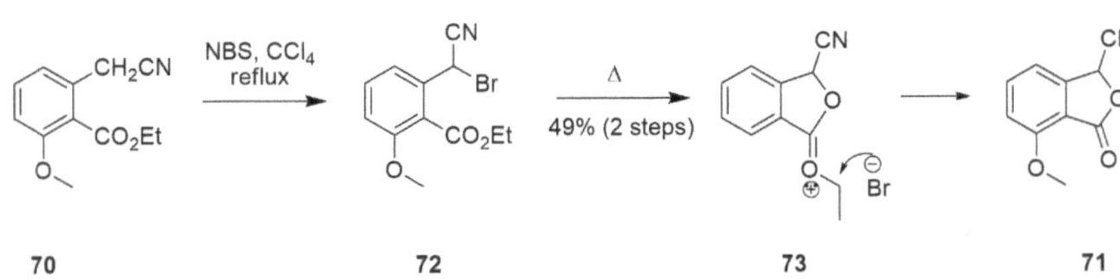

SCHEME 9.10 Kraus synthesis of 3-cyanophthalides [1983 JOC 3439 Kraus GA].

From 2-formylbenzoates: The *ortho*-formylbenzoates (**68**) can be converted into cyanophthalides (**66**) in one pot by nucleophilic addition of a cyanide ion followed by spontaneous intramolecular lactonization of the intermediate. For the synthesis of 3-cyanophthalide (**74**), a mixture of methyl 2-formylbenzoate (**75**), acetone cyanohydrin, and catalytic triethylamine is stirred at rt. to give first (**76**) and then the product (**74**).[15a] In the presence of triethylamine, acetone cyanohydrin generates $^{-}$CN under the reaction conditions. If the base is used in excess, a dimeric phthalide becomes a major product (Scheme 9.11).

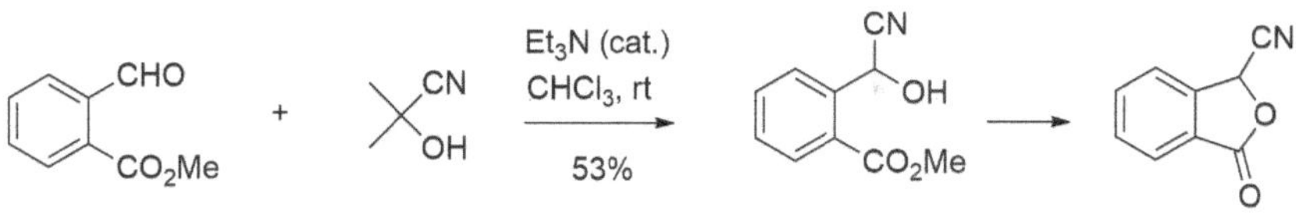

75 76 74

SCHEME 9.11 *Trans* cyanohydrin formation route [2010 THS 75 Ramstrom O].

Hauser et al. have shown the *ortho*-formylbenzoates (**68**) can be transformed smoothly to cyanophthalides (**66**) in >90% yields by the action of Et_2AlCN in DCM at room temperature. As shown in Scheme 9.12, the method encompasses a wide variety of substrates.

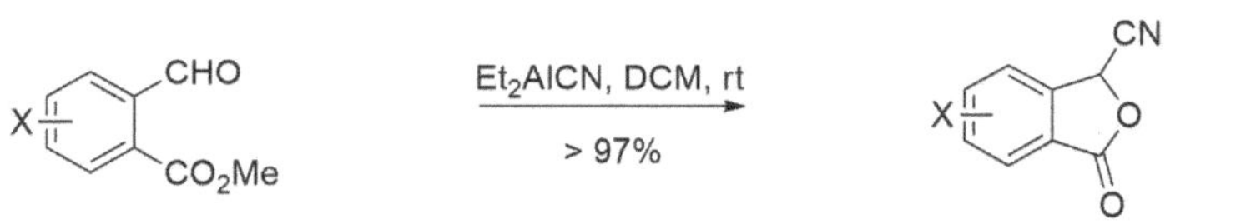

X = 2-CO_2Me, 3-CH_2SPh, 3,6-$(OMe)_2$, 3,4,6-$(OMe)_3$, naphthopyranone ring, etc.

68 66

SCHEME 9.12 Et_2AlCN-promoted synthesis of cyanophthalides from formyltoluates.

Although the method appears to be general, it did not work with tricyclic carboxylate (**77**). However, its reaction with TMSCN with catalytic ZnI_2 in DCM followed by CF_3CO_2H resulted in cyanophthalide (**78**) in good yield (Scheme 9.13).[15b]

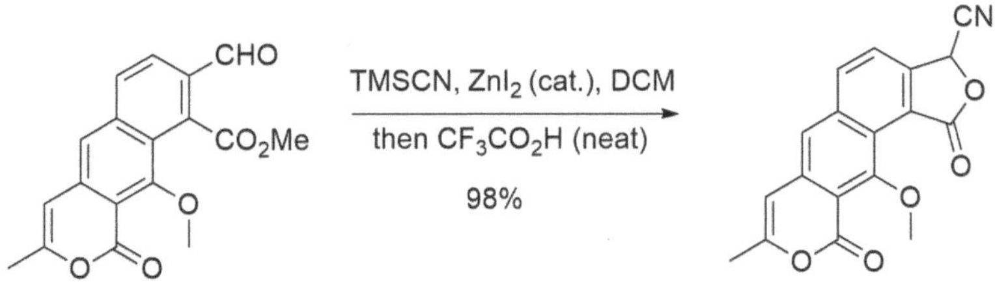

77 78

SCHEME 9.13 TMSCN-promoted synthesis of cyanophthalides [Unpublished results Hauser FM].

From Phthalaldehydic Acids

One of the attractive classes of precursors for the preparation of 3-cyanophthalides (**66**) is phthalaldehydic acids (**1**) (Scheme 9.14). These can readily be obtained by hydrolysis of 2-dialkoxymethyl benzoates, 2-dibromomethyl benzoates, 2-formylbenzoates, 2-oxazolyl benzoates, and 3-bromophthalides. Although the parent cyanophthalide (**74**) in Hauser annulation was reported by the Kraus group,

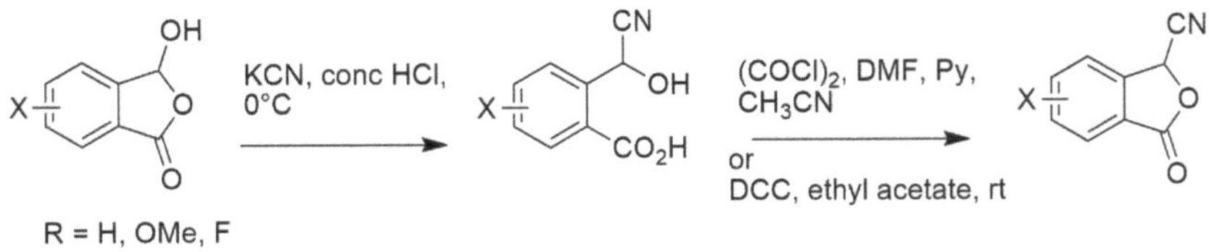

1 79 66

SCHEME 9.14 Swenton-Russell synthesis of 3-cyanophthalides.

the details of its preparation were not published until 1985, when the Swenton group disclosed the detailed study on the preparation of cyanophthalides (**66**). The reaction of KCN with phthalaldehydic acids (**1**) in the presence of aq HCl solution at 0°C furnishes cyanohydrins (**79**). Treatment of the cyanohydrins (**79**) with Vilsmeier salt (oxalyl chloride and DMF) followed by pyridine gives the corresponding cyanophthalides (**66**). Russell et al. modified the method by introducing DCC for cyclodehydration of cyanohydrins (**79**).[16]

Yields are sensitive to the substituents, especially to methoxy groups. For the synthesis of 4,7-dimthoxycyanophthalide (**80**), the additional step—i.e., cyclization of cyanohydrins by Vilsmeier reagent—is not required. It was formed in excellent yields from phthalaldehydic acid (**81**) under the conditions for formation of cyanohydrins (**79**) (Schemes 9.15 and 9.16).[17]

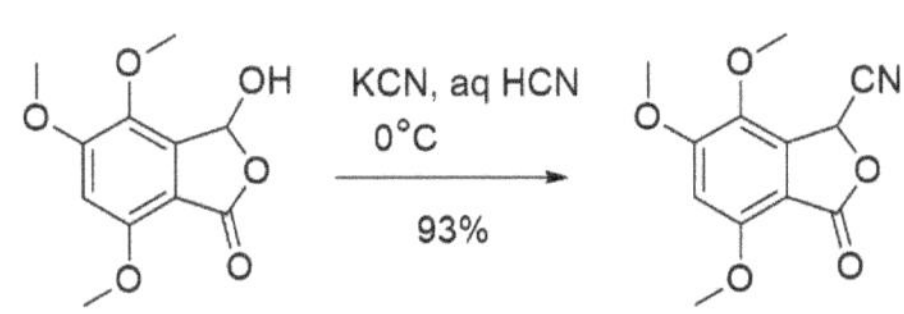

81 80

SCHEME 9.15 Direct conversion of phthalaldehydic acids to cyanophthalides [1988 JOC 5519 Bach RD].

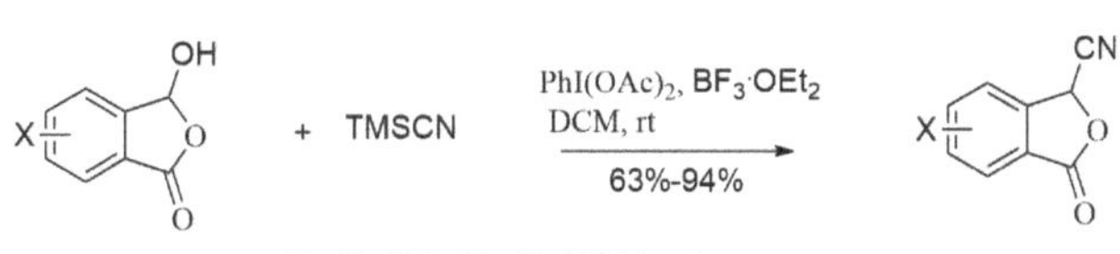

1 66

SCHEME 9.16 Mal synthesis of 3-cyanophthalides from phthalaldehydic acids [2016 CST 3097 MaL D].

The Mal protocol is an interesting addition to cyanophthalide preparation from phthalaldehydic acids. The reaction of phthalaldehydic acids (**1**) with TMSCN, phenyliodonium diacetate and BF_3 etherate in DCM at room temperature afforded cyanophthalides (**66**).[18]

From o-Formylbenzamides

One of the two most attractive methods of obtaining 3-cyanophthalides (**66**) is transformation of the readily accessible *o*-formylbenzamides (**4**) by the method of Yoshii et al. Treatment of the *N,N*-diethyl benzamides (**4**) with a mixture of TMSCN, KCN, and 18-crown-6 in DCM gives corresponding silyloxycyanides (**82**), which, upon dissolving in acetic acid, provide the desired cyanophthalides (**66**) in excellent yields (Scheme 9.17).[19] This method is adopted by most researchers.

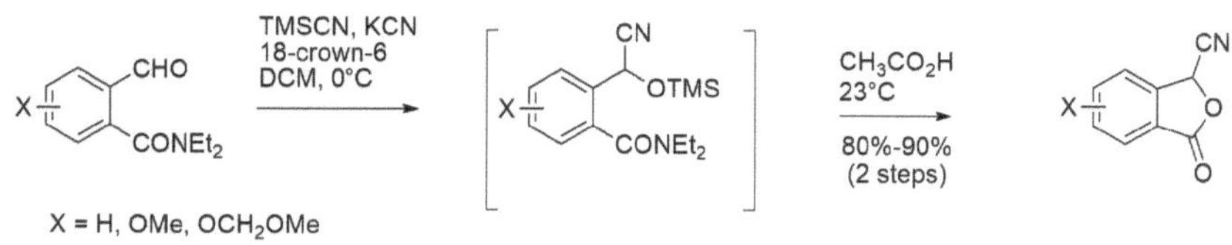

4 82 66

SCHEME 9.17 Yoshii synthesis of 3-cyanophthalides.

The conversion (**4** → **66**) can also be performed by Hauser's conditions, i.e., Et_2AlCN-DCM with comparable efficacies. Alternatively, the formylbenzamides can be converted into cyanophthalides (**66**) by treating them with an equimolar mixture of sodium cyanide and *p*-TSA in aqueous THF at 0°C to rt. While this method is excellent for the synthesis of 7-methoxy-3-cyanophthalide, for most other functionalized amides, the yields of the cyano derivatives are variable. The Yoshii method has proven very useful for the synthesis of cyanophthalides (**83**) from the commercially available 2-bromobenzaldehydes (**84**) (Scheme 9.18). The required 1,3-dioxolanes (**85**) were obtained by reacting the aldehydes (**84**) with ethylene glycol in

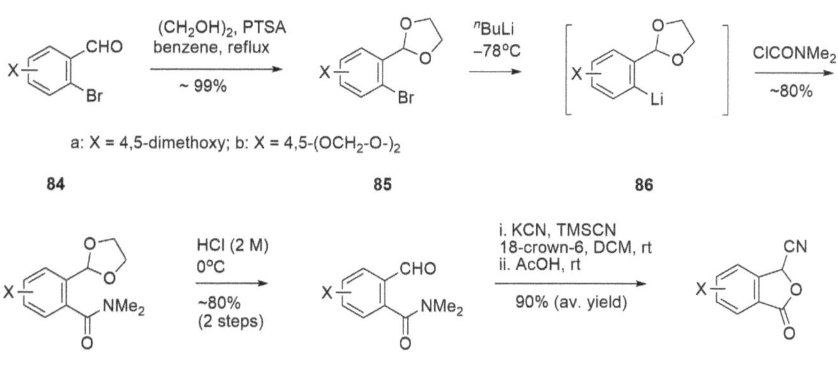

84 85 86

87 88 83

SCHEME 9.18 Preparation of cyanophthalides from *ortho*dioxolanyltoluamides [2017 EJOC 1704 Ham J].

the presence of PTSA in refluxing benzene. Nuclear lithiation of the bromoarenes (**85**) with ^{n}BuLi to lithiated benzene (**86**), followed by its amidation with *N*,*N*-dimethylcarbamoylchloride, gave amides (**87**). Deprotection of the 1,3-dioxolanes in (**87**) by HCl treatment provided 2-formylbenzamides (**88**) in good yields. Their transformation to the cyanophthalides (**83**) was achieved by the Yoshii method, which involved reaction with trimethylsilyl cyanide in the presence of KCN and 18-crown-6, followed by heating in acetic acid (Scheme 9.18).[20]

Lastly, the application of Yoshii cyanophthalide preparation to phthalide starting material is illustrated with the synthesis of 6-amino-3-cyanophthalide (**89**). The starting 6-aminophthalide (**90**) was protected as its Boc derivative (**91**) by treatment with $(Boc)_2O$. $AlCl_3$-Et_2NH-promoted lactone opening of (**91**) gave hydroxymethylbenzamide (**92**). Oxidation of (**92**) with PDC furnished formylbenzamide (**93**). Treatment of (**93**) under Yoshii conditions afforded cyanophthalide (**94**). Removal of the Boc group in (**94**) by the action montmorillonite K (**10**) furnished 6-amino-3-cyanophthalide (**89**) in the good overall yield (Scheme 9.19).[21]

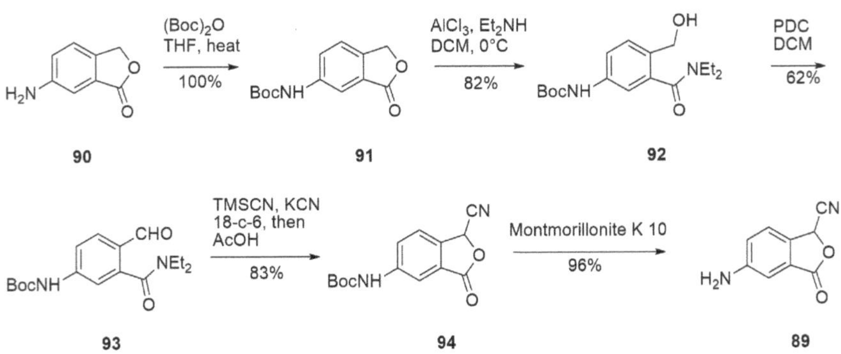

SCHEME 9.19 Synthesis of 6-amino-3-cyanophthalide (**89**) by the Yoshii method [2016 JACS 8235 Nicolaou KC].

Mal et al. have developed a three-step sequence for the conversion of phthalaldehydic acids (**1**) to 3-cyanophthalides (**66**) without using toxic cyanide. The sequence started with preparation of formylaminophthalides (**95**) by treatment of (**1**) with neat formamide. The formamides (**95**) are dehydrated with $POCl_3$-Et_3N to isocyanides (**96**), which are conveniently isomerized to cyanophthalides (**66**) with 2,6-lutidine-Tf_2O in moderate overall yields (Scheme 9.20).[22]

Phthalide-3-phosphonates are prepared from phthalaldehydic acids, 2-formylbenzamides, and 2-formylbenzoates (Scheme 9.21).[23] The reaction of 3-hydroxyphthalides (**1**) with sodium dimethyl phosphonates in methanol gave phthalide-3-phosphonates (**97**) in good yields (reaction 1). In an improved version, Watanabe and Furukawa[23] utilized phthalaldehydic amides (**4**), readily accessible from the corresponding benzamides. Phthalaldehydic

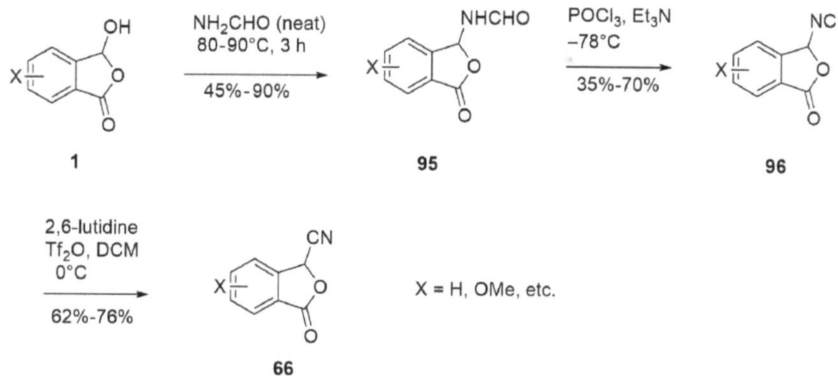

SCHEME 9.20 Synthesis of 3-isocyano and cyanophthalides [2015 SYS 2473 MaL D].

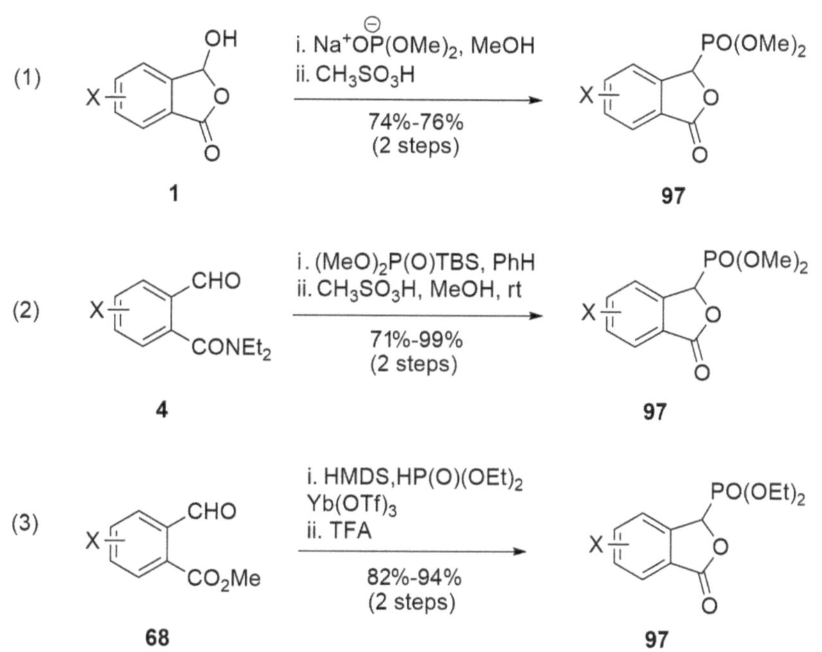

SCHEME 9.21 Synthesis of phthalide-3-phosphonates [1985 SYS 38 Napolitano E; 1991 SYT 481 Watanabe M; 2017 THS 1985 Jang DO].

amides (**4**), on reaction with *tert*-butyldimethylsilyl dimethyl phosphite followed by desilylation and cyclization using methanesulfonic acid at room temperature, were shown to give phthalide-3-phosphonates (**97**) in 71%–99% yields (reaction 2). The latest method (reaction 3) is a one-pot reaction of 2-formylbenzoates, hexamethyldisilazane, and diethyl phosphite in the presence of $Yb(OTf)_3$ to deliver (**97**).

9.1.3 Synthesis of Phthalides

The synthetic chemistry of phthalides has been exclusively detailed in the review by Mal al. There are many ways in which phthalides can be synthesized. These include preparations from monosubstituted benzenes, disubstituted benzenes, phthalaldehydic acids, and phthalic anhydrides. For the sake of simplicity, a few selected methods are highlighted in the following reaction schemes.

From 2-Bromomethyl Benzoates

Thermal decomposition of ω-bromo *o*-toluates is a classical method and convenient preparation of phthalides. For example, bromomethylbenzoate (**98**), obtained NBS bromination of (**99**) in refluxing CCl_4, cyclizes on thermolysis at 160°C and furnishes phthalide (**100**) via elimination of CH_3Br. The reaction is believed to be assisted by neighboring group participation (Scheme 9.22).[24]

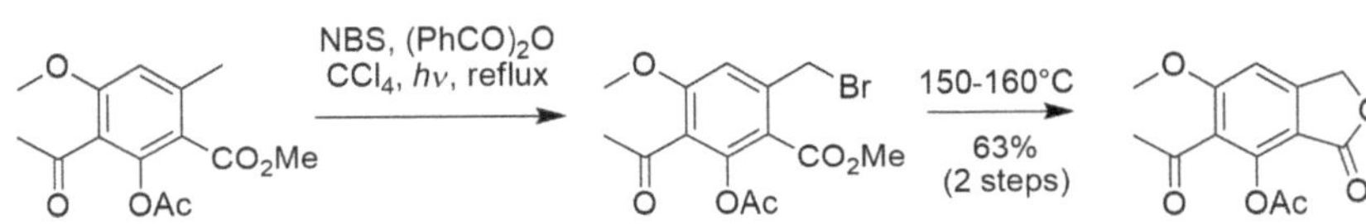

99 **98** **100**

SCHEME 9.22 Synthesis of an isobenzofuranone via thermolysis of 2-bromomethylbenzoate.

Reduction of 2-formylbenzoate is a standard technique to arrive at a phthalide. A Duff reaction of (**101**) gave (**102**), the $NaBH_4$ reduction of which furnished phthalide (**103**) (Scheme 9.23).[25]

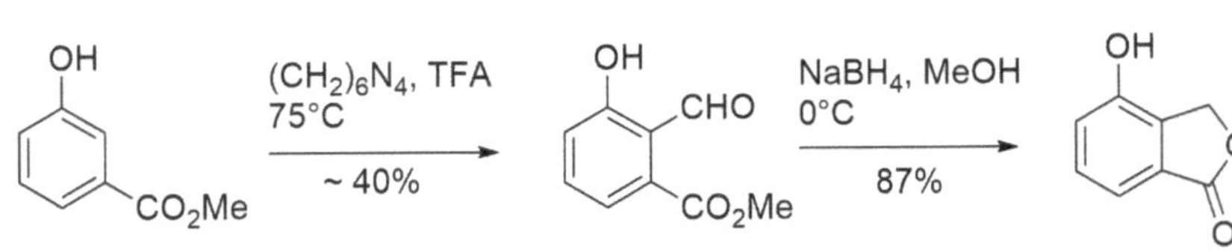

101 **102** **103**

SCHEME 9.23 Synthesis of a phthalide 3-hydroxybenzoate [2013 EJOC 4037 MaL D].

Chemoselective reduction of 1,2-diesters is suitable for preparing naphthalides. In the example cited in Scheme 9.24, the diester (**104**), available by annulation between (**105**) and (**106**), could be selectively reduced with LAH to naphthalide (**107**). The selectivity stems from hydroxy-directed delivery of hydride ions (Scheme 9.24).[26]

Likewise, selectivity can be achieved in an oxidative lactonization. Oxidation of phthalyl alcohol (**108**), obtained by LAH reduction of the corresponding diester, by manganese dioxide or barium permanganate, took place to give phthalide (**109**) selectively (Scheme 9.25).[27]

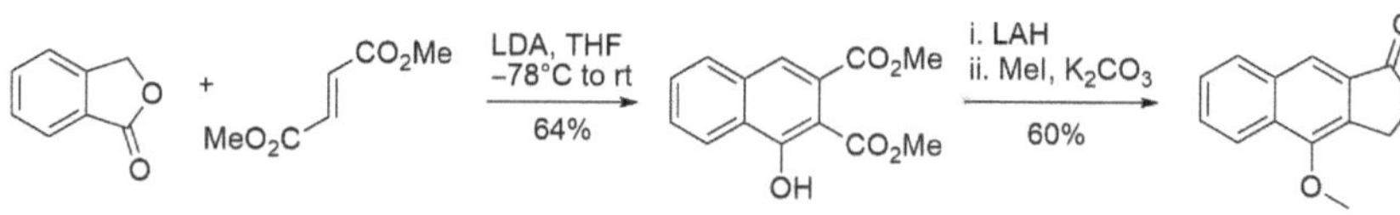

105 106 104 107

SCHEME 9.24 Synthesis of a linear naphthalide by selective reduction [2015 OBC 6344 MaL D].

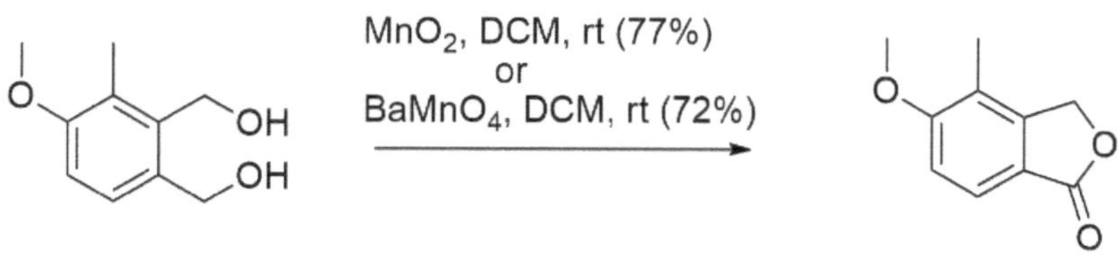

108 109

SCHEME 9.25 MnO_2 oxidation of a phthalyl alcohol to a phthalide [1980 JHC 315 Popp FD].

One of remarkable innovations in lactonizations is Yu heteroannulation. It is the direct transformation of an aromatic carboxylic acid to the corresponding lactone. In 2009, Yu et al. reported the annulation of benzoic acids (**110**) with dibromomethane leading to the formation of phthalides (**111**) (Scheme 9.26). Alkylation of C–H bonds takes place at a position *ortho* to COOH in the presence of Pd(II)-catalyst and K_2HPO_4 as the base. Dibromomethane as the alkylating agent gave higher yields and also showed greater substrate scope, compared to dichloromethane. Electron-withdrawing halogens and trifluoromethyl, keto, and ester groups are well tolerated. The exclusive monoselectivity at the less hindered *ortho*-position was explained by the rapid lactone formation of arylpalladium species like (**112**).[28]

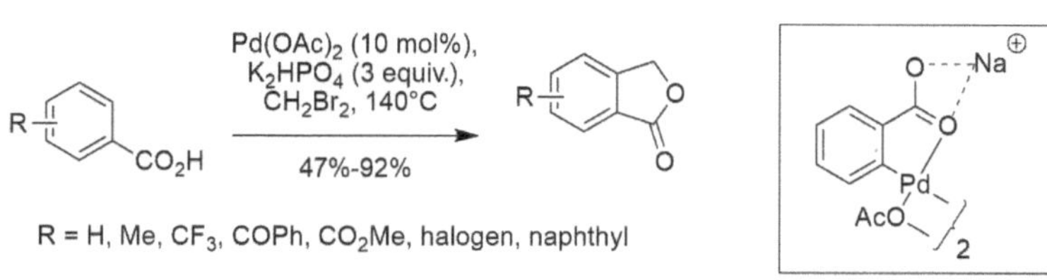

110 111 112

SCHEME 9.26 Yu lactonization route to phthalides [2009 ACIE 6067 Yu, J.-Q].

The Wong approach to phthalides makes use of reductive aromatization of oxabicyclo [2.2.1] heptadienes. For example, adduct (**113**) derived from furfurylalcohol (**114**) and dimethyl acetylene dicarboxylate was deoxygenated with $LiAlH_4$-$TiCl_4$ to produce *o*-hydroxymethylbenzoate, which, in turn, cyclized to phthalide (**115**) (Scheme 9.27).[29]

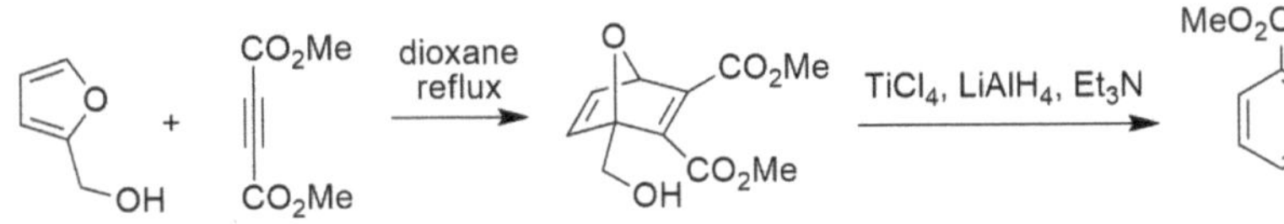

114 113 115

SCHEME 9.27 Wong Diels-Alder approach to phthalides [1984 SYS 787 Wong HNC].

The phthalaldehydic acids are useful precursors for the synthesis of phthalides. For example, phthalaldehydic acid (**116**), obtained in a two-step, one-pot operation from *o*-bromobenzaldehyde (**117**), on reduction with $NaBH_4$ in the presence of TMSOTf, furnished phthalide (**118**) (Scheme 9.28).[30]

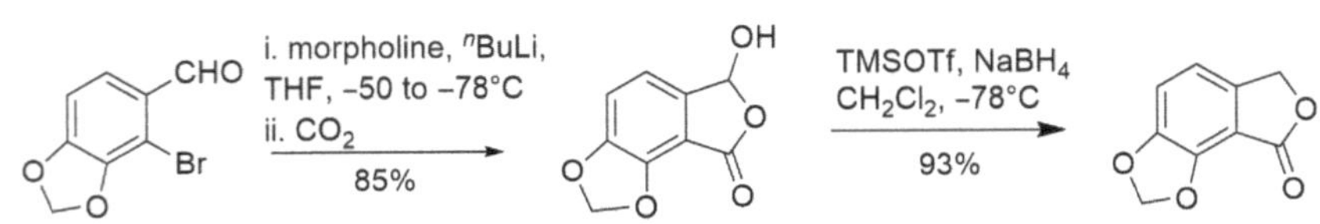

117 116 118

SCHEME 9.28 Phthalide by reduction of a phthalaldehydic acid.

Methoxy-directed *ortho*-lithiation-carboxylation of benzyl alcohols is a one-step route to phthalides. Treatment of alcohol (**119**) with ^{n}BuLi formed a lithiated intermediate (**120**), which on immediate quenching with CO_2 and acid-catalyzed lactonization produced phthalide (**121**) (Scheme 9.29).[31]

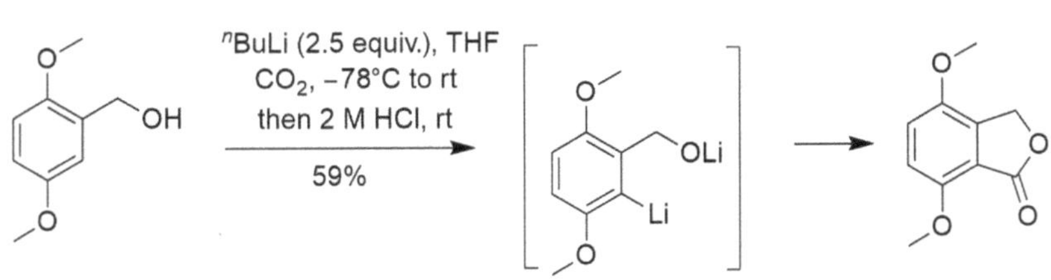

119 120 121

SCHEME 9.29 Lithiation-carboxylation of benzyl alcohols.

Solid $Mo(CO)_6$ is a popular substitute for gaseous CO. Thus, cyclocarbonylation of *o*-bromobenzyl alcohols (**122**) with $Mo(CO)_6$ in the presence of $Pd(OAc)_2$, dppf, DMAP, and DIEA yielded corresponding phthalides **123**) (Scheme 9.30). Bromobenzyl alcohols with both electron-rich and electron-deficient groups were used as substrates. Under the conditions, an angular phthalide (**124**) (Scheme 9.30) was also prepared from (1-bromo-2-naphthyl) methanol in 92% yield.[32]

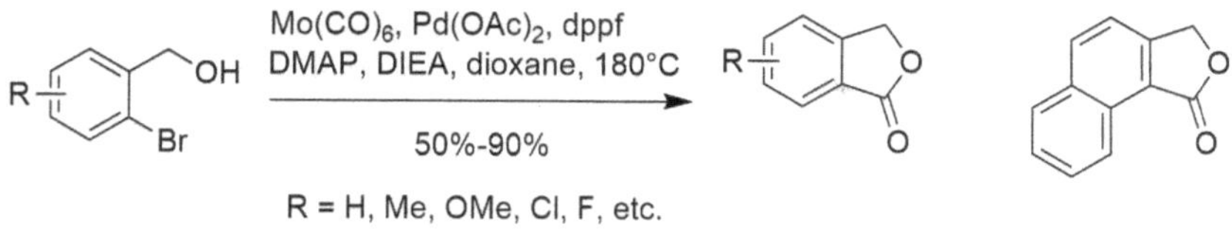

122 **123** **124**

SCHEME 9.30 $Mo(CO)_6$-mediated cyclocarbonylation of *o*-bromobenzyl alcohols to phthalides.

Transition metal catalyzed [2+2+2] cycloadditions are a common process for benzannulations. They are usually triggered by Ni, Pd, Rh, and Ru catalysts. Angular phthalide (**125**) was formed in one step by Ni(0)-catalyzed intramolecular cyclotrimerization of 2-ynyl ester of nona-2,8-diynoic acid (**126**) (Scheme 9.31).[33]

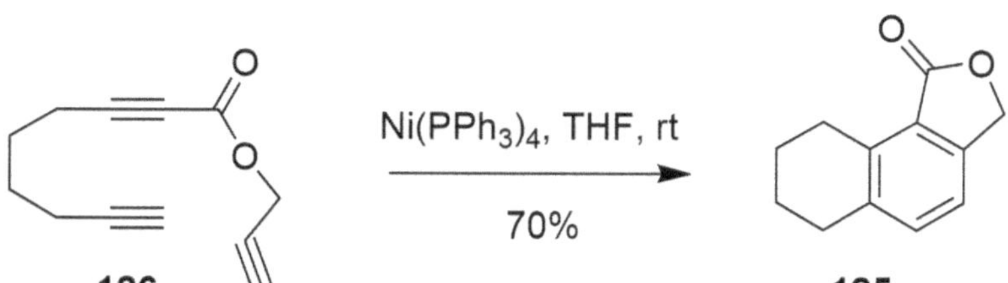

126 **125**

SCHEME 9.31 Ni(0)-catalyzed [2+2+2] cyclotrimerization to angular phthalide.

Contrary to the previous methods, the synthesis of furoindolone (**127**) starts from the preformed γ-lactone (**128**), which is readily available. The Japp-Klingemann reaction of (**128**) gave hydrazone (**129**) in 98% yield. Fischer indolization of (**129**) in the presence of TFA in acetic acid provided furoindolone (**127**) in 40% yield (Scheme 9.32).[34]

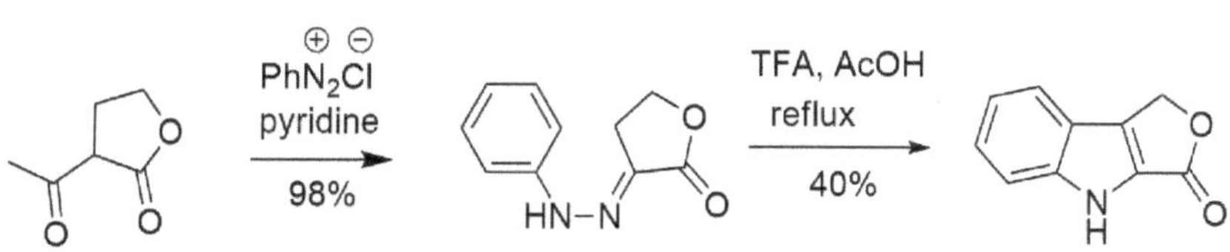

128 **129** **127**

SCHEME 9.32 Synthesis of furoindolone (**127**) by Fischer indolization.

9.1.4 Synthesis of Homophthalic Anhydrides

The parent homophthalic anhydride is prepared by oxidative cleavage of indene, followed by cyclodehydration of the resulting acid. A homophthalic anhydride may be prepared from anisole. Anisole (**130**) was selectively oxidized to *o*-toluic acid (**131**). Lateral lithiation of (**131**) with LDA followed by dimethyl carbonate quenching gave the corresponding homologated ester, which was hydrolyzed to diacid (**132**). Cyclodehydration of (**132**) with acetic anhydride furnished (**133**) (Scheme 9.33).[35]

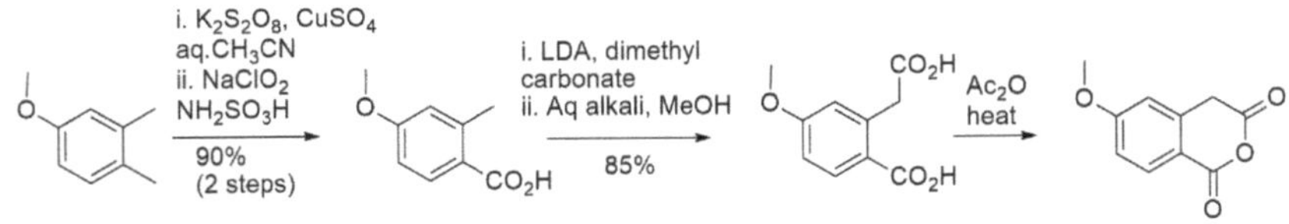

130 **131** **132** **133**

SCHEME 9.33 Homophthalic anhydride from an anisole [2002 ORS 2237 Hauser FM].

Homophthalic anhydride (**134**) was prepared by employing oxazoline-directed lithiations. Two sequential lithiations of (**135**) followed by MeI and $ClCO_2Me$ quenching afforded (**136**). Hydrolysis of the oxazolines and the ester in (**136**) produced the corresponding diacid, which was cyclodehydrated with acetyl chloride to furnish anhydride (**134**) (Scheme 9.34).[36]

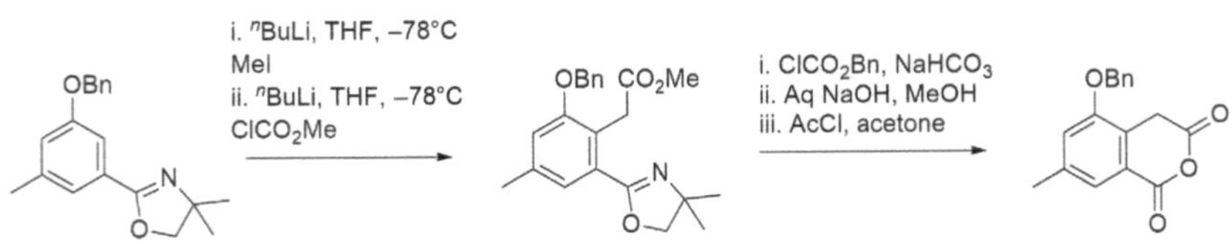

135 **136** **134**

SCHEME 9.34 Homophthalic anhydride from an oxazoline [1997 THN 16533 Suzuki K].

The Kita method for the preparation of cyclic anhydride using (trimethylsilyl)ethoxyacetylene offers an elegant alternative. It takes place under neutral conditions. Treatment of homophthalic acids with (trimethylsilyl)ethoxyacetylene in inert solvents like DCM or acetonitrile gives homophthalic anhydrides. It does not entail any aqueous work-up. The reaction of homophthalic acid (**137**) with (trimethylsilyl)ethoxyacetylene is believed to give an intermediate (**138**) on the way to homophthalic anhydride (**139**) (Scheme 9.35).[37]

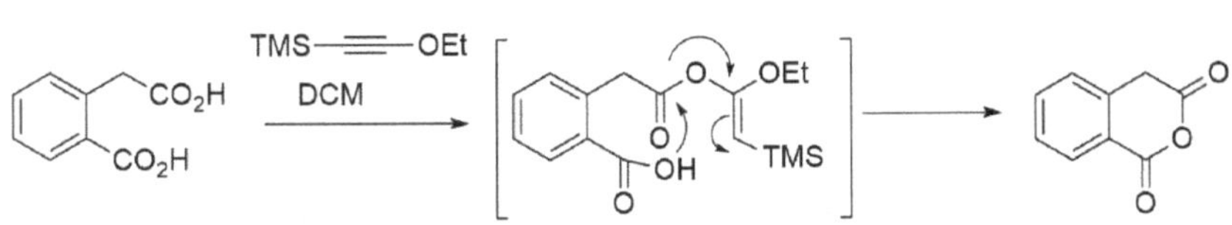

137 **138** **139**

SCHEME 9.35 Kita anhydride preparation [1986 JOC 4150 Y Kita].

Kita's method was successfully used to synthesize a pyridine analog of homophthalic anhydride. Lateral lithiation of (**140**), followed by quenching the

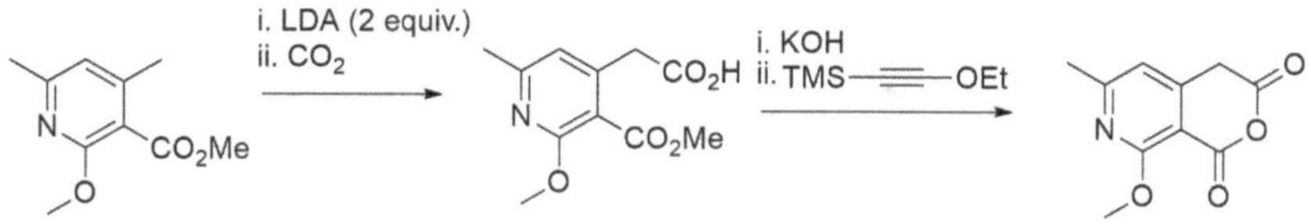

140 141 142

SCHEME 9.36 Synthesis of a pyridine analog of homophthalic anhydride [2000 JOC 89 Y Kita].

lithiated species with CO_2, gave (**141**). Alkaline hydrolysis of (**141**) gave the corresponding diacid, which was treated with the Kita reagent, to afford anhydride (**142**) (Scheme 9.36).[38]

Benzyne chemistry and Kita anhydride formation were combined to synthesize anhydride (**143**). The reaction between bromoarene (**144**) and dimethyl malonate in the presence of LTMP directly provided homophthalate (**145**). It is presumed to occur by the addition of lithium dimethylmalonate to the benzyne of (**144**), followed by intramolecular carboxylation of the resulting aryl anion. Alkaline hydrolysis of (**1450** gave diacid (**146**), which was cyclized by the Kita reagent to afford anhydride (**143**) (Scheme 9.37).[39]

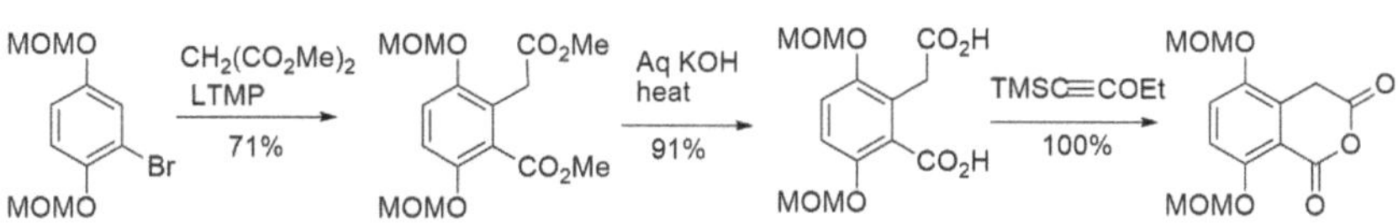

144 145 146 143

SCHEME 9.37 Homophthalic anhydride from a bromoarene [1995 ACIE 1721 Danishefsky SJ].

9.1.5 Synthesis of *o*-Toluates

For the majority of the donors of [4+2] annulations discussed in the preceding four chapters, *o*-toluates serve directly or indirectly as key precursors. Their chemistry is already highlighted. The following examples are considered special cases of toluate derivatives. The sulfoxide (**147**) was employed as the Hauser donor in the synthesis of lomaiviticin subunit. It was derived from aldehyde (**148**). Lithiation of (**148**) followed by methylation with methyl sulfate gave (**149**). Pinnick oxidation of (**149**) delivered acid (**150**). This was converted into its phenyl ester (**151**) through its acid chloride. NBS bromination of (**151**) followed by displacement of the bromine of the resultant bromide (**152**) by phenylthio group furnished (**153**). $NaIO_4$ oxidation of (**153**) provided the required sulfoxide (**147**) (Scheme 9.38).[40]

The Mootoo synthesis of *bis o*-toluate (**154**) centered on Cl_2CHOMe-mediated formylation. The required biphenyl (**155**) was prepared by Ullmann coupling of (**156**). Formylation of (**155**) with Cl_2CHOMe in the presence of $TiCl_4$ gave (**157**). Pinnick oxidation of (**157**) afforded the corresponding diacid, which was methylated with trimethylsilyldiazomethane to give (**158**).

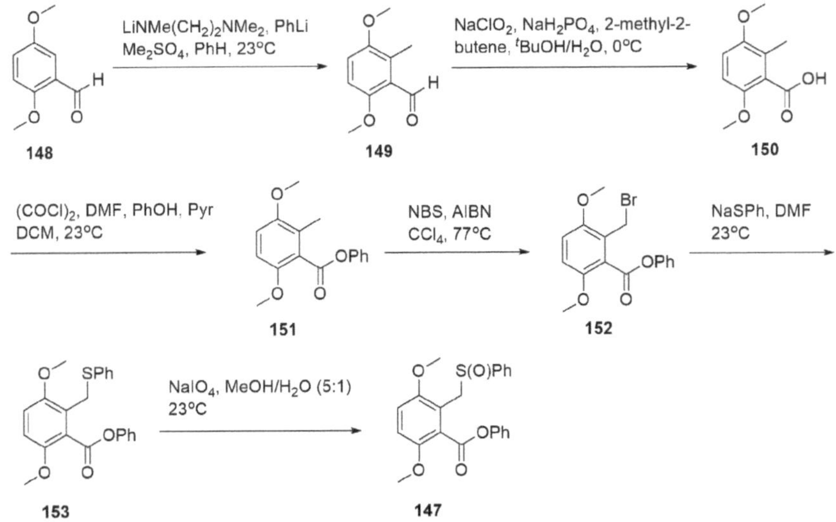

SCHEME 9.38 Synthesis of lomaiviticin precursor [2010 CEJ 13058 Shair MD].

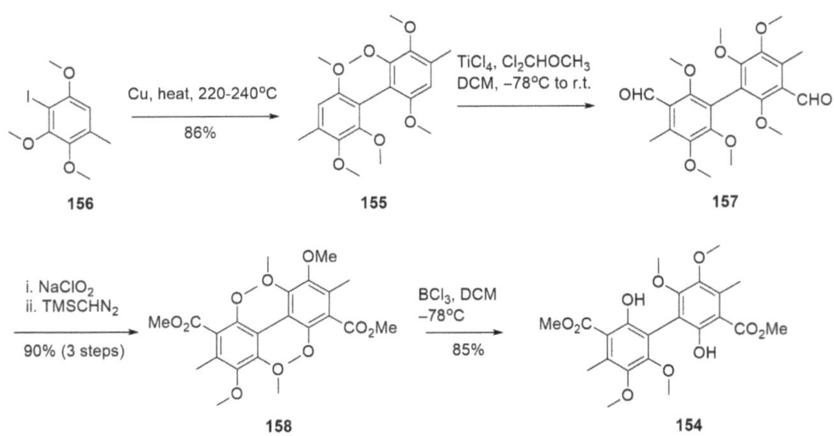

SCHEME 9.39 Mootoo bis-toluate donors [2013 SYS 2287 Mootoo DR].

BCl_3-promoted selective demethylation of (**158**) yielded bistoluate (**154**) (Scheme 9.39).[41]

The synthesis of the Shair fluoride donor (**159**) for the total synthesis of HMP-Y1 represents application of (i) oxidative dimerization of organolithiums, (ii) bromine-lithium exchange, and (iii) lateral fluorination. Directed lithiation of (**160**) and subsequent oxidation of the lithio species with $FeCl_3$ gave the dimer (**161**). Bromination of (**161**) with Br_2 followed by Br-Li exchange and $ClCO_2Me$ quenching provided (**162**). Introduction of fluorine at the lateral position of (**162**) was accomplished by lithiation with LTMP, followed by treatment with $(BrCF_2)_2$ to furnish (**159**) (Scheme 9.40).[42]

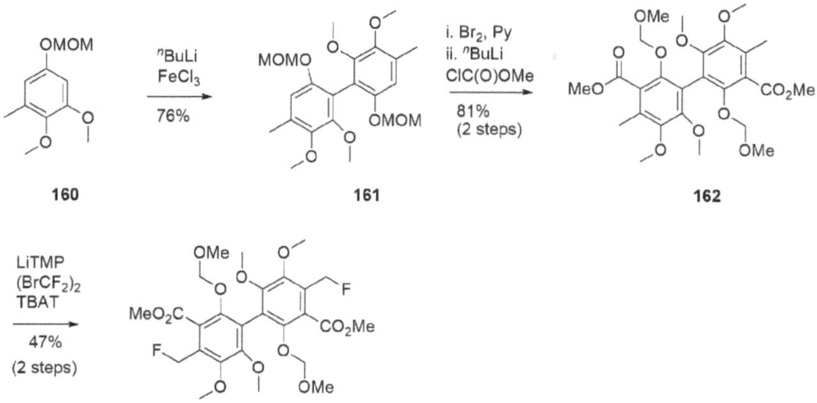

160 161 162 159

SCHEME 9.40 Synthesis of Shair fluoride donor [2012 JACS 16765 Shair MD].

9.2 REACTION CONDITIONS AND COMPARATIVE REACTIVITIES OF [4+2] ANNULATIONS

This section deals with reaction conditions, scope, and limitations of [4+2] annulations of phthalides, homophthalic anhydrides, and toluates. While Hauser phthalide annulation allows synthesis of 1,4-naphthoquinols, the Sammes annulation, Tamura annulation, and Hauser sulfoxide provide 1-naphthols. Myers annulation delivers 3,4-dihydro derivatives of 1-naphthols. Collectively, they can be classified as naphthannulation. These annulations are sensitive to one or more of four factors: (i) bases, (ii) donors, (iii) acceptors, and (iv) combination thereof.

The nature and amount of a base could be decisive factors for the success of an annulation. For Hauser phthalide annulations, the stipulated ratio of the donor to base is 1:1. Although this ratio works well to give the annulation products, the ratio of 1:3 is found to be much better for higher yields. Three-fold excess of the base is justified on the grounds of driving the reaction to completion, giving three anions: two as phenoxides and one as a leaving group. As the reaction progresses, the base is consumed to form the phenoxide salt. The choice of bases is somewhat random, arbitrary, and without logistics. Initial studies involved LDA as the only base. Later, use of LiO^{t}Bu was found to be more popular than commonly used bases like LDA, because of higher efficacies with the former. It is also true for phthalide-3-phosphonate donors. The range of reaction conditions in terms of a combination of base, solvent, additive, and temperature is wide. The reported combinations are (i) LDA-THF, (ii) LDA-HMPA, (iii) LDA-TMSCl, (iv) LDA-$ZnCl_2$, (v) MeLi-DMSO, THF, (vi) LTB, THF, (vii) LTB-LiCl, (viii) KTB, DMSO, (ix) LHMDS, (x) NaHMDS, (xi) KHMDS, (xii) NaH, (xiii) ^{s}BuLi, and (xiv) ^{t}BuLi. Although LiO^{t}Bu is most frequently used in the Hauser annulation, KO^{t}Bu is occasionally used with great success. Tso and Chen observed that there is no definitive role of the effect of counter ion of the bases used.[43] The usual reaction temperature range of the anionic annulations is −78°C to room temperature. Occasionally, the temperature is raised to 70°C. In a recent publication, Namboothiri et al. have shown that phthalide sulfone (**14**) can be condensed with nitroalkenes at room temperature in the presence of K_2CO_3 or Cs_2CO_3.[44] In their work on the annulation with methylene-indolinones, the Kesavan group also reported similar room-temperature annulation with the sulfone phthalide (**14**). Incidentally, phthalide sulfone (**14**) is superior to 3-cyanophthalide (**74**) in such annulations.[45]

Ronn et al. used two different bases sequentially for effecting the annulation between (**163**) and (**164**).[46] As the presence of excess LDA often promotes several unidentified side reactions, the procedure was modified to use LDA for the deprotonation of (**163**) and then lithium hexamethyldisilazide, a weaker base, for the deprotonation of the acidic proton in the Dieckmann step necessary to drive the equilibrium to completion to (**165**), ultimately leading to eravacycline (Scheme 9.41).

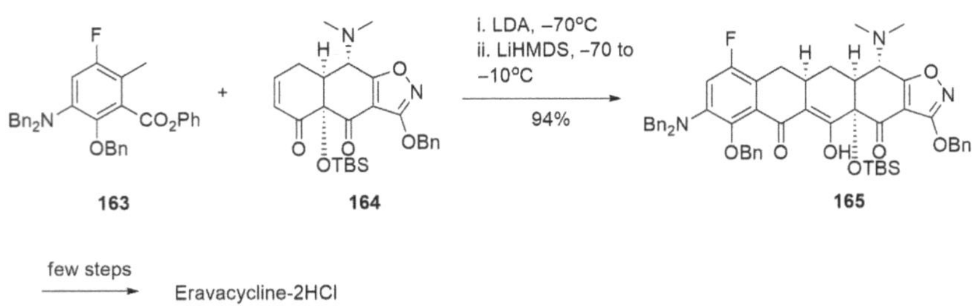

SCHEME 9.41 [4+2] Annulation with two different bases in one pot [2013 OPRD 838 Ronn M].

9.2.1 Scope and Limitations of the Hauser Type Annulations

The reactions as discussed before are very general with regard to the substrates, i.e., donors and the corresponding acceptors. They encompass a wide variety of substrates. Remarkably, many functional groups, otherwise sensitive to strong bases like LDA and LTB, survive under the reaction conditions of the Hauser annulations. They are terminal alkynes, esters, lactones, lactams, ketones, aldehydes, nitriles, and epoxides. One possible reason for the mildness of the reactions is the low reaction temperature, i.e., −78°C, at which many possible side reactions are suppressed. The 3-nucleofugal phthalides that have been engaged in Hauser annulation are shown in Fig. 9.3. The first two donors (**14** and **74**) are routinely used for condensation with α,β-unsaturated esters, α,β-unsaturated esters, and arynes.

The use of LiO^{t}Bu in the generation C3 carbanion of 3-phenylthiophthalides like (**16**) by Mal et al. in the annulation has also promoted them as a class of the effective Hauser donors. Moreover, they are readily prepared in step from the

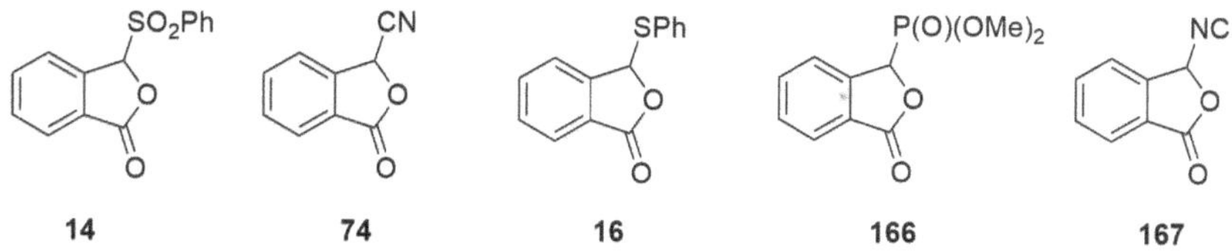

FIG. 9.3 Donors for successful Hauser annulation.

corresponding phthalaldehydic acids.[47] In their first report, Kraus et al. showed that deprotonated phenylthiophthalide gives only Michael adducts with α,β-unsaturated esters, α,β-unsaturated enones, and α,β-unsaturated cyanides. The first use of 3-phenylthiophthalide in successful naphthalene annulation in the range of 38%–70% was reported in 1983.[14] The success relied upon the use of LDA-HMPA complex as the base. The combination of phenylthiophthalide (**16**) and LiO^{t}Bu has been found to be useful in the synthesis of xanthones, pradimicinin core, and naphthofuranone natural products. Although phthalide-3-phosphonates (**97**) are found to undergo smooth annulations with different acceptors, they have not been widely used, probably due to the limited number of their preparative methods. The efficacy of the annulation with isocyanophthalide (**167**) is not prominent.[22]

There is no study available on the comparative reactivity of the donors (**14**, **16**, **74**, **166**, and **167**). From the scattered ex*peri*ments, it appears that more nucleophilic donors are needed for poorer acceptors. For example, between 3-phenylsulfonylphthalide (**14**) and 3-phenylthiophthalide (**16**), the latter is expected to generate more nucleophilic carbanion than that of (**14**), due to lesser acidity of C3-H. As a result, phthalide (**16**) is more effective for annulation with poor acceptors like vinyl sulfones. Likewise, 3-methoxyphthalide provides more nucleophilic anion than (**14**). Consequently, the anion of 3-methoxyphthalide reacts with cyclic Schiff bases. Such annulations with (**14**) are not successful. As shown in Scheme 9.42, the yields of the annulations of both (**14**) and (**16**) with enone (**168**) giving (**169**) are nearly the same under identical reaction conditions.[48]

A methoxy substituent is expected to increase nucleophilicity of the anions of phthalide sulfones (**170** and **171**) compared to (**14**). Indeed, the reaction between (**14**) and (**172**) giving (**173**) is less efficient than the reaction of (**170**) with

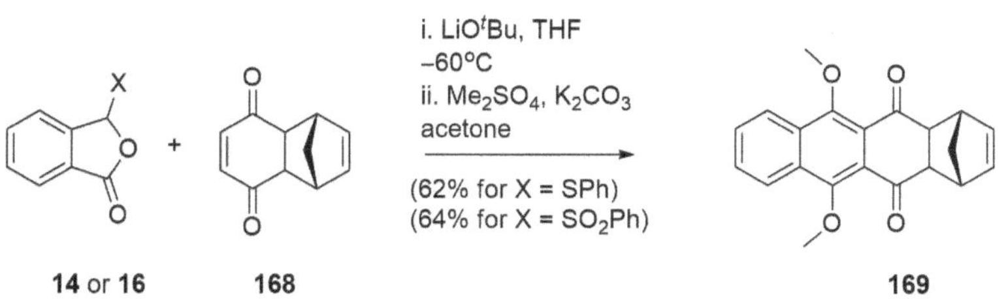

SCHEME 9.42 Comparative reactivity of phenylthiophthalide (**16**) and phenylsulfonylphthalide (**14**).

(**172**) giving (**174**). On the contrary, sulfone (**171**) gave a lesser percentage yield of (**175**) (Scheme 9.43).[49] This may be due to the steric effect of the 4-methoxy group.

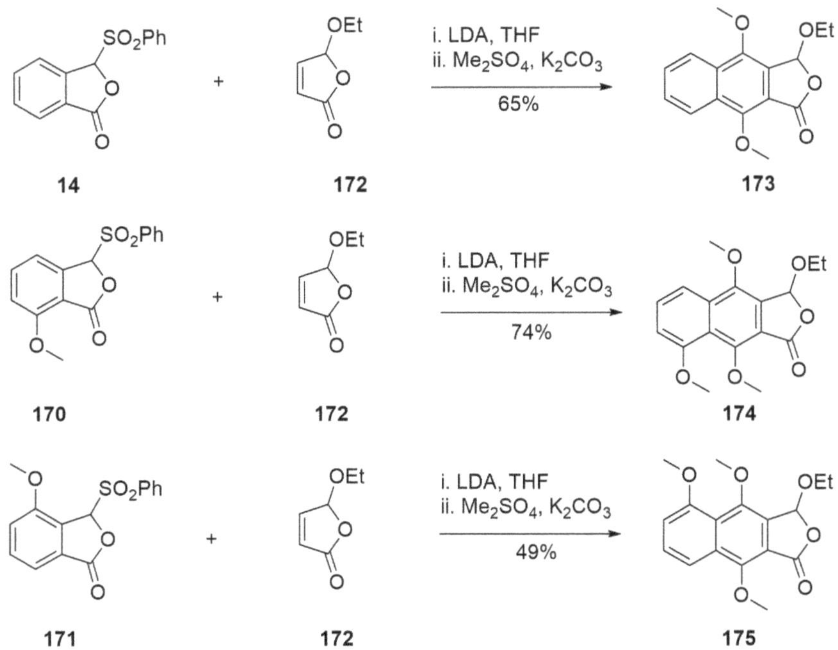

SCHEME 9.43 Hauser annulation of phthalide sulfone (**14**) with 5-ethoxyfuran-2(5H)-one.

That steric effects may be a serious impediment for the annulations is clear from the results presented in Scheme 9.44. In the reaction of (**176**) with enone (**177**), the yields of the products (**178a**), (**178b**), and (**178c**) were 85%, 52%, and 0%, respectively.[50] The failed annulation with sulfone (**176c**) is interpreted by the steric effect between the sulfone group and the methoxy group in the transition state.

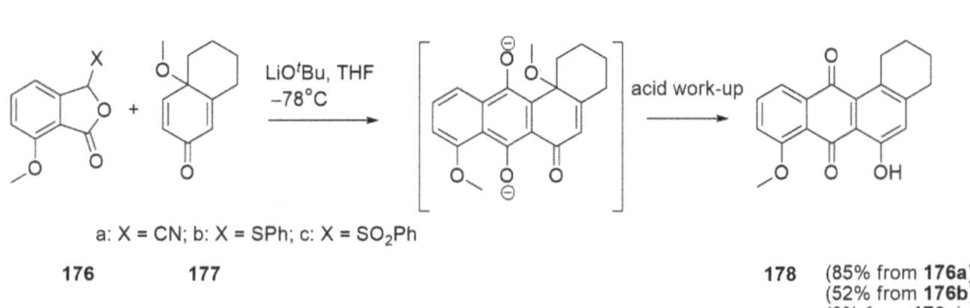

SCHEME 9.44 Annulation of different phthalides with a decalenone.

The support for the above analysis is provided by the successful annulation between (**14**) and (**179**), producing (**180**) (Scheme 9.45).[51] It is well known that a fluorine atom exerts very little steric effect. Consequently, annulation was successful.

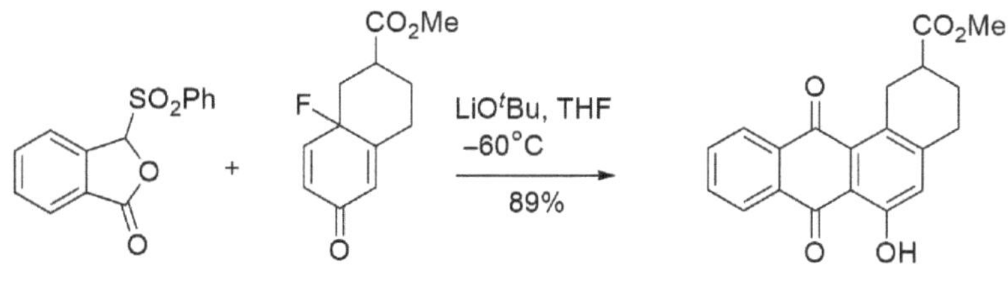

14 **179** **180**

SCHEME 9.45 Annulation of sulfone phthalide (**14**) with a fluorodecalenone.

Coumarins display interesting reactivities toward Hauser annulation. It appears that annulation is sensitive to the peri effect exerted by a substituent in the benzene ring of the coumarins, especially for the phthalide sulfones. As shown in Scheme 9.46, sulfone (**14**) reacted with coumarin (**181**) to afford the expected product (**182**). The reaction profile of sulfone phthalide (**14**) and cyanophthalide (**74**) is different in the reaction with coumarin (**183**). The sulfone phthalide (**14**) resisted annulation with dimethylcoumarin (**183**). On the other hand, cyanophthalide (**74**) gave the desired product (**184**) in 73% yield.[52]

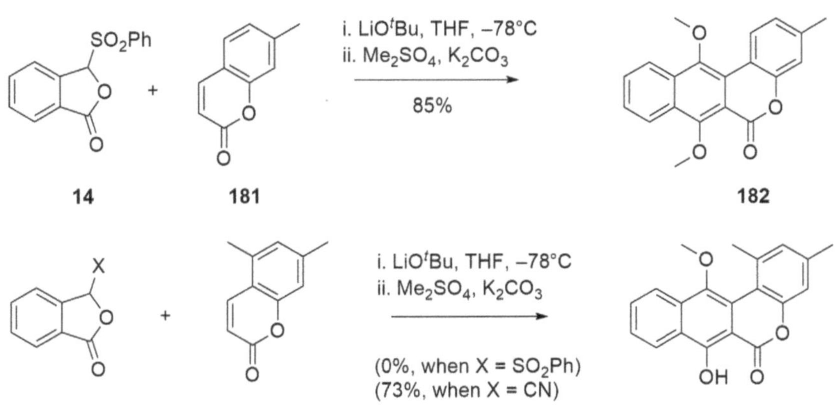

14 **181** **182**

14 or **74** **183** **184**

SCHEME 9.46 Hauser annulations with coumarins [2010 PhD Thesis Zheng R, SUNY, Albany].

A systematic study to compare the reactivity of sulfones (e.g., **185**) and analogous cyanides (e.g., **186**) was undertaken by Russell et al. (Scheme 9.47).[53] Annulations were carried out with the same acceptor (i.e., **187**). Analysis of the yields of the products (**188**) clearly indicates superiority of the cyanophthalides (**186**). There is no bearing on the electronic effect of fluorine.

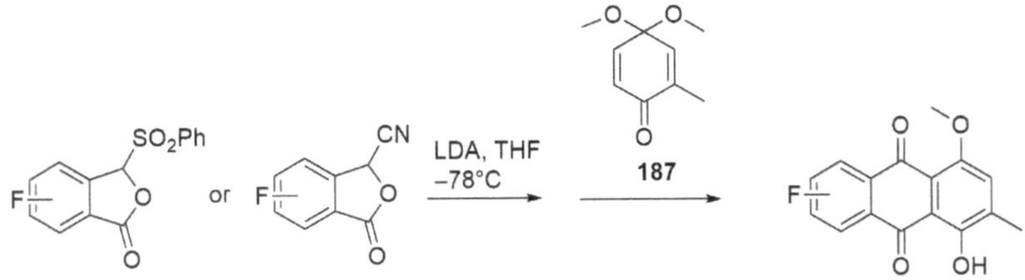

185a: 4-fluoro
185b: 5-fluoro
185c: 6-fluoro
185d: 7-fluoro

186a: 4-fluoro
186b: 5-fluoro
186c: 6-fluoro
186d: 7-fluoro

% yield of anthraquinone **188** (from)

45 (**185a**)	76 (**186a**)
42 (**185b**)	84 (**186b**)
44 (**185c**)	73 (**186c**)
74 (**185d**)	79 (**186d**)

SCHEME 9.47 Comparative efficiency of the Hauser sulfone donors and Kraus cyano donors [1987 AJC 311 Russell RA].

The reaction between (**74**) and cyclohexadienone (**189**) in the presence of LDA at −78°C did not yield the expected annulation product (**190**). This example is one of the exceptions to the otherwise general Hauser annulation (Scheme 9.48).[54]

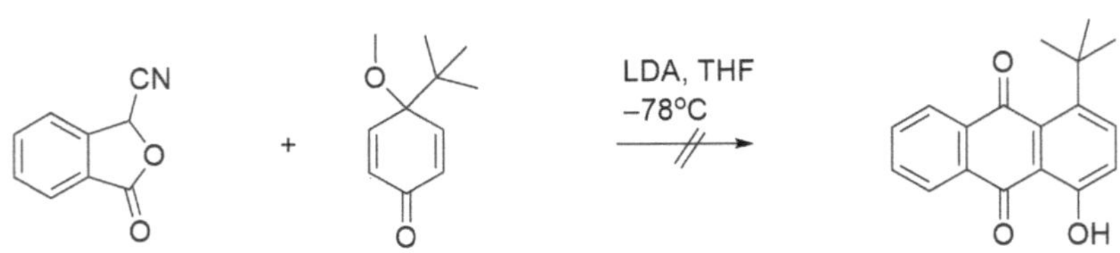

74 **189** **190**

SCHEME 9.48 Failed Hauser annulation with 3-cyanophthalide [1993 THS 545 Russell RA].

On the other hand, attempted annulation between (**74**) and cyclohexadienone (**191**) led to the formation of 3-arylated phthalide (**192**) via the initial Michael adduct (**1**) (**93**).[55] Respective annulation product was not detected (Scheme 9.49).

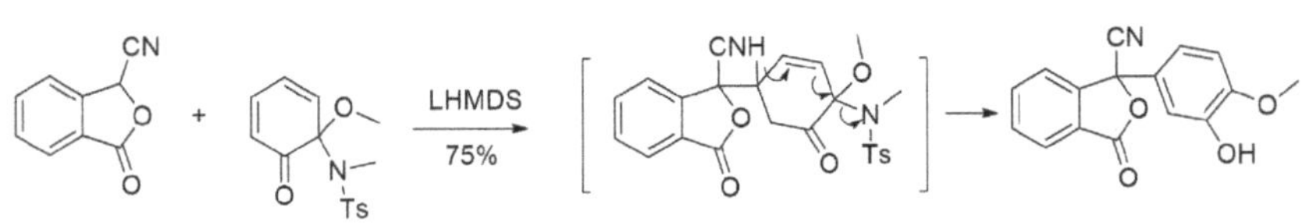

74 **191** **193** **192**

SCHEME 9.49 Failed Hauser annulation with 3-cyanophthalide [2015 ACIE 12687 Nicolaou KC].

9.2.2 Reactivity of Angular Phthalides

In the context of the total synthesis of chlorocyclinone A, Mal et al. first investigated annulation of angular phthalides having a nucleofuge in the bay region. The phthalides studied were (**194–198**) (Fig. 9.4).

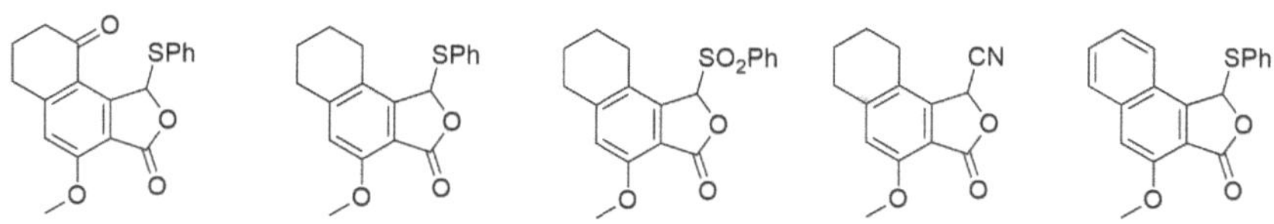

194 195 196 197 198

FIG. 9.4 Angular benzophthalides studied for annulations [2012 JOC 10235 MaL D].

Although the earlier studies with linear phthalides suggested a sure success of the annulation with the angular phthalides (**194–198**), their reaction with 2-cyclohexenone in the presence of bases (LiOtBu, LDA or LHMDS) afforded only the Michael adduct (**199**) (Scheme 9.46).[56] Annulation product (**200**) was not obtained. For the above unsuccessful annulations, one explanation could be a peri-steric effect developed during expulsion of the nucleofuge of the bay region (Scheme 9.50).

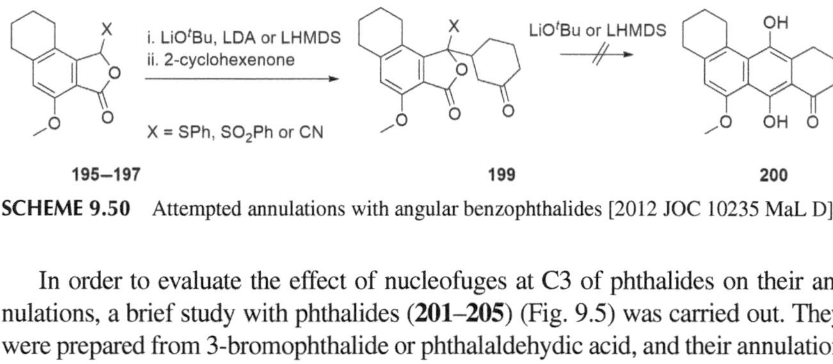

195–197 199 200

SCHEME 9.50 Attempted annulations with angular benzophthalides [2012 JOC 10235 MaL D].

In order to evaluate the effect of nucleofuges at C3 of phthalides on their annulations, a brief study with phthalides (**201–205**) (Fig. 9.5) was carried out. They were prepared from 3-bromophthalide or phthalaldehydic acid, and their annulation reactivity was examined with ethyl acrylate and cyclohexenone. Interestingly, none of them gave any annulation product. Further investigation is warranted to throw light on the effect of the nucleofuges on the reactivity of the Hauser-type donors.

9.2.3 The Sammes Annulation

Benzyne precursors react with phthalides, cyanophthalides, and sulfone phthalides under the influence of LDA or LTMP to give anthraquinone derivatives. Incidentally, all three types of donors give the same anthraquinone

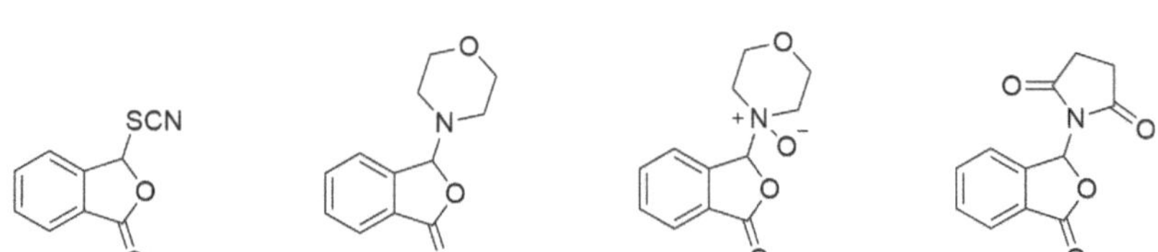

201 202 203 204

FIG. 9.5 Phthalide donors that were inert to annulations [2016 Unpublished results Mal D].

product in comparable yields. For practical purposes, simple phthalides are preferred to 3-nucleofugal phthalides, for they often give better yields. In the synthesis of calothrixin A (vide Chapter 5), the indole analog of phthalide was found to be better than the 3-nucleofugal analog in giving the corresponding annulation products. Between the phthalides and thiophthalides, the latter are slightly better annulating agents. However, the obvious annulation of thiophthalide (**206**) with 2-cyclohexenone failed to give the annulation product (**207**) (Scheme 9.51).[6]

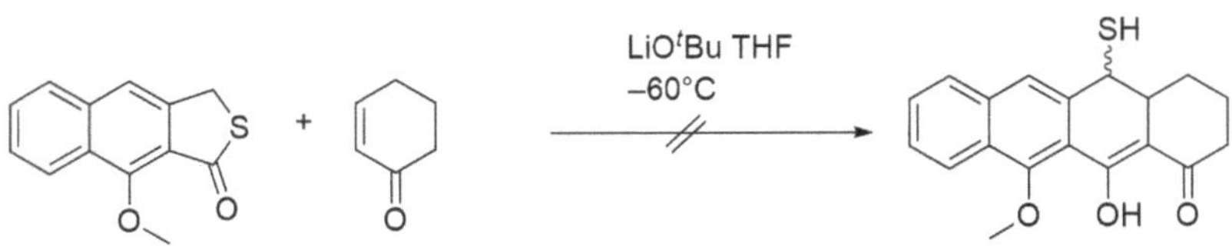

206 **207**

SCHEME 9.51 Failed annulation with a thiophthalide [1994 JCSPT1 309 MaL D].

9.2.4 Accompanying Reactions and Side Reactions

There are about seven reactions mechanistically possible besides the expected annulation when a donor and an acceptor are submitted to annulation under strongly basic conditions. These are (i) self-dimerization/decomposition of the donor, (ii) self-dimerization/decomposition of the acceptor, (iii) Michael addition, (iv) 1,2-nucleophilic addition, (v) polymerization of the acceptor, and (vi) decomposition of the products. With the Hauser synthons and Kraus donors, the expected [4+2] annulation products are 1,4-naphthoquinols. Since the quinols are susceptible to rapid aerial oxidation, the corresponding oxidized products are most often obtained.

In an attempted annulation of (**208**) with a dienoate, Br-Li exchange was tried to generate its naphthoate anion. However, the reaction gave the dimer (**209**) as the only product in 68% yield (Scheme 9.52).[57]

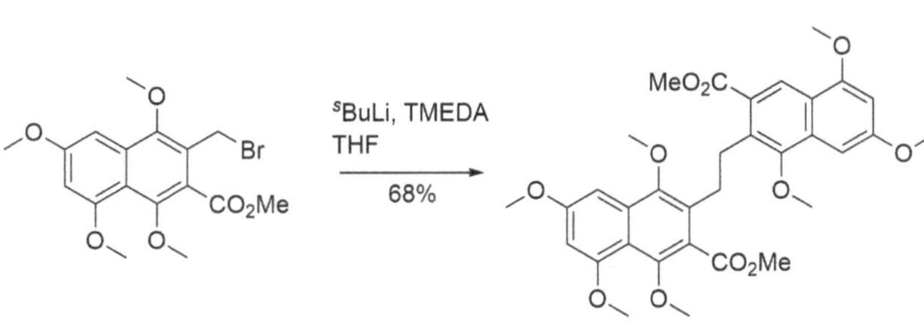

208 **209**

SCHEME 9.52 Self-destruction of the donor [2018 THN 96 MaL D].

Burnell's studies have shown that the 1,4-dipolar reagent (**210**) fails to enter into an [4 + 2] annulation with a cyclic enone due its decomposition to thiothionoanhydride (**211**) (Scheme 9.53).[58]

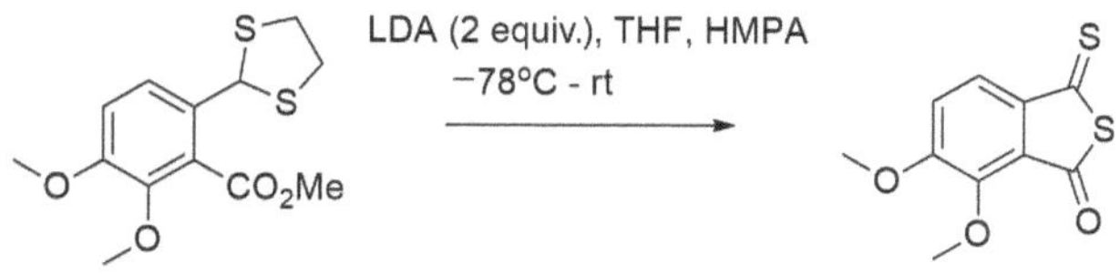

SCHEME 9.53 Decomposition of a donor [2009 THS 7021 Burnell DJ].

Likewise, an anticipated annulation may fail due to the decomposition of the acceptor. Townsend annulation between (**212**) and (**213**) in the presence of LTMP did not proceed to give (**214**), due to the fragmentation of (**213**) to benzofuran (**215**) (Scheme 9.54).[59]

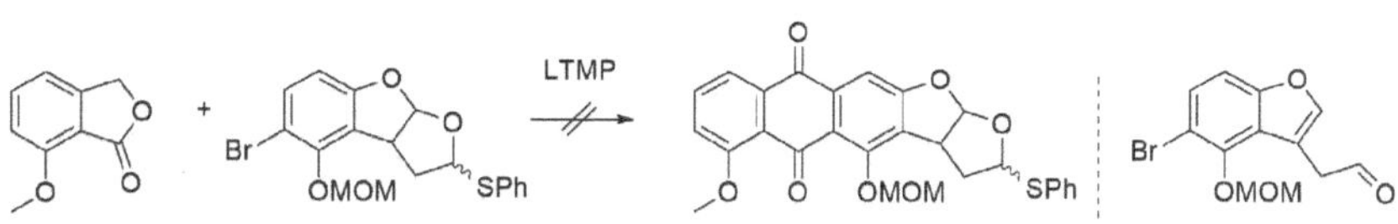

SCHEME 9.54 Decomposition of an acceptor [1999 JACS 7729 Townsend CA].

The donor of an annulation is expected to react with the acceptor in 1,4-addition mode. However, the 1,2-addition mode is sometimes observed as a competing reaction. If an isoxazole like (**216**) has no substituent at the lateral position, it would undergo 1,2-addition with quinonemonoketal (**217**) to give (**218**). When it is converted to its softer derivative (**219**) and submitted to the reaction with ketal (**217**), the Michael adduct is the end product, i.e., (**220**) (Scheme 9.55).[60] The annulation product is not obtained.

The annulation of cyanophthalides is sometimes accompanied by formation of cyanohydrins if the resulting products have a ketone substituent. In the reaction between (**221**) and (**222**) with LDA, the expected annulation product was not formed. Instead, its cyanohydrin (**223**) was obtained to a large extent (Scheme 9.56).[53]

Brimble et al. found that an unusual set of products arising out of the furanose unit of the initial annulation products under acidic conditions. The reaction between (**224**) and (**225**) in the presence of KO^tBu was supposed to deliver (**226**). However, the ultimate product was (**227**), which arose due to fragmentation of the furanose unit in (**226**) during acidic work-up (Scheme 9.57).[61]

A similar observation has been made by Chakraborty and Mal. The Hauser annulation product (**228**), obtained by the Hauser annulation, underwent acid-catalyzed fragmentation with dil. HCl to give benz[*a*]anthraquinone (**229**) (Scheme 9.58).[62]

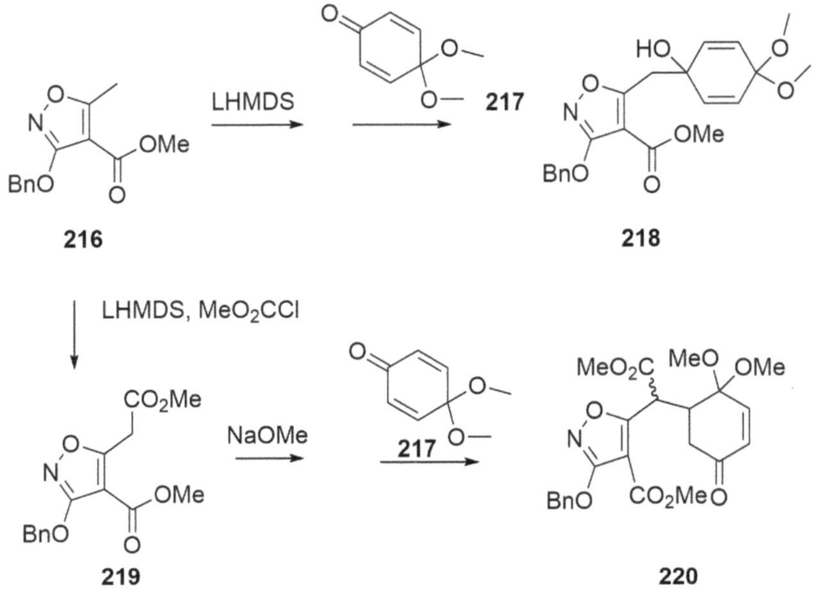

SCHEME 9.55 Competing reactions [2014 JACS 12137 Nicolaou KC].

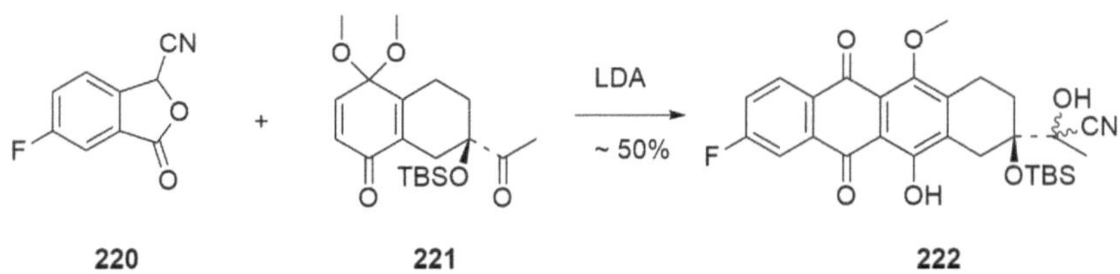

SCHEME 9.56 Cyanohydrin formation of the annulation product [1987 AJC 311 Russell RA].

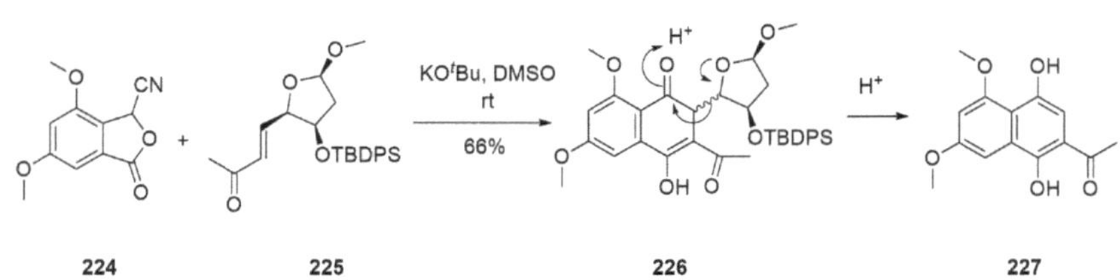

SCHEME 9.57 Elimination of furanose sugar [2008 THN 3912 Brimble MA].

From the above presentations, it appears that a 1,4-dipolar donor, an acceptor with a polarized CC multiple bond, and an alkali metal strong base would suffice to execute a successful Hauser-type annulation. However, reasonable understanding of the reaction and the possible competing reactions is warranted for real life success. There are examples that prove to be

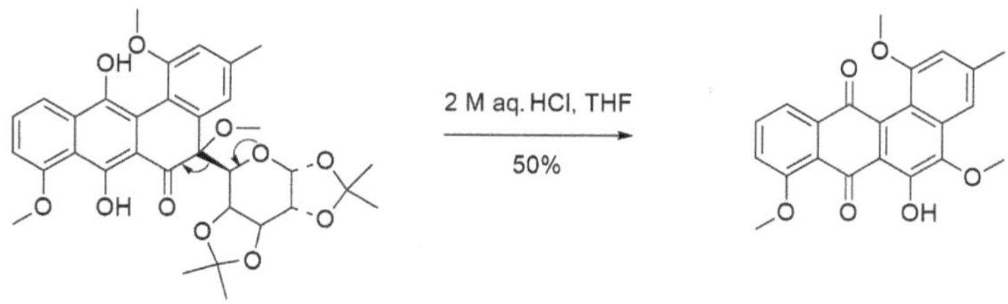

228 229

SCHEME 9.58 Elimination of a pyranose sugar [2018 JOC 1328 Mal D].

the exceptions of the generalized reaction. For example, Tamura annulations with unsaturated ketones are not often successful. This is also precluded with simple Michael acceptors such as methyl acrylate and acrylonitrile. There are more than six possible ways by which an expected [4 + 2] anionic annulation may be foiled.

REFERENCES

1. Snieckus, V. *Chem. Rev.* **1990**, *90*, 879.
2. Murty, K.V.S.N.; Pal, R.; Datta, K.; Mal, D. *Synth. Commun.* **1990**, *20*, 1705.
3. Hauser, F. M.; Rhee, R. P. *J. Org. Chem.* **1980**, *45*, 3061.
4. Tatsuta, K.; Inukai, T.; Itoh, S.; Kawarasaki, M.; Nakano, Y. *J. Antibiotics* **2002**, *55*, 1076.
5. Hauser, F. M.; Rhee, R. P.; Prasanna, S.; Weinreb, S. M.; Dodd, J. H. *Synthesis* **1980**, 72.
6. Mal, D.; Majumdar, G.; Pal, R. *J. Chem. Soc. Perkin. Trans. 1* **1994**, 309.
7. Roy, J.; Pal, R.; Mal, D. *Tetrahedron Lett.* **2015**, *56*, 6210.
8. Murty, K.V.S.N.; Hazra, N. K.; Datta, K.; Mal, D. *Ind. J. Chem.* **1997**, *36B*, 126–132.
9. Kongsriprapan, S.; Kuhakarn, C.; Deelertpaiboon, P.; Panthong, K.; Tuchinda, P.; Pohmakotr, M.; Reutrakul, V. *Pure Appl. Chem.* **2012**, *84*, 1435.
10. Hauser, F. M.; Dorsch, W. A. *Org. Lett.* **2003**, *5*, 3753.
11. Alguacil, R.; Martin, M. V.; Paredes, M. C. *Heterocycles* **2000**, *53*, 1029.
12. Matsumoto, T.; Yamaguchi, H.; Hamura, T.; Tanabe, M.; Kuriyama, Y.; Suzuki, K. *Tetrahedron Lett.* **2000**, *41*, 8383.
13. Hauser, F. M.; Gauuan, P. J. F. *Org. Lett.* **1999**, *1*, 671.
14. Kraus, G. A.; Cho, H.; Crowley, S.; Roth, B.; Sugimoto, H.; Prugh, S. *J. Org. Chem.* **1983**, *48*, 3439.
15. (a) Sakulsombat, M.; Angelin, M.; Ramstrom, O. *Tetrahedron Lett.* **2010**, *51*, 75; (b) Unpublished Results, Hauser, F. M. SUNY, Albany.
16. Freskos, J. N.; Morrow, G. W.; Swenton, J. S. *J. Org. Chem.* **1985**, *50*, 805.
17. Evans, J. C.; Klix, R. C.; Bach, R. D. *J. Org. Chem.* **1988**, *53*, 5519.
18. Ghosh, B.; Chakraborty, S.; Mal, D. *ChemistrySelect* **2016**, *1*, 3097.
19. Okazaki, K.; Nomura, K.; Yoshii, E. *Synth. Commun.* **1987**, *17*, 1021.
20. Kim, T.; Jeong, K. H.; Kang, K. S.; Nakata, M.; Ham, J. *Eur. J. Org. Chem.* **2017**, 1704.
21. Nicolaou, K. C.; Wang, Y.; Lu, M.; Mandal, D.; Pattanayak, M. R.; Yu, R.; Shah, A. A.; Chen, J. S.; Zhang, H.; Crawford, J. J.; Pasunoori, L.; Poudel, Y. B.; Chowdari, N. S.; Pan, C.; Nazeer, A.; Gangwar, S.; Vite, G.; Pitsinos, E. N. *J. Am. Chem. Soc.* **2016**, *138*, 8235.
22. Mal, D.; Ghosh, K.; Chakraborty, S. *Synthesis* **2015**, *47*, 2473.

23. (a) Napotilano, E.; Spinelli, G.; Fiaschi, R.; Marshili, A. *Synthesis* **1985**, 38; (b) Watanabe, M.; Furukawa, S. *Synlett* **1991**, 481. (c) Zhang, L.; Kim, H.; Jang, D. O. *Tetrahedron Lett.* **2017**, *58*, 1985.
24. Karmakar, R.; Mal, D. *J. Org. Chem.* **2012**, *77*, 10235.
25. Ghosh, K.; Karmakar, R.; Mal, D. *Eur. J. Org. Chem.* **2013**, 4037.
26. Roy, J.; Mal, D. *Org. Biomol. Chem.* **2015**, *13*, 6344.
27. Bhattacharjee, D.; Popp, F. D. *J. Heterocycl. Chem.* **1980**, *17*, 315.
28. Zhang, Y.-H.; Shi, B.-F.; Yu, J.-Q. *Angew. Chem. Int. Ed.* **2009**, *48*, 6097.
29. Wong, H. N. C.; Xing, Y. D.; Zhou, Y. F.; Gong, Q. Q.; Zhang, C. *Synthesis* **1984**, 787.
30. Soriano, M.D.P.C.; Shankaraiah, N.; Santosa, L. S. *Tetrahedron Lett.* **2010**, *51*, 1770.
31. Magnus, P.; Eisenbeis, S. A.; Magnus, N. A. *J. Chem. Soc. Chem. Commun.* **1994**, 1545.
32. Wu, X.; Mahalingam, A. K.; Wan, Y.; Alterman, M. *Tetrahedron Lett.* **2004**, *45*, 4635.
33. Bhatarah, P.; Smith, E. H. *J. Chem. Soc., Perkin Trans. 1* **1992**, 2163.
34. Mal, D.; Senapati, B.; Pahari, P. *Tetrahedron* **2007**, *63*, 3768.
35. Hauser, F. M.; Dorsch, W. A.; Mal, D. *Org. Lett.* **2002**, *4*, 2237.
36. Matsumoto, T.; Yamaguchi, H.; Suzuki, K. *Tetrahedron* **1997**, *53*, 16533.
37. Kita, Y.; Akai, S.; Azimura, N.; Yoshigi, M.; Tsugoshi, T.; Yasuda, H.; Tamura, Y. *J. Org. Chem.* **1986**, *51*, 4150.
38. Iio, K.; Ramesh, N. G.; Okajima, A.; Higuchi, K.; Fujioka, H.; Akai, S.; Kita, Y. *J. Org. Chem.* **2000**, *65*, 89.
39. Shair, M. D.; Yoon, T.; Danishefsky, S. J. *Angew. Chem. Int. Ed.* **1995**, *34*, 1721.
40. Lee, H. G.; Ahn, J. Y.; Lee, A. S.; Shair, M. D. *Chem. Euro. J.* **2010**, *16*, 13058.
41. Li, J.; Mootoo, D. R. *Synthesis* **2013**, *45*, 2287.
42. Liau, B. B.; Milgram, B.; Shair, M. D. *J. Am. Chem. Soc.* **2012**, *134*, 16765.
43. Tso, H.-H.; Chen, Y.-J. *J. Chem. Res. (S)* **1995**, 104.
44. Kumar, T.; Mane, V.; Namboothiri, I. N. N. *Org. Lett.* **2017**, *19*, 4283.
45. Lokesh, K.; Kesavan, V. *Eur. J. Org. Chem.* **2017**, 5689.
46. Ronn, M.; Zhu, Z.; Hogan, P. C.; Zhang, W.-Y.; Niu, J.; Katz, C. E.; Dunwoody, N.; Gilicky, O.; Deng, Y.; Hunt, D. K.; He, M.; Chen, C.-L.; Sun, C.; Clark, R. B.; Xiao, X.-Y. *Org. Process Res. Dev.* **2013**, *17*, 838.
47. Ghorai, S. K.; Roy, H. N.; Bandophthyay, M.; Mal, D. *J. Chem. Res. (S)* **1999**, 30.
48. Majumdar, G.; Murty, K.V.S.N.; Mal, D. *Tetrahedron Lett.* **1994**, *35*, 6139.
49. Hauser, F. M.; Prasanna, S. *J. Am. Chem. Soc.* **1981**, *103*, 6378.
50. Mal, D. Roy, HN. *J. Chem. Soc. Perkin Trans.* **1999**, *1*, 3167.
51. Pahari, P.; Senapati, B.; Mal, D. *Tetrahedron Lett.* **2007**, *48*, 2635.
52. Zheng, R. PhD Thesis, SUNY, Albany, 2010.
53. Russell, R. A.; Pilley, B. A.; Irvine, R. W.; Warrener, R. N. *Aus. J. Chem.* **1987**, *40*, 311.
54. Mitchell, A. S.; Russell, R. A. *Tetrahedron Lett.* **1993**, *34*, 545.
55. Nicolaou, K. C.; Lu, M.; Chen, P.; Shah, A. A. *Angew. Chem. Int. Ed.* **2015**, *54*, 12687.
56. Karmakar, R.; Mal, D. *J. Org. Chem.* **2012**, *77*, 10235.
57. Basak, S.; Mandal, S.; Mal, D. *Tetrahedron* **2018**, *74*, 96.
58. Morrison, C. F.; Stamp, C. T. M.; Burnell, D. J. *Tetrahedron Lett.* **2009**, *50*, 7021.
59. Graybill, T. L.; Casillas, E. G.; Pal, K.; Townsend, C. A. *J. Am. Chem. Soc.* **1999**, *121*, 7729.
60. Nicolaou, K. C.; Hale, C. R. H.; Nilewski, C.; Ioannidou, H. A.; ElMarrouni, A.; Nilewski, L. G.; Beabout, K.; Wang, T. T.; Shamoo, Y. *J. Am. Chem. Soc.* **2014**, *136*, 12137.
61. Andrey, O.; Sperry, J.; Larsen, U. S.; Brimble, M. A. *Tetrahedron* **2008**, *64*, 3912.
62. Chakraborty, S.; Mal, D. *J. Org. Chem.* **2018**, *83*, 1328.

FURTHER READING

63. Wheeler, D. D.; Young, D. C.; Erley, D. S. *J. Org. Chem.* **1957**, *22*, 547.
64. Sinhababu, A. K.; Borchardt, R. T. *J. Org. Chem.* **1983**, *48*, 2356.
65. Chenard, B. L.; Dolson, M. G.; Sercel, A. D.; Swenton, J. S. *J. Org. Chem.* **1984**, *49*, 318.
66. Meyers, A. I.; Avila, W. B. *J. Org. Chem.* **1981**, *46*, 3881.
67. Eliel, E. L.; Rivard, D. E.; Burgstahler, A. W. *J. Org. Chem.* **1953**, *18*, 1679.

Chapter 10

Applications of [4+2] Anionic Annulations in the Total Synthesis

The elegance, potential, and efficacy of a reaction are best judged by its applications in total syntheses.[1] Many applications of the [4+2] anionic annulations in the total syntheses have already been covered in previous chapters in conjunction with the respective reactions. In this chapter, the select total syntheses are presented to focus upon the utility and versatility of various anionic annulations in general.

10.1 HAUSER SYNTHESIS OF γ-CITROMYCINONE (1)

Anthracycline antibiotics are an important class of antitumor drugs widely used in cancer chemotherapy. They are often represented by clinical agents: daunorubicin and doxorubicin. γ-Citromycinone (**1**) is a rare anthracyclinone, isolated in minute quantities. Its B-ring hydroxy pattern is uncommon. The first total synthesis of this anthracyclinone was reported by Hauser and Mal in 1984.[2] As shown in Scheme 10.1, the founding step of the synthesis of citromycinone was annulation (LiO^{t}Bu, THF) between sulfone phthalide (**2**) and (acetoxymethy l)cyclohexenone (**3**), which provided tetrahydroanthracenone (**4**). Aerial oxidation of (**4**) in hot DMF gave anthraquinone (**5**) in 81% yield. Chemoselective *O*-alkylation of the phenolic group in (**5**) with 2-(chloromethy l)-l-butene furnished corresponding allyl ether, alkaline hydrolysis (aq NaOH, THF), which gave (**6**) in 91% overall yield. In situ reduction of the anthraquinone moiety of (**6**) with sodium dithionite in hot DMF effected a clean Claisen rearrangement. Aerial oxidation of the resultant product afforded 2-ethylpropenylsubstituted anthraquinone (**7**) in 96% yield. The phenolic group in (**7**) was methylated (Me_2S0_4, K_2CO_3, acetone) to furnish its ether, the hydroxymethyl group of which was then oxidized with barium manganate to produce aldehyde (**8**) in 90% yield. Treatment of (**8**) with stannic chloride pentahydrate regiospecifically afforded intramolecular carbonyl-ene products (**9**) as an *E,Z* mixture of 9,13-olefinic compounds in 92% yield. Since the geometry of the C-9,13 olefin in (**9**) was of no consequence for the final products, the isomeric mixture was used in subsequent steps. Sharpless

Anionic Annulations in Organic Synthesis. https://doi.org/10.1016/B978-0-12-813800-7.00010-5

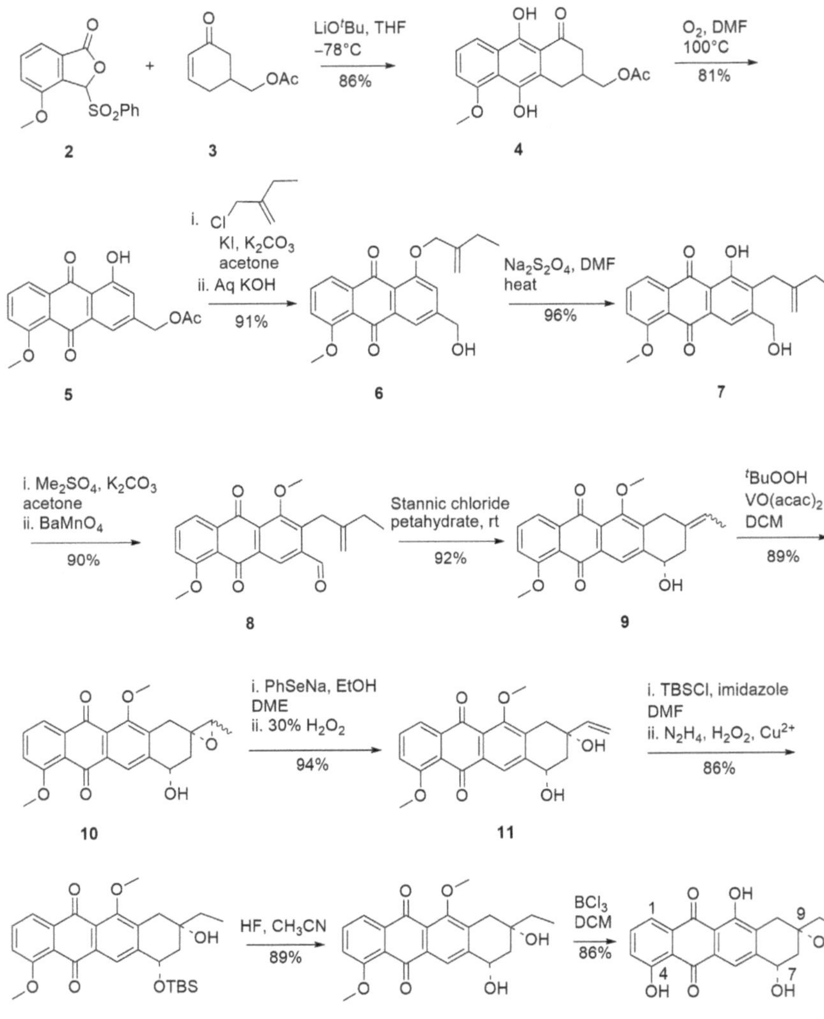

SCHEME 10.1 Total synthesis of (±)-γ-citromycinone (**1**) [1988 JACS 2919 Hauser FM].

epoxidation (^{t}BuOOH, VO(acac)$_2$) of the 9,13-olefinic moiety in (**9**) stereospecifically furnished *cis*-epoxides (**10**) in 89% yield. Opening of the epoxide moiety in (**10**) with phenyl selenide anion followed by in situ oxidation of the selenium intermediate resulted in vinyl derivative (**11**). Since the hydrogenation of the 13,14-olefinic moiety in (**11**) proved problematic, the 7-hydroxy group was protected as TBS ether (TBSCl, imidazole). Reduction of the TBS derivative by hydrazine afforded (**12**). Desilylation of (**12**) with aqueous HF in acetonitrile produced the dimethyl ether derivative of citromycinone (**13**). Finally, demethylation of (**13**) with BCl_3 furnished (±)-γ-citromycinone (**1**).

A notable feature of the synthesis is that the C7-hydroxy group was installed without resorting to conventional radical bromination pathway, which had the limitation of scale-up.

10.2 MAL SYNTHESIS OF CHLOROCYCLINONE A (14)

The angucycline family of antibiotics, characterized as benz[*a*]anthraquinone derivatives, has been a subject of growing interest due to the wide range of their structures and biological activities. New members are continually being added to the family. In 2007, four new chlorinated members, chlorocyclinones A–D, were isolated from a broth culture of *Streptomyces* sp. (DSM 17045). Seven different substituents/functionalities including chlorine at C2, have featured this unique group of angucyclines. Each of the chlorocyclinones has been reported to antagonize rosiglitazone-induced PPAR-γ activity, thus implying therapeutic potential in antidiabetic therapies, especially for the treatment of type II diabetes.

The first total synthesis of chlorocyclinone A (**14**), a natural angucycline was accomplished by Mal et al. in 28 steps from commercially available chemicals, and is depicted in Scheme 10.2–10.4.[3] It clearly demonstrates that Hauser annulation is a dependable reaction that can be utilized in the latter stages of a total synthesis. Hauser annulation between dearomatized tetralin (**15**), obtained from (**16**), and cyanophthalide (**17**), obtained from (**18**), gave tetracyclic carbon framework of chlorocyclinone, i.e., (**19**). Eight more steps applied to (**19**) completed the total synthesis of the target (**14**) (Scheme 10.2).

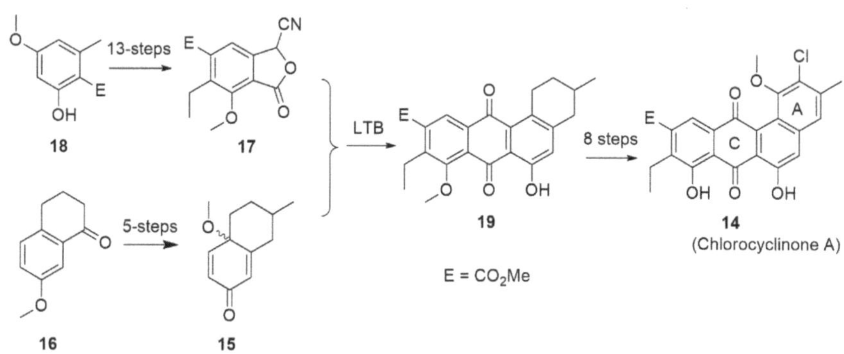

SCHEME 10.2 Outline of the total synthesis of chlorocyclinone A (**14**).

The synthesis of the phthalide (**17**) was achieved in 13 steps from (**18**), and is presented in Scheme 10.3. The synthesis of the decalenone (**15**) from tetralone (**16**) is displayed in Scheme 10.4.

Acylation of (**18**) with acetyl chloride in the presence of $TiCl_4$ provided acylsubstituted benzene derivative (**20**) in 75% yield (Scheme 10.3). To achieve selectivity in the next step, i.e., bromination, a compound (**20**) was converted

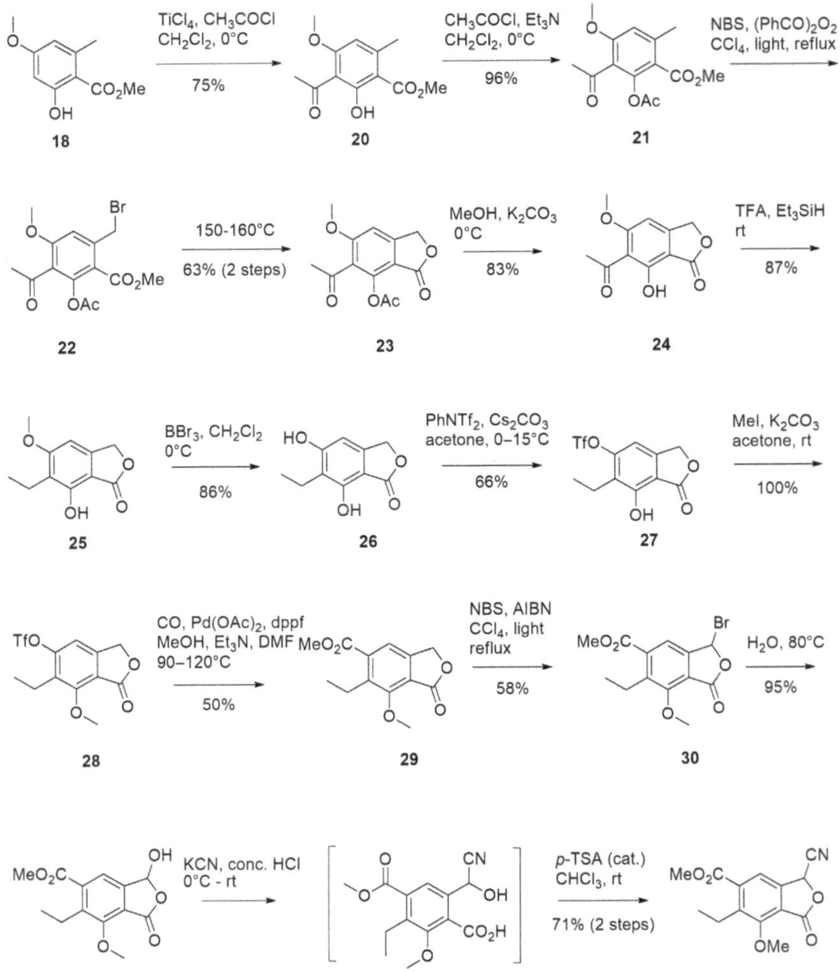

SCHEME 10.3 Synthesis of phthalide (**17**).

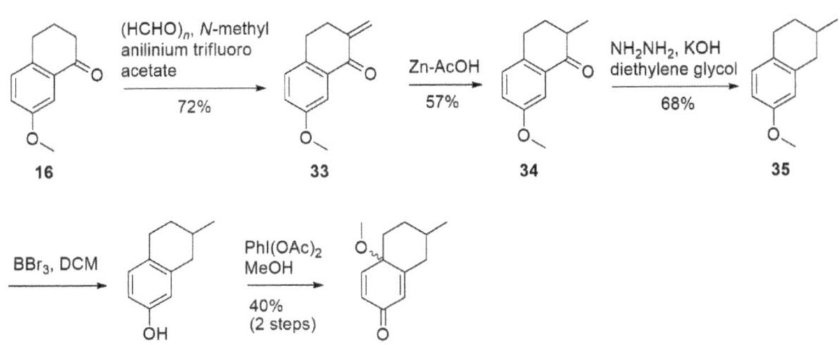

SCHEME 10.4 Synthesis of methoxy decalenone (**15**).

into its acetate (**21**). NBS bromination of (**21**) gave bromo derivative (**22**), which was thermolyzed to afford phthalide (**23**). Deacetylation of (**23**) provided compound (**24**) (52% over two steps). Reduction of the acyl carbonyl in (**24**) was performed with trifluoroacetic acid and triethylsilane to give phthalide (**25**). BBr_3-mediated demethylation of the phthalide (**25**) furnished dihydroxyphthalide (**26**). Selective protection of (**26**) using *N*-phenyltriflimide and Cs_2CO_3 led to triflate (**27**). *O*-Methylation of the triflate (**27**) produced compound (**28**) in 66% yield. Pd-catalyzed methoxycarbonylation of the triflate (**28**) with carbon monoxide afforded phthalide ester (**29**) in 50%. NBS mediated bromination of the phthalide (**29**) in refluxing CCl_4 under light selectively produced phthalide (**30**) in 58% yield as the sole product. The resulting 3-bromophthalide (**30**) was carefully heated in water at 80°C to furnish phthalaldehydic acid (**31**) in 95% yield. Treatment of the compound (**31**) with KCN in an acidic medium at 0°C resulted in the corresponding cyanohydrin (**32**), which underwent lactonization to the desired cyanophthalide (**17**) in 71% yield, upon overnight standing at rt. in chloroform solvent containing a catalytic amount of *p*-TSA. Decalenone (**15**) was prepared from commercially available 7-methoxy-α-tetralone (**16**) (Scheme 10.4). Treatment of (**16**) with paraformaldehyde in the presence of *N*-methylanilinium trifluoroacetate in THF furnished methylene derivative (**33**) in 72% yield. Reduction of (**33**) with Zn–AcOH yielded 2-methyltetralone (**34**) in 57% yield. The tetralone (**34**) was then subjected to Huang–Minlon reduction to give tetrahydronaphthol (**35**) in 68% yield. Demethylation of (**35**) with BBr3 in DCM, followed by oxidation of (**36**) with phenyliodonium diacetate in methanol, provided the methoxydecalenone (**15**) in overall 40% yield. The most striking step of Scheme 10.3 is the selective bromination of CH_2 group of the γ-lactone rather than CH_2 group of the ethyl side chain of (**29**). This is because of greater stabilization of the incipient radical by the lone pair of furan oxygen atom.

Annulation of cyanophthalide (**17**) with methoxydecalenone (**15**) in the presence of LiO^tBu at −78°C afforded benzanthraquinone (**37**) in 85% yield (Scheme 10.5). Krohn photo-oxidation of (**37**) under sunlight provided keto derivative (**38**) in quantitative yield. The phenolic hydroxy group of (**38**) was protected as methyl ether derivative (**39**). The A-ring motif was then fabricated by *gem*-α-dichlorination of (**39**), followed by aromatization via dehydrochlorination. Dimethyl ether (**39**) was *gem*-α-dichlorinated by using SO_2Cl_2 and dry HCl gas to form (**40**) in 94% yield. Dehydrochlorination of (**40**) with NaOMe in refluxing methanol gave A-ring aromatized product (**41**) in 94% yield. Free phenolic OH of (**41**) was methylated with MeI-K_2CO_3 in acetone to give (**42**). Demethylation of (**42**) with BCl_3 furnished (**43**). Further demethylation of (**43**) with $AlCl_3$ afforded trihydroxy derivative (**44**). Chemoselective methylation of C1-hydroxy group of (**44**) was achieved with MeI-K_2CO_3 in acetone to furnish chlorocyclinone A (**14**) in 39% yield.

This study of the total synthesis of chlorocyclinone A (**14**) has led to an interesting finding, i.e., annular phthalides with a nucleofuge in the bay region are not suitable donors for the Hauser annulations.

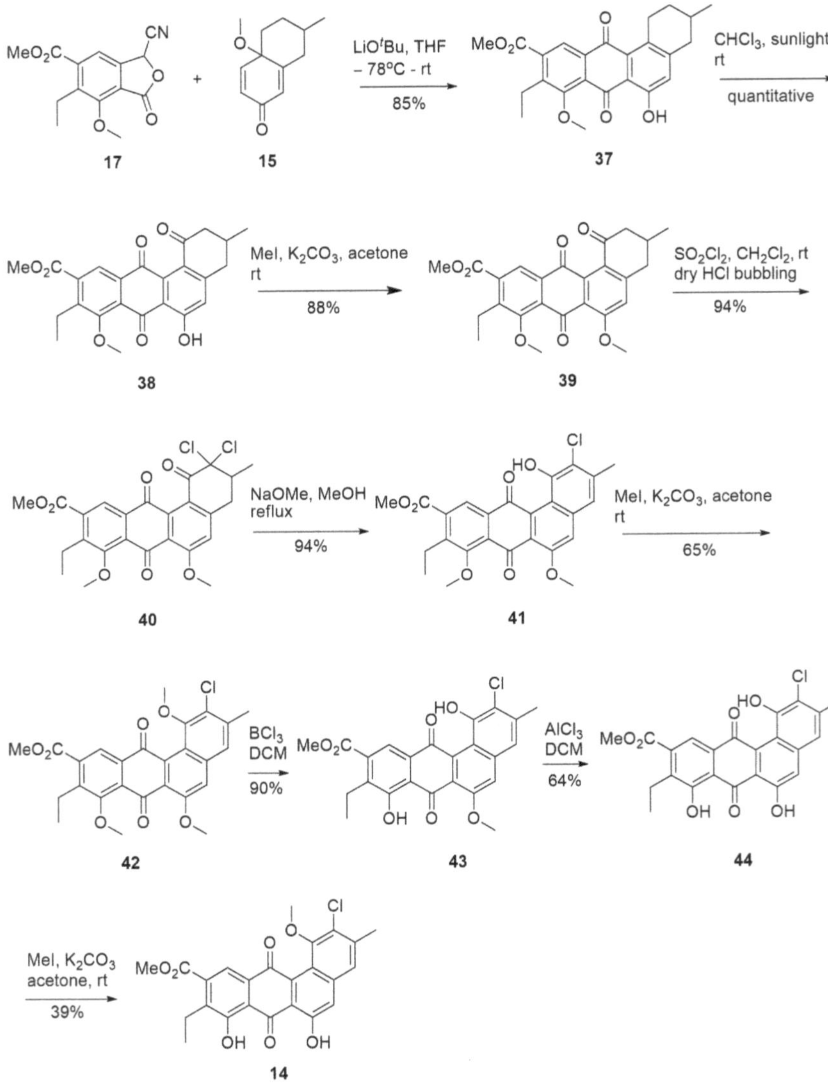

SCHEME 10.5 Last nine steps of the synthesis of chlorocyclinone A (**14**).

10.3 SUZUKI SYNTHESIS OF AQUAYAMYCIN (45)

The Hauser annulation has greatly simplified the synthesis of aquayamycin (**45**), the first ever isolated *C*-glycosidic angucycline. A combination of the Hauser reaction with Pederson intramolecular pinacol coupling led to the completion of the first total synthesis of aquayamycin by the Suzuki group.[4] It is to be noted that oxygenated angucyclines are very sensitive to acids, bases, heat, and light. More specifically, their A-rings are susceptible to rearrangements. The key step of the synthesis of aquayamycin (**45**) has been briefed in Scheme 10.6.

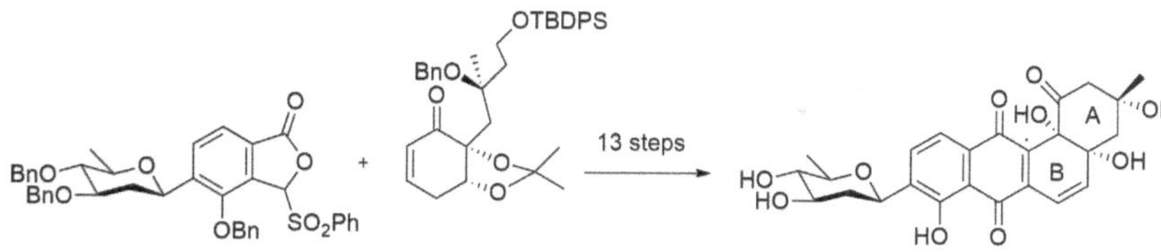

46 47 45

SCHEME 10.6 Outline of the total synthesis of aquayamycin.

It involves the union of CD fragment (**46**) and AB fragment (**47**) followed by 13 linear steps en route to the total synthesis of the target.

The synthesis of intermediate (**46**) was presented in Chapter 9. Cyclohexenone derivative (**47**) was prepared in 11 steps from enantiomerically pure cyclohexane-1,2,3-triol derivative (**48**) (Scheme 10.7). Protection of the OH group in (**48**) by PMBCl gave its PMB ether. Its semi-hydrogenation with the Lindlar catalyst followed by Hg^{2+} mediated oxidation with $PdCl_2$-$CuCl_2$ afforded ketone (**49**). The reaction of allylzinc with (**49**) produced the corresponding enantiomerically pure alcohol, which was protected as benzyl ether to deliver (**50**). Ozonolysis of (**50**) followed by $NaBH_4$ gave the corresponding primary alcohol, which was protected as its TBDPS ether. This ether was subjected to DDQ oxidation to furnish (**51**). Swern oxidation of (**51**) provided ketone (**52**). The double bond at C2 of (**52**) was introduced by Tsuji procedure of enol allylcarbonate formation and $Pd(OAc)_2$ oxidation to give (**47**).

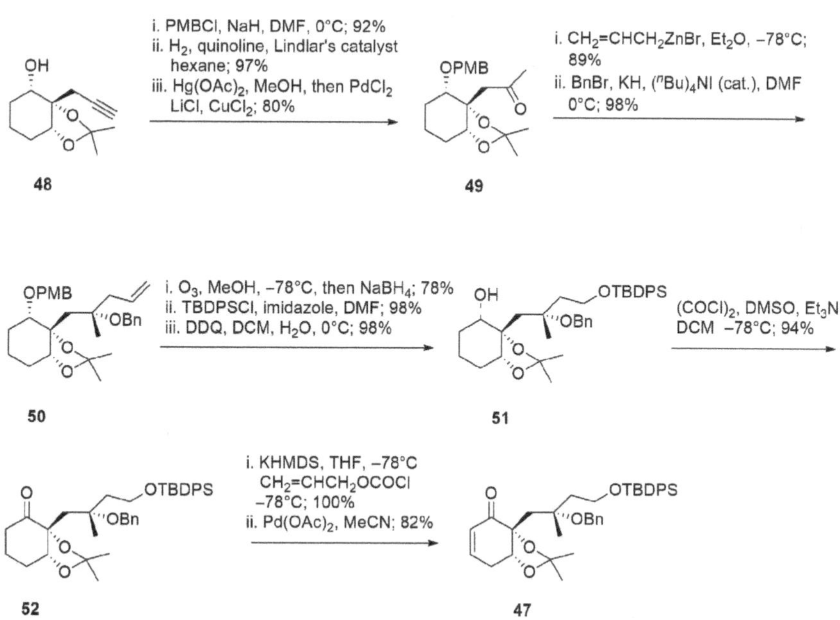

SCHEME 10.7 Synthesis of the Hauser acceptor (**47**) for aquayamycin.

Phthalide sulfone (**46**), prepared in more than 10 steps as described in Chapter 9, was condensed with highly oxygenated cyclohexenone (**47**) in the presence of LiO^{t}Bu to give the corresponding hydroquinone. This was, without purification, converted to its *O*-dimethyl derivative (**53**), the overall yield for the two steps being 73% (Scheme 10.8). The key aldehyde (**54**) was prepared from (**53**) by desilylation followed by Swern oxidation. Pedersen intramolecular pinacol coupling of (**54**) using VCl_3.(THF) and Zn followed by Swern oxidation of the pinacol product provided the tetracyclic intermediate (**55**). This was then converted to (**56**) in two steps: (i) acetonide hydrolysis and (ii) mesylate formation of the secondary alcohol group. Reductive removal of the benzyl groups of (**56**) followed by selective deprotection of C-8 phenolic group of the resultant product by benzyl bromide and Cs_2CO_3 gave (**57**). CAN oxidation of (**57**) followed by reductive debenzylation furnished tetracycle (**58**). Elimination of the mesyl group in (**58**) with *i*-Pr_2NEt completed the total synthesis of aquayamycin (**45**), the overall yield being 21% for 13 steps from sulfone phthalide (**46**) and cyclohexenone derivative (**47**) (Scheme 10.8).

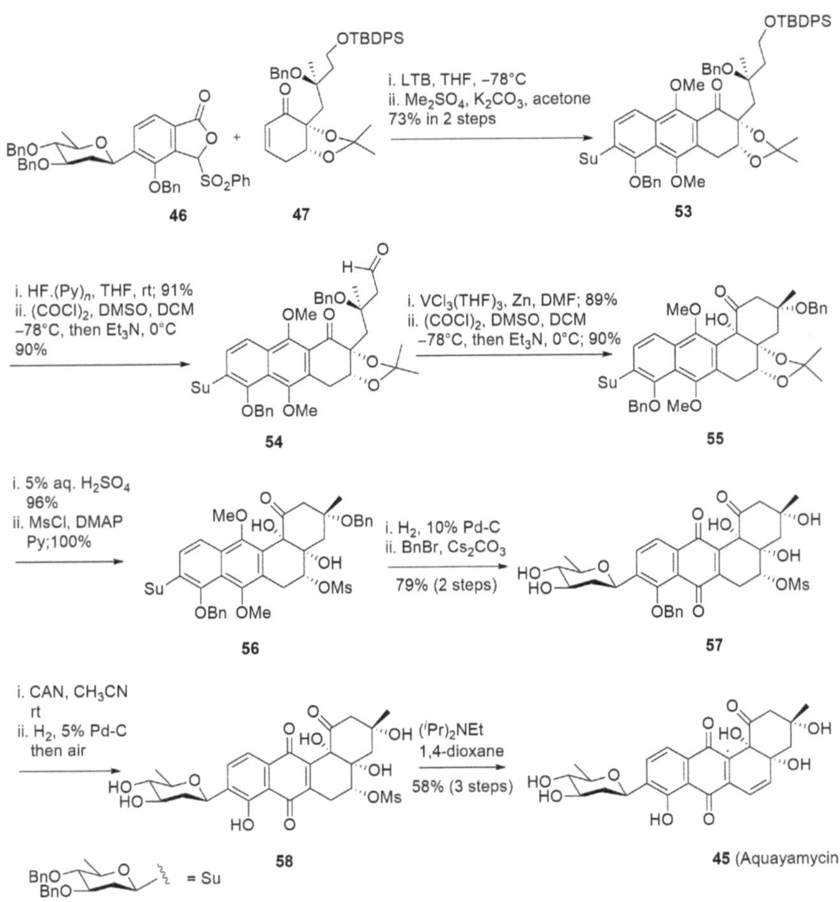

SCHEME 10.8 Last stages of the synthesis of aquayamycin (**45**).

10.4 KOERT TOTAL SYNTHESIS OF ISOQUINOCYCLINONE (59)

The only total synthesis of isoquinocyclinone (**59**), the aglycone of isoquinocycline A and B, was reported by the Koert group.[5] The heptacyclic target consists of a pyrrolo[2,3-*b*]pyrrole system joined to a tetracene quinonoid. Its synthesis was achieved in about 30 steps in three stages: (i) synthesis of decalenone (**60**), (ii) elaboration of (**60**) to tetracene subunit (**61**) using a Hauser annulation, and (iii) fabrication of pyrrolo[2,3-*b*]pyrrole unit via a Ni(0)-mediated cyanide addition, leading to the completion of the synthesis of (**59**) (Fig. 10.1).

The decalenone (**60**) was synthesized by using Diels-Alder reaction between diene (**62**) and dienophile (**63**) as the key step (Scheme 10.9). The diene (**62**) was prepared from acrolein in four steps and the corresponding dienophile (**63**) from 2-butene-1,4-ol in three steps. Lithiation of iodopropane (**64**) with ^{t}BuLi followed by addition of methacrolein gave the corresponding alcohol, which was acetylated with acetic anhydride and pyridine to furnish (**65**). Treatment of (**65**) with in situ generated Pd(0) resulted in elimination of acetic acid giving diene (**62**). The Diels-Alder reaction of the diene (**62**) with the enone (**63**) in the presence of dimethylaluminium chloride led to racemic cyclohexene (**66**). Acid-catalyzed intramolecular aldol condensation of (**66**) produced the desired

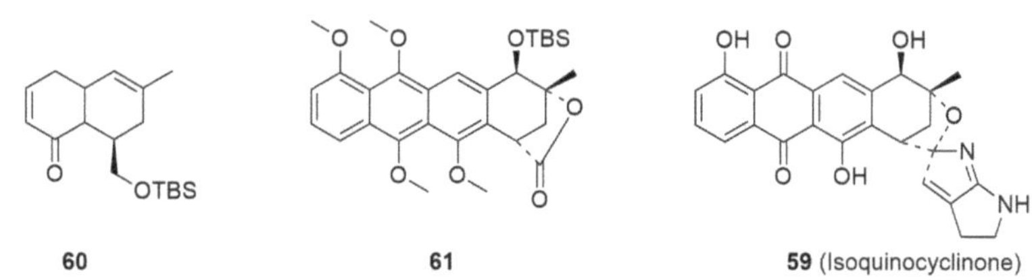

FIG. 10.1 Structures of isoquinocyclinone (**59**) and its synthetic intermediates [2014 CEJ 11300 Koert U].

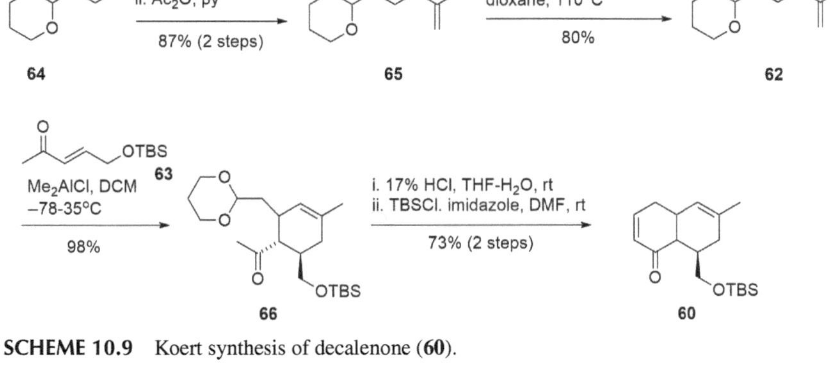

SCHEME 10.9 Koert synthesis of decalenone (**60**).

decalenone, which was derivatized as TBS-protected decalenone (**60**) by treatment with TBSCl-imidazole.

The Hauser annulation between the cyanophthalide (**67**) and the decalenone (**60**) in the presence of LDA at −78°C afforded the corresponding tetracene diol, which was aerially oxidized to give monohydroxy tetracene (**68**) (Scheme 10.10). *O*-methylation of (**68**) with methyl sulfate and potassium carbonate followed by desilylation of the resulting ether furnished tetracene alcohol (**69**). The alcohol function in (**69**) was converted into carboxylic acid by two consecutive oxidations: Dess-Martin oxidation and Pinnick oxidation to yield acid (**70**). Iodolactonization of (**70**) with I_2-KI afforded lactone (**71**). The benzylic iodide (**71**) was transformed to benzylic epoxide (**72**) by treatment with methanol-K_2CO_3. Acid-catalyzed opening of the epoxide (**72**) afforded hydroxy lactone (**73**). Reductive *O*-methylation of (**73**) with Me_2SO_4, $Na_2S_2O_4$ and KOH gave the respective tetramethoxytetracene, free hydroxy of which was protected as TBS ether to provide (**74**).

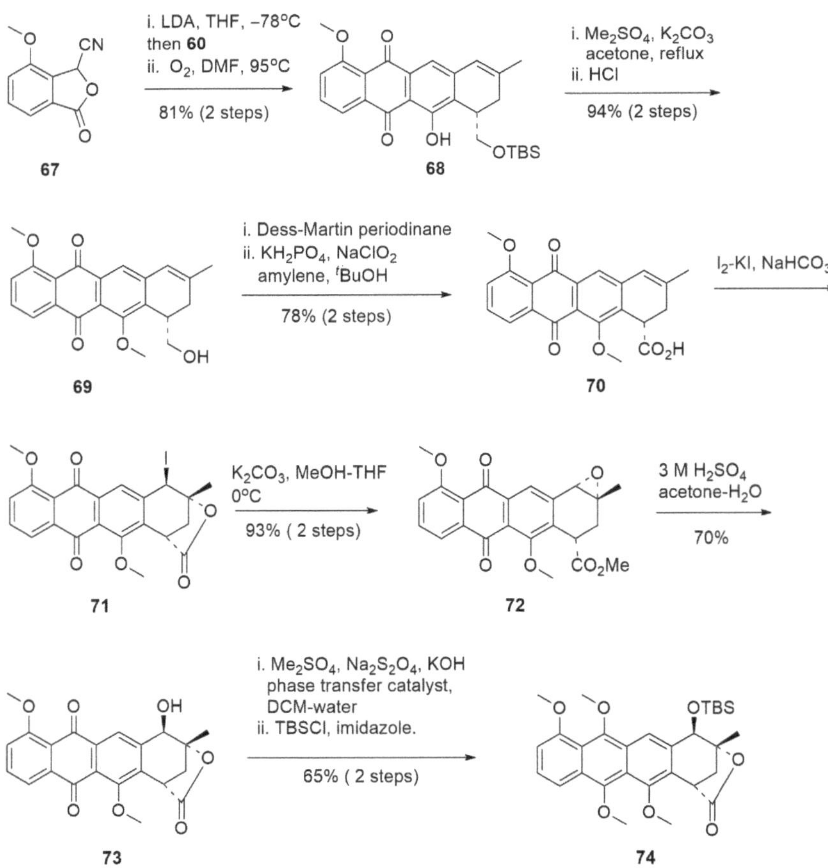

SCHEME 10.10 Synthesis of tetracene lactone (**74**).

The reaction of lithio salt of silyl ether (**75**) with the lactone (**74**) afforded alkynone (**76**) (Scheme 10.11). This was subjected to a nickel-catalyzed 1,4-addition of cyanide ion to afford cyclic imidate (**77**). It was transformed to ditosylate (**78**) in three steps. Treatment of (**77**) with TsCl-Et_3N followed by sequential reaction with HCl and TsCl-Et_3N provided (**78**). The imidate (**78**) was converted into cyclic amidine (**79**) by its exposure to NH_3-TMSCl. It may be noted the corresponding epimeric amidine was also formed in the reaction. The hexacyclic intermediate (**79**) was then was oxidized with CAN to the corresponding anthraquinone so that a thermal ring closure took place in the presence of 2,6-lutidine to give the desired pyrrolopyrrole (**80**). The protecting groups of (**80**) were removed by sequential application of TBAF, HF, and BCl_3 to give (**59**). Thus, isoquinocyclinone (**59**) was synthesized in a sequence of 29 steps in 0.68% overall yield. This total synthesis features a unique construction of a pyrrolopyrrole substructure.

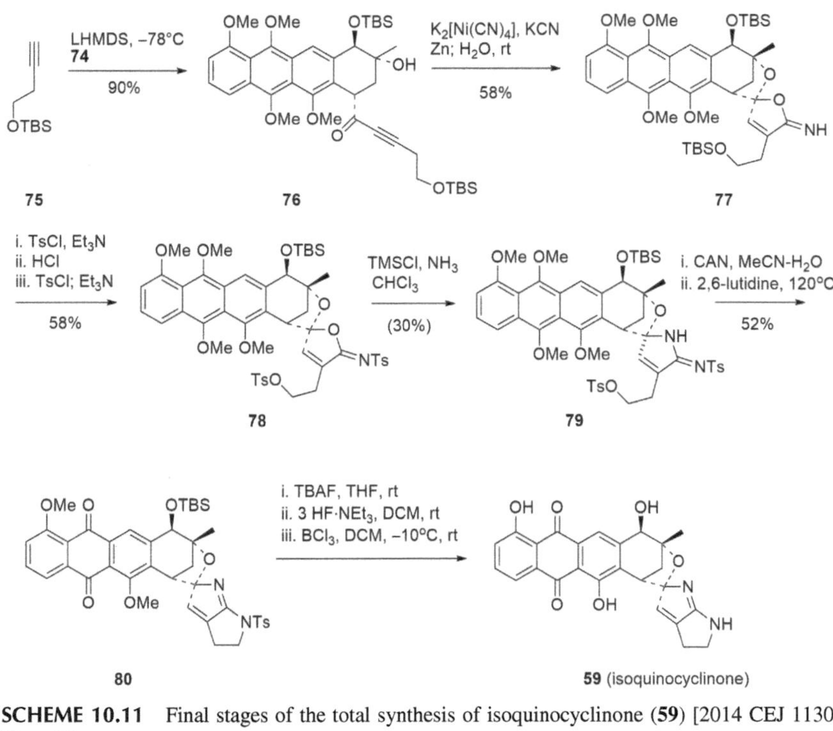

SCHEME 10.11 Final stages of the total synthesis of isoquinocyclinone (**59**) [2014 CEJ 11300 Koert U].

10.5 NICOLAOU SYNTHESIS OF UNCIALAMYCIN (81)

Presented in Schemes 10.12–10.15 is the second-generation synthesis of uncialamycin (**81**) by Nicolaou et al.[6] This synthesis was completed in 22 steps starting from hydroxyisatin in 11% overall yield. The salient feature of the new linear

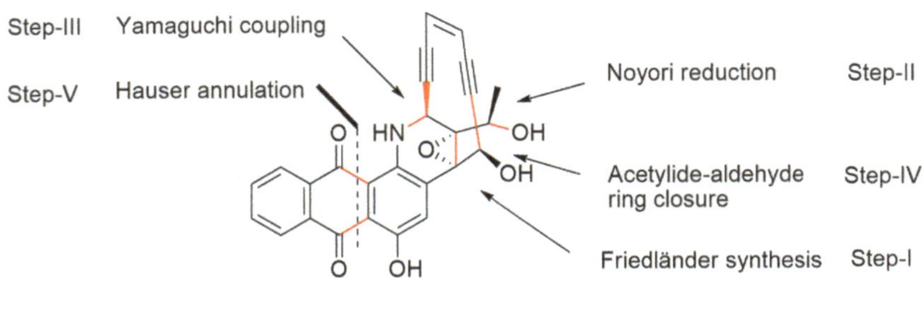

81

SCHEME 10.12 Various stages of the Nicolaou synthesis of uncialamycin [2016 JACS 8235 Nicolaou KC].

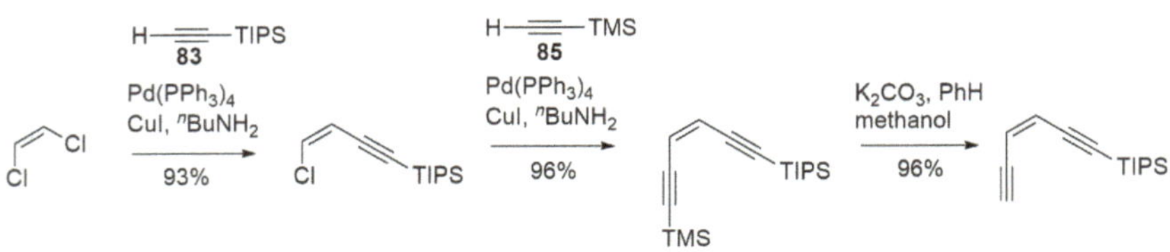

82 **84** **86** **87**

SCHEME 10.13 Preparation of enediyne fragment (**87**) [2016 JACS 8235 Nicolaou KC].

synthesis is the Nicolaou modification of the Hauser annulation, which, on this occasion, was performed with a semiquinone aminal, making the synthesis more efficient than that with iminoquinone acceptors. The other salient steps of the synthesis are graphically displayed in Scheme 10.12. These are: (i) Friedländer quinoline synthesis, (ii) Noyori asymmetric reduction, (iii) Yamaguchi coupling, and (iv) acetylene-aldehyde cyclization. Furthermore, dearomatization of hydroxyquinoline motif is also an important step of the total synthesis.

Sonogashira coupling between (**82**) and (**83**) in the presence of $Pd(PPh_3)_4$ and CuI afforded chloroenyne (**84**) in 93% yield, which was coupled, under similar conditions, with trimethylsilylacetylene (**85**) to give enediyne (**86**) 96% yield. Selective monodesilylation of TMS group of (**86**) with K_2CO_3 in MeOH furnished the desired TIPS-monoprotected *cis*-enediyne (**87**) in 96% yield (Scheme 10.13).

BBr_3-promoted demethylation of 5-methoxyisatin (**80**) at room temperature led to hydroxyisatin, which was submitted to the Friedländer reaction (aq NaOH, 25°C) with methyl ketone (**89**), to produce hydroxyquinoline ketoacid (**90**) in 91% yield via the intermediate aryl vinylogous amide (**91**) (Scheme 10.14). Bisalkylation of (**90**) with 3,4-dimethoxybenzyl bromide (DMB) in the presence of Cs_2CO_3 afforded *bis*-DMB derivative (**92**). Without purification, it was submitted to Noyori asymmetric reduction [HCO_2H, Et_3N, (*S,S*)-(**93**) cat.] and in situ lactonization to yield (**94**) in 82% overall yield and 96% ee. Next, reduction of the lactone moiety of (+)- (**94**) with DIBAL followed by silylation of the resulting lactol with TESCl gave TES-ether (**95**) in 91% yield

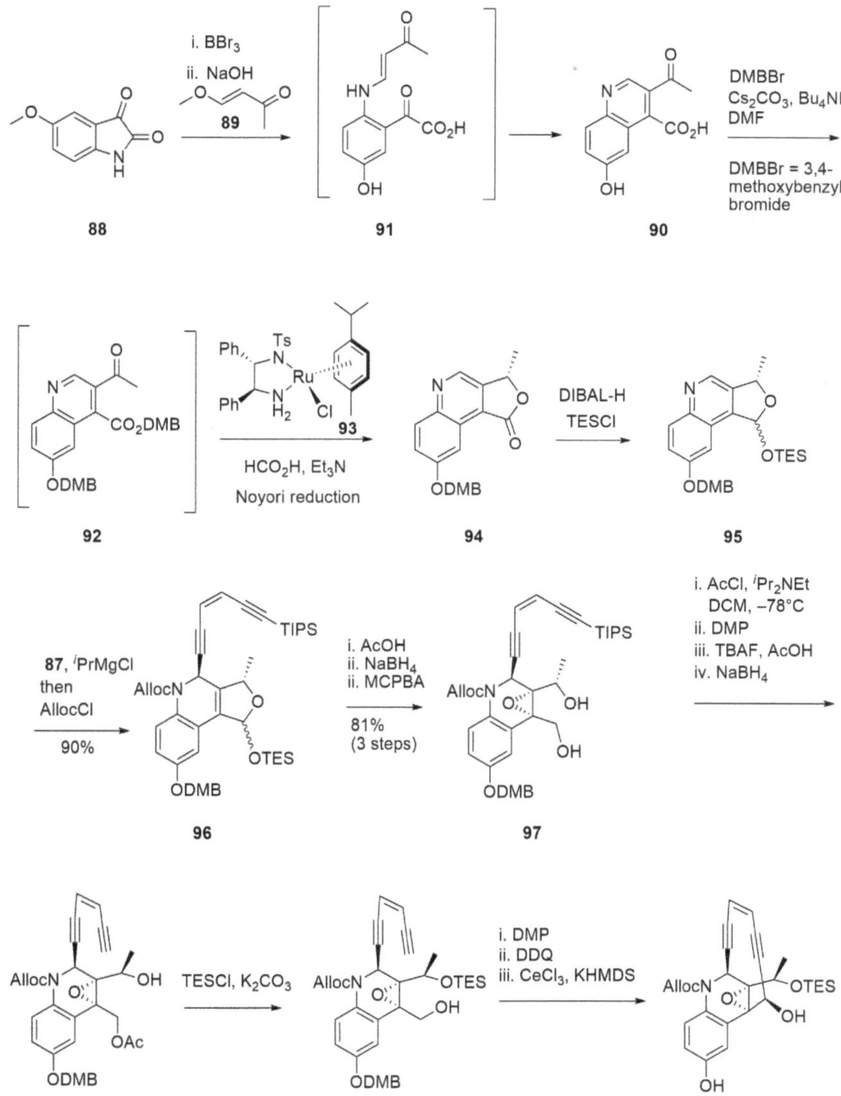

SCHEME 10.14 Friedlander route to quinoline unit (**90**) and its elaboration to intermediate (**100**) [2016 JACS 8235 Nicolaou KC].

(c.1:1 dr). The addition of (**95**) to a preformed solution of the Grignard reagent derived from enediyne (**87**) and ^{i}PrMgCl at 0°C followed by addition of AllocCl furnished an exclusively anti-addition product (+)-(**96**) in 90% yield. The antiselectivity observed in this reaction is perhaps due to the steric effect of the lactone methyl group. The TES group of (+)-(**96**) was removed with AcOH, and the resulting lactol was treated with $NaBH_4$ in MeOH to afford the corresponding

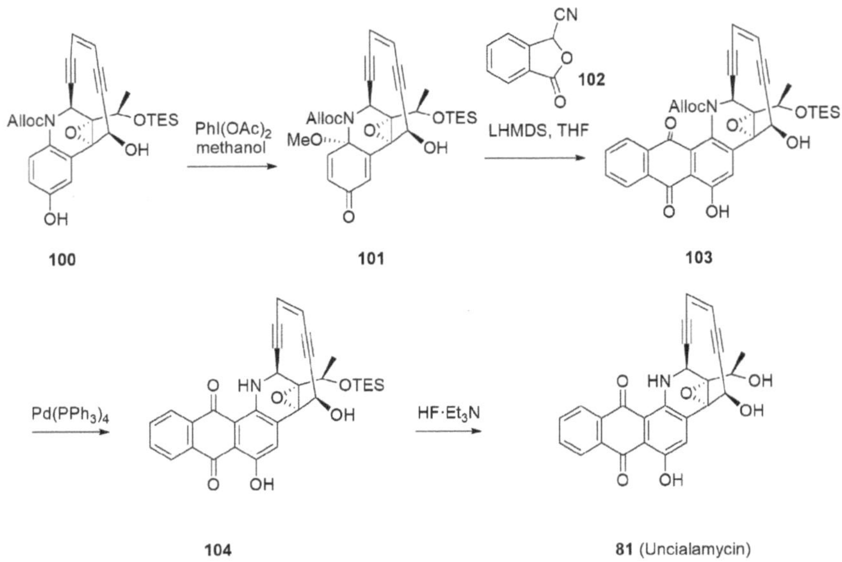

SCHEME 10.15 Final stage of Nicolaou synthesis of uncialamycin utilizing Hauser annulation [2007 ACIE Nicolaou KC].

diol, whose epoxidation with MCPBA furnished epoxide diol (+)-(**97**) in 81% overall yield. Monoacetylation of (+)-(**97**), with AcCl and Hünig's base ($^{i}Pr_2NEt$, DCM, −78°C) followed by oxidation with Dess-Martin periodinane afforded corresponding ketoacetate. Removal of the TIPS group of the ketoacetate by TBAF, AcOH followed by reduction with $NaBH_4$ led stereochemically inverted hydroxyl compound (**98**). The hydroxyl group of (**98**) was protected as the TES ether and the acetate group was cleaved with K_2CO_3, MeOH to furnish compound (**99**). DMP oxidation of (**99**) followed by DDQ oxidation of the resultant aldehyde gave the corresponding cyclization substrate, the treatment of which with KHMDS-$CeCl_3$ provided the cyclized product (**100**) in good yield and stereoselectivity.

Next, the advanced precursor (**100**) was dearomatized with $PhI(OAc)_2$-MeOH to form *p*-methoxy semiquinone animal (**101**). The reaction of cyanophthalide (**102**) with LHMDS in THF at −78°C followed by addition of the acceptor (**101**) afforded protected uncialamycin (**103**) (Scheme 10.15). Alloc and TES protecting groups of (**103**) were removed by exposure to $Pd(PPh_3)_4$ and 3HF•Et_3N, respectively, to deliver (+)-uncialamycin (**81**) via intermediate (**104**).

10.6 YOSHII SYNTHESIS OF GRANATICIN (105)

A notable application of the Hauser annulation is found in the Yoshii total synthesis of optically active natural granaticin (**105**), which can be described as a quinone-*C*-glycoside or a benzoisochromanoquinone. Granaticin (**105**), first isolated in 1957 from the culture of *Streptomyces olivaceus*, had drawn much attention from both chemical and biochemical researchers due to its unusual sugar moiety and pyran ring. As shown in Scheme 10.16, the application of the annulation identified (**106**) and (**107**) as fragments for the late-stage coupling.[7]

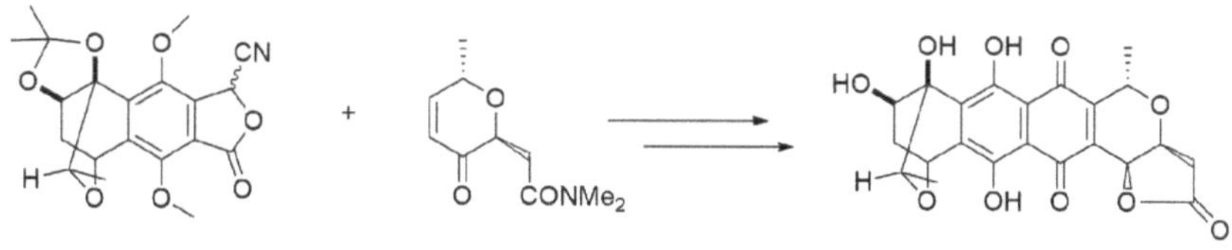

106 **107** **105** (Granaticin)

SCHEME 10.16 Key intermediates of Yoshii synthesis of granaticin [1989 CCS 354 Yoshii E].

The synthesis of the granaticin cyanophthalide fragment (**106**) is described in Scheme 10.17.[8] The reaction of tetralone (**108**) with lithiated methyl vinyl ether (**109**) followed by acid treatment of the resultant product afforded ketol (**110**) in 90% yield. This was then reduced with sodium borohydride to afford diol (**111**) as a mixture of diastereomers. Regioselective dehydration of (**111**) to allylic alcohol (**112**) was accomplished following a conventional three-step sequence: formation of monoacetate, dehydration of the acetate with thionyl chloride in the presence of pyridine, and hydrolysis of the resultant allylic acetate in 65% yield from (**111**). Resolution of the allylic alcohol (**112**) was done by (−)-*N*-(1-phenylethyl)carbamate to give (**113**). Dihydroxylation of (**113**) with OsO_4-NMO in aqueous *tert*-butyl alcohol at 60°C produced a mixture of triol (**114**) and its 1′-epimer in 80% yield, from which pure (**114**) was obtained by recrystallization in 65% yield. Treatment of (**114**) with NBS in carbon tetrachloride at 40°C in the presence of azobisisobutyronitrile (AIBN) afforded benzylic bromination product (**115**), which on treatment with 1 equiv. of silver perchlorate in THF at room temperature afforded (**116**) in 60%–67% yields. The reaction of (**116**) with 2-methoxypropene in the presence of camphorsulfonic acid furnished acetonide (**117**). Treatment of (**117**) with 1 equiv. of ^{n}BuLi at −100°C and 3 equiv. of diethylcarbamoyl chloride at −70°C afforded (**118**) in 90% yield. When the compound (**118**) was treated with 3 equiv. of ^{t}BuLi at −80°C followed by 4 equiv. of DMF, aldehyde (**119**) was obtained in 96% yield. The aldehyde (**119**) was then reacted with trimethylsilylcyanide in the presence of KCN and 18-crown-6 ether in DCM to form the respective (*O*-trimethylsilyl)cyanohydrin, which was, without purification, dissolved in acetic acid at room temperature. Extractive work-up followed by exposure of the product to 2-methoxypropene gave cyanophthalide (**106**) in 84% yield as a 1:1 diastereomeric mixture.

The acceptor (**107**) was obtained from di-*O*-acetylrhamnal (**120**). Its reaction with ketene methyl *t*-butyldimethylsilyl acetal in the presence of BF_3 etherate in toluene gave respective *C*-glycoside, which was deacetylated with KOH-methanol to form (**121**). It was then mesylated with MsCl-Et_3N to give (**122**). Silica gel promoted intramolecular rearrangement afforded lactone (**123**). The lactone (**123**) was then converted to keto carboxamide (**107**) by aminolysis with

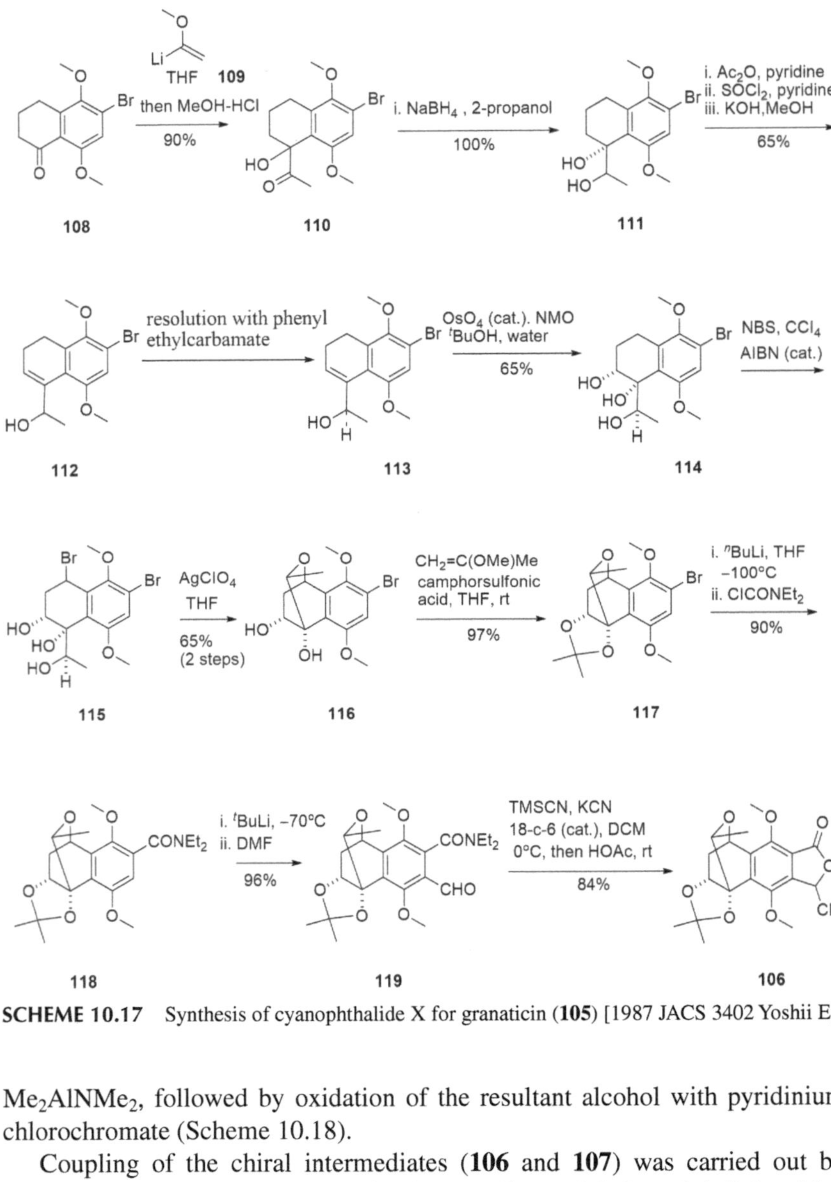

SCHEME 10.17 Synthesis of cyanophthalide X for granaticin (**105**) [1987 JACS 3402 Yoshii E].

Me_2AlNMe_2, followed by oxidation of the resultant alcohol with pyridinium chlorochromate (Scheme 10.18).

Coupling of the chiral intermediates (**106** and **107**) was carried out by $LiCH_2SOMe$ to give the respective hexacyclic naphthalene-1,4-diol, which was *O*-methylated with Me_2SO_4-K_2CO_3 to provide (**124**) in overall 50% yield (Scheme 10.19). Beyond this crucial annulation, four more steps were required to complete the synthesis of granaticin (**105**). The ketone carbonyl was reduced with $Li^{s}Bu_3BH$ to afford (**125**). Lactonization of (**125**) to (**126**) was achieved by moist silica gel. Oxidation of (**126**) with ceric ammonium nitrate to dimethoxy-1,4-naphthoquinone (**127**), followed by demethylation with $AlCl_3$-Et_2S, resulted in granaticin (**105**).

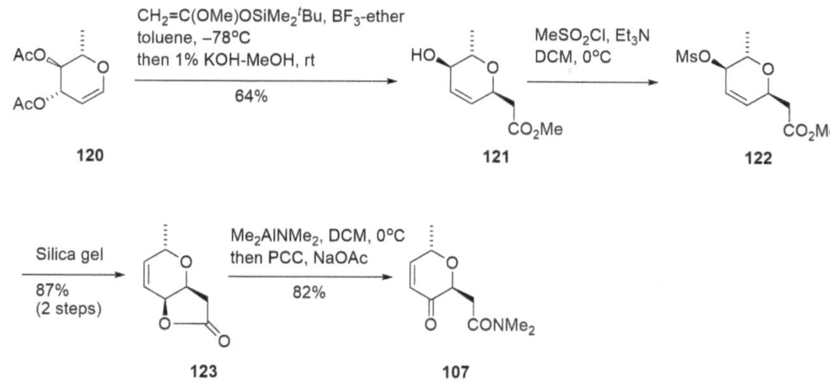

SCHEME 10.18 Synthesis Hauser acceptor for granaticin [1989 CCS 354 Yoshii E].

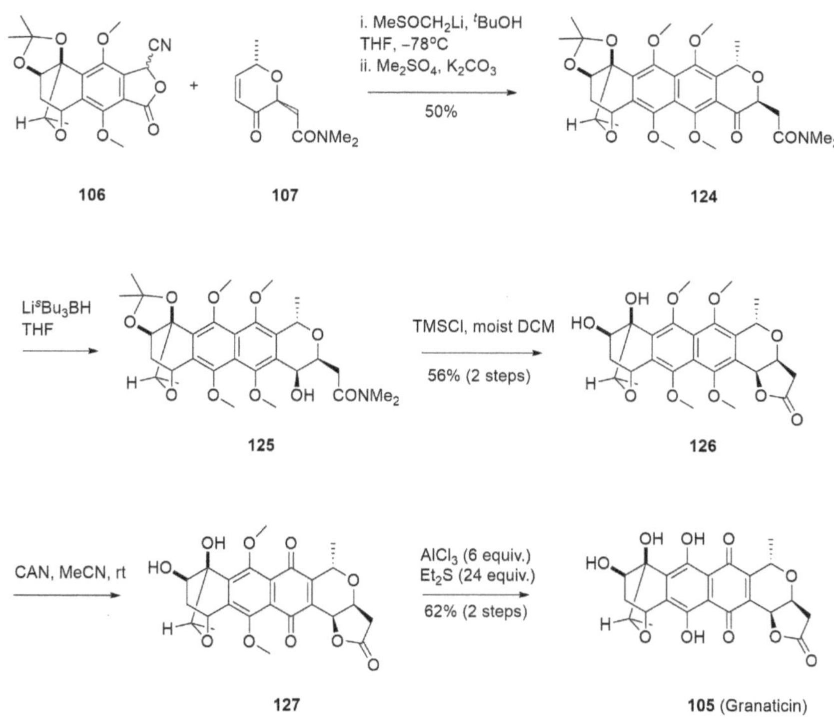

SCHEME 10.19 Final stage of the total synthesis of granaticin X (**105**).

10.7 BACH SYNTHESIS OF FREDERICAMYCIN A (128)

Both the Bach group and Julia group independently utilized Hauser annulation as the key step for the synthesis of fredericamycin A (**128**).[9] The Bach synthesis is presented here.[9a] It consisted of elaboration of indanone (**129**) to (**130**), and coupling of the latter with cyanophthalide (**131**) in the last stages of the total synthesis furnishing the core structure of fredericamycin A (**128**) (Scheme 10.20).

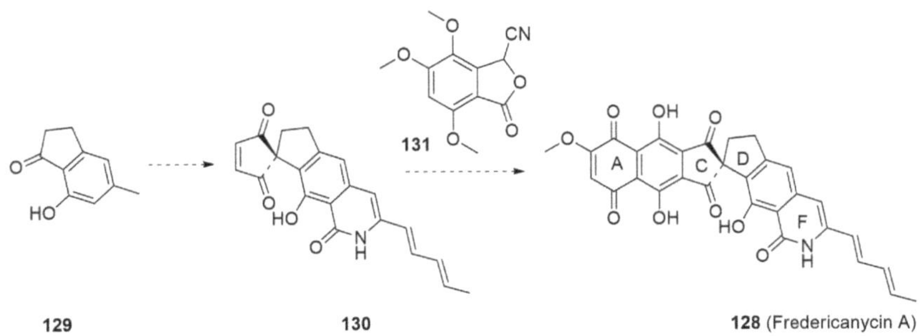

SCHEME 10.20 Outline of the Bach synthesis of fredericamycinA (**128**) [1994 JACS 9921 Bach RD].

The amide (**132**), prepared from the corresponding acid, was formylated with DMF via *ortho*-lithiation with ^{s}BuLi. The resulting formylbenzamide was hydrolyzed with 10% HCl in acetic acid to give phthalaldehydic acid (**133**). Treatment of (**133**) with KCN in aq HCl solution directly afforded the donor cyanophthalide (**131**) in 93% yield (Scheme 10.21).

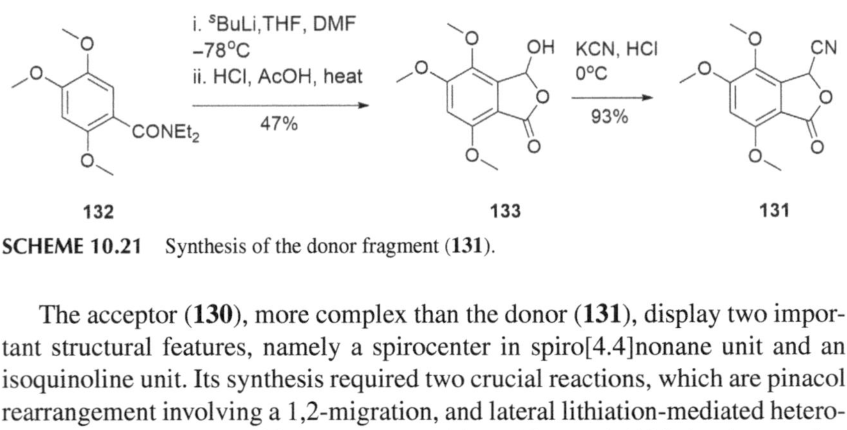

SCHEME 10.21 Synthesis of the donor fragment (**131**).

The acceptor (**130**), more complex than the donor (**131**), display two important structural features, namely a spirocenter in spiro[4.4]nonane unit and an isoquinoline unit. Its synthesis required two crucial reactions, which are pinacol rearrangement involving a 1,2-migration, and lateral lithiation-mediated heteroannulation. Indenone (**129**) was converted into thioacetal (**134**) by the reaction with ethanethiol and stannic chloride. Treatment of the thioacetal (**134**) with siloxycyclobutene (**135**) in the presence of mercuric trifluoroactete furnished the spiro compound (**136**). This conversion proceeded through nucleophilic displacement of thioether group followed by pinacol rearrangement of the intermediate (**137**). The diketone (**136**) was bisacetalated with ethylene glycol to give bisketal (**138**), which was selectively brominated with NBS. The bromo product was treated with MeI-KH to give its methyl ether (**139**). Lithiation-promoted alkoxycarboxylation with methyl chloroformate gave (**140a**). For the synthesis of isocoumarin (**141**), the substrate (**140a**) was first lithiated at the lateral methyl group with LTMP to give (**140b**), which was transmetalated with CuCN, LiCl, and then quenched with an excess of sorboyl chloride to give isocoumarin (**141**) via intermediate (**142**). Three more steps—aminolysis with ammonia, deketalization with aq HCl solution, and dehydrogenation with phenyltrimethylammonium tribromide—completed the conversion of the isocoumarin (**141**) to acceptor isoquinolone (**130**) (Scheme 10.22).

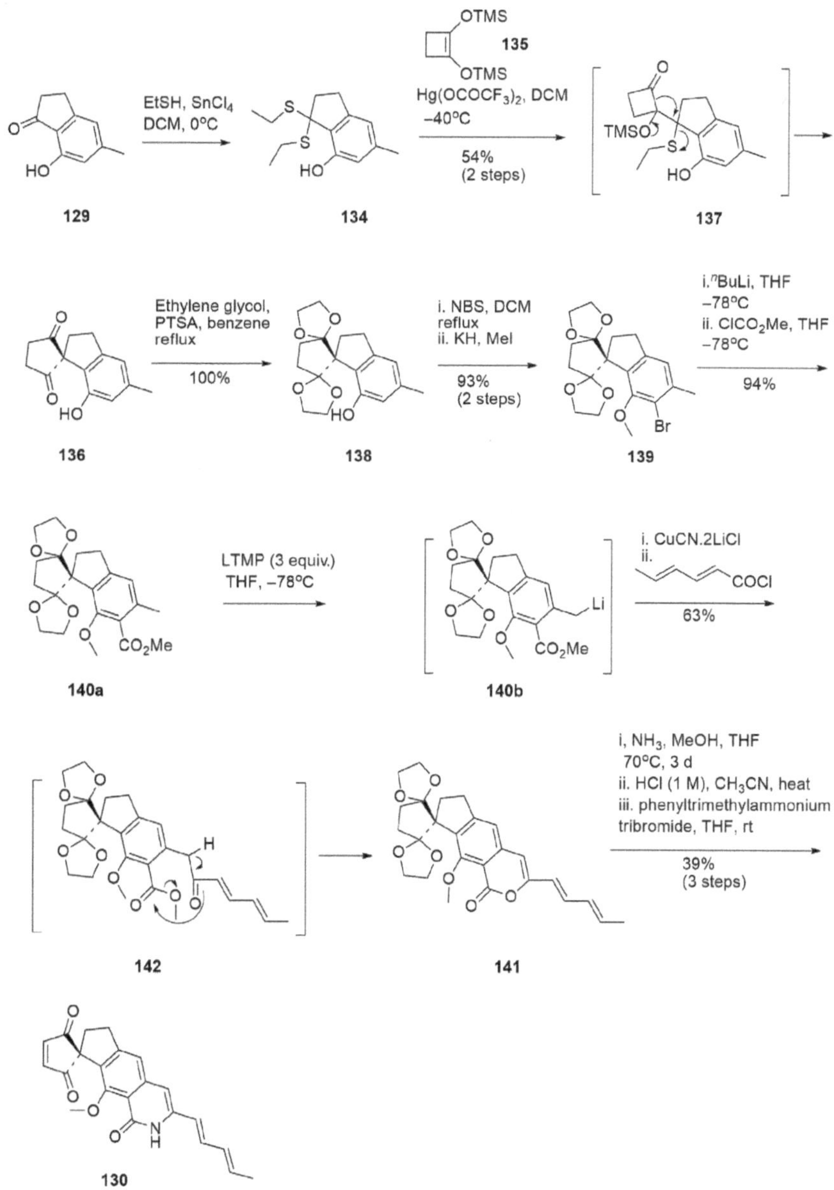

SCHEME 10.22 Synthesis of the acceptor fragment (**130**).

The final stage of the total synthesis was achieved by a modification of the Hauser annulation strategy. When the lithio anion of (**131**), generated by the action of LDA at −78°C on (**131**), was treated with enedione (**130**), a diastereomeric mixture of Michael adducts (**143**) was obtained. This mixture of isomers was then cyclized to fredericamycin derivative (**144**) with NaOEt in ethanol, the overall yield of the two-step annulation being 78%. Such a modification was previously reported by Julia et al. in 1993.[9b] Their annulation was applied to a cyclopent[*g*]isoquinoline derivative. The similar two-step strategy was applied to annulation with cyclopentenone by Clive et al. in 1994. This work of Bach et al. did not require any protection of the NH group of pyridine moiety during Hauser annulation. The final step of the total synthesis was demethylation followed by aerial oxidation of the resulting quinol. The compound (**144**) was first treated with boron tribromide (3 equiv.) and then exposed to the air for two days to provide (±)-fredericamycin A (**128**) as a violet solid. Thus, the total synthesis required a total of 12 steps from (**130**) and (**129**), the overall yield being 6% (Scheme 10.23).

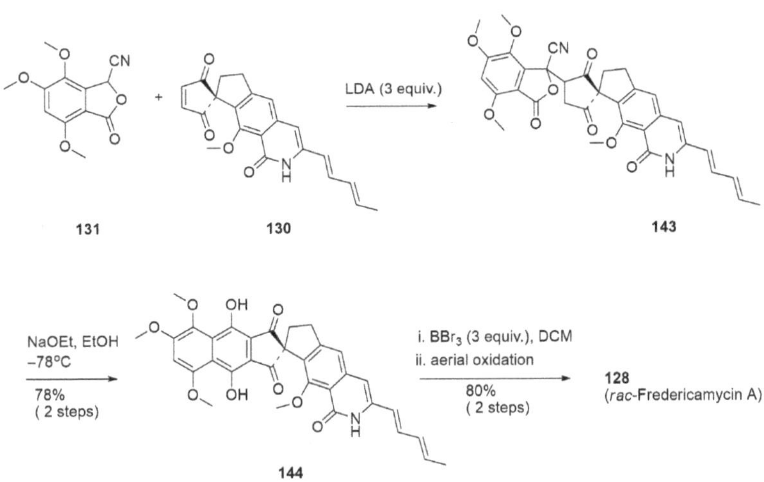

SCHEME 10.23 Last stages of the total synthesis of fredericamycin A (X) [1994 JACS 9921 Bach RD].

10.8 MYERS SYNTHESIS OF TRIOXACARCIN A (145)

In an elegant synthetic approach to trioxacarcin A (**145**), which inhibits the growth of cultured human cancer cells, Myers and co-workers made use of Hauser annulation as the key step to assemble the hydroaromatic anthracenone core (Scheme 10.24).[10] The required cyanophthalide (**146**) was synthesized from a tetrasubstituted amide in a multistep sequence. Hauser annulation of (**146**) with cyclohexenone (**147**) in the presence of LiO^{t}Bu afforded anthracenone (**148**) after selective *O*-methylation. With several more steps, the hydroaromatic scaffold (**148**) was elaborated to trioxacarcin A (**145**). The olefinic side chain in (**148**) was degraded by K_2OsO_4-$NaIO_4$ to give the corresponding aldehyde, which was reacted with bromocatecholborane for deprotection of

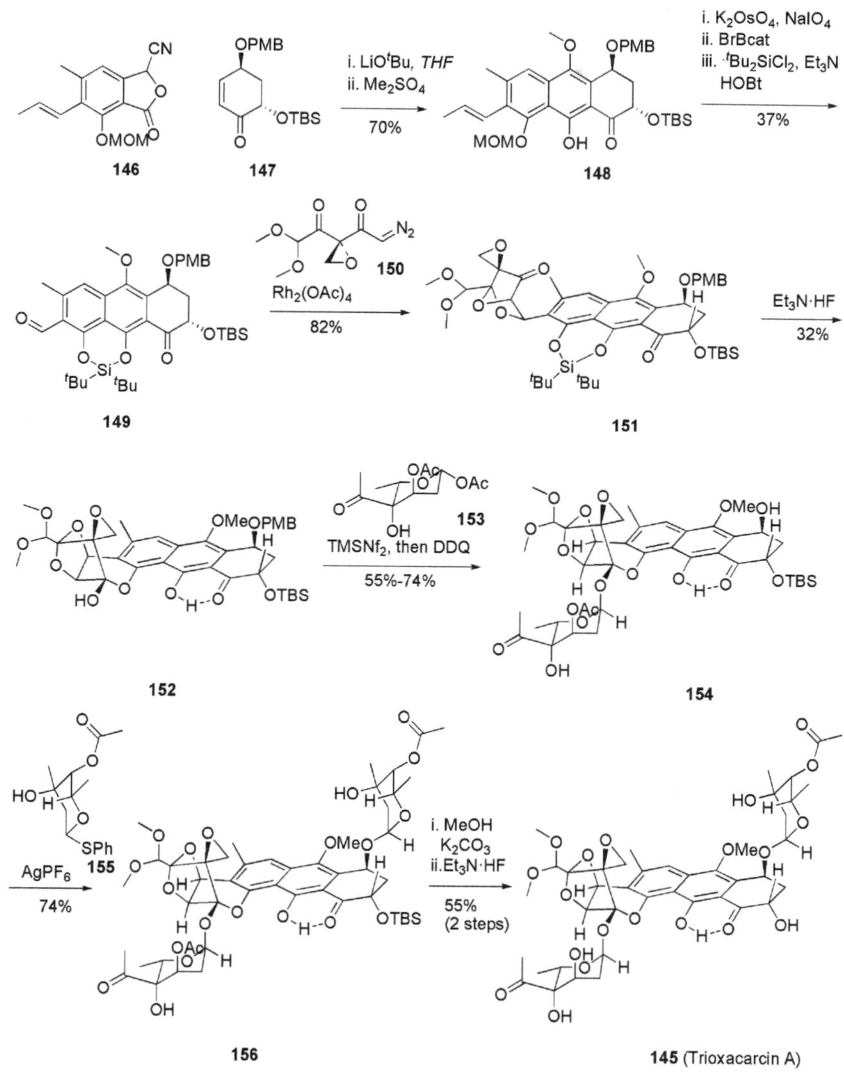

SCHEME 10.24 Myers synthesis of trioxacarcin A (**145**) [2013 NC 886 Myers AG].

the MOM group. Silylation (^{t}BuSiCl$_2$, HOBt, DIPEA) of the dihydroxy compound provided a stable di-*tert*-butylsiloxane derivative (**149**). Treatment of the anthracenone derivative (**149**) with diazoketone (**150**) in the presence of rhodium(II) acetate furnished (**151**), representing a fully oxygenated polycyclic skeleton of the trioxacarcins. The reaction presumably proceeded through cycloaddition of a carbonyl ylide. Conditions (triethylamine-trihydrofluoride) for the cleavage of the cyclic di-*tert*-butylsiloxane protecting group promoted removal of the group as well as hemiketalization giving (**152**). Glycosylation of (**152**) with (**153**) in the presence of TMSNTf$_2$, followed by removal of the

PMB group with DDQ, furnished (**154**), which was further glycosylated with (**155**) to provide (**156**). Selective deacetylation of (**156**) with methanol-K_2CO_3 followed by removal of the TBS group with triethylamine-trihydrofluoride, furnished trioxacarcin A (**145**).

Akin to Myers's approach, Nicolaou's approach (Scheme 10.25) to the trioxacarcins was founded on a Hauser annulation as one of four key steps. By employing this annulation, they could accomplish the total synthesis of trioxacarcin DC-45-A2 (**157**), which is a naturally occurring antitumor antibiotic that serves as a biosynthetic precursor to a variety of other biologically active members of trioxacarcins. The Nicolaou synthesis was founded upon hydroanthracenone (**158a**), which was regiospecifically assembled by annulation of (**158b**) with cyclohexenone (**147**).[11]

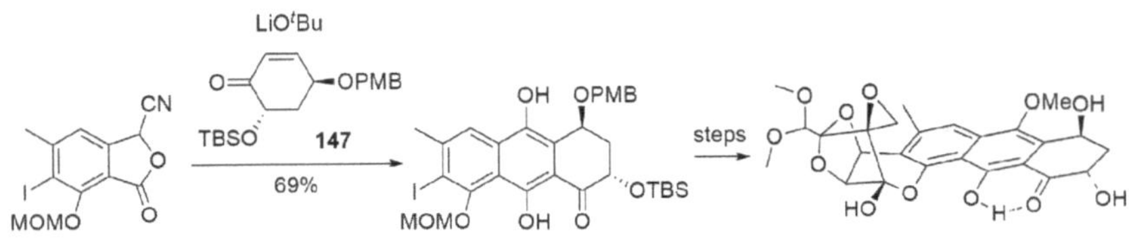

158b **158a** **157** (Trioxacarcin DC-45-A2)

SCHEME 10.25 Nicolaou synthesis of trioxacarcin DC-45-A2 [2015 ACIE 3074 Nicolaou KC].

10.9 YU SYNTHESIS OF LANDOMYCIN A (159)

The landomycins are a small but unique group of angucycline antibiotics. They feature a benz[*a*]anthraquinone aglycon, i.e., landomycinone with nonaromatic B ring, and deoxyoligosaccharide chains of various lengths attached at the C8 OH. These compounds show potent antitumor activities. The major synthetic challenges of the landomycins are: (i) construction of the unstable B ring, (ii) assembly of the deoxyoligosaccharides, and (iii) conjugation of the sugars to the poorly nucleophilic hydrogen-bonded C8 phenol of landomycinone. Yu synthesis was accomplished in two stages: synthesis of (**160**) and its conjugation with a pentasaccharide.[12]

The synthesis of (**160**) was achieved employing Hauser annulation to form the C ring and intramolecular Michael addition to form the B ring (Scheme 10.26). Cyanophthalide (**161**) and enone (**162**) were prepared from 2,5-dihydroxybenzoic acid and 3,5-dimethylphenol, respectively. Treatment of (**161**) with LiHMDS in THF at −78°C followed by addition of (**162**) delivered the desired naphthalene-1,4-diol (**163**) in 48% yield. The low yield of the annulation product (**163**) is probably due to the equilibration of the initial of Michael adduct to more stable enolate. Asymmetric reduction of ketone (**163**) with (*S*)-CBS catalyzed reduction led to the naphthodihydroquinone C6-ol derivative, which was oxidized with ceric ammonium nitrate to give naphthoquinone (**164**) in 84% yield with 93% ee. The C6 OH in (**164**) was protected as *p*-nitrobenzyl

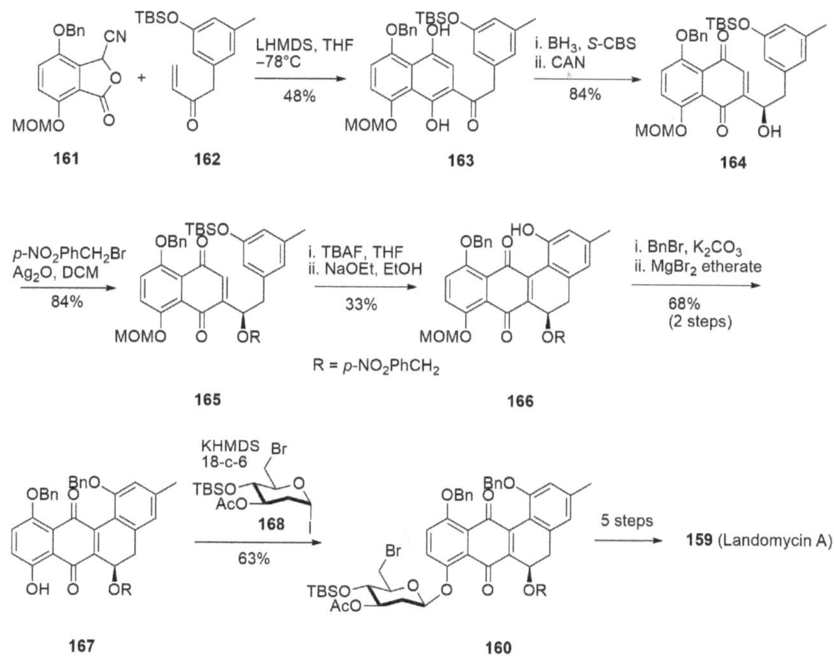

SCHEME 10.26 Yu synthesis of landomycin A (**159**) [2011 JACS 12433 Yu B].

ether to give (**165**). Removal of the 1-O-TBS protection followed by immediate intramolecular Michael addition afforded landomycinone derivative (**166**) in 33% yield. Reprotection of the C1 phenol with a benzyl group (84%) and removal of the 8-O-MOM group [$MgBr_2$ etherate, THF) provided the desired landomycinone (**167**). Treatment of (**167**) with in situ prepared iodide (**167**) in the presence of KHMDS and 18-c-6 ether gave β-glycoside (**160**) in 63% yield. Five more steps, including conjugation of a pentasaccaride, led to the total synthesis of landomycin A (**159**). Thus, this synthesis was completed in 63 steps and 0.34% overall yield (Fig. 10.2).

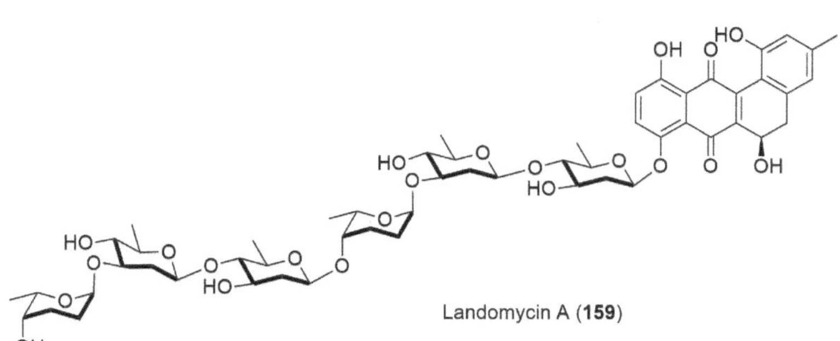

FIG. 10.2 Structure of landomycin A (**159**).

10.10 TATSUTA TOTAL SYNTHESIS OF HIBARIMICINONE (169)

Hibarimicinone (**169**), a potent v-Src tyrosine kinase inhibitor, was first synthesized by Tatsuta et al.[13] It relied upon the application of lateral lithiation initiated annulation of thiophthalides (Scheme 10.27). The dimeric tetracyclic ring skeleton of hibarimicinone (**169**) was constructed by thiophthalide annulations introduced by Mal et al. The chiral decalenone (**170**) was synthesized in several steps from commercially available L-arabinose. Bisthiophthalide (**171**) was synthesized as shown in Scheme 10.28. Under the influence of NaHMDS, the bisthiophthalide (**171**) underwent annulation with 2 equiv. of decalenone (**170**) to furnish the corresponding annulation product which was methylated to give (**172**) in modest yield. Seven more steps led to the total synthesis of hibarimicinone (**169**). The Shair group reported the synthesis of hibarimicinone (**169**) using a similar strategy (vide Chapter 7, transformation **219** → **222**).

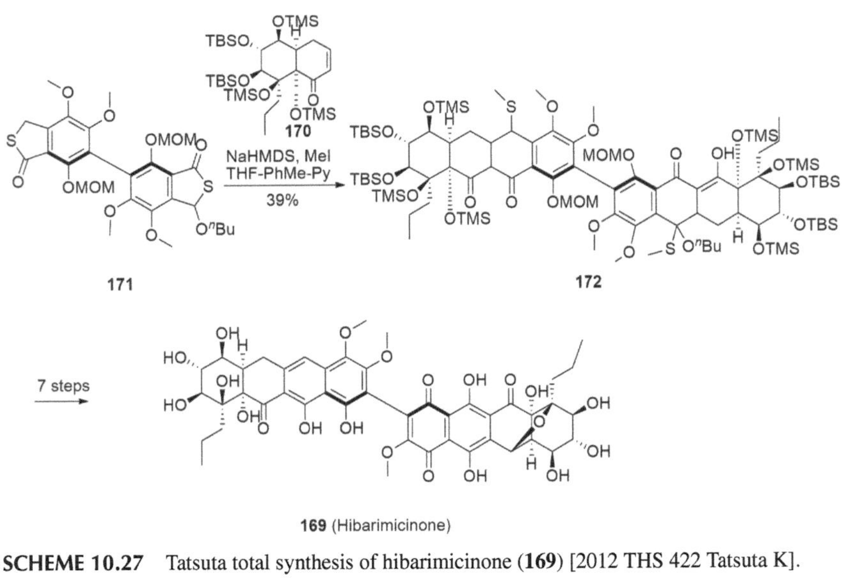

SCHEME 10.27 Tatsuta total synthesis of hibarimicinone (**169**) [2012 THS 422 Tatsuta K].

The synthesis of donor thiophthalide (**171**) was started from toluate (**173**). It was benzylically brominated to give (**174**). It was reacted with potassium thioacetate followed by NaOMe to furnish thiophthalide (**175**) containing free phenols. The racemic biaryl (**175**) was resolved with camphanic chloride to give optically pure (**176**) in good yield. It was then protected as its MOM ether (**177**). Monobromination of (**177**) with 1,3-dibromo-5,5-dimethylhydantoin, and AIBN in refluxing carbon tetrachloride, followed by treatment with ^{n}butanol, gave thiophthalide dimer (**171**) (Scheme 10.28).

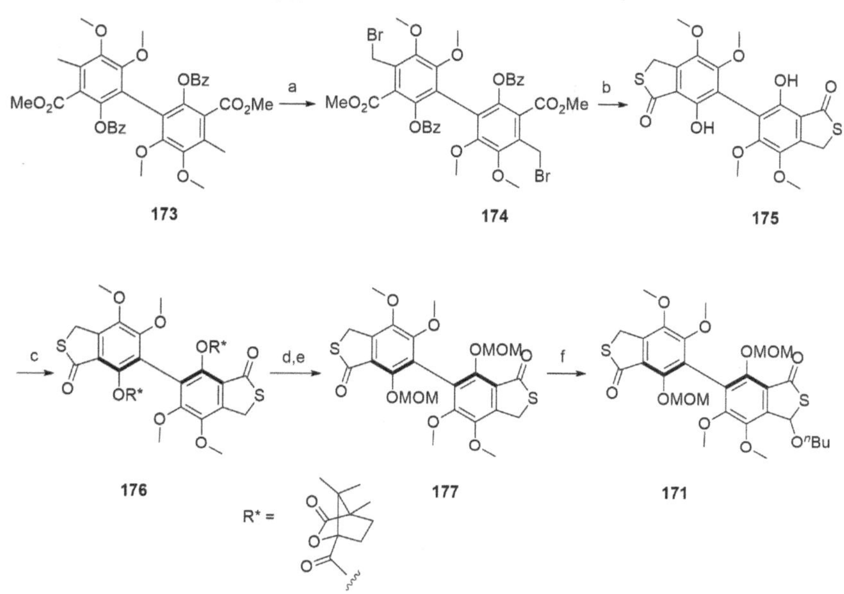

SCHEME 10.28 Synthesis of dimeric thiophthalide donor (**171**). (a) 1,3-Dibromo-5,5-dimethylhydantoin, AIBN, CCl_4, 70°C, 97%; (b) AcSK, THF, rt., then NaOMe, rt., 69%; (c) (−)-camphanic chloride, Et_3N, $ClCH_2CH_2Cl$, −78°C, then resolution, 43%; (d) NaOMe, MeOH-THF, 0°C, 95%; (e) MOMCl, $^iPr_2NEt$, DCM, rt., 93%; (f) 1,3-dibromo-5,5-dimethylhydantoin, AIBN, CCl_4, 60°C, then ^nBuOH, Et_3N, 50°C, 45%.

The synthesis of decalenone (**170**) is described in Scheme 10.29. L-Arabinose (**178**) was converted into sulfone derivative (**179**) in seven steps. It was silylated to give an open chain compound (**180**) in one step. The $SnCl_4$-promoted intramolecular Mukaiyama aldol condensation of (**180**) resulted in the formation of cyclohexenone (**181**). The Diels-Alder reaction between the enone (**181**) and 1-trimethylsiloxy-1,3-butadiene (**182**) proceeded stereoselectively to give adduct (**183**). Jones oxidation converted (**183**) to the corresponding enone, the SmI_2 reduction of which removed the phenylsulfonyl group to give respective Sm(III) enolate. This was oxidized to provide tertiary alcohol (**184**). A trisiloxy compound (**184**) was subjected to regioselective desilylation of the C10 OTBS group, and the resultant alcohol was oxidized with IBX to triketone (**185**). Regio- and stereoselective reduction of the ketone (**185**) was achieved in excellent yield by treatment with $NaBH(OAc)_3$ in EtOH to give alcohol (**186**) with the desired stereochemistry at C10. After extensive silylation of (**186**) with TMSOTf, the resultant product was reacted with allylmagnesium chloride to give tertiary alcohol (**187**). Silylation of the tertiary alcohol group in (**187**), followed by treatment with DBU in hot toluene in the presence of ^iPrOH, promoted the regioselective desilylation to afford enone (**188**). Hydrogenation of (**188**) yielded the decalin segment (**170**).

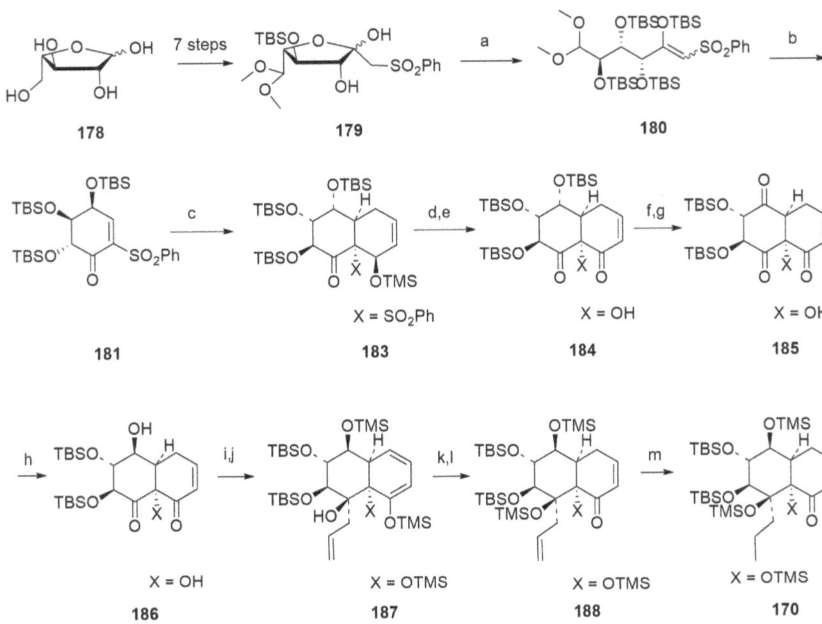

SCHEME 10.29 Synthesis of decalenone (**170**) [2012 THS 422 Tatsuta K]. (a) TBSOTf, 2,6-lutidine, DCM, 40°C, 94%; (b) $SnCl_4$, DCM, −78°C, 71%; (c) 1-trimethylsiloxy-1,3-butadiene (**182**), DTBC, PhMe, 90°C, 90%; (d) Jones reagent, acetone, rt., 92%; (e) SmI_2, THF, −78°C, then Oxone®, −78°C to rt., sat. $NaHCO_3$, rt., 75%; (f) $SnCl_4$, DCM, −30°C, 76%; (g) IBX, PhMe–DMSO, 50°C, 85%; (h) $NaBH(OAc)_3$, EtOH, 0°C, 92%; (i) TMSOTf, 2,6-lutidine, $ClCH_2CH_2Cl$, 80°C, 88%; (j) $CH_2{=}CHCH_2MgCl$, Et_2O, 0°C, 86%; (k) TMSOTf, 2,6-lutidine, $ClCH_2CH_2Cl$, 80°C, 87%; (l) DBU, ^{i}PrOH, PhMe, 80°C, 90%; (m) H_2, 10% Pd-C, PhMe, rt., 88%.

10.11 TOWNSEND SYNTHESIS OF AVERUFIN (189)

The first application of Sammes annulation is found in Townsend's brilliant synthesis of averufin (**189**), a pivotal biosynthetic intermediate of aflatoxin B_1.[14] Aflatoxin is a potent environmental carcinogen arising from the fungal infection of food grains and nuts by certain *Aspergillus* species and is one of the most highly rearranged polyketide natural products. The required bromoarene (**190**) was prepared in five steps from bis(*O*-methoxymethyl)resorcinol (**191**), involving two *ortho*-directed lithiations (Scheme 10.30). The resorcinol derivative (**191**) was lithiated with ^{n}BuLi and then treated with DMF to afford regioselectively arene aldehyde (**192**) in 63% yield. Its reaction with the Grignard reagent (**193**), derived from bromopentanone derivative, at room temperature furnished chain elongated arene (**194**). HCl-catalyzed acetolysis of the arene provided tricyclic arene (**195**). Regioselective bromination of the arene ketal (**195**) was executed through its regioselective *ortho* lithiation followed by reaction of the intermediate (**196**) with cyanogen bromide, and the desired bromo derivative (**190**) was obtained in 95% yield.

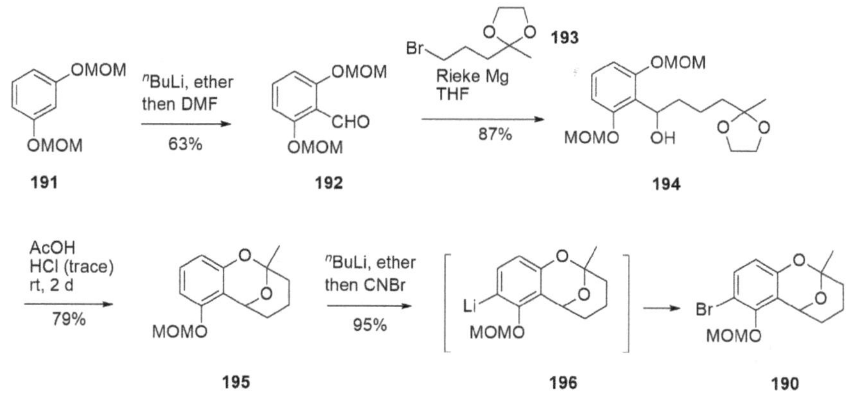

SCHEME 10.30 Townsend synthesis bromoarene precursor of averufin (**189**) [1981 JACS 6885 Townsend CA].

The required phthalide (**197**) was prepared in three steps from the benzoate (**198**). LAH reduction of the ester (**198**) gave alcohol (**199**), which was regioselectively brominated with NBS to give (**200**). Bromine-lithium exchange of (**200**) with ^{s}BuLi, followed by quenching with carbon dioxide, gave the phthalide (**197**) (Scheme 10.31). Its reaction with aryl bromide (**190**) in the

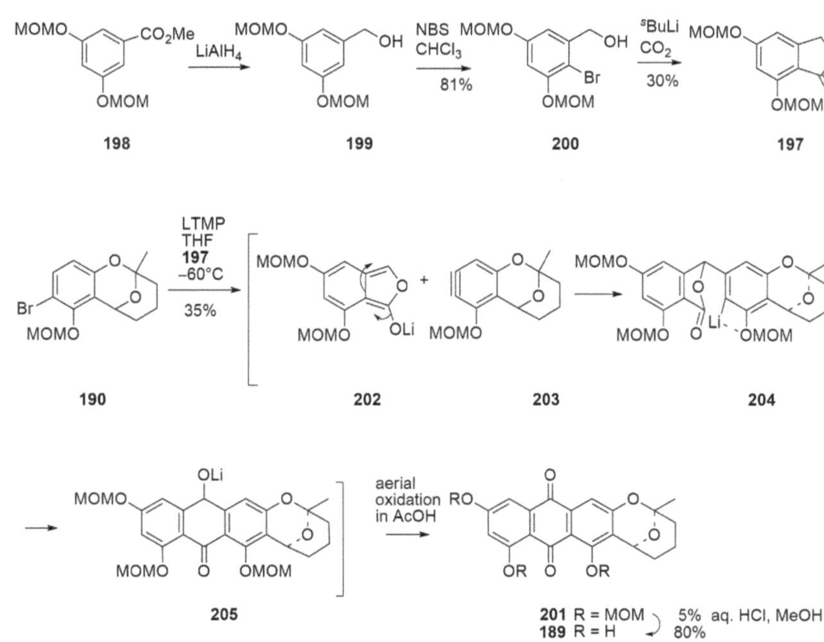

SCHEME 10.31 Townsend total synthesis of averufin (**189**) [1981 JACS 6885 Townsend CA].

presence of excess of LiTMP afforded (**201**) in 35% yield upon air oxidation. The sequence of steps leading to (**201**) is believed to involve formation of an aryne intermediate with elimination of LiBr. The phthalide anion (**202**) generated under the reaction conditions undergoes regioselective addition to (**203**), which is directed by the MOM protecting group. The resulting lithio species (**204**) is stabilized by coordination with the MOM group. Acetic acid quenching of (**205**) followed by aerial oxidation resulted in the formation of quinone (**201**). Removal of the MOM groups of (**201**) with 5% aq HCl yielded averufin (**189**).

10.12 MAL SYNTHESIS OF CLAUSEVATINE D (206) AND CLAUSAMINE C (207)

The synthesis of clausevatine D (**206**) and clausamine C (**207**), due to Mal and Jana, was based upon the LiO^{t}Bu-promoted annulation between furo-indolone (**208**) and dimethyl maleate (**209**), which resulted in the corresponding carbazole.[15] It was converted to methyl ether by treatment with MeI-K_2CO_3 to give carbazole derivative (**210**). $AlCl_3$-induced deprotection of (**210**) afforded (**211**) in 78% yield. Under very stringent conditions, the reaction of the carbazole (**211**) with K_2CO_3 and prenyl bromide at rt. afforded *O*-prenylated derivative (**212**) in 65% yield. When the carbazole (**212**) was heated in refluxing diethylaniline (DEA), *para*-Claisen rearrangement product (**213**) was obtained in about 50%. yield. Treatment of the diester (**213**) with aq KOH in refluxing MeOH provided *retro*-Kolbe-Schmitt product (**214**) in 68% yield. When the acid (**214**) was subjected to the action of MCPBA followed by aqueous $NaHCO_3$ solution, clausevatine D (**206**) was obtained in 60% yield through a unique oxidative cyclization. Selective *O*-methylation of clausevatine D (**206**) with K_2CO_3-MeI in acetone furnished clausamine C (**207**) in 77% yield. Thus, this total synthesis features three key reactions: Sammes annulation, *p*-Claisen rearrangement, and *retro*-Kolbe-Schmitt reaction (Scheme 10.32).

10.13 NICOLAOU SYNTHESIS OF VIRIDICATUMTOXIN B (215)

Viridicatumtoxin B (**215**) is a unique member of tetracycline family of antibiotics. In contrast to the majority of tetracyclines, this member features a geranyl-derived subunit in the form of a spirobicyclic EF ring system. It is active against Gram-positive bacteria, including methicillin-resistant Staphylococcus aureus (MRSA). Nicolaou's synthetic plan is depicted in *retro* synthetic manner (Scheme 10.33).[16] It consisted in the formation of three new rings, i.e., A, C, and E rings from four intermediates, namely (**216**–**219**) through two [4+2] anionic annulations: Tamura and Myers. Challenges

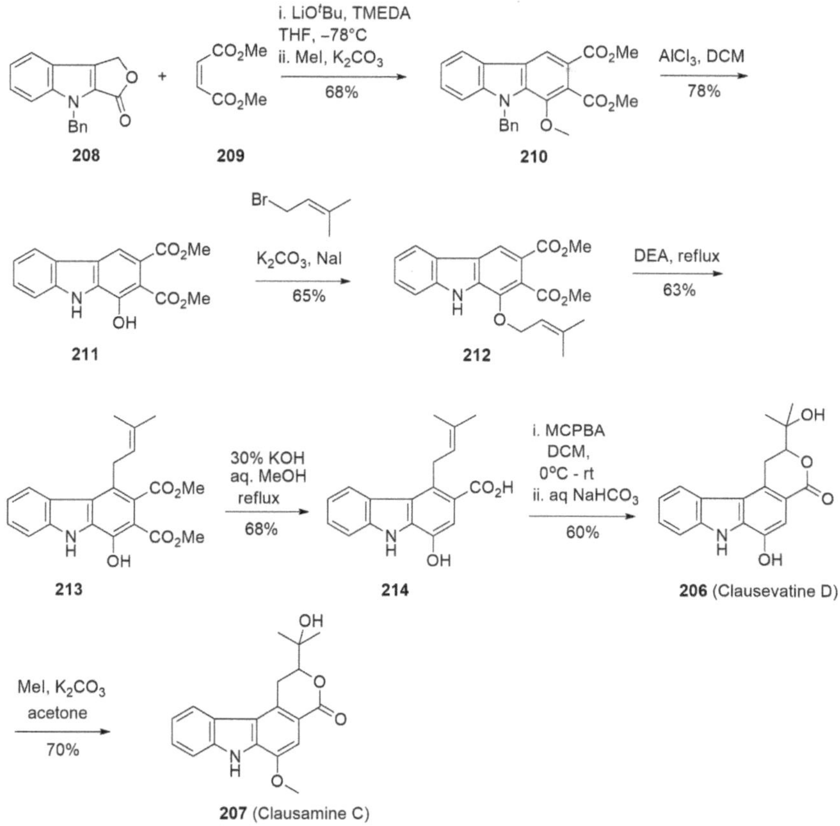

SCHEME 10.32 Mal total synthesis of clausamine C (**206**) and clausevatine D (**207**) [2010 CCS Mal D].

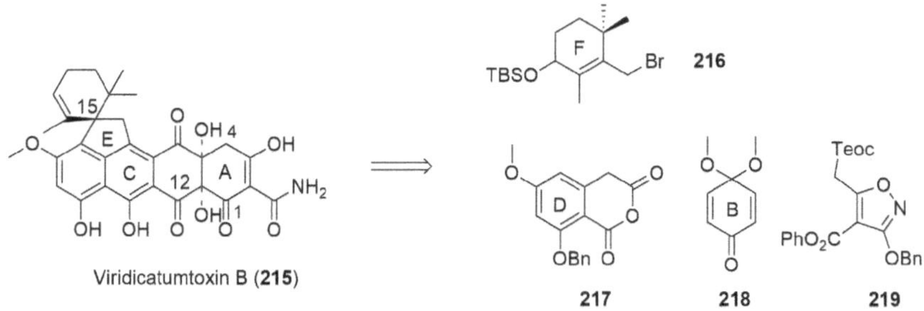

SCHEME 10.33 Retrosynthesis of viridicatumtoxin B (X).

also existed in the introduction of angular tertiary hydroxy groups, i.e., the 4a-OH and 12a-OH groups.

The synthesis of viridicatumtoxin B (**215**) began with Tamura annulation between homophthalic anhydride (**217**) and quinone monoketal (**218**). The corresponding product (**220**) was elaborated to the intermediate (**221**), as presented in Scheme 10.34.

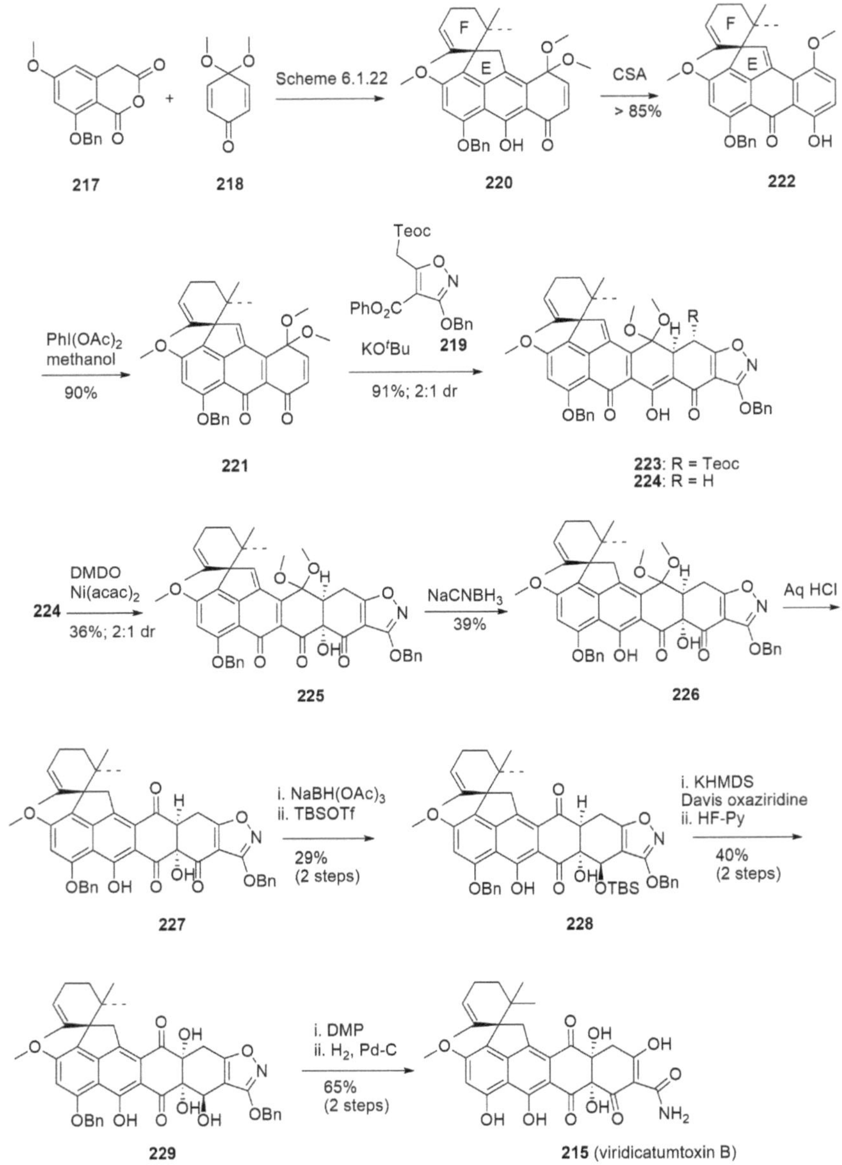

SCHEME 10.34 Synthesis of viridicatumtoxin B (**215**).

The reaction of quinone monoketal (**220**) with CSA afforded *p*-quinomethide (**222**). $PhI(OAc)_2$ oxidation of (**222**) in MeOH delivered (**221**) in 90% yield. Myers annulation of (**221**) with isoxazole phenyl ester (**219**) in the presence of KO^tBu furnished heptacyclic compound (**223**) along with its C15 epimer in a combined yield 91% combined yield. Removal of the Teoc (2-(trimethylsilyl)ethoxycarbonyl) group with TBAF-NH_4F gave intermediate (**224**), together with its C15 epimer in a combined yield of 86%. DMDO-promoted oxidation of (**224**) in the presence of $Ni(acac)_2$ afforded 12a-hydroxy derivative (**225**) with its C15-epimer, whose reaction with $NaCNBH_3$ furnished (**226**). Deketalization of (**226**) with aq HCl gave (**227**) in 100% yield. Reduction of (**227**) with $NaBH(OAc)_3$ furnished C1-hydroxycompound in 47% yield, which was then silylated with TBSOTf in the presence of 2,6-lutidine to afford (**228**) in 61%yield. Davis oxidation of (**228**), via its trianion intermediate, furnished 4a-hydroxy derivative in 65% yield, the desilylation of which with HF·Py gave (**229**) along its hemiacetal. DMP oxidation of the mixture followed by catalytic hydrogenation provided viridicatumtoxin B (**215**).

10.14 MAL SYNTHESIS OF EUPLECTIN (230)

Euplectin (**230**) is one of rare natural products with an indenone moiety fused to a 5-hydroxychromone moiety. It was reported to exhibit cytotoxicity against the growth of murine P-815 mastocytoma cells. The first total synthesis of this interesting target was accomplished by Mal and De by application of three key reactions: a Hauser sulfoxide annulation, a selective iodination, and a late-stage *retro*-Diels-Alder reaction.[17]

The sulfoxide (**231**) required for the Hauser annulation was synthesized from resorcylate (**232**). NBS bromination of (**232**) under radical conditions gave the corresponding bromomethyl derivative, which was reacted with PhSH-Et_3N followed by MeOH-K_2CO_3 to resorcylate (**233**). Its condensation with crotonic acid in the presence of $POCl_3$-$AlCl_3$, followed by acid-catalyzed ring closure, gave chromanone (**234**). $NaIO_4$ oxidation of (**234**) at room temperature yielded sulfoxide (**235**) in 63% yield. Dithianylation of (**235**) with 1,3-propanedithiol in the presence of BF_3 etherate afforded the donor (**231**) for the Hauser sulfoxide annulation. The annulation of the sulfoxide (**231**) with enone (**236**) in the presence of LiO^{t}Bu in THF-DMSO at −60°C, followed by heating the reaction mixture at reflux for 2 h before quenching, yielded annulation product (**237**) in 88% yield. The dithiane group of (**237**) was smoothly cleaved by treatment with $HgCl_2$ in aq CH_3CN to give annulated chromanone (**238**) in 84% yield. Transformation of the chromanone (**238**) to chromone (**239**) was not a trivial problem. Finally, hexacycle (**238**) was derivatized to the acetate (**239**)-controlled reaction of which with I_2-Selectfluor afforded the requisite product (**240**) in moderate yield (60%). This was then treated with DBU to give chromone (**241**) in 62% yield. Treatment of (**241**) with BBr_3 furnished (**242**) in 78% yield. Flash vacuum pyrolysis (FVP) of hexacycle (**242**), was effected at 550°C and 0.01 Torr to furnish euplectin (**230**) in 64% yield (Scheme 10.35).

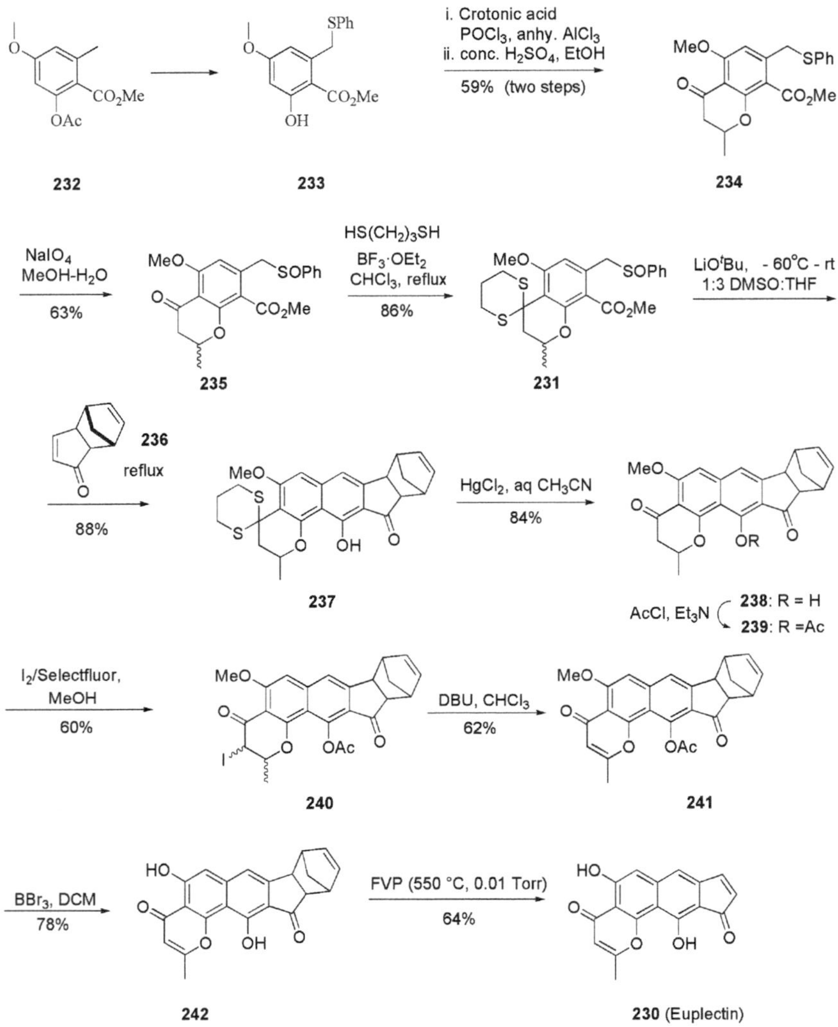

SCHEME 10.35 Synthesis of euplectin (**230**).

10.15 WEINREB TOTAL SYNTHESIS OF TRI-*O*-METHYLOLIVIN (259)

Olivomycin A (**243**) is a clinically effective member of the aureolic acid family of antitumor antibiotics, which are characterized by a complex tricyclic aglycon attached to various di- and trisaccharides. Consequently, olivin (**244**), the aglycon of olivomycin A (**243**) (Fig. 10.3), has been the subject of many synthetic approaches. Presented below is the total synthesis of tri-*O*-methylolivin by Weinreb et al.[18]

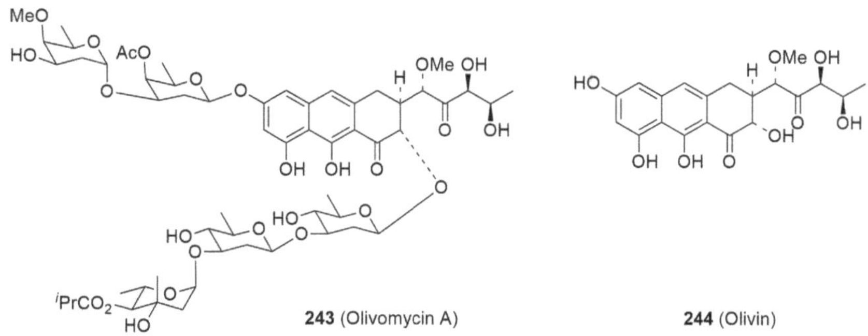

FIG. 10.3 Structures of olivomycin A (**243**) and olivin (**244**).

Ketone (**245**) was obtained in 89% yield by condensation (Scheme 10.36) of the dianion prepared from 3,5-dimethoxyphenylacetic acid (**246**) with ester (**247**). Reduction of (**245**) with sodium borohydride afforded the corresponding alcohol as a 5:3 mixture of epimers, which was subjected to Birch reduction with Li-NH_3 to give (**248**). The unstable Birch product (**248**), so obtained, was converted to the corresponding bis-ketal (**249**). Oxidation of (**249**) with CrO_3-Py gave bis-ketal ketone (**250**) (35% from **245**).

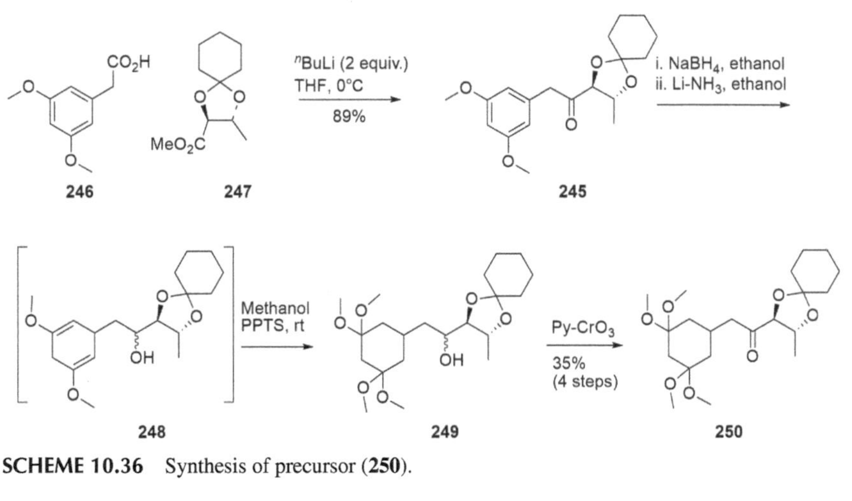

SCHEME 10.36 Synthesis of precursor (**250**).

Ketone (**250**) was kinetically deprotonated with LDA and then silylated with triethylsilyl chloride to afford enol ether (**251**) as a single isomer. MCPBA oxidation of (**251**) gave the relatively stable epoxide (**252**) in 68% yield (Scheme 10.37). Treatment of (**252**) with pyridinium *p*-toluenesulfonate in DCM at room temperature furnished ketone (**253**) as an inseparable 1:1 mixture of ring epimers in 63% yield. The side chain carbonyl group of (**253**) was selectively reduced with sodium borohydride in ethanol at −78°C to give a mixture of alcohols, from which the desired ring isomer (**254**) was obtained in 65% yield.

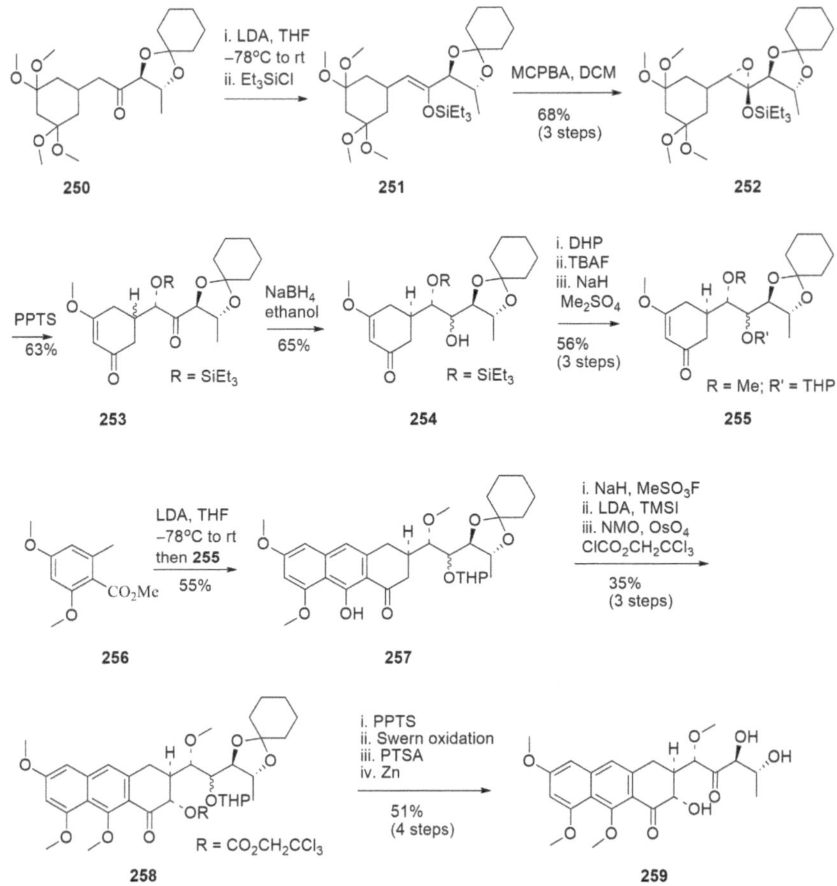

SCHEME 10.37 Total Synthesis of tri-*O*-methylolivin (**259**) [1984 JACS 1811 Weinreb].

It was protected as the tetrahydropyranyl ether and its silyl ether group was subsequently removed with tetrabutylammoniumfluoride to yield the corresponding alcohol. NaH-dimethyl sulfate-promoted methylation of the alcohol afforded β-methoxyenone acceptor (**255**) with an overall yield of 56%. Annulation of orsellinate (**256**) with β-methoxycyclohexenone (**255**) in the presence of LDA −78°C gave tricyclic phenolic ketone (**257**). This was *O*-methylated with NaH-$MeOSO_2F$ to afford its ether in 68% yield. The resulting ether was converted into its trimethylsilylenol ether, which, without purification, was subjected to oxidation with *N*-methylmorpholine N-oxide-OsO_4 to give *the trans* acyloin, which was protected as trichloroethyl carbonate (**258**). Removal of the THP protecting group in (**258**) with methanol and PPTS gave the respective alcohol. Three more steps—Swern oxidation, removal of the cyclohexylideneketal protecting group with PTSA, and cleavage of carbonate group with Zn—afforded tri-*O*-methylolivin (**259**).

10.16 CHRZANOWSKA SYNTHESIS OF (*S*)-(−)-8-OXOXYLOPININE (260)

The application of the annulation of a toluamide is illustrated with the Chrzanowska enantiospecific synthesis of (*S*)-(−)-8-oxoxylopinine (**260**), a precursor of xylopinine, a protoberberine alkaloid.[19] The only chiral center of (**260**) is created during the annulation with the help of a chiral oxazolidine auxiliary derived from (*S*)-alaninol. Acid (**261**) was converted into amide (**262**) by the reaction with thionyl chloride followed by (*S*)-alaninol. Cyclization of (**262**) to oxazolidine (**263**) was effected in 73% yield by 2,2-dimethoxypropane in the presence of PTSA. The reaction of the laterally lithiated species (**264**), generated by lithiation of (**263**) with ^{t}BuLi at −72°C, with dihydroisoquinoline (**265**), afforded oxoxylopinine (**260**) with 13% yield and 99% ee (Scheme 10.38).

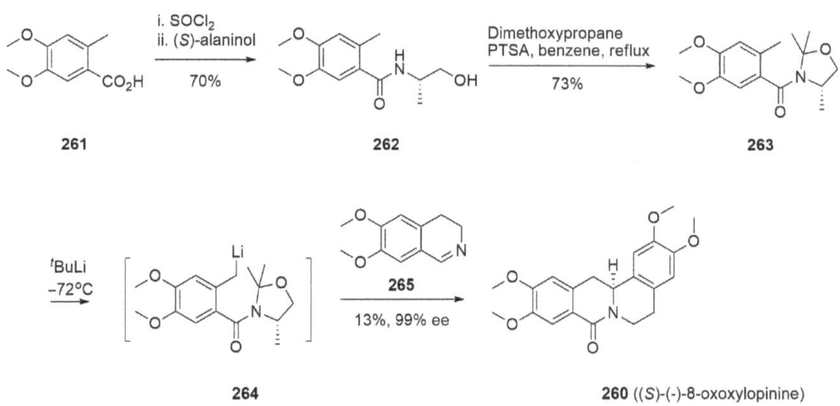

SCHEME 10.38 Enantiospecific synthesis of (*S*)-(−)-8-oxoxylopinine (**260**) [2015 THY 225 Chrzanowska M].

In conclusion, 16 randomly selected total syntheses of members of anthracyclines, angucyclines, tetracyclines, aureolic acids, carbazole alkaloids, and protoberberine alkaloids have been displayed. Attempts have been made to cover all types of anionic [4+2] annulations, namely Hauser, Sammes, Tamura, Hauser-Leusen, Staunton-Weinreb, Myers, etc. discussed in Chapters 4–8. In conjunction with the annulations, the chemistry of various donors and acceptors has been included to impress upon the reader with the associated problems and solutions.

REFERENCES

1. Nicolaou, K. C.; Sorensen, E. J. Classics in Total Synthesis; Wiley-VCH: Weinheim, 1996.
2. Hauser, F. M.; Mal, D. *J. Am. Chem. Soc.* **1984**, *106*, 1862.
3. Karmakar, R.; Mal, D. *J. Org. Chem.* **2012**, *77*, 10235.
4. Matsumoto, T.; Yamaguchi, H.; Tanabe, M.; Yasui, Y.; Suzuki, K. *Tetrahedron Lett.* **2000**, *41*, 8393.

5. Dischmann, M.; Frassetto, T.; Breuning, M. A.; Koert, U. *Chem. Eur. J.* **2014**, *20*, 11300.
6. Nicolaou, K. C.; Wang, Y.; Lu, M.; Mandal, D.; Pattanayak, M. R.; Yu, R.; Shah, A. A.; Chen, J. S.; Zhang, H.; Crawford, J. J.; Pasunoori, L.; Poudel, Y. B.; Chowdari, N. S.; Pan, C.; Nazeer, A.; Gangwar, S.; Vite, G.; Pitsinos, E. N. *J. Am. Chem. Soc.* **2016**, *138*, 8235.
7. Okazaki, K.; Nomura, K.; Yoshii, E. *J. Chem. Soc. Chem. Commun.* **1989**, 354.
8. Nomura, K.; Okazaki, K.; Hori, K.; Yoshii, E. *J. Am. Chem. Soc.* **1987**, *109*, 3402.
9. (a) Wendt, J. A.; Gauvreau, P. J.; Bach, R. D. *J. Am. Chem. Soc.* **1994**, *116*, 9921; (b) Saint-Jalmes, L.; Lila, C.; Xu, J. Z.; Moreau, L.; Pfeiffer, B.; Eck, G.; Pelsez, L.; Rolando, C.; Julia, M. *Bull. Soc. Chim. Fr.* **1993**, *130*, 447.
10. Magauer, T.; Smaltz, D. J.; Myers, A. G. *Nat. Chem.* **2013**, *5*, 886.
11. Nicolaou, K. C.; Cai, Q.; Qin, B.; Petersen, M. T.; Mikkelsen, R. J. T.; Heretsch, P. *Angew. Chem. Int. Ed.* **2015**, *54*, 3074.
12. Yang, X.-Y.; Fu, B.-Q.; Yu, B. *J. Am. Chem. Soc.* **2011**, *133*, 12433.
13. Tatsuta, K.; Fukuda, T.; Ishimori, T.; Yachi, R.; Yoshida, S.; Hashimoto, H.; Hosokawa, S. *Tetrahedron Lett.* **2012**, *53*, 422.
14. Townsend, C. A.; Davis, S. G.; Christensen, S. B.; Link, J. C.; Lewis, C. P. *J. Am. Chem. Soc.* **1981**, *103*, 6885.
15. Jana, A.; Mal, D. *Chem. Commun.* **2010**, *46*, 4411.
16. Nicolaou, K. C.; Hale, C. R. H.; Nilewski, C.; Ioannidou, H. A.; ElMarrouni, A.; Nilewski, L. G.; Beabout, K.; Wang, T. T.; Shamoo, Y. *J. Am. Chem. Soc.* **2014**, *136*, 12137.
17. Mal, D.; De, S. R. *Org. Lett.* **2009**, *11*, 4398.
18. Dodd, J. H.; Starrett, J. E., Jr.; Weinreb, S. M. *J. Am. Chem. Soc.* **1984**, *106*, 1811.
19. Meissner, Z.; Chrzanowska, M. *Tetrahedron: Asymmetry* **2015**, *26*, 225.

Chapter 11

Miscellaneous Anionic Annulations

In the preceding chapters, many annulations related to [4+2] annulations are presented. In this chapter, few more anionic annulations that have already proved to be useful in organic synthesis, or have been of interest are covered. These belong to types: [4+2], [3+3], [3+2], [2+1], and heteroannulations.

11.1 DESLONGCHAMPS [4+2] ANNULATION AND NHC-CATALYZED [4+2] ANNULATIONS

Active methylene compounds like ethyl acetoacetate or Nazarov reagents are useful donors for various annulations. The Deslongchamps [4+2] annulation is the annulation between a Nazarov reagent or γ,δ-unsaturated active methylene compound and a 2-carboxycyclohexenone in the presence of a base to give a decalin.[1] As shown in Scheme 11.1, the reaction between keto ester (**1**) and cyclohexenone (**2**) under the influence of Cs_2CO_3 followed by decarboxylation with PTSA in refluxing benzene afforded *cis,syn*-decalin (**3**) in 89% yield (reaction 1). In the related Robinson annulation (reaction 2), Nazarov reagent (**4**) reacted with cyclohexanone donor (**5**) to give decalin (**6**). The Deslongchamps annulation is viewed as a tandem Michael-Michael process or an anionic Diels-Alder reaction. On the other hand, the Robinson annulation (reaction 2) consists of a Michael addition and an aldol reaction.

The Deslongchamps annulation, discovered in 1988, has now witnessed various developments. For example, functionalized phenanthrene system (**7**) was formed in 66% yield, when the monocyclic Nazarov reagent (**8**) was submitted to the reaction with cyclohexenone (**2**) in the presence of Cs_2CO_3 and subsequently to trifluoroacetic acid (Scheme 11.2).[2]

Over the years, the substrate the scope of this reaction has steadily increased (vide Figs. 11.1 and 11.2). Some of the prominent acceptors such as (**9–13**) are shown in Fig. 11.1. Typical donors (i.e., 14–18) of the Deslongchamps annulation are given in Fig. 11.2.

The major impact of the Deslongchamps annulation is found in the construction of the complex steroid scaffolds. The example of a 13α-methyl 14α-hydroxy steroid is presented in Scheme 11.3. The standard Deslongchamps annulation between (**2**) and (**19**) initially gave (**20**), which then underwent aldol

Anionic Annulations in Organic Synthesis. https://doi.org/10.1016/B978-0-12-813800-7.00011-7

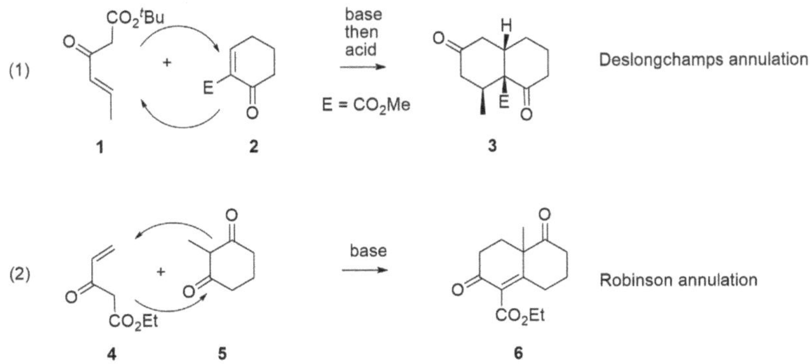

SCHEME 11.1 An example of the Deslongchamps annulation.

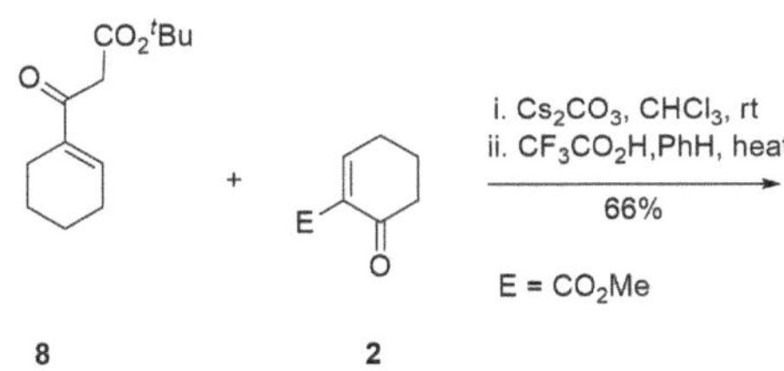

SCHEME 11.2 Synthesis of a phenanthrene by Deslongchamps annulation.

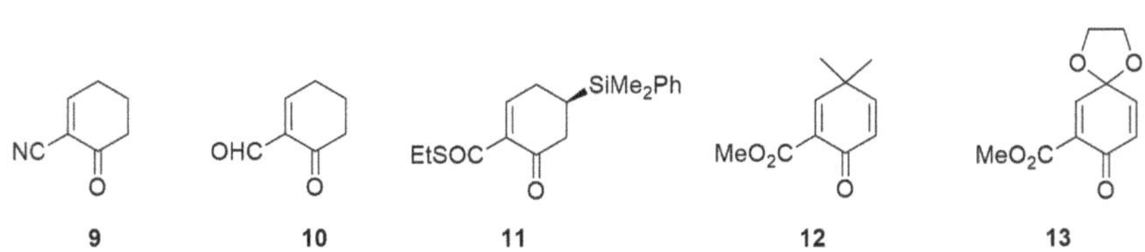

FIG. 11.1 Some Deslongchamps acceptors.

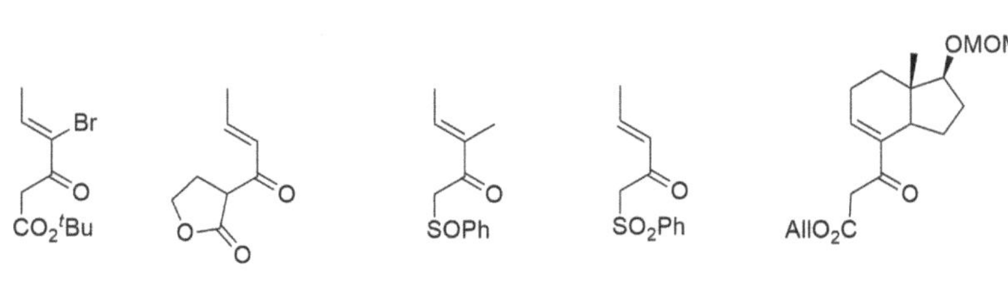

FIG. 11.2 Some Deslongchamps donors.

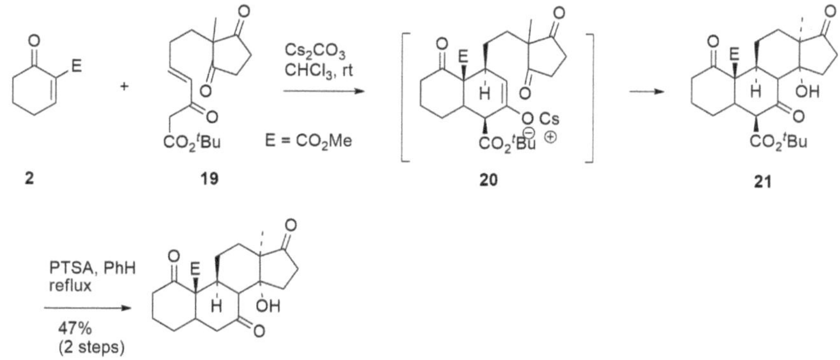

2 19 20 21 22

SCHEME 11.3 Synthesis of a steroid scaffold by Deslongchamps annulation.

reaction to furnish tetracycle (**21**). Decarboxylation of (**21**) with PTSA in refluxing benzene afforded (**22**) in 47% overall yield.

One of the finest applications of the Deslongchamps annulation is the stereoselective total synthesis of (+)-cassaine (**23**), which is described as a terpenoid alkaloid. It is an inhibitor of sodium potassium ATPase, and pharmacologically similar to digitoxin. Its synthesis involving intermediates (**24**) and (**25**) is outlined in Scheme 11.4. (+)-Carvone (**26**), was elaborated to the intermediate (**24**) in >10 steps, which included a Birch reduction and a Robinson annulation.

The intermediate (**24**) was reacted with donor (**27**) in the presence of Cs_2CO_3 to give tricycle (**28**), dealkoxycarbonylation of which gave (**29**). In a multistep sequence, compound (**29**) was converted into tricycle (**30**). Reduction of (**30**) with Mg-methanol followed by DIBAL reduction, followed by DMP oxidation furnished aldehyde (**31**). Treatment of (**31**) with Comins reagent gave enol triflate (**32**). Removal of PMB group in (**32**) followed by DMP oxidation afforded ketone (**33**). Pd(0)-catalyzed carboxylation of (**33**) provided ester (**34**). Finally, deprotection of the MOM group in (**34**) led to the synthesis of (+)-cassaine (**23**) (Scheme 11.5).

The concept of the Deslongchamps annulation was intelligently modified to fabricate cyclohexanone annulation, which may be termed as Brückner

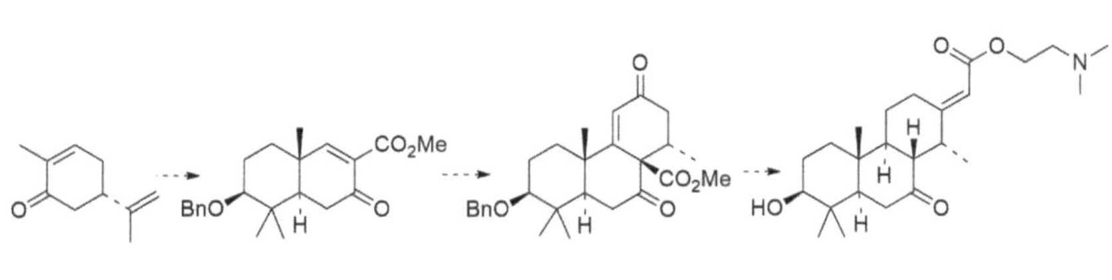

26 24 25 23 ((+)-Cassaine)

SCHEME 11.4 Outline of the synthesis of cassaine (**25**) by Deslongchamps annulation [2013 ORS 6270 Deslongchamps P].

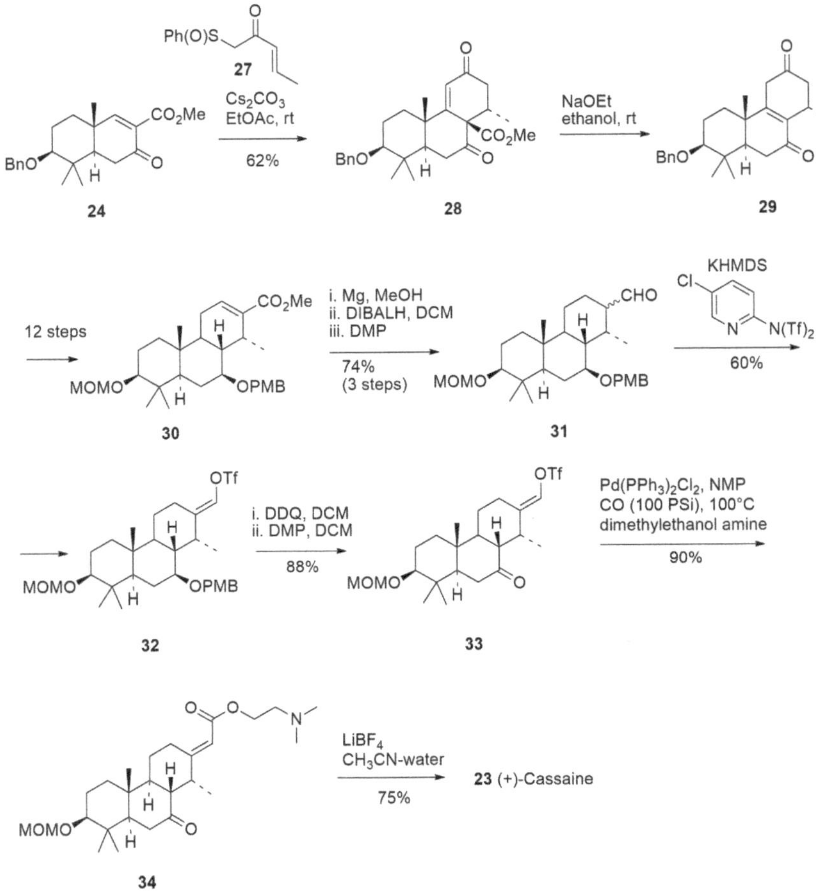

SCHEME 11.5 Synthesis of cassaine (**23**) [2013 ORS 6270 Deslongchamps P].

annulation. While the former consists of two successive Michael additions, the latter involves a sequence of Michael addition-vinyl Grignard addition-Michael addition. For both the annulations, the last step is 6-*endo*-trig cyclization. The Brückner annulation is illustrated by the synthesis of *cis*-decalin (**35**) in Scheme 11.6. In a single operation, the decalin (**35**) was prepared from (**36**) and (**37**). Their reaction gave (**38**) through enolates (**39** and **40**). Interception of the lactone (**38**) with vinylmagnesium bromide afforded (**35**) through the intermediate (**41**).

With the core structure (**35**) synthesized, functional manipulations were resorted to complete the synthesis of (**42**) (isoacanthodoral). All standard methods like Wolff-Kishner reduction or Huang-Minlon reduction failed to deoxygenate ketone (**35**). However, Shapiro olefination of tosylhydrazone (**43**) delivered dienoic ester (**44**), which was regioselectively hydrogenated using 5% Rh on charcoal as the catalyst to give (**45**), the overall yield being 86%. The ester

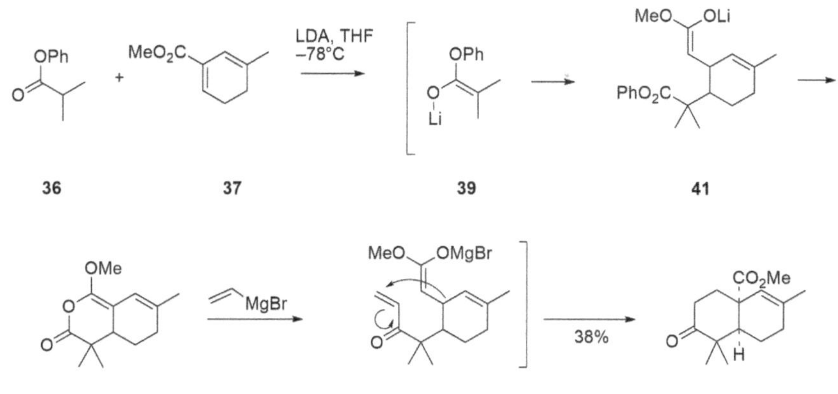

SCHEME 11.6 Brückner annulation [2009 ORS 4842 Brückner R].

(45) was reduced with DIBAL to give the corresponding neopentyl lalcohol, which was oxidized with DMP to furnish aldehyde (46). The aldehyde (46) was then homologated to cyanide (47) in 97% yield by Leusen reagent (Tosmic). Reduction of the nitrile group in (47) with DIBAL provided isoacanthodoral (42) (Scheme 11.7).

Liu annulation (Scheme 11.8) is a reaction between cyclobutenones (48) (as four-carbon 1,4-dipoles) with active methylene ketones (49) (as two carbon 1,2-dipoles) in the presence of LDA to give heavily substituted phenols (50) in a single step in regiocontrolled manner.[3] The products of the annulation are convertible to highly substituted acridine derivatives.

Mechanistically, the reaction is initiated by the deprotonation of the active methylene compounds (49) with LDA to anions (51) (Scheme 11.9). These anions undergo intermolecular aldol reaction with (48) to give cyclobutenolate intermediate (52). Then a strain-driven ring-opening of (52) takes place to lead to new anions (53). Subsequently, the anions (53) enter into intramolecular aldol

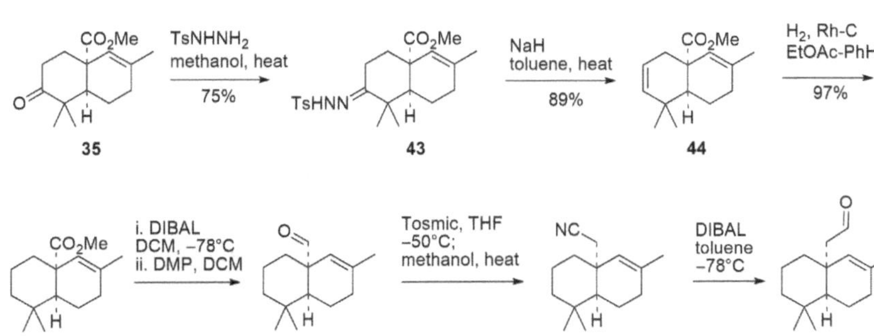

SCHEME 11.7 Brückner synthesis of isoacanthodoral (42).

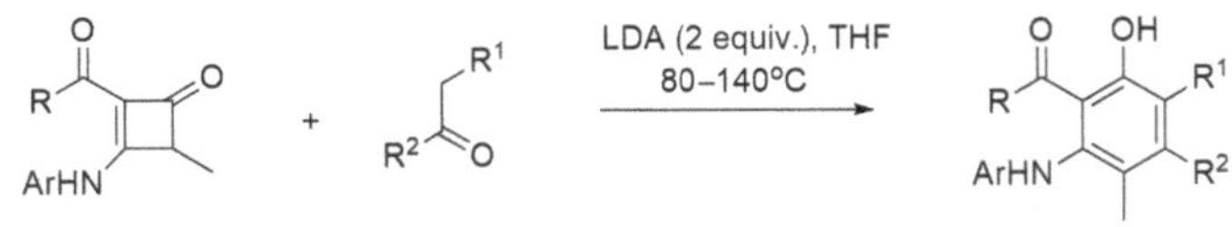

R = Me, Et, OMe. R^1 = CO_2R, CONHR, COR, etc. R^2 = Me, Et, Ph.
Ar = 4-$NO_2C_6H_4$, 4-MeC_6H_4, etc.

48 **49** **50**

SCHEME 11.8 Liu annulation of cyclobutenones with active methylene compounds [2012 JOC 5173 Liu Q].

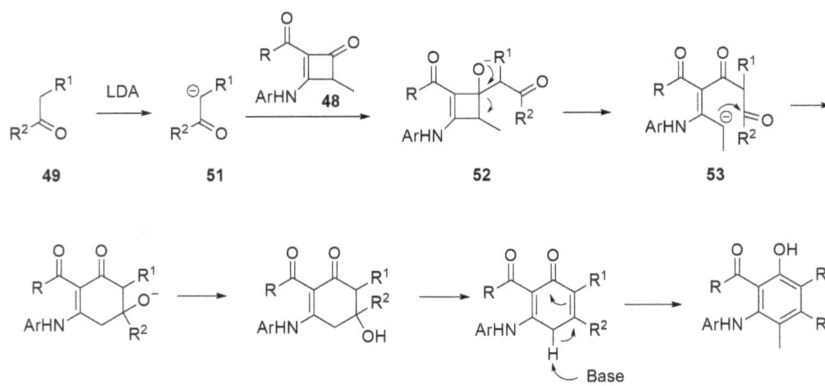

54 **55** **56** **50**

SCHEME 11.9 Mechanism of the annulation with cyclobutenones.

cyclization to afford six-membered intermediate (**54**). Finally, a sequence of protonation-deprotonation-dehydration-aromatization involving intermediates (**55** and **56**) gives the annulation products (**50**).

Another annulation involving an active methylene compound (i.e., **57**), that contains a pendant ester group is shown in Scheme 11.10. In the presence of NaH, dimethyl 2-geranyl-l,3-acetonedicarboxylate (**57**) reacted with 4-pivaloyloxy-2-butynal (**58**) to give fully substituted benzene derivative (**59**) (33% yield) in one pot with all substituents in order.[4] The reaction takes place via Michael addition of the anion of (**57**) to (**58**) to produce anion (**60**), which undergoes intramolecular regioselective Dieckmann-cyclization to give (**61**). Isomerization of carbocycle (**61**) gives resorcylate (**59**) (Scheme 11.10).

Aliphatic chloro compounds with appropriate activating groups such as cyanide can serve as versatile 1,4-dipolar reagents in the synthesis of carbocycles and heterocycles. These reagents can readily form chloroalkyl organometalics, which can enter into annulation with aldehydes to give cyano substituted pyrans and furans. As shown in Scheme 11.11, the reaction of (**62**) with benzaldehyde in the presence of KO^{t}Bu smoothly produced tetrahydropyranyl nitrile (**63**) as a mixture of diastereomers.[5] This reaction is presumed to proceed through

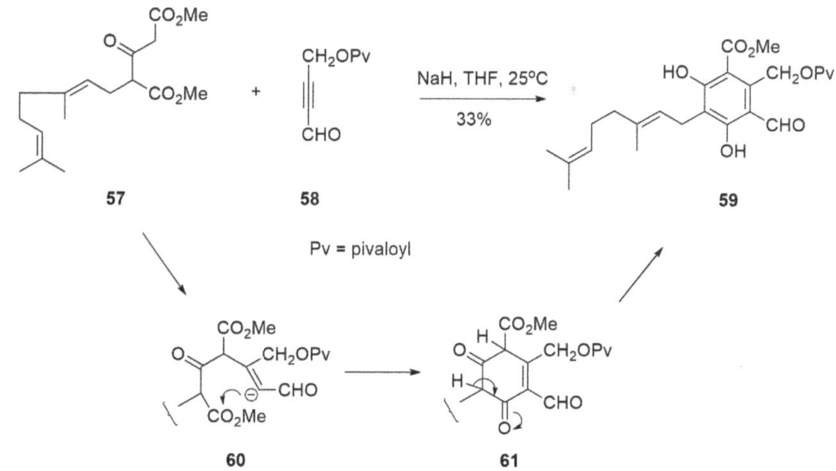

SCHEME 11.10 Synthesis of a mycophenolic acid intermediate [1998 THS 2881 Covarrubias-Zuniga A].

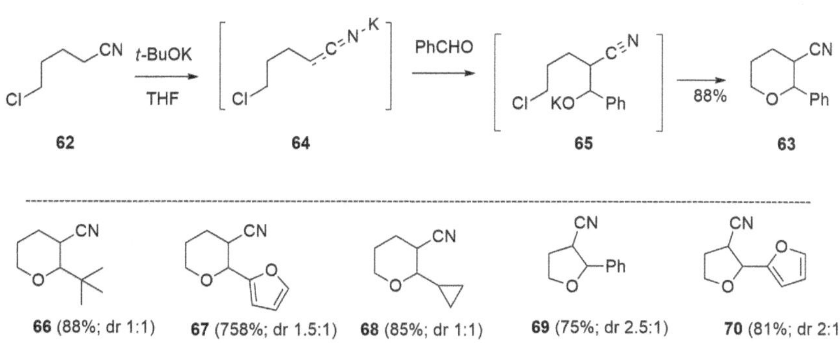

SCHEME 11.11 Fleming annulation with ω-chloroalkylnitriles [2003 JOC 3943 Fleming FF].

potassiated nitrile (**64**), which intercepts benzaldehyde to generate alkoxide (**65**), suited for intramolecular displacement to the product (**63**). This reaction was generalized with examples like (**66–70**).

ω-Chloro Grignard reagents (**71**) react with cyclic γ-hydroxy-α,β-alkenenitriles (**72**) via conjugate addition-alkylation to give cis-hydrindanes (**73**) and decalins (**74**) in a single synthetic operation. Typical examples (**75–78**) are given in Scheme 11.12.[6] Mechanistically, deprotonation and alkyl transfer from (**72**) generate the key alkylmagnesiumalkoxide (**79**). Preferential equatorial disposition of the magnesium alkoxide. (**79**) favorably positions the nucleophile for an axial delivery of the alkyl chain, generating the nitrile anion (**80**) (Scheme 11.13). Cyclization of (**80**) results in the formation of the *cis*-decalin (**74**).

NHC-catalyzed [4+2] annulation: N-Heterocyclic carbenes (NHCs) are defined as singlet carbenes, in which the carbene carbon is directly attached

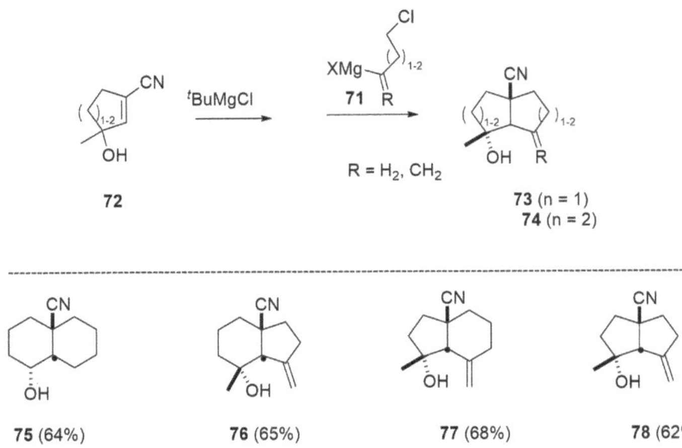

SCHEME 11.12 Fleming annulation with ω-chloro Grignard reagents [2002 ORS 2493 Fleming FF].

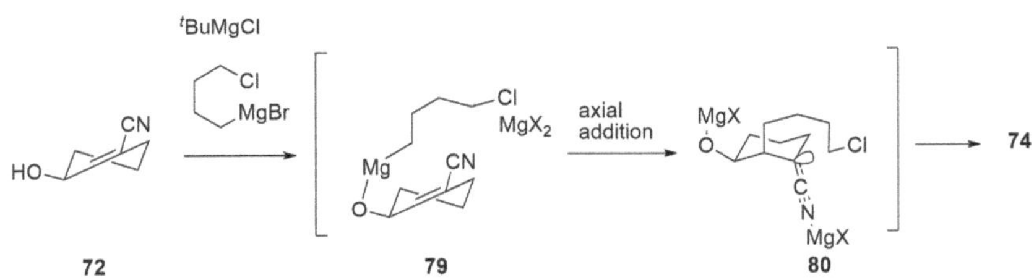

SCHEME 11.13 Probable mechanism of the Fleming annulation.

to nitrogen of a heterocyclic framework. Since the isolation of free NHC, 1,3-di(adamantyl)imidazol-2-ylidene, which was prepared by deprotonation of its imidazolium precursor, NHC-catalyzed reactions have been an attractive field of research.[7] Many nonconventional chemical syntheses involving *umpolung* of aldehydes have been realized over the last 25 years. It is to be noted all NHC-catalyzed reactions are promoted by stoichiometric amount of bases. The attractive feature of NHCs is the reversal of the polarity of aldehydes. In the following paragraphs, few NHC-catalyzed annulations are presented. Ye et al. synthesized 3-substituted-4-hydroxytetralones (**81**) from phthalaldehyde (**82**) and α,β-unsaturated ketones (**83**) in moderate yields with good *trans*-selectivities by cascade Stetter-intramolecular aldol reaction in the presence of thiazolium salt (**84**) and Cs_2CO_3 (Scheme 11.14).[8] The *trans*-selectivity of the reaction is explained by the loss of NHC unit after the aldol reaction. On other hand, the sequential Stetter reaction and aldol reaction of the above substrates gave 3-substituted-4-hydroxytetralones with good *cis*-selectivity. Mechanistically, the reactivity of one of two aldehyde groups of (**82**) is reversed due to the action of NHC. Consequently, phthalaldehyde (**82**) acts as a 1,4-dipole reagent. NHC (**85**), generated from the thiazolium salt (**84**) adds to aldehyde (**82**) to form

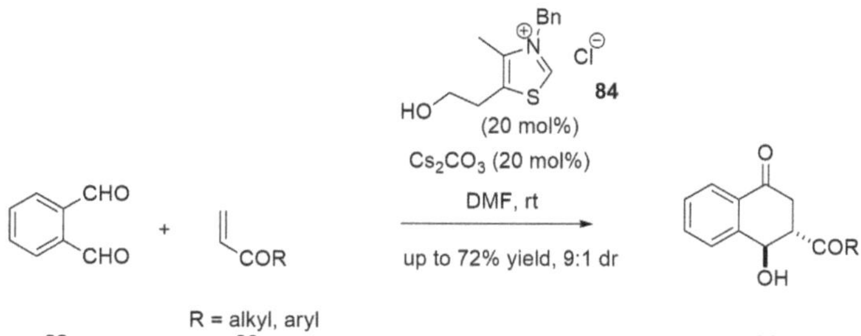

82 83 81

SCHEME 11.14 NHC catalyzed synthesis annulation of 3-substituted-4-hydroxytetralones [2010 JOC 273 Ye S].

enol (**86**), which, in turn, reacts with enone (**83**) in Michael fashion to give (**87**) (Scheme 11.15). Intramolecular aldol reaction of (**87**) generates cyclic intermediate (**88**). Finally, loss of the thiazole unit from (**89**) results in the products (**81**).

Unsaturated esters (**90**) undergo annulation with hydrazones (**91**) under the influence of NHC derived from triazolium salt (**92**) in a highly efficient and stereoselective manner to give δ-lactams (**93**) (Scheme 11.16).[9] These lactam products could readily be transformed to optically enriched pipecolic acid derivatives.

The reaction is proposed to take place via deprotonation of the γ-hydrogen of NHC activated species (**94**) to produce zwitterions (**95** and **96**) (Scheme 11.17). Nucleophilic γ-carbon addition of (**95**) to hydrazone (**91**) eventually leads to δ-lactam (**93**) with the expulsion of the NHC. The stereochemical outcome of the reaction was explained by face selective Diels-Alder reaction of the active species (**96**) with the dienophiles (**91**).

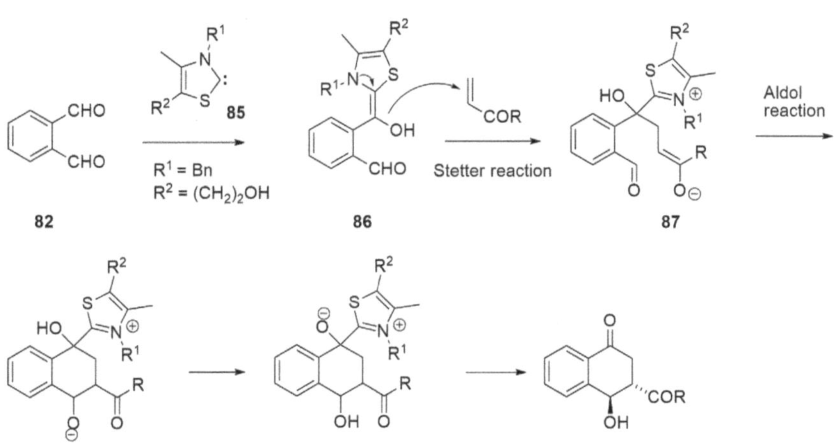

82 86 87

88 89 81

SCHEME 11.15 Mechanism of the formation of the tetralones **81**.

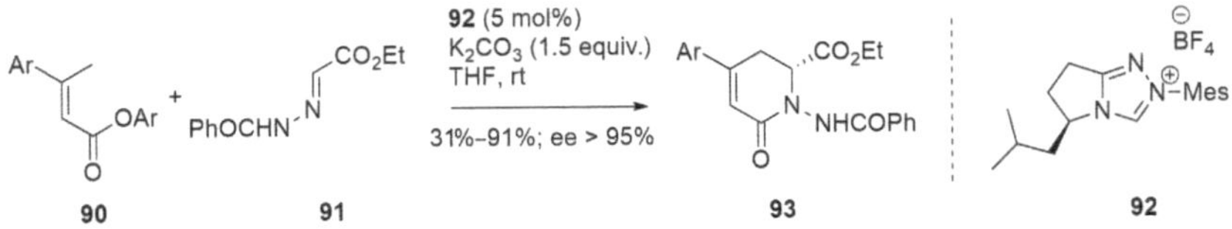

SCHEME 11.16 Chi γ-carbon activation of α,β-unsaturated esters [2013 ORS 5028 Chi YR].

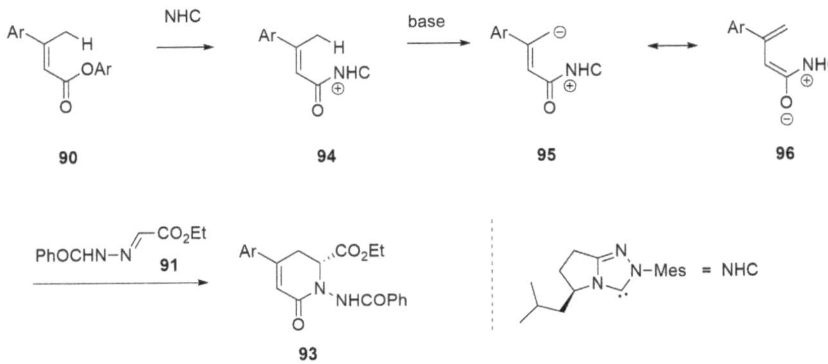

SCHEME 11.17 Mechanism of Chi γ-carbon activation of unsaturated esters.

Akin to the Chi work, Xu et al. effected NHC promoted annulation of 2-methylindole-3-carboxylates with isatin-derived ketimines to the corresponding spirocyclic adducts in good yields with excellent enantioselectivities.[10]

Recently, Glorious group reported activation of 2-bromomethylbenzaldehydes (**97**) by NHC en route to *o*-quinodimethane intermediates. Coupling of these intermediates with ketones such as phenylglyoxylates, isatins, or trifluoromethyl ketones (**98**) via [4+2] annulation provides access to functionalized 1-isochromanones (**99**) (Scheme 11.18).[11] Regarding the mechanism, the formation of NHC from salts (**100** or **101**) is considered first. Addition of NHC (**102**) to the aldehyde (**103**) results in the adduct (**104**), which loses HBr to form*o*-quinodimethide in the forms (**105**) and (**106**) (Scheme 11.19). Their reaction with ketones (**98**) in [4+2] annulation fashion followed by the loss of NHC gives isochromanone (**99**).

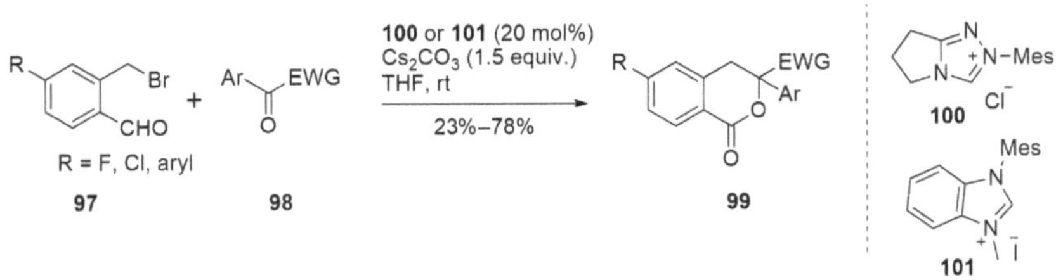

SCHEME 11.18 Glorius synthesis of isochromanones (**99**) [2016 ORS 4444 Glorius F].

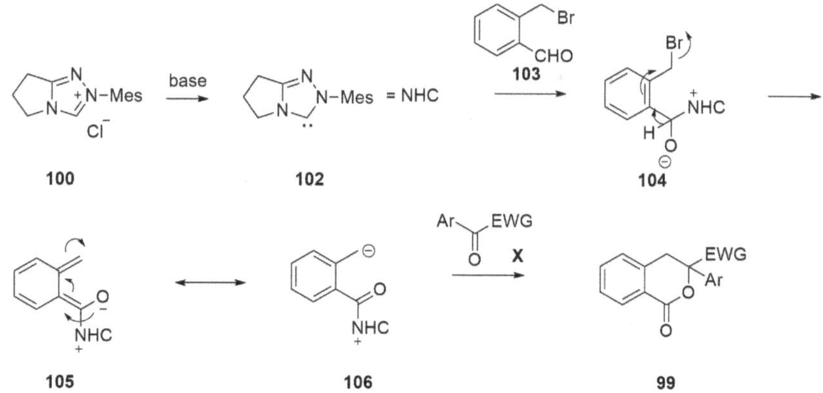

SCHEME 11.19 Probable mechanism of the Glorius annulation.

Lupton annulation: The NHC-catalyzed reaction between α,β-unsaturated acyl fluorides and silyl dienol ethers directly provides 1,3-cyclohexadienes through a [4+2] annulation. For example, the annulation between (**107**) and (**108**) gave (**109**) in the presence of NHC (**110**) in THF. Mechanistically, it involves several steps, namely, Michael addition, aldol cyclization, lactonization and decarboxylation. In this case, nucleophilic substitution of imidazolylidene NHC (**110**) with (**108**) provides α,β-unsaturated acyl azolium (**111**) and dienolate (**112**) with liberation of TMSF (Scheme 11.20). Subsequently, Michael addition and aldol cyclization between (**111**) and (**112**) takes place to give alkoxide (**113**), which cyclizes to form lactone (**114**). Thermal loss of carbon dioxide from (**114**) gives the product (**109**).[12] At temperatures below −20°C,

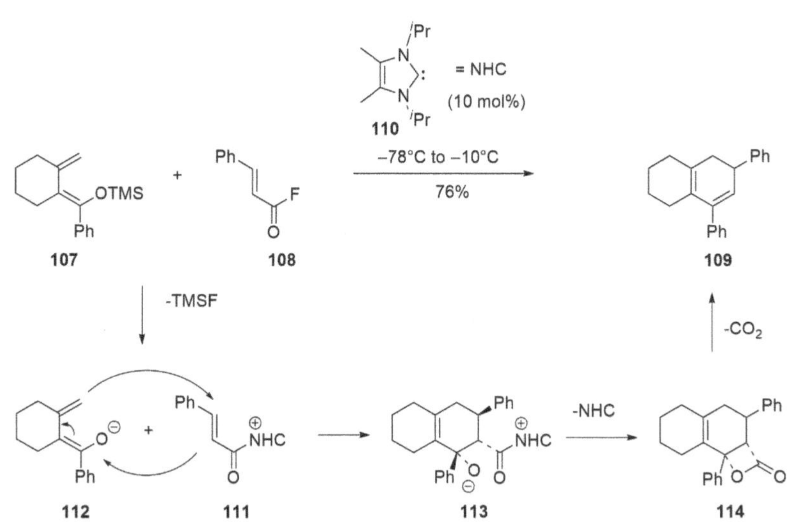

SCHEME 11.20 Lupton annulation: example and mechanism [2011 JACS 4694 Lupton DW].

it was possible to chemically intercept the β-lactone intermediate (**114**), thus supporting the above mechanism. The merit of the Lupton annulation lies in the nature of the products, i.e., 1,3-cyclohexadienes for which there are few direct methods.

11.2 ANIONIC [3+3] ANNULATIONS

Perhaps, the simplest anionic [3+3] annulation is that between methyl acetoacetate (**115**) and methyl crotonate (**116**) in the presence of NaOMe, giving cyclohexenone ester (**117**), which can readily be aromatized with iodine in refluxing methanol to furnish resorcylate (**118**) (Scheme 11.21). This cyclocondensation is believed to take place in two steps: Michael addition and Claisen condensation.[13]

Pietrusiewicz [3+3] annulation involves interaction between phosphorus ylide (**120**) and α,β-unsaturated aldehydes (**121**) in furnishing cyclohexenone esters (**122**). It is proposed to occur via Michael adduct (**123**), followed by its intramolecular Wittig reaction (Scheme 11.22).[14] Typical examples are transformations (**124**) to (**125**), and (**126**) to (**127**).

Diphenylprolinol silyl ethers have been attractive catalysts in the asymmetric synthesis over the years. In 2009, Hayashi et al. reported the synthesis of chiral cyclohexenone derivatives (**128**) by annulation of aryl substituted α,β-unsaturated aldehydes (**129**) with dimethyl3-oxopentanedioate (**130**) (Scheme 11.23) using *tert*-butyldimethylsilyl ether of diphenylprolinol

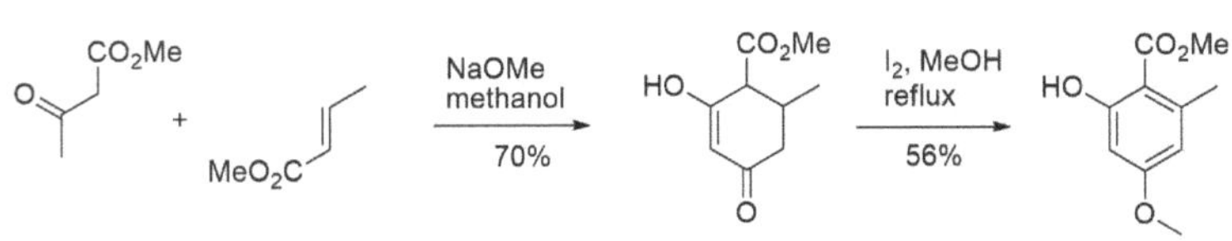

115 **116** **117** **118**

SCHEME 11.21 Condensation of methyl acetoacetate with methyl crotonate [2007 THN 11781 Mal D].

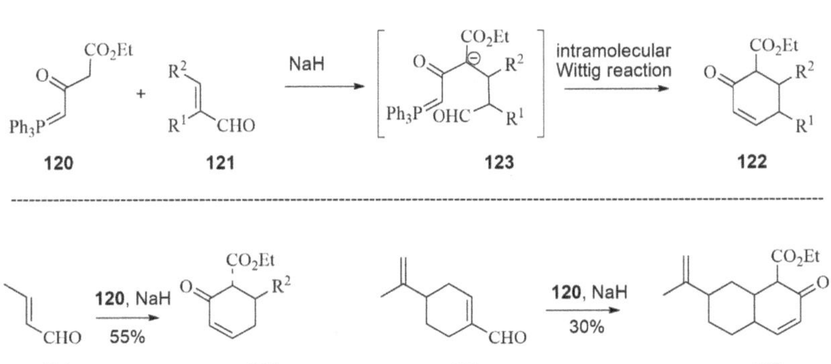

SCHEME 11.22 [3+3] annulation route to cyclic enones [1983 JOC 788 Pietrusiewicz KM].

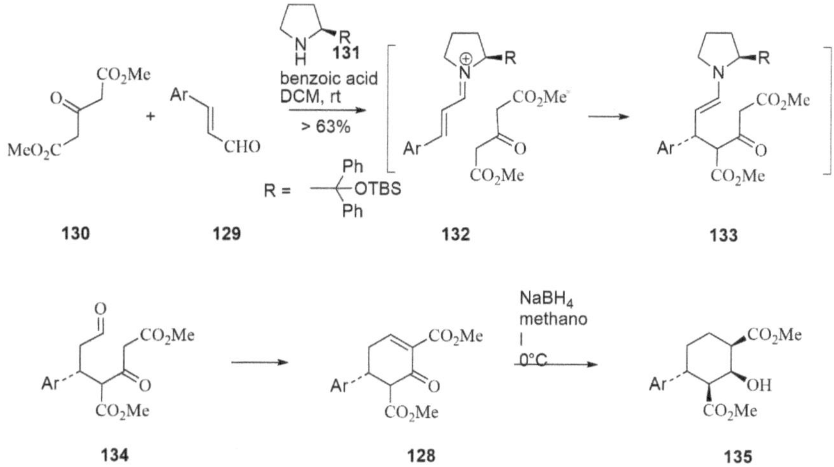

SCHEME 11.23 Hayashi annulation [2009 ORS 45 Hayashi Y].

(**131**) (Hayashi catalyst).[15] Enantioselective Michael reaction of oxopentanedioate (**130**) to the iminiumion (**132**), is proposed to generate enamine (**133**). Hydration of the enamine (**133**) occurs to afford aldehyde (**134**). Subsequently, intramolecular Knoevenagel condensation of (**134**) gives cyclohexenones (**128**). The cyclohexenones (**128**) were derivatized and characterized as saturated cyclohexanes (**135**) by $NaBH_4$ reduction.

A benzene ring can be constructed from nonbenzene precursors by a sequence of Michael addition and Claisen reaction. As shown in Scheme 11.24, sulfoxide activated methyl ketone (**136**) underwent [3+3] annulation with methylenelactone (**137**) in the presence of NaH in DME presumably to give 1,3-cyclohexadione (**138**), which on thermolysis furnished pterocarpan scaffold (**139**).[16]

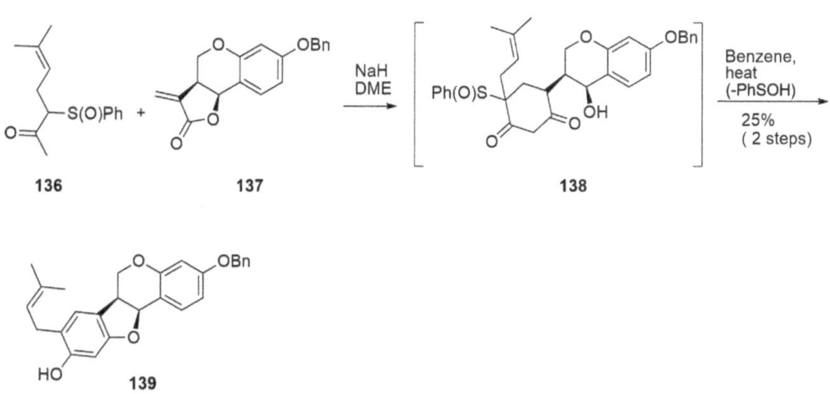

SCHEME 11.24 [3+3] Annulation with a sequence of Michael addition and Claisen reaction [1988 JCSCCS 374 Kim S-W].

Alternatively, NHC-mediated condensation can be used for benzene synthesis. When α,β-unsaturated aldehydes are reacted with NHCs, conjugated Breslow intermediates are produced. Upon oxidation, these intermediates are converted to acyl ion equivalents. Consequently, they can act as donors through γ-hydrogen activation. On the basis of this hypothesis, Chi et al. developed a one-step synthesis of tetrasubstituted benzenes, which is presented in Scheme 11.25. The reaction between enals (**140**) and enones (**141**) in the presence of NHC precursor (**142**) and oxidant (**143**) at room temperature afforded substituted benzenes (**144**).[17]

Mechanistically, Chi annulation can be interpreted in terms of Breslow intermediate (**145**) formed from enals (**140**) and NHC (**146**). Oxidation of (**145**) affords acyl equivalents (**147**), which are highly susceptible to deprotonation to give (**148**) (Scheme 11.26). Michael addition of (**148**) to enones (**150**) affords (**150**), which undergo further deprotonation to give (**151**), which, in turn, cyclize to furnish (**152**). Lactonization of (**152**) gives bicycle (**153**). Loss of CO_2 from (**153**) provides dihydrobenzene derivatives (**154**). Dehydrogenation of (**154**) then affords products (**155**).

11.3 ANIONIC [3+2] ANNULATIONS

Anionic [3+2] annulations are fairly common in organic synthesis. Quite often, they are discussed under the umbrella of 1,3-dipolar cycloadditions or the synthesis of five-membered heterocycles. Herein, selected few [3+2] anionic annulations that are not covered under cycloadditions are presented.

Padwa annulation is an annulation of (phenylsulfony1)allene with an activated olefin in the presence of a nucleophilic reagent providing a cyclopentene. Indeed, when allene (**156**), acrylonitrile (**157**), and sodiumbenzenesulfinate (cat.) are stirred in THF at ambient temperature, cyclopentene (**158**) was isolated in 73% yield (reaction 1, Scheme 11.27). Similar results were obtained when sodium cyanide or sodium nitrite were used as catalysts (reaction 2, Scheme 11.27).[18]

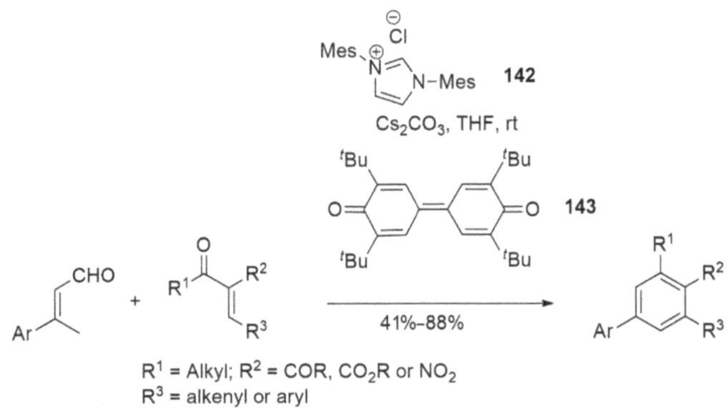

140 **141** **144**

SCHEME 11.25 Chi [3+3] annulation to tetrasubstituted benzenes [2014 NCS 5027].

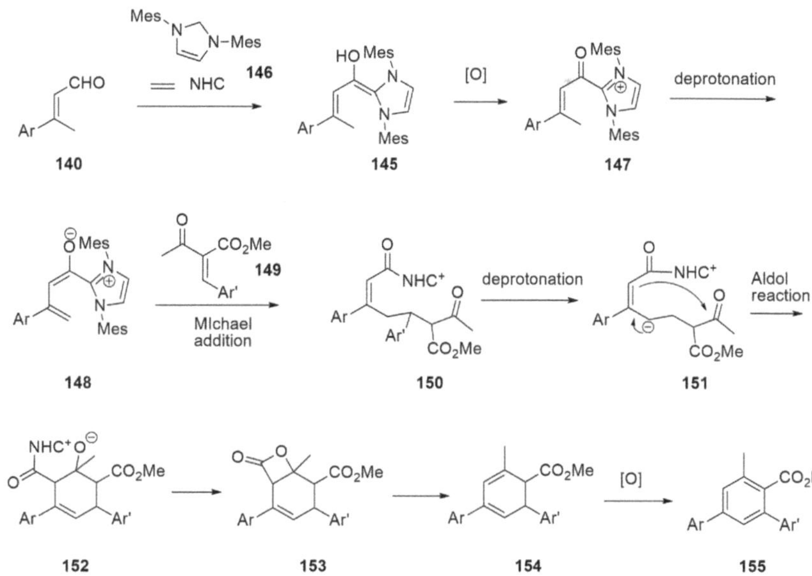

SCHEME 11.26 Possible intermediates of the Chi [3 + 3] annulation.

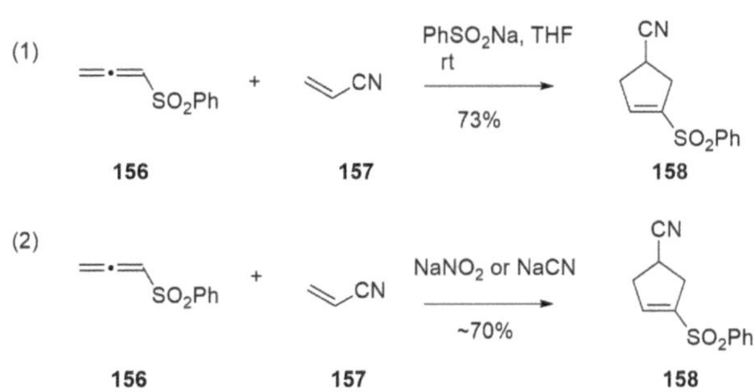

SCHEME 11.27 Padwa cyclopentene annulations [1988 JACS 1617 Padwa A].

The annulation is initiated by Michael addition of an appropriate nucleophile, namely benzenesulfinate or nitrite to allene sulfone (**156**) to form carbanion (**159**), which, in turn, undergoes Michael addition to acrylonitrile (**157**) to give (**160**) (Scheme 11.28). Subsequently, (**160**) cyclizes in 5-*endo*-trig mode to furnish cyclopentane derivative (**161**). Proton exchange reaction of (**161**) gives carbanion (**162**). Elimination of the nucleofuge from (**162**) affords cyclopentene sulfone (**158**).

A neat application of the Padwa annulation is found in the synthesis of echinosporin (**163**).[19] The annulation between enone (**164**) and allenyl sulfone (**156**) in the presence of benzenesulfinate afforded (**165**) in stereocontrolled manner (Scheme 11.29). Stereoselective reduction of the ketone in (**165**) with L-Selectride in DCM followed by *O*-silylation afforded (**166**). Dihydroxylation

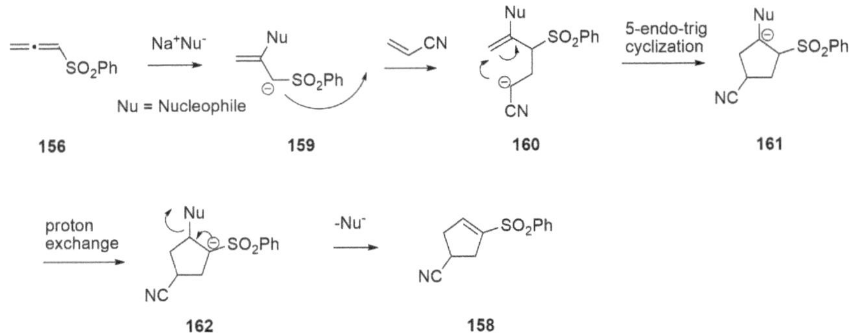

SCHEME 11.28 Mechanism of the Padwa annulation.

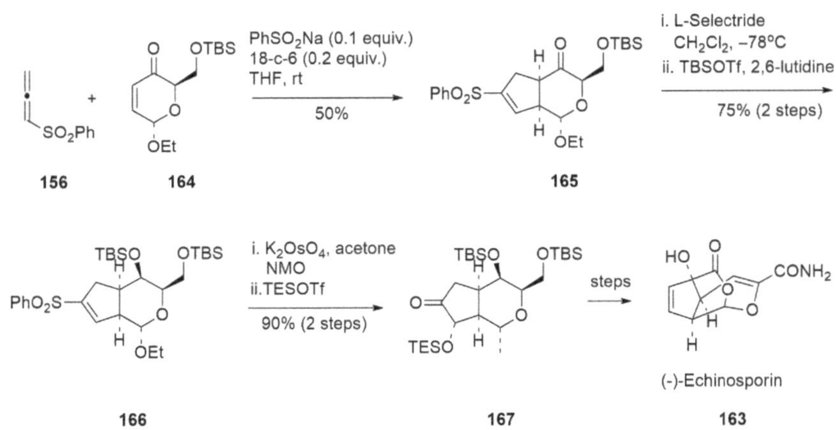

SCHEME 11.29 Synthesis of echinosporin via Padwa annulation [2012 ORS 3024 Hale KJ].

of (**166**) with potassium osmate followed by *O*-triethylsilylation with TESOTf and 2,6-lutidine furnished (**167**) in overall 90% yield, which was transformed to echinosporin (**163**) in several steps. In a further development, Hale et al. utilized the annulation product (**165**) in achieving a new synthesis of (+)-brefeldin A.

Like allene sulfone (**156**), allenoates can undergo [3+2] annulations to form cyclopentenes. The reaction of ethyl 2,3-butadienoate (**168**) with diethyl fumarate (**169**) in the presence of triphenyl phosphine at room temperature afforded cyclopentenetriester (**170**) in 67% yield (reaction 1, Scheme 11.30).[20a] Similar was the reaction between (**168**) and (**171**), giving cyclopentene leave a spacetriester (**172**). The stereochemistry of the products (**170** and **172**) reflects retention of olefin configurations. Under similar reaction conditions, the reaction between (**168**) and ethyl acrylate furnished two regioisomeric cyclopentene diesters in an unequal proportion. The mechanism of the Lu annulation with ethyl acrylate is shown in Scheme 11.31. The Lu annulation is also applicable to ethyl 2-butynoate in the place of allenoates.

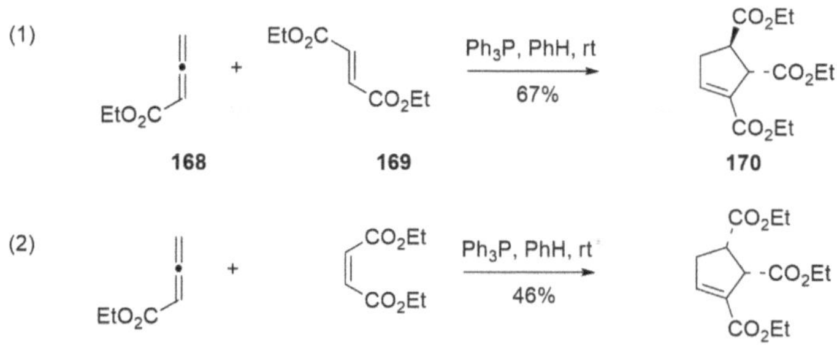

SCHEME 11.30 Lu [3+2] annulation [1995 JOC 2906 Lu X].

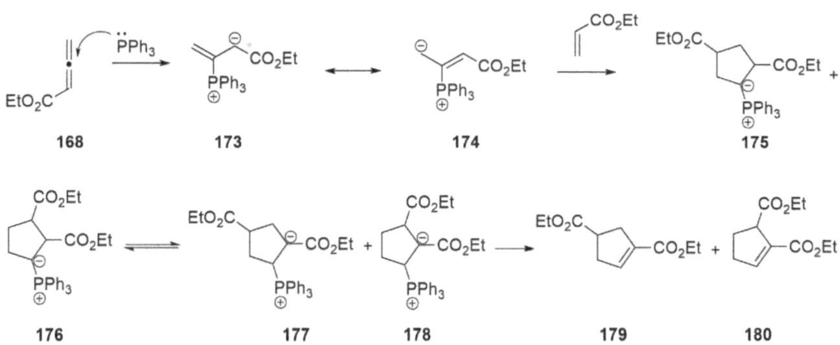

SCHEME 11.31 Mechanism of the Lu [3+2] annulation.

As to the mechanism, triphenylphosphine adds to allenoate (**168**) to form zwitterions (**173** and **174**). These ions then undergo Michael-initiated ring closure to give cyclopentene derivatives (**175** and **176**). Equilibration of these anions to zwitterions (**177** and **178**) followed by elimination of triphenylphosphine results in the formation of the products (**179** and **180**).

When a dipeptide-based chiral phosphine was used in the Lu reaction, [3+2] annulation of allenoates to α-substituted furnished functionalized cyclopentenes with quaternary stereogenic centers in high yields and with excellent enantioselectivities.[20b]

For regio- and stereoselective synthesis of substituted cyclopentenes, Brière synthesis seems to be noteworthy. The reaction between allylic sulfone (**181**) and unsaturated ketones (**182**) under phase transfer organocatalytic conditions (**183**) provide trisubstituted cyclopentenes (**184**).[21] The anion (**185**), produced under the basic conditions, undergoes Michael addition to (**182**) to form (**186**), which, in turn, cyclize to give the products (**184**) (Scheme 11.32) with excellent diastereoselectivity.

The α-selectivity of the allyl sulfone carbanions (e.g., **185**) has been harnessed by Hassner et al. to develop trimethylenemethane synthon (**187**).[22] As

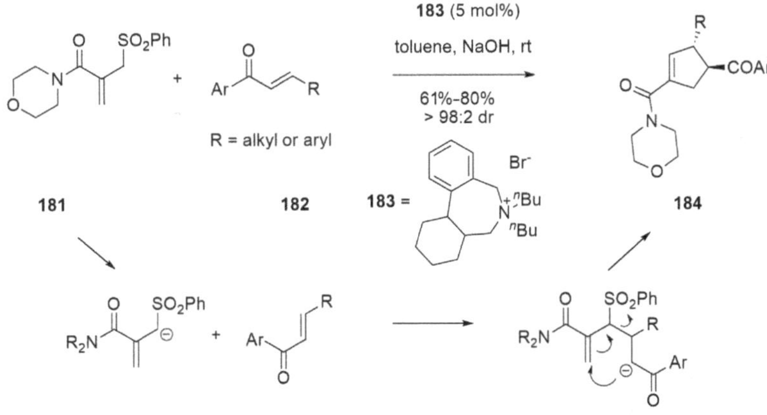

SCHEME 11.32 Brière synthesis of cyclopentenes [2011 *JOC* 4194 Brière, J-F].

displayed in Scheme 11.33, the synthon (**187**) reacts with a variety of acceptors in the presence of LDA at low temperature to furnish cyclopentene derivatives. Acceptors (**188**), (**189**), and (**190**) gave (**191**), (**192**), and (**193**), respectively, in good yields. The reactions can be described as Michael-initiated ring closures of the α-anion of the allyl sulfone (**187**).

Nair annulation: In recent years, there has been an upsurge of interest in the use of NHCs in developing new annulations. In 2006, Nair et al. discovered a novel cyclopentene annulation based upon the use of NHC precursor IMes•HCl (1,3-dimesitylimidazolium hydrochloride).[23] The reaction of enal (**194**) with enones (**195**) in the presence of IMes•HCl (**196**) and DBU in THF at afforded

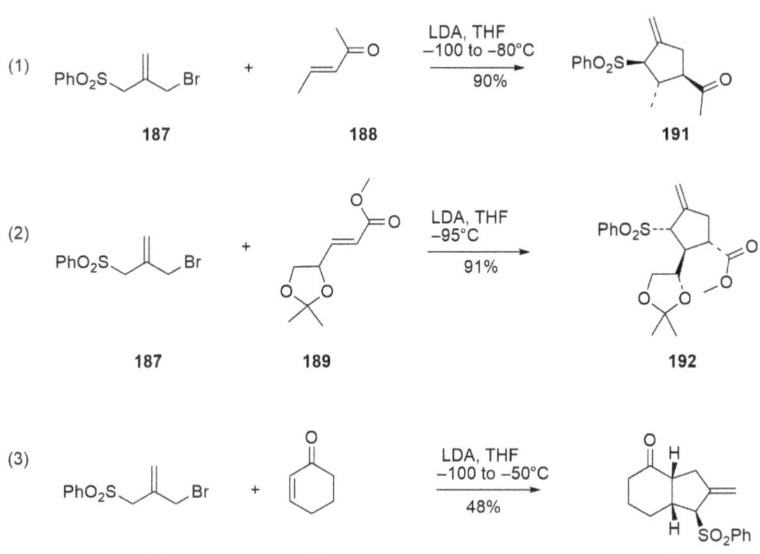

SCHEME 11.33 Hassner synthon in [3+2] annulation [2000 PAC 1671 Hassner A].

cyclopentenes (**197**) in good yields (55%–88%) and excellent diastereoselectivity, with the *trans* diastereomers being exclusive products. Importantly, aliphatic substitution is tolerated on each partner of this reaction (Scheme 11.34). In an important modification, Nair et al. used methanol as the additive to synthesize hydroxycyclopentane carboxylates.[24]

Mechanistically, deprotonation of IMes·HCl (**196**) gives IMes (**198**), which adds to the aldehyde function of cinnamaldehyde (**199**) to form alkoxide (**200**). Proton transfer of (**200**) gives (**201**), which can exist as resonance structure (**202**). Conjugate addition of (**202**) (homoenolate of cinnamate $PhCHCH_2CONHC$) to chalcone (**203**) occurs to give enolate (**204**). Proton transfer gives enolate (**205**), which undergoes an intramolecular aldol reaction to give cyclopentanealkoxide (**206**). Further cyclization of (**206**) forms β-lactone (**207**). Since substituted β-lactones are thermally unstable, β-lactone (**207**) decarboxylates to give 1,3,4-triphenylcyclopentene (**208**) (Scheme 11.35). Although both the *cis*- and *trans*-isomers are formed, the *trans* isomer predominates in this reaction.

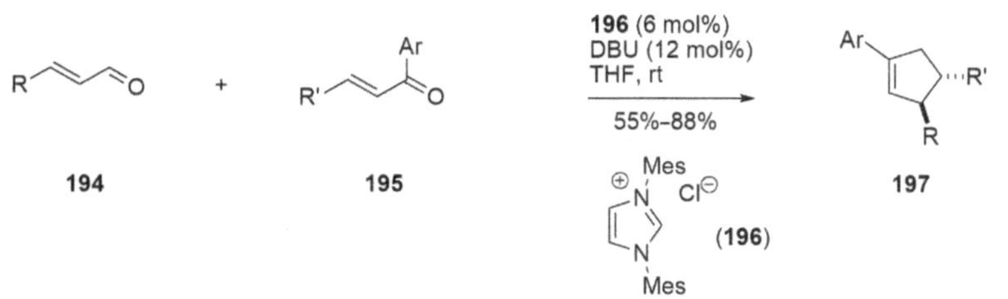

SCHEME 11.34 Nair annulation for the synthesis of 1,3,4-trisubstituted cyclopentenes.

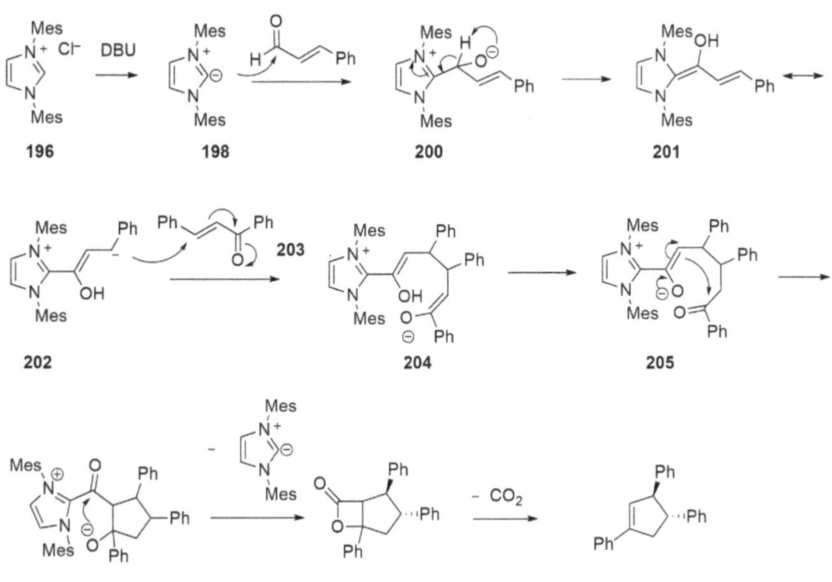

SCHEME 11.35 Mechanism of the Nair annulation.

Bode NHC-mediated cyclopentene annulation is a remarkable advancement. Unlike Nair annulation, it gave *cis*-cyclopentenes with excellent enantioselectivities (Scheme 11.36).[25] The reaction of enal (**209**) with unsaturated ketone (**210**) in the presence of catalyst (**211**) gave cyclopentene (**212**).

The stereoselectivity arises from the mechanism involving asymmetric intermolecular aldehyde-ketone crossed benzoin reaction and a NHC-promoted oxy-Cope rearrangement involving the intermediates (**213–216**) (Scheme 11.37).

Scheidt variant of the *cis* disubstituted cyclopentenes employs a Lewis acid, such as titanium isopropoxide and a *N*-heterocyclic carbene.[26] The cooperative catalysis is responsible for the observed high degree of enantioselectivities as well as diasereoselectivities.

Chi annulation: Chi annulation is a novel cyclopentene annulation.[27] The chemistry involved with the annulation is fundamentally different from the Nair

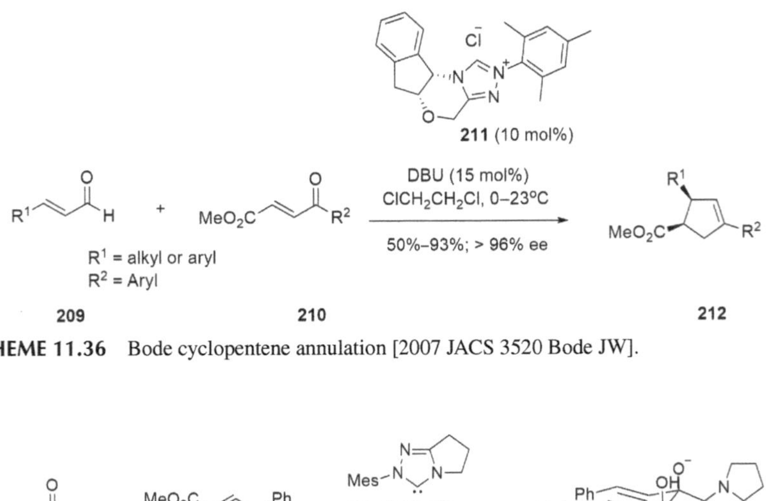

209 **210** **212**

SCHEME 11.36 Bode cyclopentene annulation [2007 JACS 3520 Bode JW].

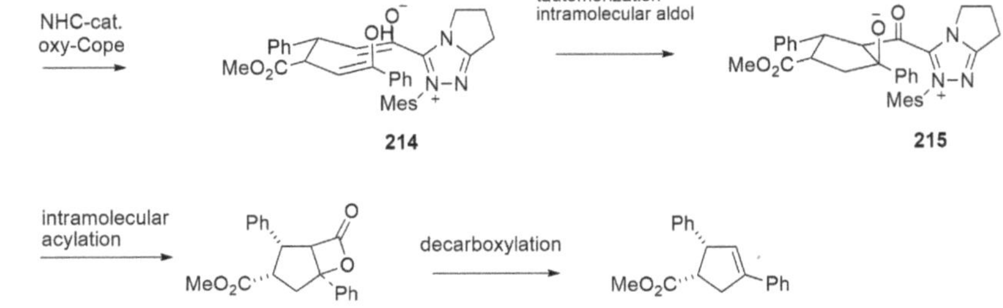

216 **217**

SCHEME 11.37 Mechanistic proposal of Bode annulation.

annulation or Bode annulation which utilizes a cinnamaldehyde. One of two reacting components of the Chi annulation is a saturated ester. The underlying principle of the reaction is direct β-carbon activation of the esters by mediation of a NHC, thereby providing their extraordinary functionalization. All three types of the Chi annulation are given in Scheme 11.38. Ester (**218**) reacted with enones (**219**) in the presence of DBU and triazolium salt (**222**) to give cyclopentenes (**223**). Under similar reaction conditions, the ester (**218**) gave lactones (**224**) in reaction with trifluoro ketones (**220**) in good yields. Likewise, hydrazones **221**) reacted with esters (**218**) to afford γ-lactams (**225**) with good yields and er.

The Chi annulation can be interpreted in terms of the formal dianion intermediate (**226**), which is formed from NHC activated species (**227**). Abstraction of α-hydrogen by DBU gives enolate (**228**) (Scheme 11.39). Subsequently, proton transfer generates ester enolate (**229**). In summary, all the annulations, namely, Nair, Bode, Scheidt and Chi annulations, entail ester homoenolates. Thus, they are versatile reactive intermediates for carbocycle and heterocycle annulations.

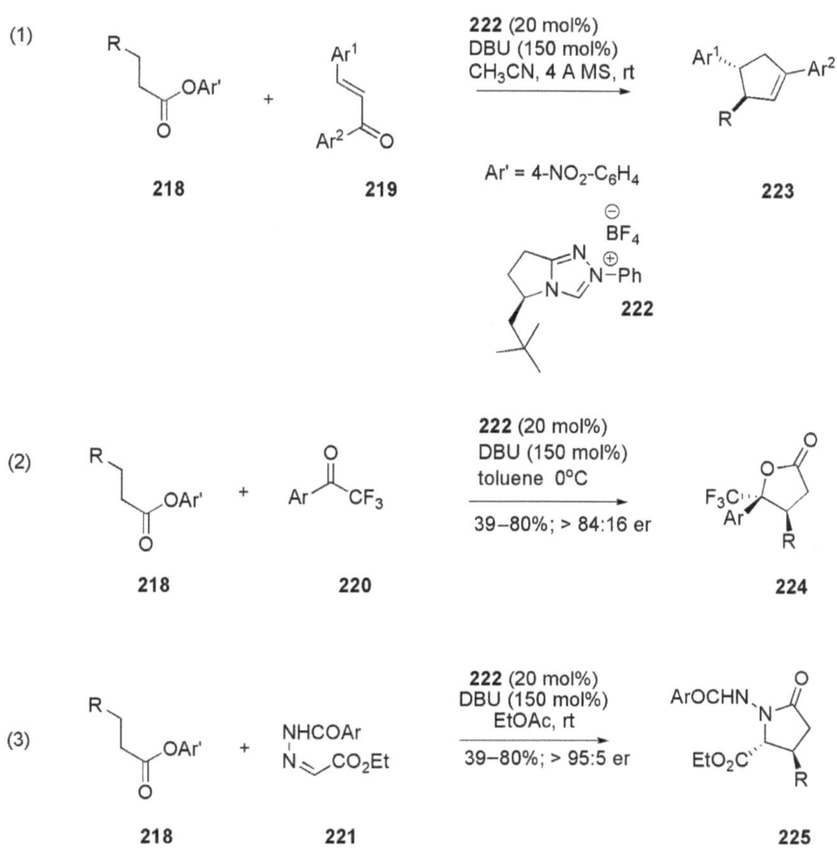

SCHEME 11.38 Chi annulation [2013 NCY 835 Chi YR].

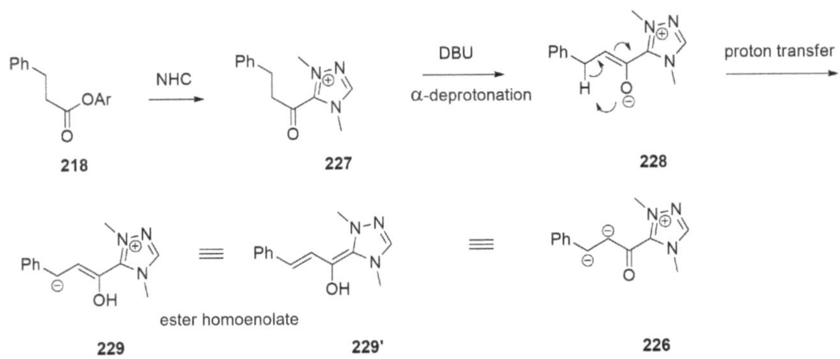

SCHEME 11.39 Proposed intermediates for Chi annulations via β-carbon activation of saturated esters.

11.4 ANIONIC [2 + 1] ANNULATION: MCCOY CYCLOPROPANATION, COREY-CHAYKOVSKY REACTION, DENNEY AND WADSWORTH-EMMONS CYCLOPROPANATION

Robinson annulation and its concept were confined to the synthesis of six-membered carbocycles till 1958, when McCoy reported an application of the concept in the synthesis of cyclopropane dicarboxylates. The year 1958 was the year for cyclopropane synthesis. In this year, the seminal report of Simmons and Smith on the reaction of alkenes with diiodomethane in the presence of zinc dust to afford cyclopropanes was published. The reaction is popular as Simmons-Smith cyclopropanation. In formal sense, it can be considered an annulation. In the same year, McCoy synthesis of cyclopropane carboxylates was published.[28] During the early 1960s, he generalized the reaction and established its stereochemical outcomes. Today, the best synthesis of cyclopropane-1,2-dicarboxylic acid, especially the *cis*-diacids, is probably that of McCoy as evident from the recent synthesis of several cyclopropane diacids.[29] As depicted in Scheme 11.40, activated esters (**230**) react with unsaturated esters (**231**) in the

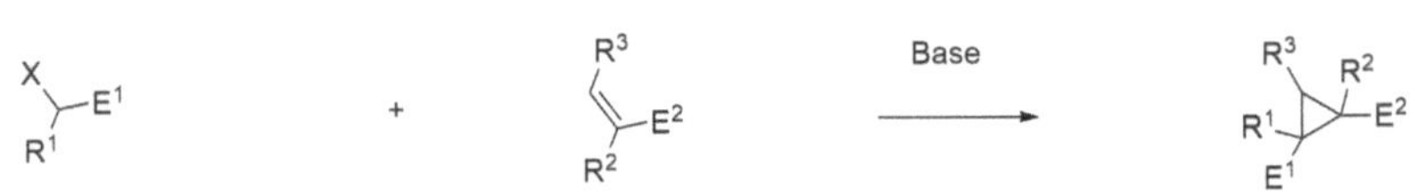

X = Cl, Br,PhS(O)NR, CN

E^1 = CO_2Me, CO_2Et, $P(O)(OMe)_2$, COR, C(O)TBS, S(O)Ph, NO_2, F

R^1 = H, alkyl, aryl, Cl, CO_2Et,

E^2 = CO_2Me, CO_2Et, $CO_2{}^tBu$, $P(O)(OMe)_2$, COR, CON(Me)O(Me), CN, NO_2

R^2 = Alkyl, CO_2Et, $OP(O)(OMe)_2$

R^3 = Alkyl, aryl, CF_3, CO_2Et

230 231 232

SCHEME 11.40 McCoy cyclopropanation.

presence of a base to give substituted cyclopropanes (**232**) in one pot. The bases used are NaH, NaOMe, KO^{t}Bu, LDA, LHMDS, LTMP, etc. Various solvents are used for the McCoy reaction. Toluene is the best solvent for setting the *cis* configuration of the ester groups.

In Scheme 11.41, few selected examples of the McCoy reaction are presented. Reaction 1 is one of earliest examples. The esters (**233** and **234**) can be condensed in the presence of KO^{t}Bu to give *cis*-diester (**235**) and its isomer (**236**). Reaction 2 demonstrates the suitability of silyl ketone (**237**) and coumarin (**238**) in giving (**239**) and (**240**). Reaction 3 shows a milder condition for carrying out cyclopropanation with (**241**) and (**242**) to produce (**243**). That a McCoy reaction can be highly diastereoselective is represented by the reaction between (**244**) and (**245**), giving (**246**) (reaction 4). The other diastereoisomer was not formed.

All of these reactions proceed through a sequence of three generic reactions, namely deprotonation, Michael addition and intramolecular nucleophilic displacement of a nucleofuge. For the example shown in Scheme 11.42, deprotonation of (**244**) takes place in the first step to form carbanion

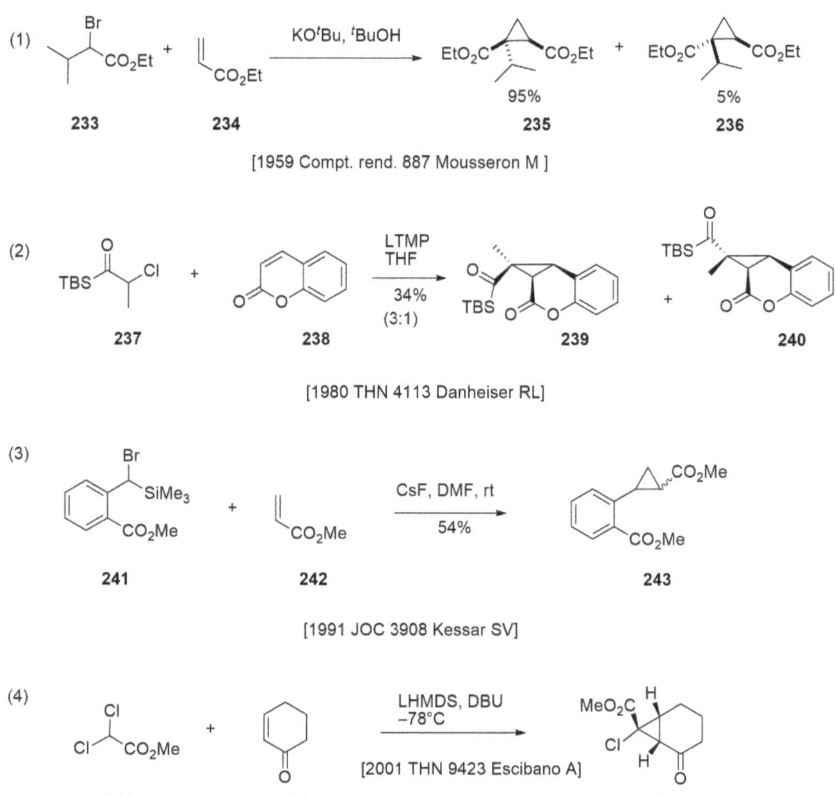

SCHEME 11.41 Selected examples of the McCoy cyclopropanation.

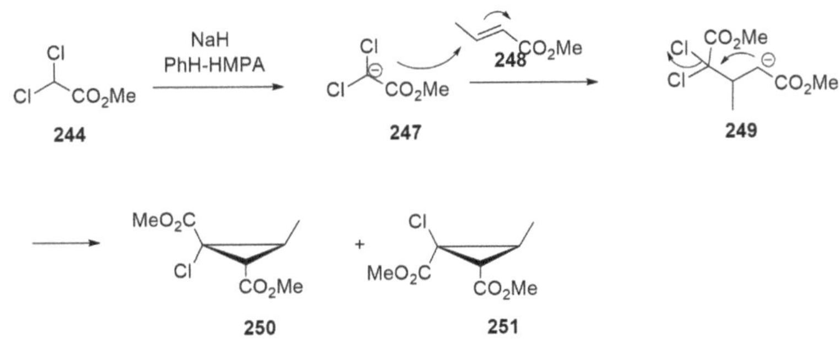

SCHEME 11.42 Mechanism of the McCoy cyclopropanation.

(**247**). Michael addition of (**247**) to methyl crotonate (**248**) generates carbanion (**249**), which undergoes intramolecular displacement of chlorine to give products (**250** and **251**).

The salient feature of the McCoy cyclopropanation is solvent-dependent stereochemical outcome.[30] The reaction between α-substituted acrylic esters or acrylonitriles with α-haloesters in the presence of base may be controlled by the solvent so as to allow the stereoselective formation of either *cis*- or *trans*-cyclopropanes. Reactions with no solvent or with hydrocarbon or ether solvents give predominantly the *cis* isomers. Those in hexamethylphosphoramide-benzene mixtures or in dimethylformamide give predominantly the *trans* isomers. These results are interpreted by electrostatic attractions between the groups in the precursor anions for the ring closure in solvents with low dielectric constant.

The McCoy annulation has found a neat application in the total synthesis of plakortide E (**252**).[31] The key dioxolane intermediate (**253**) was prepared from triethyl 2-ethyl-1,1,2-cyclopropanetricarboxylate (**254**), which was obtained from (**255**) and (**256**) by McCoy cyclopropanation. After hydrolysis with aqueous NaOH and dehydration with acetyl chloride, the corresponding anhydride was reduced with $NaBH_4$ to afford lactone (**257**). Lactone (**257**) was subsequently reduced with borane to form the corresponding alcohol (**258**). It was converted to lactone (**259**) by sequential Swern oxidation and Wittig reaction. Lactone (**259**) was then irradiated with a 300 W sunlamp at room temperature under oxygen atmosphere in the presence of catalytic amounts of Ph_2Se_2 and AIBN to give 1,2-dioxolane (**260**) in 99% yield. The lactone moiety of compound (**260**) was reduced with $LiBH_4$, and the CC double bond was reduced with diimide to afford *cis*-(**253**) (Scheme 11.43). More than 12 steps were required to elaborate (**253**) to plakortide E (**252**).

Toyota's synthesis of *cis*-2-fluorocyclopropane-1-carboxylic acid (**261**), a key intermediate for quinolone antibiotics, relied upon application of a McCoy cyclopropanation (Scheme 11.44).[32] The reaction of chloromethyl sulfoxide (**262**) with acrylate (**263**) in the presence of NaHMDS gave *cis* ester (**264**) in 51% yield. Fluorination of (**264**) with fluorine in nitrogen furnished *trans*

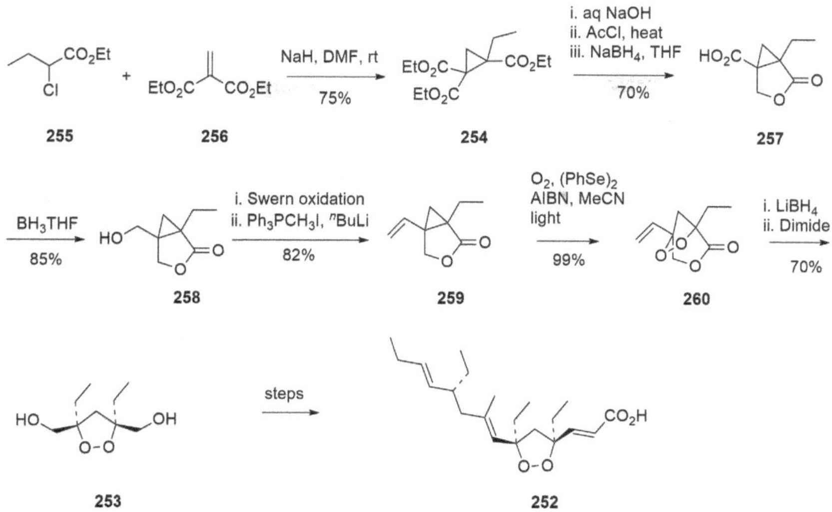

SCHEME 11.43 Total synthesis of plakortide E (**252**) [2011 CEJ 5874 Wong HNC].

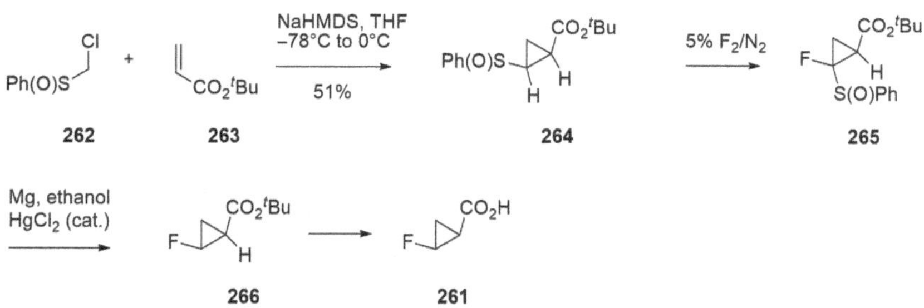

SCHEME 11.44 Synthesis of *cis*-2-fluorocyclopropane-1-carboxylic acid [1996 THS 8507 Toyota A].

ester (**265**) in 35% yield. When ester (**265**) was treated with Mg in ethanol containing a catalytic amount of $HgCl_2$, *cis* ester (**266**) was produced in 85% yield. Treatment of the ester (**266**) with 10% aq HCl quantitatively afforded the *cis* acid (**261**).

The vinylogous McCoy cyclopropanation was developed Hudlicky et al. When 2-cyclohexenone was added to a preformed lithium dienolate derived from ethyl α-bromobutanoate (**267**) with LDA and HMPA at −78°C, vinyl cyclopropanated cyclohexanone (**268**) was formed in 93% yield via (**269**) (Scheme 11.45).[33] The reaction is general for both cyclic and acyclic α,β-unsaturated ketone acceptors. It works well with α-chlorobutenoate, but the yields are lower than those with the bromo analogs. The vinylogous McCoy reaction was found useful for a total synthesis of retigeranic acid A, a complex natural product.[34]

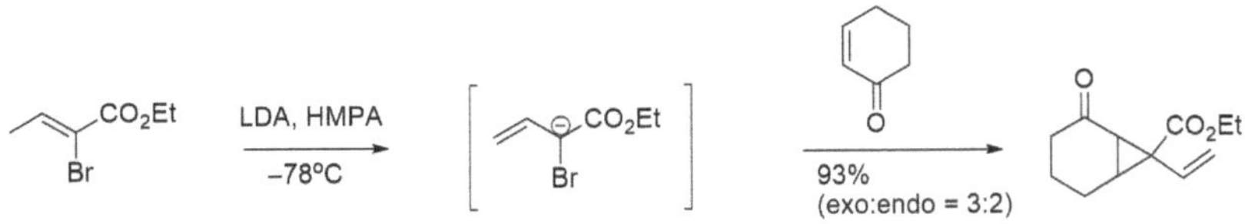

267 269 268

SCHEME 11.45 Vinylogous McCoy cyclopropanation [1986 JOC 4746 Hudlicky T].

Corey-Chaykovsky reaction: It is described as the reaction of dimethyloxosulfonium methylide, generated from trimethylsulfoxonium iodide, with α,β-unsaturated esters or α,β-unsaturated ketones giving cyclopropane derivatives.[35] In the present context, it can be described as a [2 + 1] annulation. In both the examples (reactions 1 and 2) given in Scheme 11.46, sulfur ylides, formed in situ, undergo Michael addition followed by ring closure with loss of a dialkyl sulfoxide to give the cyclopropane derivatives (**270** and **271**).

Payne cyclopropanation is a modification of the Corey-Chaykovsky in that it uses a precursor of a stabilized sulfur ylide. As shown in Scheme 11.47, the reaction between (**272**) and (**273**) selectively gave CC double bond cyclopropanated product (**274**), via the Michael adduct (**275**).[36]

Wadsworth-Emmons cyclopropanation:Denney annulation, discovered in as early as 1959, gives *trans* cyclopropane carboxylates. The reaction of

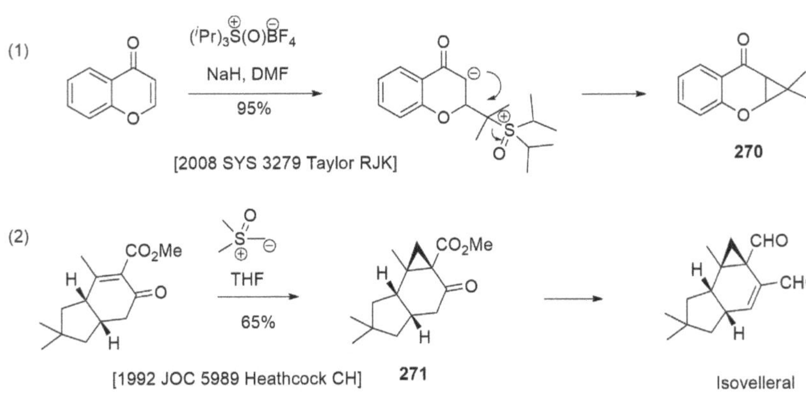

270

271

Isovelleral

SCHEME 11.46 Examples of Corey-Chaykovsky reaction.

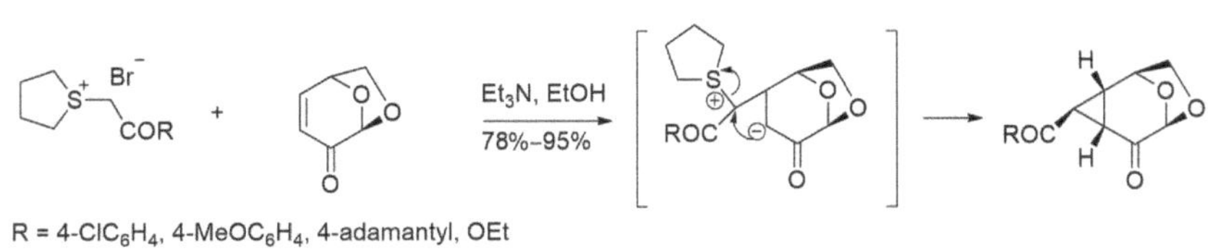

272 273 275 274

SCHEME 11.47 Payne cyclopropanation [2007 THY 1986 Samet AV].

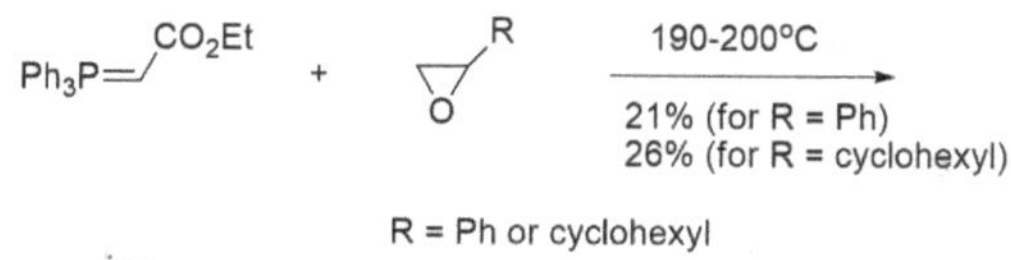

276 277 278

SCHEME 11.48 Denney annulation, the original report [1959 JACS 6330; 1962 JACS 3944 Denney DB].

phosphorane (**276**) with epoxides (**277**) at 190–200°C yielded *trans* cyclopropane carboxylate (**278**) (Scheme 11.48).[37]

In 1961, Wadsworth group discovered a similar cyclopropanation. Their reaction involves a phosphonoacetate and an epoxide under strongly basic conditions. It takes place at relatively lower temperature compared to the Denney cyclopropanation. Phosphonate (**279**) reacted with styrene oxide (**280**) in the presence of NaH to give *trans* cyclopropane **(281)** (Scheme 11.49).[38] The degree of stereoselectivity of the cyclopropanation is exceedingly high. The reaction of (*R*)-styrene oxide (**282**) with triethylphosphonoacetate (**279**) in the presence of NaH in xylenes at 135°C furnished (*S*, *S*)-2-phenylcyclopropanecarboxylate (**283**) in >95% ee, suggesting the final ring closure had taken place with complete inversion of the epoxide configuration (Scheme 11.50). The proposed mechanism involves initiation of the reaction by nucleophilic attack of the phosphonate anion (**284**) at methylene carbon of the epoxide (**282**) to give alkoxide

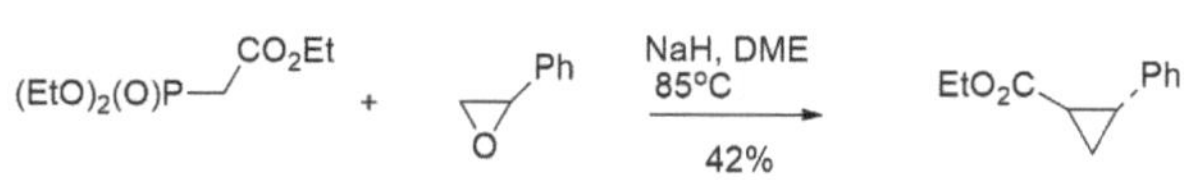

279 280 281

SCHEME 11.49 Wadsworth-Emmons cyclopropane annulation [1961 JACS 1733 Wadsworth, Jr. WS].

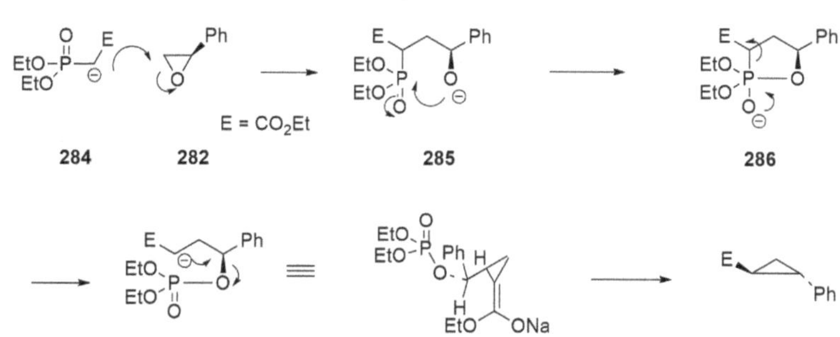

284 282 285 286

287 287 283

SCHEME 11.50 Mechanism of the Wadsworth-Emmons cyclopropanation [2003 ORS 2331 Armstrong A].

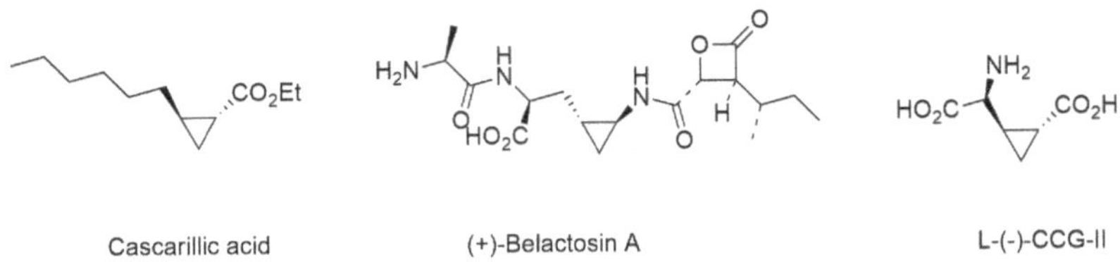

FIG. 11.3 Naturally occurring cyclopropane acids synthesized by Wadsworth-Emmons cyclopropanation.

(**285**). Intramolecular attack of alkoxide ion at phosphonate group gives intermediate (**286**), which collapses to form phosphate (**287**). Intramolecular displacement reaction of the intermediate (**287**) in transoid transition state furnishes cyclopropane (**283**).

The Wadsworth-Emmons cyclopropane annulation has been successfully applied to the enantioselective total synthesis of cyclopropane-containing natural products, such a scascarillic acid, belactosin A and L-(−)-CCG-II (Fig. 11.3). The details of the Kumar synthesis of grenadamide (**288**) are presented in Scheme 11.51. Grenadamide (**288**) was isolated from the marine cyanobacterium *Lyngbya majuscula* in 1998. Its synthesis started from commercially available epoxide (**289**). The epoxide (**289**) was converted into cyclopropyl ester (**290**) in 85% yield and >99% ee by Wadsworth-Emmons reaction with triethyl phosphonoacetate in the presence of NaH.[39] The ester (**290**) was then double homologated by Kowalski method to furnish compound (**291**) via intermediate (**292**). LiOH promoted hydrolysis of the ester (**291**) furnished acid (**293**) in 82% yield. Conversion of the acid (**293**) into its acid chloride by using $SOCl_2$ and coupling of the resultant chloride with phenylethylamine furnished grenadamide (**288**) in 65% yield.

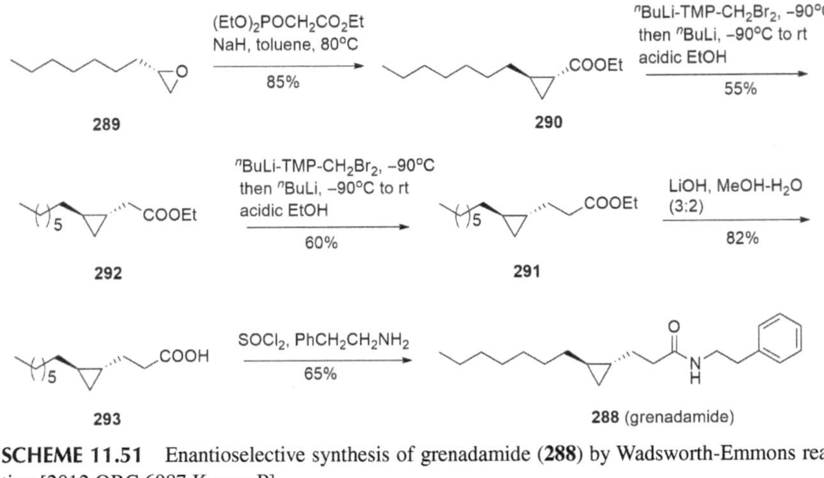

SCHEME 11.51 Enantioselective synthesis of grenadamide (**288**) by Wadsworth-Emmons reaction [2012 OBC 6987 Kumar P].

11.5 HETEROANNULATIONS: LEUSEN ANNULATION

Heteroannulations probably outnumber the carbocycle annulations. But, there are fewer heteroannulations, which are carried out under anionic conditions. The notable ones are Krohnke pyridine synthesis, Friedländer quinoline annulation and Leusen annulation.

Akin to the carbocycle annulation, Tamura et al. developed a new heteroannulation where heteroatom substituted benzoates serve as 1,4-dipole reagents. Treatment of benzoates (**294**) with allenebiscarboxylate (**295**) in the presence of NaH or KO^{t}Bu allowed formation of heterocycles (**296**).[40] Although the initial Michael adduct can be isolated, the reactions are conveniently performed in single operation to obtain the annulated products. Initial Michael addition of (**294**) with (**295**) gives anionic species (**297**), which undergo Dieckmann cyclization to form heterocyclic motifs (**298**). Isomerization of (**298**) permits the formation of final products (**296**) (Scheme 11.52). Tamura heteroannulation is applicable to *o*-aminobenzaldehyde, *o*-aminoacetophenone and *o*-aminobenzophenone (**299**). Their reactions with allenecarboxylate (**295**) in the presence of KOtBu gave quinoline carboxylates (**300**) in good yields via (**301**) (Scheme 11.53).

Friedländer annulation is another [4+2] heteroannulation, which is represented by its general (reaction 1) and classical form (reaction 2) (Scheme 11.54).[41] It is defined as the base- or acid-promoted condensation of aromatic 2-amino-substituted benzo carbonyl compounds (**302**) with appropriately substituted carbonyl derivatives (**303**) containing a reactive α-methylene group, followed by cyclodehydration to give quinolines (**304**). For example, amino aldehyde (**305**) reacted with acetone (**306**) in the presence of NaOH to give 2-methylquinoline (**307**). Although the Friedländer annulation is well established, its mechanism is not. Experimental results support two different

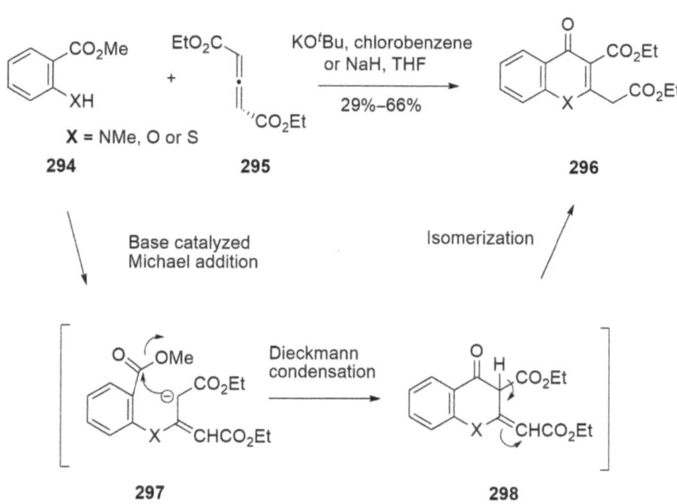

SCHEME 11.52 Tamura heteroannulation [1985 JOC 1542 Tamura Y].

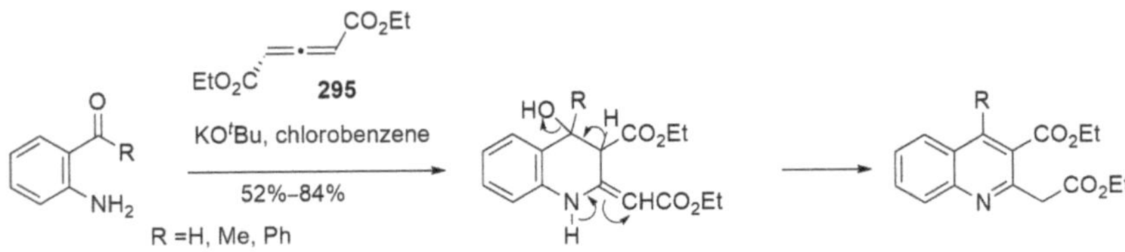

299 301 300

SCHEME 11.53 Formation of quinoline carboxylate by Tamura heteroannulation.

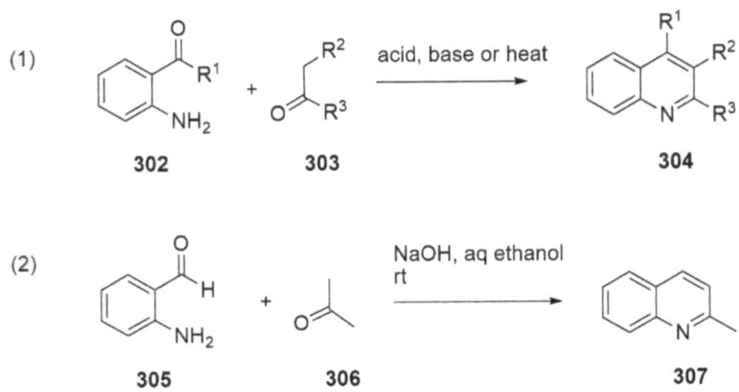

SCHEME 11.54 Friedländer annulation.

mechanistic proposals. One pathway involves formation of Schiff base followed by intramolecular aldol reaction. The other pathway could be aldol reaction followed by Schiff base formation leading to quinolines.

The Friedländer annulation has been applied to the synthesis of camptothecin analogues. Under thermal conditions, ketone (**308**) underwent reaction with (**309**) to give (**310**), featuring camptothecin scaffold (Scheme 11.55).

Leusen [3 + 2] annulation: Leusen [3 + 2] annulation was discovered in 1972 by Leusen group. Perhaps, this is the first reported annulation that builds up an aromatic ring.[42] The key reagent is to sylmethylisocyanide (Tosmic) (pK_a = 14), which is known to undergo deprotonation at the α-position under influence of strong bases like NaH, KOtBu, K_2CO_3, etc. The conjugate anion reacts with a variety of electrophiles. When tosmic (**311**) is reacted with Michael acceptors (**312**) in the presence of NaH at room temperature, pyrrole derivatives (**313**) are formed in a regiospecific manner (Scheme 11.56). Tosylmethylisocyanide (tosmic) provides CNC unit for the new pyrrole rings. With aldehydes, it forms oxazoles.

A representative mechanism for acceptor (**314**) is shown in Scheme 11.57. Tosmic anion (**315**) readily undergoes Michael addition to (**314**) to form (**316**). It then characteristically ring closes to form pyrrole nucleus (**317**). Anion dislocation of (**317**) gives (**318**). Subsequent loss of the tosyl group from (**318**) affords (**319**), which undergoes isomerization to give pyrrole (**320**).

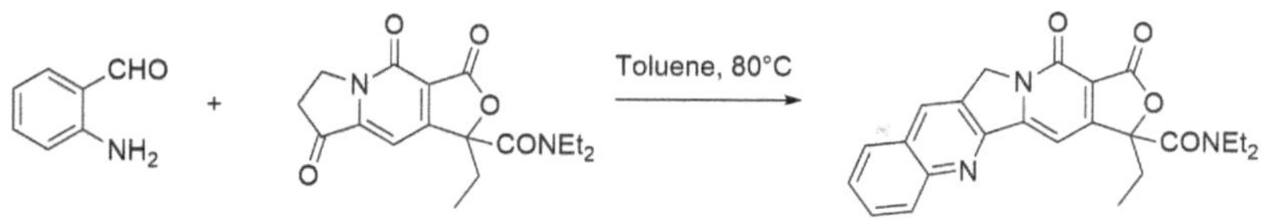

308 309 310

SCHEME 11.55 Synthesis of a camptothecin analog.

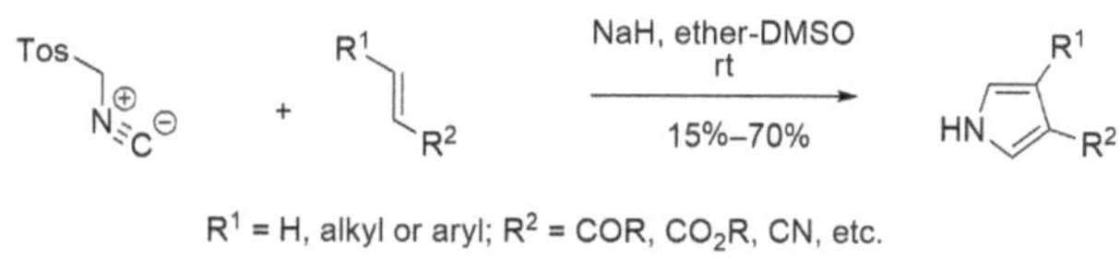

311 312 313

SCHEME 11.56 Leusen [3+2] annulation: synthesis of pyrroles [1972 THS 5337 Leusen AMV].

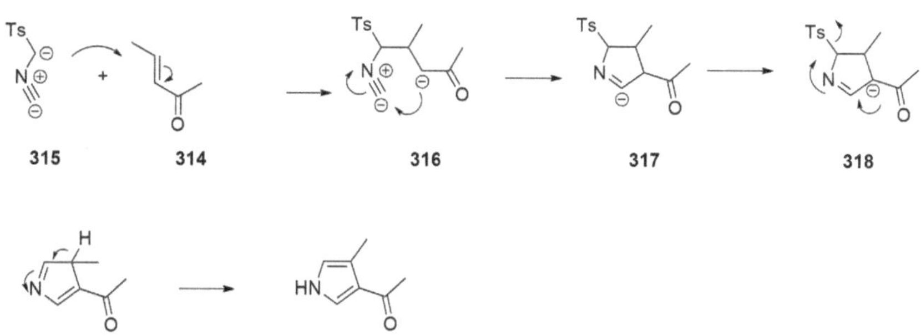

315 314 316 317 318

319 320

SCHEME 11.57 Mechanism of Leusen annulation for pyrrole synthesis.

The preparation of tosmic is given in Scheme 11.58. The reaction of sodium salt of *p*-toluenesulfinic acid (**321**) with formaldehyde in formamide gives formamide sulfone (**322**). Dehydration of (**322**) with $POCl_3$ in the presence of triethylamine gives tosylmethyl isocyanide (**311**).

The Leusen annulation has widely been studied. It is a versatile reaction. It allows synthesis of a number of classes of azole heterocycles. Scheme 11.59 gives an overview of the types of azoles that can assembled by the Leusen annulation. Reaction 1 involves interaction between β-chloro ketone (**323**) and tosmic derivative (**324**) to give pyrrole derivative (**325**). With cinnamonitriles (**326**), tosmic derivative (**327**) gave pyrrole carbonitrile (**328**). Interestingly, for cinnamaldehyde (**329**), the condensation of tosmic (**330**) took place selectively at aldehyde CO bond to furnish oxazole (**331**). For the synthesis of imidazole (**332**), imine (**333**) was expectedly condensed with tosmic (**334**).

The Leusen tosmic mediated imidazole synthesis has been employed for the synthesis of (+)-pilocarpine. It is a naturally occurring imidazole alkaloid.

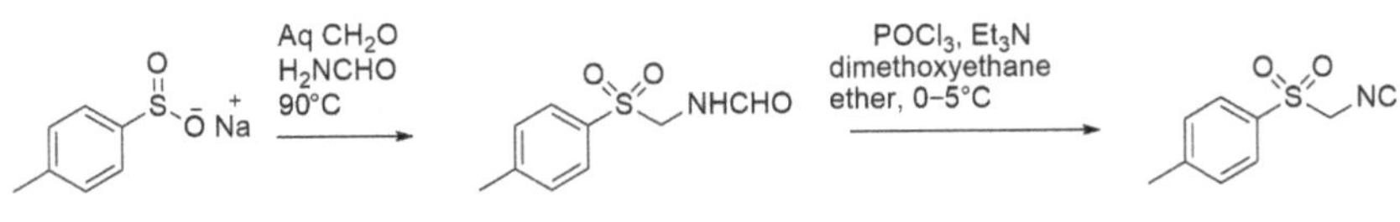

321 322 311

SCHEME 11.58 Preparation of tosylmethylisocyanide (tosmic).

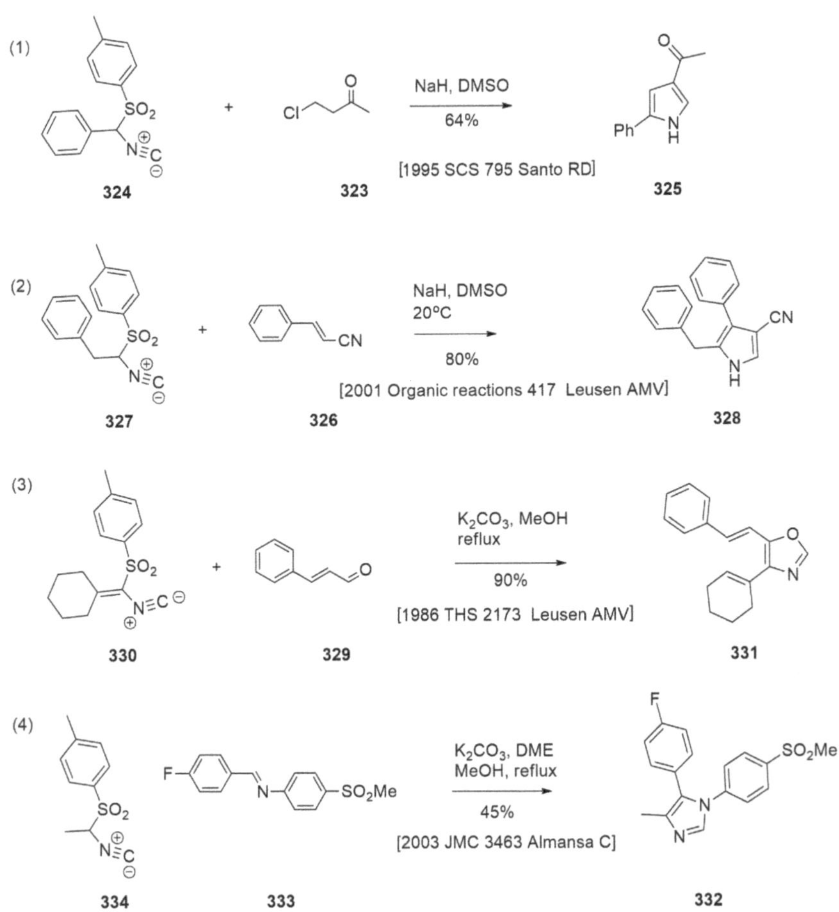

SCHEME 11.59 An overview of the scope of substrates for Leusen annulation.

Although its physiological activities are diverse, it is most commonly used for the treatment of glaucoma. Treatment of a solution of carbonate (**335**) in THF with $Pd(OAc)_2(PPh_3)_2$ under a carbon monoxide atmosphere gave γ-lactone (**336**) (92% yield, >98% de and >98% ee) via the palladium-catalyzed decarboxylation/carbonylation (Scheme 11.60).[43] Pyridinium *p*-toluenesulfonate-acetone deketalization of (**336**) furnished aldehyde (**337**) in 93% isolated yield. *N*-methyl

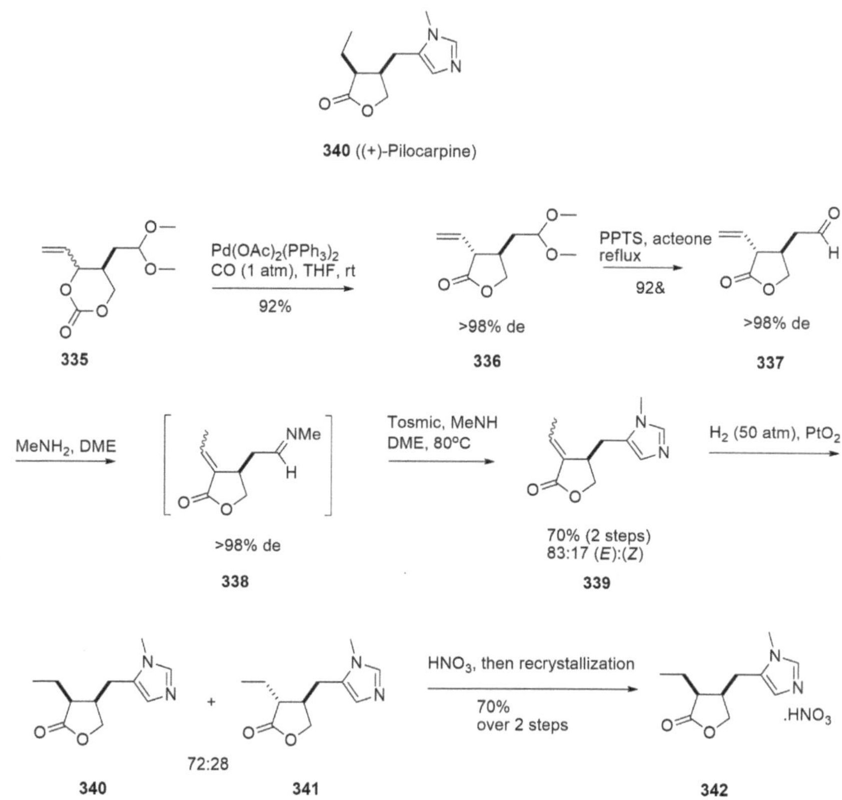

SCHEME 11.60 Total synthesis of (+)-pilocarpine [2009 THN 8283 Davies SG].

imine (**338**) was prepared from aldehyde (**337**) by passing $MeNH_2$ gas through a solution of the aldehyde (**337**) in DME in the presence of 4 Å molecular sieves at 0°C. Imine (**338**) was then treated with tosmic (**311**), K_2CO_3 and anhydrous methylamine gas in MeOH at 80°C for 24 h to give (**339**) in 70% yield as an 83:17 [(*E*):(*Z*)] mixture. Hydrogenation of the mixture over PtO_2 at 50 atm pressure in MeOH quantitatively gave a 72:28 mixture of (+)-pilocarpine (**340**) and (+)-isopilocarpine (**341**). Recrystallization of the mixture of nitrate salts from EtOH gave (+)-pilocarpine nitrate (**342**) in 70% yield.

The modification of the Leusen annulation by Magnus et al. in connection with their synthesis of CC-1065 is noteworthy.[44] It extends to the chemistry of a dienoate, i.e., ethyl sorbate (**343**). This study resulted in a new synthesis of 3,3′-bipyrroles via two successive annulations. Quite interestingly, in the first event, the annulation (NaH-DMSO-ether, 20°C) of the tosmic (**311**) took place at the 4,5-double bond of the sorbate (**343**) to give (**344**). Tosylation of (**344**) furnished (**345**). Treatment of the tosyl derivative (**345**) with tosmic (**311**) in the presence of $HN(SiMe_3)_2$-NaH-THF at 0°C afforded 3,3′-bipyrrole (**346**) (Scheme 11.61).

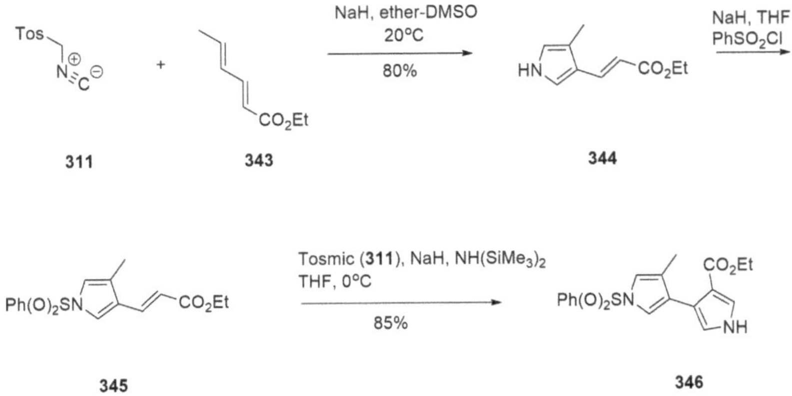

SCHEME 11.61 Leusen annulation with a dienoate [1983 CCS 26 Magnus P].

Barton-Zard pyrrole synthesis: The Barton-Zard pyrrole synthesis, first reported in 1985, is the reaction of alkyl isocyanoacetates with nitroalkenes or vinylsulfones in the presence of a base to give 2-substituted pyrroles.[45] Mechanistically, it is very similar to the Leusen synthesis. The differences between the two are the nature of the leaving groups and their positions. In Barton-Zard reaction, a nitro group is the leaving group, which is positioned in the acceptor molecule. The yields of this reaction are generally high in the range of 80%–90%. An organic strong base, such as DBU and guanidine, is generally used in this reaction. The reaction of isocyano acetate (**347**) with nitro compound (**348**) in the presence of DBU furnished trisubstituted pyrrole (**349**). It proceeded through nitroalkene (**350**) (Scheme 11.62). The Barton-Zard reaction has been useful for the synthesis of biologically active pyrroles, porphyrins, polypyrroles, and dipyrromethene dyes.

The Pathak variant (Scheme 11.63) of the above annulation involves carbohydrate-derived acceptors containing an arylsulfonyl group, the nature of the products being the same, i.e., sulfone-free pyrroles with chiral appendages.[46] For reaction 1, Barton-Zard product obtained from (**351**) and (**352**) further underwent cleavage of the sugar ring to give (**353**). The reaction of (**351**) with optically active bisvinyl sulfone (**354**) in the presence of KO^{t}Bu provided bispyrrole product (**355**).

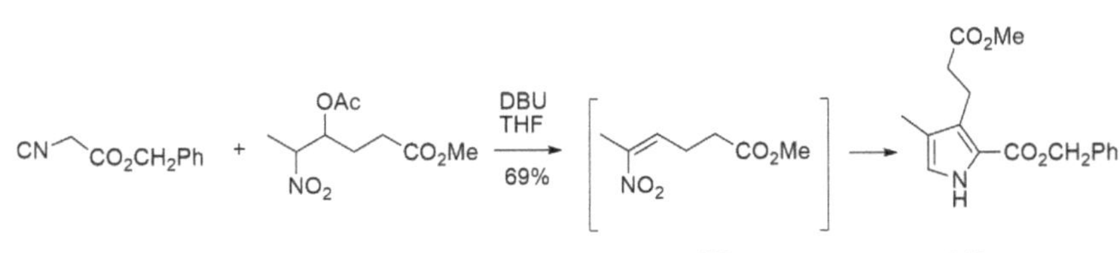

SCHEME 11.62 Example of a Barton-Zard pyrrole annulation.

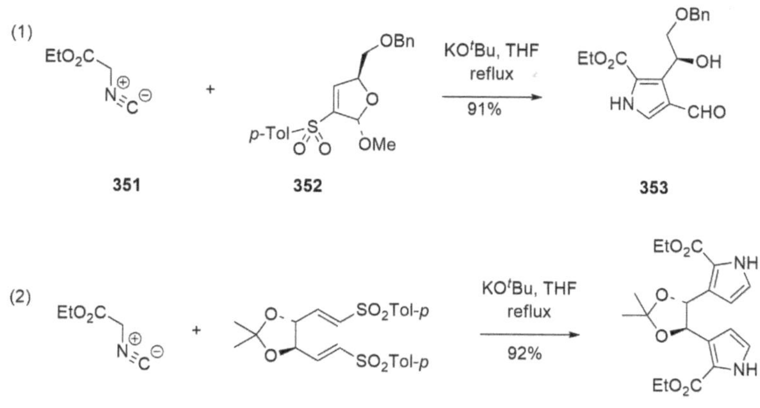

351 352 353

351 354 355

SCHEME 11.63 Annulation with conjugated sulfones [2009 JOC 669 Pathak T].

Many aromatic nitro compounds undergo annulations with ethyl isocyanoacetate (**351**). An example is the condensation of 1-nitronaphthalene (**356**) with ethyl isocyanoacetate (**351**) in the presence of (*tert*-butylimino)tris(pyrrolidino) phosphorane (BTPP), giving (**357**) (Scheme 11.64).

In Barton-Zard annulation, the nitroalkenes or vinyl sulfones acceptors can be substituted by alkyne esters. Thus, the reaction between (**351**) and (**358**) in the presence of KO^{t}Bu gave (**359**) (Scheme 11.65). The annulation between the activated isocyanides and electron deficient alkynes can be conducted with catalytic amounts of Cu2O and 1,10-phenanthroline.

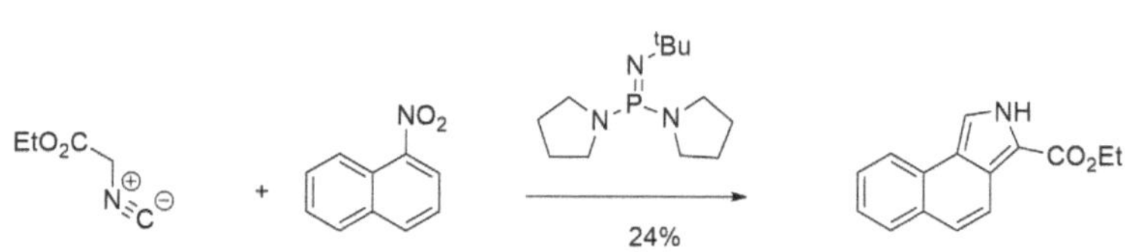

351 356 357

SCHEME 11.64 Barton-Zard annulation of a nitroarene [1996 THS 8391 Ono N].

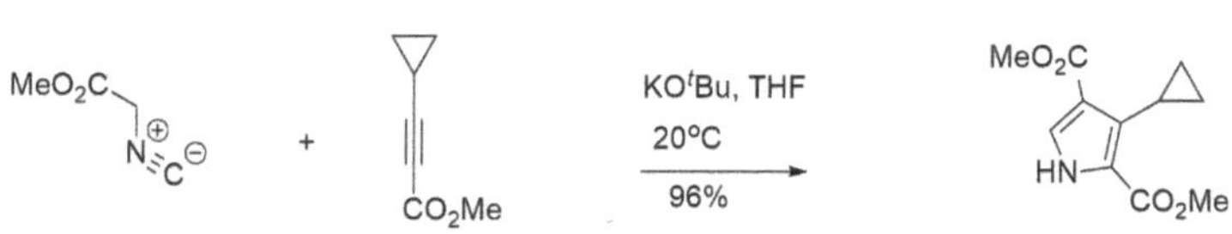

351 358 359

SCHEME 11.65 Annulation of an alkyne acceptor [2005 ACIE 5664 de Meijere A].

11.6 TRANSITION METAL-CATALYZED ANNULATIONS

Transition metal reagents and catalysts, over last years, have made enormous impact on the growth of synthetic chemistry. Numerous methods based upon the reagents have been developed. Notable are the applications of $Co_2(CO)_8$, $Ni(CO)_4$, $Fe(CO)_5$ and palladium, Rh and Ru reagents. Since a great majority of the transition metal catalyzed reactions are involved with the stoichiometric amounts of alkali metal salts or organic bases, it is likely that the anionic reactions can, in principle, be coupled with the former resulting in annulations. Alper synthesis of dihydro-α-naphthols, which can be considered as a multicomponent reaction, involves a Pd(0)-promoted arylation. The reaction of iodoarylmalonates (**360**) with Michael acceptors (**361**) and CO in the presence of $Pd(OAc)_2$-Xphos and triethylamine at 120°C provided dihydro-α-naphthols (**362**) in good yields and diastereoselectivity (Scheme 11.66).[47] Examples are the transformations: (**363** → **364**), (**365** → **366**), (**367** → **368**), and (**369** → **370**).

It is reasonable to assume that the cascade starts with Michael addition of the anion (**371**) of the malonate (**372**) with vinyl ketones (**373**) in the presence of a base, giving adducts (**374**) (Scheme 11.67). These adducts then undergo oxidative addition with palladium(0) to form intermediates (**375**), which are sufficiently labile for insertion of carbon monoxide to form (**376**). Intramolecular attack of the enolate carbon gives palladacycle (**377**). Reductive elimination of

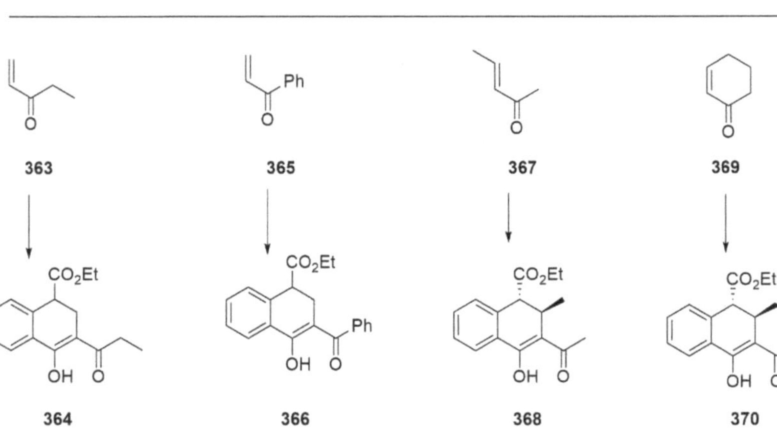

360 361 362

363 365 367 369

364 366 368 370

SCHEME 11.66 Alper [3+2+1] annulation for dihydro-α-naphthols [2009 ORS 278 Alper H].

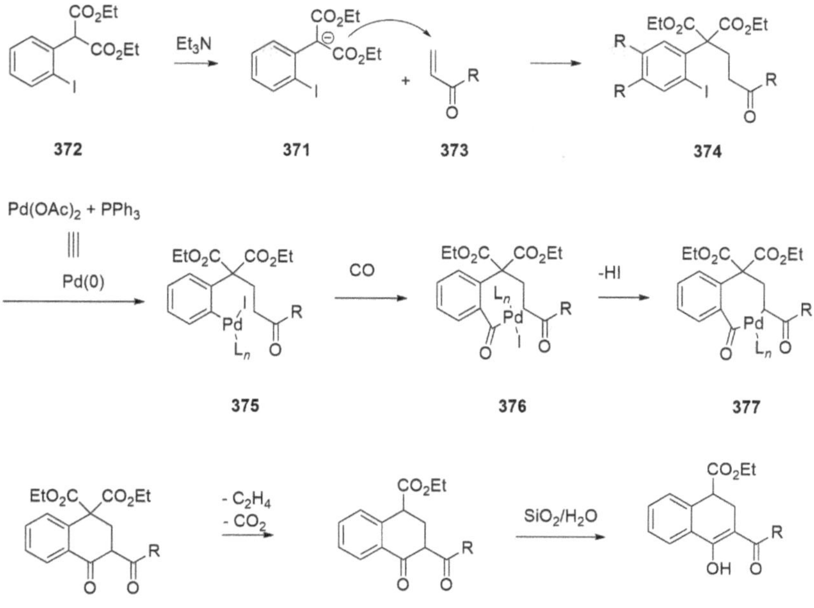

372 371 373 374 375 376 377 378 379 380

SCHEME 11.67 Mechanism of the Alper annulation.

Pd(0) from intermediate (**377**) furnishes tetralones (**378**). Triethylamine promoted demethoxycarbonylation of (**378**) gives (**379**), which isomerizes to the enols (**380**) on silica gel.

Larock annulation: Larock annulation, discovered in 1991, is a formal [3+2] annulation. It is one of the most well-established heteroannulations. It involves annulation between 2-iodoanilines (**381**) with alkynes (**382**) in the presence of a Pd(0) catalyst and a base to give indoles (**383**) (Scheme 11.68).[48] Both internal alkynes and terminal alkynes are useful. These catalytic annulation processes are exceeding useful and versatile. The are widely employed for the preparation of indole derivatives in both industrial scale as well as academic settings. This is largely due to the broad range of regioselectivity of the reaction. The Larock annulation can be extended to 2-bromoaldehyde for the indenone synthesis as well.

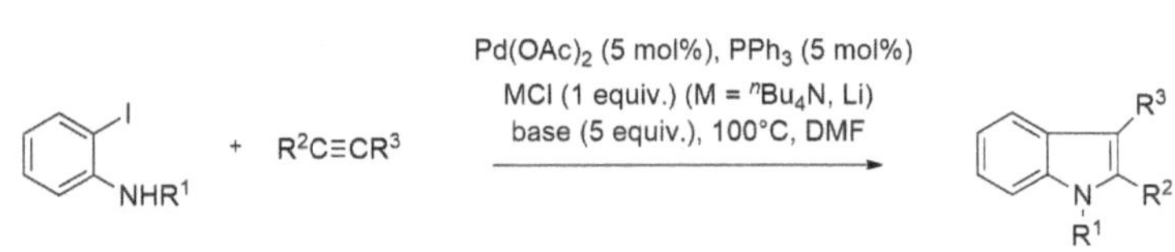

381 382 383

SCHEME 11.68 General scheme for Larock annulation.

Larock annulation presumably proceeds through six discrete steps (Scheme 11.69). These are (i) reduction of Pd(II) species to Pd(0), (ii) coordination of the chloride to form a chloride ligated palladium species, (iii) oxidative addition of Pd(0) to *o*-iodoanilines to generate arylpalladium intermediates (**384**→**385**), (iv) coordination of the alkyne to the palladium center of the intermediates (**385**→**386**), (v) regioselective *syn*-insertion into the arylpalladium to form a six membered heteropalladacycles (**386**→**387**), and (v) reductive elimination of Pd(0) (**387**→**388**). The regioselectivity of the annulation with an asymmetrical alkyne is governed by the carbopalladation step. The steric effect controls the conversion of intermediates (**386**) so as to place the smaller group next to the *peri*-hydrogen of the palladacycles (**387**).

The first report of an intramolecular Larock annulation to produce 3,4-fused indole derivatives was published by Jia group. Iodoanilines (**389**), on exposure to the standard conditions, afforded the indole derivatives (**390**). This reaction is generalized with the synthesis of (**391–396**) (Scheme 11.70).[49]

Concomitant to Jia's work, Boger group employed an intramolecular Larock annulation to assemble indole-embedded macrocyclic peptide scaffold of chloropeptin I. Thus, the reaction of (**397**) in the presence of $Pd(OAc)_2$ and DtBPF (1,1′-bis(di-*tert*-butylphosphino)ferrocene) gave (**398**) in reasonable yields (Scheme 11.71).

The applications of the Larock annulation are varied. It has been applied to the total synthesis of (+)-terreusinone, complestatin (chloropeptin II), fargesine, fuchsiaefoline, etc. The intramolecular strategy is utilized in the first synthesis of fargesine (**399**). As shown in Scheme 11.72, the synthesis of fargesine (**399**) was achieved in eight steps, starting from iodonitrobenzene (**400**). Reductive amination of (**400**) with amine (**401**) gave (**402**), which was protected with $(Boc)_2O$ to furnish (**403**). Reduction of the nitro group in (**403**) afforded iododaniline (**404**). Intramolecular Larock annulation of (**404**) provided 3,4-fused indole (**405**). Exposure of (**405**) to trifluoroacetic acid gave (**406**). Reductive methylation of the secondary amine in (**406**) with formaldehyde and $NaBH_4$

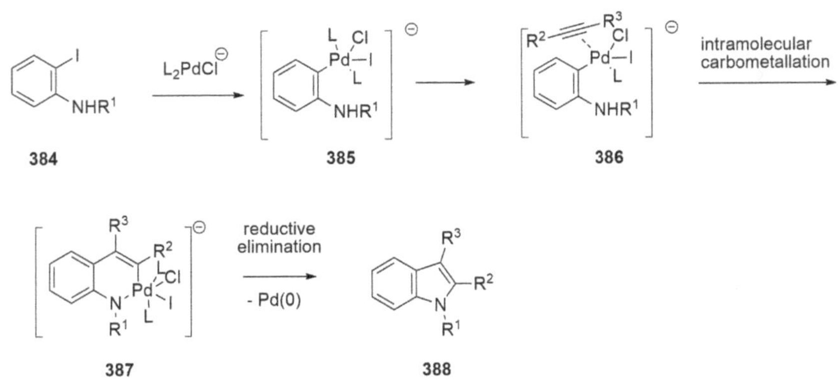

SCHEME 11.69 Mechanism of the Larock annulation.

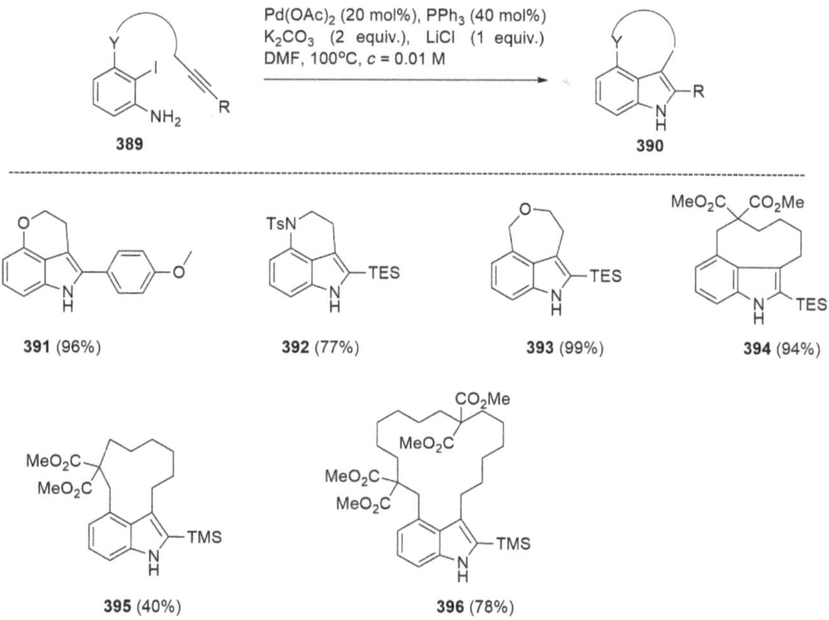

SCHEME 11.70 Intramolecular Larock annulation [2013 ACIE 4902 Jia Y].

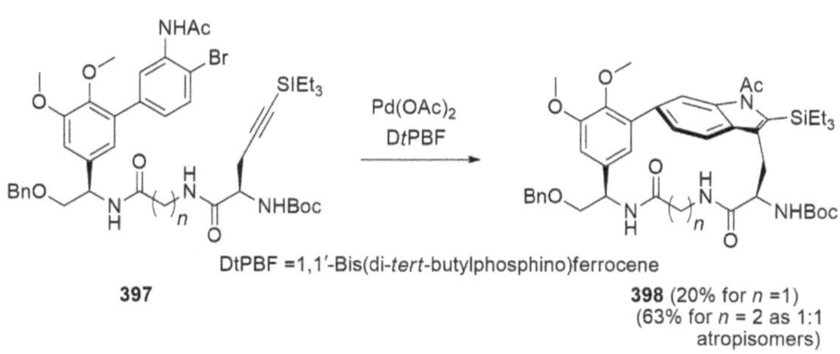

SCHEME 11.71 Boger macrocyclization [2013 JACS 1600 Boger DL].

furnished (**407**), MCPBA oxidation of which gave (**408**). Removal of the Boc group from (**408**) provided fargesine (**399**).[50]

Chen indole synthesis: Chen indole synthesis consists in a palladium-catalyzed annulation between *o*-iodoanilines and ketones with an α-hydrogen. Treatment of 2-iodoaniline (**409**) with ketones (**410**) in the presence of $Pd(OAc)_2$ in DMF at 105°C straightway gives 2,3-disubstituted indoles (**411**).[51] The use of both DABCO as the base and DMF as the solvent is crucial to the success. The generality of the reaction is tested with examples such as (**412–415**). Since this reaction is carried out with DABCO as the base, acid-sensitive substrates survive the conditions. Formation of the indole products (**411**) can be explained by an intramolecular Heck reaction of the intermediate enamines (**416**) (Scheme 11.73).

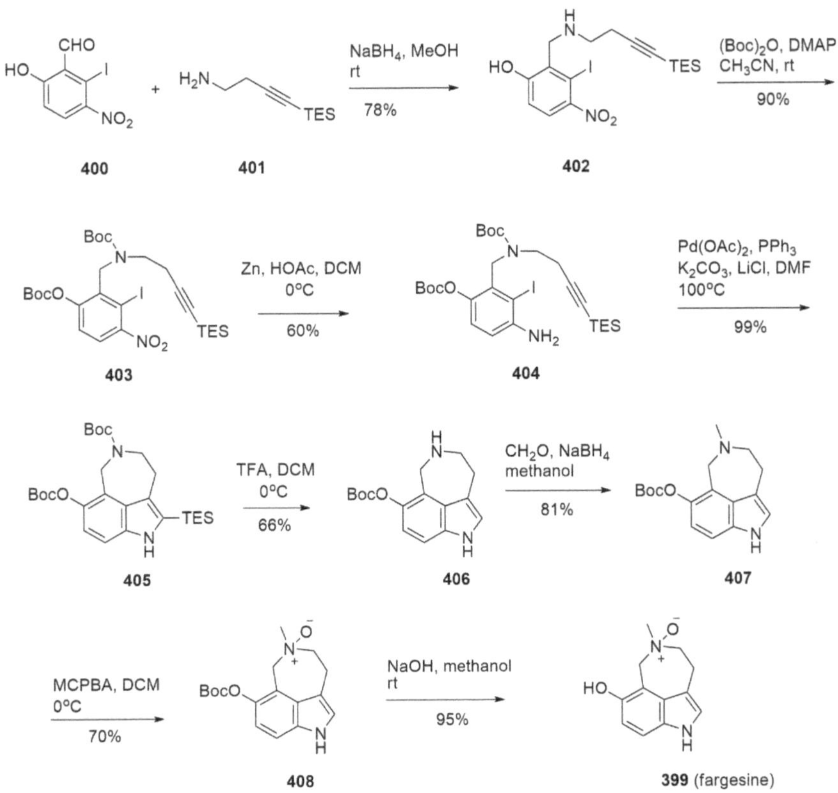

SCHEME 11.72 Total synthesis of fargesine by intramolecular Larock annulation [2014 THN 5136 Jia Y].

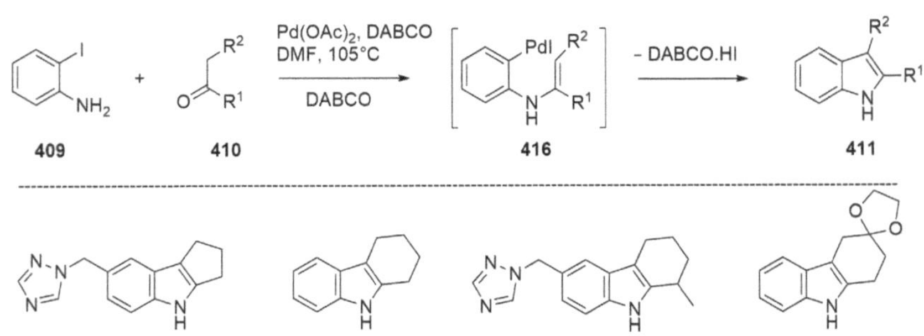

SCHEME 11.73 Chen indole synthesis [1997 JOC 2676 Chen C].

In conclusion, a host of miscellaneous anionic annulations have been presented in this chapter. The annulations are Deslongchamps annulation, Lupton annulation, Chi annulation, Friedländer annulation, Nair annulation, Padwa annulation, Leusen annulation, Barton-Zard annulation, Larock indole annulation,

Chen indole annulation, etc. All of these annulations require stoichiometric amount of bases, and form a six-membered ring or a five-membered ring in a single operation.

REFERENCES

1. Lavallée, J.; Deslongchamps, P. Tetrahedron Lett. 1988, 5117.
2. Lavallée, J.; Deslongchamps, P. Tetrahedron Lett. 1988, 6033.
3. Han, X.; Zhao, Y.; Meng, J.; Ren, C.; Liu, Q. J. Org. Chem. 2012, 77, 5173.
4. Covarrubias-Zuniga, A.; González-Lucas, A. Tetrahedron Lett. 1998, 38, 2881.
5. Fleming, F. F.; Gudipati, V.; Steward, O. W. J. Org. Chem. 2003, 68, 3943.
6. Fleming, F. F.; Zhang, Z.; Wang, Q.; Steward, O. W. Org. Lett. 2002, 4, 2493.
7. Flanigan, D. M.; Romanov-Michailidis, F.; White, N. A.; Rovis, T. Chem. Rev. 2015, 115, 9307.
8. Sun, F.-G.; Huang, X.-L.; Ye, S. J. Org. Chem. 2010, 75, 273.
9. Xu, J.; Jin, Z.; Chi, Y. R. Org. Lett. 2013, 15, 5028.
10. Xu, J.; Yuan, S.; Miao, M. Org. Lett. 2016, 18, 3822.
11. Janssen-Müller, D.; Singha, S.; Olyschläger, T.; Daniliuc, C. G.; Glorius, F. Org. Lett. 2016, 18, 4444.
12. Ryan, S. J.; Candish, L.; Lupton, D. W. J. Am. Chem. Soc. 2011, 133, 4694.
13. Pahari, P.; De, S. R.; Mal, D. Tetrahedron 2007, 63, 11781.
14. Pietrusiewicz, K. M.; Monkiewwicz, J. J. Org. Chem. 1983, 48, 788.
15. Hayashi, Y.; Toyoshima, M.; Gotoh, H.; Ishikawa, H. Org. Lett. 2009, 11, 45.
16. Ozaki, Y.; Mochida, K.; Kim, S.-W. J. Chem. Soc., Chem. Commun. 1988, 374.
17. Zhu, T.; Zheng, P.; Mou, C.; Yang, S.; Song, B.; Chi, Y. R. Nat. Commun. 2014, 5, 5027.
18. Padwa, A.; Yeske, P. E. J. Am. Chem. Soc. 1988, 110, 1617.
19. Flasz, J. T.; Hale, K. J. Org. Lett. 2012, 14, 3024.
20. (a) Zhang, C.; Lu, X. J. Org. Chem. 1995, 60, 2906. (b) Han, X.; Wang, Y.; Zhong, F.; Lu, Y. J. Am. Chem. Soc. 2011, 133, 1726.
21. Gembus, V.; Postikova, S.; Levacher, V.; Brière, J.-F. J. Org. Chem. 2011, 76, 4194.
22. Hassner, A.; Ghera, E.; Yechezkel, T.; Kleiman, V.; Balasubramanian, T.; Ostercamp, D. Pure Appl. Chem. 2000, 72, 1671.
23. Nair, V.; Vellalath, S.; Poonoth, M.; Suresh, E. J. Am. Chem. Soc. 2006, 128, 8736.
24. Nair, V.; Babu, B. P.; Vellalath, S.; Varghese, B.; Raveendran, A. E.; Suresh, E. Org. Lett. 2009, 11, 2507.
25. Chiang, P.; Kaeobamrung, J.; Bode, J. W. J. Am. Chem. Soc. 2007, 129, 3520.
26. Cardinal-David, B.; Raup, D. E. A.; Scheidt, K. A. J. Am. Chem. Soc. 2010, 132, 5345.
27. Fu, Z.; Xu, J.; Zhu, T.; Leong, W. W. Y.; Chi, Y. R. Nat. Chem. 2013, 5, 835.
28. McCoy, L. L. J. Am. Chem. Soc. 1958, 80, 6568.
29. Annunziato, G.; Pieroni, M.; Benoni, R.; Campanini, B.; Pertinhez, T. A.; Pecchini, C.; Bruno, A.; Magalhães, J.; Bettati, S.; Franko, N.; Mozzarelli, A.; Costantino, G. J. Enzyme Inhib. Med. Chem. 2016, 31(S4), 78.
30. McCoy, L. L. J. Am. Chem. Soc. 1962, 84, 2246.
31. Sun, X.; Tian, X.; Li, Z.; Peng, X.; Wong, H. N. C. Chem. Eur. J. 2011, 17, 5874.
32. Toyota, A.; Ono, Y.; Kaneko, C.; Hayakawa, I. Tetrahedron Lett. 1996, 37, 8507.
33. Hudlicky, T.; Radesca, L.; Luna, H.; Anderson, F. E., III J. Org. Chem. 1986, 51, 4746.
34. Hudlicky, T.; Fleming, A.; Radesca, L. J. Am. Chem. Soc. 1989, 111, 6691.
35. Corey, E. J.; Chaykovsky, M. J. Am. Chem. Soc. 1965, 87, 1353.

36. Samet, A. V.; Shestopalov, A. M.; Lutov, D. N.; Rodinovskaya, L. A.; Shestopalov, A. A.; Semenov, V. V. Tetrahedron: Asymmetry 2007, 18, 1986.
37. Denney, D. B.; Boskin, M. J. J. Am. Chem. Soc. 1959, 81, 6330.
38. Wadsworth, W. S.; Emmons, W. D. J. Am. Chem. Soc. 1961, 83, 1733.
39. Kumar, P.; Dubey, A.; Harbindu, A. Org. Biomol. Chem. 2012, 10, 6987.
40. Tamura, Y.; Tsugoshi, T.; Mohri, S.; Kita, Y. J. Org. Chem. 1985, 50, 1542.
41. Marco-Contelles, J.; Pérez-Mayoral, E.; Abdelouahid Samadi, A.; María do CarmoCarreiras, M. D. C.; Elena Soriano, E. Chem. Rev. 2009, 109, 2652.
42. Van Leusen, A. M.; Siderius, H.; Hoogenboom, B. E.; van Leusen, D. Tetrahedron Lett. 1972, 5337.
43. Davies, S. G.; Roberts, P. M.; Stephenson, P. T.; Storr, H. R.; Thomson, J. E. Tetrahedron 2009, 65, 8283.
44. Magnus, P.; Or, Y.-S. J. Chem. Soc., Chem. Commun. 1983, 26.
45. Barton, D. H. R.; Zard, S. Z. J. Chem. Soc., Chem. Commun. 1985, 1098.
46. Bhattacharya, R.; Atta, A. K.; Dey, D.; Pathak, T. J. Org. Chem. 2009, 74, 669.
47. Zheng, Z.; Alper, H. Org. Lett. 2009, 11, 3278.
48. Larock, R. C.; Yum, E. K. J. Am. Chem. Soc. 1991, 113, 6689.
49. Shan, D.; Gao, Y.; Jia, Y. Angew. Chem. Int. Ed. 2013, 52, 4902.
50. Gao, Y.; Shan, D.; Jia, Y. Tetrahedron 2014, 70, 5136.
51. Chen, C.; Liebermn, D. R.; Larsen, R. D.; Verhoeven, T. R.; Reider, P. J. J. Org. Chem. 1997, 62, 2676.

FURTHER READING

1. Ravindar, K.; Caron, P.; Deslongchamps, P. Org. Lett. 2013, 15, 6270.
2. Hampel, T.; Brückner, R. Org. Lett. 2009, 11, 4842.

Chapter 12

Concluding Remarks

The term annulation is seemingly synonymous with Robinson annulation. A discussion on "annulation" often starts with Robinson annulation, discovered in 1935, and stops there too. But today, it has grown to be a vast subject. Since this is the first book in the field of the annulations, it should have ideally covered all types of annulations: anionic, cationic, radical, thermal, photocatalytic, and transition metal catalytic. A future edition should probably be titled *Annulations in Organic Synthesis*. However, for the reasons cited in the introduction, it has been confined to the anionic annulations and as such titled *Anionic Annulations in Organic Synthesis*. The term "annulation" is different from the cycloadditions by number of steps and/or loss of a small molecule. Cycloadditions are reserved for interactions of pi-systems. This book is largely devoted to the Hauser-type [4+2] annulation. It presents a lopsided view of uncommon annulations and annulations covered in other books. In this chapter, some annulations of the author's choice are briefly presented.

Aryne annulation: Since the structural assignment of benzynes by J. D. Roberts in 1953, the aryne chemistry has continued to entice the synthetic organic chemists for its possible applications.[1] In this paragraph, the cycloadditions of arynes formed under anionic conditions, such as $NaNH_2$, LDA, BuLi, CsF, etc., are highlighted. Typically, annulations with arynes proceed through either [4+2] cycloadditions or [2+2] cycloadditions. Alternatively, arynes can undergo stepwise nucleophilic additions. Condensation of (**1**) and (**2**) under aryne-forming conditions (CsF) afforded benzocycloheptenone (**3**) featuring amurensinine (Scheme 12.1).[2] The reaction is believed to proceed through the cyclobutene intermediate (**4**).

Papaverine is a clinically used nonnarcotic antispasmotic agent. Its synthesis was accomplished by an aryne annulation. The reaction between silyl aryl triflate (**5**) and *N*-acyl enamine (**6**) in the presence of tetrabutylammoniumdifluorotriphenylsilicate furnished isoquinoline (**7**) in good yield (Scheme 12.2).[3]

Buszek indolyne annulation has paved the way for the synthesis of annulated indole natural products. In Scheme 12.3, their synthesis of (±)-trikentrin A (**8**) is presented. The key reaction is the condensation between dibromoindole (**9**) and cyclopentadiene (**10**) in the presence of ^{n}BuLi, giving tetracycle (**11**) in good yield. As was conceivable, this reaction proceeded through an indolyne intermediate. Three more steps yielded (±)-trikentrin A (**8**).[4]

Anionic Annulations in Organic Synthesis. https://doi.org/10.1016/B978-0-12-813800-7.00012-9

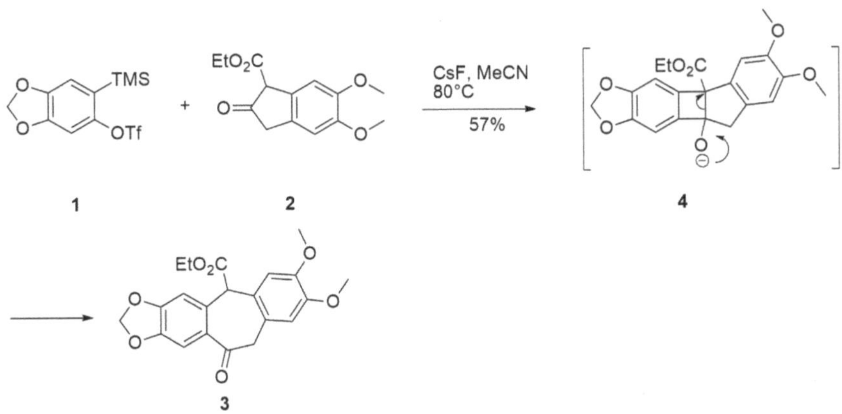

SCHEME 12.1 Aryne annulation for the synthesis of a benzocycloheptenone.

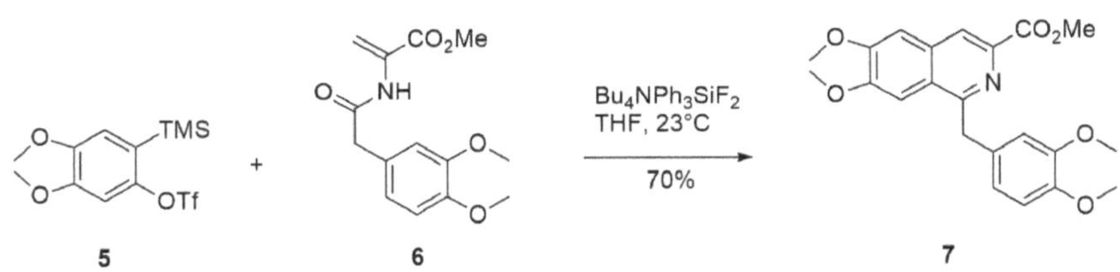

SCHEME 12.2 Synthesis of papaverine precursor [2008 JACS 1558 Stoltz BM].

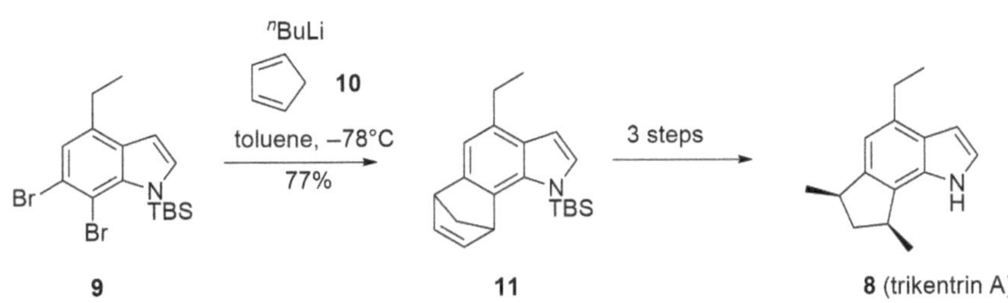

SCHEME 12.3 Synthesis of (±)-trikentrin A (**8**) via aryne annulation.

Aryne annulations are extendable to [2+2] cycloaddition mode to produce a wide range of benzocyclobutenes. One of the simple examples of such annulations is the reaction between dibromobenzene (**12**) and dimethoxyethene (**13**) in the presence of sodamide giving benzocyclobutene (**14**) via benzyne (**15**) (Scheme 12.4).

Among the thermal annulations, the Alder-Rickert reaction can be classified as an annulation since it involves the loss of a small molecule. The reaction of diene (**16**) with acetylene carboxylate (**17**) in refluxing xylenes afforded phenol (**18**) through the loss of ethylene from the intermediate (**19**) (Scheme 12.5).[5]

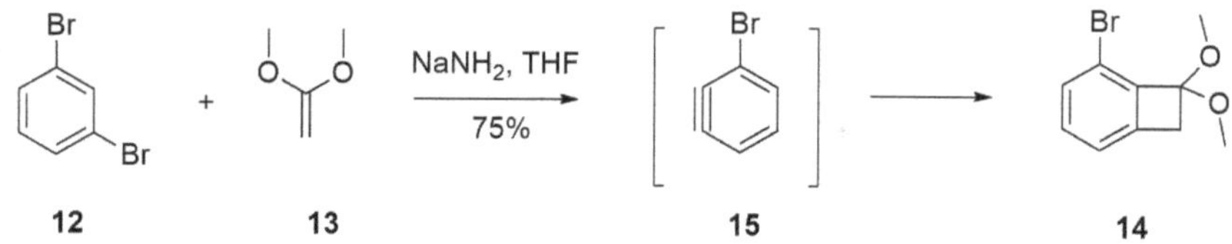

SCHEME 12.4 Aryne [2+2] annulation.

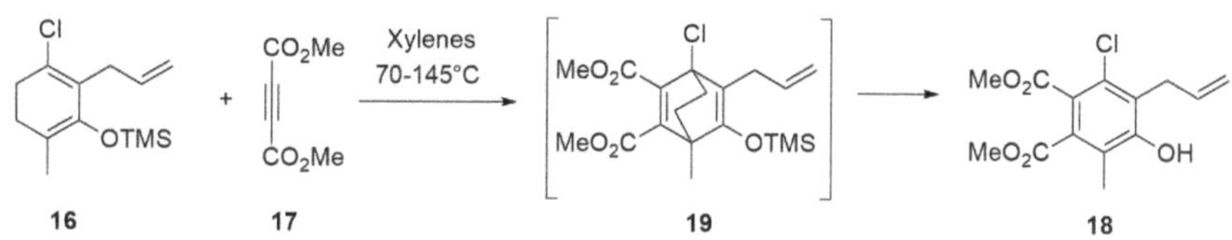

SCHEME 12.5 Example of an Alder-Rickert reaction.

Danheiser benzannulation: This is an intermolecular thermal reaction between a cyclobutenone and a heteroatom substituted alkyne, resulting in regioselective formation of a substituted phenol. The reaction of cyclobutenones (**20**) with alkynes (**21**) gives resorcinol derivatives (**22**) in a one-pot operation.[6] In the alkynes (**21**), OMe can be replaced by a OTIPS, SMe, NHMe, NMe_2, or N(Boc) Me. This annulation is believed to proceed through a cascade of four thermally allowed pericyclic reactions, as shown in Scheme 12.6. Mechanistically, cyclobutenones (**20**) undergo ring-opening to give vinylketenes (**23**), which undergo [2+2] cycloaddition with alkynes (**21**) to form vinylcyclobutenones (**24**).

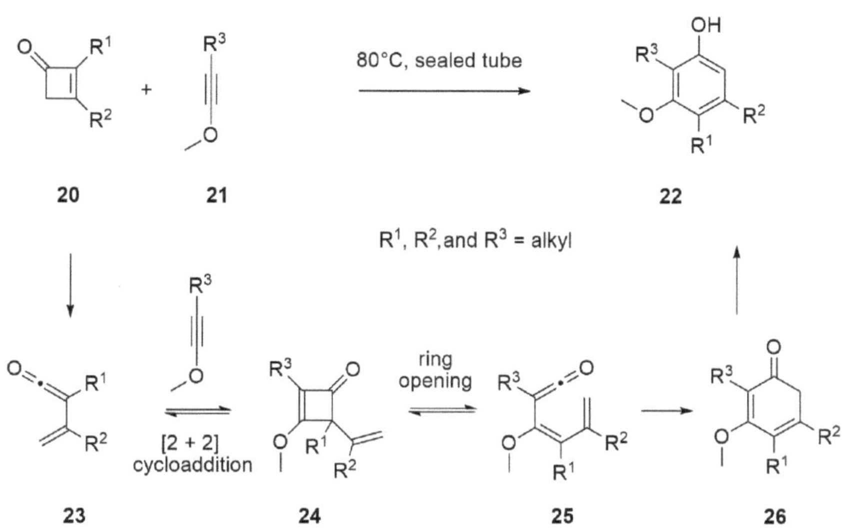

SCHEME 12.6 An example of Danheiser benzannulation.

Ring-opening of (**24**) generates extended vinylketenes (**25**). Six π-electron cyclization of (**25**) gives (**26**), which in turn, tautomerizes to give resorcinols (**21**).

Danheiser annulation has been found useful for direct and regioselective synthesis of cannabinols. Heating a 1:1 mixture of silyloxyacetylene (**27**) and 3-pentylyclobutenone (**28**) in toluene at 80°C afforded the resorcinol (**29**) (Scheme 12.7). Without purification, the disilyl derivative (**29**) was treated with refluxing acidic ethanol to give Δ-6-tetrahydrocannabinol (**30**).[7]

Dötz annulation: This is a reaction between a carbene complex (e.g., **31**) with an alkyne (e.g., **32**) to produce the arene complex (**33**) with the chromium tricarbonyl group attached to the newly formed benzene ring. Aerial oxidation of (e.g., **33**) results in naphthoquinol derivative (**34**).[8] It is formally considered as a reaction among carbene (**35**), alkyne (**32**), and carbon monoxide, as shown in Scheme 12.8. The detailed mechanism is beyond the scope of the book.

One of the finest applications of the Dötz annulation is the synthesis of juglomycin (**36**) by Fernades et al. As shown in Scheme 12.9, the annulation between (**37**) and (**38**) provided naphthol (**39**) in 78% yield, which was further elaborated to juglomycin (**36**) in six steps (Scheme 12.9).[9]

APEX reaction: The single-step annulative π-extension (APEX) reaction has recently become a valuable tool for efficient and rapid access to π-extended fused arenes from small arenes in a step/atom economic manner. It is defined

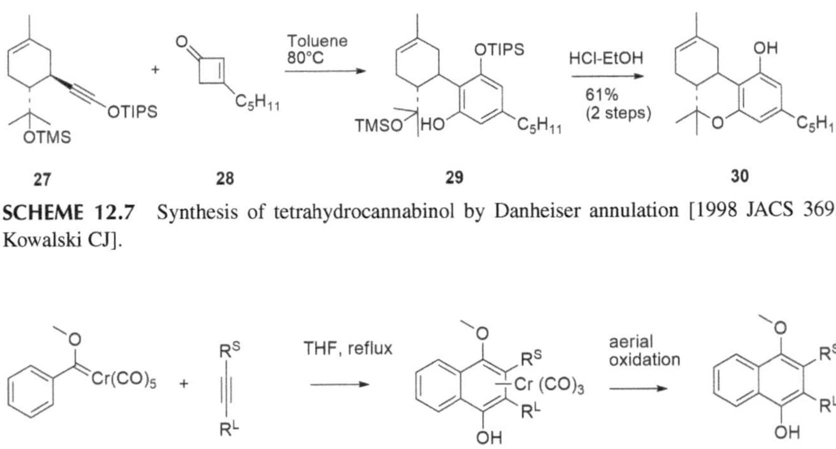

SCHEME 12.7 Synthesis of tetrahydrocannabinol by Danheiser annulation [1998 JACS 3693 Kowalski CJ].

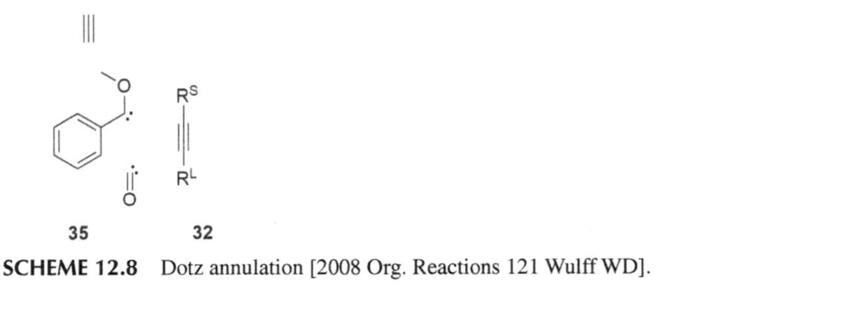

SCHEME 12.8 Dotz annulation [2008 Org. Reactions 121 Wulff WD].

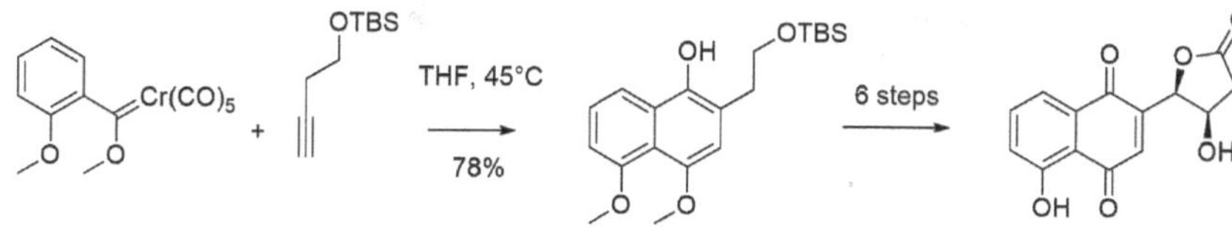

37 38 39 36 (-) Juglomycin

SCHEME 12.9 Dötzannulation route to juglomycin (**36**) [2008 THS 3899 Fernades RA].

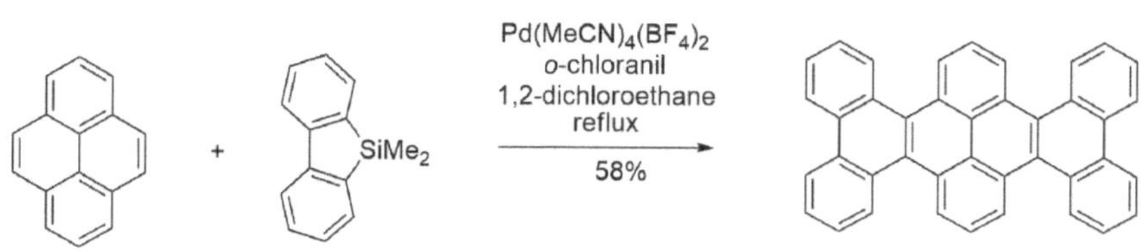

40 41 42

SCHEME 12.10 Synthesis of hexabenzo[*a,c,fg,j,i,op*]tetracene [2016 SYT 2081 Itami K].

by Itami et al. as a single-step π-extension reaction of a template arene to construct one or more new fused aromatic rings.[10] In the field of material science, annulative π-extension reactions have received attention as a specialized class of reaction. In a recent communication, a Pd-catalyzed annulation has been described for K-selective π-extension. When a mixture of pyrene (**40**) and 5,5-dimethyldibenzo[*b,d*]silole (**41**) was heated in refluxing 1,2-dichloroethane in the presence of $Pd(MeCN)_4(BF_4)_2$ and *o*-chloranil, hexabenzo[*a,c,fg,j,l,op*] tetracene (**42**) was formed in 58% yield (Scheme 12.10).[11]

Transannulation: Although the extension of a conventional annulation is perceptible in terms of conceivable mechanisms, its success depends greatly on suppressing the competing reactions such as Michael addition and 1,2-addition leading to dimerization. It is foreseeable that transition metal-mediated annulations would gain popularity in the near future because of the significance of the Pd-catalyzed cross-coupling reactions and the CH activation-directed annulations. In recent years, transannulations (annulations at the expense of a preexisting ring) have risen to prominence due to the pioneering work of Fokin et al. and Gevorgyan et al.[12] For example, triazole (**43**), obtained from tosyl azide and phenylacetylene by a Cu-catalyzed click reaction, reacted with benzonitrile (**44**) at 140°C in the presence of $Rh_2(Oct)_4$ to give imidazole (**45**) (Scheme 12.11).[13] Similarly, a palladium-catalyzed transannulation reaction of *N*-aroylbenzotriazoles with alkynes gave indoles.[14]

A probable mechanism of the Fokin-Gevorgyan annulation is given in Scheme 12.12. Under the influence of Rh catalysts, ring-opening of triazole (**43**) takes place to form Rh-iminocarbenoid (**46**), which, upon reaction with benzonitrile, leads to zwitterion (**47**). 5-Endo cyclization of (**47**) gives (**48**). Subsequent loss of Rh metal from the intermediate (**48**) produces imidazole (**45**).

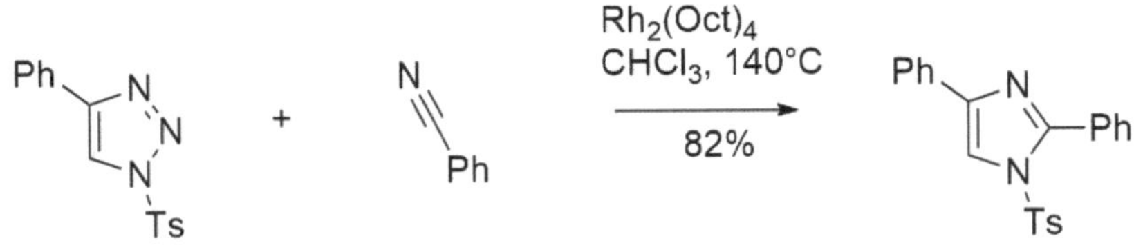

43 **44** **45**

SCHEME 12.11 Fokin-Gevorgyan annulation.

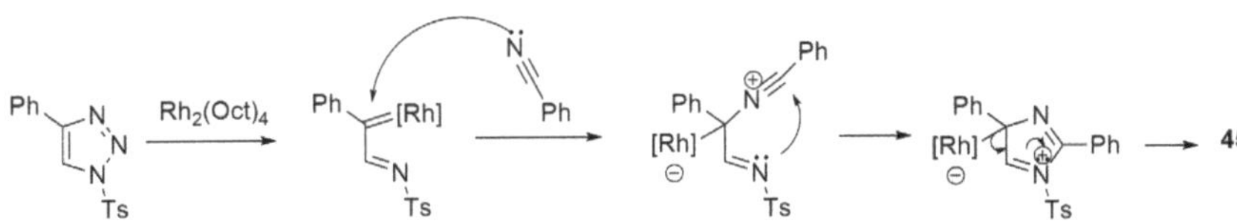

43 **46** **47** **48**

SCHEME 12.12 Mechanism of the Fokin-Gevorgyan annulation.

In 2014, the Murakami group discovered an intramolecular transannulation of a triazole, wherein a benzene nucleus acted as a nucleophilic dipolarophile, resulting in dearomatization of the benzene ring.[15] Treatment of triazole (**49**) with rhodium(II) pivaloate dimer in 1,2-dichloroethane at 80°C gave 3,4-fused dihydroindole (**50**) in excellent yield. The reversible ring-chain tautomerization of the triazole moiety of (**49**) generates α-diazo imine (**51**), which, in turn, forms rhodium carbenoid (**52**) with the expulsion of nitrogen (Scheme 12.13).

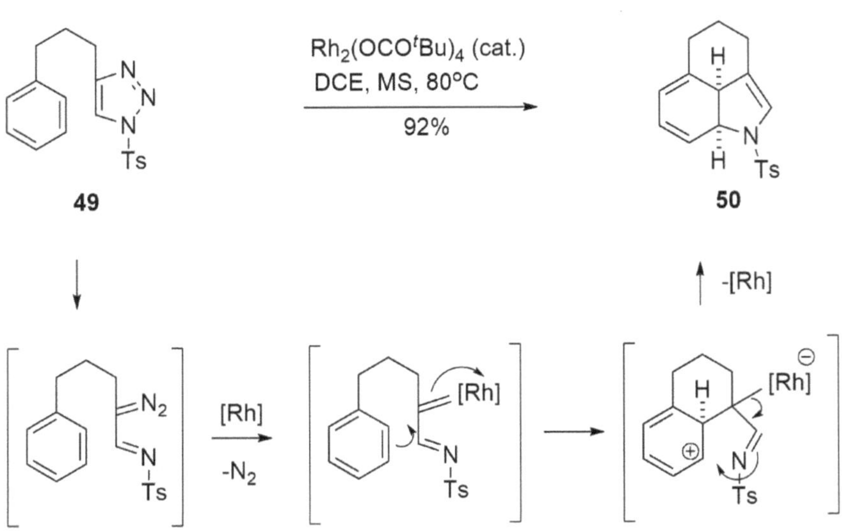

51 **52** **53**

SCHEME 12.13 Murakami-Miura transannulation [2014 JACS 2272 Murakami M].

The carbenoid carbon of (**52**), being electrophilic in nature, induces the intramolecular attack of the phenyl ring to furnish the zwitterionic intermediate (**53**). The anionic rhodium then causes cyclization through the imino nitrogen to produce (**50**).

Murakami-Miura transannulation manifested in the synthesis of Uhle's ketone (**54**), which been a key intermediate in the synthesis of *Ergot* alkaloids. The triethylsilyl group of the dihydroindole (**55**), obtainable by a Murakami-Miura transannulation, was deprotected by treatment with tetrabutylammonium fluoride (TBAF) to give (**56**). Oxidation of the resultant alcohol (**56**) with manganese dioxide afforded the corresponding ketone, which was converted into Uhle's ketone (**54**) by detosylation with potassium hydroxide (Scheme 12.14).

Lysergol (**57**), an *Ergot* alkaloid, is implicated in emotional, neuropsychiatric, and neurodegenerative disorders. The synthesis of the alkaloid (**57**) was started with (**58**), which, in turn, was obtained from (*R*)-Garner's aldehyde (Scheme 12.15).[16] It was elaborated to (**59**), which was then subjected to a copper-catalyzed click reaction with tosyl azide to furnish *N*-sulfonyl-1,2,3-triazole (**60**). Subsequently, a rhodium-catalyzed Murakami-Miura

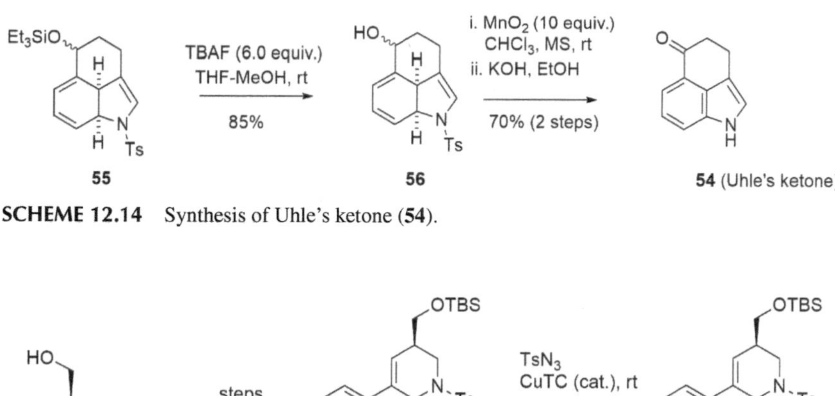

SCHEME 12.14 Synthesis of Uhle's ketone (**54**).

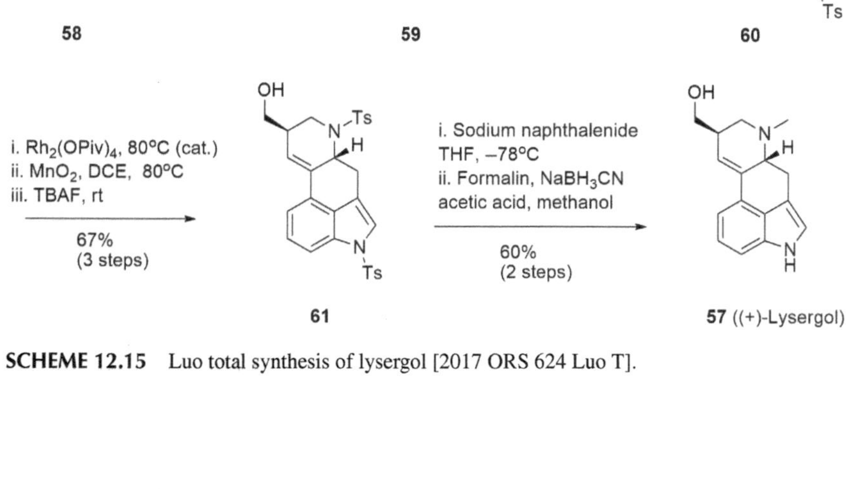

SCHEME 12.15 Luo total synthesis of lysergol [2017 ORS 624 Luo T].

transannulation reaction followed by oxidative aromatization and deprotection of the TBS group furnished compound (**61**) in 67% yield over two steps. Finally, deprotection of Ts groups in (**61**) with sodium naphthalenide and reductive amination of the resultant amine completed the synthesis of (+)-lysergol (**57**).

In summary, varied annulations have been covered in this book, although it was intended to cover only anionic annulations. The annulations included are Darzens condensation, Robinson annulation, Hauser annulation, Sammes annulation, Tamura annulation, Hauser sulfoxide annulation, Staunton-Weinreb annulation, Myers annulation, Deslongchamps annulation, Padwa annulation, Hassner annulation, Chi annulation, Nair annulation, Lupton annulation, McCoy annulation, Wadsworth-Emmons annulation, Friedländer annulation, Larock annulation, Leusen annulation, Barton-Zard annulation, Chen annulation, aryne annulation, Danheiser annulation, Dötz annulation, Fokin-Gevorgyan transannulation, and Murakami-Miura transannulation. Many of the above annulations are predictable and robust. Their widespread use in constructing polycyclic compounds is testimony to the importance of these reactions. Industrial utility of some of the anionic annulations seems to be brighter. For example, justicidin lignans are poised to be manufactured by the employment of Hauser annulation.[17] The Mal synthesis of vitamin K has the potential to be used as an industrial method.[18] Lastly, the author would like to see a future graduate student practicing a retrosynthesis based upon an annulation as the transform.

REFERENCES

1. Tadross, P. M.; Stoltz, B. M. *Chem. Rev.* **2012**, *112*, 3550.
2. Tambar, U. K.; Ebner, D. C.; Stoltz, B. M. *J. Am. Chem. Soc.* **2006**, *128*, 11752.
3. Gilmore, C. D.; Allan, K. M.; Stoltz, B. M. *J. Am. Chem. Soc.* **2008**, *130*, 1558.
4. Buszek, K. R.; Brown, N.; Luo, D. *Org. Lett.* **2009**, *11*, 201.
5. Patterson, J. W. *J. Org. Chem.* **1995**, *60*, 560.
6. Danheiser, R. L.; Gee, S. K.; Perez, J. J. *J. Am. Chem. Soc.* **1986**, *108*, 806.
7. Kowalski, C. J.; Lal, G. S. *J. Am. Chem. Soc.* **1988**, *110*, 3693.
8. Marcey, W.; Wulff, W. *Org. React.* **2008**, *70*, 121.
9. Fernades, R. A.; Chavan, V. P. *Tetrahedron Lett.* **2008**, *49*, 3899.
10. Ito, H.; Ozaki, K.; Itami, K. *Angew. Chem. Int. Ed.* **2017**, *56*, 11144.
11. Yanoa, Y.; Itoa, H.; Segawaa, Y.; Itami, K. *Synlett* **2016**, (27), 2081.
12. Chattopadhyay, B.; Gevorgyan, V. *Angew. Chem. Int. Ed.* **2012**, *51*, 862.
13. Horneff, T.; Chuprakov, S.; Chernyak, N.; Gevorgyan, V.; Fokin, V. V. *J. Am. Chem. Soc.* **2008**, *130*, 14972.
14. Nakamura, I.; Nemoto, T.; Shiraiwa, N.; Terada, M. *Org. Lett.* **2009**, *11*, 1055.
15. Miura, T.; Funakoshi, Y.; Murakami, M. *J. Am. Chem. Soc.* **2014**, *136*, 2272.
16. Yuan, H.; Guo, X.; Luo, T. *Org. Lett.* **2017**, *19*, 624.
17. Kim, T.; Jeong, K. H.; Kang, K. S.; Nakata, M.; Ham, J. *Eur. J. Org. Chem.* **2017**, 1704.
18. Mal, D.; Ghosh, K.; Jana, S. *Org. Lett.* **2015**, *17*, 5800.

Appendix

Problems and Exercises

1. Complete the following reaction sequences, giving appropriate structures of compounds represented by bracketed alphabets.

(a) [1987 JOC 1273 Kraus GA]

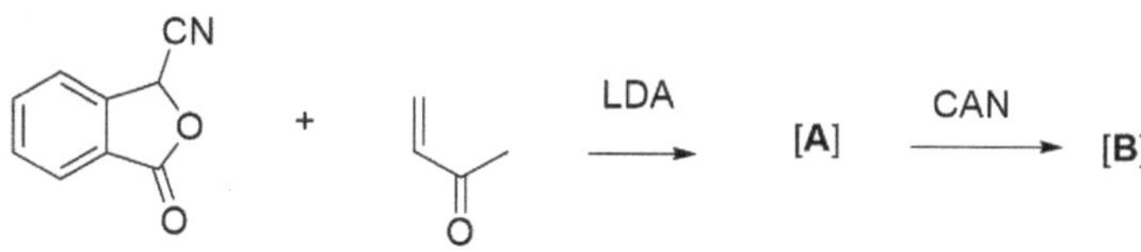

(b) [1994 THS 6139 Mal D]

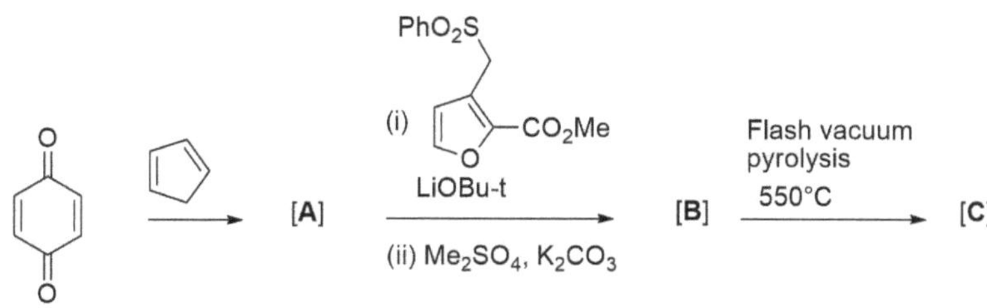

(c)

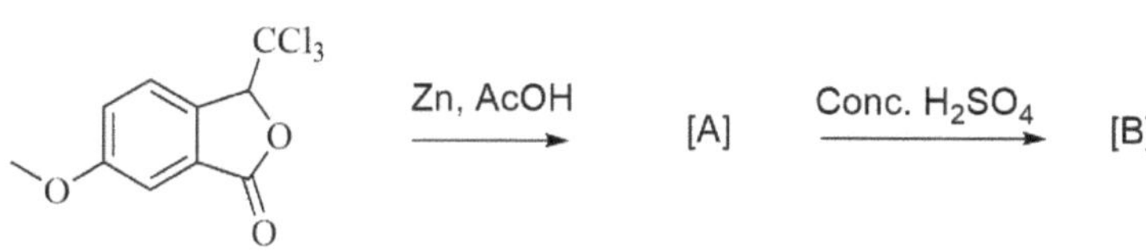

(d) [1989 CPB 304 Kim S-W]

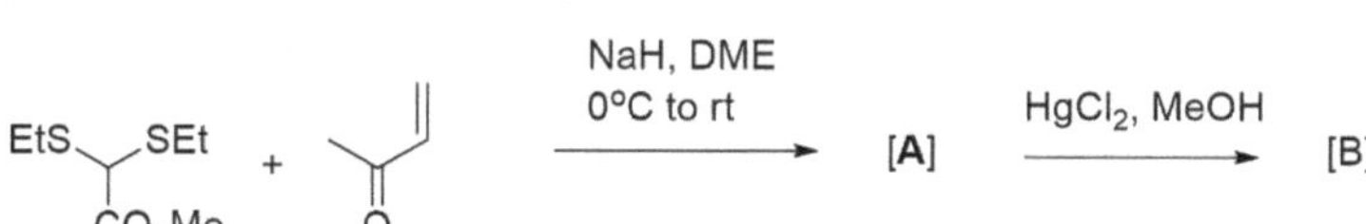

(e)

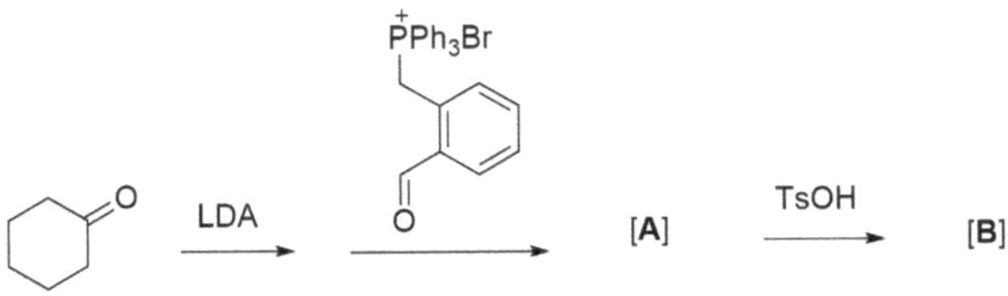

(f)

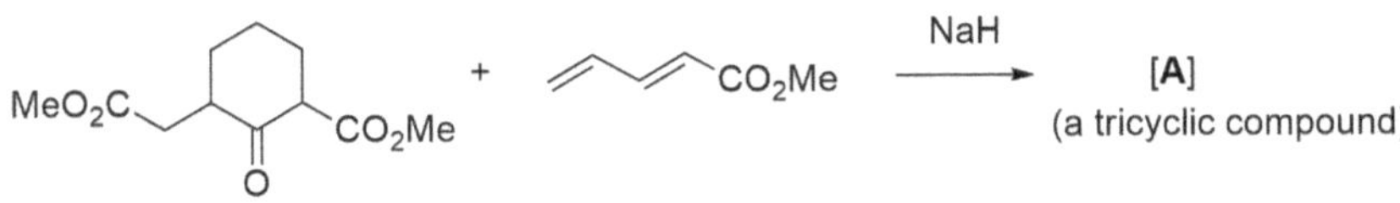

(g)

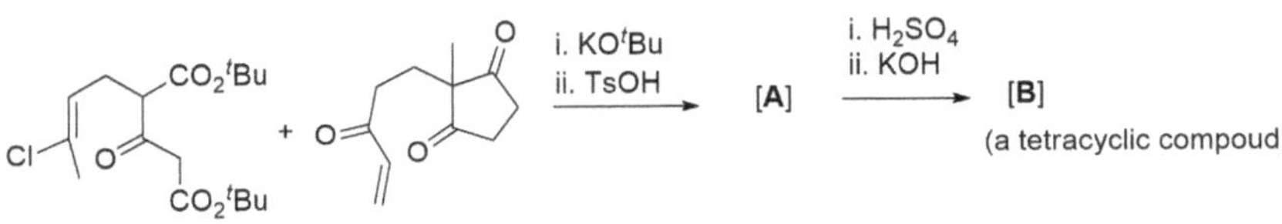

(h) [1990 SCS 1273 Rai KML]

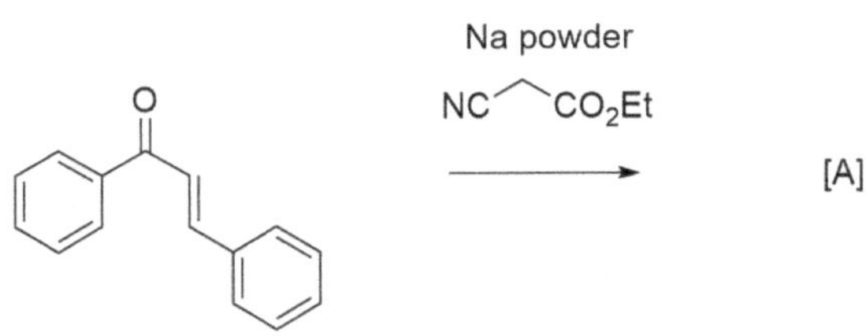

(i) [1998 THS 7657 Afonso A]

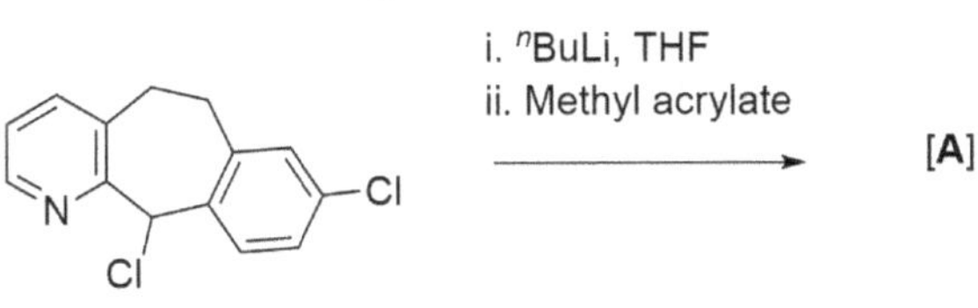

(j)

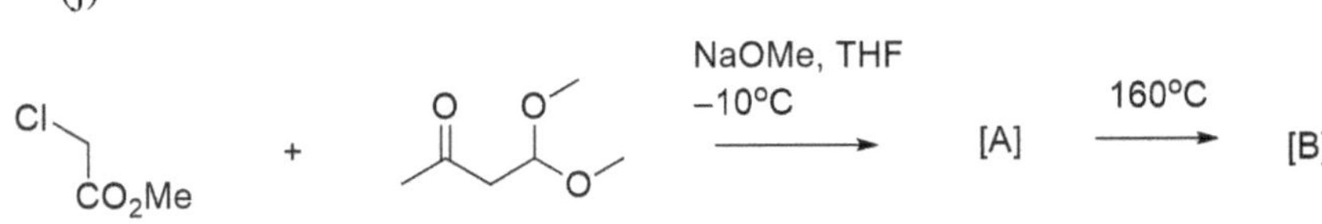

^{1}H NMR data of [B]: δ 7.45 (d, 1H, J = 2), 6.35 (d, 1H, J =2), 3.91 (s, 3H), 2.35 (s, 3H).

(k)

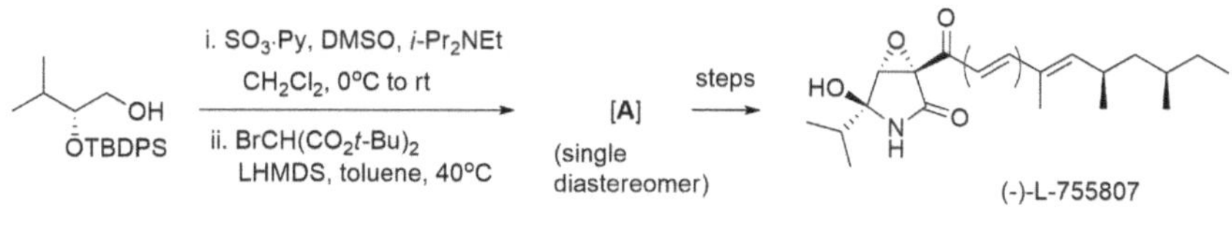

[2016 ORS 1920 Kogen H]

2. Predict the structures of the possible major products of the following annulations:

(a)

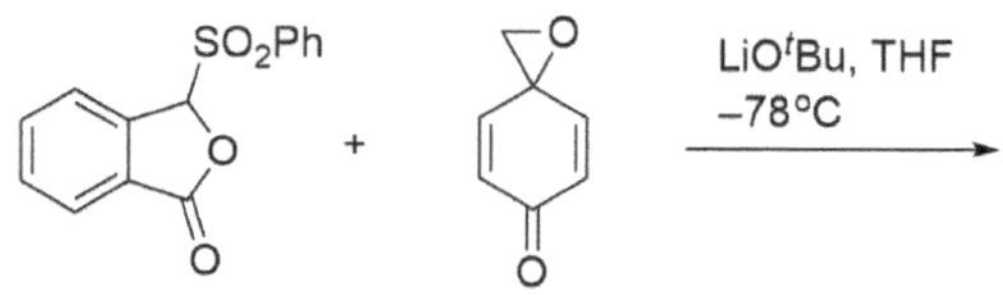

(b)

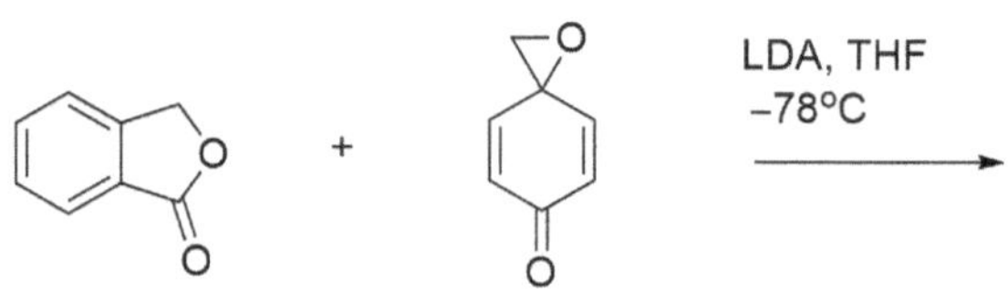

(c)

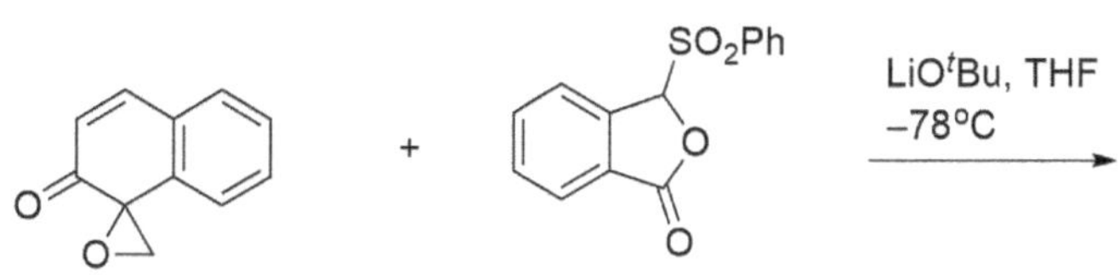

(d)

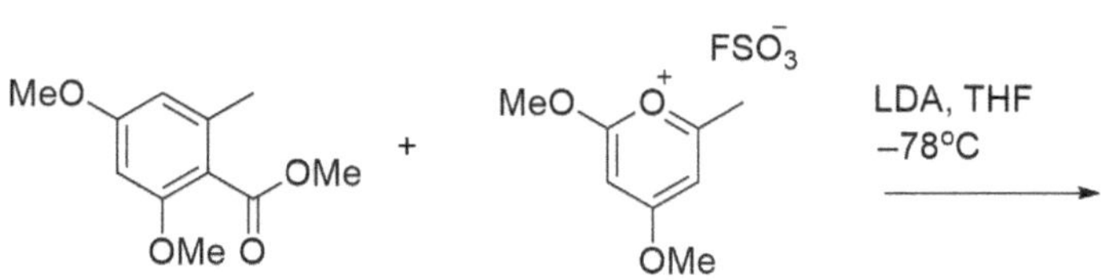

(e) [2015 EJMC 583 Cho W-J]

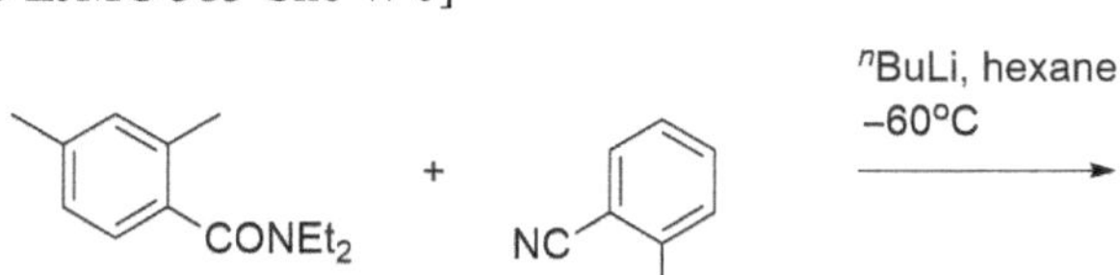

(f) [1993 Bull Soc. Chim. Fr. 447 Julia M]

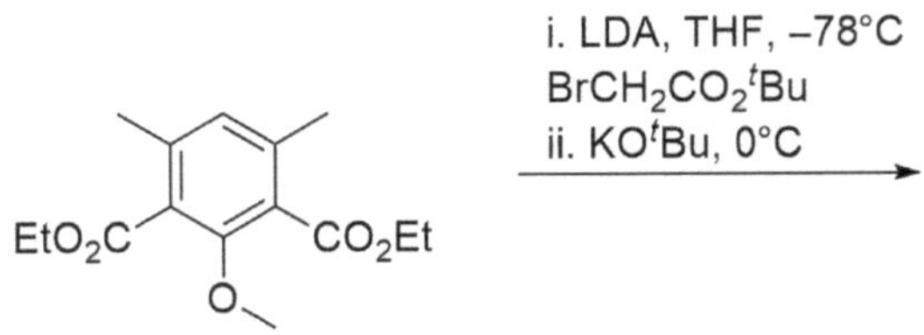

(g) [2011 EJOC 6544 Occhiat EG]

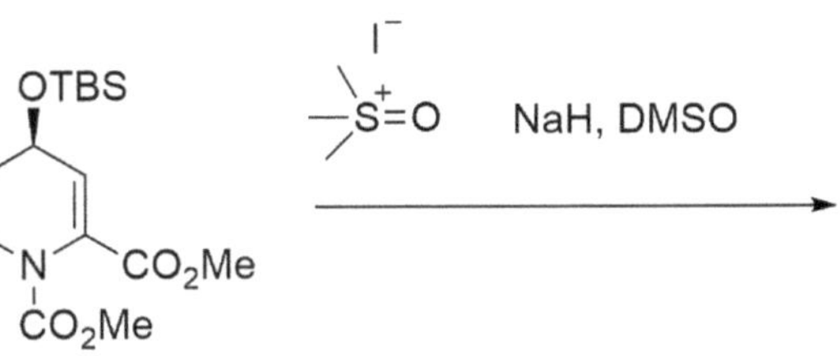

(h) [1974 JACS 1607 Fuchs PL]

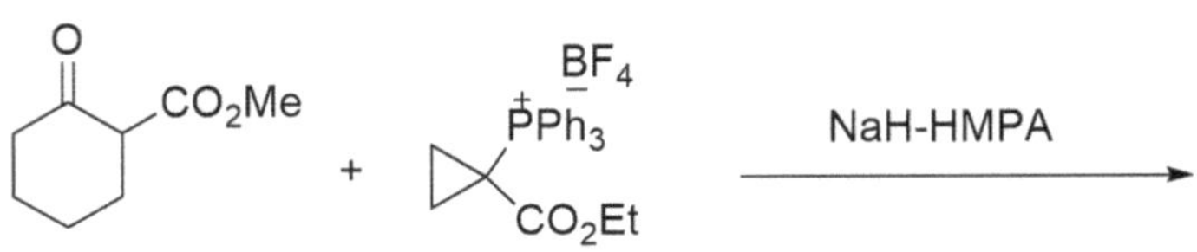

(i) [1979 JCSCCS 449 Koreeda M]

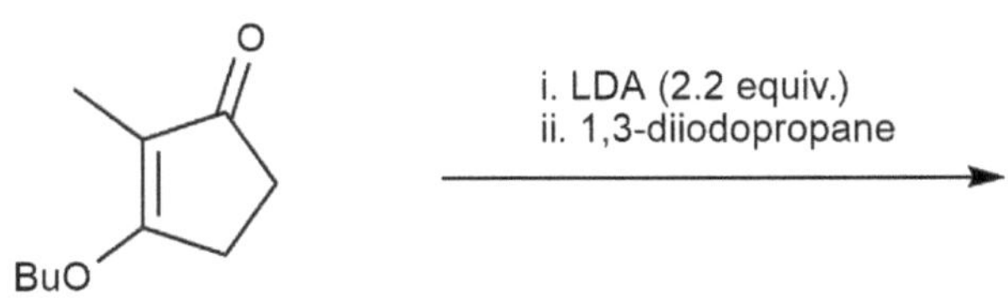

(j)

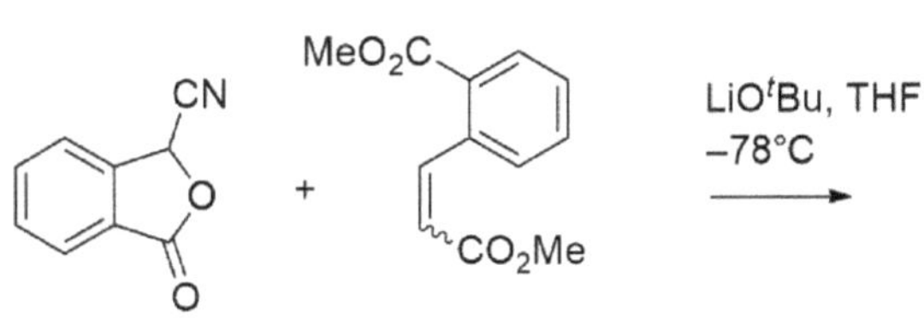

(k) [2017 ORS 3099 Kawabata T]

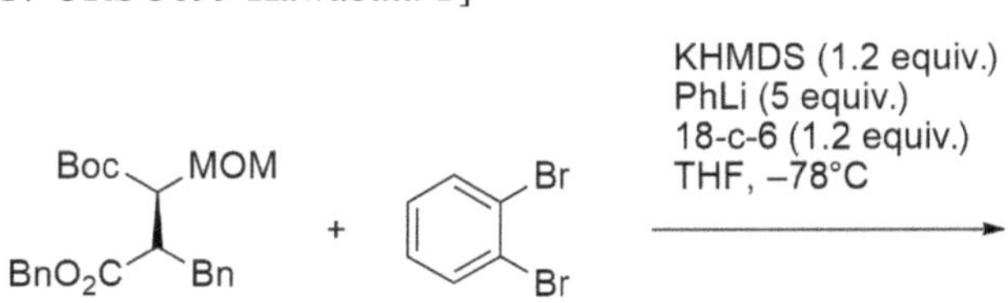

(l)

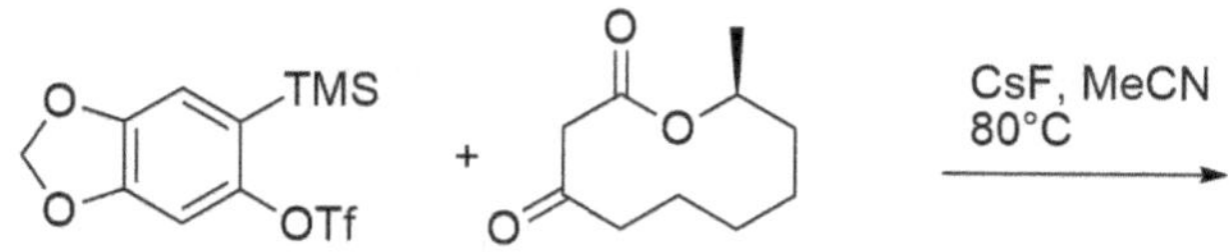

(m)

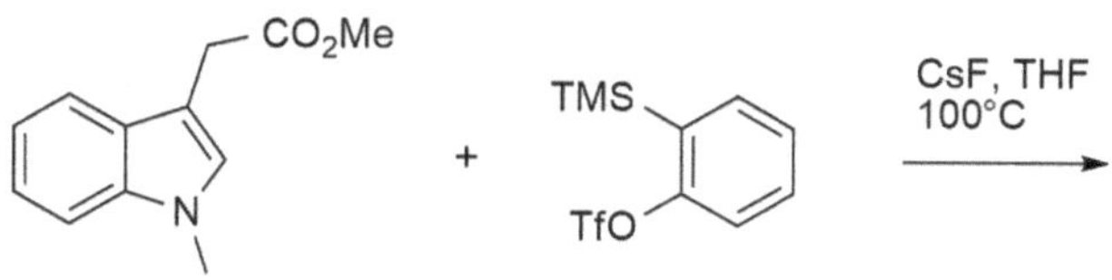

(n)

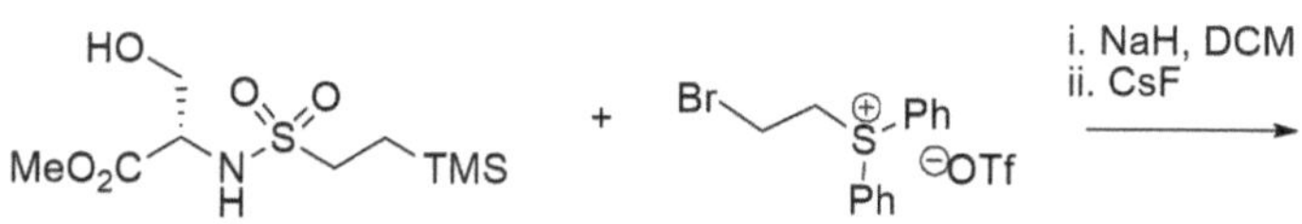

(o) [2004 JACS 14370 Bode J]

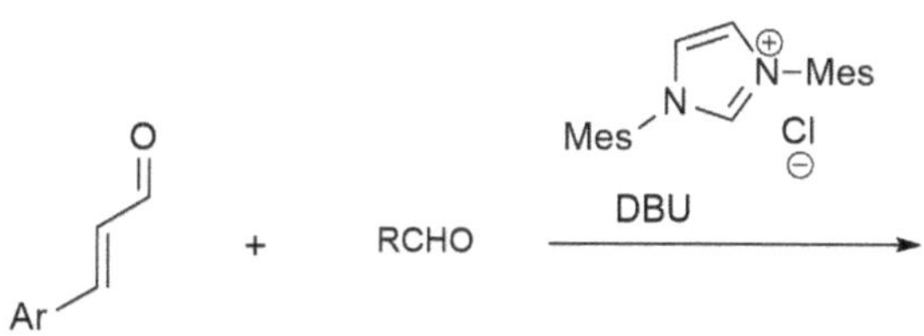

(p)

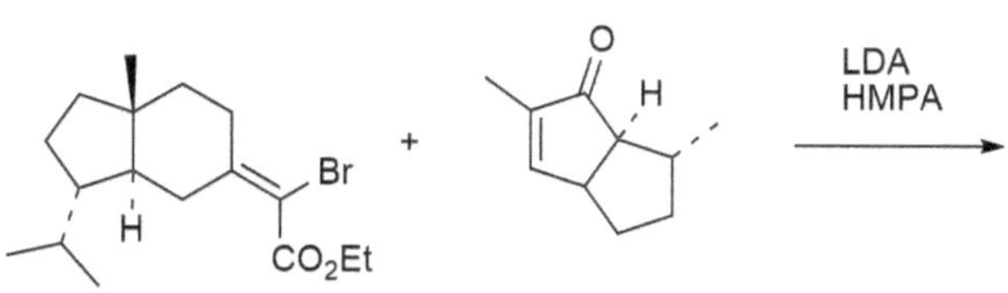

3. Suggest a route to synthesize the following by an anionic [4 + 2] cycloaddition:

(a)

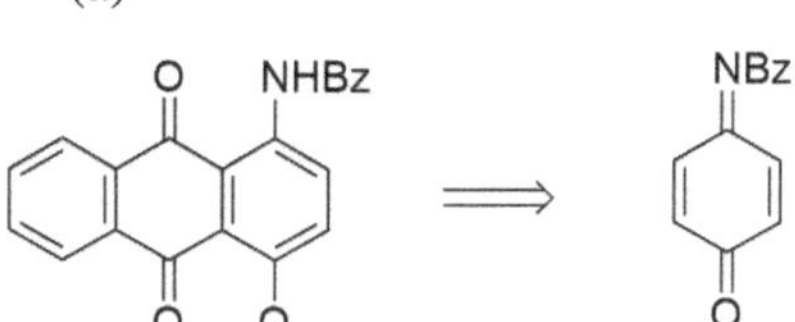

[Key 1990 JOC 2027 Swenton JS]

(b)

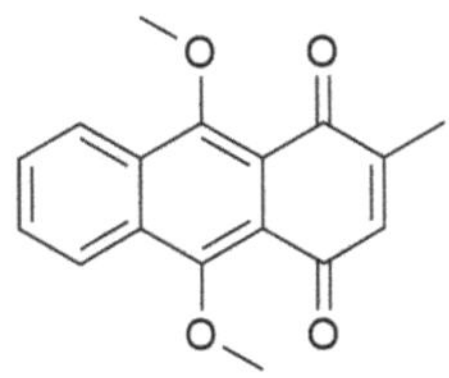

[Key: 2008 EJOC 3014 Mal D]

(c)

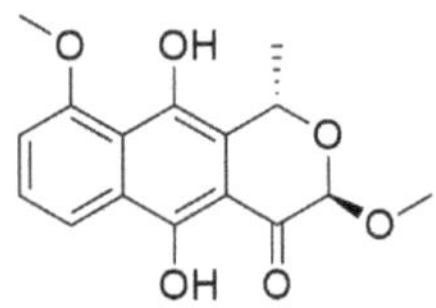

(d)

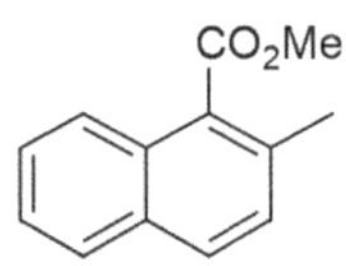

(e)

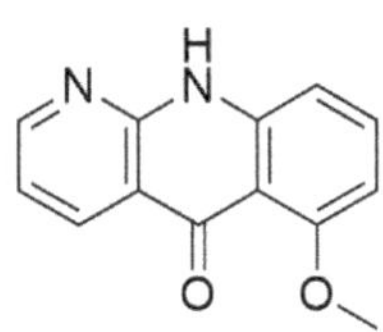

(f)

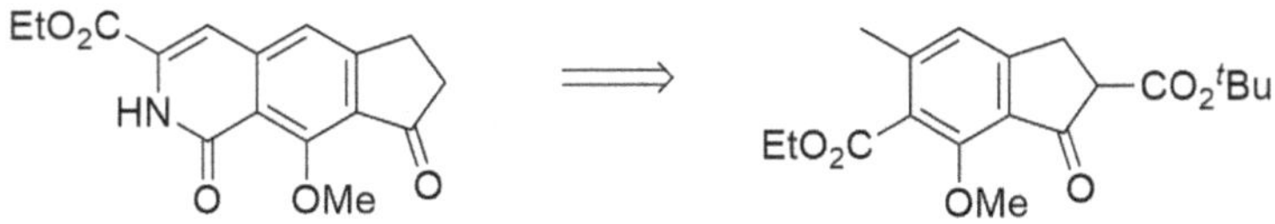

(g)

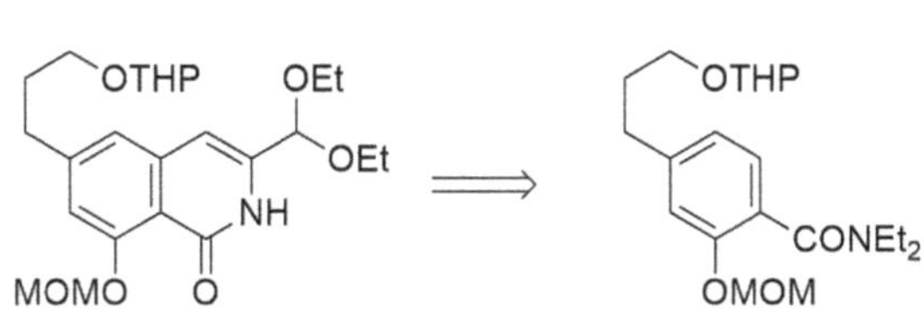

(h)

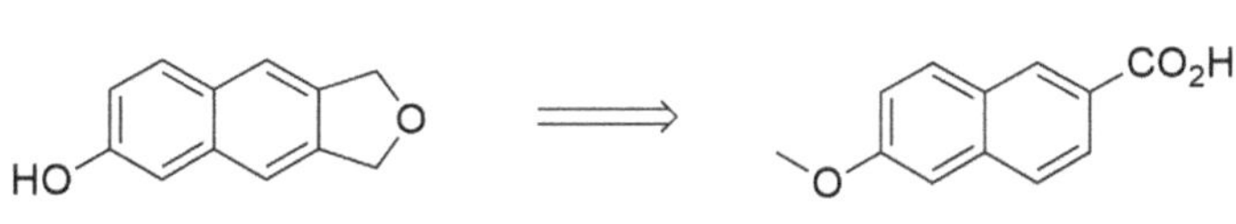

4. Provide a suitable mechanism for each of the following transformations:

(a)

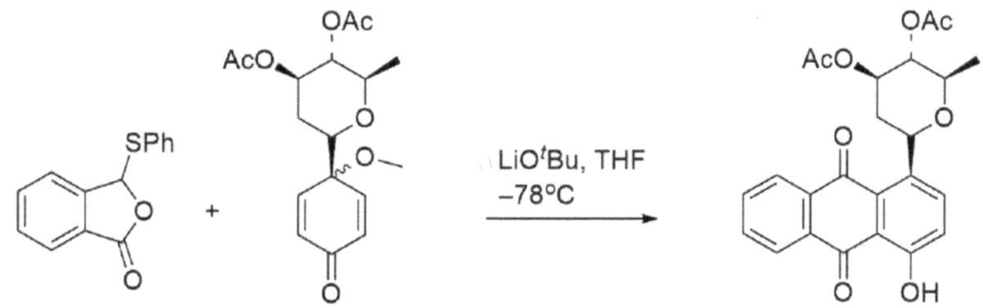

(b) [2014 THN 8879 Pratap R]

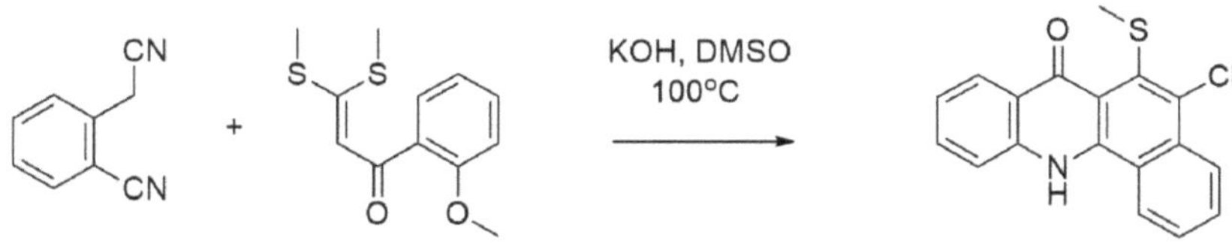

(c)

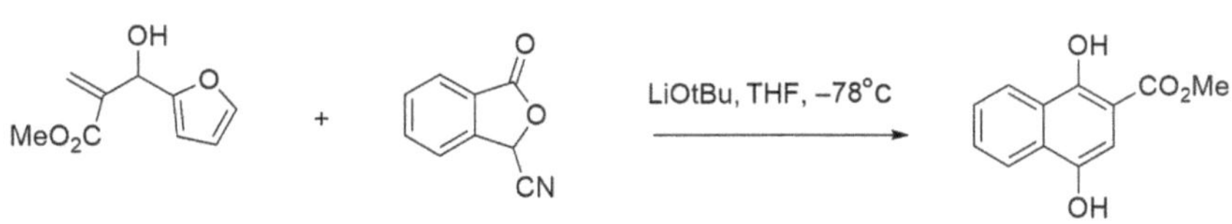

(d)

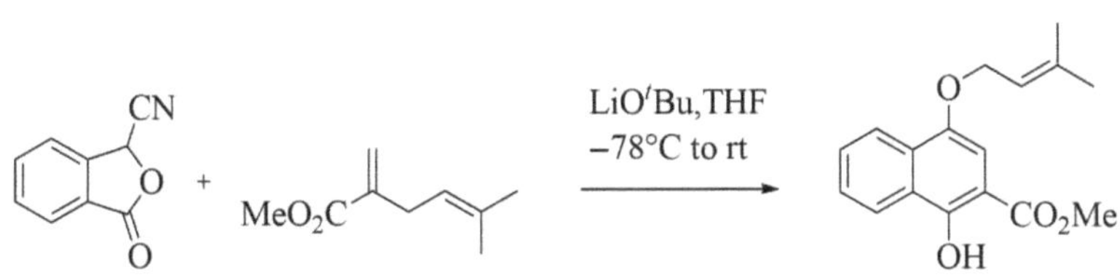

(e) [1989 Chem. Commun. 281 Marchand AP]

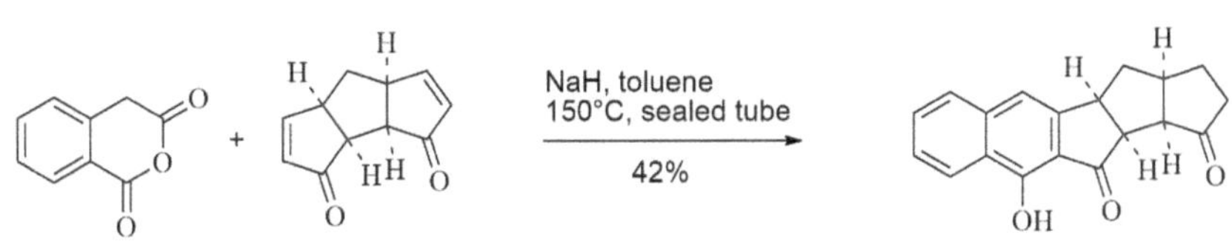

(f) [2007 JOC 3965 Huang X]

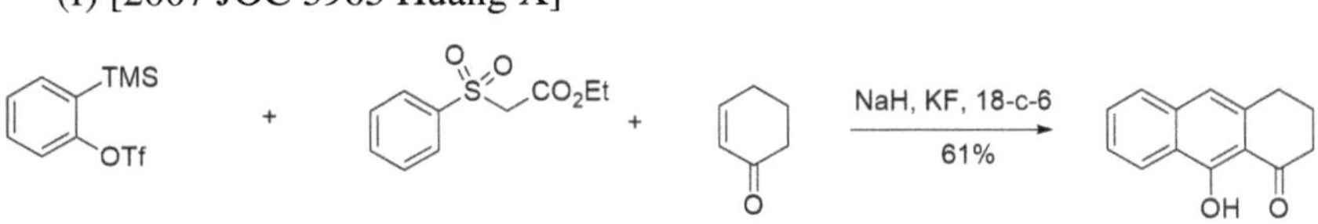

(g) [2011 SYT 1005 Ye S]

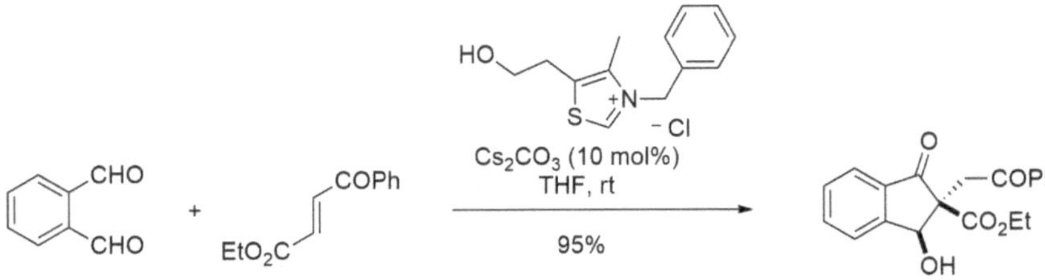

(h) [2014 Scheidt KA]

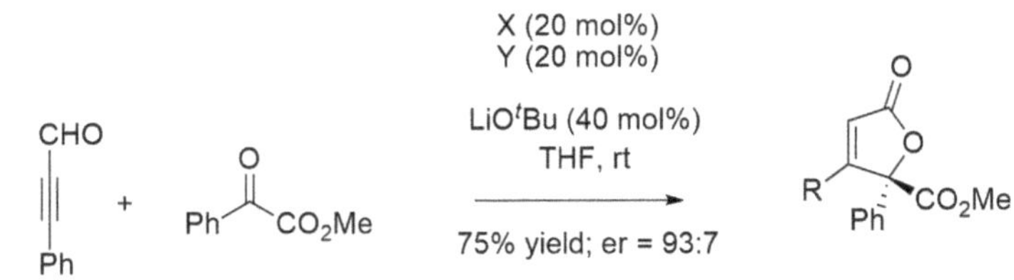

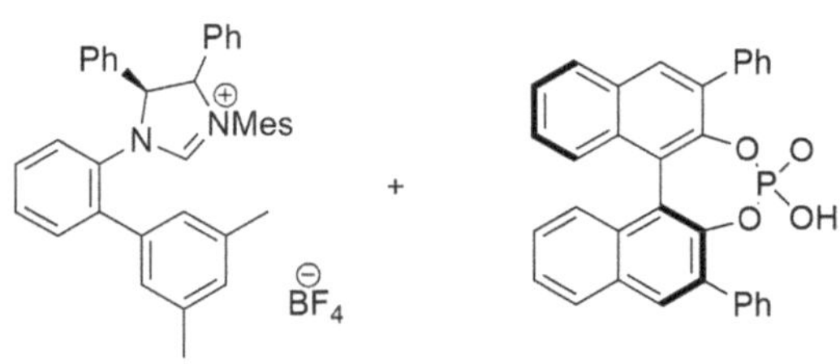

X Y

(i) [1984 JOC 1800 Swenton JS]

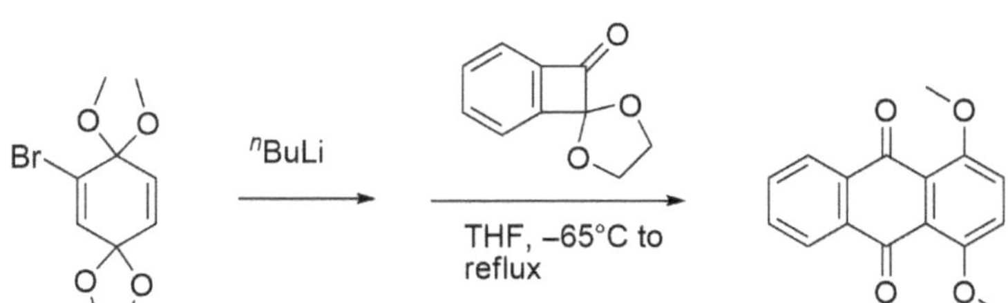

(j) [2003 JOC 2572 Jung ME]

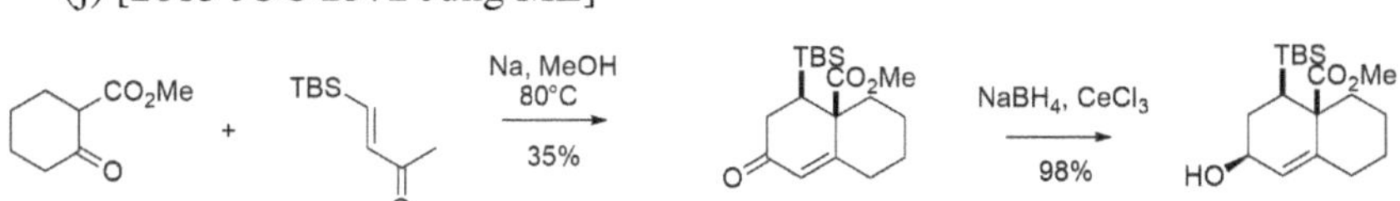

Major product

(k) [2018 JOC 2131 Ghorai MK]

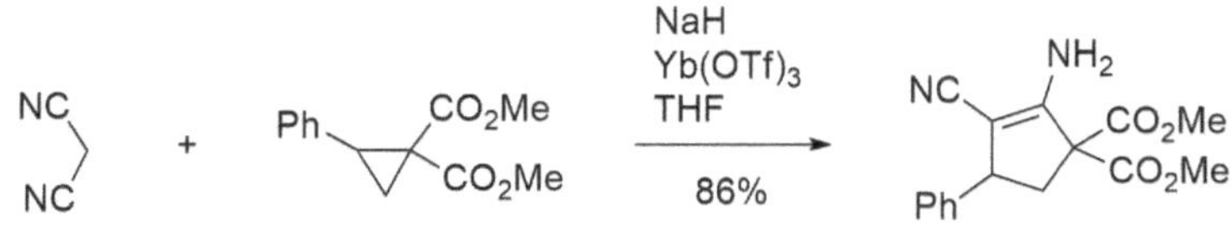

(l)

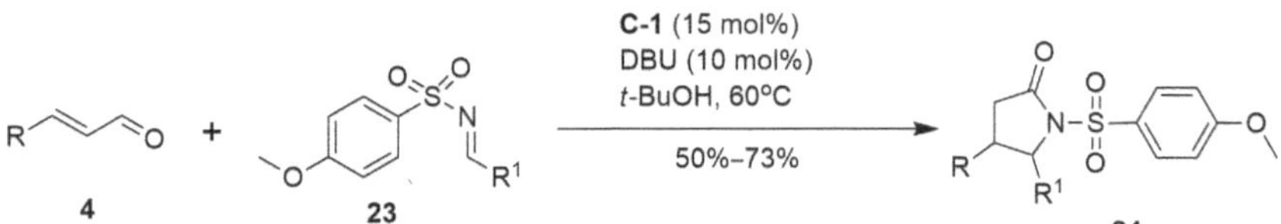

Key: S. Ye and J. Bode

(m) [2017 SCS 1250 Le Lagadec R]

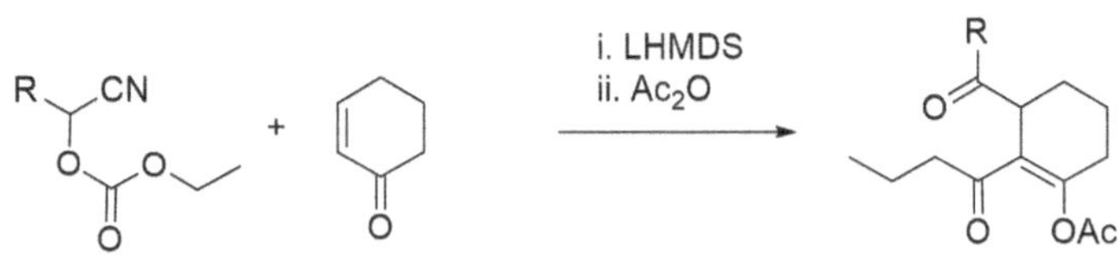

(n) [1996 CCS 1181 Mal D]

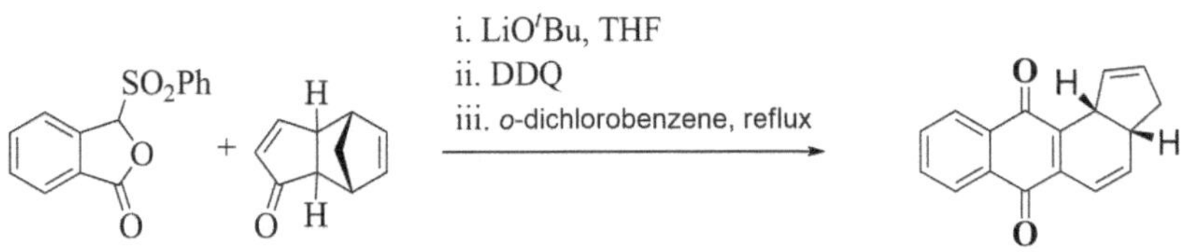

(o) [2015 THN 1247 Mal D]

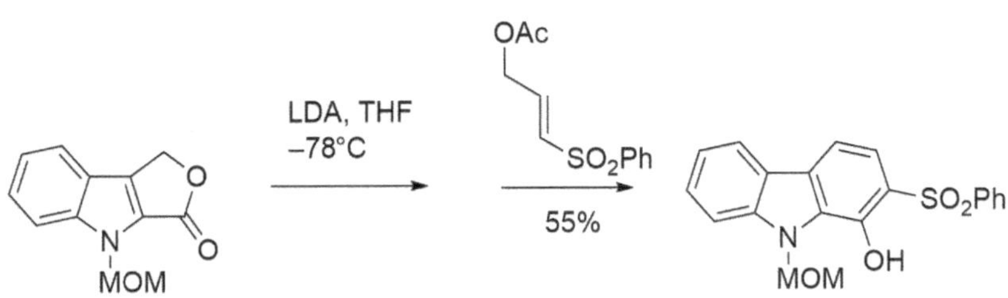

(p)

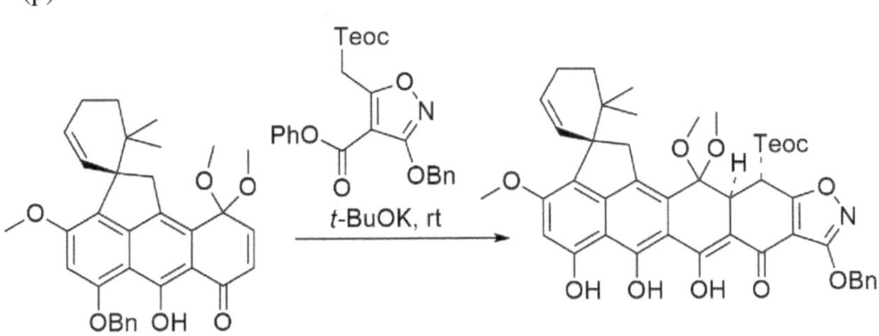

(q)

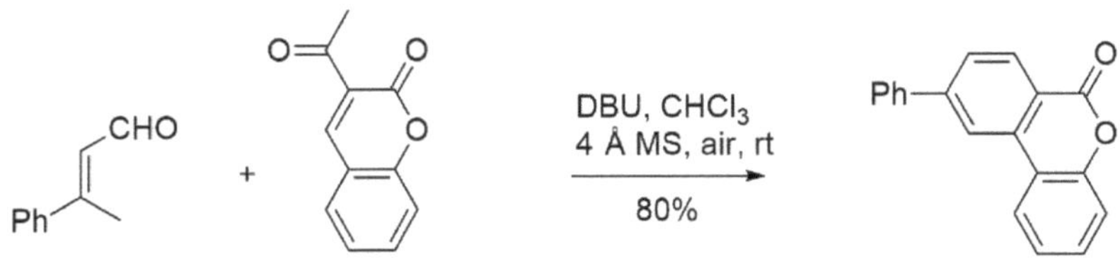

[2016 Adv.Synth. Catal. 707 Chi YR]

(r)

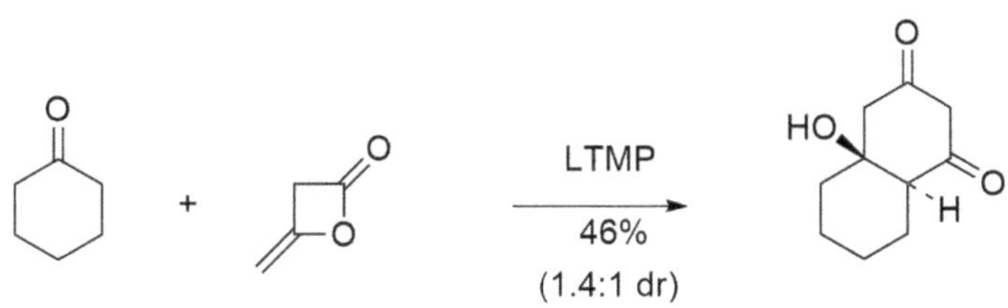

[2016 JACS 14868 Maimone TJ]

(s)

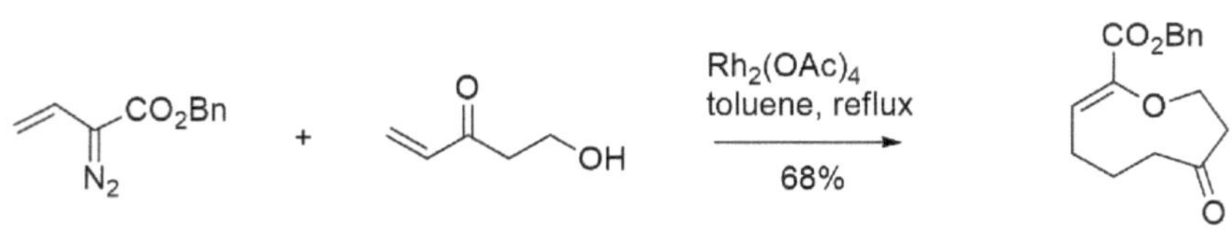

[2016 ORS 6340 Sharma I]

(t)

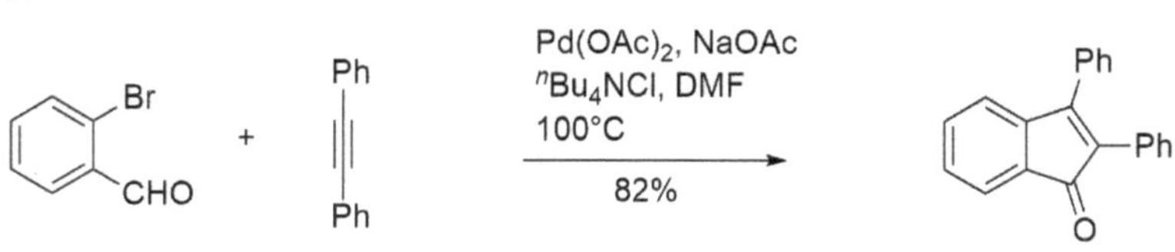

[1993 JOC 4579 Larock RC]

(U)

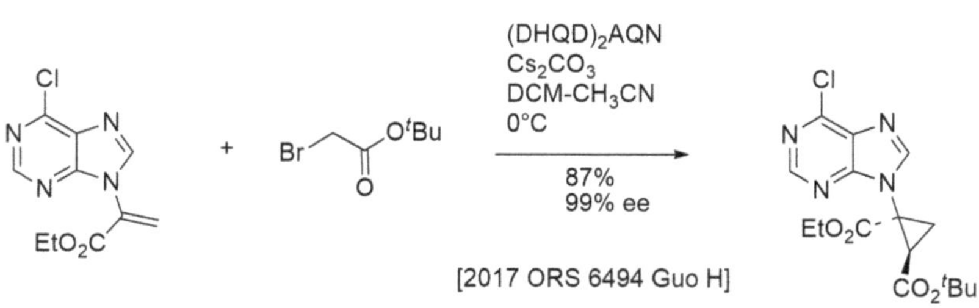

[2017 ORS 6494 Guo H]

(v) [2014 JOC 8428 Anbarasan P]

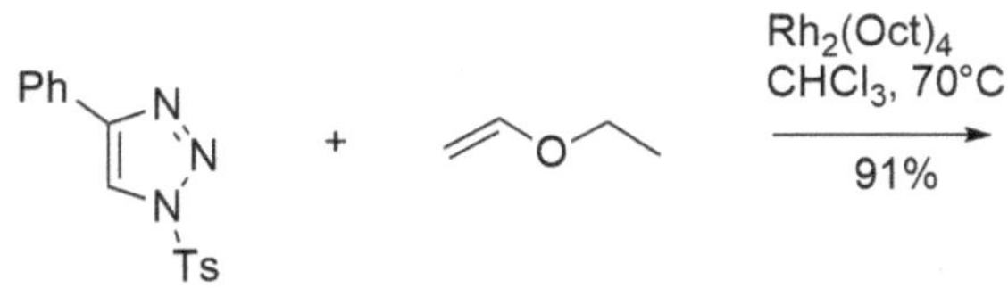

5. Suggest a synthetic route for each of the following targets, using at least one annulation:

(a) Parvinaphthol B (**1**) was isolated in 2016 from the root of *Pentas parvifolia.* It was shown to be active against the triple-negative human breast cancer cell line. It has been synthesized by Han et al. [2017 THS 4779 Han YT].

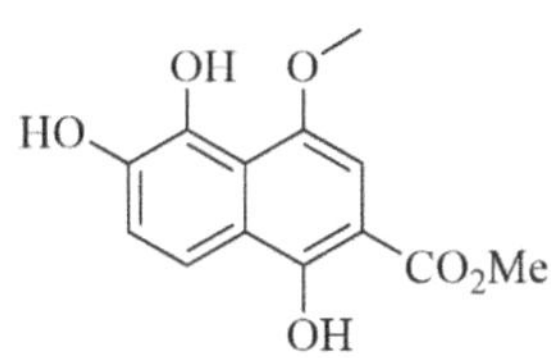

Parvinaphthol (**1**)

(b) Thalictrifoline (**2**) was isolated from a species of Corydalis. Its stereochemistry was determined by a synthesis by Cushman et al. [1981 THS 2333 Cushman M].

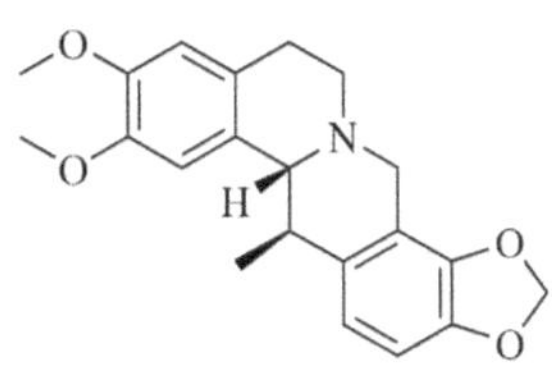

Thalictrifoline (**2**)

Index

Note: Page numbers followed by *f* indicate figures, *t* indicate tables and *s* indicate schemes.

B

C

D

E

F

G

H

I

N

O

P

Q

R

S

W

X

Y

Z

Biography

Dipakranjan Mal

is a former professor of the Department of Chemistry of the Indian Institute of Technology Kharagpur, India. He attended R. K. Mission Vidyamandira and University College of Science for his B.Sc. (Hons) and M.Sc. degrees, respectively, from Calcutta University. He received his Ph.D. degree (1981) from the University of Missouri at Kansas City, having worked under Professor Layton L. McCoy on intramolecular hydrogen bonding. After a postdoctoral stint with Professor Frank M. Hauser at Oregon Graduate Center, he joined Bose Institute, Kolkata as a lecturer in 1984. During the period 1987–2017, he worked in the Department of Chemistry, Indian Institute of Technology, Kharagpur. He has also been a visiting professor at the State University of New York, Albany. His research interests are focused on the development of anionic and thermal domino strategies, and total synthesis of angucyclines, anthracyclines, carbazoles, furocoumarins, and phthalide quinonoid natural products. His notable discoveries are: (i) a decarboxylative [4+2] annulation; (ii) an anionic [4+3] annulation; and (iii) an olefination based on thiophthalides. He has supervised 30 doctoral students and published over 130 research publications, including two reviews in *Chemical Reviews*. He has developed a 40-lecture video course on "Heterocyclic Chemistry," which is available in YouTube. He is a recipient of the honor certificate of Phi Kappa Phi, USA. He is a Fellow of the West Bengal Academy of Science and Technology as well as the Royal Society of Chemistry (UK).

FSC
www.fsc.org
MIX
Papier | Fördert gute Waldnutzung
FSC® C083411

Kolibri 360 GmbH
Ferdinand-Jühlke-Straße 7
99095 Erfurt, Deutschland
produktsicherheit@kolibri360.de